MW00568089

SCIENCE PROBE 10

Second Edition

Science Probe 8 (second edition)
Science Probe 9 (second edition)
Science Probe 10 (second edition)

John Wiley & Sons Canada Limited
Junior Secondary Science Series
with accompanying Teacher's Resource Guides

SI Metric

The illustration on the cover is an artist's representation of **Science Probe 10, Second Edition**. The coloured bands represent the six units of the book: Exploring Space, Harnessing Energy, Investigating Matter, Examining the Earth, Exploring the Basis of Life, and Understanding Ecology. As you use the book, you will have an opportunity to become actively involved in learning more about each of these topics. Knowing this, what do you think the hand on the cover represents? Why are the bands woven through the fingers of the hand?

Look at the back cover. What happens to the coloured bands as they are extended onto the back cover? Why do you think the artist did this? See if you can interpret the figures on the back cover as well.

Cover illustration by Michael Reinhart

SCIENCE PROBE 10

Second Edition

Author Team

Jean Bullard
Frank Cloutier
Nancy Flood
Gordon Gore
Eric S. Grace
Bruce Gurney
Alan J. Hirsch
Dennis Hugh
Chandra Madhosingh
Gerry Millett
Allen Wootton

John Wiley & Sons
Toronto • New York • Chichester • Brisbane • Singapore

Canadian Cataloguing in Publication Data

Main entry under title:

Science probe 10

2nd ed.

For use in secondary schools.
Includes index.
ISBN 0-471-79534-8

1. Science. I. Bullard, Jean, 1938–

Q 161.2.S35 1992 500 C91-093495-9

Printed and bound by Arcata Graphics Book Group

2 3 4 5 AG 98 97 96 95

Science Probe 10 Project Team

Acquisitions Editor: Wilson Durward

Project Editor: Nancy Flight

Developmental Editors: Nancy Flight, Anne Norman, John F. Ricker, Barbara Tomlin, Yvonne Van Ruskenveld

Copy Editor, Proofreader, Indexer: Claudette Reed Upton

Editorial Assistant: Margalo Whyte

Editorial Consultant: Lee Makos

Pedagogical Consultant: Robin Hansby

Content Consultant: Jean Bullard

Safety Consultant: Margaret Redway (Fraser Science Awareness Incorporated)

Contributing Writers: Kathleen Gibson, Eric S. Grace, Don Plumb, John F. Ricker

Design: Michael van Elsen Design Inc.

Picture Research: Rosemary Wurts

Word Processing: Carol Herter, Joy Woodsworth

Production Co-ordinator: Deborah Starks

Illustrations: Jim Loates Illustrations, Dorothy Siemens Illustrations & Design, Jane Whitney Medical & Scientific Illustrator

Typesetting, Assembly, Film: Compeer Typographic Services Ltd.

Acknowledgements

Science Probe 10 (second edition) is the result of the efforts of a great many people. The publisher would like to thank those who assisted in the development of this learning resource.

Focus Group Members
Anand Atal, Burnaby South Secondary; Patricia Dairon, Delta Secondary; Dr. Gaalen Erickson, University of British Columbia; Lloyd Esralson, Burnaby North Secondary; Roger Fox, Premier's Advisory Council on Science and Technology; Lon Mandrake, Seaquam Secondary; Tony Williams, Dr. Charles Best Junior Secondary; Sandy Wohl, Hugh Boyd Junior Secondary

Teacher Reviewers
Anand Atal, Burnaby South Secondary (Unit I); Kim Bondi, Maple Ridge Senior Secondary (Unit IV); Frank Cloutier, Delta Secondary (Unit V); David Crampton, D.P. Todd Secondary (Unit VI); Darrell Ferrill, Johnston Heights Junior Secondary (Unit VI); Stephen J. Formosa, College Heights Secondary (Unit V); Roger Fox, Premier's Advisory Council on Science and Technology (Unit II and Unit III); Ralph George, Delview Junior Secondary (Unit II); Rick Guenther, South Peace Secondary (Unit IV); Herb Johnston, Howe Sound Secondary (Unit III); Gerry Millett, Crofton House (Unit II); Bert Parke (Unit I); W.A. Vliegenthart, Correlieu Secondary (Unit II and Unit VI); Sandy Wohl, Hugh Boyd Junior Secondary (Unit I); Dick Zarek, Duchess Park Secondary (Unit III)

Content Reviewers
Dr. Tom Dickinson, Caribou College (Unit VI); Carlo Giovanella, University of British Columbia (Unit IV); Dr. Mel Heit, Simon Fraser University (Unit III); Dr. Peter Matthews, University of British Columbia (Unit II); Dr. Nancy Ricker, Capilano College (Unit V)

We would also like to thank those people who provided information or reviewed portions of this book in manuscript form:

Dr. F.M. Anglin (Unit IV); Sheila Backeland (Unit II); Ann Barrett (Unit V), Hilda Ching (Unit II), Sarah Dench (Unit II), Karen Ewing (Unit VI), Dr. J.M. Friedman (Unit V), Monique Frize (Unit II), Geological Survey of Canada staff (Unit IV), Rick Heemskerk (Unit IV), Don Howes (Unit VI), Jeanne Johnson (Unit IV), Jeff Johnson (Unit III), Bob LaBelle (Unit II), Don Luney (Unit II), Dave Luttman (Unit V), Barbara McGillivray (Unit V), Kerry McKenzie (Unit II), Maps B.C. staff (Unit VI), Dr. Richard G. Mathias (Unit V and Unit VI), Gordon Rawlings (Unit II), Shawn Robins (Unit VI), Susan Smily (Unit I), Rod Stratton (Unit V); Kenneth P. Stubbs (Unit VI), Brent Tymchuk (Unit II), Bob Williamson (Unit VI), Mel Williamson (Unit II), Susan Winther (Unit II), Joe Wisniewski (Unit II), Anna Wong (Unit V), Dr. R.G. Worton (Unit V)

We would especially like to thank the staff and students of the following schools for taking part in photograph sessions: Caledonia Senior Secondary in Terrace, Cloverdale Junior Secondary in Surrey, and Hugh Boyd Junior Secondary in Richmond.

Table of Contents

To the Teacher

This is an exciting time to be teaching science. It is also a very demanding time. Today, science teachers all over the world are being challenged to rethink the goals, purposes, and processes of science education. Teaching approaches are changing as the importance of scientific literacy for all citizens becomes increasingly clear. Teachers are being challenged to demonstrate the relationships between science and technology and the societal decision-making process.

Science Probe 10, Second Edition, responds to these challenges. This program gets students involved in the processes of science, in problem solving, and in discussions about science, technology, and social issues. The program helps students become active participants in the learning process. In each chapter of *Science Probe 10*, students are encouraged to consider what they already know about certain topics or concepts. They are then given opportunities to build upon this knowledge and develop critical thinking abilities. The information, visual materials, activities, and questions in *Science Probe 10* all help students acquire basic learning skills and scientific knowledge. Throughout *Science Probe 10*, students are encouraged to identify and gather appropriate data, and to draw inferences from these data. They are also encouraged to reason and think independently by identifying different viewpoints and evaluating evidence. This acquisition of knowledge, skills, and attitudes will help students become educated citizens, aware of their rights and prepared to exercise their responsibilities as individuals within the world community.

The dedicated team of authors, reviewers, and editors that has produced *Science Probe 10* is proud to present these balanced, solid, and very teachable materials. With these resource materials, science teachers will be able to provide the kind of creative program that meets the challenges and demands of science education today and into the twenty-first century.

Features of Science Probe 10

- *Science Probe 10* has an introductory chapter about the nature of science followed by six units, each representing a major area of science. These units can be studied in any sequence.
- The text includes a variety of activities designed to foster the development of scientific skills and processes. The diverse formats include step-by-step experiments, reading activities, informal discussions, and student-designed investigations. All activities are designated by the same type of heading, for example, *Activity 2C*. The first activity in each chapter is designed to help students consider what they already know about the topic of the chapter and what they may want to find out.
- Within an activity, the word CAUTION! followed by a few words or sentences indicates that the activity contains potentially dangerous procedures (such as those that call for the use of scalpels) or materials (such as acids and bases). More information on safety precautions can be found in the Safety Rules section that follows.
- Enrichment activities in the margin under the heading *Extension* are intended to challenge students or to reinforce new concepts.
- Enrichment material in the margin under the heading *Did You Know?* includes interesting facts and scientific discoveries.
- Boxed features examine special topics related to the chapter. There are four types of features: (1) Profile, (2) Career, (3) Computer Application, and (4) Science in Our World.
- Terms considered essential for understanding new concepts are printed in **boldface**, defined in the text, included in the *Vocabulary* at the end of the chapter, and listed in the *Glossary*.
- At the end of each numbered section, questions are listed under the heading *Review*. These questions allow students to check their understanding of the content as they progress through the chapter. Recall questions cover the range of concepts in the section, and other questions require reflection, evaluation, comparison, and application.
- Each chapter concludes with a five-part chapter review. *Key Ideas* lists the main ideas of the chapter. *Vocabulary* lists the boldface words and suggests two activities using these key terms. *Connections* provides questions requiring skills such as comparison, evaluation, prediction, and problem solving. *Explorations* suggests projects, research, and group activities. *Reflections* includes questions that allow students to consider how the chapter has changed their ideas or to reflect in an open-ended way on what they have learned.
- The *Glossary* includes definitions of all key concepts (shown in boldface print in the chapters) along with other important scientific terms.
- The *Appendices* provide additional information of general use and technical material related to a specific unit.
- The *Index* helps students to locate important topics and to find connections between units.
- The accompanying *Science Probe 10 Teacher's Resource Guide* provides teaching strategies, lesson plans, practical tips, answers to questions, additional questions and activities, sample tests, blackline masters (some for overhead projection, some for student use), and other useful information for planning and teaching each unit.

Safety Rules

Your school laboratory, like your kitchen, need not be dangerous. In both places, understanding how to use materials and equipment and following proper procedures will help you avoid accidents.

The activities in this textbook have been tested and are safe, as long as they are done with proper care. Take special note of the instructions accompanying the word CAUTION! whenever it appears in an activity. These instructions alert you to potentially dangerous equipment (such as Bunsen burners and scalpels), procedures (such as those involving electricity or chemicals), and materials (such as acids and bases, or other toxic or flammable materials).

Follow the guidelines and general safety rules listed below. Your teacher will give you specific information on additional safety routines. You will also be informed about the location and proper use of all safety equipment.

1. Learn the location and proper use of the safety equipment available to you, such as safety goggles, protective aprons, heat-resistant gloves, fire extinguishers, fire blankets, eye-wash fountains, and showers. Find out the location of the nearest fire alarm.
2. Inform your teacher of any allergies, medical conditions, or other physical problems you may have. Tell your teacher if you wear contact lenses.
3. Give your teacher your complete attention when listening to instructions.
4. Read through the entire activity before you start. Before beginning any step, make sure you understand what to do. If there is anything you do not understand, ask your teacher to explain.
5. Do not begin an activity until you are instructed to do so.
6. If you are designing your own experiment, obtain your teacher's approval before carrying out the experiment.
7. Clear the laboratory bench of all materials except those you are using in the activity.
8. Follow your teacher's instructions regarding the No Crowding Zone.
9. Wear protective clothing (a lab apron or a lab coat) and closed shoes during activities involving equipment or materials. Long hair should be tied back.
10. Wear safety goggles when using hazardous or unidentified materials and when heating materials.
11. Do not taste or touch any material unless you are told to do so by your teacher.
12. Do not chew gum, eat, or drink in the laboratory.
13. Read and make sure you understand all safety labels.
14. Label all containers.
15. When taking something from a bottle or other container, double-check the label to be sure you are taking exactly what you need.
16. If any part of your body comes in contact with a chemical or specimen, wash the area immediately and thoroughly with water. If your eyes are affected, do not touch them but wash them immediately and continuously for at least 15 minutes and inform your teacher.

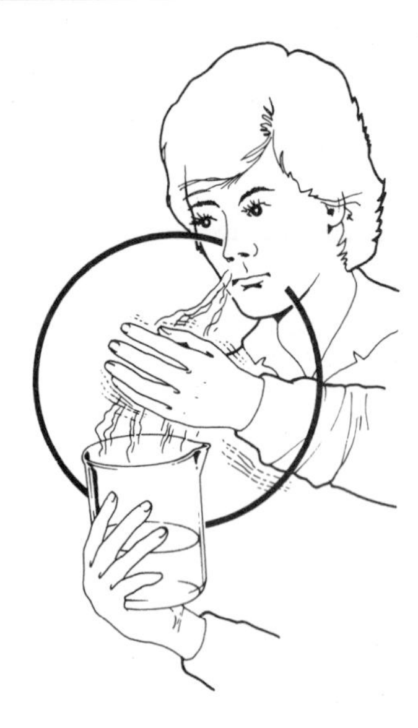

17. Handle all chemicals carefully. When you are instructed to smell a chemical in the laboratory, follow the procedure shown here. Only this technique should be used to smell chemicals in the laboratory. Never put your nose close to a chemical.
18. Do not pour liquids while holding the container(s) close to your face.
19. Place a test tube in a rack before pouring liquids into it.
20. Clean up any spilled water, chemicals, or other materials immediately, following instructions given by your teacher.
21. Do not return unused chemicals to original containers. Do not pour them down the drain. Dispose of chemicals as instructed by your teacher.
22. Whenever possible, use electric hotplates. Use flames only when instructed to do so. Read the special Bunsen burner safety procedures listed under safety rule number 41.
23. When heating materials, always wear safety goggles and use hand protection if required.
24. When heating glass containers, make sure you use clean Pyrex or Kimax. Do not use broken or cracked glassware. Always keep the open end pointed away from yourself and others. Never allow a container to boil dry.
25. When heating a test tube over a flame, use a test tube holder. Holding the test tube at an angle and facing away from anyone, move it gently through the flame so that the heat is distributed evenly.
26. Handle hot objects carefully. Hotplates can take up to 60 minutes to cool off completely. Hot and cold hotplate burners can look the same. If you suffer a burn, immediately apply cold water or ice.
27. Make sure your hands are dry when handling electrical cords and plugs. Keep water away from electrical cords, plugs, and sockets.
28. Always unplug electrical cords by pulling on the plug, not the cord. Report any frayed cords or damaged outlets to your teacher.
29. Make sure electrical cords are not placed where anyone can trip over them.
30. When using a scalpel, always cut away from yourself and away from others.
31. When walking with or passing a scalpel, a pair of scissors, or any sharp or pointed object, keep the pointed surface facing the floor away from people.
32. Watch for sharp or jagged edges on all equipment.
33. Place broken or waste glass only in specially marked containers.
34. Follow your teacher's instruction when disposing of waste materials.
35. Report to your teacher all accidents (no matter how minor), broken equipment, damaged or defective facilities, and suspicious-looking chemicals.
36. Be sure all equipment is shut off when not in use. Be ready to shut off equipment quickly if it breaks down or if an accident occurs.
37. Clean all equipment before putting it away.
38. Wash your hands thoroughly using soap and warm water after any activity where you are handling materials. This practice is especially

important when you handle chemicals, biological specimens, or micro-organisms.

39. Do not take any equipment, materials, chemicals, or biological specimens out of the laboratory.

40. Do not practise laboratory experiments at home unless directed to do so.

41. If a Bunsen burner is used in your science classroom, make sure you follow the procedures listed below. (Note: Hotplates should be used in preference to Bunsen burners whenever possible.)
    - Do not wear scarves or ties, long necklaces, or earphones around your neck. Tie back long hair, and roll back and secure loose sleeves before you light a Bunsen burner.
    - Obtain instructions from your teacher on the proper method of lighting and using the Bunsen burner.
    - Do not heat a flammable material (for example, alcohol) over a Bunsen burner.
    - Be sure there are no flammable materials nearby before you light a Bunsen burner.
    - Do not leave a lighted Bunsen burner unattended.
    - Always turn off the gas at the valve, not at the base of the Bunsen burner.

42. If a fire occurs, make sure you follow the procedures listed below.
    - Do not panic. Remain calm.
    - Notify your teacher immediately. Since every second is vital, act quickly to provide help in an emergency.
    - Shut off all gas supplies at the desk valves if it is safe to do so.
    - Pull the fire alarm.
    - Follow your teacher's instructions if your assistance is required.
    - If your clothing is on fire, roll on the floor to smother the flames. If a fellow student's clothing is on fire, use a fire blanket to smother the flames.
    - Avoid breathing fumes.
    - Do not throw water on a chemical fire.
    - If the fire is not quickly and easily put out, leave the building in a calm manner.

43. Become familiar with the warning symbols that are placed on potentially dangerous substances. You should be able to identify and understand each of the labels shown here. The warning symbols on household products were developed to indicate exactly why and to what degree a product is dangerous. The Workplace Hazardous Materials Information System (WHMIS) symbols were developed to standardize the labelling of dangerous materials used in all workplaces, including schools. Pay careful attention to any warning symbols on the products or materials that you handle.

Hazardous Household-Product Symbols

poison

flammable

explosive

corrosive

danger

warning

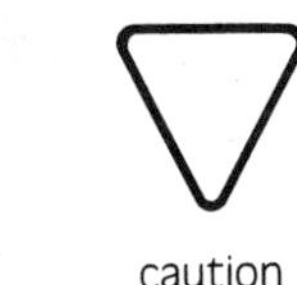
caution

Workplace Hazardous Material Information System (WHMIS) Symbols

Class A
compressed gas

Class D Division 1
poisonous and infectious material causing immediate and serious toxic effects

Class D Division 3
biohazardous infectious material

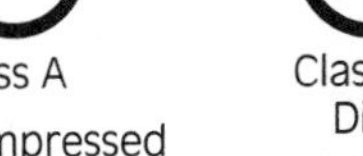

Class B
flammable and combustible material

Class E
corrosive material

Class C
oxidizing material

Class D Division 2
poisonous and infectious material causing other toxic effects

Class F
dangerously reactive material

To the Student

This book is about science. It contains information and ideas about how science works and about past scientific discoveries. There is information on household electricity, chemical substances, earthquakes, reproduction, global problems such as pollution, and much, much more. In addition, this book contains activities and other suggestions to help you learn science. A few of the many things you might do are described in this section.

Consider What You Already Know

At the beginning of each chapter in this book, you will find an activity that helps you consider what you already know about a particular topic. For instance, Activity 1A in Chapter 1 asks you to think about the process of science. As you read Chapter 1, you may find that you want to change or add to some of your original ideas. You may also find that you know a lot more about the way scientists work than you thought you did.

Write in Your Learning Journal

Throughout this book, you will find references to "your learning journal." A learning journal is a notebook where you write your ideas and comments about your learning as you work through a course. Writing down your ideas will help make them clear to you and to others, should you choose to share your writing. As well as helping clarify your thoughts, your learning journal will be a record of your learning in science.

Read Actively

Some of the activities in this book ask you to do something as you read. For instance, in Activity 8D in Chapter 8, you are asked to make two-column notes while you read. The first column contains what you know before the reading, and the second column contains information you gain from the reading. There are also reading activities that ask you to construct charts and tables, make a flow chart of a process, or draw diagrams. These reading activities will help you learn by allowing you to organize the ideas in a useful way.

Make a Concept Map

After the vocabulary list at the end of each chapter in this book, you will find a concept mapping activity. You will be asked to express your understanding of a topic by arranging some of the chapter's key ideas, or concepts, in a meaningful way. For example, this concept map was created to express an understanding of the word "water."

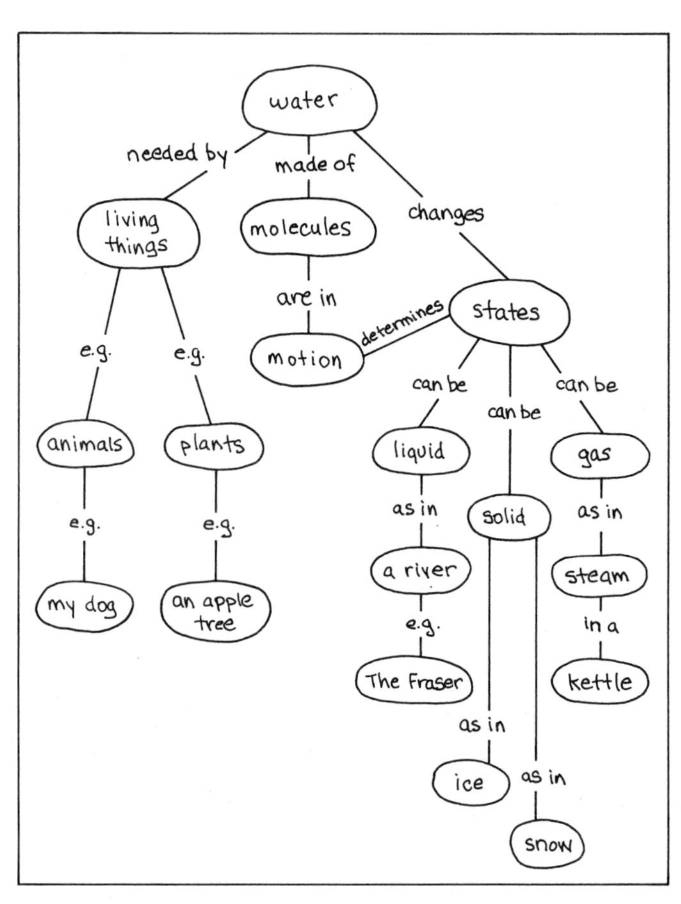

Making a concept map begins with choosing a topic and selecting words connected with the topic. These words are then arranged in a way that suggests how they are connected with the topic. For instance, if you wanted to consider what "chair" means, you might create something like this concept map.

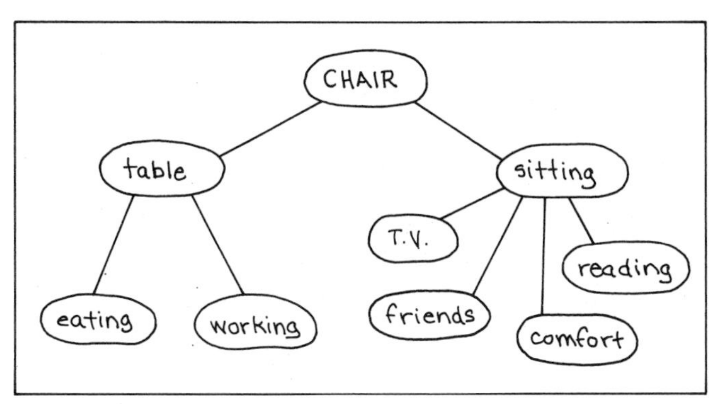

As you can see in this example, the topic is placed at the top of the map. Underneath come concepts related to "chair," and under these are more concepts. The concepts become more specific, or more descriptive, as you move down the map away from the topic word. The lines show connections, or relationships, between concepts.

To show relationships more clearly, connecting words are then placed on or across the lines.

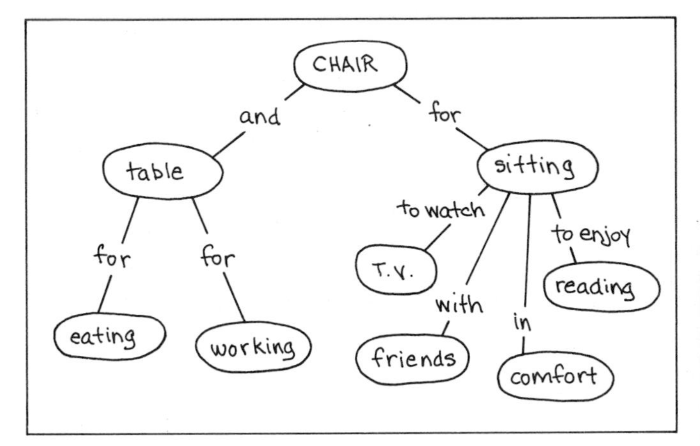

Any two concepts and the words between them make a short statement about the connection or relationship between the two concepts. The statement "sitting to enjoy reading" is an example of this. The connecting lines show that the two concepts ("sitting" and "reading") are related and the connecting words ("to enjoy") show how they are related.

Sometimes there are relationships between concepts that are on opposite sides of the map.

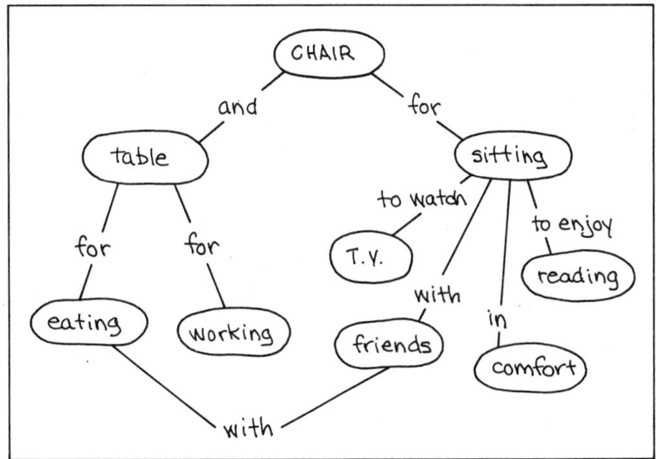

These relationships can also be shown with lines and connecting words. For example, "eating with friends" shows a relationship between concepts on opposite sides of this map. You will often find this kind of relationship once your map is constructed.

Constructing a concept map allows you to explore relationships between ideas and to discover new relationships. Since learning science involves understanding new ideas, concept mapping can be a very useful exercise. But remember, there is no "right answer" when it comes to concept mapping. There are many different ways of showing the relationships between ideas.

ACTIVITY A Making a Concept Map

In this activity, you will construct a concept map. What do you think will be easy about constructing a concept map? What do you think will be difficult?

PROCEDURE

1. Write the following words on small pieces of paper: plants, animals, organisms, daisy, mouse, photosynthesis, eat, maple tree, insect. Decide which word will be the topic of your concept map or choose another word to be the topic of your map.
2. Arrange your pieces of paper on your desk. Remember to place your topic word at the top of your map and to arrange the other words in some meaningful relationship to your topic and to each other.
3. Once you complete an arrangement, copy it onto a large piece of paper. Draw lines to connect concepts that you think are related. On the lines, write one to three connecting words that make a logical statement about the two concepts.
4. Look for other connections you can now see between the concepts on your map. Draw lines and write connecting words. As you work on your map, you may find you wish to change the topic you chose. Feel free to do so. This is your concept map and it should represent what you think or know about the various concepts. You may be able to think of other concepts that you could add to this map. Feel free to do this as well until your concept map represents what you know about the topic.
5. Compare your concept map with maps drawn by two or three of your classmates.

DISCUSSION

1. What is a concept?
2. What is a concept map?
3. What did you find easy about concept mapping? What did you find difficult?
4. How did you decide on a topic for your concept map?
5. Do you think that concept mapping will help you to learn? Why or why not?

Look Back at Your Learning

This book provides many opportunities to review and reconsider things you have learned and thoughts you have had. The Reflections section at the end of each chapter contains questions that ask what caused you to change your mind or that ask you to add to your own ideas. For example, in Chapter 21 you are asked: "How have your ideas changed since reading this chapter? Why have your ideas changed?"

Share Your Ideas

Many of the activities in this book involve taking risks. You are asked to hold out your ideas to others, to express what you think you know, and to hear other ideas in return. For example, Activity 1A in Chapter 1 asks you to express your ideas about science in a concept map, a letter, or a cartoon strip. You are then asked to share what you have made with others. If you are afraid to do this because you might be "wrong," then you will miss the opportunity to see whether or not you can explain to another person what you think or know. So share your ideas. Say what you think. Then listen to the responses and the ideas of others and compare them with your own. You will not only learn more about science, you will also learn how to learn. You will develop a skill that will serve you long after you have finished with this book.

CHAPTER

Science, Technology, Society, and You

Only one scientist appears in this photograph, but putting scientists and their cameras into orbit actually took decades of effort by thousands of scientists, technologists, and others from all over the world. We have learned many things from the work done by these men and women. Photographs taken from space have shown that Earth's biosphere (the layers of air, soil, and water) is really very thin. Photographs of the biosphere have also helped us realize that all of Earth's inhabitants truly share a single environment. More and more often today, we hear that our environment is in trouble. Some people say that all environmental problems arise from science and technology. Others say the problems can only be solved through science and technology.

What is science? What is technology? How do scientists and technologists solve problems and make discoveries? What effects, good and bad, may science and technology have on human beings and on the other living things that share the Earth with us? What action should society take to protect this shared environment? In this chapter and later in this book, you will explore these questions.

ACTIVITY 1A Thinking about Science

Take a moment to think about what you know about science. Not the facts you know, but what you know about the process of science itself. How do scientists learn about our world? How do they express their ideas to others? How do scientists solve problems and make discoveries? How do they know when they have discovered something new? Write some of your thoughts about these questions in your learning journal. Then, after you have thought and written about the process of science, express your ideas in any one of the following forms:

- a concept map (see page xii)
- a letter to a classmate
- a cartoon strip

No matter how you choose to express your ideas, share them with others—your classmates, teacher, friends, or family.

1.1 What Is Science?

Many dictionaries describe **science** as a body of knowledge — all the facts, explanations, and ideas that scientists use to describe the world we live in and the universe beyond it. But "science" actually means much more. Science is both a body of knowledge and an organized method, or process, for gaining more knowledge about the world in which we live.

The process of science is not restricted to scientists. For example, suppose some amateur detectives discover several mysterious objects on a lawn and take a photograph (Figure 1.1). As the detectives try to solve the mystery, they will be using some of the methods of science: they will be making observations and trying to explain them. The objects on the lawn are puzzling, but the detectives can use past knowledge and ask questions to solve the mystery. In the same way, a scientist faced with a problem asks questions, predicts possible answers, performs experiments, and examines information to look for causes and effects.

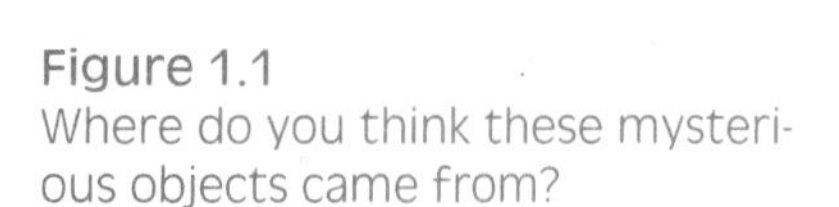

Figure 1.1
Where do you think these mysterious objects came from?

A scientist, like a detective, searches for explanations. The process of science can be described as an organized search for structure and patterns in the world around us.

Both the knowledge and the process of science are equally important. No detective can solve a mystery without knowing the facts and no scientist can search for explanations without a broad base of factual knowledge. You already have a great deal of scientific knowledge. For example, it would not take you long to figure out why the paper clip in Figure 1.2 is pulling upward on the thread. Just a little knowledge of science is enough. Yet, 500 years ago, this sight would have mystified most observers. Many important discoveries may seem simple to us now, but were surprising and important at the time they were made. The following story will show how one curious person used science to solve a "mystery."

Figure 1.2
Can you guess how the paper clip is held in position?

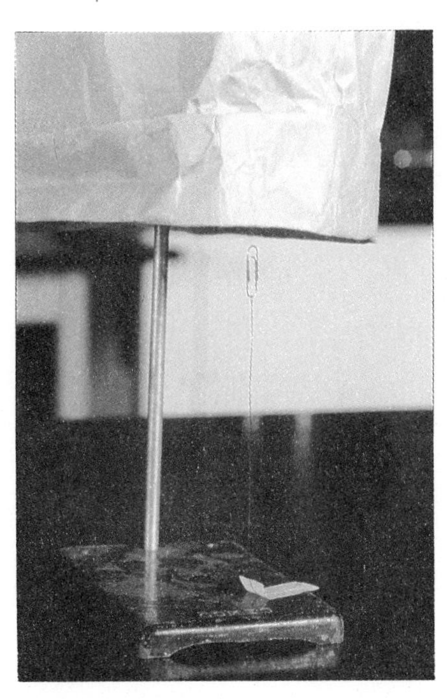

REDI AND SPONTANEOUS GENERATION

For centuries, people saw that most animals produced offspring that were like themselves. However, they also thought they saw some creatures being produced from non-living matter: frogs and eels emerging from mud, insects appearing in animal hair. After observing that maggots appeared in meat left for several days, some observers made two assumptions. They assumed the decaying flesh to be non-living material. Then they assumed that the decaying flesh had been transformed into living maggots through some unknown process. They referred to this as spontaneous generation.

Francesco Redi (1626-1697) was an Italian scientist who was not sure that living things could be produced from non-living matter. He wondered if maggots were the offspring of houseflies, and used a scientific approach to test his idea. He designed an experiment to show that if houseflies were kept away from the decaying flesh, no maggots would appear. As you can see in Figure 1.3, he placed meat and fish in three

DID YOU KNOW?

The housefly has highly specialized mouthparts that allow it to eat decaying food and waste. The housefly's mouth is like a tube with a wide sponge on the end—a very efficient device for sucking up liquid.

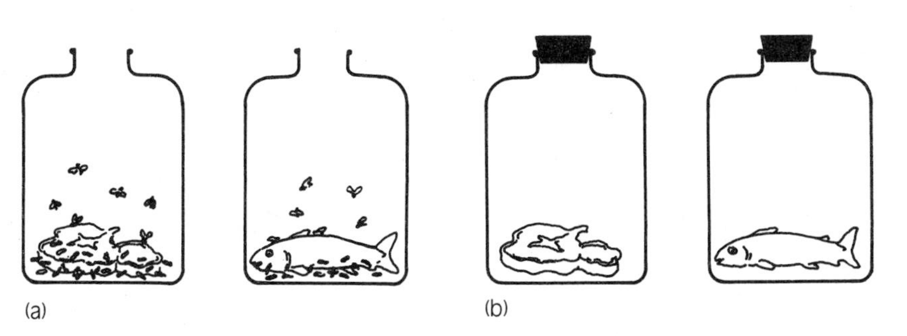

Figure 1.3
Redi's experiment: (a) The first set of flasks contained meat and fish left open to the air. These became covered in maggots. (b) The second set of flasks was sealed. No maggots appeared. (c) The third set of flasks was covered with fine net. No maggots appeared. Redi's understanding of how food could be contaminated led, much later, to the development of techniques for sealing and heating food to destroy decomposing organisms.

EXTENSION

Francesco Redi was a poet as well as a scientist. Find out more about Redi's life and his poetry.

sets of flasks. By later comparing the first two sets of flasks, he was able to conclude that the flesh of dead animals could not spontaneously generate maggots. Redi used the third set of flasks to prove that spontaneous generation was not prevented in the second set by some other factor, such as the lack of fresh air.

By asking questions, experimenting, and observing, Redi was able to reach a conclusion and discredit the notion of spontaneous generation. His work and the work of others later led to methods and products that could be used to stop food from being spoiled by houseflies and other organisms. Redi's "solution" to a mystery eventually helped society.

REVIEW 1.1

1. "Science is a process." Explain what this statement means.
2. "Science is a body of knowledge." Explain what this statement means and provide an example of your own scientific knowledge.
3. Define "science."
4. Describe the experiment that discredited the notion of spontaneous generation.

1.2 What Is Technology?

The work of Francesco Redi in the 1600s ultimately led to some food-preserving techniques. Often scientific discoveries do lead to processes and products that help us get a practical job done. This application of scientific knowledge is referred to as **technology**. Technology can be seen as something that enables us to make use of the natural world. Humans were using technology long before formal science began. Fire-making techniques, farming methods, and labour-saving devices such as waterwheels were developed without any scientific understanding at all. But nowadays, scientific discoveries frequently lead to new technology. For example, discoveries about the nature of light led to the invention of the laser and the compact disc player. In Activity 1B, you will look at some other examples of technology.

ACTIVITY 1B *Technology around You*

Examples of technology are all around you, at school, in the workplace, and in your home. In this activity, you will consider some of the products of technology in a common household scene.

PROCEDURE

Examine the kitchen scene in Figure 1.4. In your notebook, classify the things you see in the illustration. Which things in this scene involve energy or changes in energy? Which items involve life or once-living components? (Classify these items further as different types of living organisms.) Which items are synthetic—that is, made from artificial materials rather than from natural plants, animals, or minerals?

DISCUSSION

1. What would have been different if this picture had been drawn 200 years ago? Which items would have been missing? Why would these items have been missing? What might have been used instead?
2. Identify the product of technology that you think has affected your life most. Explain.

Figure 1.4
Use the information in this illustration to complete Activity 1B.

MODERN TECHNOLOGY—BURDEN OR BENEFIT?

Technology has produced great changes in our way of living. Communication systems have changed dramatically, as shown in Figure 1.5. Messages that once took months, years, or hours to reach their destination can now arrive in seconds. Transportation systems have also changed. You may travel from home to school on a bus, instead of walking. And most wage-earners in our society drive or ride to work, often travelling long distances with little effort.

Technology has changed our daily lives, too. Many household tasks are now done by machines. Dishes can be cleaned in a dishwasher and food can be prepared in a microwave oven. Work that once took a great deal of effort can now be done easily. Computers are commonly used to store and sort large amounts of information.

Many aspects of human health have been improved by technology, as well. More babies survive at birth, more children grow into adults, and more adults live long lives now than in the past. Sewage treatment, food-preserving techniques, and a variety of drugs and vaccines have all helped eliminate disease. Clearly, technology has many benefits.

But unfortunately, there are also problems associated with the technology that provides these benefits. Working at a computer can cause back or eye problems, and life-saving drugs can have unpleasant side effects. Many of technology's products — the automobile especially —

require huge amounts of Earth's resources. Irreplaceable fossil fuels are needed to make and run the vehicles that allow us to travel so easily. Even more disturbing, the use of automobiles is polluting the Earth's atmosphere. Technology has given us longer, healthier, easier lives, but it has also given us life-threatening air pollution. Because applying scientific knowledge can have unexpected social consequences, there are many questions that must be asked and decisions that must be made when we develop and use technology.

(a)

(c)

(d)

(b)

Figure 1.5
Technology has changed the way people communicate.

(a) Around 1500 B.C., the Egyptians used pictures (hieroglyphics) as a form of written communication.
(b) Around A.D. 1500, the Inca Indians of Peru communicated by means of *quipus,* which coded their messages in a series of knots on ropes connected to one another.
(c) In the 1800s, rapid long-distance communication was made possible by the invention of the telegraph.
(d) Now many devices, including computers, make rapid communication even easier.

REVIEW 1.2

1. Explain the difference between science and technology, using an example of each.
2. Give an example of technology not mentioned in this chapter that can have both good and bad effects. Explain.
3. Suggest at least one environmental problem that has been caused by technology.

1.3 Science: A Way of Solving Problems

Since science and technology can bring us both benefits and burdens, every citizen needs to understand how scientists "do" science in order to make effective decisions about such issues. Many people assume that science is an unchanging collection of facts. This is not true. Science is actually an ever-changing body of knowledge and an active process that thousands of workers around the world engage in every day.

To scientists, questions about nature are the everyday problems they must solve. Their approach to solving these problems is almost always orderly, but it is not always exactly the same. For example, questions about living things may require a different approach than questions about the universe. However, certain basic steps are common to most scientific research. You will consider these steps in Activity 1C.

ACTIVITY 1C A Deck of Science Cards

In this activity, you will see how a scientist would go about answering a specific question.

MATERIALS

paper
scissors

PROCEDURE

1. Make a set of cards by cutting sheets of paper into pieces about 10 cm by 4 cm. On each of the cards, write one of the following phrases:
   - assembling equipment
   - carrying out the experiment
   - checking results against the original prediction
   - collecting equipment
   - designing new experiments to study the problem further
   - designing the experiment
   - identifying the problem
   - interpreting observations
   - looking for patterns or regularities in observations
   - making conclusions
   - making observations
   - predicting an answer to the problem
   - recording observations
2. Place all of the cards face up on the desk in front of you. Move the cards around on the desk until you have created a list of steps that you think a scientist would follow in solving a problem (Figure 1.6).
3. Compare your list with at least two other students' lists.
4. Compare your list with at least two other lists from current science textbooks. Look in the index of each text for "scientific method," "scientific process," or "research methods." If possible, compare your list with one in an older science textbook. Try to find science textbooks on two different subjects, perhaps physics and chemistry.

DISCUSSION

1. How was your list like and unlike other lists—your classmates' lists and the lists in science textbooks?
2. Which of these steps do you think are the four most important ones? Why?
3. What other steps could be added to this method of problem solving?

4. What other logical sequence of steps might be used to solve a problem?
5. Design an experiment to answer the question: "Does temperature affect how quickly a sugar cube dissolves in water?"

Figure 1.6
A scientist follows certain steps when trying to answer a question. You can consider the order of these steps by arranging some "science cards" on your desk.

EXTENSION

When scientists communicate with each other, they often use specialized language that is difficult for non-scientists to understand. Fortunately, a few scientists have written understandably about their own work or about scientific research in general. Read a book by or about a scientist, then share what you have learned with others in a discussion group. Any of the following books will give you new insights into the nature of science, scientists, and their research: *The Double Helix* by James D. Watson, *What Little I Remember* by Otto Frisch, *Madame Curie* by Eve Curie, or any book by Loren Eisley. Your librarian will be able to suggest other books.

SCIENCE PROCESS SKILLS

The procedures used in scientific research are worth considering in more detail. These procedures are often called science process skills or simply **process skills**. Non-scientists also use process skills, and you are likely familiar with observing, measuring, and many other of the process skills listed alphabetically in Table 1.1. For example, you are *identifying a problem* when you wonder why your hair dryer will not start. You are *classifying data* when you decide to store both your science lab reports and your English essays on a computer disk labelled "schoolwork." And you are *communicating* when you draw a map for a lost motorist. The meanings and uses of less familiar process skills will become clearer as you continue your science studies.

THE SCIENTIFIC METHOD

Most prehistoric people studied nature in order to put resources to practical use. By 400 B.C., some Greek scholars studied nature simply to learn more about it. And around A.D. 1500, some European naturalists began to study nature by means of the orderly approach we now call the **scientific method**. Table 1.1 lists process skills in alphabetical order. The scientific method uses the skills in a logical order. In general, the first step is to identify a specific research problem. Then scientists must design an experiment that will help answer the question. To do this, a scientist often makes an "educated guess" at the result before starting. Such a prediction is called a **hypothesis** (plural: hypotheses).

Whenever possible, scientists test their hypotheses by conducting experiments. Some fields of science, such as astronomy and earth science, do not lend themselves to experimental methods of investigation.

Table 1.1 Science Process Skills

Name of skill	How scientists use the skill
Classifying	Sorting objects or ideas into groups on the basis of observations
Communicating	Presenting observations in an organized way (often involves rearranging data or putting them in a new order)
Constructing models	Using a familiar object or system to describe or explain an unfamiliar object or system
Experimenting	Performing activities to test an inference, prediction, hypothesis, or model (involves recognizing what conditions might affect outcome)
Hypothesizing	Suggesting a possible solution to a problem, based on present knowledge and experience (often called making an "educated guess")
Identifying a problem	Recognizing a problem and asking a meaningful question about it that can be answered through research
Inferring	Drawing a conclusion from facts or observations
Interpreting	Analysing results to look for a pattern; making generalizations about observations
Measuring	Using numbers or instruments to obtain information about an object, for example, how big, small, hot, or cold it is
Observing	Using the senses to obtain information about objects or events in the natural environment or in an experiment
Predicting	Making a reasonable guess about the future based on what has been observed in the past and the present
Recording	Keeping records of all observations, regardless of whether or not they seem to support the hypothesis

But other fields do. For instance, Franceso Redi had a question that was biological in nature. He thought that maggots found in decaying meat might be the offspring of houseflies. He hypothesized that the houseflies laid their eggs on the meat. He then planned and performed extensive experiments to test this hypothesis. He kept records at every stage of his experiment and finally determined that the decaying meat did not spontaneously produce maggots. He communicated his findings to other scientists by presenting his observations in an organized way. Further research eventually confirmed and expanded on Redi's hypothesis.

A hypothesis supported by observation and repeated experimentation may come to be called a **theory.** The kinetic molecular theory, which you studied in earlier grades, is an example of a theory that has stood the test of time. Spontaneous generation is an example of a theory that has not stood the test of time. If the results of experiments contradict a

DID YOU KNOW?

Different kinds of models are widely used in science. For example, computer models are used to design parts for automobiles. Miniature airplanes are used to study the principles of flight. Science teachers use models to help students understand extremely small living things. And physicists study the motion of billiard balls to understand the behaviour of particles too small to be seen.

theory, it has to be modified or discarded. This process makes the scientific method self-regulating. Scientists themselves check each other's ideas and make sure that theories stay up to date.

Table 1.2 shows one way to describe the scientific method. How does the order of the steps in Table 1.2 compare with what you decided on in Activity 1C? Most scientists would agree that they use most of these steps at some time. But most would also admit that they do not always use them all and do not always use them in the same order. Nevertheless, the order shown here provides a good starting point, as long as you remember that science is a creative process and different people will use the method in different ways.

EXTENSION

All our senses have limitations. For example, bats make sounds that humans cannot hear. So do dolphins and whales. But how do scientists know this? How can scientists recognize and study sounds that they cannot hear? Find out what technology scientists use to extend the range of human hearing.

Table 1.2 The Steps of the Scientific Method

1. Identify the problem.
2. Make a hypothesis.
3. Design and perform an experiment to test the hypothesis.
4. Interpret and analyse results.
5. Communicate procedures, observations, and conclusions to other scientists.

THE LIMITATIONS OF SCIENCE

Science has limits. Scientific knowledge is not absolute truth. Rather, it is the total of our present understanding about nature. This understanding is subject to constant revision and updating. Redi's work is an example of this. He designed an experiment to test a theory that was accepted by most people at the time. He soon discovered that the theory of spontaneous generation did not apply to houseflies.

It is tempting to think that scientists of the past must have been very foolish to accept such a theory in the first place. After all, modern scientists seldom accept ideas or explanations until they have been tested by experiment. But nothing can ever be proved for certain. Next week, next year, or next century, at least some of the ideas we accept today will surely suffer the same fate as the theory of spontaneous generation. Those who understand the nature of science will not be dismayed, for they know that this is how science progresses.

REVIEW 1.3

1. What does the term "process skills" mean? Name three process skills and describe how they can be used in your daily life.
2. What does the term "scientific method" mean? List the usual steps of the scientific method in their usual order. Explain why this order might not always be followed.
3. When you place money in a soft-drink machine, how does a can of pop appear? Hypothesize to answer the problem. Compare your hypothesis with a classmate's.

1.4 Why Study Science?

There are many reasons to study science, but perhaps the most important one involves you as a member of society. Science and technology have always had an impact on society and always will. When used properly, science and technology can enhance life for humans and other living things. When not used properly, science and technology can threaten life.

An understanding of science is needed today to help all citizens make informed decisions. Some of these decisions are individual and personal in nature: "I am going to exercise more to improve my health." Some decisions involve important social issues: "Our town will reduce automobile exhaust to control air pollution." Issues related to drugs, disease, famine, population growth, and natural disasters such as earthquakes face us every day. Perhaps the most important issue of this decade relates to Earth itself. Many people believe our beautiful, irreplaceable planet is in great trouble. Is this true? If so, what can and should be done?

In just a few years, you will be voting to choose people to make decisions on your behalf. What decisions do you want them to make? For example, should the government encourage industries to spend money creating more jobs or to spend it on pollution controls? Answers to such questions are never as easy as they may seem at first glance. And when decisions that relate to science and technology must be made, it is helpful to know enough science to evaluate the arguments on both sides. This is especially true of environmental issues.

SCIENCE AND YOU

An understanding of science and technology is also needed for your personal future. Keeping up with scientific developments is a natural part of professions such as engineering, medicine, law, and education. But waiters, police officers, cashiers, car mechanics, bus drivers, and actors need to keep up too. Why? Because all these people work in a rapidly changing world, and much of the change is linked to science and technology. Jobs of today differ from those of the past, and jobs of the future will differ from those of today. Most of today's high school students will probably work at six or more different *kinds* of jobs before retirement, each one requiring a significant amount of job retraining. Understanding what is going on in science and technology will help workers plan for the future and take new training.

In a complex world where jobs change and difficult social problems must be solved, no one can afford to stop learning science. By becoming your own teacher, you can keep learning science throughout life. This will help you adapt to change and take part in the societal decision-making process. A good way to become your own teacher is to develop independent study skills. Activity 1D will allow you to practise some of these skills.

DID YOU KNOW?

Machines keep changing the way we work. Someone using a manual typewriter can type 30 000 keystrokes an hour. Someone using a word processor can type 80 000 keystrokes an hour!

EXTENSION

Throughout your life you will have to make decisions about whether to use tobacco, alcohol, medicines, and pesticides. Understanding some biology, chemistry, and ecology will help. Design a poster that shows how an understanding of science can help you make an important decision.

ACTIVITY 1D A Research Project

What do you think of when you hear about scientific research? Test tubes? A smelly laboratory? Expensive equipment? You might be surprised to learn that scientists nearly always begin their research in the library. There they can find out what has already been learned about the topic they are researching. In this activity, you will conduct library research on a scientific topic of your choice. Later you may choose to prepare a report using the information you gather.

PROCEDURE

1. Begin your research by making a short list of possible topics. Professional scientists usually choose topics they find interesting. Any topic you find interesting probably involves science in some way. The following examples show how many possible topics there are in subjects and activities that are not usually considered to be "scientific."

   Art. Interesting research topics connected with art include the science behind metal casting, oil painting, computer art, neon art, and mechanical art.

   Cars. There are many scientific topics connected with automobiles of the past, present, and future. Possible areas of research include aerodynamic design, ceramic engines, and electric cars. What kind of car *should* we be developing for the twenty-first century?

   Food. Research can be done on how we produce food and how we choose a nutritious diet. Some researchers today are questioning the effect of cattle farming on the atmosphere.

   Music. Whether you like to listen to music or play it, there are dozens of possible science research projects (Figure 1.7). All modern playback equipment has strong links to physics. Computers are used in the recording process. The construction of large church organs requires an extensive understanding of acoustics—the science that explains why sound quality is so bad in most large domed sports arenas.

   Sports. There is a wide range of possible research topics involving sports. Pitchers, place kickers, skateboarders, and tennis players all need to understand the physics of objects in motion. Swimmers and runners need to understand the biology and chemistry of gas exchange.

2. After making a short list of topics, visit the library and identify scientific links to these topics.

3. Over the course of a week, try to choose one main topic and two or three subtopics. Write down a few questions you would like to answer.

4. Before you settle on a topic, make sure you will be able to get information to answer your questions. Scientists spend a lot of time and effort identifying researchable topics. For example, scientists might agree that the "greenhouse effect" is important enough to research, but this topic is wide-ranging and must be narrowed down. Good researchers identify and clarify problems by asking questions: what? why? and how? Asking questions will also help you identify a specific topic that has enough available information for your library research project.

5. If possible, make this a long-term project. Gather information on your topic whenever you can. You might see something on television or in a magazine that clears up a puzzling point, or that raises a new question. A little time spent each week on your chosen topic will soon make you a well-informed amateur, if not an actual expert.

Figure 1.7
The sound of an electric guitar depends partly on the properties of the materials used to make the guitar. A library research project could involve finding out what materials are used to make an electric guitar and how these materials affect the instrument's sound.

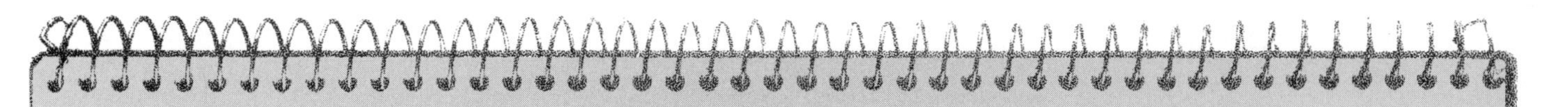

Profile

VERENA TUNNICLIFFE

Peering a few metres ahead, Verena Tunnicliffe sees what few humans have seen before—the fascinating and unusual animals that cluster around hot-water vents on the ocean floor. To do her research, Tunnicliffe squeezes into a small underwater vehicle, called a submersible, and takes a journey down 2 km beneath the waves. In this dark, deep-water world, Tunnicliffe studies giant tube worms, crabs, clams, and vent fish. As a marine zoologist, she wants to discover more about these animals. Turnnicliffe wonders, "How do these animals survive under the special conditions of their environment—the high pressure and continuous darkness?"

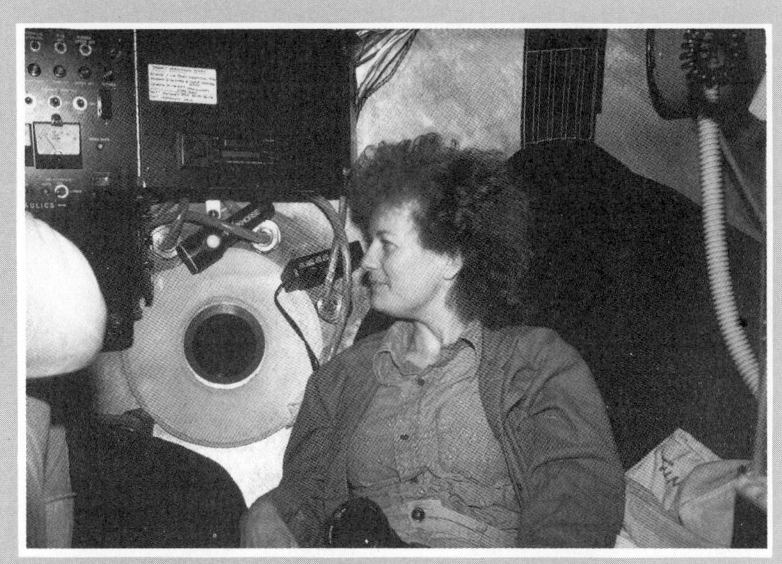

Verena Tunnicliffe is shown here inside a submersible, 2000 m below sea level.

Tunnicliffe has always been fascinated by the ocean. Her university studies in geology and biology, and a previous job as a biologist, were good preparation for her present work. Today, the dives in the submersible are part of her job as a teacher and researcher at the University of Victoria.

The technology that allowed the building of submersibles has made Tunnicliffe's work possible. Until recently, human beings were unable to travel to the deepest parts of Earth's oceans. Now scientists have a new world to explore. But even with the help of submersible technology, planning for a dive still takes a lot of time and a lot of care. "It can take up to two years of planning to get one or two dives," says Tunnicliffe. "Using a research vessel and crew costs thousands of dollars a day, so everyone must know exactly what they will be doing." Usually several scientists share a research voyage that may last three to four weeks.

As the submersible descends for a 10-hour dive, the sunlight rapidly fades. Soon the researchers can see only what is in their beam of light, a small area about 6 m × 3 m. "It's like exploring the Rockies at night with a flashlight. We could pass within metres of something exciting and never know it." The researchers use more than just their eyes, though. They have a long list of chores and experiments to carry out. Using remote-controlled claws, they manipulate rocks and collect samples. They pay careful attention to data displayed on onboard computers. The data are collected in a variety of ways. Sensors on the outside of the vessel measure temperature and analyse water chemistry. Hydrophones pick up sounds. Other devices map peaks and valleys on the ocean floor.

This kind of work is an example of "pure research." Tunnicliffe's aim is to find out more about the world we live in. "Nothing," she says, "is quite like experiencing the unknown and trying to understand it. It's a lot of fun, but science should be fun."

REVIEW 1.4

1. Give two examples of science-related issues that have been in the news lately.
2. "Science and technology have changed the way we do things." Explain the meaning of this statement with some examples from your daily life.
3. How might the study of science be useful if you wanted to be a radio announcer, carpenter, secretary, or tree planter?
4. How could an understanding of the scientific method allow you to help your school or your community?

CHAPTER • REVIEW

Key Ideas

- Science is both a body of knowledge and an organized method for gaining more knowledge and understanding of the natural world. Both aspects of science are equally important.
- Technology involves designing and testing processes and products to get a practical job done. Technology often arises from scientific knowledge or methods.
- Technology provides many benefits, but problems may be associated with these benefits. Questions must be asked and decisions must be made by society regarding the use of technology.
- The process skills used in scientific research include classifying, communicating, constructing models, experimenting, hypothesizing, identifying problems, inferring, interpreting, measuring, observing, predicting, and recording.
- The scientific method is an orderly way of solving problems that can be used by non-scientists as well. The following list describes the scientific method.
  1. Identify the problem.
  2. Make a hypothesis.
  3. Design and perform an experiment to test the hypothesis.
  4. Interpret and analyse results.
  5. Communicate procedures, observations, and conclusions to other scientists.
- A scientific theory is a hypothesis that has been supported by repeated experiment. All scientific theories and knowledge are subject to constant revision and updating.
- Studying science can help individuals adapt to changes in the workplace. Understanding science and technology can also help citizens make decisions about issues that affect society.

Vocabulary

science
technology
process skill
scientific method
hypothesis
theory

1. Read about concept maps on pages xii and xiii. Use the words from the vocabulary list and any other words you need to create a concept map about science.
2. Write a clue for each word in the vocabulary list. Give your list of clues to a classmate to solve.

Connections

1. For each of the following scientific facts, identify a technology that developed as a result.
   (a) Some gases are "lighter" (less dense) than air.
   (b) Falling water has energy.
   (c) The Earth behaves like a giant magnet.
   (d) Diamond is one of the hardest substances.
   (e) Some chemicals change colour when they are exposed to light.
   (f) Light bends when it passes from air into glass or from glass into air.
2. You are a scientist in one of Canada's food-testing laboratories. One day, an inventor visits your laboratory and declares that she has discovered a remarkable new product she calls "Magic Sweet." She has spent three years working on it and makes the following claims:
   - It sweetens foods and drinks.
   - It has no food energy.
   - It is perfectly safe for human consumption.
   - It is very inexpensive to produce, as it comes from a commonly available source.

   Outline the steps your laboratory would take to test these claims.

3. Look back at the photograph on page 1. What evidence do you see of technology? What other technology must have been involved to make this photograph possible? How do you think this technology could be used to solve environmental problems? Are there any problems in your community that science and technology could help solve?

Explorations

1. Some scientific discoveries are said to be the result of luck or chance. The great French scientist Louis Pasteur (1822–1895) said that "chance favours the prepared mind." Investigate the "lucky discoveries" of William Perkin, Wilhelm Roentgen, Alexander Fleming, or Barbara McClintock. Do you agree that their discoveries were the result of "chance"? Design a cartoon using Pasteur's words and showing what you think Pasteur meant.

2. Find out about Canadian scientists who have won Nobel prizes (Figure 1.8). Prepare a presentation or report describing the achievement of one Canadian Nobel Prize winner.

Figure 1.8
Alfred Nobel (1833–1896), the founder of the Nobel Prize. Nobel was a Swedish chemist, engineer, and industrialist. He invented dynamite in the 1860s and became a wealthy man. He left the bulk of his fortune in trust to establish the Nobel prizes, now awarded in the fields of physics, chemistry, literature, physiology or medicine, peace, and economic science.

Reflections

1. Has the information in this chapter changed the way you think about science?

2. What else would you like to know about how scientists work?

UNIT I

Exploring Space

CHAPTER 2
Working in Space

What do the words "space technology" make you think of? Perhaps you think of rockets or space shuttles. Do you also think of digital watches, hand-held calculators, or the hard plastic in your bicycle helmet? Items such as these were developed as by-products of space technology. Now they are part of your daily life.

Space exploration is an example of the link between science and technology. Scientific knowledge is used to develop the technology to build and launch spacecraft such as the space shuttle and the proposed space station in the illustration. Once the spacecraft are in use, they provide information that helps scientists discover things they did not know before. These discoveries lead to new applications in technology.

Space exploration is also an excellent example of international co-operation. The biggest and most expensive projects are shared by several countries in Europe, as well as the United States, the Soviet Union, Japan, China, and Canada. Canada's role in space is an important one.

As you study the exciting topic of space exploration, you will discover the uses of satellites, space stations, space probes, and future space colonies. You will also learn about Canada's contributions to space exploration. As you read this unit, you will find that there are problems that must be overcome to get into space and live there. Is the cost of exploring space worth the expense? How does space exploration affect our society? You will consider these and other questions as you learn about space exploration.

CHAPTER

2 Working in Space

Some spacecraft, such as space stations, are designed to have people living and working aboard them. However, other spacecraft, such as satellites, are not designed for human travel. The satellite in this photograph is being retrieved by two astronauts. Satellites like this one travel in paths around the Earth, high above the atmosphere. Every time you watch a television program "brought to you by satellite," you are using space technology.

Canada is a leading nation in the construction and use of satellites. We are also involved in the human aspects of space exploration. Canadian astronauts have travelled in space and are in training for future space missions.

ACTIVITY 2A Questions about Space Exploration

You may already know some facts about space exploration and spacecraft. Perhaps you have thought of some advantages and disadvantages of space exploration. This activity will help you compare your current ideas with those you have after studying this unit.

In your learning journal, divide two pages into three columns. Title the columns "Questions about Space Exploration," "Predictions before Starting Chapter 2," and "Answers after Completing Chapter 2." In the first column, make a list of questions that you have about space technology and exploration. Your teacher may ask you to work together with one or more other students. In your group, make a list of questions on space technology and exploration using everyone's ideas. Record these questions in column 1 of your learning journal, leaving out those that are the same as your own. In the second column, write your prediction of what the answer will be for each question. You will complete the third column after studying the unit.

2.1 Exploring Space

A basic characteristic of humans is the need to explore. Explorers were the first people to cross the oceans, the first to climb the Earth's highest mountains, and the first to walk on the moon. After the explorers achieved their goals, other people followed them, often with greater ease and safety.

The biggest single frontier left to explore is **space**, which is the region above the Earth's atmosphere. The atmosphere consists of air and everything it holds, such as water vapour. The atmosphere extends upward from the Earth's surface, gradually becoming less dense until it finally becomes almost nothing at an altitude of about 150 km.

EXTENSION

List events and possible dates of space events that you think might occur in the next 50 years.

COMPARING AIRCRAFT AND SPACECRAFT

Can an aircraft travel in space? Can a spacecraft travel in the Earth's atmosphere with no engines operating, as it can in space? The answer to both questions is no. To find out why, let us look at the characteristics of each type of craft.

An **aircraft** is a vehicle that travels through the air; examples include jet airplanes, propeller airplanes, and helicopters. The engines of these vehicles must operate continuously to keep them above the ground. The engines use oxygen in the air to burn the fuel they need to travel. Most aircraft do not travel higher than about 20 km above the surface of the Earth. If they go too high, there is not enough oxygen to support the burning of the fuel.

DID YOU KNOW?

The altitude record for a jet aircraft is about 38 km, and for a helicopter it is about 12 km.

A **spacecraft** is a vehicle designed to travel in space, usually 200 km or more above the Earth's surface. Once the spacecraft rises above the atmosphere and is travelling fast enough, it no longer needs to operate its engine full time in order to travel around the Earth. The spacecraft must be above the atmosphere; otherwise air resistance would slow it down. Although many spacecraft travel in paths (called **orbits**) around the Earth, some go to the moon or to other parts of the solar system (Figure 2.1).

DID YOU KNOW?

Most aircraft travel at speeds lower than 1000 km/h. Many spacecraft travel at speeds of about 30 000 km/h.

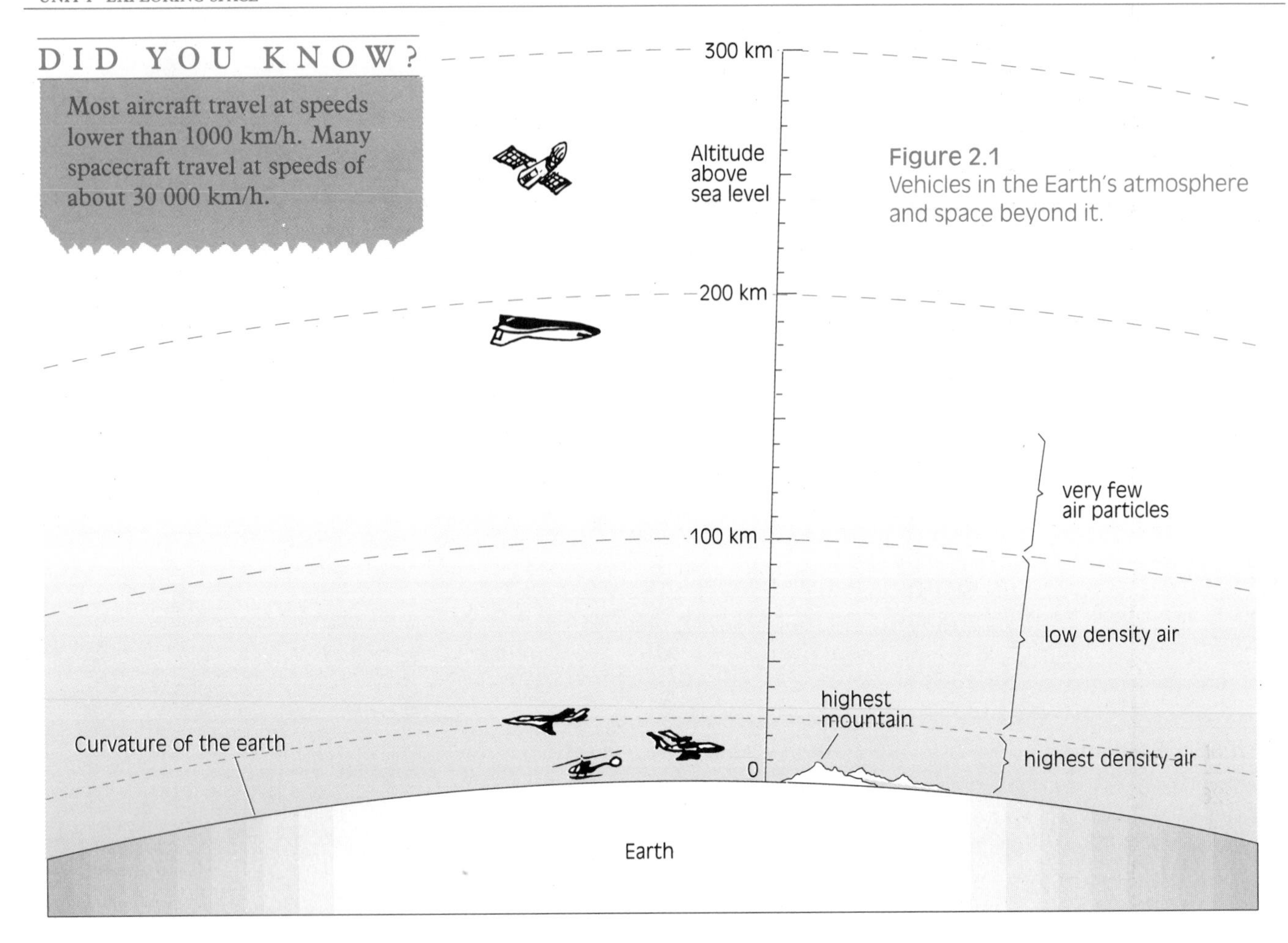

Figure 2.1
Vehicles in the Earth's atmosphere and space beyond it.

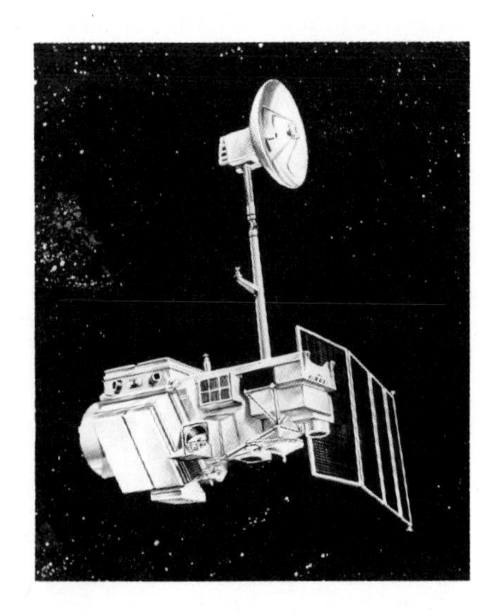

Figure 2.2
This satellite, called *Landsat D*, is one type of unpiloted spacecraft.

UNPILOTED AND PILOTED SPACECRAFT

Spacecraft can be piloted or unpiloted: an **unpiloted spacecraft** has no people on board. Satellites are the most common type of unpiloted spacecraft. A **satellite** is an object that travels in an orbit around another object; for example, the moon is a natural satellite of the Earth. The first artificial satellite was launched into space in 1957 by the Soviet Union. Since then, hundreds of satellites have been sent into space, mostly in orbits around the Earth (Figure 2.2). Another important type of unpiloted spacecraft, called a **space probe,** is one with many instruments on board that is sent to discover more about planets, comets, the sun, and other parts of the solar system.

The first **piloted spacecraft** (that is, one with people on board) was sent into space in 1961 by the Soviet Union. Piloted craft are not sent into space as often as unpiloted ones.

One type of piloted spacecraft is a **space station**, a large spacecraft in which humans live and work. It is designed to stay in space for many years. Figure 2.3 shows an example of another piloted spacecraft, the space shuttle built by the United States.

Figure 2.3
The U.S. space shuttle is an example of a piloted spacecraft.

HISTORICAL OVERVIEW OF SPACE EXPLORATION

Table 2.1 summarizes many of the main events of the history of space exploration.

Table 2.1 Highlights in the history of space exploration

Year	Event
1926	First liquid-fuelled rocket is tested by Robert Goddard (Figure 2.4).
1957	First artificial satellite, *Sputnik 1*, is launched by the Soviet Union.
1961	Yuri Gagarin, a Soviet cosmonaut, is first human to orbit the Earth (cosmonaut is the Soviet term for astronaut).
1962	Canada's first satellite, *Alouette 1*, is launched.
1963	Valentina Tereshkova, a cosmonaut, is first woman to travel in space.
1966	First probe lands on the moon without crashing.
1969	Neil Armstrong, an American astronaut, is first human to walk on the moon (Figure 2.5).
1971	First Soviet space station, *Salyut 1*, is placed in orbit.
1972	First Canadian communications satellite, *Anik A-1*, is placed in orbit.
1973	First American space station, *Skylab*, is placed in orbit.
1977	Space probes *Voyager 1* and *Voyager 2* are launched to explore the outer solar system.
1981	First space shuttle is launched.
1984	First retrieval and repair of a satellite is accomplished using the Remote Manipulator System called Canadarm; first Canadian astronaut, Marc Garneau, travels in space.
1986	Soviet space station *Mir* is launched.
1988	Soviet cosmonaut is first human to remain in space for one year.
1990	*Hubble Space Telescope* is placed in orbit.

Figure 2.4
Robert Goddard is seen ready to launch the world's first liquid-fuelled rocket in 1926.

Figure 2.5
This footprint was placed on the moon over two decades ago. Would it still be there today?

REVIEW 2.1

1. In the phrase "space exploration," what does the word "space" mean?
2. Why do humans want to explore space?
3. Describe the main differences between an aircraft and a spacecraft.
4. List two examples of
   (a) unpiloted spacecraft,
   (b) piloted spacecraft.
5. Make an illustrated time-line of the major events in space exploration using Table 2.1 and reference books. (Reference books and encyclopedias will help you design interesting illustrations.)

2.2 Getting Into Space

What happens if you try to throw a ball straight upward? It soon falls back down, of course, pulled by gravity. Large bodies, such as the Earth, exert a strong force of gravity on nearby objects. A vehicle can be launched into space only by overcoming this strong force of gravity. This job is done by rocket engines.

DID YOU KNOW?

The force of gravity on an astronaut on the moon is only 1/6 as much as the force of gravity on the same astronaut on Earth.

PAYLOADS AND LAUNCHERS

If you were to throw a ball to a friend, your arm could be called the launcher, and the ball, the payload. In space exploration, a **payload** is any satellite, piloted spacecraft, or cargo launched into space. The **launcher** is the device that carries a payload into space. Its main part is a rocket engine.

A toy balloon is a simple model that helps explain how a rocket engine works. When an inflated balloon is released, it flies quickly and uncontrollably around the room. As the air under pressure inside the balloon escapes from the open neck, the balloon is forced in the opposite direction. The force that causes an object to move is called **thrust.** In the case of the balloon, if the air is escaping to the left, the thrust on the balloon is to the right (Figure 2.6).

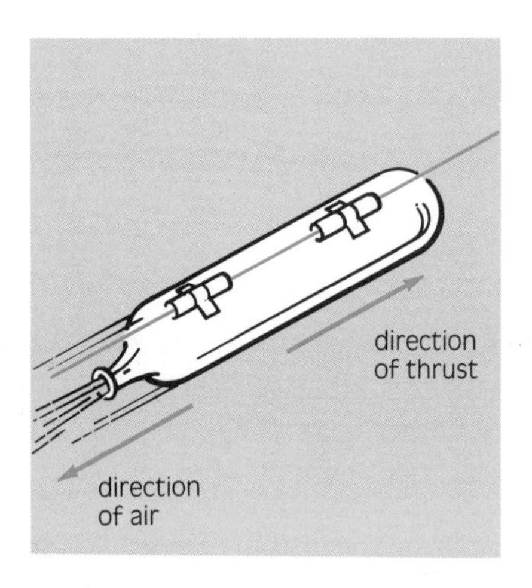

Figure 2.6
Using a balloon to demonstrate thrust. Fishline is fed through two pieces of a drinking straw, which are in turn taped to a blown-up balloon. When the balloon is released, air leaves the open neck rapidly in one direction, creating a thrust on the balloon in the opposite direction.

Like a balloon, a rocket engine is enclosed except at the nozzle end. Unlike a balloon, a rocket engine operates through the action of two chemicals. One is the fuel and the other is an oxygen-containing chemical that causes the fuel to burn quickly. The burning of the fuel produces large amounts of hot, expanding gases. These exhaust gases leave rapidly through the engine's nozzle. The exhaust gases (mostly water vapour) travelling rapidly in one direction cause the thrust on the rocket in the opposite direction (Figure 2.7).

The launch of a common spacecraft, the space shuttle (Figure 2.8), shows how rocket engines are used in space travel. The shuttle has three main rocket engines. Attached to the shuttle is the huge external tank that stores the chemicals used by these engines during launching. Also attached are two large booster rockets, used only during the first stage of the launch.

When a shuttle is launched, the two booster rockets assist the main engines for about two minutes, until their fuel is used up. The booster rockets then separate from the shuttle and parachute down to the ocean, where they are recovered to be used again. About six minutes later, just before the shuttle reaches the final orbit, the large external tank, now empty, separates from the shuttle and falls back to Earth. The tank breaks into pieces that are later recovered but not used again. At this stage, the speed of the shuttle is about 28 000 km/h or about 8 km/s. At that speed, the shuttle would take only about 12 minutes to travel all the way across Canada. This high speed is needed to put the shuttle into its orbit around the Earth and keep it there (Figure 2.9).

Once the external tank leaves the shuttle, the shuttle's three main engines no longer operate. To get the shuttle into its final orbit, smaller engines, called orbital maneuvering engines, are used.

Figure 2.7
Basic design of a liquid-fuelled rocket engine. The fuel and an oxygen-containing chemical are pumped into the combustion chamber where the fuel burns rapidly, producing gases under high pressure. The gas molecules speed up as they escape through the nozzle neck; then they speed up even faster when they get to the nozzle mouth. These escaping gas molecules exert an upward thrust on the rocket engine.

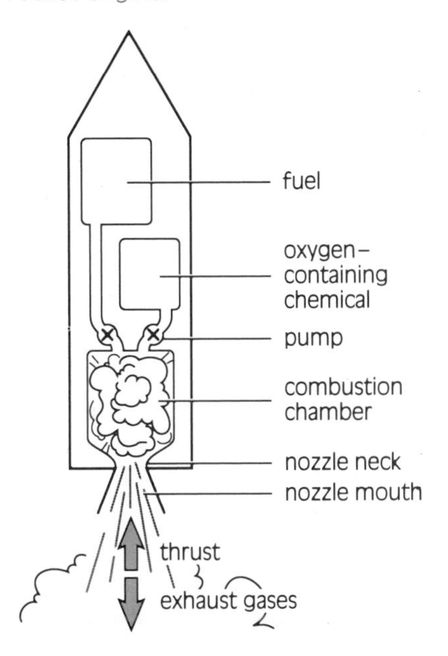

Figure 2.8
Launch of the space shuttle *Discovery*.

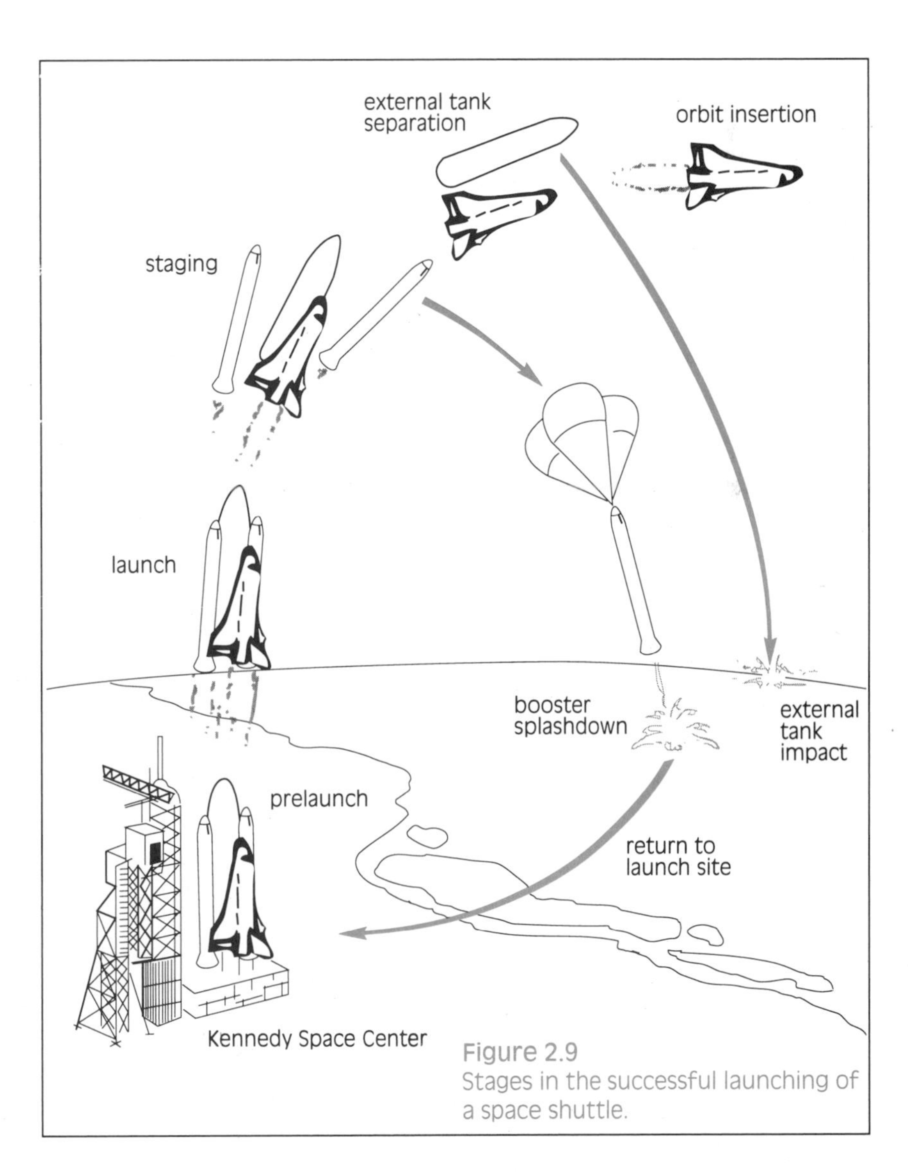

Figure 2.9
Stages in the successful launching of a space shuttle.

ACTIVITY 2B *Using a Toy Water Rocket*

A toy water rocket is a safe and easy way to study how rockets work. Although it is powered only by water, it works on the same principles as a rocket powered by chemical fuels.

MATERIALS

toy water-rocket kit
clean water
safety goggles

PROCEDURE

CAUTION!
Operate the rocket outdoors only and do not aim it at people or buildings.

1. Examine the water rocket and describe its features, including the size and shape of the nozzle. (A diagram will help.)
2. Put on your safety goggles.
3. Hold the rocket upside down and insert the funnel from the kit. Add water up to the level of the maximum fuel line.
4. Attach the pump rod to the rocket and lock them together using the lock ring.
5. Turn the rocket upright and pump air into it. Count the number of strokes up to a maximum of 20.
6. Aim the rocket straight upward and release the lock ring. Try to determine the maximum height reached by the rocket and record it.
7. Launch the rocket two or three more times, changing the amount of water each time, but keeping the number of strokes of the pump at 20. Try to determine the maximum height reached by the rocket with each launch and record it.
8. Launch the rocket two or three more times, keeping the amount of water the same, but changing the number of strokes of the pump used each time. Try to determine the maximum height reached by the rocket with each launch and record it.

DISCUSSION

1. How is thrust obtained in a water rocket?
2. Explain the factors that affect the thrust.
3. Compare the operations of a water rocket and a rocket fuelled by chemicals. Include both similarities and differences.

REVIEW 2.2

1. (a) Name the force that must be overcome in order to launch a vehicle into space.
   (b) Why is this force greater on the Earth than on the moon?
2. Classify each of the following as a payload or a launcher:
   (a) a weather satellite
   (b) a rocket
   (c) an astronaut
3. Name the force that is exerted by the exhaust gases on a rocket engine as they leave the engine's nozzle. Explain how this force is created.
4. A firecracker at a fireworks display is launched by applying the same principle used in launching real rockets. Using a diagram, explain how a firecracker is launched vertically upward.
5. Space scientists suggest that, in the future, space probes sent to distant planets such as Saturn and Jupiter could be launched from the moon. Describe the advantages of this suggestion.
6. When the space shuttle returns to Earth, it encounters tremendous heat caused by air resistance in the Earth's atmosphere. To reduce this heat, the shuttle must slow down. Describe how the shuttle's rocket engines are used to slow the shuttle.
7. What will happen to a spacecraft in orbit around the Earth if its speed becomes too slow? Explain.

2.3 Earth-Orbit Satellites

On a clear evening, just after the sun sets, you may observe very small shiny objects travelling quickly across the sky. These are satellites moving in orbits around the Earth. Earth-orbit satellites have many uses, some of which affect your daily life.

SATELLITES IN LOW EARTH ORBIT

In 1962, Canada became the third country in the world to send a satellite into space. This satellite, *Alouette 1*, was used to study particles in the Earth's upper atmosphere. Thus, its function was similar to that of many satellites: scientific research.

Alouette 1 was an example of a satellite in **low Earth orbit,** an orbit just above the Earth's atmosphere, at an altitude of about 200 km to 1000 km. These orbits may be around the equator, around the poles, or at any angle between the equator and the poles (Figure 2.10). At speeds of about 28 000 km/h, such satellites take only about 1.5 h to travel once around the Earth.

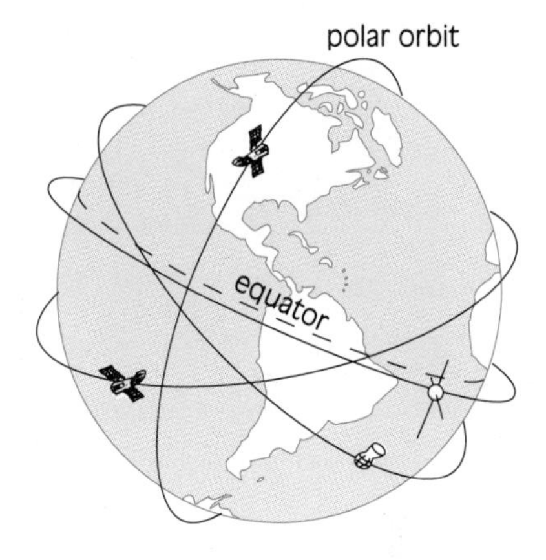

Figure 2.10
Examples of low Earth orbits.

Besides carrying out scientific research, the main function of low Earth-orbit satellites is **remote sensing**. This means making observations from a distance using imaging devices. One type of image is a photograph taken with a camera. Other types of images are created from infrared (heat) waves, and other invisible energy waves. The satellites detect these waves by using special cameras or antennas similar to television and radio antennas. The information gathered by the satellites is beamed to Earth using radio waves (Figure 2.11).

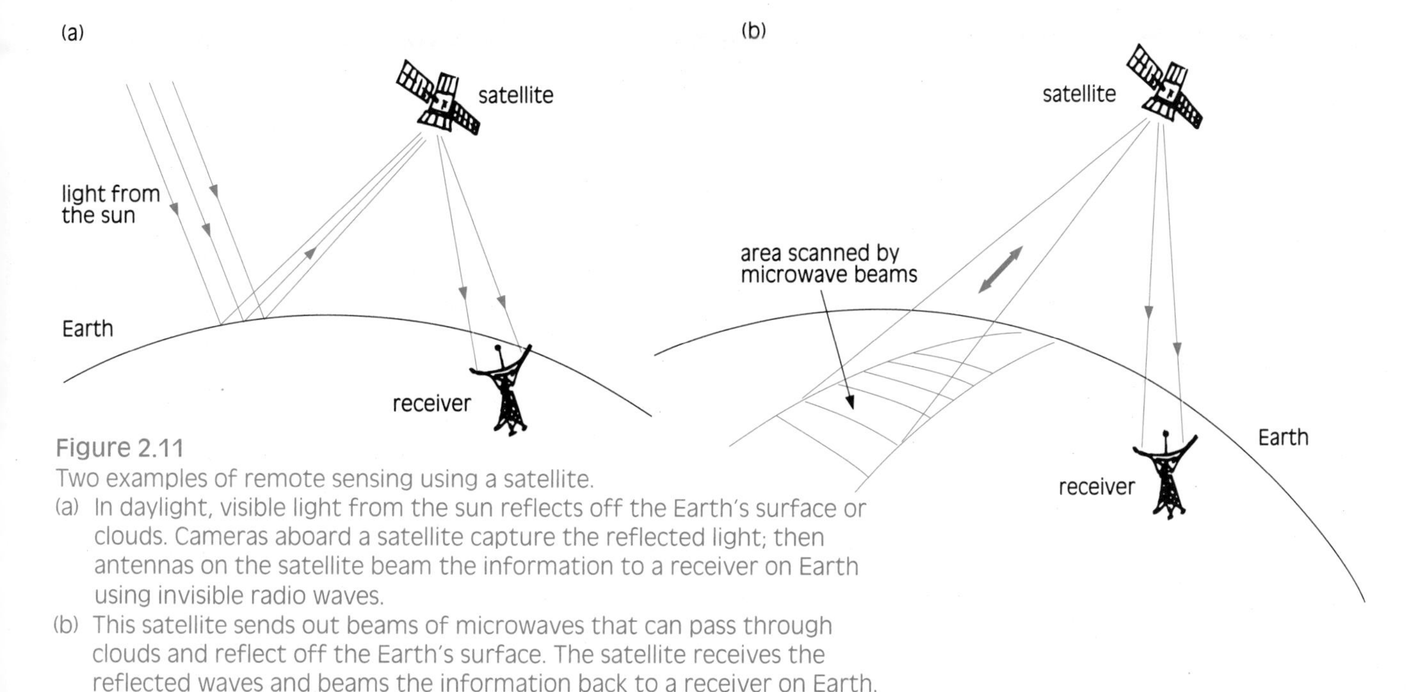

Figure 2.11
Two examples of remote sensing using a satellite.
(a) In daylight, visible light from the sun reflects off the Earth's surface or clouds. Cameras aboard a satellite capture the reflected light; then antennas on the satellite beam the information to a receiver on Earth using invisible radio waves.
(b) This satellite sends out beams of microwaves that can pass through clouds and reflect off the Earth's surface. The satellite receives the reflected waves and beams the information back to a receiver on Earth.

Weather forecasting relies heavily on remote sensing using photographs of cloud patterns around the world. These pictures help predict what the weather will be like in local areas. More importantly, however, they help save lives by providing weather forecasters with information on approaching storms, such as hurricanes (Figure 2.12).

Figure 2.12
Whirling storm clouds are obvious in this image provided by a weather satellite.

Remote sensing is also used by the military. For example, some nations have spy satellites that obtain images of other nations' activities on the Earth's surface.

Scientists also use images from satellites to study water resources, food crop management, forests, insect damage, pollution, and fault lines on the Earth's surface.

A great advance in remote sensing is the use of images created by reflecting radio waves and microwaves off objects on the Earth's surface. A Canadian satellite, *Radarsat*, will be able to do this beginning in the mid-1990s. (Radar uses radio waves to detect the features of objects.) Radio-wave images of the Earth's surface have advantages over photographs and infrared images because they can be taken at night and through clouds as well as on clear days.

DID YOU KNOW?

The cameras aboard some spy satellites are so sensitive that they are able to read letters and numbers about the size of a person's hand from hundreds of kilometres away.

ACTIVITY 2C Looking at a Satellite Image

Satellite images show you the Earth in a new way. In this activity, you will look at a photograph created by a computer that combined different types of images received from a satellite. At first, such a photograph is more difficult to understand than an ordinary photograph. Practice helps.

MATERIALS

Figure 2.13: Satellite image of Vancouver and its surroundings

PROCEDURE

Look at the image carefully and describe what you see in it. Consider such things as parks, rivers, the ocean, built-up areas, forests, and airports.

DISCUSSION

1. How is this image different from an ordinary photograph?
2. Describe how the image might have looked if it had been taken 100 years ago.
3. Describe how an image of this area taken 100 years in the future may look. (Consider changes to the delta caused by the river as well as changes caused by humans.)
4. How would each of the following experts use this satellite image of Vancouver and its surroundings:
   a) an environmental expert for the large park (Stanley Park)?
   b) a government official in charge of airplane traffic at the international airport?
   c) an engineer in charge of building and maintaining bridges?
   d) a housing developer?
   e) an official of a shipping company?

Figure 2.13
Satellite image of Vancouver and its surroundings.

SATELLITES IN GEOSYNCHRONOUS ORBIT

Figure 2.14
A typical satellite dish for home use.

Our lives would be very different if we did not have telephones, televisions, and radios, and if we could not transmit computer data. These products of the telecommunications industry help us communicate over small and large distances. (The prefix "tele" is from the Greek word *tele*, meaning far.)

Many of the features of modern telecommunications would be impossible without satellites, but satellites in low Earth orbits have limited use as communications satellites. To discover why, imagine a television satellite dish, such as the one shown in Figure 2.14, trying to receive signals from such a satellite. The dish would have to follow the satellite as it moved quickly across the sky. Then suddenly the satellite would go below the horizon, and all signals from the satellite would be lost. In order for the satellite dish to receive signals constantly, the satellite would have to remain in the same location above the Earth's surface.

The orbit of such a satellite is called a **geosynchronous orbit**. "Geo" means Earth and "synchronous" means taking place at the same rate. Since it takes the Earth 24 h to turn once on its own axis, a satellite in geosynchronous orbit must take 24 h to orbit the Earth. The easiest place to control such an orbit is directly above the Earth's equator, at an altitude of 36 000 km above sea level. To keep its position at this altitude, the satellite must travel at a speed of 11 060 km/h. (See Figure 2.15.)

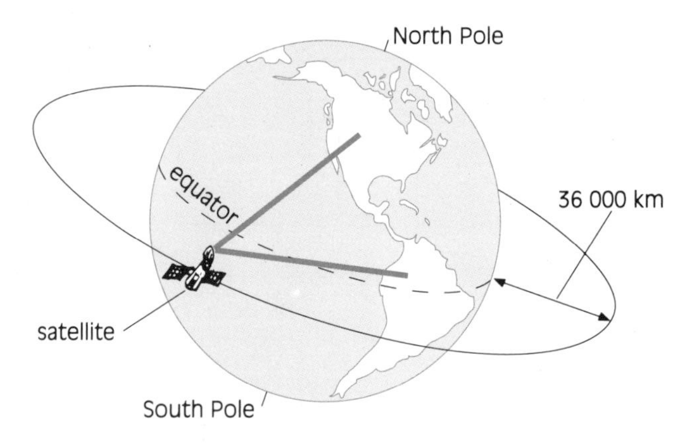

Figure 2.15
A geosynchronous orbit lies just above the equator. A satellite there receives signals from Earth-based transmitters and sends signals back to Earth-based receivers.

Canada has become a leader in telecommunications because of its huge size and the remote location of many communities. The first satellite for domestic communications put into geosynchronous orbit was Canada's *Anik 1*, launched in 1972. The most recent technological advances in the *Anik* series include a satellite that sends and receives signals from mobile telephones in cars, trucks, and ships. Similar satellites provide radio service to remote areas of Canada.

Satellites in geosynchronous orbit also assist in search and rescue operations. When a ship or an aircraft is lost or involved in an accident, a device on board sends signals that a satellite can receive and send back to Earth. These signals tell the rescuers where the ship or aircraft is so they can try to get to that location.

EXTENSION

Many resources are available to help you discover more about Canada's use of satellites for telecommunications. Do research on one aspect of satellite telecommunications, and prepare a report on it. Possible topics include the history of Canadian satellites, related jobs and industries, applications, geosynchronous orbits, polar orbits, economics, and new technologies.

ORBITING ASTRONOMICAL OBSERVATORIES

To astronomers looking out into space from the Earth's surface, the atmosphere acts like a fog. It reduces their ability to see the stars, galaxies, and other parts of the universe clearly. Astronomers have known for a long time that they could improve their understanding of the universe if they could observe it from above the atmosphere. Telescopes aboard orbiting satellites now enable them to take observations from space. Astronomers use computers to interpret the thousands of images that have been obtained this way.

EXTENSION

The most famous orbiting observatory as of the writing of this book is called the *Hubble Space Telescope*. When it was put into orbit in 1990, it was expected to provide visible images of the stars 10 times as clear as those provided by equal-sized telescopes on Earth. Scientists soon discovered, however, that the main mirror was flawed. In reference materials, read the story of the *Hubble Space Telescope*, including how scientists tried to find ways of correcting the flaw.

REVIEW 2.3

1. What are the features of a low Earth orbit?
2. Name four uses of satellites that travel in low Earth orbits.
3. (a) What is a geosynchronous orbit?
   (b) What type(s) of satellite must use this orbit? Explain why.
4. How has Canada contributed to the use of Earth-orbit satellites?
5. Will Canada's newest contribution to remote-sensing satellites, *Radarsat*, be useful for weather forecasting? Why or why not?
6. What is the advantage of having an observatory in space?

2.4 Space Probes and Space Stations

A space probe is a payload sent to explore distant parts of the solar system. The probe carries instruments that collect information (including photographs) and transmit it back to Earth. The first space probes, launched in 1959, went to the moon where they took close-up photographs. They showed us the first views ever seen of the far side of the moon. Since then, probes have been sent to every planet in the solar system except Pluto, as well as to the sun and even to a few comets. Although nine flights to the moon between 1968 and 1972 were piloted (including six in which astronauts landed on the moon's surface), all space probes since then have been unpiloted.

An example of a successful probe is the American *Voyager 2* craft (Figure 2.16), launched in 1977 to study four of the five outer planets in the solar system: Jupiter, Saturn, Uranus, and Neptune. The timing of the launch allowed the craft to travel from one outer planet to the next in the least possible time because the planets were aligned as closely as possible, an event that occurs only once every 176 years!

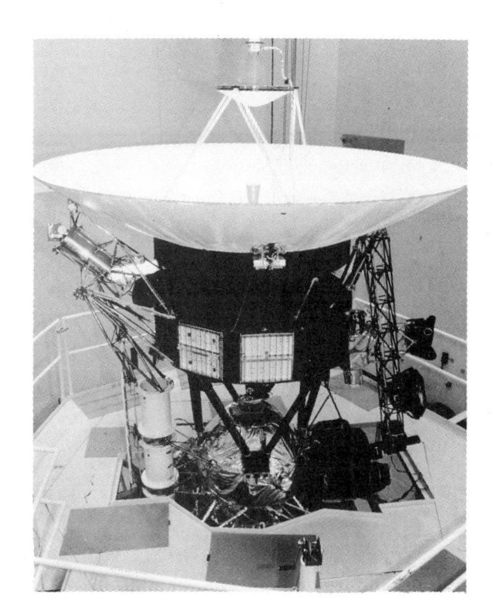

Figure 2.16
The *Voyager 2* space probe.

The *Voyager 2* craft carried several imaging devices, instruments to detect particles and electric and magnetic forces, antennas to send and receive signals, and small engines needed to correct the path of the craft from time to time. Throughout its years of travel, *Voyager 2* sent back clear images of the planets and their moons, some of which are shown in Figure 2.17, along with detailed information on their atmospheres.

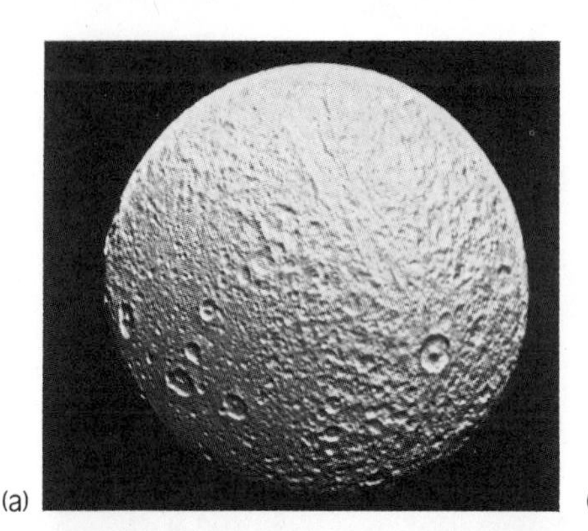

(a)

(b)

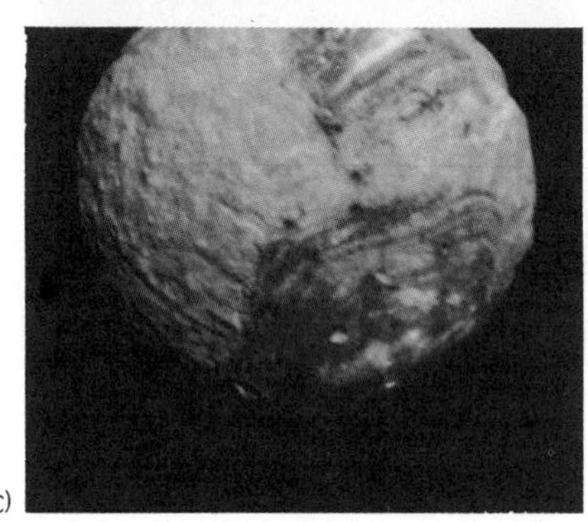

(c)

Figure 2.17
Images provided by the *Voyager* space probes.
(a) Tethys, a moon of Saturn
(b) Saturn and its rings
(c) Miranda, a moon of Uranus.

It has given scientists a huge quantity of material that will take years to analyse. By the time *Voyager 2* passed its last viewing spot, the planet Neptune, it had travelled for 12 years, covering a distance of 4.5 billion kilometres.

Space probes such as *Voyager 2* have revealed a great deal of new information about our solar system. This information helps us better understand the origins and characteristics of the universe, the solar system, and the Earth. For example, learning about the greenhouse effect in the thick atmosphere of Venus will help scientists understand the greenhouse effect on Earth. Information from space probes will also help space scientists plan when and how to mine minerals on other planets and moons. Table 2.2 lists some space probes that have been or soon will be carried into space.

Table 2.2 Important space probes

Name of Probe	Points of Interest
Giotto	Built by European countries; came within 1400 km of Halley's comet in 1986
Galileo	Will explore Jupiter and its moons; most sophisticated probe yet built; to be launched in the early 1990s
Magellan	Will be sent to Venus to map its surface using radar waves; to be launched in the early 1990s
Observer	Will be sent to Mars to take more detailed photographs than ever before; will help plan future exploration of Mars; to be launched in the early 1990s
Ulysses	First probe intended to look closely at the poles of the sun; joint European and American mission launched in 1990
Cassini	Will study an asteroid, then Jupiter, then one of Saturn's moons; to be launched in 1996

SPACE STATIONS

Long before humans can plan piloted trips to planets such as Mars or build colonies on other planets or our moon, more has to be learned about how the human body can survive in space. Much of the research on living in space can be carried out in space stations, where people can live for months at a time. The only space stations that have been used or are planned are those that orbit the Earth. Some of their uses include scientific research and experimentation, materials research, manufacturing large, pure crystals, and retrieving and repairing satellites. Some applications of the research include better medicines, better crops, and better liquid fuels. Eventually, interplanetary probes could be launched from a space station in orbit either around the Earth or around the moon. Such a launch would require much less energy than launching the probe from the Earth's surface, because the force of gravity is much less on a space station than it is on Earth.

The first space station, *Salyut 1*, was put into low Earth orbit by the Soviet Union in 1971. Six more *Salyut* stations followed, but none is in use any longer. A more recent Soviet space station, launched in 1986, is called *Mir* (meaning "peace"). Using their space stations, the Soviets have learned a great deal about how space travel affects humans. Many cosmonauts have had long stays in space, some as long as a year.

In 1973, the United States placed its first space station, called *Skylab*, in orbit. Astronauts went up four times to perform experiments during *Skylab's* short lifespan (less than one year).

Major plans are under way by several nations to build and use an orbiting space station called *Freedom*. The European Space Agency, the United States, Japan, and Canada are cooperating in this large venture.

Canada's contribution to the space station is important. Canada has a reputation for building sophisticated robots, such as the Remote Manipulator System called Canadarm, used on the U.S. space shuttle (Figure 2.18). The technology of the Canadarm is being extended to develop a stronger manipulator arm (able to move masses up to 100 000 kg) and two smaller arms, called end effectors, to perform more sensitive tasks. These devices, together with servicing tools, would be placed on a sliding platform on the space station's main frame. The entire Canadian construction, called the Mobile Servicing Centre (MSC), would be used by visiting astronauts to construct the space station.

EXTENSION

The Canadian Solar Sail Project is Canada's first attempt to send a craft beyond an Earth orbit. In 1992, the project will enter a race to Mars called the Columbus 500 Space Sail Cup, in honour of the 500th anniversary of Christopher Columbus's first voyage from Europe to America in 1492. The project involves building a craft that will be launched into Earth orbit using a rocket. From that orbit, the craft will sail to Mars powered only by particles streaming out from the sun. Look for newspaper and magazine articles about the project. Prepare a report describing how the craft will travel to Mars.

Figure 2.18
The Canadarm, officially called the Remote Manipulator System, is a robotic manipulator with many uses. In this photograph, it is supporting an astronaut who is performing an experiment on constructing a truss in space, a process that will be applied when the space station *Freedom* is built. The Canadarm is also used to move cargo and to help deploy and retrieve satellites. The arm is 15 m long, and is controlled by an astronaut inside the shuttle. It has a shoulder, an elbow, and a wrist, similar to a human arm. Small cameras follow the arm's movements. On Earth the arm cannot even support its own mass of 364 kg, but in space it can manipulate masses up to 30 000 kg.

After helping in the construction of the station, the MSC would assist in maintaining the station, moving supplies and equipment into and out of the station, docking visiting shuttles, sending satellites out from the station and retrieving them, and assisting astronauts during activities outside the station.

REVIEW 2.4

1. Describe the uses or functions of space probes.
2. Name 10 bodies in the solar system that have been explored by space probes.
3. Explain how the information from *Voyager 2* helps scientists to understand our solar system.
4. Why will space probes eventually be launched from space stations rather than from Earth?
5. The space probe *Ulysses*, expected to pass over the poles of the sun in 1994 and 1995, developed a slow wobble shortly after it was launched in October 1990. What are some problems that might result if such a wobble could not be corrected?
6. Describe what Canada's role will be in the proposed space station *Freedom*.

ACTIVITY 2D *Thinking about Space*

The following section describes what it would be like to live and work on a space station. As you read Section 2.5, think about the kinds of work that astronauts do and the way a space station is designed for astronauts to be able to survive in space. In your learning journal, write down the things you did not know about living and working in space before you read this section.

2.5 Living and Working in Space

Whenever an American space shuttle is launched, the people aboard stay in space for about seven days. In the future, when the multinational space station *Freedom* is operating, crew members will stay in space for several weeks or months.

Imagine that you are one of Canada's astronauts about to be taken in a space shuttle to the space station. As you wait in the shuttle, you think back to the day you applied to become a crew member of a space mission. There were three types of positions available:

- Pilot astronauts who control and command the spacecraft
- Mission specialist astronauts who co-ordinate the flight operation, including crew activities, scientific experiments, and payload operations
- Payload specialist astronauts who are scientists, engineers, or technicians with skills needed to do tasks for a specific mission

On this flight, you are a payload specialist. Your job is to replace the fuel tank and replace a damaged part of a Canadian communications

EXTENSION

Look for information on recent progress (or problems) in the development of the space station *Freedom*. Has the original design been altered? If so, why? Is the Canadian Mobile Servicing Centre still an important component? What else can you discover?

satellite. As the shuttle blasts off from the launch pad, the sound is tremendous, and you are pushed back in your seat with a force that makes you feel three times as heavy as normal. Within two minutes, you feel a jolt as the solid rocket boosters separate from the shuttle. About six minutes later, you feel another jolt as the main fuel tank separates. Then you feel as if you are floating; only the seat-belt keeps you in your seat.

The sensation of floating in space has led to the terms "weightlessness" and "zero gravity." These terms give the impression that there is no gravity acting on an astronaut in space. This is false because gravity is needed to keep a spacecraft in an Earth orbit. The weightless feeling occurs because the spacecraft and its contents are continuously falling toward the Earth, as they follow the Earth's curvature at some altitude above the Earth's surface. In the rest of this chapter, we will use the term **microgravity** to describe the situation where objects in orbit are falling toward the Earth and thus seem weightless (Figure 2.19).

Figure 2.19
Pilot Robert Crippen appears to be weightless. Actually he is taking advantage of microgravity to perform some acrobatics.

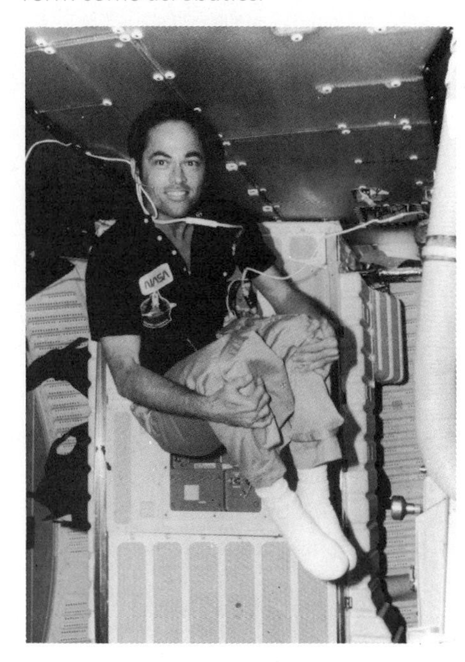

Soon the shuttle approaches the station and docks. As you enter, you are greeted by other crew members who have been on the space station for several months. With them, you eat your first space meal, part of which is prepared by adding water to a mixture in a plastic pouch (Figure 2.20). The other part of the meal consists of fresh vegetables grown on the station as an experiment to help absorb carbon dioxide from the air and provide oxygen for the crew. Water, both hot and cold, is available as a by-product of the fuel cells that produce some of the energy aboard the station.

EXTENSION

Do research in the library to learn about the construction and operation of a fuel cell.

You do your share of work to help keep the conditions on the station clean. Sanitation is important in space because micro-organisms can grow easily in conditions of microgravity and an enclosed system like a spacecraft. The eating equipment, dining area, toilet, and sleeping facilities must be cleaned regularly. Garbage and worn clothing are sealed in airtight bags to be returned to Earth later.

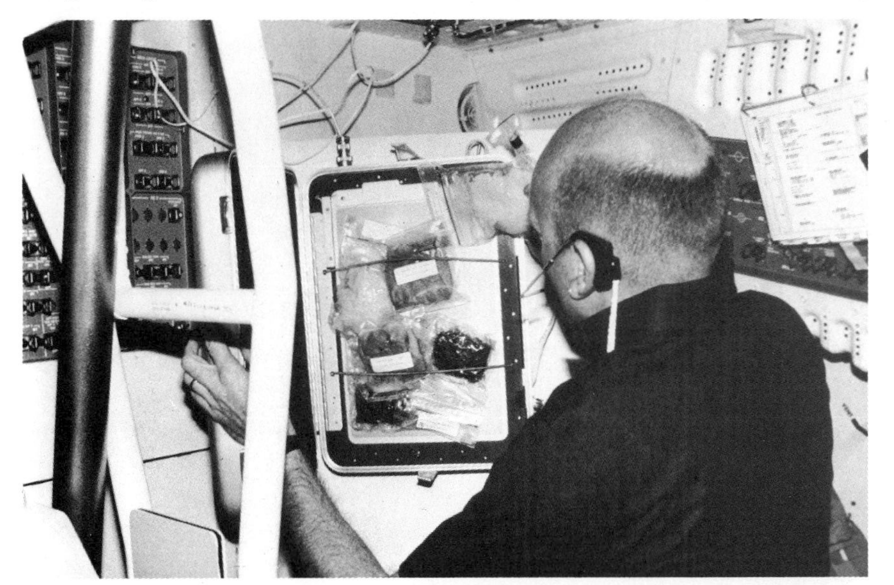

Figure 2.20
Food in sealed plastic bags can be heated in a specially designed food warmer.

After working at a computer terminal for some time, you talk to your family on Earth and spend a long time exercising on the specially designed treadmills, cycles, and rowing equipment (Figure 2.21a). You must exercise regularly to keep your muscles, bones, and blood from deteriorating in microgravity conditions. Then, after another meal and some relaxation, you fasten yourself into the sleeping hammock tied to one wall (Figure 2.21b).

You spend much of the next day working at computers and practising the use of the manipulator arms. You are not allowed to leave the station to carry out your main tasks until you have been in space at least three days. This time is needed to reduce the chances of motion sickness.

On the fourth day, you put on a space-walking suit (Figure 2.22) to protect you from the vacuum of space and the sun's harmful radiation. You go outside the station and attach yourself to the Canadian Mobile Servicing Centre. Using the manipulator arm, you grab the satellite that has moved close to the station and pull it close to you. You then replace the satellite's used fuel tank with a fresh one, and use the small manipulator arms to replace the damaged part. After testing the satellite, you use the large manipulator arm to send the satellite away from the space station. Earth-based controllers then ignite the satellite's engines long enough to get it back into its proper orbit.

On the following days, you perform other required tasks as well as various experiments. When your mission is complete, you board the shuttle for your trip back to Earth. When the shuttle lands, you step out and try walking under the influence of gravity. You feel very heavy at first, but soon you get used to being back on Earth.

Figure 2.21
(a) Astronaut Guion Bluford runs on a treadmill while wired to monitors that record his body's reaction to exercise.
(b) Astronaut Anna Fisher is seen in a sleeping hammock attached to a wall.

(a)

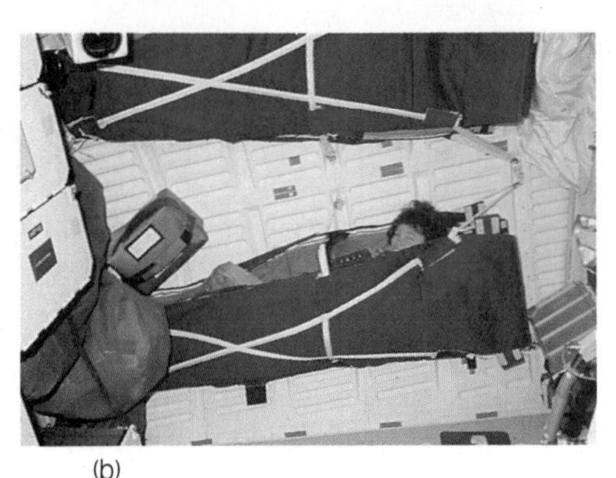

(b)

Figure 2.22
The space-walking suit is equipped with its own maneuvering unit strapped to the astronaut's back. The maneuvering unit acts like a small rocket engine, although it uses compressed nitrogen gas for propulsion rather than any dangerous fuel.

Profile

DR. ROBERTA BONDAR

If you ever experience the thrill of hearing Roberta Bondar speak about being an astronaut, you will understand why all Canadians should be proud of her. As one of Canada's six astronauts, she will travel as a payload specialist on the U.S space shuttle.

Even as a little girl, Bondar dreamed of being an astronaut. When she was a teenager, she built rocket and spacecraft models and covered her bedroom walls with space posters. Through an aunt who worked at the National Aeronautics and Space Administration (NASA) in the United States, she was able to collect the sew-on badges from all the U.S. space missions. Bondar says she never watched a NASA launch without wanting to be on board.

Bondar went to elementary and secondary school in Sault Ste. Marie, Ontario. She loved high school and got good marks, but she was not always at the top of her class. She was too busy with other activities, from sports to running the science club.

In university, Bondar earned a bachelor of science degree in zoology and agriculture, followed by a master of science degree in pathology, which is the study of the nature and causes of diseases. She went on to obtain a doctor's degree in neurobiology, the study of the nervous system. Later, she earned a doctor of medicine degree. She has the most

degrees of all the Canadian astronauts.

As soon as Bondar saw the advertisement for astronauts in 1983, she applied. Six months later, she was chosen as one of Canada's six astronauts. To gain this honour, she had competed with 4300 other people. She and the other astronauts were chosen because of their academic backgrounds, technical knowledge, and physical fitness.

When she became an astronaut, Bondar began training programs to prepare her for living and working in space. One of the many experiences she recalls vividly was the training to adapt to microgravity. A special aircraft climbs to a high altitude, then aims toward the ground. Everyone inside feels weightless for less than a minute. Then the aircraft curves back up to repeat the motion several times. Bondar described this experience as "like being on the wildest roller coaster you can imagine." Like most people, she suffered motion sickness at the beginning of the training.

In 1990, Bondar was chosen to be a payload specialist on the first mission of the International Microgravity Laboratory. At the time this book was written, she had already spent more than a year in special training for this mission. During the mission, which will be on board the space shuttle, she will conduct more than 40 experiments for 13 countries. In one set of experiments, she will investigate the flow of blood to the brain in microgravity. This research relates to her study on Earth of the flow of blood to the brains of stroke victims. Another set of experiments involves taste sensitivity in microgravity. The results may help in understanding how the nervous system adapts to various situations.

Bondar does not spend all her time being an astronaut. She shares her knowledge by serving on committees and giving speeches across Canada. Her interests include flying aircraft, hot-air ballooning, canoeing, target shooting, fishing, biking, cross-country skiing, and squash. Bondar's advice to young people is to enjoy a variety of interests and keep physically fit. She followed her own advice and achieved her dream.

REVIEW 2.5

1. The amount of time it takes a space shuttle to reach its orbit after launching can be measured in (choose one) minutes/hours/days.
2. For which type of crew position in a space mission would you apply if you were
   (a) a doctor specializing in ear/nose/throat problems?
   (b) a military airplane pilot?
   (c) a chemical engineer?
   Explain your answers.
3. Assume a food expert has asked you what you would like to eat during the first two days you are aboard a space station. What would your answer be?
4. Why is exercise important in space?
5. Most of us will never travel in space. The closest we can come to feeling increased and decreased forces of gravity acting on us is on rides at an amusement park.
   (a) Which rides would cause you to feel much heavier than normal?
   (b) Which rides would cause you to feel nearly weightless for a few seconds?
   Explain your answers.

2.6 The Present Impact and Future Prospects of Space Exploration

How important will space exploration be in the 21st century? People who oppose more space missions argue that the large amounts of money spent on space exploration would be better spent cleaning up the Earth's environment and reducing poverty throughout the world. People who promote space exploration agree that these issues are important, but argue that the money spent on space missions is only a small portion (perhaps 2 per cent or less) of any nation's budget. Furthermore, they point out, all of society benefits from space exploration.

Many benefits, such as telecommunications, come directly from using spacecraft. Other benefits, called **spinoffs**, are the indirect results of space exploration. Examples of space spinoffs include:

- microelectronics (such as digital watches, home computers, pacemakers, and hand-held calculators)
- new materials, such as Velcro (used to hold objects together), Teflon (the non-stick coating), flame-resistant materials (Figure 2.23), metal alloys with a memory (used for dental braces), and hard plastics (used in safety helmets)
- robotics used in industry, mining, forestry, and offshore oil exploration (Figure 2.24)
- reading machines for the blind
- vehicle controllers for the handicapped
- safety devices such as smoke detectors
- water recycling processes
- cheaper and more efficient solar cells and chemical batteries
- convenience foods, such as freeze-dried foods
- anti-nausea medication, in the form of a patch stuck behind the ear to prevent motion sickness (Figure 2.25)

Figure 2.23
A firefighter's safety equipment.

Figure 2.24
Robotic manipulation was first designed for space use, but it has been adapted for the nuclear industry to handle dangerous radioactive materials on Earth.

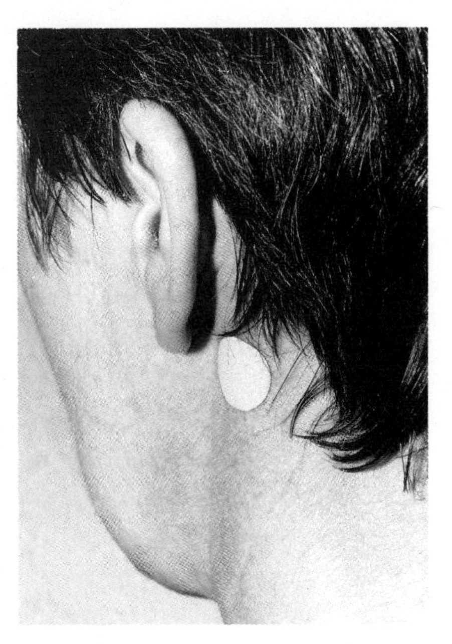

Figure 2.25
The patch placed behind the ear contains a drug (called scopolamine) that helps prevent motion sickness. The patch allows the drug to pass slowly through the skin and into the bloodstream.

OUR FUTURE IN SPACE

If space stations are successful, the next stages of space travel will likely be exploration of the Earth's moon, and the planet Mars and its moon, Phobos. These bodies will be mined and the materials used to build and maintain structures in space. Eventually, human communities may be built on the moon and on Mars.

Many problems must be overcome before humans can travel to, and live on, Mars. One of the biggest problems is the reaction of the human body to microgravity. Motion sickness, weakened muscles, reduced calcium in bones, and loss of blood cells have all been observed in astronauts. Researchers are trying to discover why these reactions occur and how to reduce their effects on long space trips. A feature of any piloted trip to Mars will probably be artificial gravity, created by rotating the spacecraft. On arriving on Mars, the astronauts will find a gravitational force only 38 per cent of that on Earth. This may cause problems for the human body over long periods of time.

Visitors to Mars will require food, oxygen, and warm shelter. Oxygen and small amounts of water could be extracted from the thin atmosphere, which consists mostly of carbon dioxide. Water could also be extracted from the permafrost on Mars and from the recycling of plant and human wastes. Large greenhouses could be used to grow fruits and vegetables. Living quarters would have to be buried beneath the Martian soil to protect the humans from the harmful solar radiation.

Some scientists predict that by the time humans have settled on Mars, we will have propulsion systems powerful enough to allow spacecraft to reach extremely high speeds. Perhaps such a craft, if unpiloted, would be sent to explore neighbouring stars in our galaxy.

REVIEW 2.6

1. What is meant by the term "spinoff"?
2. Which spinoffs listed in this section are most closely linked to Canada's contributions to space exploration?
3. Use your imagination to predict some possible spinoffs of future space exploration.
4. Design a "Table of Contents" for a magazine titled *This Year in Space Exploration*, published in the year 2020. Include titles of both the regular features of the magazine and articles written by specialists.

CHAPTER • REVIEW

Key Ideas

- Space travel includes travelling around the Earth above the atmosphere as well as travelling to other bodies in the solar system. Most spacecraft are unpiloted, but some have people on board.
- The common method of launching payloads into space is to use chemically propelled rocket engines. The engines provide thrust as expanding gases leave in the opposite direction.
- Putting a spacecraft into orbit around the Earth takes only about 10 minutes. Once there, the spacecraft travels at high speed, following the Earth's curvature.
- Earth-orbit satellites are used for scientific research, remote sensing, weather forecasting, telecommunications, and search and rescue operations. For the latter two uses, satellites are usually placed in geosynchronous orbits.
- Many of Canada's greatest contributions to space exploration have been in the use of Earth-orbit satellites.
- Space probes, which have many instruments and measuring devices, have been sent to the moon, the sun, some comets, and seven of the eight other planets in the solar system. These probes have revealed much about the solar system.
- Space stations are designed for humans to live and work in space.
- Canada's contributions to the space shuttle program and the future space station include Remote Manipulator Systems and our own astronauts.
- Future space exploration will probably involve exploration by robots, and human colonies on the moon, Mars, and perhaps other solar system bodies.

Vocabulary

space
aircraft
spacecraft
orbit
unpiloted spacecraft
satellite
space probe
piloted spacecraft
space station
payload
launcher
thrust
low Earth orbit
remote sensing
geosynchronous orbit
microgravity
spinoff

1. Create a concept map to show what you have learned about space exploration.
2. Write a science fiction story about space exploration using as many of the words in the vocabulary list as you can.

Connections

1. Compare an aircraft engine and a rocket engine, explaining why an airplane cannot fly in space.
2. A space shuttle has three main engines and other smaller ones.
   (a) At what times are the three main engines used?
   (b) What purpose(s) do the smaller engines serve?
   (c) Are there times during a space shuttle mission when no engines are operating? Explain.
3. A sports event takes place in Australia, and at almost the same time you are able to watch that event on television. Describe how this form of telecommunication works.
4. Compare the uses of space probes with the uses of space stations.
5. (a) Why is the expression "zero gravity" misleading for an astronaut inside a spacecraft in an Earth orbit?
   (b) Under what condition(s) would it not be misleading?
6. During the last few days, in what ways have you benefited from space exploration? (Include both direct benefits and spinoffs.)

Explorations

1. One of the problems of sending humans to explore Mars is that a lot of fuel is required to launch a spacecraft from Mars to bring them back to Earth. How would you design a space mission to minimize the amount of fuel that the spacecraft would need to carry when it takes off from Earth?
2. The U.S. space shuttles are useful vehicles, but each time one is launched, the large external fuel tank is discarded. Suggest changes to the shuttle design that would allow the fuel tank to be used again or not needed at all.
3. Robotic rovers designed to roam across the irregular surface of Mars will have to be "smart"; in other words, they will have to make their own decisions about whether or not it is safe to move forward. Explain why such rovers could not be adequately controlled by Earth-based controllers.
4. Compose a letter to NASA (the National Aeronautics and Space Administration) describing details of experiments you would like astronauts to carry out during a mission aboard a space station.

Reflections

1. Find Activity 2A in your learning journal. Complete column 3 ("Answers after Completing Chapter 2"). Compare this column with column 2. How have your answers changed after studying this chapter?
2. Criticize or defend one of the following statements:
   (a) Space exploration benefits the human race.
   (b) Money spent on space exploration would be better spent cleaning up the global environment.
3. In December 1990, news reports indicated that the last of the older Soviet space stations, *Salyut 7*, was losing speed and slowly dropping in altitude. Predictions were that it would crash to Earth within several months. State your opinion regarding the clean-up of the pieces that crash to the Earth. Should the country where the pieces land be responsible? Should the Soviet government be responsible? Would you advise setting up an international committee to take care of materials that fall from space? How would your answers be affected if you learned that there was a nuclear power plant aboard the falling vehicle?

UNIT II

Harnessing Energy

Have you ever stopped to think about how much electrical energy you use every day? Perhaps you are awakened each morning by an electric alarm clock. Next you probably turn on a light in your room. For your breakfast, you might help yourself to some orange juice from the refrigerator and pop a couple of slices of bread into the toaster. Maybe you listen to the radio while you are eating. Later in the day, you might watch a show on television.

These are just a few examples of the many ways you use electrical energy every day. In our society, few things are as vital as electricity. How to produce electricity inexpensively, without harming the environment, is a growing concern. Scientists are experimenting with new technology for producing electricity. They are also looking at alternatives to electrical energy, such as solar energy and nuclear power.

In this unit, you will investigate how electrical energy is produced, used, measured, and safely controlled. You will also explore ways of reducing your use of electrical energy and evaluate different sources of electricity and alternatives to electrical energy. In addition, you will find out about the relationship between electricity and magnetism. Finally, you will investigate radiation, including both its harmful effects and its many beneficial applications.

B.C. Hydro

CHAPTER

3

Static and Current Electricity

If you have ever experienced a lightning storm, you can probably understand why people have always found these storms awe-inspiring and frightening. In the past, people explained lightning through legends or myths. Ancient Greeks, for example, believed that lightning bolts were hurled by Zeus, the king of their gods, when he was angry. In North America, the myths of the Tlingit and the Cree say that lightning flashes when the thunderbird winks.

Scientists have also tried to find explanations for lightning. The American scientist, writer, and politician Benjamin Franklin (1706–1790) noticed that lightning was similar to electric sparks and guessed that it was caused by electricity. To test his idea, he fastened a key to one end of a long piece of silk and tied a wire to the other end. Then he attached the wire to a kite and flew the kite during a lightning storm. The key sparked in the same way it did when Franklin produced electricity in his indoor experiments. Franklin had the evidence he needed.

Franklin was lucky that lightning did not strike his kite directly, or (like the next two people who tried the experiment) he might have been killed. Fortunately, he survived the experiment and went on to investigate lightning and electricity more fully and in safer ways. You can too.

ACTIVITY 3A What I Already Know about Electricity

In this activity, you will summarize what you know and what you would like to know about electricity. Divide a page in your learning journal into four columns. Put one of the following questions at the top of each section as a heading:

- What do I already know?
- What do I think I know (but am not sure)?
- What would I like to know?
- What do I need to know?

Write your questions and ideas under each of the headings. Then get together with one or more other students and compare what you have written. You may wish to add to or subtract from your list after you have discussed it with others. Keep your list handy; you may want to refer to it throughout the chapter as you investigate electricity.

3.1 Static Electricity

Have you ever walked across a rug on a dry day and then felt a small shock when you touched an object in the room? Do you know why this happens?

As you walk across a rug, your body can pick up a **static electric charge**, an electric charge that stays in place for some length of time. Static electric charges are often produced when different materials, like your shoes and a rug, are rubbed together. You are usually not aware of these charges until you touch an object; then the electric charge flows from your body and you experience a slight shock.

Fortunately, you can detect static electric charge, or static electricity, as it is also known, without experiencing a shock. You can do this with an **electroscope**, a device that detects static electric charges. One type of electroscope can be made by suspending a small Styrofoam ball that is coated with graphite (the material in pencil lead) on a thread.

With this type of electroscope, it is easy to find materials that produce static electric charges when they are rubbed together. If a charged object is brought near the ball of the electroscope, it either **attracts** or **repels** the ball (Figure 3.1). If two objects attract, they try to pull one another together. If two objects repel, they try to force one another apart.

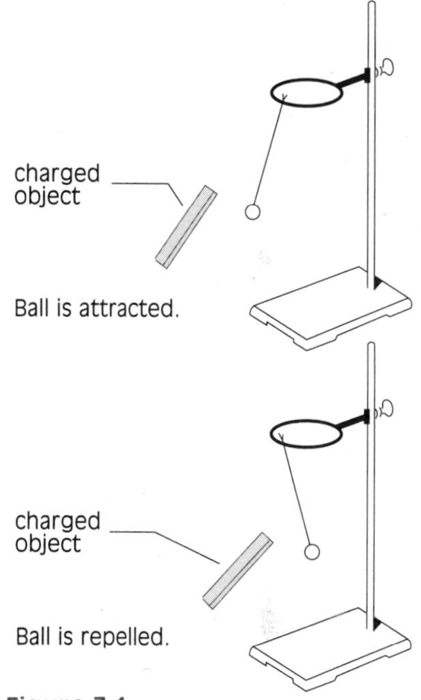

Figure 3.1
A charged object attracts or repels the ball of an electroscope.

ACTIVITY 3B Producing Static Electric Charges

In this activity, you will find materials that produce static electric charges when they are rubbed together.

MATERIALS

electroscope
For other materials, see Procedure.

PROCEDURE

1. Work with a group of students to make up a list of materials that can be rubbed together and tested for static charges. Some materials that you could try are plastic combs (Figure 3.2), pens, and rulers; wooden pencils; metal rods; and other similar objects. You could rub these objects with a paper towel, different types of fabric, and plastic bags, for example. Use your imagination.
2. Prepare a chart on which you can record the materials you rub together and the results you obtain when you test them with an electroscope.

→

Figure 3.2
This student ran a comb through his hair. Has the comb become charged?

3. Rub these materials together and use your electroscope to test them for the presence of a charge.

DISCUSSION

1. Of all the combinations you tried, which pairs of materials seemed to produce the most static electricity when they were rubbed together?
2. Besides the static charge, were there any other changes in the materials that you used?
3. With which materials were you unable to produce a static charge?
4. If you experience a lot of small "shocks" at home or in school (when you touch something metal or another person), what pairs of materials do you think might be responsible for producing the static electricity?

TWO TYPES OF CHARGE

In the early 18th century, a French scientist named Charles Du Fay (1698–1739) discovered that there are two types of static electric charge. Later in the century, Benjamin Franklin named these two types of electric charge positive and negative.

A **positive charge** is developed on acetate plastic (the clear plastic your teacher may use on an overhead projector) when it is rubbed with cotton or a paper towel. A glass rod rubbed with silk also develops a positive charge. A **negative charge** is developed on vinyl plastic (the plastic from which some raincoats are made) when it is rubbed with wool. Amber rubbed with fur also develops a negative charge. An object without any charge is **neutral**.

Why did Du Fay conclude that there are only two types of static electric charge? You can find out for yourself in the following activity.

ACTIVITY 3C *Effects of Static Charges*

In this activity, you will find answers to the following questions:

- What is the effect of one positively charged object (+) on another?
- What is the effect of one negatively charged object (−) on another?
- What is the effect of a positively charged object on a negatively charged object?
- What is the effect of a charged object on a neutral object (0)?

MATERIALS

two acetate strips
ring stand
paper towel or cotton cloth
two vinyl strips
wool cloth
watchglass
wooden metre stick

PROCEDURE

1. Make a copy of Table 3.1 in your notebook.
2. Tape an acetate plastic strip to a ring stand so that it hangs freely (Figure 3.3).
3. Charge the acetate strip by rubbing it with cotton or a paper towel. Then charge a second acetate strip in the same manner. Bring the second acetate strip near the first. Do the strips attract or repel one another? Record the result in your table.

Figure 3.3
Tape an acetate strip to a ring stand like this.

Table 3.1 Sample Data Table for Activity 3C

Suspended object	Charge	Object brought near	Charge	Result (attracts or repels)
Acetate	+	Acetate	+	
Vinyl	–	Vinyl	–	SAMPLE ONLY
Vinyl	–	Acetate	+	
Metre stick	0	Acetate	+	
Metre stick	0	Vinyl	–	

Figure 3.4
What is the effect of a charged strip on a metre stick?

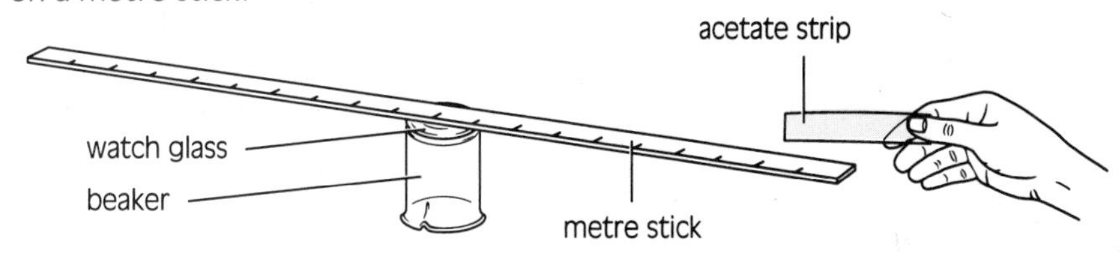

4. Repeat the experiment using two vinyl strips and rubbing them with wool. Record whether the strips attract or repel one another.
5. Bring a charged acetate strip near a suspended vinyl strip. Record what happens.
6. Neutralize a metre stick by running your hands along its length. Then balance the stick on a watchglass as shown in Figure 3.4. Bring a charged acetate strip near one end of the metre stick but do not let it touch the stick. Is the metre stick attracted or repelled? Record the result.
7. Bring a charged vinyl strip near the end of the metre stick without touching the stick. Again, record what happens on your table.

DISCUSSION

1. Why was it necessary for the suspended acetate or vinyl strips to be free to move in this experiment? Why was the metre stick balanced on a watchglass?
2. Now that you have done the experiment, what are your answers to the questions that you considered at the beginning of the activity?
3. Suppose you suspended a plastic spoon with a piece of tape so that it was free to move, and then charged the spoon by rubbing it with a paper towel. If you charged a second plastic spoon in the same manner and brought it near the first spoon, would you expect the two spoons to attract or repel? Explain your answer.

RULES OF STATIC ELECTRICITY

The effects of charged objects on one another and on neutral objects can be summarized in two rules:

1. Objects with like charges repel one another, whereas objects with unlike charges are attracted to each other.
2. Charged objects attract neutral objects.

Charles Du Fay concluded that there were two kinds of charge because he could summarize his experimental observations with these rules. But the rules do not explain why an object can gain an electric charge. That explanation was not discovered until near the beginning of this century.

EXTENSION

Will a charged rod affect a stream of water from a tap? Bring a negatively charged rod near (but not touching) a fine stream of water. Now try a positively charged rod. Does a positively charged rod have the same effect on the water as a negatively charged one?

REVIEW 3.1

1. Describe two ways in which static charges can be detected.
2. What is the effect of
   (a) two objects with like charges on one another?
   (b) two objects with unlike charges on one another?
   (c) a charged object on a neutral object?
3. An acetate strip charged by rubbing it with cotton is brought near a charged plastic ruler. The two repel one another. What can you conclude about the charge of the ruler? Give the reasons for your answer.
4. An acetate strip charged by rubbing it with cotton is brought near a balloon. The balloon is attracted to the acetate strip. What can you conclude about the charge of the balloon? Give the reasons for your answer.
5. If a vinyl strip is charged by rubbing it with wool, would you expect a charge to develop on the wool? If so, what kind of charge?
6. How could you find out what charge is on your comb after you comb your hair? Design a test and carry it out. What is the charge? Explain how you obtained your results and why you are certain of the charge.

DID YOU KNOW?

Words such as "electricity," "electron," and "electronics" all come from the same Greek word, *elektron*. *Elektron* is the Greek word for "amber." As long ago as 600 B.C., the Greeks knew that amber, a fossilized resin from trees, would pick up small bits of dust, lint, and other light-weight materials when the amber was rubbed.

3.2 Static Electricity and the Atom

Benjamin Franklin and other scientists of his time made many correct predictions about static electricity, even though their knowledge of the subject was incomplete. In the 1700s, scientists did not know that all matter is made up of tiny particles called **atoms** and that static electricity is due to the structure of the atom.

THE MODERN EXPLANATION OF STATIC ELECTRICITY

Atoms have two main parts, one positively charged and the other negatively charged (Figure 3.5). Both of these parts are made of smaller particles. Since these particles are smaller than atoms, they are said to be **subatomic**.

Figure 3.5
The main parts of an atom are the positively charged nucleus and the negatively charged electrons.

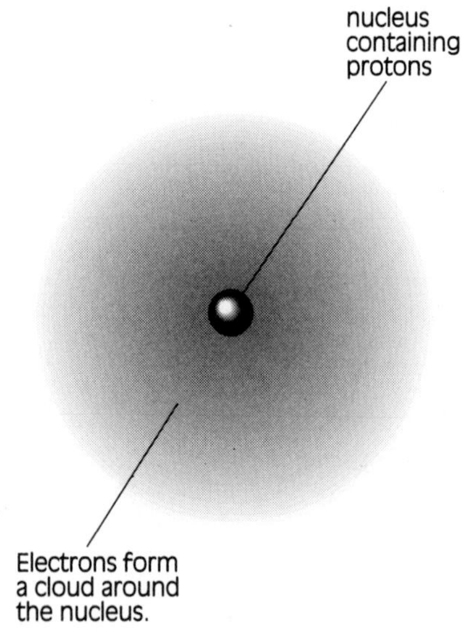

The negatively charged part of an atom is made up of subatomic particles called **electrons**. The existence of these particles was demonstrated in 1897 by the English scientist Sir Joseph John (J. J.) Thomson (1856–1940). A modern and modified version of the apparatus that Thomson used in his experiments is found in television sets. The picture on the screen is produced as a result of the tiny, negatively charged electrons striking the screen. Figure 3.6 shows how a television picture tube works.

The positively charged part of an atom is concentrated in a very dense central part called the **nucleus** (plural: nuclei). The nucleus was discovered in 1911 by Ernest Rutherford (1871–1937). In 1914, Rutherford showed that the nucleus contained tiny, positively charged particles, which he called **protons**. Although a proton's charge is opposite to that of an electron, the amount of charge on each is the same. Protons are about 2000 times more massive than electrons, however.

Static electricity can be explained using the model of the atom shown in Figure 3.5. Neutral atoms contain equal numbers of protons and electrons. Huge numbers of atoms are joined together to make objects big enough to see. Overall, these objects are neutral, like the atoms of which they are composed. When two objects are rubbed together, some electrons may be transferred from the atoms of one object to the atoms of the other object (Figure 3.7). If this happens, a static electric charge results on each object. The object that gains electrons has more electrons than protons, so it is negatively charged. The object that loses electrons has more protons than electrons; it is positively charged. Thus, the transfer of electrons causes objects to gain static electric charges.

Only electrons are transferred in this process because they are removed from atoms relatively easily. Electrons are located outside the nucleus, and small amounts of energy can remove them or add them to some of the atoms in an object.

Adding or removing protons is much more difficult. All atoms of the same element have the same number of protons in their nuclei. This number of protons is the **atomic number** of the element. If a proton is taken from or added to the nucleus of an atom, an atom of a different element is produced. Changes like this occur only in nuclear reactions, since they require very large amounts of energy, much more than can be made available to single atoms by rubbing things together.

The attraction between charged and neutral objects results from the movement of electrons within the neutral object. The electrons move towards a nearby positively charged object and away from a negatively charged one. This movement produces an **induced charge**, an electric charge that results from the presence of a charged object nearby, rather than from direct contact. Overall, the neutral object remains neutral, since no electrons are added to or removed from it.

Figure 3.8 illustrates the induced charges on a neutral ball. Two induced charges are produced. Near the positively charged rod, there is an induced negative charge. Farther away, the induced charge is positive. The negative side of the ball (left) is closer to the charged rod than is the positive side of the ball (right). Thus, the attraction between the rod and the ball's left side is stronger than the repulsion between the rod and the ball's right side. As a result, the neutral ball is attracted to the charged rod.

Figure 3.6
A television picture tube is based on the principles of the experimental apparatus used by J. J. Thomson. As electrons travel through the tube, they are deflected electrically and magnetically. When electrons hit the screen, they produce a glow of light in the screen's chemical coating.

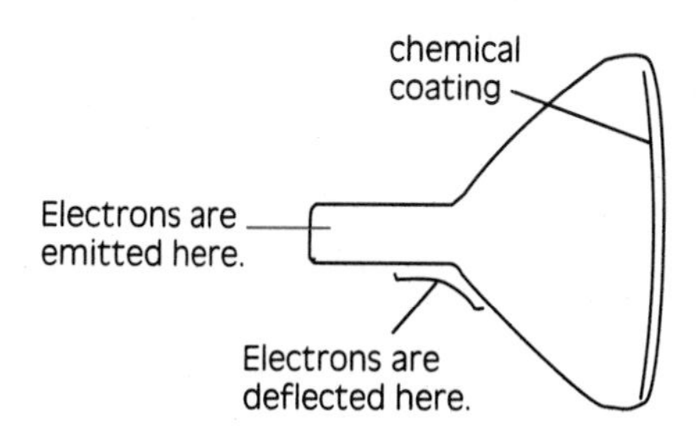

Figure 3.7
When objects become charged with static electricity, electrons are transferred. The object receiving electrons becomes negatively charged; the object losing them becomes positively charged.

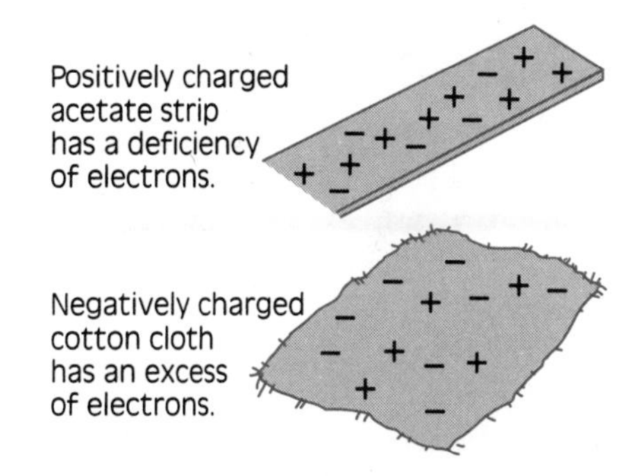

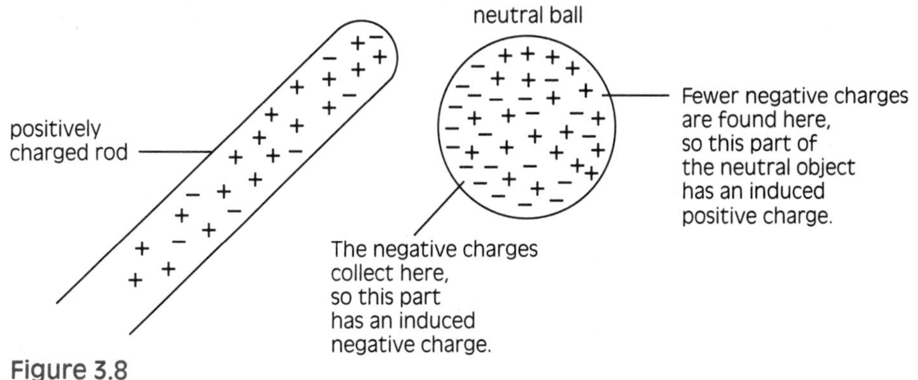

Figure 3.8
Two induced charges result when a charged object is brought near a neutral object.

Profile

ERNEST RUTHERFORD

"The most incredible event that has ever happened to me in my life." That is what Ernest Rutherford said about his discovery that led him to suggest that an atom has a nucleus. His discovery of the nucleus was essential for an understanding of the atom.

Ernest Rutherford developed his love of science and discovery when he was growing up in New Zealand. Born in 1871, he was the second of 12 children. His father was a farmer and a wheelwright (someone who makes and repairs wheels), and the family did not have much money. As a boy, Rutherford learned to be inventive. He was good with his hands and often rebuilt household tools and equipment.

At the age of 15, Rutherford won a scholarship to a New Zealand university, and in 1889, he won another scholarship to Cambridge University in England. He graduated from Cambridge with a first-class degree in physical sciences and mathematics.

At the beginning of his scientific career, Rutherford worked in England with J. J. Thomson, who discovered the electron. From 1898 to 1907, Rutherford lived in Canada, teaching physics at McGill University in Montreal and doing experiments with radioactive elements. In 1907, he moved back to England, where he continued his work on radiation. As a result of this work, he won the 1908 Nobel Prize in chemistry.

Ernest Rutherford (right) conferring with J. J. Thomson.

Rutherford's 1911 model of atomic structure was the basis for the model that was developed in the 1920s by Niels Bohr, which is still in use today. From 1919 until his death, Rutherford directed the Cavendish Laboratory in Cambridge. The research he conducted there was the beginning of today's field of experimental nuclear physics.

It has been said that Rutherford had the knack of knowing what was likely to be important in an experiment. One colleague said that he was "like an enormously powerful battleship ploughing majestically through the storms and seas." Another remembered that he frequently made simple mistakes in arithmetic. His genius was due to his imagination, his determination, and his enthusiasm for his subject.

In the lab, Rutherford could be found rushing about reading electroscopes and electrometers, or casually chatting with his colleagues. If something difficult or interesting came up, he would pull up a stool, sit down, and throw out helpful comments or suggestions. He was generous in his acknowledgement of others' work and was ready to help the scientific advancement of women as well as men.

It is rare that one person can be said to create an entire field of science. But Rutherford's contributions to science so consistently broke new ground that he is often compared with Sir Isaac Newton. When Rutherford died in 1937, he was honoured by being buried near Newton, in Westminster Abbey in England.

THE UNIT OF STATIC CHARGE

The amount of negative or positive charge on an object is measured in **coulombs (C)**, named after the French scientist Charles Augustin de Coulomb (1736–1806), who did many experiments on the nature of static electricity. In 1911, many years after the definition of the coulomb, the American scientist Robert Millikan (1868–1953) found that one coulomb of negative charge equals the charge of 6.24×10^{18} electrons. Since the charge of a proton is equal in size to that of an electron, one coulomb of positive charge is equal to the charge of 6.24×10^{18} protons. The static charges with which you are familiar typically are much smaller—in the range of tenths of microcoulombs (1 microcoulomb is 10^{-6}, or 0.000 001, coulomb).

EXTENSION

Static electric charges can have hazardous or unwanted effects. For example, the huge charges associated with lightning can easily kill a person. Conveyor belts in factories may generate static electricity. If static charges become large enough, an electric arc may result. Such an arc can ignite nearby flammable materials, causing an explosion.

Static electric charges can also have beneficial uses. Photocopiers, sandwich wrapping, soaps, electrostatic spray painting, and concentrators—in which valuable metal ores are separated from rock—all use static electricity. Choose one of these examples and find out how it makes use of static electricity.

REVIEW 3.2

1. What kind of charge is on an object if the object gains an excess of electrons?
2. If an object loses electrons, what kind of charge is on the object?
3. Why must a neutral object have an equal number of protons and electrons?
4. Explain why protons are not transferred when an object gains a charge.
5. (a) What is a coulomb?
   (b) How many electrons make a coulomb of negative charge?
   (c) What is the charge of one electron in coulombs?
6. Figure 3.9 shows a picture of a student holding onto the dome of a machine called a Van de Graaff generator. When it is in operation, the dome of the machine becomes charged. Explain why the student's hair stands on end.
7. A student combed his hair and then brought the comb near some sawdust. At first, the sawdust was attracted to the comb. Then, after touching the comb, it was repelled. Explain these observations.

Figure 3.9
You will get a charge from a Van de Graaff generator.

3.3 Current Electricity

Imagine pushing two electrons together. You would have to push, because electrons repel one another. Pushing the electrons would be something like compressing a spring. You would have to work to push them together, and when you let go, they would push apart. When you push the electrons together, you increase their stored energy; but when you let go, the stored energy decreases.

Separating an electron from a proton would be like stretching a spring, because the two charged particles attract each other. You would have to work to separate the electron and proton, and you would increase their stored energy as you separated them.

The spring and the charged particles in these examples contain stored energy. Stored energy is called **potential energy**. Because the potential energy of the electrons and protons results from their electrical charge, it is called **electrical potential energy**. When you rub two neutral objects together to produce static electric charges on them, you are "pushing" electrons together and separating electrons from protons. As a result, you increase the electrical potential energy of the objects.

Figure 3.10
You can make a chemical cell like this one.

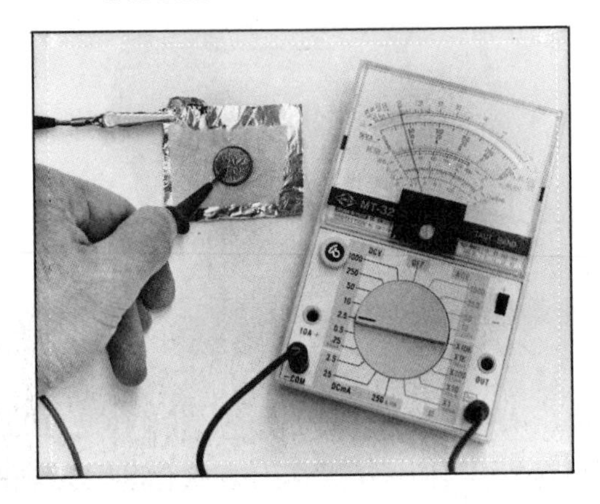

The objects do not maintain the increased electrical potential energy forever. Electrons are transferred to and from the air surrounding the objects until the charges of the objects are again neutral. Then the objects have the same amount of electrical potential energy that they began with.

Under some circumstances, a fairly constant difference in electrical potential energy can be maintained. A **chemical cell**, for example, uses the potential energy of chemicals to produce a difference in electrical potential between the two electrical connections, or **terminals**, of the cell. Through chemical action, one of the terminals is supplied with electrons, making it negatively charged. Electrons are removed from the other terminal, making it positively charged. You can even make a chemical cell yourself. To do this, sandwich a piece of paper towel soaked in vinegar between a clean penny and a small piece of aluminum foil (Figure 3.10).

VOLTAGE

Voltage is the amount of electrical potential energy that is supplied per coulomb of charge transferred. The unit of voltage is the **volt (V)**, named after the Italian scientist Alessandro Volta (1745–1827), who experimented with chemical cells like the one shown in Figure 3.10.

MEASURING VOLTAGE

The voltage of a cell can be measured with a device called a **voltmeter**. Often a meter that measures voltage and several other electrical quantities is combined into a multi-range meter. Such meters are handy for measuring various electrical quantities. You may be using a multi-range meter to make measurements in Activities 3D, 3E, and 3F. If you use a meter at home, it is most likely to be a multi-range meter.

In multi-range meters, one dial does the job of several separate meters. For this reason, the dial shows several different scales. The scales and the method of selecting the quantity you wish to measure may vary from meter to meter. Some meters, for example, may have a lowest voltage scale of 0–10 V, and you select the scale you want by turning a switch.

On another meter, the lowest voltage scale may be 0–15 V, and you select the scale you want by connecting one of the meter leads at a terminal marked with the scale (Figure 3.11). Generally, the black meter lead is connected to the common, or negative, terminal, and the red lead is connected to the terminal corresponding to the desired scale.

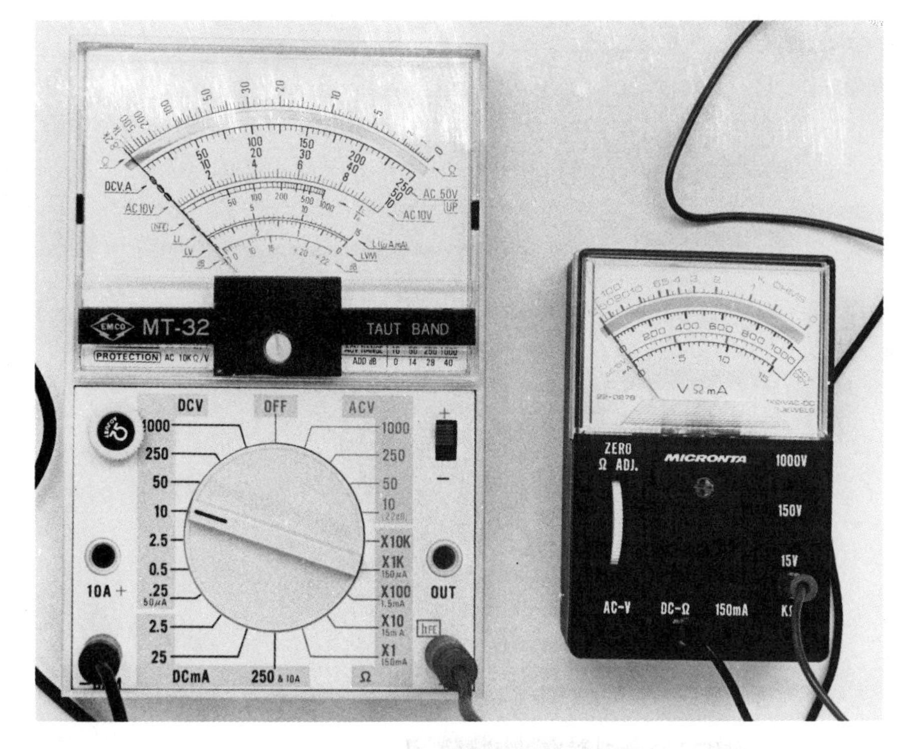

Figure 3.11
The multi-range meter on the left uses a switch to select the meter scale. The meter is set to measure voltage on its 0–10 V scale. To select the scale on the right-hand multi-range meter, you must connect the meter leads to the correct terminals. This meter is set to measure voltage on its 0–15 V scale.

For each scale, the meter is connected differently inside the meter housing. The interpretation of the numbers at which the meter needle points depends on the scale that was selected. When the meter leads are connected as shown in Figure 3.12a, the meter indicates a voltage of 6 V. In Figure 3.12b, the choice of scale has changed and so has the reading; it is now 60 V. The reading in Figure 3.12c indicates a voltage of 150 V. What is the voltage of the electrical cell shown in Figure 3.10?

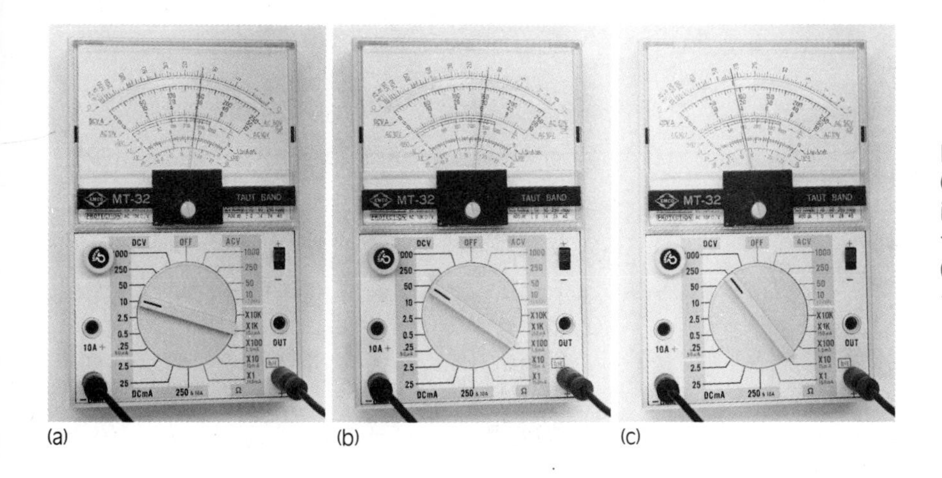

(a) (b) (c)

Figure 3.12
(a) On the 0–10 V scale, the meter indicates 6 V. (b) On the 0–50 V scale, the meter indicates 30 V. (c) On the 0–250 V scale, the meter indicates 135 V.

BATTERIES

The chemical cell shown in Figure 3.10 is not very good at producing electricity. A much better chemical cell is shown in Figure 3.13a. You probably call this cell a battery. In fact, a **battery** consists of two or more chemical cells connected together.

Cells can be connected in a **series circuit** or a **parallel circuit** to make a battery. In a series circuit, the positive terminal of one cell is connected to the negative terminal of a second cell. In a parallel circuit, the positive terminals of all the cells are connected together and so are all the negative terminals (Figure 3.14).

(a)

(b)

Figure 3.13
(a) This is a chemical cell. Inside it contains all the materials needed to produce electricity. (b) This is a battery. It is made up of several cells.

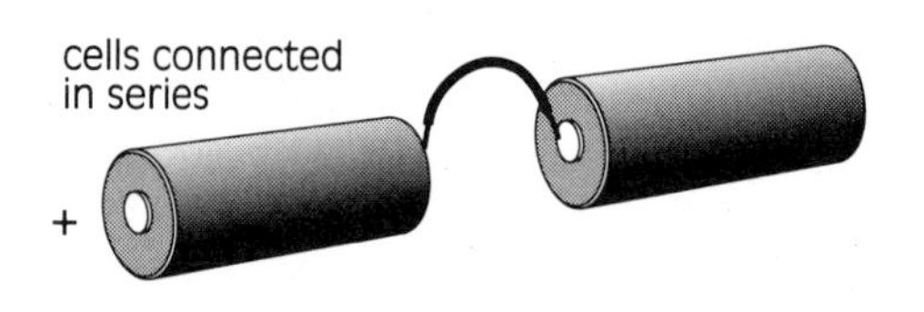

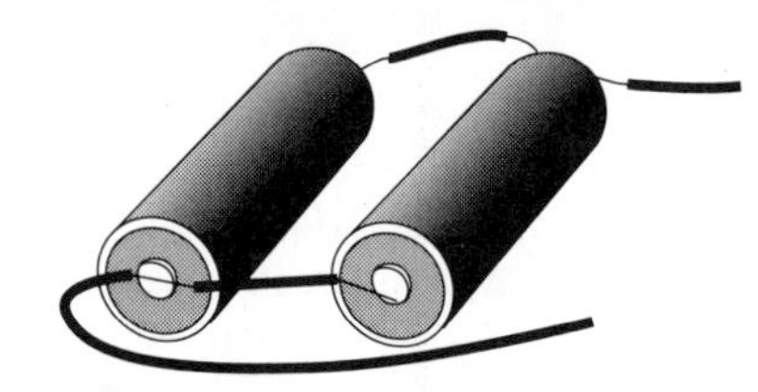

Figure 3.14 ►
Chemical cells can be connected in series or in parallel.

ACTIVITY 3D *Voltage of Electrical Cells in Series and Parallel*

In this activity, you will make batteries by connecting chemical cells in series and parallel. Then you will measure the voltages of the batteries.

MATERIALS

four chemical cells and holders
multi-range meter
connecting wire
alligator clips

Table 3.2 Sample Data Table for Activity 3D

Cell connection	Number of cells	Voltage (V)
—	1	
Series	2	SAMPLE ONLY
Series	3	
Series	4	
Parallel	2	
Parallel	3	
Parallel	4	

PROCEDURE

1. Prepare a table like Table 3.2 in your notebook.
2. Connect a multi-range meter so that it is on the scale nearest to 0–10 V. You will be using the meter for your voltage measurements.
3. Touch the black negative terminal of the meter to the negative terminal (bottom) of a chemical cell and the red positive terminal to the positive terminal (top) of the cell. Record the voltage of the cell. Then measure and record the voltage of each of the other cells to make sure that they all have the same voltage.
4. Now connect two cells in series. Measure the voltage of the two cells by connecting the multi-range meter from the negative terminal of the first cell to the positive terminal of the second cell. Record the voltage of the battery.
5. Predict the voltage you will measure with three and four cells connected in series. Connect the cells and test your prediction. Record the voltage in your table in each case.
6. Arrange two cells so that they are connected in parallel. Measure and record the voltage of this battery.
7. Predict the voltage of three and four cells connected in parallel. Connect the cells and test your prediction. Record your results in your table.

DISCUSSION

1. What is the voltage of a single cell?
2. State a general rule that allows you to find the voltage of a battery of cells connected in series.
3. State a general rule that allows you to find the voltage of a battery of cells connected in parallel.
4. Figure 3.13b shows the interior of a 1.5 V battery in which four cells are connected in parallel. What advantage would such a connection have over a single chemical cell?
5. Common non-rechargeable electrical cells have a voltage of 1.5 V. What minimum number of these cells is required for a 9 V battery? How must these cells be connected? Is this what you see in the 9 V battery shown in Figure 3.13b?

ELECTRIC CURRENT

If a wire is connected between the terminals of a chemical cell, electrons will flow from the negative terminal to the positive terminal because of the difference in voltage. This flow will continue until the chemicals in the cell run out of energy.

When scientists were experimenting with electricity in the late 1700s and early 1800s, they did not know about electrons. They thought that electricity resulted from a flow of fluid from a positively charged object, where there was an excess of the fluid, to a negatively charged object, where there was less of the fluid. They called this flow an **electric current**. By the time it was discovered that an electric current in wires results from the flow of electrons from the negative terminal to the positive terminal, the convention was well established that current flows from positive terminal to negative terminal. **Conventional current** is, by definition, the current that flows from a positively charged object to a negatively charged one, even though electrons move in the opposite direction (Figure 3.15).

Electric currents flow in other things besides wires. For example, electric currents can flow in solutions and gases. In these materials, the current can result from the movement of positively charged particles as well as negative ones. When current results from the movement of positively charged particles, conventional current and the particles move in the same direction.

Figure 3.15
Electrons move through the wire from a negatively charged terminal to a positively charged terminal. This movement results in a transfer of charge between the terminals. Conventional current is defined as flowing from the positive terminal to the negative terminal.

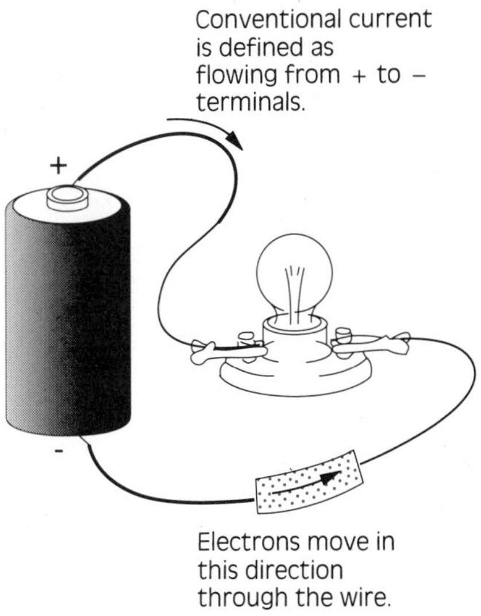

The amount of electric current is measured in **amperes (A)**, named in honour of the French scientist André Marie Ampère (1775–1836). A current of one coulomb of electric charge per second is called one ampere. For a current of one ampere to flow through a wire, 6.24×10^{18} electrons must move through any cross-section of the wire each second.

A device called an **ammeter** can be used to measure the electric current flowing in a wire. To do this, an ammeter must be inserted into the wire so that the current passes through the meter. You can see how this is done in Figure 3.16.

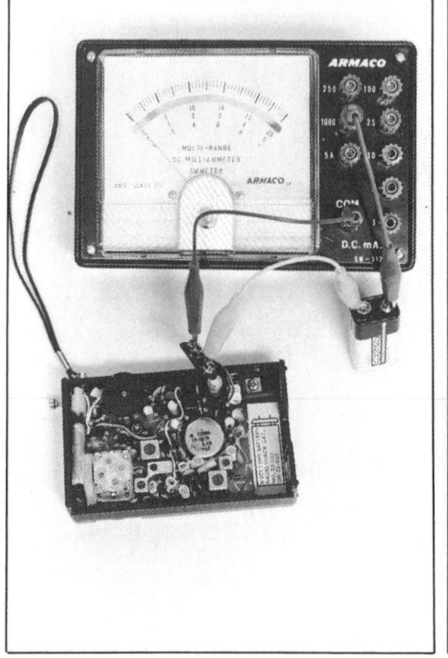

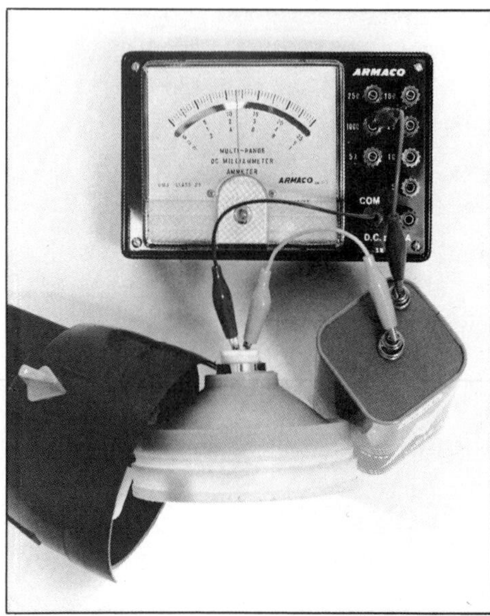

Figure 3.16 ➤
These ammeters are connected to measure current. The flashlight uses much more current than the transistor radio.

Figure 3.17
Current flows easily in conductors (a). Current does not flow easily in insulators (b).

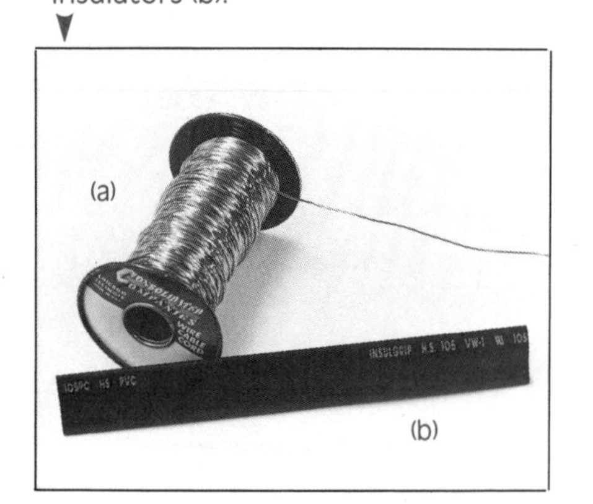

Ammeters often include several scales so that they can measure both small and large currents. For example, a large flashlight may use a current of several amperes, whereas a transistor radio may use only a small fraction of an ampere. For this reason, a smaller unit of current, the milliampere (mA) or 0.001 A, is often convenient.

Large numbers of electrons can flow only when materials allow their easy passage. Materials in which electrons can move easily are called **conductors**. All metals are conductors, for example. However, some, like silver and copper, are better conductors than others, such as lead, iron, and mercury. Materials in which electrons do not move easily are called **insulators**. Glass, most plastics, and rubber are good examples of insulators (Figure 3.17).

ELECTRIC CIRCUITS

In a typical flashlight, two cells are connected to a switch and a light bulb. When the switch is closed, there is a complete pathway for an electric current to flow from the cells to the light bulb, through the light bulb, and then back to the cells. A complete pathway for electric current, from the source and back again, is called an **electric circuit**.

Figure 3.18 shows one way to draw the electric circuit in a flashlight. A drawing like this takes time to complete, and the electric circuit does not stand out from other, less essential details of the flashlight. For these reasons, **schematic diagrams** are often used to represent electric circuits. Schematic diagrams use standard symbols to depict circuit components. The standard symbols that you will need to know are shown in Figure 3.19. Try using these symbols to make a schematic drawing of the flashlight shown in Figure 3.18.

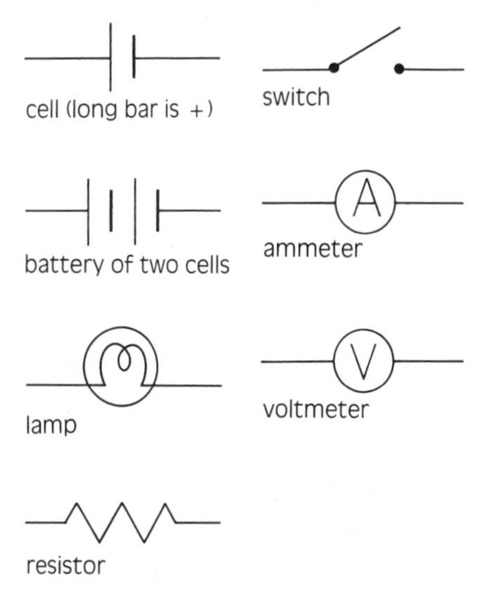

Figure 3.19
These symbols are used to represent the parts of an electric circuit.

Figure 3.18
The electric circuit in a flashlight.

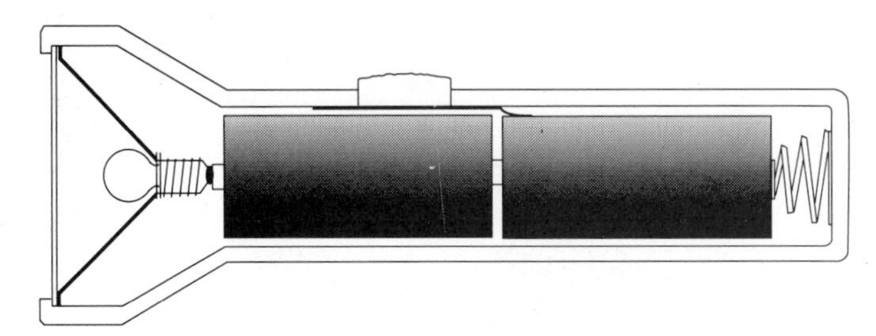

ACTIVITY 3E *Investigating Electric Current*

This activity will give you an opportunity to use an ammeter and will also introduce you to some simple electrical circuits.

MATERIALS

multi-range ammeter or multi-range meter
two chemical cells and holders
push-button switch
three lamps and sockets
connecting wire
alligator clips

PROCEDURE

1. Figure 3.16 shows one ammeter connected so as to measure the current used by a large flashlight, and a second ammeter connected to measure the current used by a transistor radio. Make a schematic drawing to show how the ammeter is connected in the flashlight circuit. The ammeter is connected in series so that the current passes through the meter.
2. With your partner, decide how you should connect a simple circuit consisting of a battery of two cells in series, a switch, an ammeter, and a small lamp. Make a schematic drawing of your circuit. Construct the circuit you have drawn schematically and measure the current registered on the ammeter. Also record the brightness of the lamp. Record these observations on your schematic diagram.
3. How can you change the circuit so that the current has to go through two lamps along the same conducting path? Discuss this problem with your partner and see if you can devise a way in which this can be done. Draw a schematic diagram for your solution. Now construct the circuit and measure the current (Figure 3.20). Record both the current and the brightness of the lamps on your schematic diagram.
4. Predict what will happen if you add another lamp, connected so that the current has to go through all three lamps along the same conducting path. Draw a schematic diagram for the circuit in which this would happen. Then connect the circuit and test your predictions. What was the current and how bright were the lamps? As before, record your results on your circuit diagram. Were your predictions correct?
5. How can two lamps be connected in a circuit so that current coming from the battery can go through either one lamp or the other? How must the switch be connected to turn on both lamps at once? See if you and your partner can come up with a way to do this and then draw a schematic diagram of the circuit. In the schematic diagram, show the ammeter connected so that all the current that leaves the →

battery must flow through the meter. Predict how the current in the ammeter will be different from the circuit you constructed in step 2. How bright will the lamps be? Now connect the circuit and test your predictions. Record your results on your schematic diagram.

6. Predict what will happen if you arrange three lamps in the same way as they were connected in step 5. Test your prediction and record your results.

DISCUSSION

1. Draw a diagram to show two lamps connected to a battery of two cells in series so that there is only one pathway for the current.
   (a) Add arrows to the diagram to indicate the direction of current flow.
   (b) If one of the lamps in this circuit burned out, what would happen to the other one?
   (c) If the ammeter were connected in different parts of the circuit, would it indicate different amounts of current? Give the reasons for your answer.
2. Draw a diagram to show two lamps connected to a battery of two cells in a series so that there are two pathways for the current.
   (a) Add arrows to the diagram to indicate the direction of current flow in the two pathways.
   (b) If one of the lamps in this circuit burned out, what would happen to the other one?
   (c) If the ammeter were connected in different parts of this circuit, would it indicate different amounts of current? Give the reasons for your answer.
3. Devise a hypothesis to explain why the current leaving the battery decreases as lamps are added in the series circuit, whereas this current increases as more lamps are connected in parallel.

Figure 3.20
Connect your ammeter to measure the current leaving the battery.

SERIES AND PARALLEL CIRCUITS

Circuits connected so that there is only one pathway for the electric current are called series circuits. In a parallel circuit, the current can flow through two or more pathways. In some circuits, there are both series and parallel connections. These circuits are called **series-parallel circuits** (Figure 3.21).

When you turn off a lamp in your home, you use a switch that is connected in series to the lamp. When you break this circuit, or any other series circuit, no more current will flow. In a parallel circuit, however, current can still flow in spite of a break in part of the circuit. That is why you can turn off the bathroom light while the kitchen light is still burning. Each lamp is connected in parallel to the source of electricity.

Figure 3.21
Circuits can be connected in series, parallel, or combinations of series and parallel.

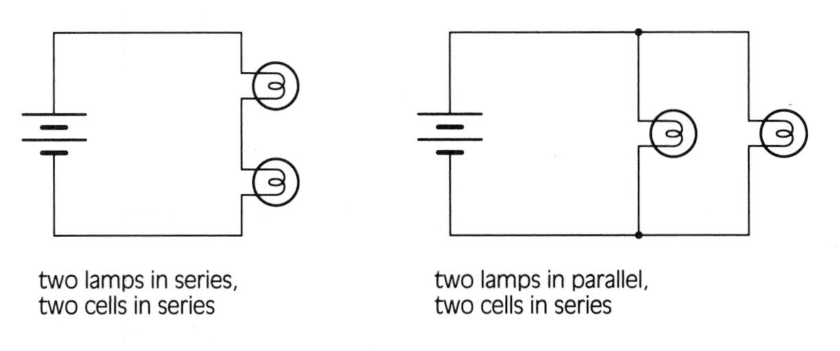

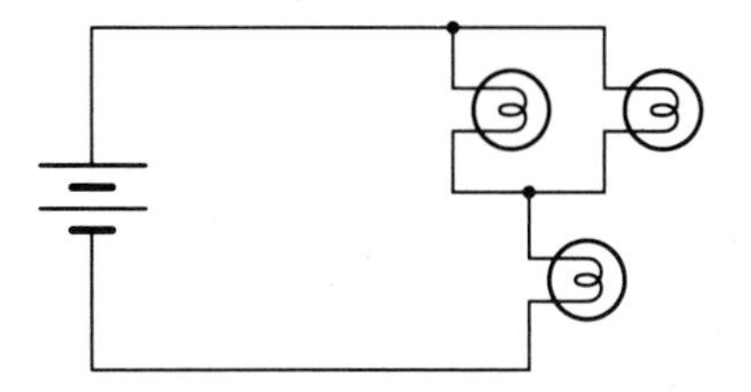

As you know, electric cells can also be connected in series or parallel. When cells are connected in series to make a battery, the voltage of each cell is added to give the total voltage of the battery, just as each individual step in a flight of stairs adds to the total height of the stairs (Figure 3.22). All the current must pass through each of the cells, however, because there is only one pathway.

Batteries made by connecting cells in parallel have a total voltage equal to the voltage of an individual cell. Each of the cells in the battery does less work, and thus lasts longer than a single cell, because the current from the battery is provided by all the cells together. Connecting cells in parallel is like widening a step; the height remains the same, but the load on any part of the step is lessened because there are alternative pathways.

EXTENSION

Draw a schematic diagram to show a circuit with three lamps, a switch, and a battery of two cells in series, connected so that there are three alternative pathways for electrons. On the circuit diagram, draw the locations where ammeters would have to be placed to measure the current through each of the lamps and the current leaving and returning to the battery. Predict the readings on each of the ammeters. Then connect the circuit and test your predictions.

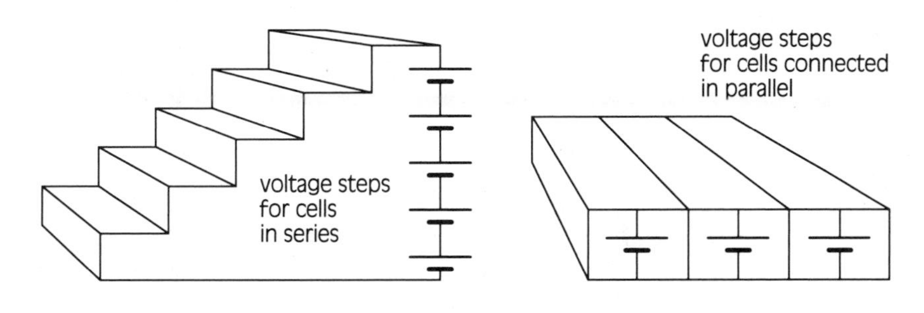

Figure 3.22
Cells connected in series are like the steps in a flight of stairs. All the traffic goes up each step to reach the top, just as all the current must pass through each cell. Cells connected in parallel are like steps placed side by side. The height, or total voltage, is not increased, but there are alternative pathways for the traffic and for the current.

RESISTANCE

If a voltage is applied to a circuit, current will flow. But the same voltage does not always produce the same amount of current. The amount of current in the circuit depends on a third quantity, the circuit's **resistance**. Resistance is a measure of how hard it is for an electric current to flow through a material.

▲
Figure 3.23
All these devices act as resistors in electrical circuits.

Devices made of materials that offer resistance to the flow of current in a circuit are called **resistors**. The tungsten filaments in the lamps used in Activity 3E are resistors. The heating elements in electric stoves, toasters, and kettles are resistors. Television sets and radios contain many resistors. Some of these are made of high-resistance wire coiled tightly on a form, but most are made of short lengths of compressed carbon (Figure 3.23).

Variable resistors, or rheostats, can be used to control current to lights or motors, or to control the volume of sound from a radio or stereo. Figure 3.24 shows how a length of high-resistance wire made of nickel and chromium can be used as a variable resistor. As the connecting alligator clip is slid back and forth along the wire, the resistance to current increases or decreases. This action decreases or increases the current, thus controlling the speed of the motor. The volume control of a radio works in much the same way, except that it has a more compact shape and is more convenient to use.

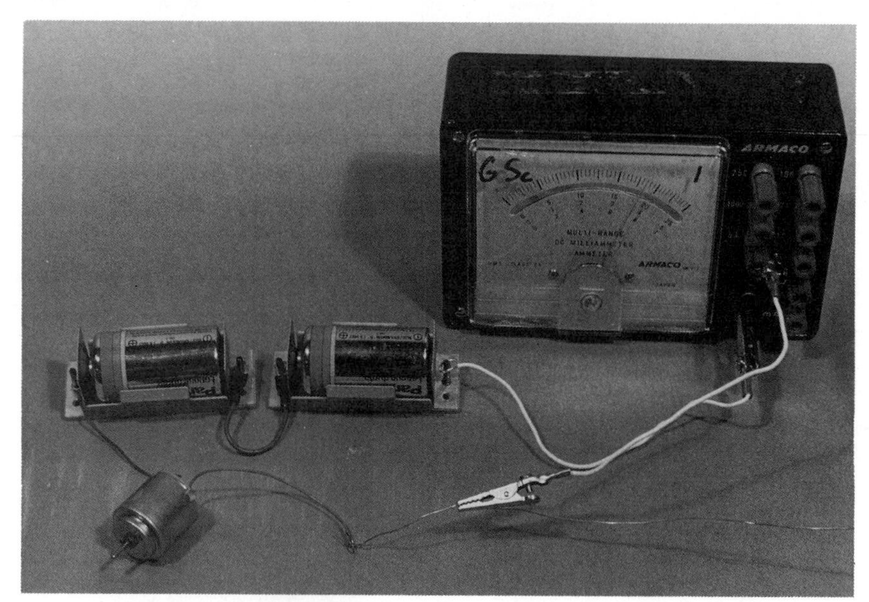

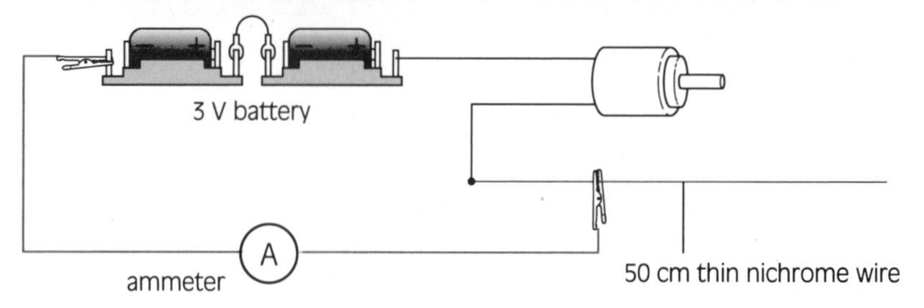

Figure 3.24 ►
This length of high-resistance wire acts as a rheostat, or variable resistor. The wire is comparable in its action to the rheostat from a radio.

Resistance is measured in **ohms**, named after the German physicist Georg Ohm (1789–1854), who experimented with resistance. The symbol for an ohm is the Greek letter omega, Ω. Most multi-range meters include a scale for measuring resistance, and you can use this scale to measure the resistance of many different resistors. By holding on to the ends of the meter leads, you can even measure your own resistance (Figure 3.25).

Figure 3.25
You can measure your resistance by setting a multi-range meter to its resistance scale and holding the tips of the meter leads in your hands. Is your resistance constant?

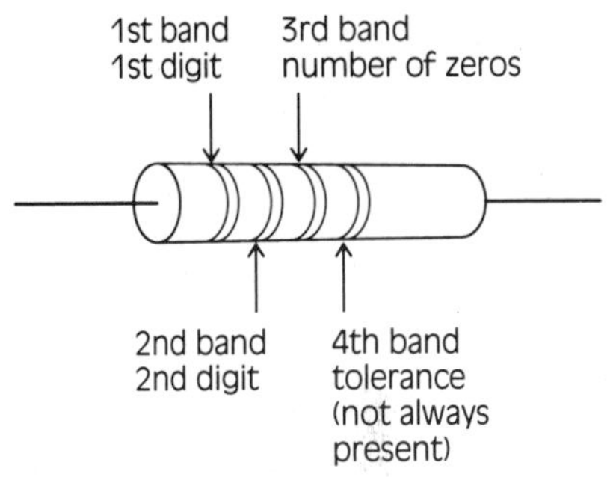

Figure 3.26
The meaning of the coloured bands on a carbon resistor.

Resistors made of compressed carbon use coloured bands to indicate their resistance. The coloured bands begin closer to one end of a resistor than the other (Figure 3.26). To read the resistance, the resistor must be held so that the bands are closest to the left-hand side; then the resistance can be read from left to right.

Each colour stands for a different number. The colours and numbers are shown in Table 3.3.

Table 3.3 Resistor Colour Code

Colour of band	Number
Black	0
Brown	1
Red	2
Orange	3
Yellow	4
Green	5
Blue	6
Violet	7
Grey	8
White	9

The first coloured band nearest the end of the resistor tells you the first digit of its resistance. The second digit is given by the second band. The third band gives the number of zeros.

Many resistors have a fourth band that gives the uncertainty, or tolerance, of the resistance. A gold fourth band means that the resistance is accurate to within 5 per cent of the indicated value, whereas a silver band means that the accuracy is within 10 per cent. If there is no fourth band, the resistor is accurate within 20 per cent.

Here is an example of how you find the resistance of a carbon resistor. The coloured bands are brown, black, red, and silver (Figure 3.27).

The first digit is 1, the second digit is 0, and the number of zeros after the second digit is 2. This resistor has an indicated resistance of 1000 Ω. Since the fourth band is silver, however, the resistance is accurate to within 10 per cent, or 100 Ω, of this indicated value. Thus, the actual resistance of the resistor is between 900 and 1100 Ω.

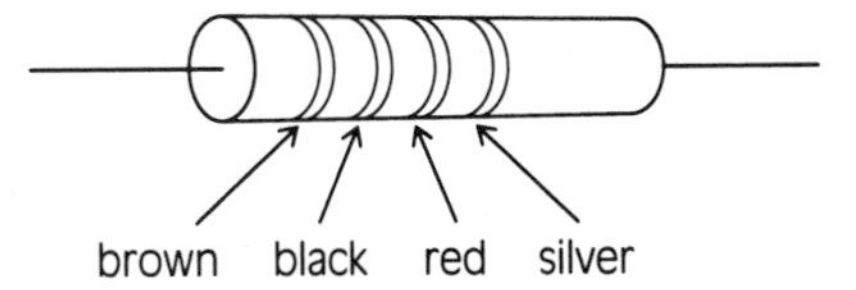

Figure 3.27
This resistor has a resistance of 1000 Ω.

ACTIVITY 3F Voltage and Current for a Carbon Resistor

In this activity, you will find how the current flowing through a resistor is affected by the applied voltage.

MATERIALS

four chemical cells and holders
multi-range meter
two resistors of different values
connecting wires
alligator clips

PROCEDURE

1. Make two copies of Table 3.4, one for each resistor, in your notebook.
2. Use the colour bands of one of your resistors to determine its resistance. Write this resistance, and the tolerance, in the space at the top of one of the tables you drew in step 1. Then do the same for the second resistor.
3. Use your multi-range meter set to a resistance scale to measure the resistance of your two resistors. You should obtain results that are within the uncertainty of the value you found from the colour bands. Record your measured values in your tables.
4. Use your multi-range meter to measure the voltage of each of the four cells you are using. Be sure that the voltage of each cell is the same. Then, knowing that you will connect the cells in series, you can fill in the voltage column of your tables.
5. Set up the circuit shown schematically in Figure 3.28. In this diagram, the arrow through the battery indicates that it has a variable voltage. Use your multi-range meter to measure current. Begin with the meter set to its highest amperage scale.
6. With one cell in the circuit, measure the current through the first of your resistors. If you get a small reading on your ammeter, use a more sensitive scale. Record the current (in amperes) in your table. (To change from milliamperes to amperes, move the decimal point three places to the left.)
7. Predict what will happen to the current if you use two cells in series. Connect the circuit and measure and record the current. Repeat this procedure with three and four cells in the circuit.
8. Repeat steps 5, 6, and 7 with the second resistor and record the results you obtain on the second table.

Table 3.4 Sample Data Table for Activity 3F

Marked resistance ______ Ω Tolerance ______
Measured resistance ______ Ω

Number of chemical cells	Voltage (V)	Current (A)	Ratio (V/A)
1			
2			
3			
4			

SAMPLE ONLY

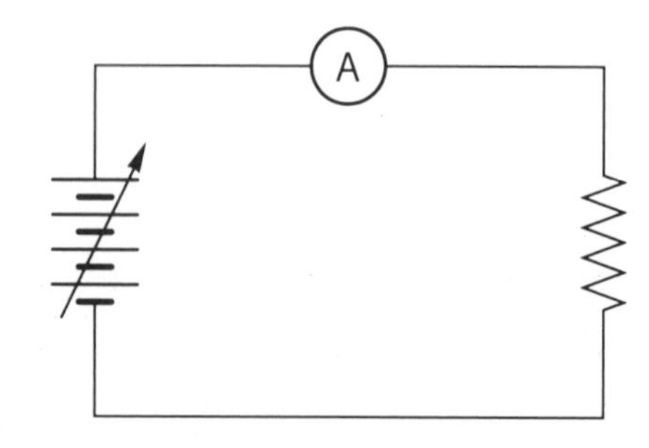

Figure 3.28
Connect your apparatus like this to measure the current and voltage for a carbon resistor.

DISCUSSION

1. What happens to the current in the carbon resistor when the battery voltage
   (a) doubles,
   (b) triples,
   (c) quadruples?
2. Predict what the current in the resistor would be if you used a battery of
   (a) 9 V,
   (b) 12 V,
   (c) 3.2 V.
3. Complete the ratio column of the tables by dividing the voltage applied to the resistor (in volts) by the current through the resistor (in amperes). For each table, does the ratio remain constant or nearly constant? How does the ratio compare with the marked resistance and the resistance you measured with the multi-range meter?

OHM'S LAW

Early in the 1800s, Ohm experimented with wires and found that the current through a wire is proportional to the applied voltage. If, for example, the voltage applied to a wire was doubled, the current through it would double as well. Ohm's discovery that the current through a material is proportional to the applied voltage is called **Ohm's law.**

Ohm's law applies to wires and many other materials as long as the electric current does not cause some change in the material. For example, the filament of a light bulb is made of wire. At its operating voltage and current, however, the wire filament of a light bulb gets very hot, increasing its resistance. Because of this change, Ohm's law no longer applies.

Ohm defined the resistance of the wire as the ratio of voltage to current. Thus,

$$\text{resistance} = \frac{\text{voltage}}{\text{current}}$$

This definition is often written as

$$R = \frac{V}{I}$$

where V stands for voltage, I for current, and R for resistance. If the voltage is measured in volts and the current in amperes, the units of resistance are volts/amperes. Thus, an ohm is one volt/ampere.

Using some algebra, you can rearrange the equation like this:

$$V = IR$$

or like this:

$$I = \frac{V}{R}$$

The equation $R = V/I$ can be applied to parts of the complex electronic circuits found in radios and television sets, radar sets, and computers. There are also some simpler applications, as you can see in the following examples.

1. A small portable stove operates from a 120 V outlet. When it is in use, 6 A flow through it. What is the resistance of the heating element of the stove?

$$R = \frac{V}{I} = \frac{120\,\text{V}}{6\,\text{A}} = 20\,\Omega$$

2. The filament of a certain light bulb has a resistance of 96 . To operate at its proper brightness, it needs a current of 1.25 A to pass through it. What voltage must be applied to this light bulb to make it work?

$$V = IR = 1.25\,\text{A} \times 96\,\Omega = 120\,\text{V}$$

3. If a kettle is operated from a 120 V outlet and its heating element has a resistance of 10 Ω, what current is in the heating element?

$$I = \frac{V}{R} = \frac{120\,\text{V}}{10\,\Omega} = 12\,\text{A}$$

REVIEW 3.3

1. How does a chemical cell produce a voltage difference between its terminals? Do you produce a voltage difference between a piece of acetate plastic and a paper towel when you rub them together?
2. What advantages are there to connecting cells in (a) series and (b) parallel?
3. What is conventional current? How does it differ from a flow of electrons?
4. What units are used to measure current? What is a milliamp?
5. In a series circuit of Christmas tree lamps, what would happen to the lamps if one burned out? Would this happen if the lamps were arranged in parallel?
6. Give three examples of resistors that you would find in your home.
7. Where would you find a rheostat in your home?
8. A 3 V battery sends 0.10 A through a light bulb. What is the resistance of the filament of the bulb?

CHAPTER • REVIEW

Key Ideas

- Static charges result when an object has an excess or a deficiency of electrons.
- Objects with like charges repel one another, whereas objects with unlike charges attract each other.
- Charged objects attract neutral objects.
- Each element has a unique number of positively charged protons in the nuclei of its atoms. The number of protons is indicated by the element's atomic number.
- Electrons are negatively charged and are found outside the nuclei of atoms.
- Voltage is the amount of work done per coulomb of charge transferred. It is measured in volts.
- An electric current is a continuous flow of charged particles that results from a voltage difference in an electric circuit. Electric current is measured in amperes.
- A series circuit provides only one pathway for current, whereas in a parallel circuit there is more than one pathway.
- Resistors are devices that limit the flow of current in a circuit. Their resistance is measured in ohms.
- Ohm's law states that the current through a material is proportional to the voltage applied to it. Ohm's law applies as long as the resistance of the material does not change.
- Resistance is defined by the equation $R = V/I$.

Vocabulary

static electric charge
electroscope
attract
repel
positive charge
negative charge
neutral
atom
subatomic
electron
neutron
nucleus
proton
atomic number
induced charge
coulomb (C)
potential energy
electrical potential energy
chemical cell
terminals
voltage (V)
volt
voltmeter
battery
series circuit
parallel circuit
electric current
conventional current
ampere (A)
ammeter
conductor
insulator
electric circuit
schematic diagram
series–parallel circuit
resistance
resistor
ohm
Ohm's law

1. Make a concept map linking 12 to 15 words of your choice from the vocabulary list.
2. (a) In your notebook, draw the parts of Figure 3.29 that relate to static electricity. Then label these parts with

Figure 3.29
This student is holding an acetate strip charged by rubbing with a paper towel near the electroscope.

terms from the vocabulary list.

(b) Draw a schematic diagram for the circuit shown in Figure 3.30. Then label the diagram appropriately with as many of the terms from the vocabulary list as you can.

Connections

1. If you rub a balloon against your clothing and then touch the balloon to a wall, it will often stick there. Explain why this happens.

2. (a) Compare the mass of an electron with that of a proton.
   (b) How does the space occupied by these particles in an atom compare?

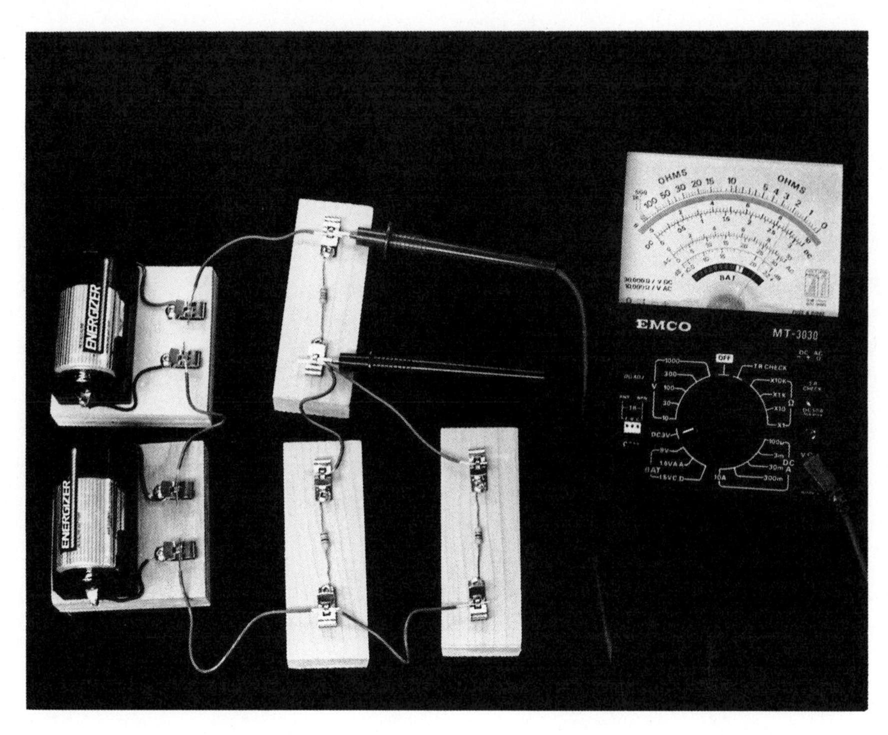

Figure 3.30
Draw and label the schematic diagram of this circuit.

3. In Activity 3B, you may have tried charging a metal bar by rubbing it with some other material. If so, you were most likely unsuccessful in this attempt. Why?

4. If lamps are connected in parallel to a 3 V battery, what voltage would you measure between the terminals on each lamp? Draw a diagram to show the circuit and how you would connect a voltmeter to measure the voltage for each lamp.

5. Figure 3.31 shows schematic diagrams for three circuits. In which circuit will the ammeter reading be (a) the greatest and (b) the least? Give reasons for your answers.

6. Calculate the current that would be indicated by the ammeter in Figure 3.31a.

7. A current of 10 A flows through a red-hot stove element when the voltage applied to the element is 240 V.
   (a) Calculate the resistance of the element under these conditions.
   (b) A student removed the element from the stove and connected a 6 V battery to its terminals. She found that the current through the element was greater than she had predicted on the basis of Ohm's law. Explain why.
   (c) Using Ohm's law, calculate the current the student would predict in (b).

Figure 3.31
In which circuit will the current be the greatest? In which circuit will it be least?

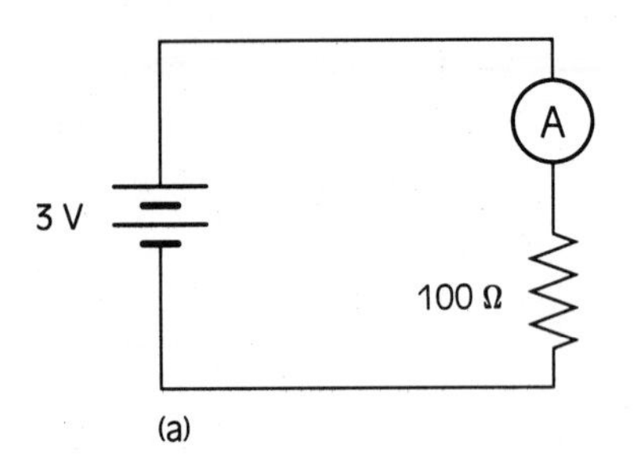

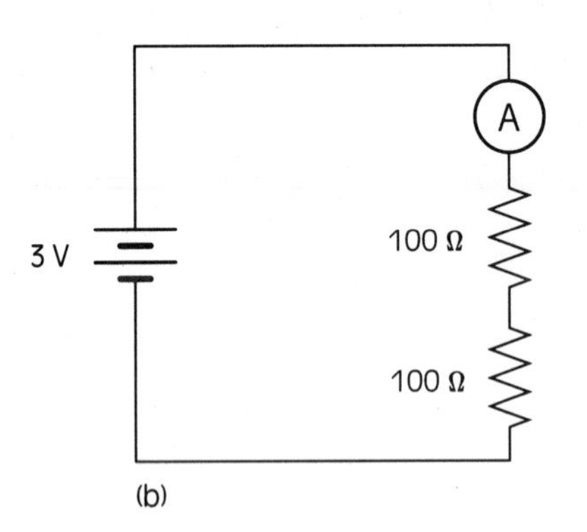

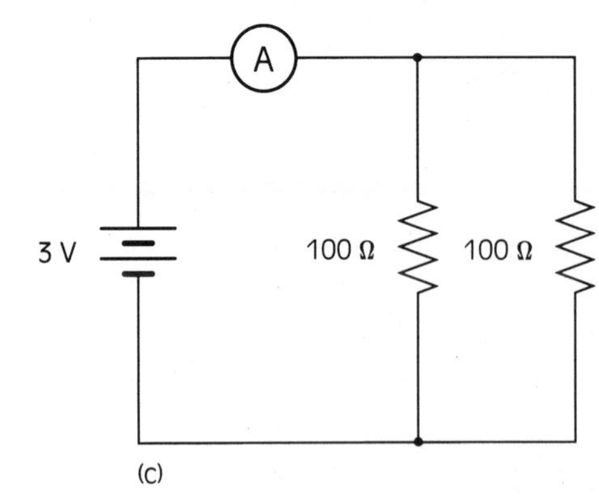

Explorations

1. Make a poster that explains how to use and read a multi-range meter.
2. Use a reference book to find out how an ammeter can be converted into a voltmeter. Then, with your teacher's permission, see if you can make a 0–10 V voltmeter from an ammeter.
3. What is the total resistance of two or more resistors connected in series? Devise a simple experiment that would allow you to find the answer to this question. When you know the answer, write a rule for the total resistance of resistors in series.
4. What is the total resistance of two or more resistors connected in parallel? See if you can find out by carrying out a simple experiment. If you can, write a rule for the total resistance of resistors in parallel.

Reflections

1. Refer to Activity 3A in your learning journal. Which ideas about electricity do you still hold? Which ideas have changed? Why or how did they change? Make a diagram to show new ideas that you have learned.
2. Benjamin Franklin's explanation of electricity was incorrect, even though it allowed him to make many correct predictions. Was Franklin a poor scientist because of errors in his explanation? Could there be errors in the modern theory of electricity? Are there some things about electricity that you cannot yet explain?

CHAPTER 4

Magnetism and Electricity

The aurora borealis, or northern lights, can occasionally be seen in many parts of British Columbia. Auroral displays occur at unpredictable times—often late at night or early in the morning. City lights can block out the night sky and make all but the brightest aurora invisible. The aurora may also flicker as if someone were waving a searchlight at the sky. For these reasons, many British Columbians have never seen an auroral display. Others may have seen one without realizing what it was.

If you are fortunate enough to see the aurora borealis, you may see shimmering curtains of light that fill the sky from the horizon to a point high overhead. The light can sizzle and slash across the sky in hues of pink and green. This beautiful phenomenon is the result of the interaction between the Earth's magnetism and electrically charged particles in the Earth's upper atmosphere. It is just one example of the relationship between electricity and magnetism.

ACTIVITY 4A What Do I Know about Magnetism?

In this activity, you will try simple experiments with magnets. Then you will combine your present knowledge of magnetism with any additional information you learn from the experiments to write brief summary statements.

MATERIALS

two small magnets like those used to hold messages on refrigerators
assorted materials such as paper clips, bits of paper and Styrofoam, thumbtacks, pennies, nickels, pencils, pens, and rulers

PROCEDURE

1. Observe what happens when you bring your two magnets together in different ways. Record your observations in your learning journal.
2. Find how different materials are affected by a nearby magnet. Record your observations in your learning journal.

DISCUSSION

In two or three statements, summarize
(a) the effect of magnets on one another,
(b) the effect of magnets on different materials.

4.1 Magnetic Forces and Fields

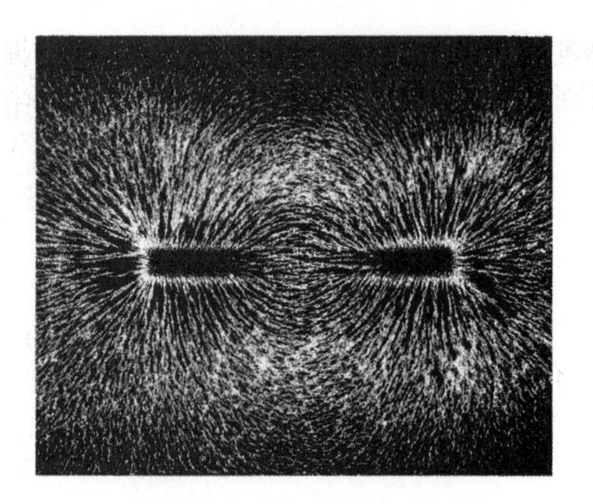

Figure 4.1
Iron filings show the shape of the magnetic field around a magnet.

A **magnet** is an object that can exert a pull, or force, on a piece of iron. This magnetic force is different from many of the other forces with which you are familiar, however. For example, to move a book, you have to touch the book and pull it. A magnet can attract a piece of iron without touching it.

The English scientist Michael Faraday (1791–1867) invented the idea of a **magnetic field** to help people understand how magnets could exert forces on objects without touching them. Faraday imagined that a magnetic field surrounds a magnet, and perceived that any nearby piece of iron, or a second magnet, would be affected by its interaction with the magnetic field. Thus, a magnetic field may be defined as a region in space where magnetic effects can be detected.

Because of their small size and mass, tiny pieces of iron called iron filings are especially affected by a magnetic field. If iron filings are sprinkled carefully around a magnet, they tend to collect in areas where the magnetic field has the greatest effect (Figure 4.1). In these areas, the magnetic field is considered to be the strongest. Fewer iron filings collect in other areas. There the magnetic field is weaker. The iron filings are randomly distributed where the magnetic field of the magnet is too weak to affect them. Thus, iron filings can be used to indicate the shape of a magnetic field.

ACTIVITY 4B Mapping Magnetic Fields

In this activity, you will use iron filings to map the magnetic field around one or two bar magnets.

MATERIALS

two bar magnets
paper
salt shaker filled with iron filings

PROCEDURE

1. Place a magnet beneath a piece of paper as shown in Figure 4.2. Then use a salt shaker to sprinkle iron filings on top of the paper. Tap the paper gently as you do this, and the magnet's magnetic field will affect the arrangement of the iron filings. Draw the appearance of the iron filings' arrangement in your notebook, showing just enough detail to indicate the general shape of the field.

   When you have finished your drawing, return the iron filings to the salt shaker so that you can reuse them. Lift the paper straight up from the magnet and form it into a V shape so that the filings collect in the bottom of the V. Pour the filings back into the salt shaker.

2. Place two bar magnets end to end (but not touching) so that they attract. Then use paper and iron filings to map the magnetic field, as you did in step 1. Draw the shape of the magnetic field in your notebook.

3. Place two bar magnets end to end so that they attract and are touching. Then use paper and iron filings to map the magnetic field, as you did in step 1. Draw the shape of the magnetic field in your notebook.

4. Place two bar magnets end to end (but not touching) so that they repel. Then use paper and iron filings to map the magnetic field, as you did in step 1. Draw the shape of the magnetic field in your notebook.

5. Use iron filings to find the shape of the magnetic field around two bar magnets arranged in the three different ways shown in Figure 4.3. As before, draw the shapes in your notebook.

DISCUSSION

1. Where is the magnetic field strongest near a single bar magnet? How do you know?

2. What happens to the strength of the magnetic field around a single bar magnet as the distance from the magnet increases? How do you know?

3. Describe the shape of the magnetic field around two magnets that are
   (a) attracting one another,
   (b) repelling one another.

4. Compare the drawing you made in step 3 above (when you placed two bar magnets end to end so that they attracted and touched one another) with your drawing for a single bar magnet. What are the similarities between the two drawings?

5. Predict the shapes of the magnetic fields around other arrangements and shapes of magnets—for example, a single horseshoe magnet or a circular ceramic magnet. Test your predictions.

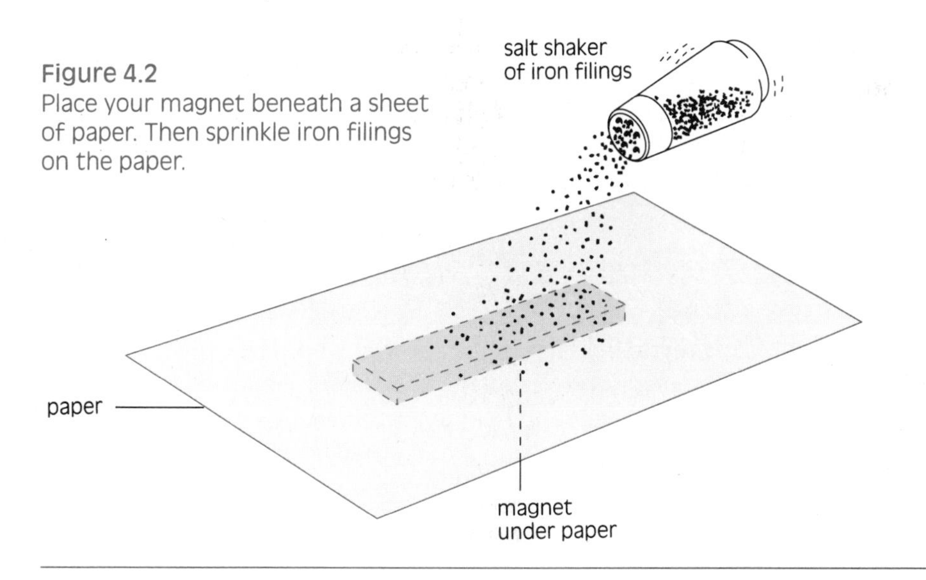

Figure 4.2
Place your magnet beneath a sheet of paper. Then sprinkle iron filings on the paper.

Figure 4.3
Find the shape of the magnetic field for each of these arrangements of magnets.

magnets side by side and attracting

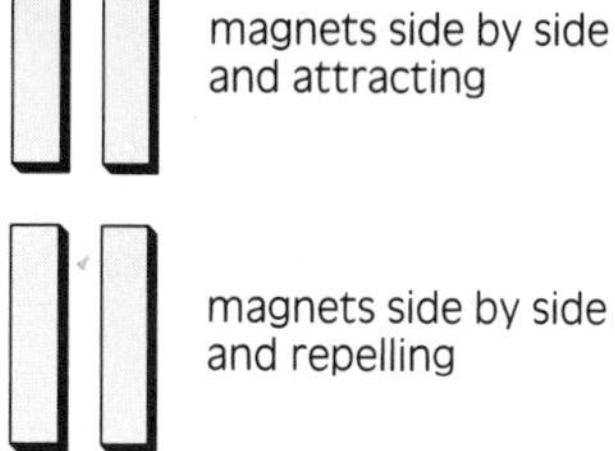

Figure 4.4
This suspended bar magnet acts as a compass. The end of the magnet that points northward is the magnet's north pole.

MAGNETIC POLES

The regions of a magnet where the magnetic field is strongest are called the **magnetic poles** of the magnet. On a bar magnet, these poles are located at either end of the magnet. Since the ends of two magnets can either attract or repel one another, the poles of each magnet must be of two types.

The two types of poles can be distinguished from one another by suspending a bar magnet so that it is free to move (Figure 4.4). When the magnet is released, it will turn so that one end points northward and the other end points southward. The end of the bar magnet that points northward is called a **north pole**, and the end that points southward is a **south pole.**

The suspended bar magnet is a type of compass; it can be used to indicate direction. A compass used in orienteering or by a hiker has a suspended bar magnet too; it is just in a more convenient form.

The needle of a compass placed near the north pole of a bar magnet will turn so that its south pole points at the bar magnet. The needle of a compass placed near the south pole of the bar magnet will turn so that its north pole is directed at the magnet. Thus, you can see that unlike magnetic poles attract one another.

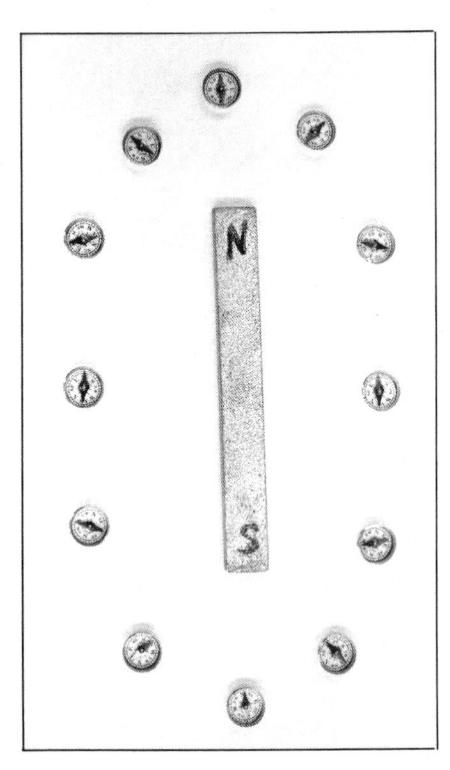

Figure 4.5
These small compass needles point in the direction of a magnetic line of force.

MAGNETIC LINES OF FORCE

The needles of a number of small compasses placed near a bar magnet line up as you can see in Figure 4.5. You can imagine a line connecting one magnetic pole to the other in these compass needles. The arrowhead on the line shows the direction of the north poles of the compass needles.

Michael Faraday used lines like these to depict magnetic fields. He called the lines **magnetic lines of force**. Magnetic lines of force are most concentrated where the magnetic field they depict is the strongest. Where the magnetic field is weaker, the lines are farther apart. The needle of a tiny compass placed anywhere along a magnetic line of force indicates the direction of the line at that point; the needle's north pole points in the direction of the magnetic line of force.

In the presence of a magnet, iron filings act like tiny compass needles. That is why the patterns they form near a magnet resemble drawings that show the magnetic lines of force of a magnetic field.

THE EARTH'S MAGNETIC FIELD

The Earth and several other planets are surrounded by magnetic fields. The Earth's magnetic field has the shape that would be expected if there were a giant bar magnet deep inside the Earth. As a result, there are two places where the magnetic field is most concentrated. These places are called the Earth's **geomagnetic poles**; they are not in the same places as the Earth's geographic North and South poles. Also, the positions of the geomagnetic poles are not constant. At present, the north geomagnetic pole is on Bathurst Island in northern Canada, about 1500 km from the Earth's geographic North Pole. The south geomagnetic pole is on the edge of Antarctica, just inside the Antarctic Circle and about 2600 km from the Earth's geographic South Pole (Figure 4.6). Since a compass needle points at these geomagnetic poles, there usually is a difference in direction between **magnetic north** (north as indicated by a compass) and "true" or geographic north (the direction to the Earth's geographic North Pole).

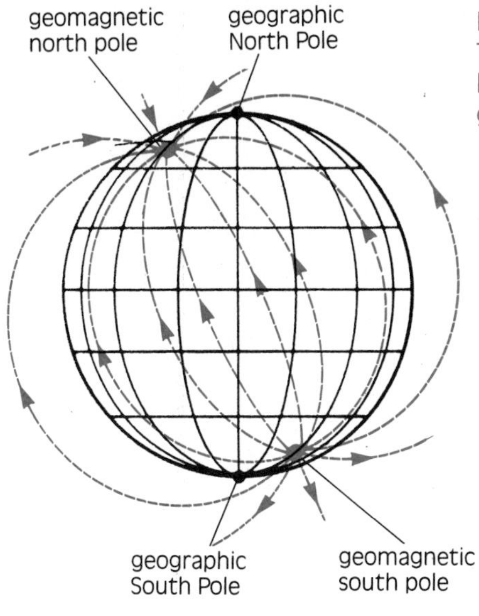

Figure 4.6
The Earth's geomagnetic poles are located fairly near the Earth's geographic poles.

DID YOU KNOW?

Although the magnetic pole near the north is called the geomagnetic north pole, it attracts the north end of a compass. Thus, since opposite magnetic poles attract, the geomagnetic north pole is actually a south pole. Likewise, the south geomagnetic pole is actually a north pole, since it attracts the south end of a compass needle.

REVIEW 4.1

1. Figure 4.7 shows the arrangement of iron filings around three bar magnets. If the end of the bar magnet marked "A" is a north pole, what kinds of poles are B and C?
2. In your notebook, trace the positions of the bar magnets shown in Figure 4.7. Then draw some of the lines of force around the magnets. Use arrows to show the direction of these lines of force.
3. How did you determine the directions of the magnetic lines of force in question 2? How would you determine the directions of the magnetic lines of force if you were given a magnet whose poles were not labelled N or S?
4. (a) If you followed a magnetic compass north from where you live, where would you end up?
   (b) If you followed the lines of longitude north from where you live, where would you end up?
5. Explain why a compass needle points to the east of true north when it is used in British Columbia. Would a compass needle point to the east of true north in Nova Scotia?
6. At the Earth's geomagnetic north pole, the north end of a compass needle is pulled strongly downward. Explain why.

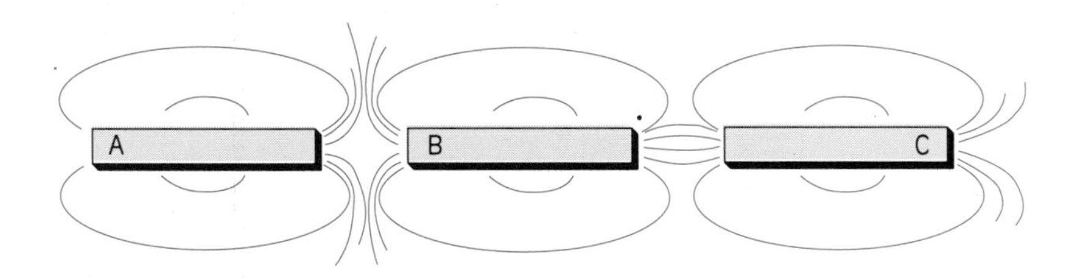

Figure 4.7
What types of poles are A, B, and C?

4.2 Magnets and Magnetism

Any materials that are affected by a magnetic field are said to be **magnetic**. Only a few of the **elements** (substances made of just one type of atom) are magnetic. Often they can be combined with non-magnetic elements and still retain their magnetism. For example, magnetic iron that contains a small amount of carbon makes steel. Steel is also magnetic.

ACTIVITY 4C Magnetic Substances

In this activity, you will use iron filings to show the magnetic field around magnets and a magnetic material.

MATERIALS

two bar magnets
paper
iron filings
soft iron bar

PROCEDURE

1. Cover a soft iron bar with paper and sprinkle iron filings on the paper as in Activity 4B. Do you observe any regular pattern to the iron filings?

Figure 4.8
Arrange your magnets and iron bar like this.

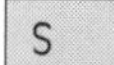

2. Place the iron bar you used in step 1 between two bar magnets as shown in Figure 4.8.

   Again cover the bar with paper and sprinkle iron filings on top. Do you observe any regular pattern to the iron filings now? Make a sketch in your notebook to show the arrangement of the filings.

3. Without disturbing the paper and iron filings, remove the bar magnets that you placed near the magnetic material in step 2. Then tap the paper gently so that the iron filings can rearrange themselves. Do the filings indicate any regular magnetic field pattern?

DISCUSSION

1. Is there a definite magnetic field around an unmagnetized iron bar? Explain how you know.
2. When an iron bar is placed in a magnetic field, is there a definite magnetic field around the bar? How do you know?

EXTENSION

Magnets and magnetism are used in many places. You can find them in telephones, can openers, loudspeakers, generators, tape recorders, screwdrivers, and refrigerator and cupboard doors, for example. List all the uses of magnets or magnetism you can find in your home. Compare the uses you find with those of your classmates.

WHAT CAUSES MAGNETISM

Magnetism originates inside atoms. There the movement of electrons around the nucleus produces a magnetic field. As a result, there is some magnetism associated with every atom. In atoms of most elements, however, the magnetic fields produced by the various movements of the electrons nearly cancel each other out. These elements, therefore, do not show strong magnetic properties.

Only four elements are strongly magnetic. Iron, nickel, cobalt, and gadolinium have atoms in which, it is believed, there are more electrons moving one way than any other way. Thus, the magnetic effects of the moving electrons do not cancel each other out. Atoms of these elements tend to arrange themselves into groups in which all the individual magnetic fields are lined up in the same direction. Groups of aligned atoms are called **magnetic domains**. Each magnetic domain acts like a small magnet with its own north and south poles.

If the magnetic domains in a substance are aimed in many different directions, as in Figure 4.9a, the substance is magnetic. It is not itself a magnet, however, because normally the magnetic effects of the domains tend to cancel each other out. If a piece of a magnetic element such as iron is placed in another, external magnetic field, most of the iron's domains line up with the lines of force in this field. The iron is then temporarily magnetized and acts like a magnet. This effect is shown diagrammatically in Figure 4.9b.

Figure 4.9
These drawings show how magnetic domains are arranged in (a) unmagnetized iron and (b) partially magnetized iron.

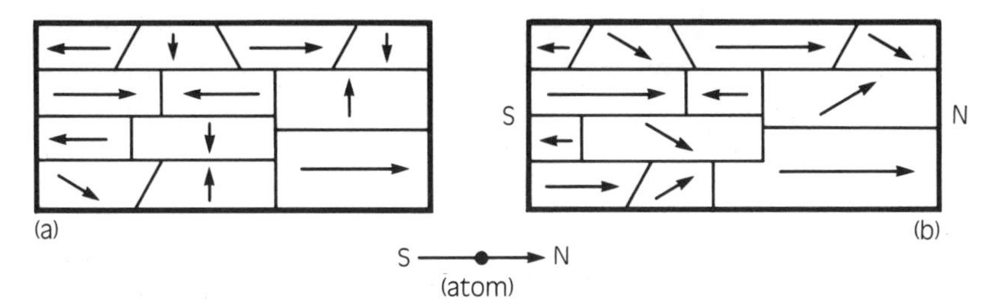

DID YOU KNOW?

Above a certain temperature called the Curie point, the strongly magnetic elements iron, nickel, cobalt, and gadolinium become non-magnetic. The very rare element gadolinium becomes non-magnetic at 16°C, whereas the other three elements must be heated to hundreds of degrees Celsius. Can you think of a reasonable explanation for why these elements become non-magnetic at temperatures above their Curie points?

PERMANENT MAGNETS

When most magnetic materials are removed from a magnetic field, the magnetic domains revert to the orientation they had before they were placed in the magnetic field. As a result, the materials no longer act like magnets. One such material is called soft iron—the kind of iron found in nails, for example.

In some materials, it is hard for the domains to change their positions. Once they have done so, they tend to remain in that new position. These materials remain magnets for a long time, even though no other magnet is nearby. They are called **permanent magnets**. Good permanent magnets can be made by combining iron with other metals. Two examples are cobalt steel, made of cobalt and iron, and alnico, which contains aluminum, nickel, and cobalt, as well as iron.

Permanent magnets are not really permanent, since they lose their magnetism if the domains are allowed to become disordered again. The domains can become disordered if the magnets are dropped or banged together, if they are heated, or if they are improperly stored so that like magnetic poles are together (Figure 4.10).

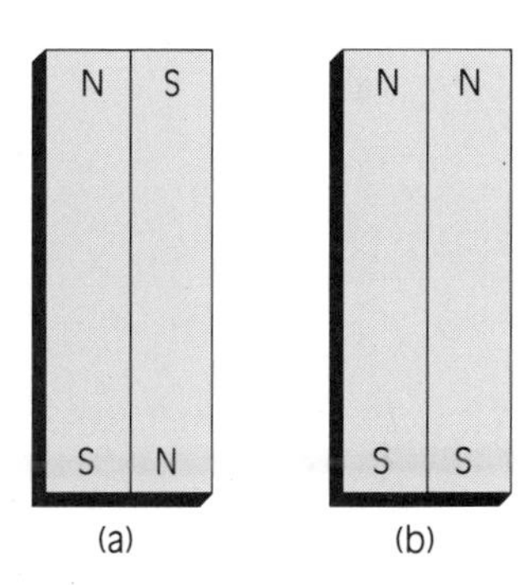

Figure 4.10
If magnets are stored with their unlike poles together, as in (a), the magnetic fields of both magnets help to keep the magnetic domains aligned. As a result, they retain their magnetism for a much longer time than if they are stored with their like poles together, as in (b).

REVIEW 4.2

1. What is a magnetic domain?
2. How are magnetic domains arranged inside
   (a) a permanent magnet?
   (b) an unmagnetized piece of iron?

   Illustrate your answers with labelled diagrams.
3. Use a diagram to show what happens to the domains inside a magnetic object when the object is placed in a magnetic field.
4. (a) Explain why "permanent" magnets are not really permanent.
   (b) How can magnets be made to lose their magnetism quickly?

→

(c) How should magnets be stored so that they will retain their magnetism as long as possible?
(d) Explain why your storage methods would help maintain the magnetism of the magnets.

5. Use what you have learned about magnetic domains to explain why iron filings line up along magnetic lines of force. Use a diagram to help with your explanation, if necessary.

6. If you cut a bar magnet in half, you do not obtain separate north and south poles. Use what you have learned about the causes of magnetism to explain why not.

4.3 Electromagnetism

Imagine yourself as a science student at the University of Copenhagen almost 200 years ago. Your physics instructor is Professor Hans Oersted (1777–1851). Professor Oersted is explaining that there may be a connection between electricity and magnetism, but no one has yet discovered what that connection is. Professor Oersted gives a demonstration with a battery, some wire, and a compass. All at once he becomes very excited. Electric current is flowing through the wire and something is happening to the compass needle, even though the wire is not made of iron or any other strongly magnetic substance. When the lecture is over and you and your fellow students are leaving, you notice that the professor is still excitedly experimenting with his apparatus.

This demonstration, which occurred in a lecture at the University of Copenhagen in 1819, was the first evidence for the connection between electricity and magnetism. Oersted had good reason to be excited by the discovery. It was a major scientific breakthrough that has led to the present belief that all magnetic fields are caused by the movement of charged particles, like the electrons moving around atomic nuclei, or the electrons moving through wires.

ACTIVITY 4D *The Magnetic Field around a Current-Carrying Wire*

How does a wire carrying an electric current affect a magnet? You will find out in this activity.

MATERIALS

PART I

50 and 20 cm lengths of stiff bare copper wire (14-18 gauge)
alligator clips
electric cell or rechargeable nickel-cadmium cell
push-button switch
small compass

PART II

materials from Part I
paper cup
three more small compasses

PROCEDURE

PART I

1. Connect the two pieces of wire, the cell, and the push-button switch to make the circuit shown in Figure 4.11. When the switch is closed, why will there be a large current flowing in the wires? The push-button switch will ensure that the current is not left on between observations. Why is this important?

CAUTION!
Keep wires away from your face and other students' faces.

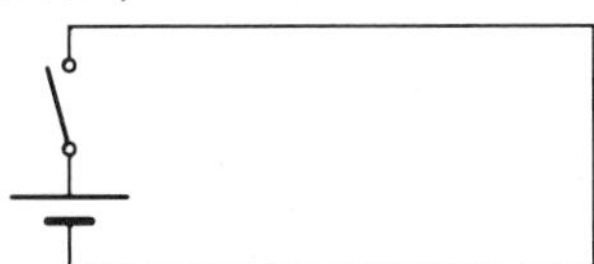

Figure 4.11
Connect your circuit like this.

2. With the switch disconnected, bring the compass near the longer piece of wire. Does the wire affect the compass needle?

3. Line up the longer section of copper wire so that it points in the same direction as a compass needle placed underneath it (Figure 4.12).

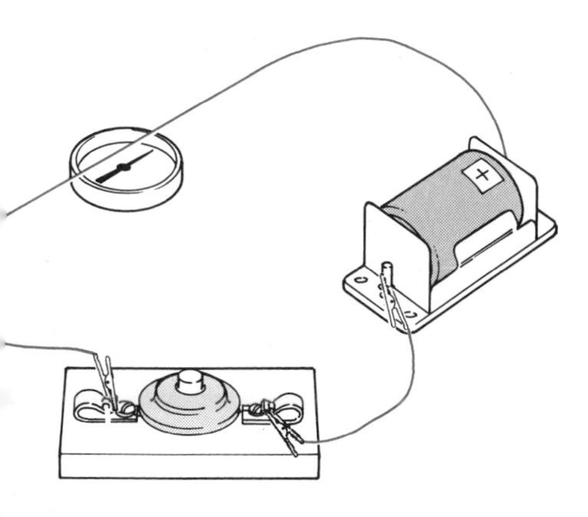

Figure 4.12
Arrange the compass needle so that it is parallel to the wire.

4. Draw a diagram like Figure 4.13a in your notebook. The " + " represents the end of the wire that is connected to the positive terminal of your cell. Push the button just for an instant and observe the direction of the compass needle. Draw an arrow to show where the compass needle came to rest when the push button was depressed.

5. Predict what will happen if you reverse the connections to the cell. Try it and see if you were correct. Record your observations on the second part of the diagram.

Figure 4.13
Draw these diagrams in your notebook and use them to show which way the compass needle points in each of your experiments.

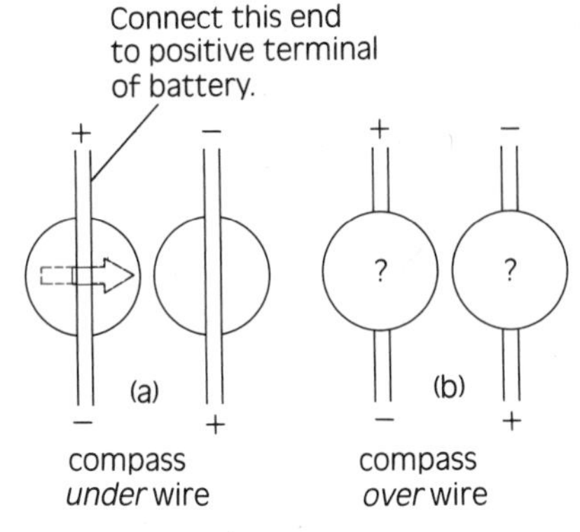

6. In your notebook, draw a diagram showing the compass above the wire, as in Figure 4.13b. Predict what will happen to the compass needle if you place the compass in this position. Then put the compass above the wire and see if you were correct. Again, record your results on your diagram.

7. Predict what will happen when you reverse the connections to the cell. Then test your predictions and record your results on the second part of your copy of Figure 4.13b.

DISCUSSION

1. Would you notice any effect on the compass needle due to the magnetic field around the wire if the compass and the wire were first set up as shown in Figure 4.14? Explain your answer.

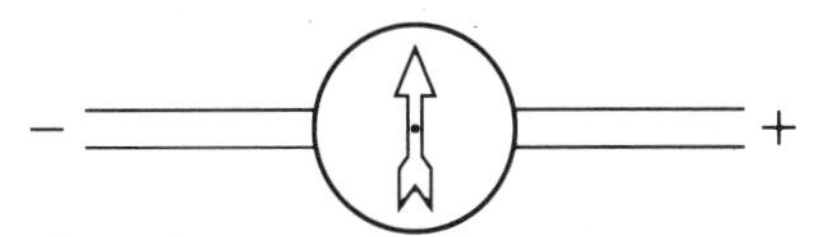

Figure 4.14
If the compass needle and wire are arranged like this when no current is flowing in the wire, will you notice any effect on the compass needle when current flows?

2. Are the magnetic lines of force near a current-carrying wire aligned in the same direction as the current flow or at right angles to it? Explain how you know.

PART II

Now that you have some idea about the direction of the magnetic lines of force above and below a wire carrying an electric current, you may be able to predict how the lines of force are arranged all around a wire. If you set up your apparatus so that it is arranged as shown in Figure 4.15, what should happen when the current is flowing?

Figure 4.15
Arrange the long wire so that it passes vertically through the paper cup. Use the paper cup to support your compasses.

1. Predict which direction the compass needles will point when the wire shown in Figure 4.15 is carrying a current. Record your prediction in a diagram like that shown in Figure 4.16.

Figure 4.16
Use diagrams like these to show which way the compass needles point in each of your experiments. Show which end of the cell is connected to the top end of the wire with a + or — as shown.

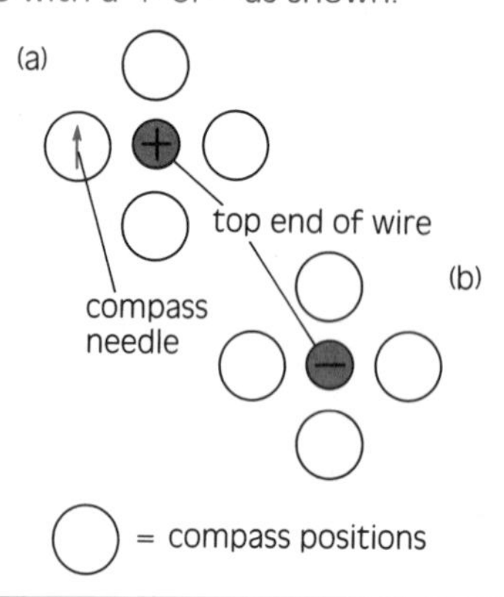

2. Now connect your wire, push button, and cell exactly as shown in Figure 4.15. Be sure that you have connected the cell correctly. Now press the button and see if your predictions were correct. If the results were different from those you predicted, show the results with a diagram.

3. If you reverse the connections to the cell, what should happen? Draw a diagram to show your prediction, as you did in step 1. Then test your prediction and record the results.

DISCUSSION

1. Are the results you obtained in Part II of this experiment consistent with those you obtained in Part I? Explain how you know.

2. Does the magnetic field around a wire in which a current is flowing have two distinct poles? Explain your answer.

3. How does reversing the direction of current in a wire affect the magnetic lines of force around the wire?

THE RIGHT-HAND RULE

Within one week after Oersted reported his discovery in 1820, the French scientist André Marie Ampère (1775–1836) discovered the results you may have found in Activity 4D. Ampère developed a rule, the **right-hand rule**, for predicting the direction of the magnetic lines of force around a current-carrying wire. The right-hand rule states that if you point the thumb of your right hand in the direction of the conventional current flow in a wire, the curl of your fingers will then indicate the direction of the magnetic lines of force around the wire (Figure 4.17).

As for permanent magnets, it is possible to show the shape of the magnetic field around a current-carrying wire. According to the right-hand rule, the lines of force around a current-carrying wire circle the wire. This pattern can be demonstrated by using fine iron filings, but you need a large current for the procedure to work well (Figure 4.18).

EXTENSION

In Activity 4D, you found magnetic fields around bare copper wires when current was flowing through them. Would the results be different if the wires were insulated? Devise an experiment that would allow you to answer this question. Then try it out.

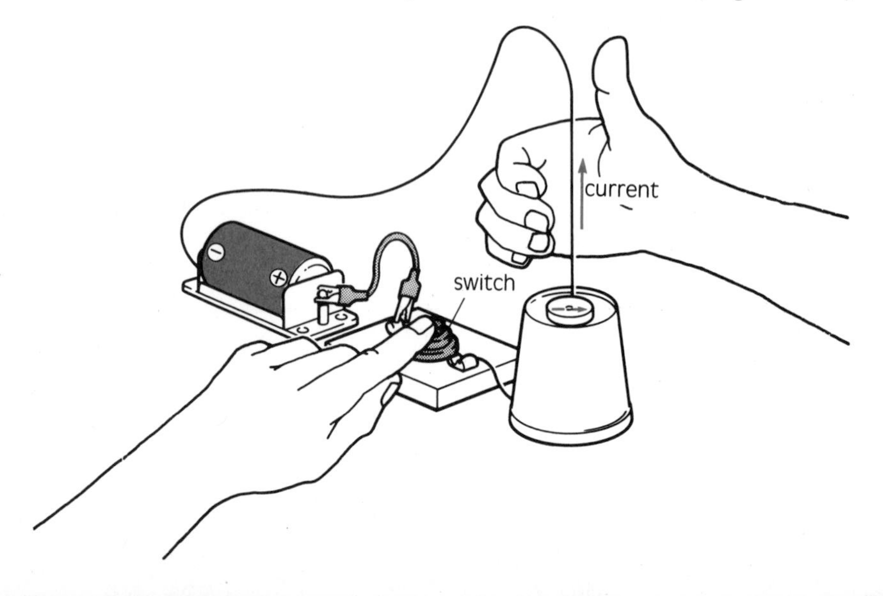

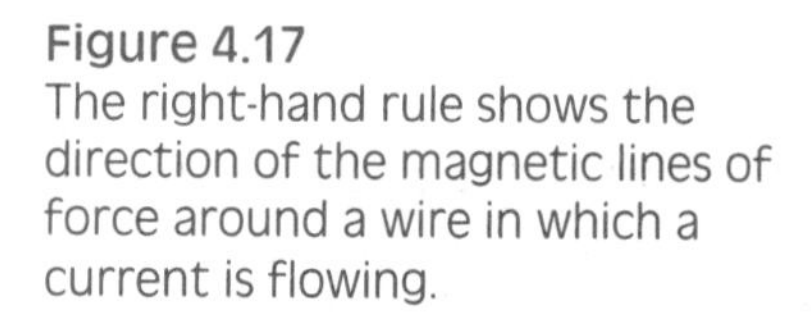

Figure 4.17
The right-hand rule shows the direction of the magnetic lines of force around a wire in which a current is flowing.

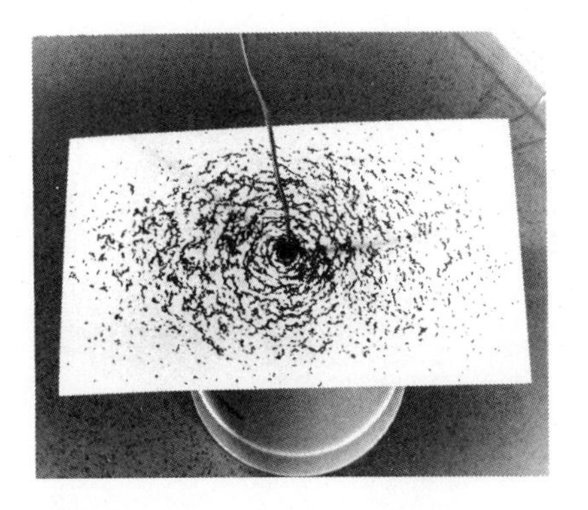

Figure 4.18
Iron filings show the shape of the magnetic field around a single current-carrying wire.

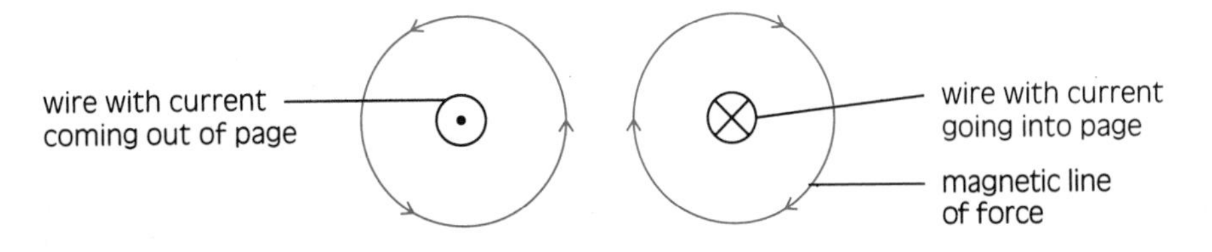

Figure 4.19
You can use these symbols to show the direction of current in a wire and the resulting magnetic lines of force around it. Imagine that the direction of current is indicated by an arrow. A dot in the middle of the wire shows the current arrowhead coming towards you, out of the page. An x shows the current arrowhead going into the page. The coloured arrow shows the direction of the magnetic lines of force, as predicted by the right-hand rule.

Filings show the shape of the magnetic field around the wire; a compass needle indicates its direction. The current direction and the resulting magnetic field are shown symbolically in Figure 4.19.

Magnetic fields around current-carrying wires have many consequences, which can be significant. For example, in an aluminum smelter, the large electric currents in wires produce undesired magnetic fields, as shown in Figure 4.20.

In many other situations, electric currents in wires can be used to produce magnetic fields where and when they are desired. In these situations, the wire is often wound into a coil.

Figure 4.20
At the Aluminum Company of Canada smelter at Kitimat, B.C., very large electric currents pass through wires leading to the containers in which aluminum is produced. As a result, there is a very strong magnetic field inside the smelter, strong enough to magnetize coins.

ACTIVITY 4E *A Coil's Magnetic Field*

You have seen the shape of the magnetic field around a straight piece of wire in which a current is flowing. In this activity, you will see what happens to the magnetic field if the wire is wound into a coil.

MATERIALS

- 2 m length of insulated copper wire (20–24 gauge)
- clear plastic pill bottle
- alligator clips
- electrical cell
- push-button switch
- small compass

→

PROCEDURE

1. Write down what you think will happen to the magnetic field of a wire that is made into a coil, or draw a diagram to show what you think will happen.
2. To make a coil, carefully wind about 20 turns of wire, spread evenly, along the length of a plastic pill bottle. Use tape to hold the turns in place (Figure 4.21).

CAUTION!
Keep wires away from your face and other students' faces.

3. Make a drawing similar to the one in Figure 4.22. Then connect your coil to the cell in the same way as it is shown in the diagram.
4. Place the compass at each of the positions in the drawing and use arrows to show which way the compass needle points when current is flowing through the coil.
5. Predict what will happen if you reverse the connections of your coil to the cell. Show your predictions on a diagram similar to Figure 4.22 but with the direction of the current reversed.
6. Test your predictions by repeating step 3 of the procedure. Were your predictions correct? If not, why?

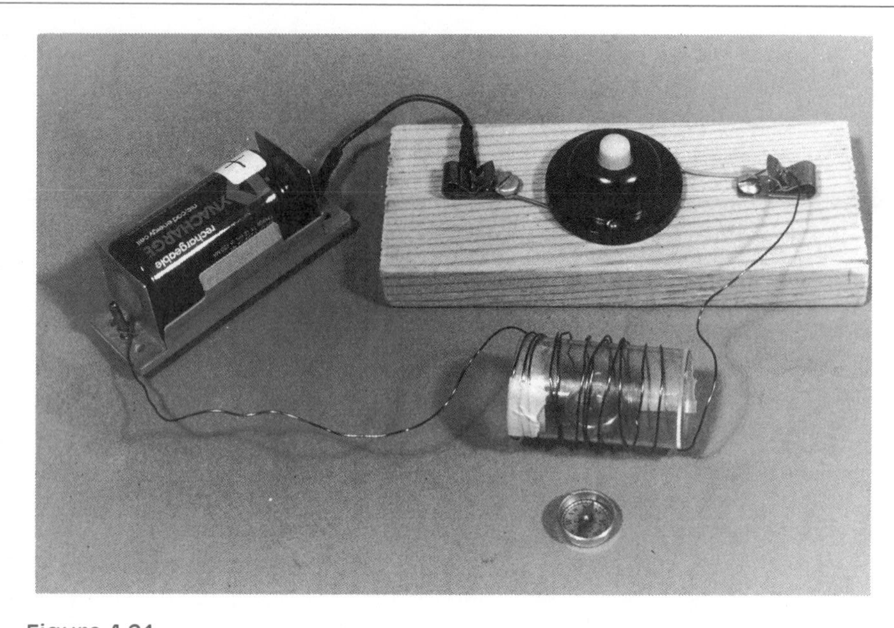

Figure 4.21
Which way will the compass needles point when current is supplied to the coil of wire?

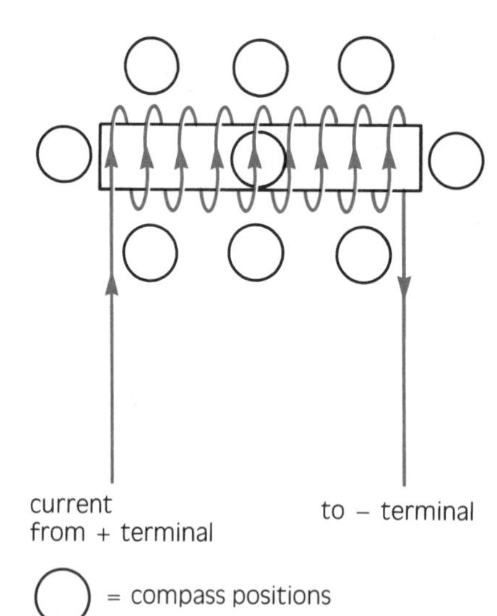

Figure 4.22
Copy this diagram into your notebook. Then connect your coil as shown in the diagram.

DISCUSSION

1. When the same amount of current is flowing through a straight wire and a coil, which seems to have the greater effect on a compass? Explain the reasons for your answer.
2. Apply the right-hand rule to a single turn of wire of the coil. Does the right-hand rule predict the direction of the magnetic lines of force near the coil? Does the right-hand rule predict the end of the coil that is the north pole?
3. Compare your drawings showing the directions of the lines of force around a coil with the diagram showing lines of force around a bar magnet in Figure 4.5. Are the drawings similar?
4. Does a coil have magnetic poles? Explain your answer. Use a labelled diagram if necessary.
5. Figure 4.23 shows how iron filings arrange themselves around (a) a single turn of wire and (b) many turns of a coil. A large current is necessary to obtain clear patterns like these.

   In your notebook, make drawings of both elements of Figure 4.23a and b. Then use the right-hand rule to find the direction of the magnetic lines of force in (a) and the direction of the current in (b).

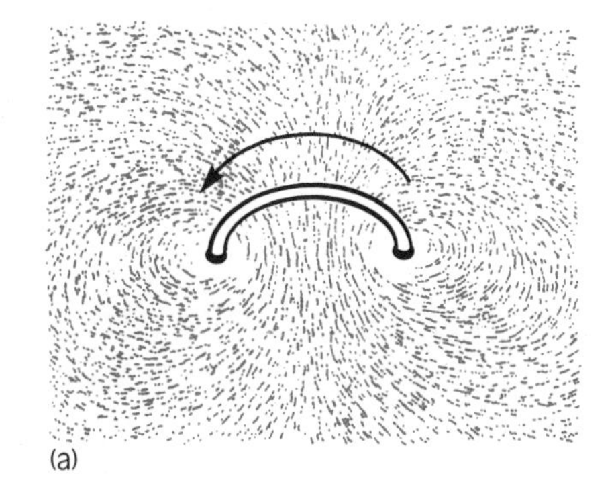

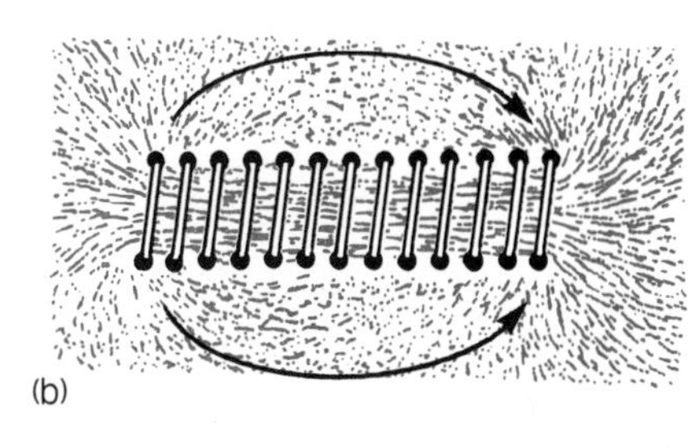

Figure 4.23
(a) The magnetic field around a single turn of wire. (b) The magnetic field around a coil. The coloured arrows indicate the direction of current flow in (a) and the direction of the magnetic field in (b).

ELECTROMAGNETS

Magnets that produce magnetic fields only when an electric current flows through them are called **electromagnets**. These magnets are usually constructed by winding a coil around a core made of a magnetic material such as soft iron. When current flows in the coil, the magnetic domains in the iron line up with the magnetic field produced in the coil.

Because electromagnets can be controlled electrically, they have many common uses. For example, they are used in doorbells, buzzers, telephones, electric motors, the "heads" of tape decks, and in the speakers contained in radios and television sets. Electromagnets control the water valves in dishwashers and washing machines, and if you drive a car, you use one every time you turn the key to start the engine (Figure 4.24).

Figure 4.24
Electromagnets are used in dishwashers, washing machines, and other appliances that you can find in your home.

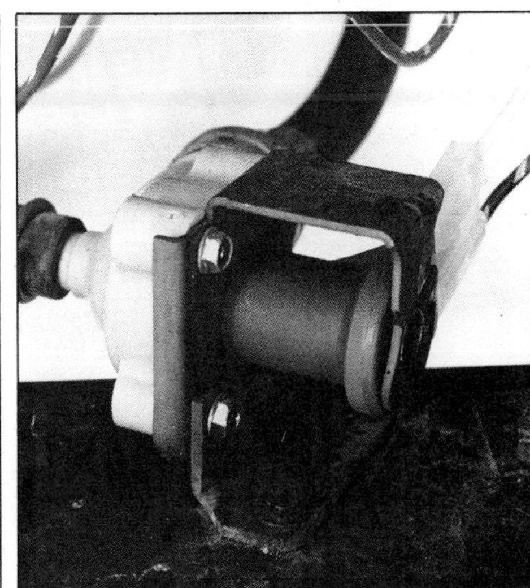

SCIENCE IN OUR WORLD

PRACTICAL USES OF ELECTROMAGNETS

Electromagnets are an important part of many complex pieces of equipment. When a fire alarm rings, for example, an electromagnet is working. Look at the figure on the left below to see what the inside of a bell looks like. When the switch is closed, current flows through the wire coil, creating an electromagnet. The electromagnet attracts the arm of the alarm and the hammer strikes the bell. However, the movement of the arm creates a gap between the arm and the contact screw, and the circuit is broken. Current stops flowing, so there is no longer a magnetic field either. As a result, the arm is no longer attracted and goes back to its resting position. Now the circuit is once again complete, and the process is repeated as long as the switch remains closed.

Tape recorders and tape players also rely on electromagnets. The actual tape on which a recording is made is a thin piece of plastic on which tiny magnetic particles are scattered. When you speak into the microphone, your voice is changed into electrical signals. Because your voice consists of sound waves that vary in frequency, intensity, and pitch, the electrical signals also vary. These signals produce a changing current that is sent to an electromagnet in the recording head of the tape recorder. Since the current going through the magnet changes, the magnetic field that is produced also changes. As a result, the tiny magnetic particles on the tape are rearranged in specific patterns. These patterns depend on the strength of the magnetic field at the time the tape passes by the recording head. The result is that the particles are reorganized in a way similar to that shown in the figure on the right below.

The same principle applies to videotapes, though they are much larger than audio tapes.

The floppy disks used in computers are also made of plastic with a magnetic coating. The magnetic patterns are arranged in circles, instead of in a straight line as in an audio tape. The rotating disk drive is able to "read" these patterns quickly to bring back the information stored on them.

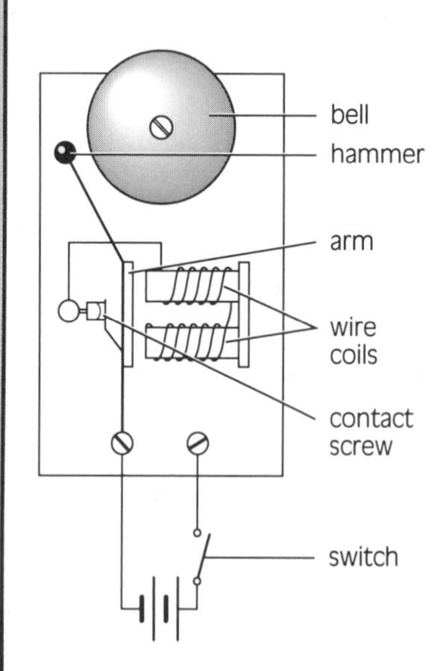

An electromagnet is used to make the bell ring in a fire alarm.

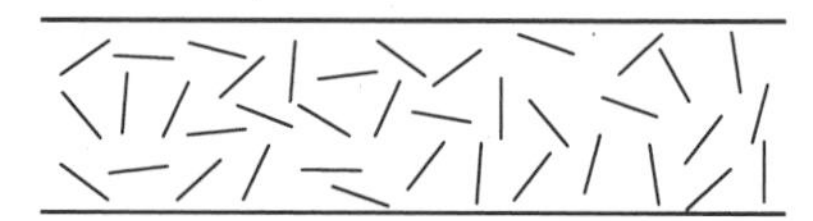

The magnetic particles on a blank tape are scattered at random.

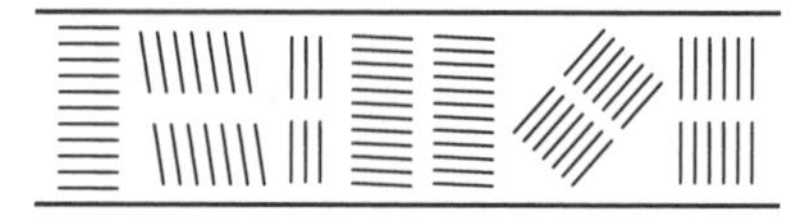

The magnetic particles on a recorded tape are arranged in various patterns, depending on the input from the microphone.

ACTIVITY 4F A Strong Electromagnet

How can you make a strong electromagnet? In this activity, you will explore some of the ways to increase the strength of an electromagnet.

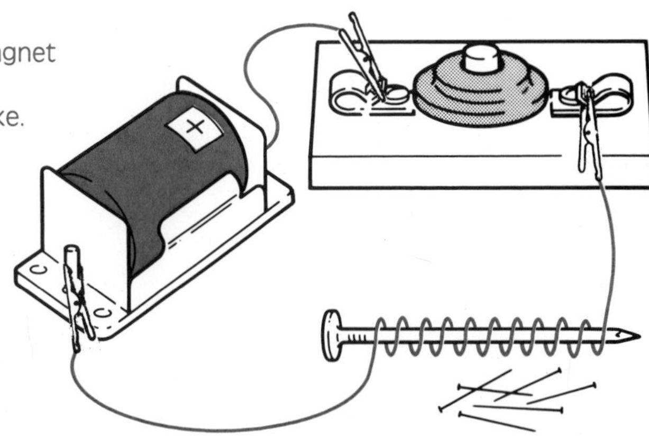

Figure 4.25
Make an electromagnet by winding wire around an iron spike.

MATERIALS (per group)

iron spike
2 m length of insulated wire
alligator clips
push-button switch
about 50 small finishing nails
four electrical cells
wooden pencil or small wooden dowel
aluminum rod

PROCEDURE

1. Start with one cell in your circuit, as shown in Figure 4.25.

> **CAUTION!**
> Keep wires away from your face and from other students' faces.

2. Wrap 10 turns of wire around the flat end of the spike. A lot of wire will be left hanging loose, but do not cut it.
3. Push the button and see if your simple electromagnet can pick up some nails.
4. Add more turns in the same direction (10 at a time). See what effect this has on the lifting strength of the magnet.
5. Predict what will happen if you use two cells in series. Test your prediction. Then try a battery of three and four cells.
6. Predict whether a core of aluminum or wood will work as well as an iron spike. Then conduct a test to see if you were correct.

DISCUSSION

1. Give at least three ways in which you can increase the strength of an electromagnet.
2. When you increase the number of cells connected to your electromagnet, what happens to the current flowing in it? Why should the cells be connected in series?
3. Is it necessary for a coil to have a magnetic core if it is to function as an electromagnet? Explain your answer.

REVIEW 4.3

1. Use the right-hand rule to predict the direction of a compass needle in each of the examples shown in Figure 4.26.
2. Use the right-hand rule to predict the positive terminal of the battery in each of the examples shown in Figure 4.27.

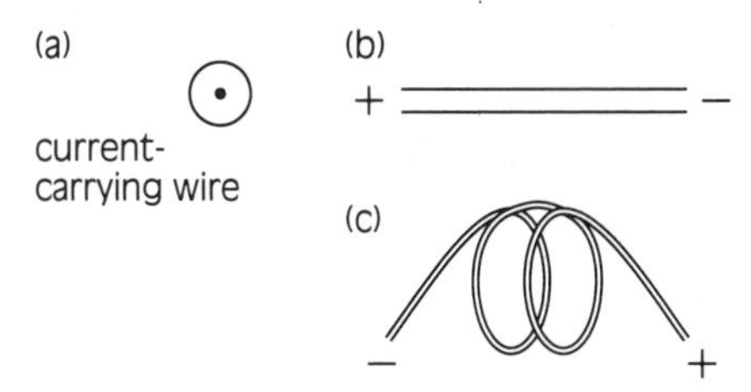

Figure 4.26
Find the direction of the magnetic lines of force.

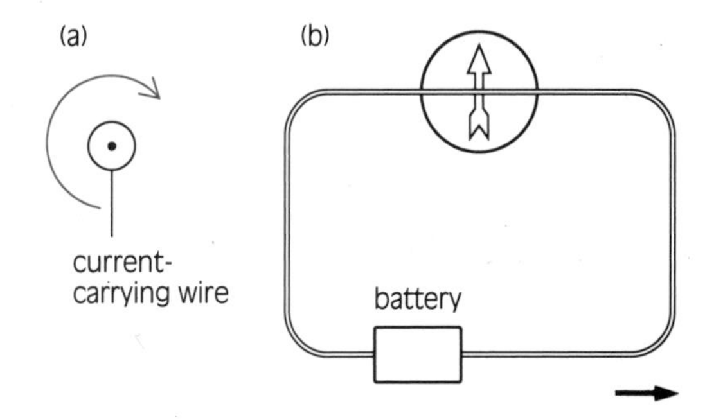

Figure 4.27
Which way does current flow in these wires? Which is the positive terminal of the battery?

3. Figure 4.28 shows an electromagnet made by wrapping a few turns of wire around a steel nail. Apply the right-hand rule to one turn and figure out which end of the electromagnet, A or B, is the north pole.
4. Suggest two ways in which you could strengthen the electromagnet in Figure 4.28.
5. Sketch Figure 4.28 in your notebook and draw a few lines of force around it, showing what its magnetic field would look like.

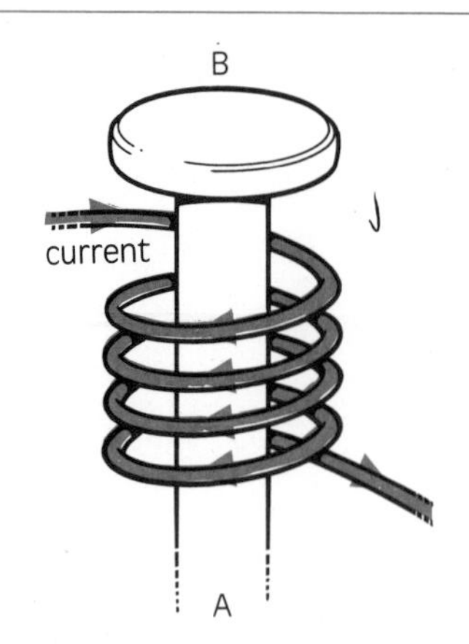

Figure 4.28
Which end of this electromagnet is north? What would the magnetic field around it look like?

4.4 Electric Generators

Oersted's discovery that an electric current flowing in a wire produces a magnetic field around the wire stimulated scientists to search for other connections between electricity and magnetism. Some scientists thought that if a current produced a magnetic field, perhaps a magnetic field could produce an electric current. In 1831, Michael Faraday and the American scientist Joseph Henry (1797–1878) showed that this was possible.

Faraday and Henry worked independently to produce an electric current with a magnetic field, but the methods they used were very similar. A battery connected to one of the coils in the apparatus (Figure 4.29) produced a magnetic field around the coil and ring. When the battery produced a steady flow of current in the coil, no current was detected in the second coil. At the instant the battery was connected or disconnected, however, an electric current was detected in the second coil, even though the two coils were not electrically joined.

In their first experiments, Faraday and Henry used electromagnets. Later, in a different experiment, Faraday demonstrated that permanent magnets could be used to produce electric current in wires. You can do this too. In your experiments, a special type of ammeter that responds to very small electric currents will make it easier for you to detect any currents that are produced. Such an ammeter is called a **galvanometer**. Its needle is centred when there is no current flow, so it can measure current that travels through the meter in either direction.

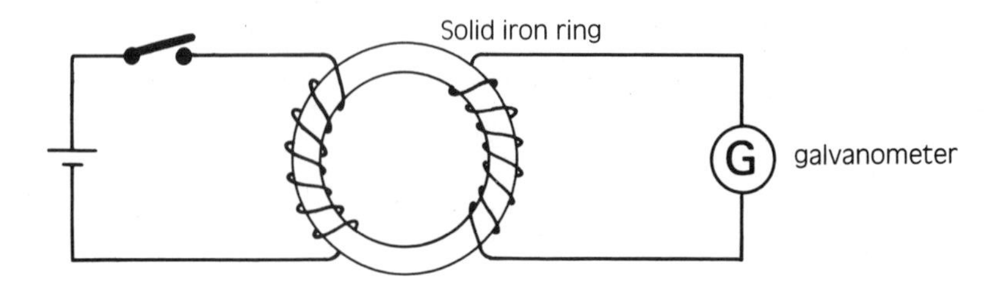

Figure 4.29
Michael Faraday and Joseph Henry used apparatus like this for their experiments.

ACTIVITY 4G Producing Electricity

In this activity, you will find out how a magnet can be used to produce an electric current in a wire.

MATERIALS

2 m length of insulated wire
alligator clips
galvanometer
two bar magnets

PROCEDURE

1. Make a coil by wrapping a length of insulated wire around your fingers. Tape the wire to hold the coil together, then connect it to a galvanometer (Figure 4.30).

> **CAUTION!**
> Keep wires away from your face and from other students' faces.

2. Move the north pole of a bar magnet into the coil. Does the galvanometer's needle move? If so, does it move to the left or the right?
3. Move the north pole of the bar magnet out of the coil. Which way does the galvanometer needle move?
4. Predict what will happen to the galvanometer needle if you move the north pole of the magnet in and out of the coil slowly and then quickly. Try the experiment and record your results. Do they match your predictions?
5. Predict in which direction the galvanometer needle will move if you move the south pole of the bar magnet in and then out of the coil. Try the experiment and record your result on diagrams.

Figure 4.30
Your coil should look like this when you are finished.

6. Predict what will happen to the galvanometer needle if you hold either pole of the magnet stationary within the coil. What do you find when you do this?
7. Increase the strength of the magnet you use by placing two magnets side by side, north pole to north pole and south pole to south pole. What do you think will happen to the galvanometer needle when you move this stronger magnet in and out of the coil? After doing the experiment, record your results.
8. Decrease the strength of the magnet you use by placing two magnets side by side so that their unlike poles are together. Predict what will happen to the deflection of the galvanometer needle when you move this weaker magnet in and out of the coil. Test your prediction.

DISCUSSION

1. A student observes that if a bar magnet is inserted into a coil, the needle of a galvanometer connected to the coil moves to the left.
   (a) In what direction will the needle move when the magnet is removed from the coil?
   (b) What does the change in direction of the galvanometer needle indicate about the current flow through the coil and the galvanometer?
2. What does the amount of deflection of the galvanometer needle indicate about the amount of current produced in the coil?
3. Describe the relative amounts of current produced in a coil when
   (a) a magnet is held stationary within a coil,
   (b) a magnet is moved slowly in and out of a coil,
   (c) a magnet is moved quickly in and out of a coil,
   (d) the strength of the magnet is increased,
   (e) the strength of the magnet is decreased.
4. According to the law of conservation of energy, energy cannot be created; it can only be changed from one form to another. Electricity is a form of energy. What source provided the energy that was converted to electrical energy in this activity?

EXTENSION

Suggest other ways in which you could increase the amount of current you can produce with a coil and magnets. Devise methods to test your predictions and try them out.

GENERATORS

Any device that can change mechanical energy into electrical energy is called an **electric generator**. The bicycle generator shown in Figure 4.31 is an example. It uses the mechanical energy of the moving bicycle wheel to turn the generator. This mechanical energy is converted into electrical energy inside the generator.

Michael Faraday made the first electric generator in 1831. Even though there were few uses for the electricity it produced at that time, the construction of this generator was an important accomplishment. Until then, all current electricity was produced using chemical cells, which were expensive and produced only small amounts of electricity. Generators, in contrast, can produce electricity in large quantities.

Generators produce electricity by moving a conductor through a magnetic field or by changing the magnetic field surrounding the conductor. Either procedure exerts a force on the electrons in the conductor, and as a result, the electrons move through it.

A coil and a magnet moving back and forth act as a generator of electricity because the movement of the magnet changes the magnetic field around the coil. An improved generator can be made by rotating a magnet near coils of wire. The changing magnetic field produced by the turning magnet is more effective in producing a current in the coils, and the magnet's motion is much smoother. The bicycle generator shown in Figure 4.31 is constructed in this manner.

In many generators, electromagnets are used instead of permanent magnets to provide a changing magnetic field. The electromagnets are made by winding coils of wire onto a steel core to make a part called an armature (Figure 4.32). (Only one coil is shown in the figure.) The ends of these armature coils are fastened to metal rings. Electricity is supplied to them through flexible electrical contacts called brushes, which slide over the surface of the rings as the armature turns. In this way, a current flows through the armature coils continuously, even though the armature is turning.

Figure 4.31
In this bicycle generator (a), a cylindrical permanent magnet revolves about a stationary coil of wire (b). AC electricity is generated in the coil.

(a)

(b)

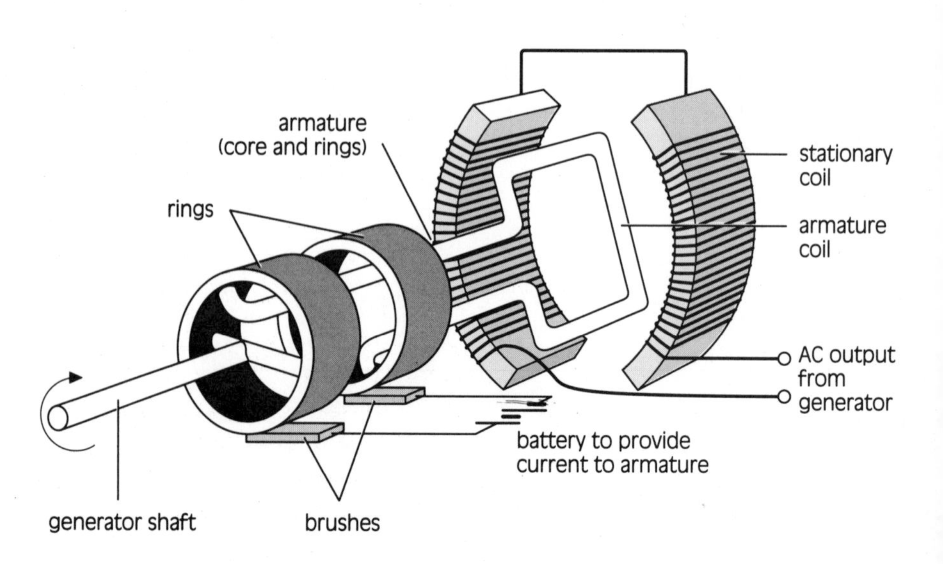

Figure 4.32
In this generator, a spinning electromagnet produces a changing magnetic field around the coils. The current for the electromagnet is supplied through the brushes and rings.

ALTERNATING CURRENT

When a magnet is inserted into a coil, the current produced in the coil flows in one direction. When the magnet is removed, the current flows in the opposite direction. Because of the changing direction of the current, this type of current is called **alternating current,** or **AC**.

The generator shown in Figure 4.32 produces alternating current. You can see why if you consider a simplified diagram of its operation. In the diagram, the armature is drawn as a bar magnet and only one stationary coil is shown. As a pole of this magnet approaches the stationary coil, current flows in one direction through the coil and galvanometer. As the pole leaves the coil, current flows in the opposite direction, just as it did in your experiment with magnets and coils (Figure 4.33).

Generators used by B.C. Hydro and other electric power companies generate AC, so the household circuits that they supply are also AC circuits. The AC changes its direction 120 times each second, much too fast to notice if a galvanometer is connected. A device called an oscilloscope can be used to show the changing direction of the current, however (Figure 4.34).

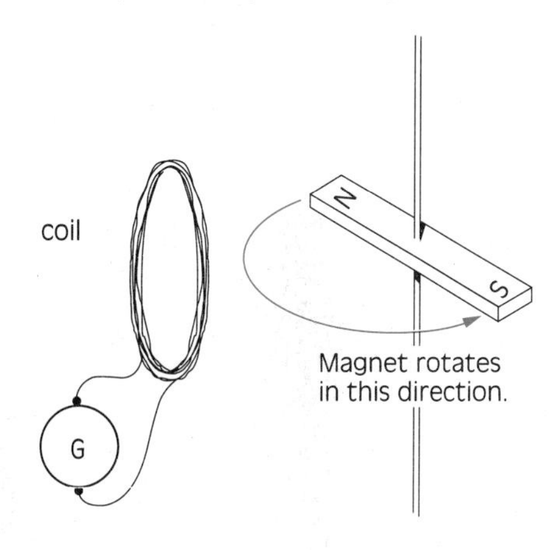

Figure 4.33
A simplified diagram of a generator.

Figure 4.34
On an oscilloscope screen, alternating current looks like this (a). The horizontal axis of the screen represents time (b). As the current increases to a maximum value, the line produced on the screen rises from zero to a maximum. Then, as the current decreases, the line goes back to zero. As the line goes below zero, the current reverses its direction. It goes to a maximum negative value, and then becomes less negative as the line goes back to zero. The current reverses again as the line rises above zero. This cycle is repeated 60 times each second for the AC electricity that is supplied to your home.

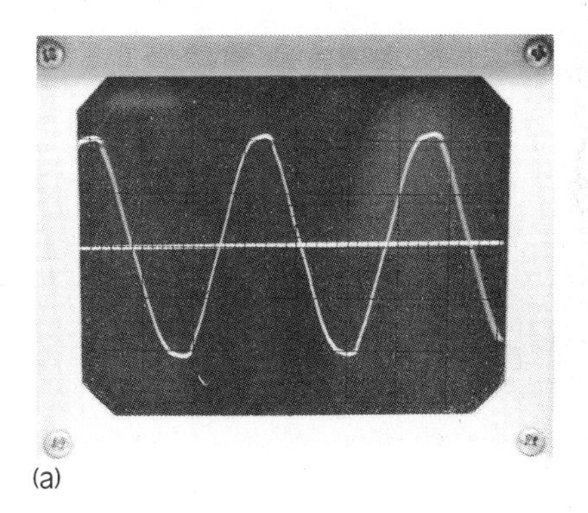

(a)

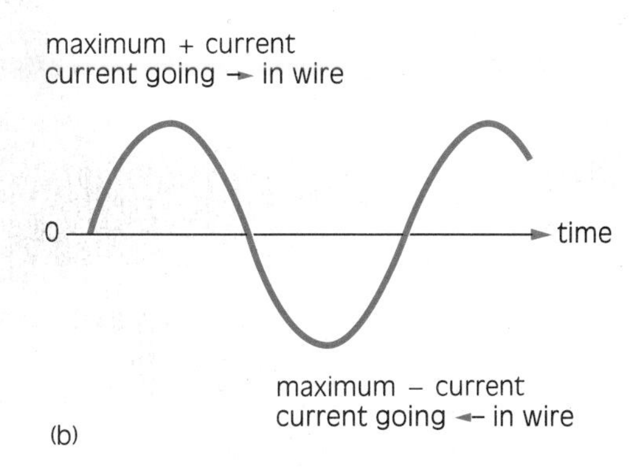

(b)

DIRECT CURRENT

Current that flows in only one direction in a circuit is called **direct current**, or **DC**. You have already had experience with direct current; it is the kind of current produced by chemical cells. That is why the ammeter needle did not change direction when you made a measurement in a battery circuit. Direct current is also used in electronic appliances such as television and stereo sets. The AC with which they are supplied is converted to DC inside these appliances.

Generators that produce direct current can be made, though the DC they produce is not constant like that from a battery. Figure 4.35 shows what DC electricity from a generator looks like on an oscilloscope screen.

Figure 4.35
The DC electricity produced by a generator looks like this on an oscilloscope screen (a). The DC always goes in the same direction in the wire—that is why the line on the oscilloscope is always above zero (b). You can see that the amount of current produced by the generator is not constant but goes up and down slightly as the generator turns. Would the DC produced by a battery fluctuate like this?

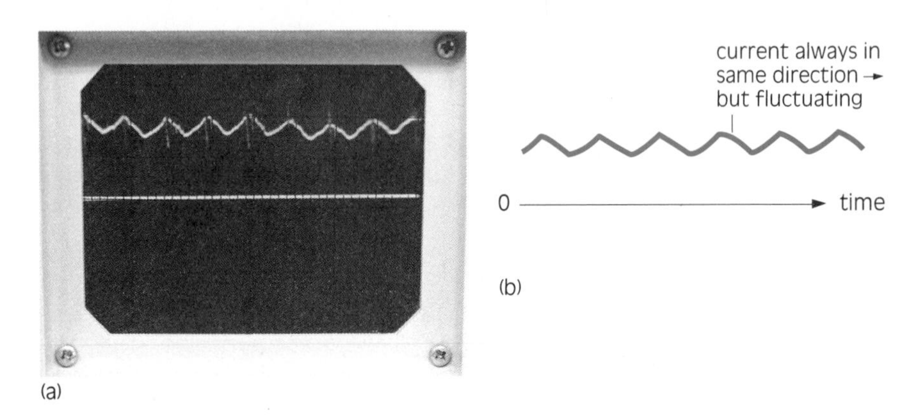

EXTENSION

You can use a motor as a generator. Obtain a small motor like those used to run battery-powered toys. See if you can devise a way to turn or spin the shaft of this motor so that it will act as a generator. Can you use your generator to recharge a rechargeable cell? What type of electricity must the motor produce if it is to recharge the cell, AC or DC?

GENERATORS OR MOTORS?

Many generators can be operated in two ways. If mechanical energy from flowing water or wind or a bicycle wheel is used to turn a generator, it converts the mechanical energy into electrical energy. If the generator is supplied with electrical energy, however, it can change the electrical energy into mechanical energy. In other words, the generator will then act like a motor. An **electric motor** has the same structure as a generator but has the opposite function. A motor changes electrical energy into mechanical energy.

In some ways, motors and generators are symbolic of a tremendous technological change that occurred in the late 1800s and the early part of this century. The electricity that generators produce made possible electrical appliances and lights, as well as products, like aluminum, that require large amounts of electricity for their manufacture. Electric motors made use of electricity to produce mechanical energy for moving the parts of machinery and home appliances such as refrigerators, freezers, and washing machines. Electricity thus became common in workplaces and in homes.

The technological change caused by electric motors and generators resulted from an understanding of the fundamental principles of magnetism and electricity, a study that began with scientists' curiosity about phenomena such as lightning, the aurora, and the Earth's magnetism.

REVIEW 4.4

1. List three ways in which the amount of electricity from an electric generator can be increased.
2. (a) What is the meaning of DC?
   (b) List three devices that use DC.
3. (a) What is the meaning of AC?
   (b) Name two sources of AC.
   (c) Draw a picture that shows the appearance of AC on an oscilloscope. Then draw a picture that shows what you predict the appearance of the DC from a battery would look like on an oscilloscope.
4. (a) Electricity is supplied to the armature of a generator through brushes and rings. Explain why continuous wires cannot be used for this purpose.
   (b) All large generators, like those used by B.C. Hydro, use electromagnets, whereas bicycle generators use permanent magnets. See if you can think of a reason for this difference.
5. If you had a generator supplying electricity to some appliances connected in parallel, would you expect any change in the generator as more appliances are turned on?

CHAPTER • REVIEW

Key Ideas

- Around a magnet there is a magnetic field, a region in which the magnet has a detectable effect. The field can be represented by magnetic lines of force.
- The direction in which a compass needle points shows the direction of the magnetic lines of force in a magnetic field.
- The magnetic field of a magnet is strongest at regions called magnetic poles.
- Two like magnetic poles repel one another; two unlike poles attract.
- An object can be magnetic without being a magnet. In a magnet, magnetic domains are lined up north to south, whereas in a magnetic material, magnetic domains are disordered.
- All magnetic fields are produced by the movement of charged particles.
- The direction of the magnetic lines of force around a current-carrying wire can be predicted by using the right-hand rule.
- Electricity can be produced by moving a conductor through a magnetic field or by changing a magnetic field near a conductor.
- Electricity may be direct current, DC, in which the current always goes in the same direction, or alternating current, AC, in which the current changes direction.

Vocabulary

magnet
magnetic field
magnetic poles
north pole
south pole
magnetic lines of force
geomagnetic pole
magnetic north
magnetic
element
magnetic domain
permanent magnet
right-hand rule
electromagnet
galvanometer
electric generator
alternating current (AC)
direct current (DC)
electric motor

1. Create a concept map relating as many of the terms in the vocabulary list as you can.
2. Make a sketch of Figure 4.36 in your notebook. Then label your sketch with as many of the vocabulary list terms as you can.

Figure 4.36
How many words in the vocabulary list apply to this drawing?

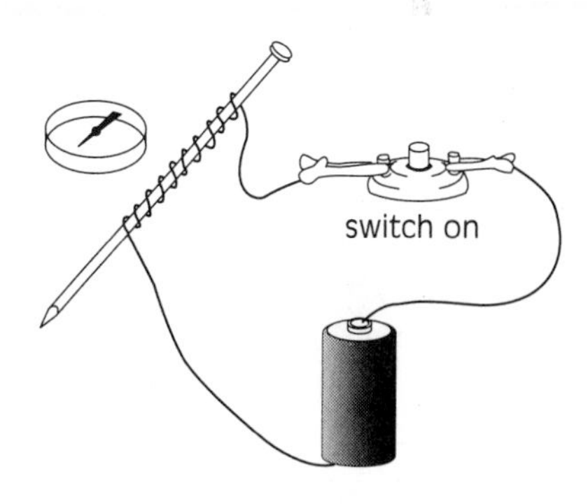

Connections

1. Information is stored magnetically on computer disks. Explain why magnets should be kept well away from them.

2. Explain why dropping or heating a permanent magnet can decrease the magnet's strength.
3. Why are electromagnets, rather than permanent magnets, used for lifting scrap steel in junkyards?
4. Could you distinguish between a bar electromagnet and a permanent bar magnet according to the patterns either magnet would produce with iron filings? Explain your answer.
5. A car engine runs a generator that keeps the car's battery charged. As the car engine turns faster, the generator turns faster. As the engine speeds up, what would you expect to happen to the amount of current produced by the generator? Why?

Explorations

1. Examine the electric motor from a toy. How many magnets are in it? (Count both permanent magnets and electromagnets.) Try to discover how the motor works.
2. Figure 4.37 shows two devices with which you are probably familiar. What is the purpose of the magnets in each of these devices?
3. Many animals look for landmarks and use the position of the sun and the stars to guide them when they migrate. It is now fairly certain that birds and other animals, such as dolphins and tuna, also have a magnetic sense. This sense may help guide them, especially at times when they cannot see landmarks. Design an experiment that would allow you to determine if a species of animal has a magnetic sense. Then look in reference books to see how others have tested for this type of sense.

Reflections

How have your ideas about magnets changed as a result of your investigations in this chapter?

Figure 4.37
See if you can explain the purpose of the magnets on (a) a lawnmower and (b) a triple-beam balance.

(a)

(b)

CHAPTER

Using Electricity in Your Home

The students lifting weights are doing a lot of work. They are using energy to do this work.

Electrical energy also does work. In your home, electrical energy does the work of operating your stove, refrigerator, toaster, and other electrical appliances.

The faster that any kind of energy is used, the more power is required. Both a toaster and the students in the photograph, for example, use energy quickly, so they use a lot of power. In fact, although each of the students shown in the picture is quite powerful, a toaster can be even more powerful. An electric water heater, another common and powerful home appliance, may be equal in power to this whole group of athletes.

The electrical power used in your home can be considerable. In this chapter, you will see how it is used and distributed.

ACTIVITY 5A What Do I Know about Work, Energy, and Power?

In your learning journal, prepare a table with three columns. In the first column, write the terms "work," "energy," and "power." In the second column, write down what you think is the scientific meaning of each term. Leave the third column blank. After you have read section 5.1, you may use the blank column for any changes or additions you need to make to your explanations.

5.1 Work, Energy, and Power

You have probably used the terms "work," "energy," and "power" many times. Each one has a common, everyday meaning as well as a stricter scientific definition. For example, you could go off to a job where you work hard and yet do very little work in the scientific sense. You can feel low in energy even when you have a lot of it, and you could be a powerful politician though you produce only a small amount of power. How does the scientific meaning of each term differ from its everyday usage?

WORK

The quickest way to show the scientific meaning of work is with an equation:

$$\text{work (joules)} = \text{force (newtons)} \times \text{distance (metres)}$$

Force, measured in newtons (N), indicates how hard an object is pushed or pulled. The distance, in metres (m), refers to the distance an object is moved by the force. The amount of **work** done is the product of these two quantities, so its units are newton metres, or **joules** (J).

Figure 5.1
This student is studying hard, but she is doing very little work.

The girl shown in Figure 5.1 is a good student. When she studies, she accomplishes her objectives, but she applies only small forces over small distances to move her pencil or to turn the pages of her books. As a result, she does only a few joules of work.

You can see this same girl training in the introductory photograph of the team of athletes. There, she is lifting a 200 N weight vertically through a distance of 0.5 m. Each time she lifts the weight, she does 100 J of work (200 N × 0.5 m = 100 J). Would she do any work if she tried to lift a weight that was too heavy for her to move?

ENERGY

Energy is defined as the ability to do work. When the girl lifts the 200 N weight, it gains an amount of potential, or stored, energy equal to the amount of work the girl did in lifting it. At its highest point, the 200 N weight has an additional 100 J of energy, though *it* is doing no work. When the weight comes back down, this 100 J of potential energy could be used to do 100 J of work. This amount of work is equivalent to the weight's loss of potential energy. Or it might simply be changed into other forms of energy, mainly heat.

As you can see from this example, there is a direct relationship between work and energy. The amount of energy corresponds to the amount of work that can be done by it. Since work is measured in joules, energy is measured in joules too.

INSTANT PRACTICE

1. A student lifts a hammer, which has a weight of 5 N, a vertical distance of 0.8 m. Calculate
   (a) the work the student does on the hammer,
   (b) the increase in the hammer's energy.
2. A student whose weight is 500 N walks up some stairs. If the height of one step is 0.2 m, calculate
   (a) the work done when the student goes up one step,
   (b) the work done when the student goes up 10 steps,
   (c) the increase in the student's potential energy from the bottom of the 10 steps to the top.
3. A student holds an 8 N science textbook stationary a distance of one metre above the floor. Calculate the amount of work done by the student while holding the book. Does the energy of the book change? Explain why or why not.

POWER

The rate at which energy is produced, absorbed, or transferred is called **power**, and this can be expressed by the equation

$$\text{power (watts)} = \frac{\text{energy (joules)}}{\text{time (seconds)}}$$

When energy is measured in joules and time in seconds, power is measured in joules per second. A joule per second is called a watt (W), in honour of James Watt (1736–1819), the Scottish engineer who developed the steam engine. Energy produced, absorbed, or transferred at the rate of one joule per second corresponds to a power of one watt.

The basketball players in Figure 5.2 are producing quite a lot of power. As they run up the stairs, they increase their potential energy by an amount equal to the work done to raise their bodies. This work equals their weight in newtons times the vertical distance they move. The sideways distance does not matter, because motion in this direction has no effect on their potential energy.

Figure 5.2
As these students run up the stairs, they produce a lot of power. This is another way of saying that they do work quickly.

ACTIVITY 5B How Powerful Am I?

In this activity, you will get an idea of the amount of power involved when you go up stairs. If you go as quickly as you can up the stairs, you will measure your maximum power. If you climb at your usual rate, you will find the power you normally produce when you climb stairs.

MATERIALS

(per class)
one or two bathroom scales

(per group of students)
stopwatch
metre stick

PROCEDURE

1. With a partner, decide on a method you can use to measure the height of a flight of stairs with a metre stick. Then determine the height of the flight of stairs and record this value in your notebook. This is the distance you will raise your body.
2. Most bathroom scales give your mass in kilograms. For this activity, you need your weight in newtons. Stand on the scale and find your mass. Record this number. Then change your mass to weight by multiplying your mass in kilograms by the force with which the Earth's gravity pulls on each kilogram, 10 N/kg. If your mass is 45 kg, for example, your weight is 450 N. Record your weight. This figure is the amount of force you will use to raise your body up the height of the stairs.
3. To find your maximum power, run up the stairs as fast as you can while your partner times you with the stopwatch. To find your normal power, go up the stairs at your usual rate while your partner times you with the stopwatch. Record the time it takes for either method (or for each method).
4. Calculate the amount of work you did in climbing the stairs by multiplying your weight by the height of the stairs (work = force × distance). This amount of work is also the increase in your body's potential energy.
5. Calculate the power you produced by dividing the increase in your body's energy by the time taken (power = energy/time).

DISCUSSION

1. When you climb the stairs, you increase your energy. Do you increase your energy more if you climb the stairs quickly than you do if you climb slowly? Explain your answer.
2. Would you produce more power if you climbed the stairs quickly than if you climbed slowly? Explain your answer.
3. A typical light bulb uses 60 W of power. Can you produce this much power? Can you produce this much power for 10 s? Would you like to produce 60 W of power steadily for an hour? Explain your answers.
4. On the back of a television set, a label like that shown in Figure 5.3 indicates how much power the TV uses. From the information shown in Figure 5.3 or the information on the back of the television set in your home, answer the following questions.
   (a) How much power does the TV use?
   (b) How much energy does the TV use in 1 s? (HINT: Remember that power = energy/time.)
   (c) If the TV is operated for 1 min, how much energy does it use? How much energy does it use in an hour?
   (d) If you had to provide the power to operate this TV by your own physical effort, how much TV would you choose to watch?
5. Most single fluorescent lighting tubes require 40 W of power for their operation. Count the number of these tubes in your classroom.
   (a) How many watts of electric power are used to light your classroom?
   (b) How much energy is used to light your classroom for 1 min? For 1 h?
6. A typical toaster uses 950 W of power. Is the toaster more powerful than you? Explain your answer.

Figure 5.3
The labels on appliances usually show their power requirements. What is the power required by this TV set?

ELECTRICAL ENERGY AND POWER

So far in this chapter, you have seen how energy, work, and power are related for objects you can see. But the same principles apply to objects that you cannot see, like electrons moving in wires. A battery, for example, exerts a force on electrons to make them move. Because a force is applied over a distance, work is done on the electrons. The more electrons that are moved, the more work is done. The more quickly the electrons are moved, the more power is used.

The voltage of a battery indicates how much work is done in joules per coulomb of charge transferred, or

$$\text{voltage} = \frac{\text{energy}}{\text{charge transferred}}$$

Thus, the work done, or energy, is given by the expression

$$\text{energy} = \text{voltage} \times \text{charge transferred}$$

Usually it is easiest to find the amount of charge transferred by measuring the current and the time over which it flows. From these data, the charge transferred can be calculated by rearranging the formula that defines current,

$$\text{current} = \frac{\text{charge transferred}}{\text{time}}$$

to

$$\text{charge transferred} = \text{current} \times \text{time}$$

Thus, the energy supplied by a battery can be calculated from the formula

$$\text{energy} = \text{voltage}\ (V) \times \text{current}\ (I) \times \text{time}\ (t)$$

Figure 5.4 shows the labels on a saw and on an electric heater. Using the information on the labels, you can calculate the energy used by either appliance. For example, if a carpenter uses the saw shown in Figure 5.4 for 10 seconds, the amount of energy used by the saw will be

$$\begin{aligned}\text{energy} &= V \times I \times t \\ &= 120\ \text{V} \times 13\ \text{A} \times 10\ \text{s} \\ &= 15\,600\ \text{J}\end{aligned}$$

DID YOU KNOW?

A commonly used unit of power is the horsepower. This unit dates back to 1783, near the start of the Industrial Revolution. At that time, steam engines were taking the place of horses as a source of power. For this reason, James Watt compared the power of his steam engines with that of a typical work horse. He called the power of one horse a horsepower.

This book uses an international system of units based on the metre, kilogram, and second. As you know, in this system the unit of power is the watt. One horsepower is equal to 746 W.

If you did Activity 5B, you found your power in watts. What is your horsepower? How does the power of a work horse compare with that of a 950 W toaster?

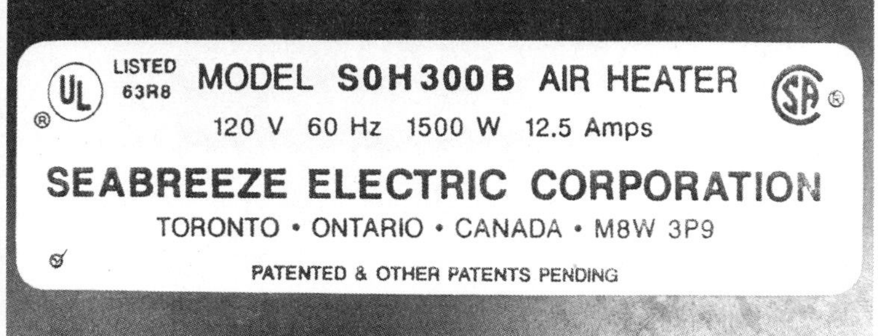

Figure 5.4
The saw's label (left) gives the saw's voltage and current requirements. You can calculate its power from the formula $P = V \times I$. The voltage, current, and power requirements for a heater are shown in the picture on the right. If you divide the power by the voltage, do you get the current? Why should you expect this result?

If the saw were used steadily for one minute, the energy used during that time would be

$$\begin{aligned}\text{energy} &= V \times I \times t \\ &= 120\,\text{V} \times 13\,\text{A} \times 60\,\text{s} \\ &= 93\,600\,\text{J}\end{aligned}$$

To calculate the power of an appliance, you do not need to know how long it is used. The reason is that

$$\begin{aligned}\text{power} &= \frac{\text{energy}}{\text{time}} \\ &= V \times I \times \frac{t}{t}\end{aligned}$$

Since the time cancels out of this equation,

$$\text{power} = V \times I$$

For the saw shown in Figure 5.4,

$$\begin{aligned}\text{power} &= V \times I \\ &= 120\,\text{V} \times 13\,\text{A} \\ &= 1560\,\text{W}\end{aligned}$$

The saw uses power at this rate for the entire length of time it is used. The energy it uses, however, is equal to the power (the rate at which it uses energy) multiplied by the time it is in operation.

The label for the heater shown in Figure 5.4 indicates both the power of the heater and its current and voltage requirements. If you divide the heater's power by the voltage, do you get the current that is indicated on the label?

INSTANT PRACTICE

1. A toaster uses 10 A of current at 120 V. Calculate
   (a) the power (in watts) used by the toaster,
   (b) the energy (in joules) used by the toaster if it is used for one minute,
   (c) the energy (in joules) used by the toaster if it is used for five minutes.
2. An electric drill uses a current of 3.5 A. If the drill operates on 120 V, what is the drill's power rating?
3. A battery-operated radio has a 6 V battery and uses 0.1 A of current. What is the power of the radio?
4. A light bulb is rated at 60 W, 120 V.
   (a) How many amperes of current does the bulb require?
   (b) How much energy does the bulb use if it is used steadily for two hours?

MEASURING ELECTRICAL ENERGY

Since the amount of energy used in a home is large compared with a joule of energy, a larger unit is required. This unit is the **kilowatt hour (kW·h)**.

You can see how kilowatt hours compare with joules by doing some calculations. For example, a 60 W (0.060 kW) light bulb operated for an hour uses the same amount of energy regardless of the units used to measure the energy. In kW·h, this energy is

$$\begin{aligned} \text{energy} &= \text{power} \times \text{time} \\ &= 0.060\,\text{kW} \times 1\,\text{h} \\ &= 0.060\,\text{kW}\cdot\text{h} \end{aligned}$$

In joules, this same energy is

$$\begin{aligned} \text{energy} &= \text{power} \times \text{time} \\ &= 60\,\text{W} \times 3600\,\text{s} \\ &= 216\,000\,\text{J} \end{aligned}$$

Thus you can calculate the number of joules in one kilowatt hour as follows:

$$\begin{aligned} \text{kW}\cdot\text{h} &= \text{kilowatts} \times \text{time (h)} \\ 1\,\text{kW}\cdot\text{h} &= 1000\,\text{W} \times 1\,\text{h} \\ 1\,\text{kW}\cdot\text{h} &= 1000\,\text{W} \times 3600\,\text{s} \\ &= 3\,600\,000\,\text{J} \end{aligned}$$

B.C. Hydro and other electric utility companies use **electric meters** to measure and record the energy used in homes. On houses, these meters are located outside at about eye level so that a meter reader can periodically record the energy used in the house. In apartment buildings, these meters are centrally located, but just as there is one meter per house, there is usually one per apartment (Figure 5.5).

Figure 5.5
(a) The electric meter for a house.
(b) The electric meters for an apartment building.

(a)

(b)

The electric meter for your home contains an electric motor that turns as long as power is being used. The motor drives the meter dials, and the longer and faster it turns, the greater the change in the reading of the dials. Figure 5.6 shows how these dials change.

Today's reading

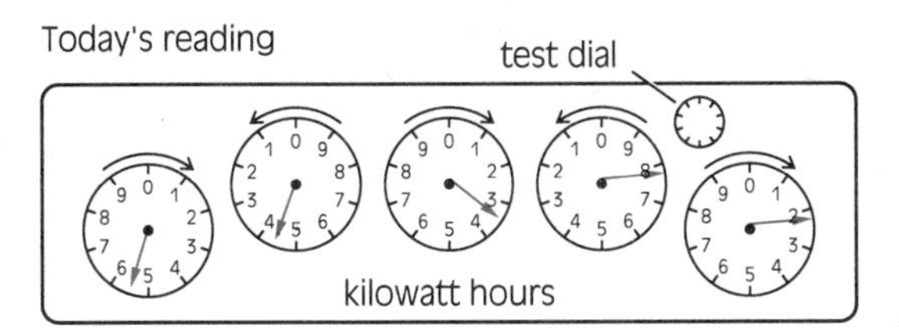

Previous day's reading

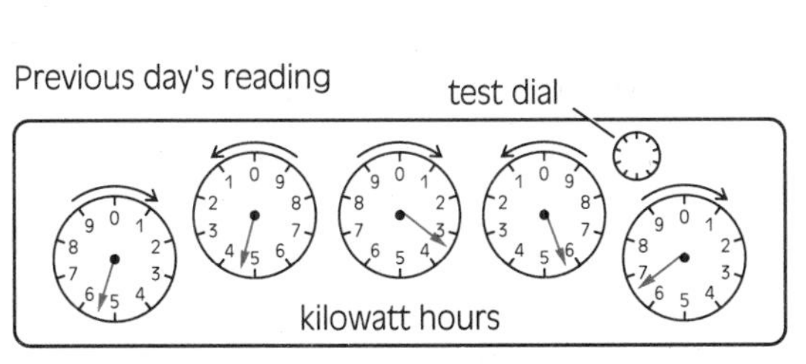

Figure 5.6
Reading an electric meter.

today's reading	54 372 kW·h
previous day's reading	−54 356 kW·h
energy used	16 kW·h

ACTIVITY 5C Energy Use in Your Home

In this activity, you will keep track of the amount of electrical energy your home uses in one week. If you live in a house, you will have to find the electric meter on the outside of your house. If you live in an apartment, ask the superintendent of the building if you can inspect your apartment's meter at daily intervals for a week. Try to make your meter readings at the same time each day.

Before you begin the activity, think about how events during the week may affect the amount of energy used. For example, would weather, doing laundry, or having baths affect energy consumption? Keep a record of the week's events in your notebook so that you can see how they affect energy consumption in your home.

PROCEDURE

1. Copy Table 5.1 into your notebook.
2. Read the dials of your electric meter and record the numbers in order from left to right. Mark down the number to which the hand on each dial points. If the dial hand is between two numbers, write down the number that the dial hand has already passed (the lower number). Refer to the examples shown in Figure 5.6 to be sure that you are doing this correctly.
3. Record the present kilowatt hour reading on the top row of your data table. Then each day for the next seven days, record the meter reading in the space for the day.
4. Each day, subtract the previous day's reading from that of the present day to ascertain the amount of energy consumed that day. Record this amount in the table. For example:

   Reading day 1 86830 kW·h
   Reading day 0 86817 kW·h
   Energy use day 1 13 kW·h

5. Each day, subtract the first reading (day 0) from that of the present day to ascertain the amount of energy used to date. Record this amount in the table. For example:

   Reading day 5 86891 kW·h
   Reading day 0 86817 kW·h
   Energy used to day 5 74 kW·h

6. In the last column of your table, record any events that could affect energy consumption in your home.

Table 5.1 Sample Data Table for Activity 5C

Day number	Date and time	Meter reading (kW·h)	Energy consumed this day (kW·h)	Energy consumed to date (kW·h)	Events that could affect energy consumption
0			0	0	
1					
2					
3					
4					
5					
6					
7					

SAMPLE ONLY

DISCUSSION

1. (a) Why do the first readings on the table start at day 0?
   (b) Why does the time of measurement matter?
2. How much energy was consumed in your home for the whole week?
3. Are there small variations in the amount of energy consumed in your home from day to day? If so, try to explain them. Your records of events for the week will help you.
4. Find out the present cost of each kW·h of energy consumed and then calculate your home's weekly costs for electrical energy.
5. Electricity bills are sent out every two months. Use your data for one week to estimate your bimonthly (two-month) energy expense for electricity.
6. Compare your estimate with your family's last electricity bill. Was your estimate close?
7. What would you change around your home if you paid the electricity bills?

REVIEW 5.1

1. Explain the difference between energy and power.
2. A TV set uses 96 W, and a hair dryer uses 1400 W. Could you provide the power they require through your own physical effort? Explain.
3. Calculate the energy used by a 3000 W water heater that operates continuously for 30 min in
   (a) joules,
   (b) kilowatt hours.
   Use these answers to explain why home energy use is measured in kilowatt hours.
4. An electric iron uses 8.3 A of current at 120 V. Calculate
   (a) the electrical power supplied to the iron,
   (b) the electrical energy used by the iron in kilowatt hours if it is operated continuously for 1 h. What happens to the electrical energy used by the iron?
5. A radio that uses a small 9 V battery requires 0.05 A from the battery. A large flashlight that has a 6 V battery requires 1.0 A from its battery. Explain why the battery for the flashlight needs to be a lot bigger than the battery for the radio.
6. Figure 5.7 shows the dials of an electric meter.
   (a) Record the meter readings and then calculate the energy consumed between the readings.
   (b) If energy costs \$0.05/kW·h, calculate the cost of the energy consumed.
7. Is it possible that appliances that use quite a lot of power use quite small amounts of energy? Explain your answer.
8. Look back over your work in this section. Decide if your ideas about energy, work, and power have changed. If they have, record these changes in the table you made in Activity 5A.

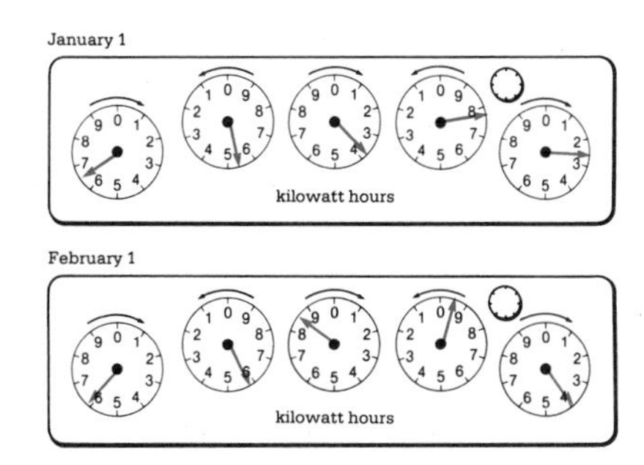

Figure 5.7
Record the energy used and calculate its cost for this example.

5.2 Conserving Electrical Energy

Think of how different your life would be without electrical energy. Imagine, for example, a day without electric lights or any electric appliances in your home. In the summer you might not miss the lights too much, but what about the refrigerator?

It can be easy to take electrical energy for granted because it is easy to use and is readily available in British Columbia. Nevertheless, it does not come without cost, and it may not always be so available.

EFFICIENCY

In schools, homes, and industry, much more electrical energy is used than is necessary. Where does the extra energy go?

Most of it goes to the environment in the form of heat, and this lost energy cannot be recovered. By changing the way electrical devices are used, we can decrease the amount of energy that is wasted. But in spite of our best efforts at conserving it, energy will be lost because some electrical devices are not very well suited to converting electrical energy into the desired form of energy.

The ability of a device to convert electrical energy to another desired form of energy is called its **efficiency**. The efficiency of electrical devices can be calculated with the following equation:

$$\text{efficiency} = \frac{\text{useful energy obtained from device}}{\text{energy used to operate device}}$$

or

$$\text{efficiency} = \frac{\text{energy output}}{\text{energy input}}$$

Often efficiency is expressed as a percentage. Percentage of efficiency is the efficiency multiplied by 100 per cent.

For example, a 100 W incandescent light bulb produces 5 J of light energy for every 100 J of electrical energy it uses. The efficiency of one of these bulbs can be shown as

$$\begin{aligned}\text{efficiency} &= \frac{5\ \text{J output}}{100\ \text{J input}} \\ &= 0.05\end{aligned}$$

Or, the bulb's percentage efficiency = 0.05 × 100% = 5%.

A device that converts electrical energy to another form of energy always produces heat. Producing unwanted heat decreases the efficiency of the energy conversion. Devices that are intended to produce heat can be very efficient, since heat is not an unwanted by-product.

Figure 5.8
A sample Energuide label. Energuide labels show the energy required for a typical month of operation for an appliance.

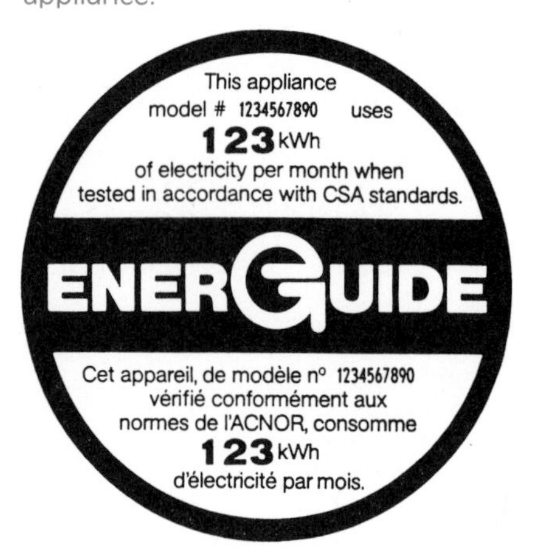

ENERGUIDE LABELS

The relative efficiency of new electrical appliances such as stoves, dishwashers, refrigerators, freezers, and clothes dryers can be checked by looking for the **Energuide label** attached to the appliance. This label, required by federal law, states how many kilowatt hours of energy per month the appliance will use (Figure 5.8). The lower the Energuide number, the more efficient the appliance. How can knowing the efficiency of appliances help you conserve energy?

ACTIVITY 5D Conserving Electrical Energy in Your Home

Can you conserve electrical energy in your home? Why should you conserve energy there? In this activity, you will try to answer these questions.

PROCEDURE

1. Copy Table 5.2 into your notebook.
2. List in your table at least six electrical devices that you have in your home that use large amounts of electrical energy.
3. In the centre column of your table, outline one or more ways you could change the way these electrical devices are used so that you would conserve energy. For example, how could you use less hot water from an electric hot water heater than you do at present? Could you dry clothes outside instead of using an electric clothes dryer?
4. Using information from labels on the electrical devices, estimate the amount of energy you think will be conserved in one week by the changes in the use of these devices that you suggest. For example, if you think that a clothes dryer could be used for 1.5 fewer hours per week, then the energy saved for a 4600 W dryer is

$$\begin{aligned} \text{energy} &= \text{power} \times \text{time} \\ &= 4.6\,\text{kW} \times 1.5\,\text{h} \\ &\quad (\text{remember, } 1000\,\text{W} = 1\,\text{kW}) \\ &= 6.9\,\text{kW}\cdot\text{h} \end{aligned}$$

5. Add up the total energy you think can be conserved by following the procedures you suggest. Record this value in your table.

Table 5.2 Sample Data Table for Activity 5D

Electrical device	How I plan to change the way I use this device	Energy I expect to conserve in one week (kW·h)
	SAMPLE ONLY	
Estimated total energy conserved		

DISCUSSION

1. What changes in your use of electrical devices would allow you to conserve the most energy? Are there some changes that would have relatively small effects on energy consumption? If so, explain why.
2. If you changed the way you used electrical devices so that you conserved electrical energy, you would make a monetary saving for your household. Suggest why any monetary savings may be less important than the fact that energy would be conserved.

EXTENSION

Try changing your use of electrical devices in your home in the ways you suggest in Table 5.2. If you did Activity 5C, you can compare the amount of energy that was used while you did that activity with the energy used during a week when you were trying to conserve energy. To do this, record the reading on the electric meter to your home at the start of an energy-conserving week and again at the end. Subtract the two readings so that you can find the energy used in your home during the week. When you tried to conserve energy, did you use less in your home?

REVIEW 5.2

1. The efficiency of a fluorescent lamp is approximately 20 per cent. What percentage of its electrical energy is changed into light? What percentage is converted to heat?
2. An electric motor is supplied with 15 kW·h of energy. If 12 kW·h of mechanical energy is obtained from the motor, what is the efficiency of the motor? What happened to the remaining energy that was supplied to the motor?
3. Could the efficiency of an electrical device ever be greater than 100 per cent? Explain your answer.
4. If electricity costs $0.05/kW·h, how much does it cost to run each of the following appliances for the time indicated? (Remember, 1000 W = 1 kW.)
   (a) a 1200 W dishwasher for 1 h
   (b) a 200 W colour TV for 6 h
   (c) a 400 W block heater for 10 h
5. Figure 5.9 shows two electric water heaters. Although both hold the same amount of water, one is more efficient than the other.
   (a) Which water heater is the more efficient?
   (b) Suggest a reason why one heater is more efficient than the other.
   (c) Suggest possible ways in which an existing water heater can be modified to make it more efficient.
6. Outline some methods by which energy can be conserved in your home and in your school.
7. Are some things that conserve electrical energy already being done in your home or school? If so, explain these procedures and how they conserve energy.

Figure 5.9
Can you explain why one of these water heaters is more efficient than the other?

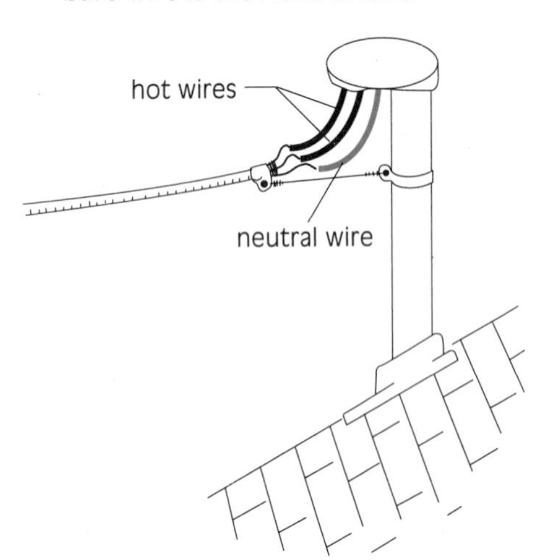

Figure 5.10
Three wires bring the electricity to a house or apartment building. The bare wire is the neutral wire.

5.3 Household Circuits

Household circuits are much like the circuits you studied in Chapter 3 except for two important differences: (1) they use alternating current instead of direct current, and (2) they operate at much higher voltages.

Alternating current first flows in one direction, then reverses, and then reverses again to its original direction. These two changes of direction make up one cycle of alternating current. The number of cycles in a second is called the **frequency** of the alternating current. Throughout North America, household alternating current has a frequency of 60 cycles per second. This frequency is written as 60 **hertz (Hz)**. A hertz is the standard unit for frequency in cycles per second.

The electricity for your home is supplied via three wires (Figure 5.10). One of these is said to be a **neutral wire** because it is connected to the ground. Thus, there is no voltage difference between it and the ground. The other two wires are **hot wires;** there is a 120 V difference between either of these wires and the neutral wire. Between the two hot wires, there is a voltage difference of 240 V. For this reason, household circuits can operate at either 120 V or 240 V (Figure 5.11).

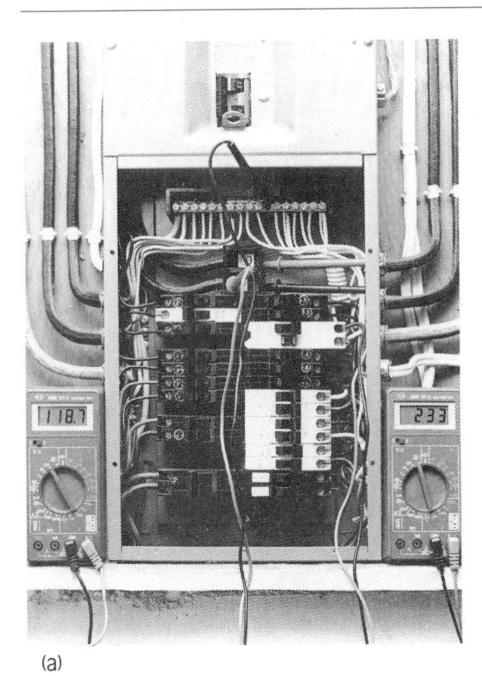

(a)

Figure 5.11
(a) The voltmeter on the left is connected between the hot wires and the neutral. It shows a voltage difference of nearly 120 V. The second voltmeter is connected between the hot wires. The voltage difference is almost 240 V. (b) The analogy of building height may help to explain the differences in voltage. The difference between the top of a building 10 m tall and ground level is the same as the difference between ground level and the bottom of a hole 10 m deep. Between the top of the building and the bottom of the hole, however, the difference is 20 m.

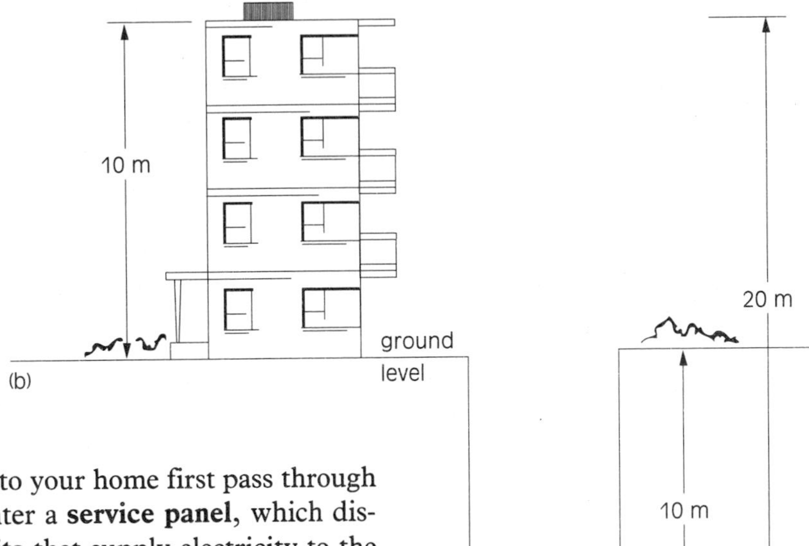

(b)

THE SERVICE PANEL

The three wires that bring electricity into your home first pass through the electric meter. From there, they enter a **service panel**, which distributes electricity to the **branch circuits** that supply electricity to the different parts of a home (Figure 5.12). In a house, the service panel is located inside. It is usually near the electric meter, but on the other side of the wall. In an apartment, the service panel is located at some convenient, but not too obvious, location.

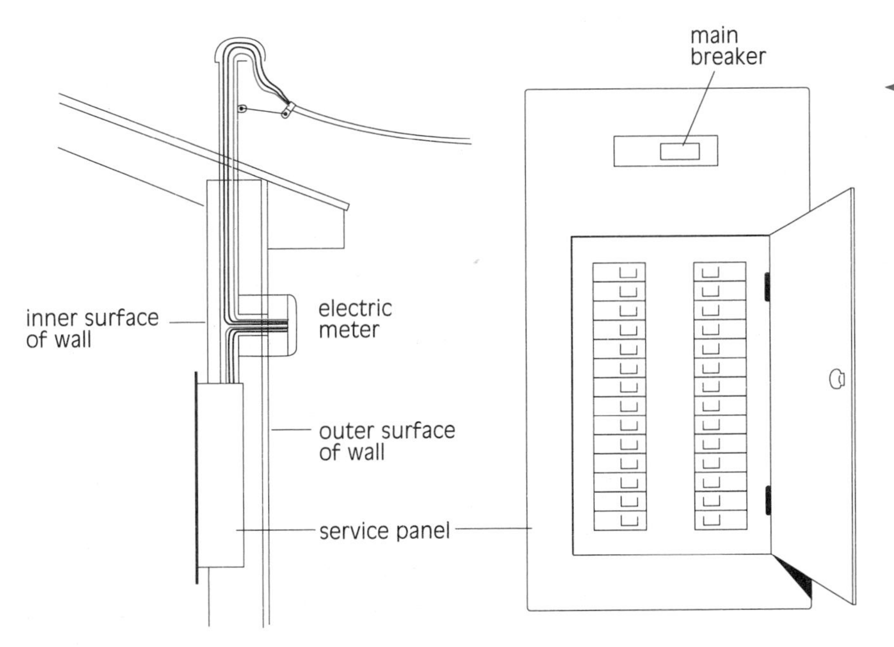

Figure 5.12
This is how electricity comes to the service panel in a house.

CAREER

B.C. HYDRO LINE WORKER

Don Luney and Brent Tymchuk are B.C. Hydro line workers in Victoria. Don is a journeyman and has been in the line trade since 1976. Brent is an apprentice in the company's Lineman Training Program.

What does a line worker do?
Don: We look after the power lines up to the meter base on your house. After that, it's your responsibility.

What is a typical day's work for a line worker?
Brent: It depends what crew you're on. All line workers have to be able to do all jobs—we move around from one crew to another.
Don: We may spend the whole morning setting a pole. Then we may have to hang a transformer on it. After that, we may get a call to go somewhere and repair something that's broken. Or we might be building an underground electrical system for a new community.

What kind of skills are useful in your work?
Don: You have to enjoy working outside. We work in all kinds of weather. Physical strength is important. We lift and place heavy objects when we're working from a bucket or the top of a pole. And we use a variety of tools to make connections; you need good eye-hand coordination for that.
Brent: The difficult part is working from a pole. It can take years and years to be really good at it.

Once you get to the top of the pole, you're in your "office." You're supported by your belt and the spurs on your boots and you're using both hands, often reaching around behind you and up over your head.

Don: You also need to be able to improvise.
Brent: During a storm, we can be 80 kilometres away from our headquarters and run out of a certain part. There are various ways of getting the line back up, so you use what you can. Then you go back later and do permanent repairs.

What's the best part of your work?
Don: Working together as a team. When you're working during a big storm, you work with the same person for 16-hour shifts with 8-hour breaks. You're depending on your partner to watch out for you, and you're watching out for him—there's a good feeling there.
Brent: During a storm, you may come out and find live wires on the ground. You have to survey the whole situation and make a decision and go to work to get people back in power. There's real satisfaction in that.

Is your work dangerous?
Brent: There is an ever-present element of danger, whether it's the danger of falling from heights or the danger of electrocution from handling live electrical wires. Coming to terms with it is a question of attitude.
Don: You have to be comfortable with the person you're with and feel confident in yourself and your knowledge. We have safety rules for everything we do. When people get hurt, it's because they lost their concentration and did something unsafe.

How does a person get into this type of work?
Brent: Grade 12 science and some education after secondary school are important. I had a couple of years of college before B.C. Hydro accepted me in construction. From there I applied to the Lineman Training Program. It's a three- to three-and-a-half-year program. To learn all the skills, I spend a few months in each of the different parts of the line trade.
Don: You have to be the kind of person who is willing to learn. That carries on into the job and for the rest of your life.

CIRCUIT BREAKERS AND FUSES

All the branch circuits are connected to the main supply wires by **circuit breakers** in the service panel (Figure 5.13). Circuit breakers are switches that turn off automatically if the current flowing through them exceeds a preset limit. Depending on how the circuit breakers for the circuits are connected to the hot and neutral supply wires, branch circuits have voltages of either 120 V or 240 V. The main supply wires themselves are connected to the service panel by a single large circuit breaker called the **main circuit breaker**. This circuit breaker can shut off all electricity to the service panel and thus to the house.

Circuit breakers prevent the wires of a circuit from overheating when too much current flows through them. **Circuit overload** and **short circuits** cause too much current to flow through the wires of a circuit. Circuit overload occurs when too many small appliances are on the same circuit and are all used at the same time. A short circuit occurs when bare wires touch and provide an undesired low-resistance pathway for electricity. As you have learned, low resistance allows a great deal of current through a wire.

Here is how a circuit breaker works. Inside the circuit breaker, there is a bar made of two different metals joined together (Figure 5.14). If too much current flows through this bi-metallic bar, it heats up and the metals expand. Since they are different metals, they expand at different rates and the bar bends. When the bar bends, it trips a switch and disconnects the circuit. When the bi-metallic bar cools, the circuit breaker can be reset manually by flipping it back to its original position.

Figure 5.13
A service panel. Note the bare ground wire leading from the top right of the panel.

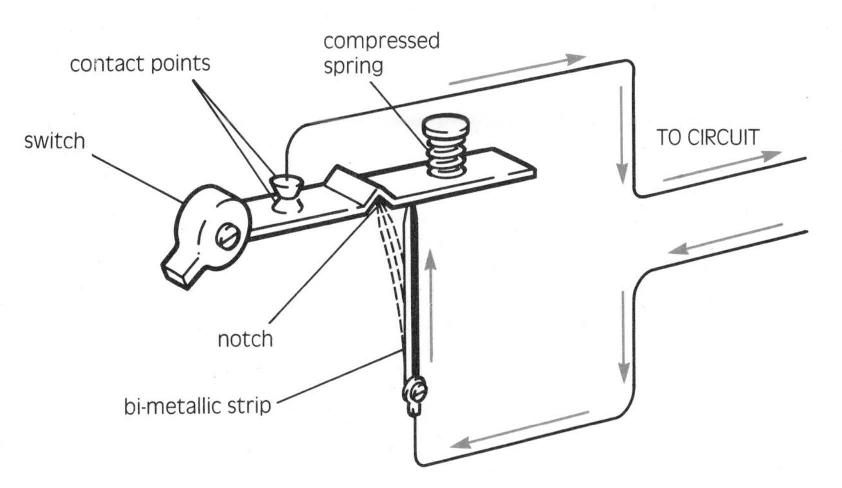

Figure 5.14
Too much current heats the bi-metallic strip, which bends to the left. The notch in the spring-loaded strip drops down over the end of the bi-metallic strip. This makes the contact points move apart, shutting off current. When the bi-metallic strip cools down, the breaker can be reset manually.

In a typical 120 V branch circuit, the maximum current for a circuit breaker is 15 A. Some 120 V branch circuits, like those intended for electric motors, are designed to carry larger currents. These circuits usually have 20 A circuit breakers and are connected with wire that is larger in diameter than that used for 15 A circuits. In this way, they can safely carry the greater current without overheating.

All 240 V branch circuits have double circuit breakers, one for each of the hot wires to which they are connected in the service panel. These

Figure 5.15
The wire on the left is used for 15 A household circuits. The wire on the right is intended for a 40 A stove circuit. The larger diameter of the wire for the stove allows more current to flow without overheating the wire.

circuit breakers often have much higher current limits so that greater power can be delivered to the circuits they supply. The circuit breaker for an electric stove, for example, is typically 40 A. The circuit breaker for an electric clothes dryer is 30 A. Because the wires in these circuits must carry more current, they are large in diameter too (Figure 5.15).

All the current for a home must pass through the main circuit breaker. Its maximum current capacity must be large enough to supply current to all the branch circuits in the home (though not with all the branch circuits at maximum current at the same time). Typical values for the maximum current of main circuit breakers are 60, 100, 125, and 200 A.

In some older homes and in electric stoves, cars, and electronic equipment, **fuses** are used in place of circuit breakers (Figure 5.16). To protect a circuit, a fuse is connected so that the current to the circuit must flow through a small piece of easily melted wire inside the fuse. If the current flowing through this wire exceeds the maximum current of the fuse, the wire will melt and the circuit will be broken before the rest of the circuit is damaged by the excess current.

Unlike circuit breakers, fuses cannot be reset once they are burned out or "blown." A blown fuse must be replaced by a fuse of the same rating. If a blown fuse is replaced by a fuse with a higher maximum current capacity, more current can flow than the circuit was designed to carry. This can cause wires to overheat or equipment to be damaged.

Figure 5.16
(a) Electric stoves have fuses located behind a front panel. (b) Fuses like these are used in cars, radios, and other appliances.

(a)

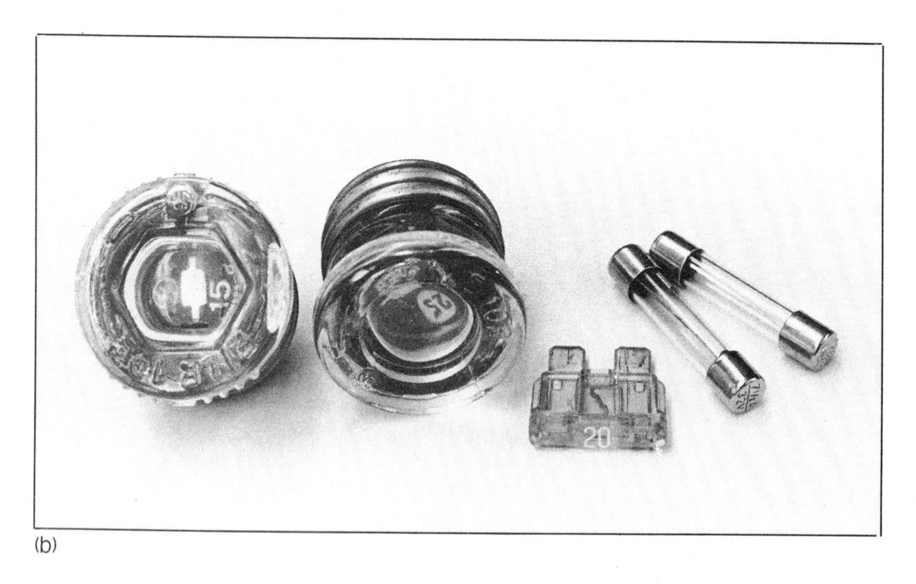

(b)

A HOUSEHOLD CIRCUIT

Most of the circuit breakers in a service panel are connected to circuits that operate several lights and outlets. For example, a 120 V, 15 A circuit breaker could supply current to two bedroom lights, a light in a hallway, and several outlets in the bedrooms.

Figure 5.17 shows how such a circuit is connected. The circuit begins at the circuit breaker that is connected to one of the hot supply wires.

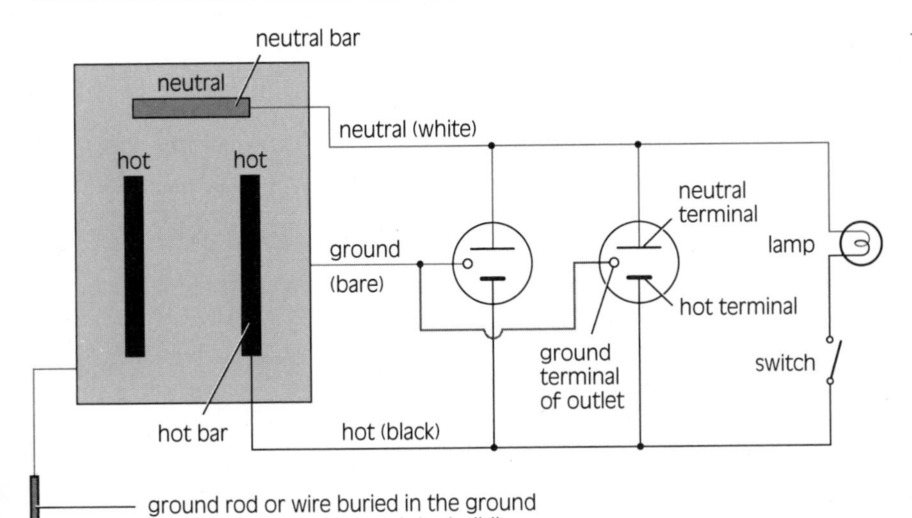

Figure 5.17
This schematic diagram shows how lights and outlets are connected to the service panel.

The neutral wire completes the circuit and makes a 120 V difference between the conductors of the branch circuit. As you can see from the diagram, all the lights and outlets are connected in parallel. For this reason, they all operate at 120 V. In addition, the parallel connection means that a light or appliance can be turned off without affecting the rest of the circuit, as would happen in a series circuit. The circuit breaker is connected in series; thus, if its maximum current capacity is exceeded, the whole circuit is shut off.

Figure 5.18
An appliance that uses 120 V may require a three-pronged plug and a three-holed outlet.

A HOUSEHOLD OUTLET

A household outlet is a socket that allows electrical appliances to be easily connected to a branch circuit. A typical household outlet is shown in Figure 5.18. This outlet is said to be **polarized** because each of its terminals is connected differently. The hot conductor is connected to the narrower of the two slot-shaped terminals in the outlet. The neutral conductor is connected to the wider one. The third terminal in the outlet is the **ground terminal**. It connects appliances to the ground electrically.

Figure 5.19
A polarized plug has one terminal that is wider than the other. It can be inserted into an outlet in only one way.

POLARIZED PLUGS

Because the hot and neutral terminals of outlets are of different sizes, it is possible to make **polarized plugs**, or plugs that have one prong that is wider than the other (Figure 5.19). These plugs can be inserted into an outlet in only one way, thus determining which is the hot wire within an appliance. On a lamp, for example, a polarized plug ensures that the lamp switch operates on the hot wire and that there will be no voltage difference in the lamp socket when the appliance is turned off. Polarized plugs also ensure that the centre terminal of the lamp socket is the hot terminal; it is much safer for a lamp to be connected in this way (Figure 5.20).

Figure 5.20
It is easy to touch the outer terminal of a light bulb while changing the bulb. If this outer terminal is connected to the neutral wire, you will not get a shock even if you touch the terminal. You could get a very severe shock if this terminal were connected to the hot wire, however.

GROUNDING

The ground terminal of an outlet is connected to the ground circuit (Figure 5.21). This circuit is connected to a long length of wire buried beneath a building's foundation, or to a metal rod, called a **ground rod**, driven deep into the ground. In this way, the ground circuit is connected to the ground.

Figure 5.21
The ground circuit carries current directly to the ground.

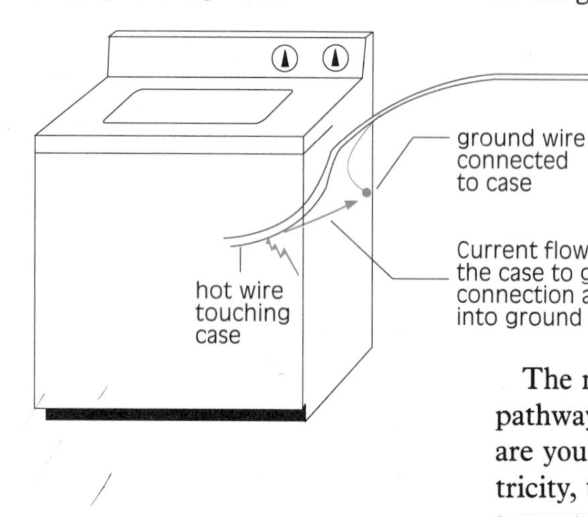

The neutral terminals of outlets, which form part of the conducting pathway for household electricity, are also connected to the ground. So are you, when you touch the ground. Since the ground conducts electricity, this means that you are electrically connected to the ground and to neutral terminals as well! This connection has no effect on your body, because there is no voltage difference between you and these terminals. If you accidentally touched the hot terminal of an outlet, however, there would be a voltage difference across your body. As a result, a current would flow through you to the ground. If this current exceeded a few milliamperes, it could be fatal (Figure 5.22).

Normally you would never touch a hot wire, but under some circumstances you might touch one by accident. Suppose, for example, that insulation is rubbed off a hot wire inside a washing machine so that the wire touches the case of the machine. If you touch the case, you will be indirectly connected to the hot wire and current will flow through you. The better your electrical connection with the machine and the ground, the more current will flow through you (Figure 5.23). You could be electrocuted.

Figure 5.22
As long as you are touching the ground in some way, you are electrically connected to it. If you touch a hot wire at the same time, a current will flow through you.

Figure 5.23
You make a poor electrical connection with the ground when you are standing on a dry wooden floor of a house. When you touch a water tap or stand barefoot in a pool of water, however, the electrical connection between you and the ground is good. For this reason, use extreme caution if you use electrical appliances near water.

This danger can be avoided by connecting the case of an appliance to the ground terminal of an outlet with the round ground terminal of a three-prong plug. This terminal is connected to the case of the appliance. If a hot wire touches the case, a current will flow from the hot wire through the case and into the ground circuit. Because the ground circuit offers a low resistance to the flow of electricity, a current that exceeds the maximum capacity of the circuit breaker will flow through this circuit to the ground (Figure 5.24). As a result, the circuit breaker will trip and the unsafe circuit will be shut off.

A grounded circuit is only as good as the connections that are made to it. If it is faulty in any way, it no longer provides protection. For example, the three-pronged plugs that include a ground connection do not fit in some extension cords. People sometimes remove the third prong entirely or bend it out of the way so that an extension cord can be used. These are unwise practices because they eliminate the protection of the ground circuit.

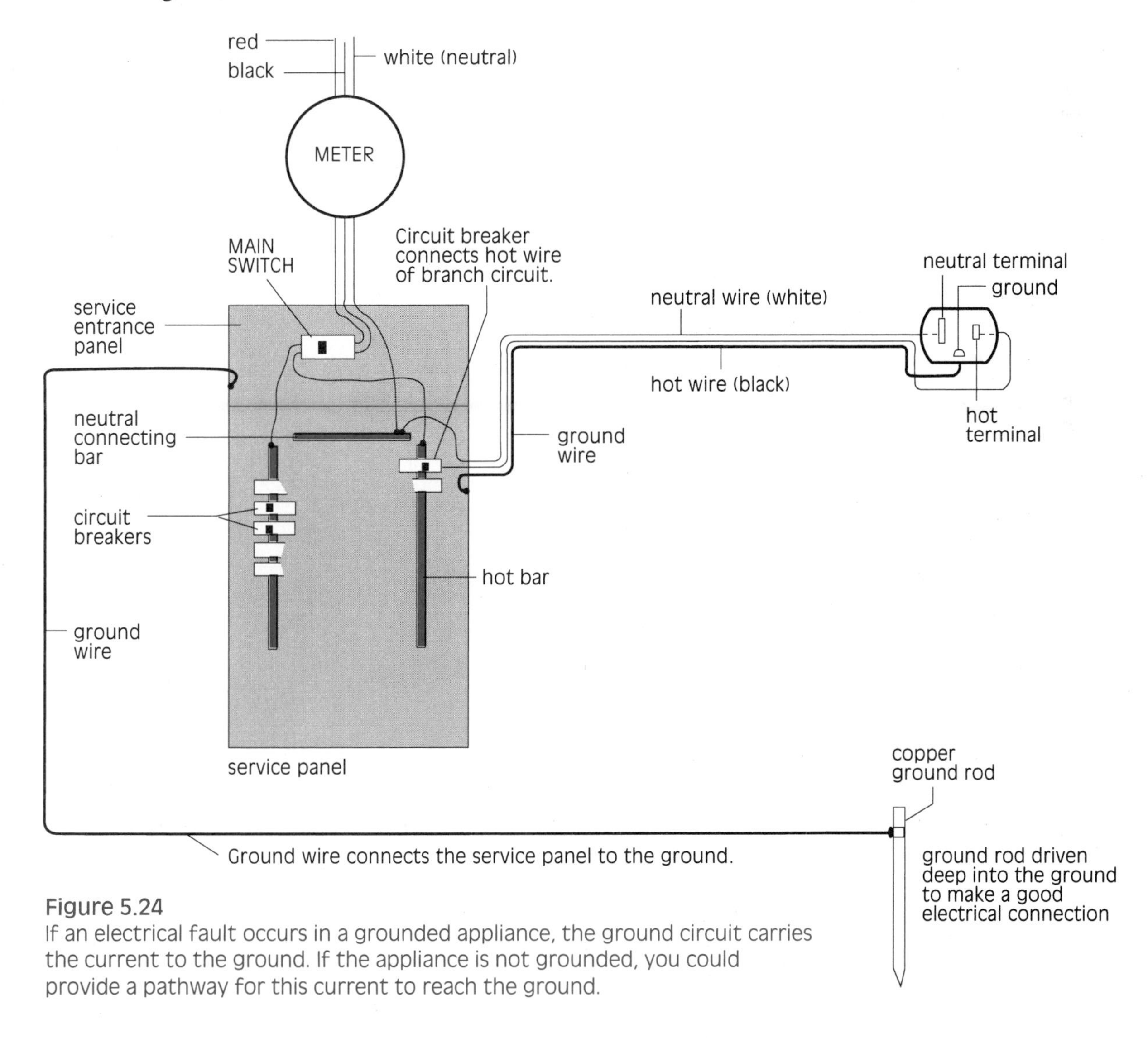

Figure 5.24
If an electrical fault occurs in a grounded appliance, the ground circuit carries the current to the ground. If the appliance is not grounded, you could provide a pathway for this current to reach the ground.

Figure 5.25
This is a GFCI breaker installed in an outdoor household outlet.

A **ground fault circuit interrupter (GFCI)** extends the protection offered by normal circuit breakers. As well as turning off if too much current flows, a GFCI circuit breaker can detect extremely small differences between the current in the hot wire and the current in the neutral wire in a circuit. In a properly operating circuit, the current in both wires should be exactly the same. If some current is flowing from the hot wire to the ground wire, however, the current in the hot wire will be greater than that in the neutral wire. If this happens, a GFCI circuit breaker breaks the circuit immediately.

GFCI circuit breakers are often installed in circuits that include outdoor outlets or electric razor outlets in bathrooms. GFCI circuit breakers may be installed in the service panel in place of a normal circuit breaker. You can see one installed in the service panel shown in Figure 5.13. Sometimes when a GFCI circuit breaker is added to a circuit after the original wiring was completed, it can be installed in an outlet as a second circuit breaker for the circuit (Figure 5.25).

DID YOU KNOW?

Specifications for electrical wiring in homes are contained in the *Electrical Code for British Columbia*. Pamphlets that outline the requirements for electrical wiring in homes are available from your local B.C. Hydro office or B.C. government office. Another source of information, though more complicated and detailed, is the book *Electrical Code Simplified* by P. S. Knight. It is available in many stores specializing in electrical hardware.

PREVENTING OVERLOAD

Because the lights and outlets on a branch circuit are all connected in parallel, the current that flows through the circuit breaker for the branch circuit is the sum of the currents required by each of the lights and outlets (Figure 5.26). For a 15 A breaker, this sum must be less than 15 A or the circuit breaker will trip.

Often people are unaware of the current requirements of their home appliances. But most appliances do have a power rating printed somewhere on them. This rating can be used to determine whether a circuit may be overloaded. For example, a 15 A circuit breaker can handle a power of $15\ A \times 120\ V = 1800\ W$. The total power used in the branch circuit must be less than this value.

If a 1200 W electric kettle and two 60 W light bulbs are used on the same circuit, will the circuit be overloaded? You can quickly calculate that the total power required is $1200\ W + (2 \times 60\ W) = 1320\ W$. This power is less than the 1800 W maximum for a 15 A, 120 V circuit, so the circuit is not overloaded.

Suppose a 900 W toaster is plugged into this same circuit at the same time. What will happen then?

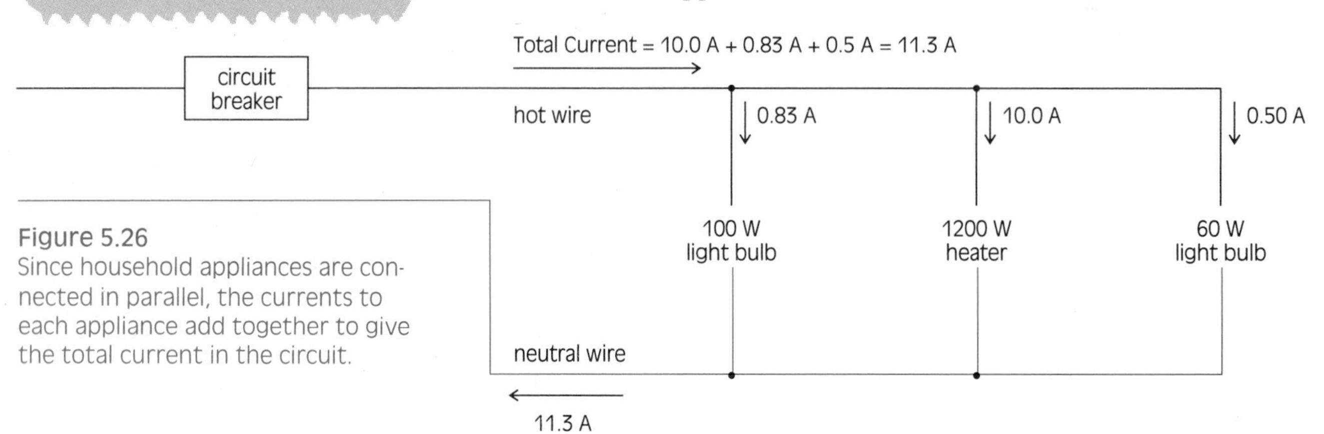

Figure 5.26
Since household appliances are connected in parallel, the currents to each appliance add together to give the total current in the circuit.

SAFETY CHECKLIST

Carelessly used, electricity can be a hazard to you and your family. Here are a number of suggestions that will help you use electricity more safely.

- Check the cords on any appliances you use. Old cords may have worn-out insulation or loose connections inside their plugs. These can cause overheating and lead to fire.
- If there are small children in your home, insert plastic plugs in unused outlets to prevent the children from pushing objects into the terminals of the outlet.
- Do not overload extension cords. There is a limit to the amount of current they can carry safely without overheating. The same is true of outlets.
- If a fuse blows or a circuit breaker trips, find out what is causing the current overload. Perhaps you have simply plugged too many appliances into one circuit. Perhaps there is a frayed cord somewhere or a short circuit in an appliance. Only when the problem has been found and corrected should you replace a fuse or reset a circuit breaker.

- Be especially careful if you are using electrical appliances near water. If you are in contact with water (if you are in the bath, touching a sink, or standing on a wet surface), or if you are touching a metal water pipe, the electrical connection between you and the ground can be especially good. In such circumstances, accidental contact with a hot wire can lead to fatal amounts of current passing through you to the ground.
- Before making any electrical repairs, be sure you understand how to make the repair correctly.
- If you must do electrical repairs, be sure to shut off current to the circuit on which you are working. (Remove the fuse or shut off the circuit breaker.)

- Never remove the third prong from a three-pronged plug.
- Use a three-wire extension cord with a three-pronged plug.
- Working outdoors? Keep ladders (particularly aluminum ones) and tools well away from power lines.
- Do not use light bulbs that exceed the maximum power for which a lamp fixture was designed.

ACTIVITY 5E *Electrical Safety*

In this activity, you will find out if electricity is used safely in your home.

PROCEDURE

1. Read over the safety checklist that precedes this activity. If you can think of additional electrical safety rules, write them in your notebook.
2. Use the safety checklist in the textbook and your own additions to it to inspect your home. In your notebook, record any unsafe electrical situations that you find.

DISCUSSION

1. Are there any unsafe electrical situations in your home? If so, what are they, and what can be done to correct them?
2. Explain the reason for any rules you added to the safety checklist.

REVIEW 5.3

1. What are the voltage, the maximum current, and the frequency of the current in a typical household circuit?
2. What is a circuit breaker and what is its function? How is it similar to a fuse and how is it different?
3. Why do major appliances such as electric stoves and clothes dryers require 240 V circuits that have large current ratings?
4. Make a sketch of an electrical outlet. Identify the hot terminal, the neutral terminal, and the ground terminal on your drawing. What is the function of each terminal?
5. How is the service panel in a home connected to the ground? Why is this connection so important?
6. Why is it unwise to disconnect the ground prong from a plug?
7. What is a polarized plug? Why are polarized plugs often used on appliances such as lamps and television sets?
8. How much current will these appliances draw from a 120 V circuit?
   (a) a 1200 W steam iron
   (b) a 600 W, two–cup coffee maker
   (c) a 60 W stereo
   (d) a 40 W lamp
9. What will happen if all the appliances listed in question 8 are used on the same 15 A circuit at the same time? Explain your answer.

CHAPTER • REVIEW

Key Ideas

- Power is the rate at which energy is produced, absorbed, or transferred. Electrical power = voltage × current.
- Electrical energy = power × time, and is measured in joules (J) if the quantities are small or in kilowatt hours (kW·h) if the quantities are large.
- Efficiency = $\dfrac{\text{useful energy obtained from an appliance}}{\text{energy used to operate the appliance}}$
- Consumption of electrical energy can be reduced by turning off unused lights and appliances and by using more efficient appliances. The less electrical energy that is used, the fewer harmful effects there will be to the environment.
- Three wires supply the electricity to a home. Two of these are hot. The third is a neutral wire.
- The main circuit breaker shuts off all circuits in a home at once.
- Circuit breakers connect branch circuits to the electricity supply at the service panel.
- Branch circuits consist of lamps and outlets connected in parallel to a hot and a neutral wire. A ground wire is included as a safety feature.
- Proper use of electricity gives many benefits. Used improperly, electricity can be hazardous.

Vocabulary

work
joule (J)
energy
power
kilowatt hour (kW·h)
electric meter
efficiency
Energuide label
frequency
hertz (Hz)
neutral wire
hot wire
service panel
branch circuit
circuit breaker
main circuit breaker
circuit overload
short circuit
fuse
polarized
ground terminal
polarized plug
ground rod
ground fault circuit interrupter (GFCI)

1. Create a concept map using terms related to work, energy, power, and electrical safety.
2. Make a labelled drawing that shows a service panel and its connections to household appliances.

Connections

1. How much current is drawn by a "tri-lite" reading lamp when each of the following settings is used? The voltage of the lamp is 120 V.
   (a) 100 W
   (b) 200 W
   (c) 300 W
2. How can an Energuide label help a person who is buying an appliance?
3. Suggest three ways in which you could save energy by using the appliances in your home more wisely.
4. (a) Why should only low-power appliances (radios, lamps, stereos) be connected to ordinary extension cords?
   (b) A certain extension cord is designed to handle no more than 6 A. What is the maximum power rating of a device that can be connected to the cord if the voltage is 120 V?
5. (a) What is meant by a circuit overload?
   (b) What is a short circuit?
   (c) Why does too much current flow in a circuit if either condition exists?
6. (a) With the aid of a diagram, explain why household circuits must be connected so that lights and outlets are in parallel with one another.
   (b) Is a circuit breaker connected in series or in parallel with the circuit? Explain your answer.
   (c) How is a switch that controls a light connected in a circuit?
7. GFCI circuit breakers are required by the B.C. electrical code for outdoor outlets and bathrooms.
   (a) How does a GFCI circuit breaker differ from a normal circuit breaker?
   (b) Why would a GFCI circuit breaker be especially important for outdoor outlets and bathrooms?

Explorations

1. The price per kilowatt hour of electrical energy to a home depends on the amount of energy used in the home. How does B.C. Hydro set this price? Does B.C. Hydro's policy on energy pricing encourage conservation of electric energy? Discuss this topic with your classmates and suggest ways in which energy pricing could be modified to make it more effective in encouraging energy conservation.
2. Make a poster that illustrates some aspect of electrical safety.
3. Make a poster that encourages conservation of electrical energy.

Reflections

1. Estimate the amount of electrical energy you use during one day. If you had to produce this amount of energy through your own physical effort, how would your life be changed?
2. Think about the ideas regarding power, work, and energy consumption and conservation you had when you began this chapter. Are your ideas different now? Why or why not?

CHAPTER

6 Generating and Transmitting Electricity

Dams like the one in this picture store water for hydroelectric generating stations, which provide most of British Columbia's electricity. These dams have changed rivers like the Peace and the Columbia forever. The once free-flowing waters of these rivers have been used to fill lakes. The forests that lined the river banks, as well as some rich agricultural areas, have been flooded. In addition, building large dams and hydroelectric generating stations requires a great deal of money and many materials, such as concrete, steel, and fuel for machinery. The use of electrical energy has freed people from much of the back-breaking and unpleasant labour of past times, but as you can see, this energy does not come without costs.

ACTIVITY 6A Costs of Generating and Transmitting Electricity

You know how electricity is used in your home, but how is this electricity produced and how is it transmitted to your home? What is the original source of the electrical energy you use? What are the costs of producing and transmitting this electricity?

1. Divide a page in your learning journal into three columns. Label the first column "Questions," the second column "Answers," and the third column "Costs."
2. In the first column, write the following questions:
   - How is electricity produced?
   - How is electricity transmitted?
   - What is the original energy source for the electrical energy?

   Leave enough space after each question for the most complete answer you can come up with.
3. In the second column, fill in the answers to the questions as best you can.
4. In the third column, record the costs you can think of for each answer.

6.1 Generating Electricity

In Chapter 4, you learned that as the shaft of a generator rotates, it causes a change in the magnetic field surrounding a coil of wire. The changing magnetic field generates a current in the wire. Thus, a generator converts the mechanical energy of the rotating shaft into electrical energy in the coil of wire.

Many different energy resources can be used to provide the mechanical energy to turn the shaft of a generator. In Canada, the most common energy resource for this purpose is moving water.

HYDROELECTRIC GENERATING STATIONS

In a typical hydroelectric generating station, water that has been stored behind a dam is allowed to fall down to turbine wheels through large tubes called penstocks. The rapidly moving water at the end of the penstock turns the turbine wheels, which spin shafts. The rotating shafts drive the generators (Figure 6.1).

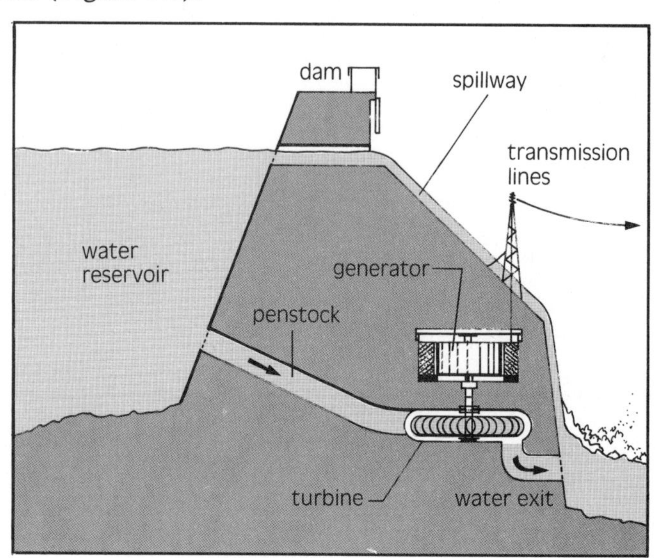

Figure 6.1
The water behind a dam has potential (stored) energy. As the water flows through the penstock, it loses potential energy and gains kinetic energy, the energy of moving things. This kinetic energy is transferred to the turbine and so to the generator. Because the turbine and generator are machines, they are said to have mechanical energy. The generator converts this mechanical energy into electrical energy.

The power produced by a hydroelectric generating station can be controlled by varying the amount of water turning the turbine wheels. If the demand for power increases, the water flow through the turbines is increased (Figure 6.2).

Figure 6.2
These generators are under construction beneath a dam. They will convert mechanical energy into electrical energy.

FOSSIL FUELS AS AN ENERGY RESOURCE

Fossil fuels are fuels that have been formed by the decomposition of once-living material over millions of years. Natural gas, petroleum, and coal are fossil fuels. When a gasoline or diesel engine burns a fossil fuel, it transforms the energy stored in the fuel into mechanical energy. This mechanical energy is transformed into electrical energy when the motor turns the shaft of a generator (Figure 6.3a). If you look under the hood of a car, you will see a generator turned by the engine. This generator keeps the car's battery charged. Some small towns in British Columbia obtain all their electric power in a similar way. In these towns, large diesel engines turn generators to produce electricity (Figure 6.3b).

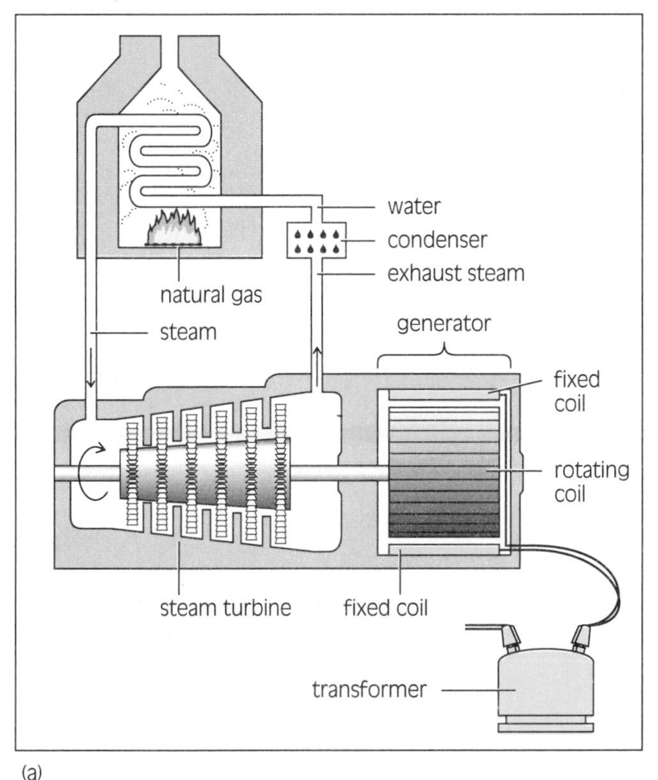

(a)

(b)

Figure 6.3
(a) Natural gas heats water, creating steam. The steam turns a turbine that runs a generator, which produces electricity. The transformer raises the voltage. (b) This diesel-electric generator converts the energy stored in diesel fuel into mechanical energy and then into electrical energy.

In most electric generating stations that use fossil fuels, the fuel is burned to heat water to boiling. Then the steam that is produced can be used to turn turbines connected to generators. In British Columbia, the Burrard Thermal Plant near Coquitlam uses natural gas for this purpose. This electric generating station is put into operation when water supplies run low and hydroelectric generating stations cannot meet the demand for electric power (Figure 6.4).

(a)

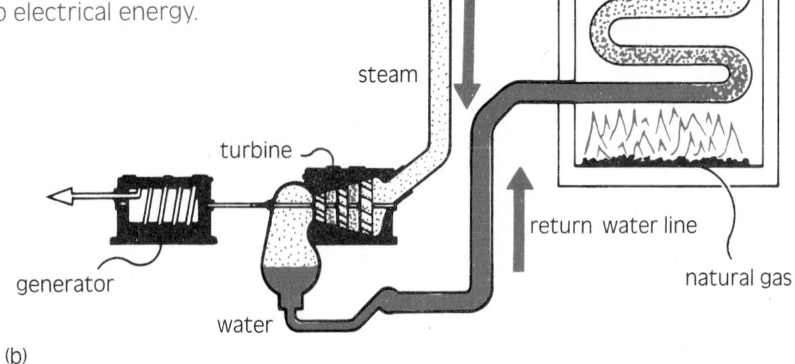

(b)

Figure 6.4
(a) The Burrard Thermal Plant near Coquitlam can produce 912 500 000 W of power. (b) In the Burrard Thermal plant, chemical energy stored in natural gas is changed to thermal energy. This energy boils water, which changes to steam. The steam, under pressure, forces the turbines to spin. The generator changes the mechanical energy of spinning into electrical energy.

OTHER ENERGY RESOURCES

In some parts of the world, the action of waves is used to turn generators; in others, wind energy is used (Figure 6.5). In France, a tidal hydroelectric generating station is in operation on the River Rance in Brittany. A similar smaller station generates electricity from the tides in Nova Scotia's Bay of Fundy. In California, a generating station uses heat from the sun to generate electricity. In New Zealand and Iceland, generating stations transform heat from the Earth into electricity. And in many places, nuclear reactors transform nuclear energy into heat, which is then converted into electricity. All these generating stations use different sources of energy, but they all use generators to produce the electrical energy.

(a)

(b)
(c)

Figure 6.5
Energy to produce electricity can be obtained from waves, wind (a), heat within the Earth (b), tides (c), and the sun.

RENEWABLE AND NON-RENEWABLE ENERGY RESOURCES

The water behind a dam, waves, wind, the sun, and heat from inside the Earth are examples of **renewable energy resources**, which constantly replenish themselves for continuous use. **Non-renewable resources,** however, cannot be replaced during a human lifetime. The fuel oil used by a diesel engine and the natural gas used by the Burrard Thermal Plant are non-renewable energy resources. Why?

ACTIVITY 6B *Sources of Energy for Generating Electricity*

Many different sources of energy can be used for the generation of electricity. In this activity, you will work with two or three other students to prepare either

(a) an attractive display about an energy resource or

(b) an oral report on an energy resource.

Some resources that supply the energy needed to turn electric generators are

- natural gas
- coal
- uranium
- wind
- heat from within the earth
- geothermal energy
- tides
- gasoline and diesel fuel
- running or falling water
- the sun
- biomass fuel (for example, wood)

PROCEDURE

1. Select an energy resource from the list. Or, if you and your group know of another energy resource that can be used to turn a generator and you are interested in learning more about it, check with your teacher to determine whether the topic is suitable before proceeding further.
2. Before you begin collecting information for a display or report, it is helpful to prepare a list of key questions about your topic. Here are some suggestions:
   (a) What are the advantages of this resource?
   (b) What are the disadvantages of this resource? Are there environmental problems associated with it?
   (c) How much electrical power can be provided by this resource? Enough for one home, a small town, or a city?
   (d) Are hazardous wastes produced by this resource?
   (e) Will the resource continue to be available for a long time?
   (f) Is this resource a renewable or a non-renewable source of energy?
   (g) In what parts of the world is this energy resource used? Could it be used in British Columbia? Is it likely to be used in British Columbia?

   Discuss your topic with the other students with whom you will be working on the display or report. From your discussion, modify and add to the above questions so that they suit your topic and your interests in it. Then use these questions to guide your research.
3. When you have finished your research, prepare your display or report and share it with your class.

DISCUSSION

1. Of the energy resources considered by your class, is there a "best" and a "worst" source? Is it necessary to explain what you mean by a best or worst source?
2. Are any of the energy resources without disadvantages?
3. Of the energy resources considered by your class, which would you prefer as a source of electricity for your home? Explain the reasons for your choice.

COSTS OF ELECTRICAL ENERGY

Any electric generating station that can produce large amounts of power for extended periods of time is expensive to build. The operation of an electric generating station can be costly too. Resources such as natural gas, oil, coal, and uranium are expensive in the quantities necessary for producing electricity.

But it is not just the monetary costs of materials and resources that must be taken into account. Consider a hydroelectric generating station, for example. The sun provides the energy to evaporate water from oceans, lakes, soil, and the leaves of plants. When the water returns to the Earth, some of it goes to rivers, where its energy is used to generate electricity (Figure 6.6).

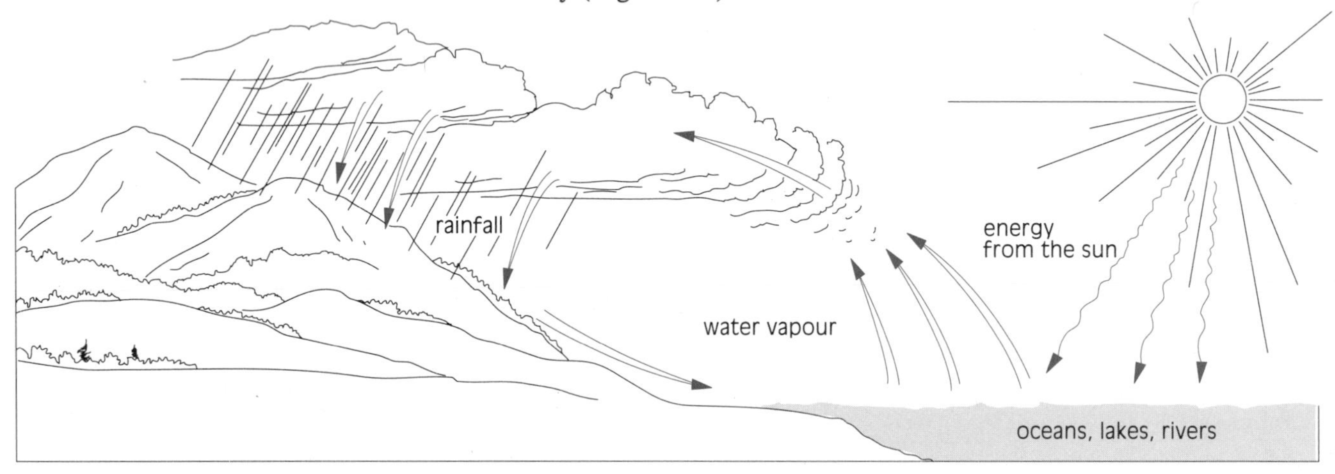

Figure 6.6
The energy used by a hydroelectric power station comes originally from the sun. Energy from the sun evaporates water, which returns to the Earth as rain or snow. This water fills the reservoirs for hydroelectric dams.

Using the sun as a source of energy might seem to be inexpensive. But to use the energy of moving water effectively, land that could be used for other purposes must be flooded. Fish stocks that once spawned in the river and its tributaries above a dam are destroyed forever. Silt carried by the river may fill the lake behind the dam and decrease its water capacity. Without its supply of silt, the delta at the river's mouth may erode. Flooding may result, making the land uninhabitable. In addition, the dammed river becomes harnessed and controlled, and all living things have to adjust to these changes.

After hydroelectric energy, fossil fuels and materials that can be used in a nuclear reactor, such as uranium, are the most important resources in Canada for producing electrical energy. Besides their initial costs, both have significant disadvantages:

1. They are non-renewable, meaning they can be used only once.
2. Nuclear fuels produce hazardous waste products that are hard to get rid of, about which you will learn more in Chapter 7.
3. Carbon dioxide gas is produced when fossil fuels are burned. In the past 100 years, the amount of this gas in the Earth's atmosphere has been steadily increasing, and this increase may cause warming of all the Earth's climates.

4. When some fossil fuels are burned, they produce gases such as sulphur dioxide and nitrogen oxides. These can combine with water to make acids. These acids kill the leaves of trees and smaller plants and fish in lakes and streams.

What are the true costs of any source of energy? You can see that they go beyond the cost of building and operating a power generating station and include the effect on the environment. Unfortunately, it is difficult to assess the cost to the environment when all the effects of the use of an energy resource are not known until after the resource is being used. In addition, different people put different values on environmental changes.

The production of electrical power for towns and cities always has some harmful environmental effects. Some sources of energy are less harmful to the environment than others. If there is a choice of energy resource, a less harmful resource should be selected. But by conserving the electrical energy you use in your home and school, you can help decrease the environmental harm caused by the production of electricity. After all, if there is less demand for electricity, there will be less reason to build more power generating stations.

REVIEW 6.1

1. Explain the difference between a renewable and a non-renewable energy resource.
2. (a) Give three examples of renewable energy resources.
   (b) Give three examples of non-renewable energy resources.
3. (a) What three energy resources supply most of the electricity used in Canada?
   (b) Only two of these resources are used in British Columbia. Which are they?
4. If electricity is to be produced for a city, what are three costs that should be considered in providing the energy for generators from each of the following sources?
   (a) coal-fired steam turbines
   (b) windmills
   (c) moving water in a tidal hydro-electric power station
5. Scientists say that refrigerators could be made that use about 25 kW·h of energy per month instead of the 125 kW·h of energy per month now used by most refrigerators.
   (a) Estimate how much electrical energy could be saved in one month in the town or city in which you live if all the refrigerators were of the more efficient type.
   (b) Estimate how much electrical energy could be saved in all of British Columbia during one month if the more efficient refrigerators were used.

6.2 Alternative Methods of Producing Electricity

The methods of producing electrical energy that have been discussed so far all rely on generators. There are also methods by which electrical energy can be produced without a generator.

PHOTOVOLTAIC CELLS

Photovoltaic cells are devices that convert the energy of light directly into electrical energy. They are often called **solar cells**. Photovoltaic cells are used extensively in space to provide the energy for operating spacecraft. They can also be used on Earth to produce electrical energy for homes. On a smaller scale, they can take the place of batteries in calculators and similar devices that have low power requirements (Figure 6.7).

DID YOU KNOW?

Photovoltaic cells are one of a number of types of electronic devices that are affected by light. Other related devices are photodiodes and phototransistors. When light strikes photodiodes and phototransistors, the flow of current through them changes with the light that strikes them.

Photodiodes and phototransistors are used for a number of specialized applications. For example, in the bar code reader at a supermarket check-out, a photodiode detects the changing light levels caused by the pattern of the bar code. Compact disk players use a similar method to obtain information from the compact disk. The music on a compact disk is encoded as a series of pits that circle the disk. As the disk spins, a laser beam scans its surface. The laser light is reflected by the high parts of the disk, but not from the pits, so the photodiode receives a pattern of reflected light from the disk. Because this pattern of light affects the current flow in the photodiode, the photodiode converts the changing light into a changing current flow that is used to produce sound.

Figure 6.7
(a) Photovoltaic cells are used to supply electrical energy for many satellites. (b) These solar panels collect light energy from the sun and convert it into electrical energy. (c) Small photovoltaic cells provide electricity for calculators.

ACTIVITY 6C Power from Photovoltaic Cells

In this activity, you will see how the amount of power available from a photovoltaic cell is affected by the intensity of light striking it.

MATERIALS

photovoltaic cell
100 Ω resistor
voltmeter
ammeter
connecting wires
alligator clips

PROCEDURE

1. Connect a photovoltaic cell, a 100 Ω resistor, a voltmeter, and an ammeter as shown in Figure 6.8. The voltmeter and ammeter should both be set to their most sensitive scales. The resistor acts like an appliance connected to the photovoltaic cell.
2. Copy Table 6.1 in your notebook.
3. Shine a constant moderate source of light on the photovoltaic cell and record the voltage and current registered on the meters. Calculate the power produced by multiplying the voltage by the current.
4. Increase the intensity of light striking the photovoltaic cell by moving the cell into direct sunlight or by moving it closer to a light source. What happens to the voltage and current measured by the meters? Again calculate the power produced.
5. If the intensity of light striking the photovoltaic cell decreases, what is likely to happen to the voltage and current that is produced by the cell? Test your prediction and record your results. Calculate the power produced as before.

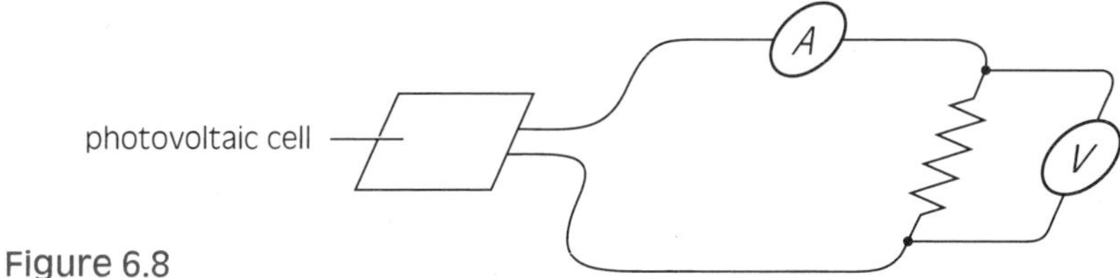

Figure 6.8
Connect your photovoltaic cells like this.

Table 6.1 Sample Data Table for Activity 6C

Light Intensity	Voltage (V)	Current (A)	Power (W) (voltage × current)
Moderate			
Bright			
Dim			

SAMPLE ONLY

DISCUSSION

1. How does the intensity of the light striking the photovoltaic cell affect
   (a) the voltage produced by the cell?
   (b) the current produced by the cell?
   (c) the power produced by the cell?
2. While light strikes the photovoltaic cell, it produces power. In this experiment, what happens to the power produced by the cell?
3. Does the photovoltaic cell produce AC or DC? How do you know?

→

4. Estimate how many photovoltaic cells similar to the one you used would be required to provide the power necessary to operate a 120 W television set on
   (a) a cloudy day when light intensity is low,
   (b) a sunny day when light intensity is high.
5. Could you operate the electrical appliances in your home on photovoltaic cells only? Explain your answer using some estimates of the amount of power used in your home and the amount of power that could be produced from an arrangement of photovoltaic cells that could be situated near, or on, your home. How could you provide electric power at night?
6. One of the disadvantages of photovoltaic cells is that they are expensive. If the photovoltaic cell you used in this activity, for example, cost four dollars, estimate the cost of the photovoltaic cells you would require to provide the power requirements of your home. Are there environmental costs to using photovoltaic cells?

CHEMICAL CELLS

Chemical cells convert stored chemical energy directly into DC electricity. There are many different types of chemical cells. The device most people call a battery is a type of chemical cell called a carbon-zinc dry cell. (You will find out how dry cells work in Chapter 11.) Other types of chemical cells work in a similar manner to carbon-zinc dry cells but use silver and zinc, zinc and mercury oxide, or lithium metal as their major ingredients.

Some chemical cells are rechargeable. These cells produce electricity when they convert chemical energy into electrical energy, and they store chemical energy when electricity is used to recharge them. Examples of rechargeable cells are the nickel-cadmium cells that can be used to replace dry cells and the lead-acid batteries that are used in cars.

Chemical cells provide energy for special purposes. For example, batteries are good for operating flashlights, portable radios, and cameras. Batteries cannot be used for long-term energy supplies, however. The chemicals they contain have a fixed operating lifetime and it requires much more energy to manufacture these chemicals than the cells are able to deliver.

Another type of chemical cell is called a **fuel cell** (Figure 6.9). A common kind of fuel cell uses the energy stored in pure hydrogen and oxygen. Hydrogen gas is pumped into the cell at one electrode and oxygen is pumped in at the other. Inside the cell, the hydrogen and oxygen are combined to make water, but this reaction is carried out in such a way that no burning occurs and no flame is produced. Instead, electrical energy is produced between the two electrodes.

So far, fuel cells have had relatively limited uses. One of these uses is to provide energy for the space shuttle. Fuel cells may have other potential uses, however. For example, hydrogen can be produced from coal. If this hydrogen is supplied continuously to fuel cells, they can produce large amounts of electricity with efficiencies that may be as high as 50 to 60 per cent. In this way, the energy from coal can be converted to electricity more directly and with greater efficiency than if the coal were burned in a conventional coal-fired electricity generating station.

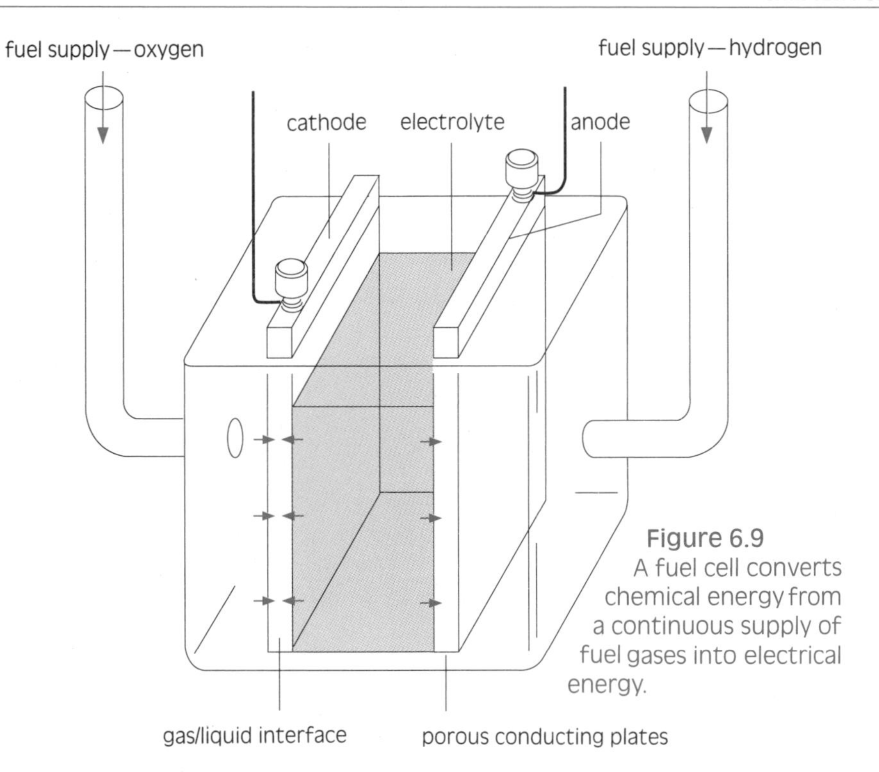

Figure 6.9
A fuel cell converts chemical energy from a continuous supply of fuel gases into electrical energy.

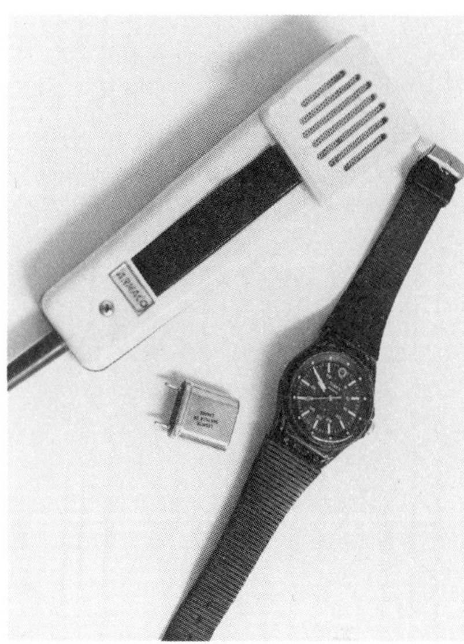

Figure 6.10
Piezoelectric crystals are used in all these devices.

THE PIEZOELECTRIC AND THERMOELECTRIC EFFECTS

When certain types of crystals are subjected to stress, they produce small amounts of electricity. This effect is called the **piezoelectric effect**, after the Greek word *piezein*, meaning "to press." Quartz crystals are piezoelectric. In a "quartz" watch, a small piece of quartz vibrates at a constant rate. As the crystal vibrates, it produces small amounts of AC electricity, which is used to regulate the watch. Quartz crystals are used in a similar way in computers and in radio transmitters and receivers.

The piezoelectric effect has many other practical uses. For example, some phonograph cartridges contain piezoelectric crystals. In these cartridges, a crystal is connected to the stylus, which makes contact with the record. When the phonograph stylus moves up and down on the record grooves, it presses against the crystal. As a result, the crystal produces varying amounts of piezoelectricity, which are amplified by the record player. Some microphones contain piezoelectric crystals as well. In these crystal microphones, the sound waves from a person's voice compress the crystal so that it produces piezoelectricity, which is electronically amplified. Figure 6.10 shows some items that use the piezoelectric effect.

The **thermoelectric effect** also produces small amounts of electricity. When two different kinds of metals are joined together at the ends and these junctions are at different temperatures, a voltage difference is produced.

Devices that make use of the thermoelectric effect are called **thermocouples** (Figure 6.11). Thermocouples can be used to measure temperature. One junction may be placed in ice water (0°C), and the other is placed where the temperature is to be measured. The greater the temperature difference, the greater the voltage registered on a voltmeter. If the scale of this voltmeter is calibrated in degrees, it will show the temperature directly.

Although both the piezoelectric effect and the thermoelectric effect have many practical uses, they are of no value for providing electrical power for homes or cities. The quantities of power they produce are far too small.

DID YOU KNOW?

A unique use of piezoelectricity is being researched in the Geophysics Department at the University of British Columbia. Gold is often found associated with quartz crystals. Explosive charges are used to cause stress in the ground, and sensitive detectors attached to the ground some distance away record any electrical signals coming from piezoelectric quartz crystals beneath the ground. These signals help to locate quartz veins—and perhaps gold as well.

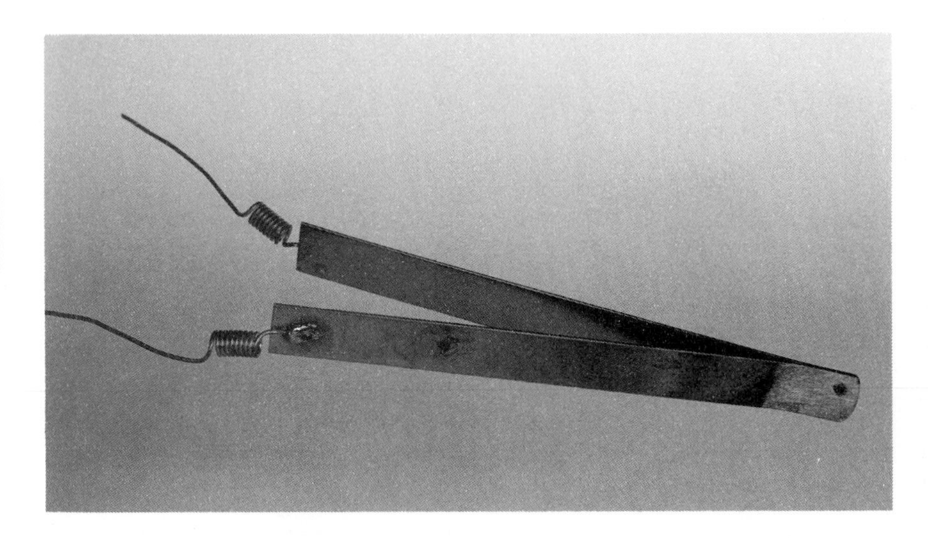

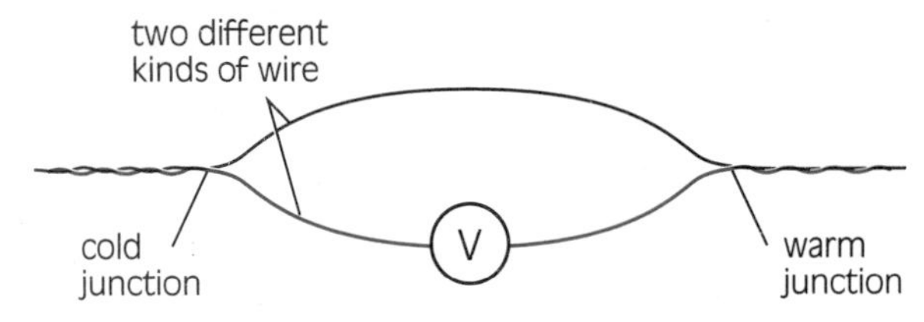

Figure 6.11
A thermocouple is made of two different metals joined together. A flow of current is produced as a result of the difference in temperature between the junctions.

REVIEW 6.2

1. What kind of electricity, AC or DC, is produced by photovoltaic cells and dry cells?
2. (a) Would clouds ever interfere with photovoltaic cells used to provide electricity for a spacecraft? Explain your answer.
   (b) As a spacecraft powered by photovoltaic cells moved away from the sun, what would happen to the power available to the spacecraft? Explain your answer.
3. Give three advantages of using photovoltaic cells to produce electricity. What are three disadvantages of their use?
4. Why is a fuel cell potentially more efficient at converting the energy in hydrogen into electricity than a power plant that uses hydrogen to run a motor that turns a generator that produces electricity?
5. Explain why a hydroelectric power station can be used to provide electricity to a city, whereas batteries cannot.
6. (a) Give two examples that show an application of the piezoelectric effect.
   (b) Give one example that shows an application of the thermoelectric effect.
   (c) Explain why neither the piezoelectric effect nor the thermoelectric effect can be used to produce electricity for a home.

SCIENCE IN OUR WORLD

WOMEN IN ENGINEERING

Engineering is one of the largest professions in North America, and engineers are among the top income earners in Canada. Engineers work in many fields and perform a wide variety of activities. They may design or develop power transmission equipment, lighting and wiring, computers, telephones or television sets, business machines, or robots, among many other things. In fact, every human-made object you see around you was designed by an engineer.

Despite these opportunities, only 3 per cent of professional engineers are women. The enrolment of women in engineering programs, though increasing, is still low; the national average in 1988–89 was 14 per cent.

Why do so few women go into engineering? Engineering has traditionally been viewed as a male activity. Some men have not welcomed women into the field. Many young women in secondary school have not been given information about engineering as an exciting, rewarding career. More women than men drop math before the end of secondary school and so do not have the courses they would need to enter engineering programs. Some women who do enter engineering programs find themselves a minority, without much support or encouragement.

According to current estimates, Canada will be facing a shortage of about 30 000 engineers by the year 2000. Already in British Columbia, more than a third of all employers in advanced technology firms report difficulties finding the engineers and scientists they want. "We must look at recruiting the untapped 50 per cent of our population — women," says Monique Frize of the University of New Brunswick.

Frize is an electrical engineer. She holds the position of Women in Engineering Chair, established in 1989. The University of New Brunswick has taken the lead in promoting engineering as a career for women and addressing the problems they face.

Frize says an important part of her job is to provide accurate information about what engineering programs and work are really like. For instance, marks of 70 or higher in high school math and science are adequate. You don't have to have straight A's.

There is growing support for women in engineering programs. Says Frize, "We will have the best technology in this country when we have both men and women engineers solving our problems.''

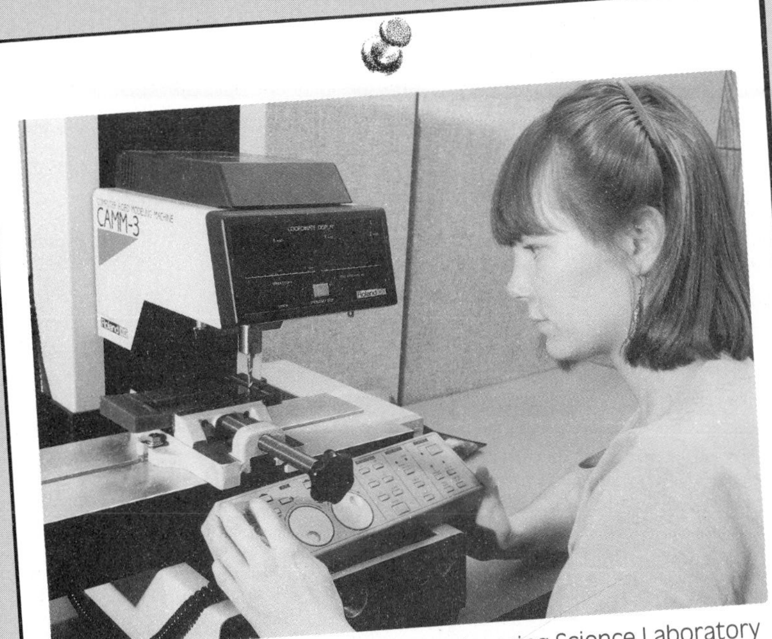

An undergraduate student in the Engineering Science Laboratory at Simon Fraser University.

6.3 Power Transmission

Power transmission lines transmit the electrical power to operate appliances in your home. They also transmit electrical power to schools, hospitals, office buildings, and factories (Figure 6.12).

Figure 6.12
Power transmission lines like this carry electricity from its source to the places where it is used.

ACTIVITY 6D *A Model Power Transmission Line*

In this activity, you will calculate the power input and output of a demonstration power transmission line.

MATERIALS

6 V battery
10 m of two-conductor speaker wire
two voltmeters
two ammeters
six lamp sockets
six 6 V bulbs
alligator clips

PROCEDURE

1. Your teacher will set up a demonstration of a power transmission line like that shown in Figure 6.13.
2. Copy Table 6.2 in your notebook.

Table 6.2 Sample Data Table for Activity 6D

Number of bulbs lit	Input			Output			
	Current (A)	Voltage (V)	Power (W)	Current (A)	Voltage (V)	Power (W)	Efficiency
0							
1			SAMPLE ONLY				
2							
3							
4							
5							
6							

Figure 6.13
A model transmission line like this will help you to understand the principles of the transmission line shown in Figure 6.12.

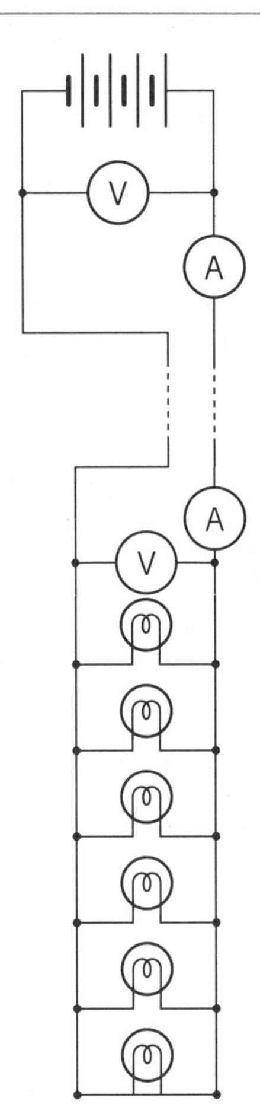

Figure 6.14
Tighten the light bulbs in their sockets one at a time and record the current and voltage input and output as each bulb is connected.

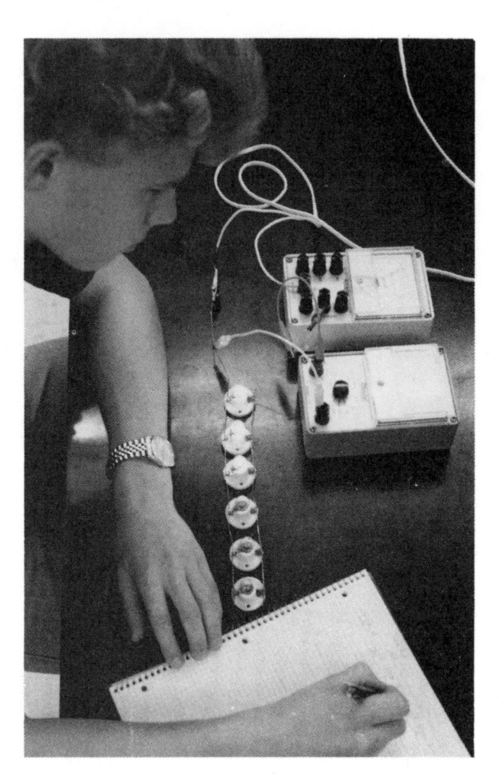

3. With the wires and meters connected, but with all the bulbs loose in their sockets so that none of them are making contact, measure the current and the voltage at the input end of the transmission line. This is the end where the battery is located. Record these measurements in your table. Calculate the power input by multiplying the current by the voltage ($P = I \times V$).

4. With all the bulbs loose in their sockets, record the current and the voltage at the output end of the transmission line. This is the end where the light bulbs are located. Record these measurements and calculate the power output.

5. Connect the light bulbs one at a time by tightening them in their sockets. Record the input and output current and voltage each time as you do this (Figure 6.14). Then calculate the input and output power.

6. Calculate the efficiency of the power transmission line by dividing the power output by the power input. Record your results in the last column of your table.

DISCUSSION

1. Answer the following questions to compare the model that you used in your classroom with the electric power system that supplies electric power to your home.
   (a) A generator supplies the power for your home. What takes the place of the generator in the model?
   (b) What do the light bulbs in your model represent?

2. How does the power input to the model power transmission line compare with the power output?

3. What happens to the efficiency of the power transmission line as
   (a) more bulbs are connected?
   (b) the current input to the transmission line increases?

4. When bulbs were connected, why was there a difference between the energy input to the transmission line and the energy output?

HIGH-VOLTAGE POWER TRANSMISSION LINES

A transmission line with a low voltage and a high current can carry the same amount of power as a transmission line with a high voltage and lower current. In practice, power in a transmission line is always transferred at very high voltages, typically 250 000 to 500 000 V, and relatively low currents of a few thousand amperes. High voltages and low currents minimize the energy losses that occur in the transmission line itself. Electrical energy is lost as a result of the heat produced as electrons move along the conductors of a transmission line. With a higher voltage and lower current, there is much less heating of the transmission line and the surrounding air, increasing the efficiency of the transmission line. Typically, a high-voltage transmission line is about 80 to 90 per cent efficient, depending on the amount of power it is carrying.

The high voltages used on transmission lines are often expressed in kilovolts (kV). One kV = 1000 V. Thus, a transmission line with a voltage of 500 000 V is a 500 kV transmission line. Since the power transferred by a transmission line is very large too, megawatts (MW) are convenient for measuring this power. One MW = 1 000 000 W.

At high voltages such as those used on transmission lines, current electricity is extremely dangerous. At these high voltages, electricity can flow from the transmission lines to anything nearby that will conduct the electricity to the ground. You can understand why if you remember Ohm's law from Chapter 3. Ohm's law states that the current $I = V/R$. If the voltage, V, is large, then significant amounts of current will flow through even a large resistance. A few examples may help you to visualize this. Suppose that you accidentally touch the positive and negative terminals of a car battery. The voltage difference between these terminals is 12 V. Under these conditions, your electrical resistance would be about 500 000 Ω. Using Ohm's law,

$$
\begin{aligned}
I &= V/R \\
&= 12\,\text{V}/500\,000\,\Omega \\
&= 0.00002\,\text{A or } 20\,\mu\text{A (microamperes)}
\end{aligned}
$$

About one milliampere (1 mA, or 0.001 A) of current must flow through you before you will notice it. Twenty microamperes of current is much less than this smallest detectable amount, so even though this current would flow through your body, you would not notice it.

In contrast, if you were to touch a high-voltage transmission line with a voltage of 500 000 V and your resistance remained the same, the current flowing through you would be

$$
\begin{aligned}
I &= V/R \\
&= 500\,000\,\text{V}/500\,000\,\Omega \\
&= 1\,\text{A}
\end{aligned}
$$

This current would have two effects. First, it would stop your heart. Second, since power = current × voltage, 500 000 W of power would be supplied to your body. Your body would convert this power into heat with terrible results.

DID YOU KNOW?

Fatal amounts of electricity vary. Ten milliamperes of current flowing through a person's body may cause the person to stop breathing. Currents of about 70 mA through the chest cause irregular heartbeat, called fibrillation. If treatment is not given immediately, death will result.

Surprisingly, if a current of about one ampere flows through a person, the effect can be less severe than for lower currents. The reason is that currents of about one ampere cause the heart to stop completely. Often it will resume beating normally when the flow of current is stopped.

Sometimes a person's heart will begin to fibrillate, or beat irregularly, as a result of a heart attack or an electric shock. If this happens, a device called a defibrillator can be used to supply a brief high current to the heart so that it will stop. After this treatment, the heart may again begin beating with a normal rhythm.

SCIENCE IN OUR WORLD

SUPERCONDUCTORS

Something is about to happen at the front of the room. A Styrofoam cup sits upside down on a table. A small black disk rests on the bottom of the cup. Liquid nitrogen is poured out of a beaker over the disk, engulfing it in a cold fog. Plastic forceps then hold a tiny chunk of dark material just above the disk. The forceps are gently removed and—the unexpected happens. The tiny chunk floats in the air like a levitating person in a magic show.

How does this happen? The black disk is a specially made ceramic. When it is cooled to an extremely low temperature, it becomes a superconductor. That means that it has no resistance to electric current and is a superior conductor of electricity. Thus, when liquid nitrogen, which has a temperature of −196° C, is poured on the disk, the disk becomes a superconductor. The chunk of material is a magnet, and like all magnets, it has a magnetic field of force around it. Placing this magnet near the superconductor causes a supercurrent within the superconductor. This super-current generates its own magnetic field. Then the two opposing fields, one from the chunk of magnet and the other in the superconductor, cause them to repel each other, just as the opposite poles of two magnets do. The magnetic force of the superconducting ceramic disk repels the magnet enough to keep it up in the air.

Superconductors are not new. In 1911, it was discovered that mercury would become a superconductor if cooled to the temperature of liquid helium (−269° C). Because it is expensive to keep temperatures this low, there were not many applications of superconductors. Over the years, more and more materials have been found to be superconductors at very low temperatures. Then, in the late 1980s, ceramic materials were prepared that would conduct at the temperature of liquid nitrogen (−196° C). This still is very cold, of course, but it is a lot warmer than liquid helium. Liquid nitrogen is less expensive to prepare and store, so more applications are possible.

Scientists predict that superconductors could revolutionize our transportation. By winding superconductors into coils, they could produce powerful, efficient magnets. These could be used to levitate trains, increasing speed even more. The electromagnetic "bullet" trains in Japan operate in a similar way.

Because superconductors offer no resistance to electric current, none of the electrical energy is transformed into heat, which can be lost to the surrounding environment or cause overheating. Thus, superconductors could be major energy savers.

Scientists are now trying to find materials that will superconduct at room temperature, or close to it. Then the number of possible applications will be enormous.

For a few moments, the dark object hangs suspended in the air above the flat disk.

Figure 6.15
Long insulators are required to keep high-voltage power transmission lines away from the towers that support them.

Because of the high voltages of transmission lines, even a kite string touching a transmission line can carry a fatal current from the line to a person and through the person to the ground. For this reason, transmission lines are supported by high towers that keep them away from the ground and prevent anyone from getting too close. Transmission lines are suspended from the towers with large insulators that keep the lines away from the support towers (Figure 6.15).

A high-voltage power transmission line requires a great deal of space to keep the wires away from buildings, trees, and people. In populated areas, the space required is impractical. In addition, within a community, the distances that power must be transmitted are relatively short, so power losses from the transmission line are not as significant as they are for a long-distance power transmission line. For these reasons, the voltage of the power transmission line is lowered before the electricity is distributed within a community. This is done at a **sub-station** like that shown in Figure 6.16.

At a sub-station, devices called **transformers** lower the voltage. Transformers can raise voltages or lower them, but they do not increase the amount of power. If the voltage is lowered, as in this case, the current increases so that the power remains the same. You can see this from the formula for power, $P = I \times V$. If the voltage is decreased by half, the current will be doubled to maintain the same power.

In fact, the electrical power output from a transformer is slightly less than the electrical power input, because transformers are not perfect. Transformers get hot while they are in operation, because some of the electrical power input is converted to heat. You can see a transformer in Figure 6.16. It is the large object in the centre of the picture. The large fan on the side of the transformer is needed to cool the transformer during hot weather.

Figure 6.16
At a sub-station, the voltage of the electricity received from a power transmission line is lowered by transformers. The electricity is transferred to distribution stations at this lower voltage.

Without transformers, the voltage of electricity cannot be easily raised for transmission over long distances or lowered for use within a community. Transformers depend on the changing voltage and current of AC electricity for their operation. Transformers will not operate on DC electricity. This is the single most important reason that electric power stations generate AC electricity instead of DC.

Electricity arrives at the sub-station via the wires on the left of Figure 6.17. Notice the large insulators that support the wires. These wires lead into transformers that convert the high voltages of the transmission lines to lower voltages, typically 60 kV. At this lower voltage, the electricity is transferred to **distribution stations** that are placed at central locations within a community (Figure 6.17). At distribution stations, transformers lower the voltage again, this time to 25 kV. This is the voltage at which the distribution station supplies electricity to the streets of a community.

Figure 6.17
A distribution station looks much like a sub-station, and it performs a similar function. At a distribution station, the voltage of electricity is lowered and then distributed to the streets of a community.

Because a voltage of 25 kV is still far too high to be safely used in a home, this voltage must be dropped further by another transformer. Many of these transformers are located along the streets of a commmunity. They may be found on power poles (Figure 6.18a) or in metal boxes in areas where underground wiring is used (Figure 6.18b). These transformers lower the voltage for your home to 240 V.

The diagram in Figure 6.19 summarizes the path of electricity from where a transmission line ends at a sub-station to its distribution to a transformer along a street.

(a)

(b)

Figure 6.18
A pole transformer or a transformer in a metal box (for underground wiring) lowers the voltage of electricity to the 240 V supplied to your home.

Figure 6.19
The distribution of electricity from a power generating station to your home is shown in this diagram.

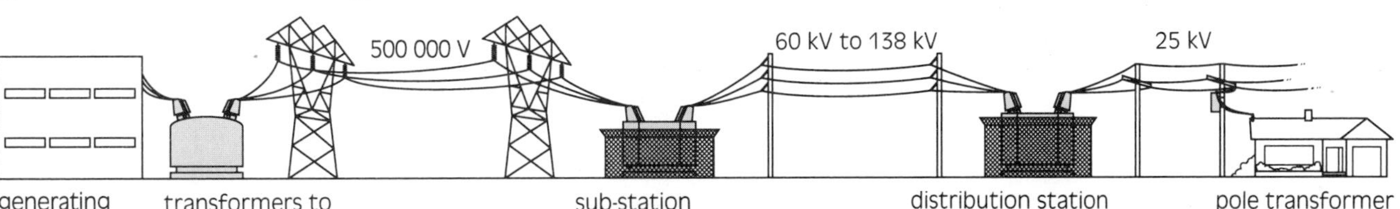

COMPUTER APPLICATION

COMPUTERS IN GENERATING AND TRANSMITTING ELECTRICITY

B.C. Hydro's system control centre in Burnaby controls all the electricity that the company generates and transmits. More than 40 generating stations with a combined total capacity of 10.5 million kW contribute to the system. Power is transmitted all over British Columbia and, at off-peak times, into Alberta and as far south as Mexico. Computers play an important role in controlling the generation and transmission of all this electricity.

The control centre is staffed by two or three people around the clock. The dispatcher who is in charge of generating electricity uses one computer console, and the dispatcher who is in charge of transmitting electricity uses another one. Each dispatcher has to know what the other is doing. Their job is to make sure that British Columbia's power needs are met, maintain enough reserves for emergencies, and keep the whole system in balance.

Maintaining balance in the B.C. Hydro electrical system is a challenging task. Demand goes up and down all the time, depending on how many people need power. Periods of peak demand are fairly predictable. Demand is usually greatest when people get up in the morning and when they turn lights on in the evening. Supply is affected by several factors, however, one of the most important being the weather. Lightning can knock out transmission lines. Heavy rain raises lake levels behind dams; to prevent overflow, more water must be sent through the turbines or (only if absolutely necessary) released down the spillways.

If monitors in the control centre show that the system is out of balance, the dispatchers must act quickly to fix the problem. When parts of the system fail (for example, when a transmission line is interrupted), the generation load goes up on other parts, causing overloads and more blackouts. If not corrected, this problem can shut down large parts—or all—of the system.

To operate the system, dispatchers must have accurate, up-to-date information. They need to know about the weather, water levels, status of each generating station, status of transmission lines, needs of B.C. users, needs of other power companies using the transmission lines, and much more. As B.C. Hydro's system has grown, and more and more electricity is used, computers have become increasingly necessary to gather, analyse, and display the increasing volume of information.

The control centre and the computer technology it uses are being redesigned, a process that will take three to four years. The senior dispatchers are part of the team working on software design. Their new consoles will have display screens with up to 24 "windows," which can show many activities going on at the same time. The new system will be able to respond faster and show a lot more information than the present system.

B.C. Hydro control centre in Burnaby, B.C.

The software team is also looking at what computers do best, what humans do best, and how the two can best be combined. Computers are better than humans at storing, organizing, and providing large amounts of information—a computer would have no problem listing all the options for any potential decision, for instance. But humans are better than computers at creative thinking and at predicting what other humans may do—so a person would be more effective at making certain (especially non-routine) decisions. The role of humans versus the role of computers is becoming a more and more important issue. Will people who use computers feel they control the technology, or will they feel it is controlling them?

REVIEW 6.3

1. A power transmission line operates at 500 000 V and has a current of 1000 A flowing through it. Calculate the power transferred by the transmission line.
2. A power transmission line has an input power of 100 000 000 W and an output power of 89 600 000 W.
   (a) Calculate the efficiency of the power transmission line.
   (b) What happenes to the 10 400 000 W of power lost between input and output?
3. Two power transmission lines have an input power of 100 MW. One of them operates at 500 kV, and the other operates at 50 kV. Which power transmission line will be more efficient? Explain your answer.
4. (a) If you hold a chemical cell between your thumb and forefinger, does current flow through your hand (Figure 6.20)?
   (b) Would you touch your thumb and forefinger across the terminals of a 150 V battery? Explain your answer.
   (c) Which would be more hazardous, to touch the thumb and forefinger of one hand to the terminals of a 150 V battery, or to touch one terminal with one hand while you touch the other terminal with the other hand?
5. What is a transformer?
6. Explain why electric power companies supply AC electricity to their customers.
7. Explain the function of
   (a) a sub-station,
   (b) a distribution station,
   (c) a power pole transformer.

Figure 6.20
Does current flow through your hand when you hold a chemical cell like this?

EXTENSION

High voltages on transmission lines cause electromagnetic fields. These fields are a combination of electric fields and magnetic fields. Often you can notice their effects on a car radio when you drive under power transmission lines. Do these electromagnetic fields cause harmful effects on people and other animals living near the power lines?

You can obtain information on the biological effects of electromagnetic fields from science and technology encyclopedias and from B.C. Hydro. Obtain as much information on the topic as you can. Then, on the basis of this information, decide if electromagnetic fields cause harmful biological effects.

CHAPTER • REVIEW

Key Ideas

- In a hydroelectric power plant, the energy of moving water is transformed into electrical energy by a generator. Any other source of energy could be used to turn the generator instead.
- Chemical cells and photovoltaic cells produce energy in an entirely different manner from that of a generator. They produce DC electricity directly from chemicals or from the light energy of the sun.
- All methods of producing large amounts of electricity have some harmful effects on the environment. Conservation of electrical energy can help minimize these harmful effects.
- Power transmission lines carry electricity from the place where it is produced to the places where the power will be used.
- Power transmission lines are operated at very high voltages and low currents so that they can supply a large amount of power with minimal loss of power along the transmission line.
- Transformers are used to raise or lower the voltage of AC electricity. Transformers operate only on AC.

Vocabulary

fossil fuel
renewable energy resource
non-renewable energy resource
photovoltaic cell
solar cell
fuel cell

→

piezoelectric effect
thermoelectric effect
thermocouple
superconductor
sub-station
transformer
distribution station

1. Create a concept map relating as many of the terms in the vocabulary list as you can.
2. Prepare a list of devices that can be used to produce electrical energy. Beside each device, indicate the type (or types) of energy it transforms into electrical energy.

Connections

1. Which sources of energy for producing electricity do you consider the least harmful to the environment? Explain why.
2. If you take into account all its advantages and disadvantages, what do you consider to be the best source of energy from which electrical power can be obtained? Give reasons for your choice.
3. The electrical energy from a hydroelectric generating station or from a wind generator ultimately comes from the sun. Explain why.
4. The piezoelectric effect is used in lighters that are included in many gas barbecues. A brief compression of the crystal in the lighter causes a high-voltage spark that ignites the propane gas supplied to the barbecue. Why is the piezoelectric effect suitable for this use but unsuitable for providing the electrical energy for your home?
5. Many household appliances (radios and television sets, for example) require DC electricity for their operation. Others, such as toasters or heaters, would work equally well on either DC or AC. Nevertheless, AC electricity is provided by all but one of the electric power companies in North America. Explain why.
6. Use the relationship $I = V/R$ to help explain why high-voltage current electricity is hazardous.
7. (a) Explain why it is extremely hazardous to fly a kite near a power transmission line.
   (b) In severe storms, power transmission lines are sometimes blown down so that they touch the ground. Why does this phenomenon create a short circuit?
8. Power cords to appliances such as electric kettles and irons become warm when the appliance is in operation. Power transmission lines also become warm when they are in operation. Explain why the heat produced in either case represents a loss of efficiency in the transmission of electrical energy.

Explorations

1. Describe how you would make a solar cell battery charger for rechargeable batteries. Include a diagram to show your apparatus and how it would be connected.
2. Make a small generator by using a magnet and some wire or obtain a bicycle lamp generator. Then devise some way to turn your generator so that it produces a supply of electricity.
3. Prepare a poster that encourages people to conserve electrical energy.
4. In the northeastern United States, many electric power generating plants use coal as a source of energy. What is the connection between this energy source and
   (a) acid rain?
   (b) the greenhouse effect?
5. Find out what a Rankine-cycle heat engine is. How have scientists suggested that this type of engine could be used to obtain energy from the ocean?
6. Find out what a magnetohydrodynamic generator is. What source of energy could be used with a magnetohydrodynamic generator?

Reflections

1. Refer to Activity 6A. Have your opinions changed after studying this chapter? If so, outline the changes.
2. Imagine that someone discovered a cheap and virtually unlimited source of energy. In addition, suppose that the source of energy could be operated without polluting its surroundings. After considering as many of the effects of such a source as you can, decide if the source would be of overall benefit or overall harm to the world.

CHAPTER

Radiation and Nuclear Energy

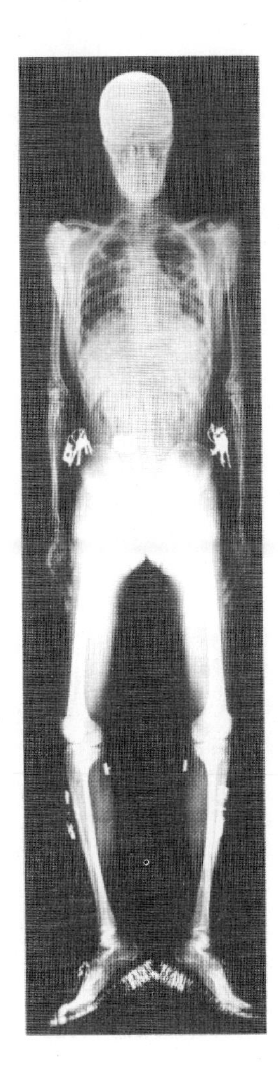

The photograph below was made with X rays shortly after they were discovered by the German scientist Wilhelm Roentgen (1845–1923) about 100 years ago. It shows a fully clothed man. The X rays were stopped by the keys in his pocket, the metal clips holding up his socks, the nails in his shoes, and the thicker bones of his skeleton, but they passed through his clothes and skin to reach the photographic film.

X rays are a type of radiation. Roentgen called this radiation X radiation because he had no idea of what it was. Since then, scientists have learned about the nature of X rays, and they have found many practical uses for them. They have also learned of the dangers of X rays.

There are other kinds of radiation besides X rays. For example, a lamp, the sun, a heater, and a uranium sample all emit radiation. In this chapter, you will learn about these different types of radiation, and you will see how they are connected to an understanding of the structure of atoms.

ACTIVITY 7A *What Is Radiation? What Is Nuclear Energy?*

Draw a line down the middle of a page in your notebook. Draw a second line across the middle of the page. With the help of two or three classmates, write down at least five things you know about radiation in the space on the top left-hand side of the page. In the space on the bottom left-hand side of the page, write down five things you know about nuclear energy. After you have completed this chapter, you will fill in the right-hand side of the page.

7.1 Electromagnetic Radiation

The term "radiation" is general and unspecific. It can be used in a number of ways. For example, radiation from the sun warms the Earth, and radiation is emitted from a sample of uranium. The radiation from a sample of uranium is probably what most people associate with the term "radiation." But they are likely to be much more familiar with the sun's radiation or that from a heater.

There are several types of radiation. Radiation from the sun is one example of **electromagnetic radiation**, or energy in the form of electromagnetic waves sent out in all directions. One form of electromagnetic radiation from the sun is light. Raindrops can separate sunlight into different colours. The colours of a rainbow make the **visible spectrum** of light (Figure 7.1).

Figure 7.1
A rainbow results when white light is separated into its component colours by drops of rain.

EXTENSION

Most photographic film is not sensitive to infra-red radiation. Special infra-red film is available, however. This film shows the world quite differently from the way in which it appears in visible light, so some photographers use it to emphasize these differences. If you have a special interest in photography and have access to a 35 mm camera, you can try taking infra-red photographs with infra-red film. To take these photographs, you will need a deep red filter that fits the camera. You can expect results like those shown in Figure 7.2.

Figure 7.2
Infra-red radiation, which is invisible to your eyes, affects special infra-red film. Infra-red photographs have an unreal appearance because the effect on this film of infra-red radiation from common objects is different from the effect of visible light on regular film.

The sun also gives off infra-red and ultraviolet rays, radio waves, microwaves, and X rays. You have probably had experience with these forms of electromagnetic radiation. For example, much of the heat from a fire results from infra-red radiation. A sunburn is the effect of ultraviolet radiation on skin. You can detect radio waves with a radio, and if you have used a microwave oven, you know that microwaves produced in the oven can heat food, even though the microwaves are invisible to you. And, as you know, invisible X rays can be used to make photographs that show whether bones are broken. Light, infra-red and ultraviolet rays, radio waves, microwaves, and X rays are examples of electromagnetic radiation.

Electromagnetic radiation can be pictured as a series of waves that spread outwards from a source, just as waves travel outwards in a pond. There is an important difference between the two kinds of waves, however. Water waves are made of matter; the waves of electromagnetic radiation are not. Remember from Chapter 4 that a magnetic field surrounds a magnet. In the same way, an electric field surrounds a charged object. An electromagnetic wave is composed of changing electric and magnetic fields that travel at the speed of light (Figure 7.3). An electromagnetic wave like that shown in Figure 7.3 is produced by a vibrating charged particle. The rate at which the charged particle vibrates is called its frequency. The electromagnetic waves that are produced by the vibrating particle have this same frequency.

If the frequency of the waves increases, their **wavelength** decreases. As you can see in Figure 7.4, the wavelength is the distance from a point on one wave to the same position on the next wave.

Different names are given to electromagnetic radiation that has different frequencies and wavelengths. Radio waves, for example, have low frequencies and long wavelengths. In contrast, X rays have high frequencies and short wavelengths. When the types of electromagnetic radiation are arranged from low frequency to high frequency, they show the **electromagnetic spectrum**, a diagram that indicates the range of electromagnetic radiation (Figure 7.5).

The energy of electromagnetic radiation is related to its frequency. As the frequency of the electromagnetic radiation increases, so does its energy. Low-frequency radio waves have low energy, light waves have higher energy, and high-frequency X rays and gamma rays have the highest energy. For this reason, X rays and gamma rays are hazardous to living things; the high energy of their electromagnetic radiation damages cells in ways you will learn more about later in this chapter and in Unit V.

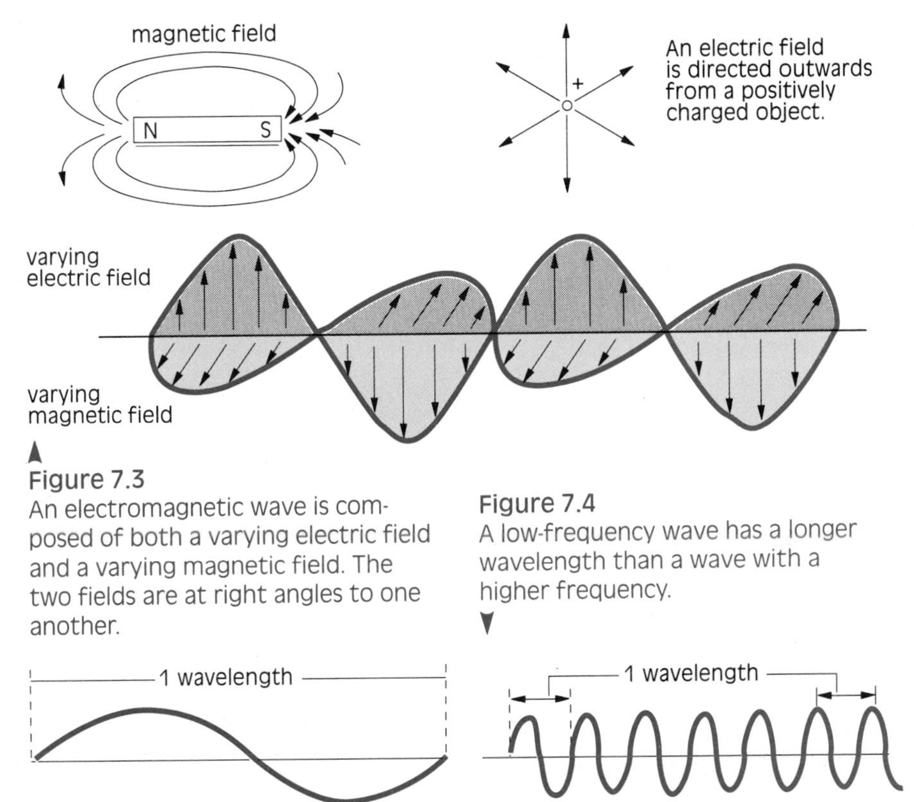

Figure 7.3
An electromagnetic wave is composed of both a varying electric field and a varying magnetic field. The two fields are at right angles to one another.

Figure 7.4
A low-frequency wave has a longer wavelength than a wave with a higher frequency.

DID YOU KNOW?

Colour television sets (more than black and white ones) give off small amounts of X rays. The reason for this is that fast-moving electrons hit the television screen and are stopped at its surface. The electrons give up a lot of energy when they are stopped, and some of it is converted to X rays that travel outward from the screen's surface. For people sitting back from a TV at a comfortable viewing distance, exposure to these X rays is very small. Young children should be discouraged from sitting on the floor right in front of the screen, however.

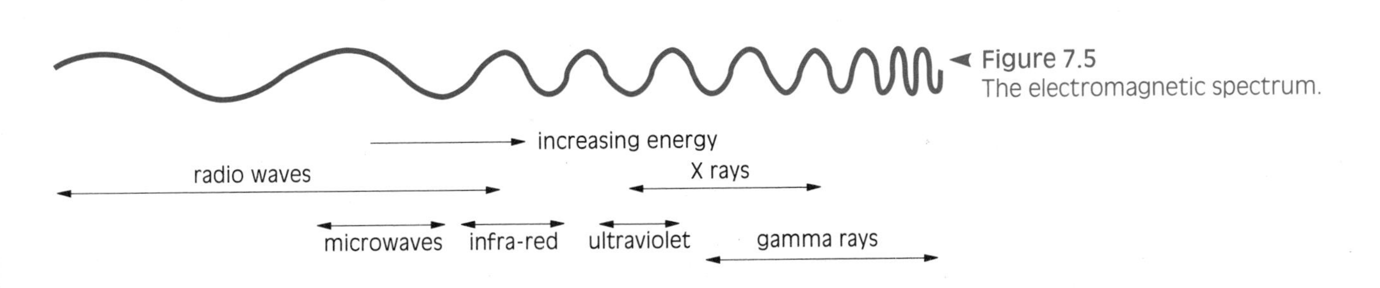

Figure 7.5
The electromagnetic spectrum.

REVIEW 7.1

1. Give evidence that shows that sunlight contains all colours of light. What colour of light is produced when these colours of light are mixed with one another?
2. If you stand near a campfire, much of the heat you receive is from electromagnetic radiation. What is the name of the kind of electromagnetic radiation that warms you?
3. As the frequency of electromagnetic radiation increases, what happens to
   (a) its energy?
   (b) its wavelength?
4. How does visible light compare with X rays in
   (a) wavelength?
   (b) frequency?
   (c) energy?
5. Why are X rays harmful to living things?
6. Many photographic materials are not sensitive to red light. This fact is convenient for photographers because they can work with these materials in the presence of red light without affecting the photographic materials. Blue light and green light do cause changes in these materials, however. Why do different colours of light have different effects?

7.2 Radiation from Atoms

In 1896, shortly after Roentgen's discovery of X rays, the French scientist Henri Becquerel (1852–1908) discovered a previously unknown type of radiation. This radiation was given off from a mineral Becquerel was studying even though Becquerel had done nothing to the mineral. Soon after Becquerel's discovery, other substances were found that also gave off radiation. Within a few years, scientists discovered that this type of radiation resulted from unstable nuclei of some atoms. Atoms that emit radiation from their nuclei are said to be **radioactive**.

Most atoms are stable; they do not break apart under normal circumstances. Yet in some ways, the stability of atoms is surprising. For example, in the nucleus of atoms, positively charged protons are crowded together in spite of the repulsive force that exists between objects with the same charge. If this repulsive electrical force were the only force acting on the protons, the nucleus would quickly disintegrate. But most nuclei do not disintegrate. In these nuclei, there is a stronger force that opposes the repulsive electrical force. This opposing force holds the nuclei together and is called the **nuclear force** (Figure 7.6). It is a short-range force that acts only within nuclei.

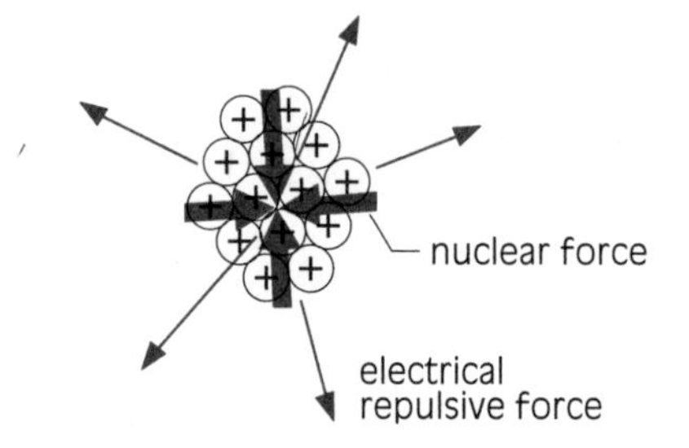

Figure 7.6
The electrical force between the protons of a nucleus pushes the protons apart. The nuclear force holds them together.

All elements except hydrogen include neutrons in their nuclei. Without neutrons, the repulsive electrical force between positively charged protons would destroy the nucleus. Hydrogen remains stable without a neutron because its nucleus contains only one proton.

RADIATION FROM RADIOACTIVE NUCLEI

Naturally occurring radioactive nuclei can emit three types of radiation: (a) **alpha radiation**, (b) **beta radiation**, and (c) **gamma radiation**. Alpha, beta, and gamma radiation have different properties. It can be difficult to study the properties of something that cannot be seen, but with the aid of a **Geiger counter**, a device that detects radiation, some of the properties of these types of radiation can be determined.

ACTIVITY 7B

Penetrating Ability of Alpha, Beta, and Gamma Radiation (Teacher Demonstration)

In this investigation, your teacher will use a Geiger counter and a set of weak radioactive sources to test the ability of alpha, beta, and gamma radiation to penetrate through materials.

As you watch this demonstration, remember that radiation can be harmful. Radioactive samples should always be handled with care and should not be used any more than necessary.

MATERIALS

one set of radioactive samples (alpha emitter, beta emitter, and gamma emitter)
Geiger counter
10 sheets of paper
10 pieces of aluminum foil
10 pieces of lead sheet

PROCEDURE

PART I

1. Your teacher will hold the detector of a Geiger counter over a source of alpha radiation. The distance between the source and the detector is then gradually increased. What happens to the reading on the Geiger counter?
2. Next, the detector of a Geiger counter is held over a source of alpha radiation (Figure 7.7). A sheet of paper is then inserted between the source and the Geiger counter. What happens to the reading on the Geiger counter? What is the effect of adding more sheets of paper?
3. With the Geiger counter detector held over the source of alpha radiation as before, a piece of aluminum foil is inserted between the source and the detector. What happens to the reading on the Geiger counter? What is the effect of adding more sheets of aluminum foil?
4. Again the Geiger counter detector is held over the source of alpha radiation. A piece of lead sheet is inserted between the source and the detector (Figure 7.8). What happens to the reading on the Geiger counter? What happens as more sheets of lead are added?

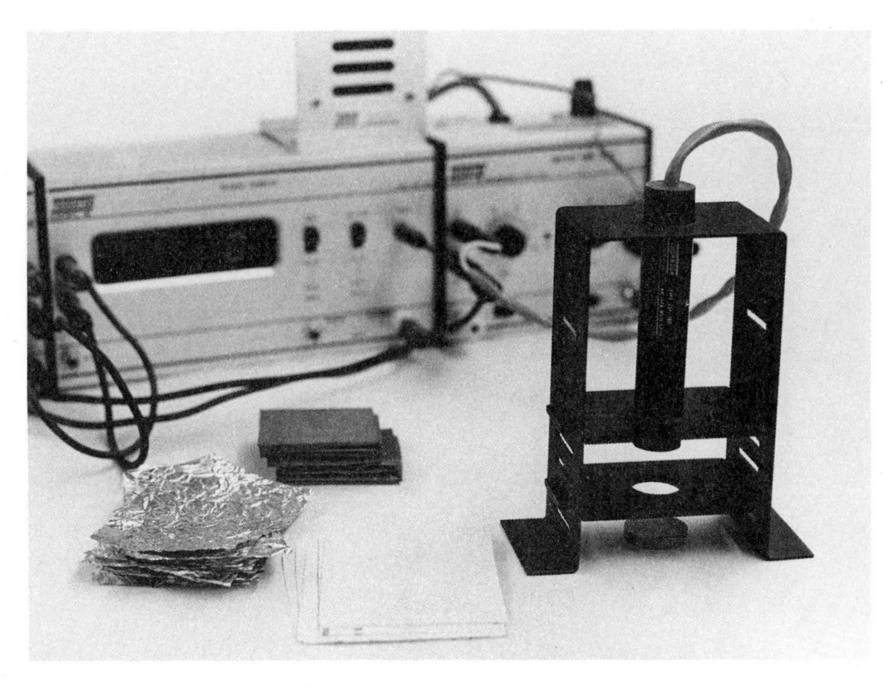

Figure 7.7
The detector tube of the Geiger counter is placed over the radiation source.

PART II

In this part of the demonstration, steps 1–4 from Part I are repeated, but this time a beta radiation source is used instead of the alpha radiation source used in Part I.

Figure 7.8
Does the radiation penetrate paper, aluminum foil, or lead placed between the source and the detector?

PART III

Steps 1–4 from Part I are repeated, but this time a gamma radiation source is used instead of the alpha radiation source used in Part I.

DISCUSSION

1. (a) Of the three types of radiation you studied in this activity, which is the most penetrating? Which is the least penetrating?
   (b) Give reasons for these conclusions.
2. Prepare a table like Table 7.1 in which you summarize the results of your observations. For each type of radiation, record whether it penetrated each material a great deal, very little, or not at all.

Table 7.1 Sample Data Table for Activity 7B

Type of radiation	Ability to penetrate			
	Air	Paper	Aluminum foil	Lead sheet
Alpha				
Beta				
Gamma				

SAMPLE ONLY

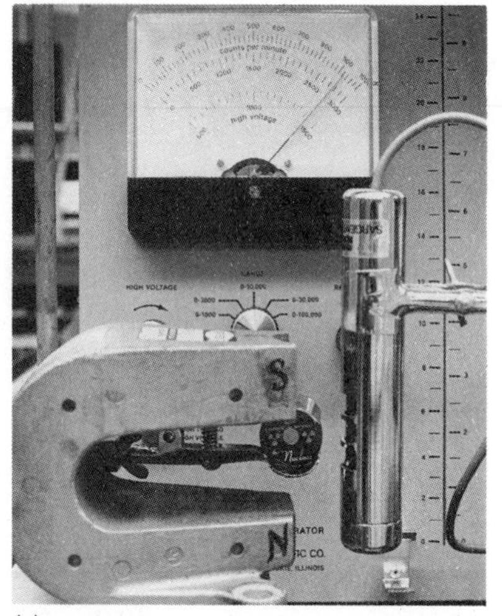

(a)

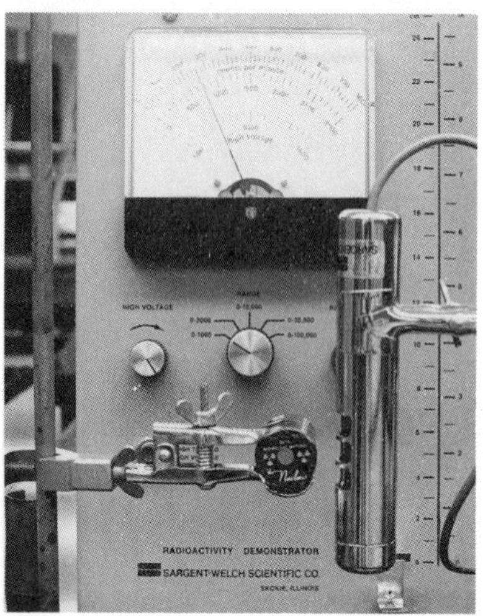

(b)

CHARACTERISTICS OF ALPHA, BETA, AND GAMMA RADIATION

Through many different types of experiments, scientists learned about the nature of alpha, beta, and gamma radiation. It was found that alpha and beta radiation consist of particles. We now know that alpha particles are helium nuclei. They are composed of two protons and two neutrons, so they have a charge of +2. Beta particles are high-energy electrons. Each beta particle thus has a charge of −1.

Because of their charges, both alpha and beta particles are affected by electric and magnetic fields. Beta particles are attracted towards positively charged objects and away from negatively charged objects. You can see the effect of a strong magnet on beta particles in Figure 7.9. In Figure 7.9a, the beta particles are deflected by the strong magnetic field of the magnet, and the Geiger counter shows a high reading. In Figure 7.9b, the beta particles are not deflected, and the reading on the Geiger counter is much less.

In contrast, gamma radiation is not made of particles. It is a form of electromagnetic radiation. It is a much more penetrating form of radiation than alpha or beta radiation, and since it is not charged, it is not deflected by electric and magnetic fields. In fact, except that its source is radioactive nuclei, it is indistinguishable from other high-energy electromagnetic radiation, such as X rays.

Figure 7.9
Alpha and beta particles are deflected by magnetic fields. When the beta particles pass the magnet, they are deflected so that they reach the detector (a). Without the magnet, fewer beta particles reach the detector, and so the Geiger counter shows a lower count (b).

ISOTOPES

The number of protons in the nuclei of an element's atoms is indicated by the atomic number of the element. The sum of the number of protons and neutrons in a nucleus is called the **mass number** of the nucleus. From the mass number, the number of neutrons in the nucleus can be calculated by subtracting the atomic number from the mass number.

All atoms of the same element have the same atomic number, but their mass numbers can differ. This is because atoms of the same element can have different numbers of neutrons in their nuclei. Atoms of the same element that have different mass numbers are called **isotopes** of that element.

The number of electrons of an isotope can be ignored because only the nucleus is considered in determining the protons and neutrons in an isotope. In a neutral atom, however, the number of electrons equals the number of protons in the nucleus.

There are a number of isotopes of the common element carbon. Three of these isotopes are carbon 12, carbon 13, and carbon 14 (Figure 7.10). Since all of them are carbon, they all have six protons. Carbon 12 nuclei have six neutrons and thus have a mass number of 12. Carbon 13 nuclei have seven neutrons, so their mass number is 13. Carbon 14 nuclei have eight neutrons and a mass number of 14.

Figure 7.10
The nuclei of three isotopes of carbon are shown in these diagrams. Each nucleus contains six protons, but the number of neutrons changes.

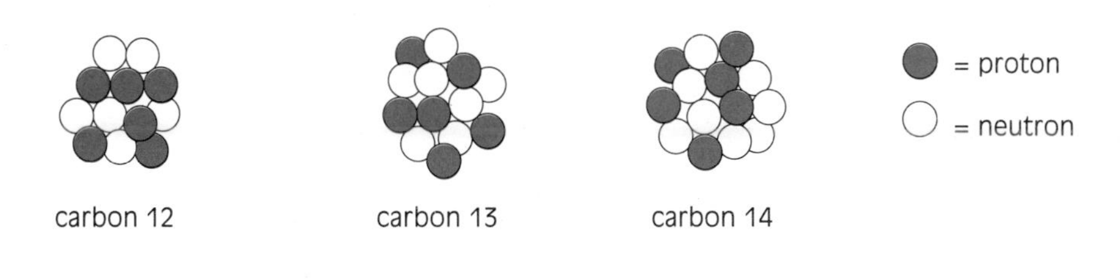

There is a shorter way to write the names of these isotopes of carbon. In this method, the above isotopes of carbon are written as

Carbon 12	$^{12}_{6}C$
Carbon 13	$^{13}_{6}C$
Carbon 14	$^{14}_{6}C$

The subscripts indicate the atomic number, and the superscripts show the mass number.

There are three isotopes of hydrogen. The most common one is $^{1}_{1}H$. This isotope has one proton and no neutrons; that is why its atomic number and mass number are the same. The other isotopes of hydrogen are called deuterium ($^{2}_{1}H$) and tritium ($^{3}_{1}H$). Both these isotopes have one proton in the nucleus, but deuterium has one neutron and tritium has two neutrons (Figure 7.11).

EXTENSION

In total darkness, place a radioactive source in contact with a strip of unexposed black and white photographic film. The source should be placed against the dull, light-sensitive side of the film. Wrap the film and the source carefully in black paper or in a black plastic bag and place it in a dark place. Leave the film and radioactive source there for a few days; then develop the film (or have the film developed—perhaps a member of your school's photography club can help you).

Write down everything you do during the experiment. Draw a neat, labelled diagram to show how the experiment was set up. State how your film was developed. But instead of calling the result a photograph, call it a radiograph because it is produced by radiation other than light.

Try this experiment with different types of radioactive sources—for example, sources that emit alpha, beta, and gamma radiation. Do these different sources produce different results? Can you explain any differences?

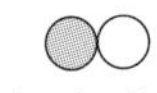

tritium
(hydrogen 3)

= proton
= neutron

Figure 7.11
These diagrams show the nuclei of the isotopes of hydrogen. Each nucleus contains one proton.

Uranium has many naturally occurring isotopes. Two of them are uranium 235 ($^{235}_{92}U$) and uranium 238 ($^{238}_{92}U$). They each have 92 protons, but uranium 235 has 143 neutrons and uranium 238 has 146 neutrons.

INSTANT PRACTICE

You can practice calculating the number of protons and neutrons in different isotopes by copying and completing Table 7.2 in your notebook. (Use the periodic table in Appendix B to find the symbol and the atomic number.)

Table 7.2 Sample Data Table

Isotope	Symbol	Number of protons	Number of neutrons
Iron 56			
Iron 54			
Aluminum 27			
Sulphur 32			
Oxygen 16			

SAMPLE ONLY

RADIOACTIVE DECAY

Different isotopes of an element have the same chemical properties. The nuclei of some isotopes are stable, however, whereas others are not. In unstable nuclei, the nuclear force is not enough to overcome the repulsive electrical forces within the nucleus. As a result, the nuclei are radioactive. When an unstable nucleus emits radiation, it undergoes a **radioactive decay**.

ALPHA DECAY

In an alpha decay, the nucleus of a radioactive isotope emits an alpha particle. Since an alpha particle is a helium nucleus with two protons and two neutrons, the nucleus formed as a result of the decay has two fewer protons and neutrons than the original nucleus. The following **nuclear equation** shows what happens when uranium 238 undergoes change by alpha decay:

$$^{238}_{92}U \longrightarrow {}^{4}_{2}He + {}^{234}_{90}Th$$

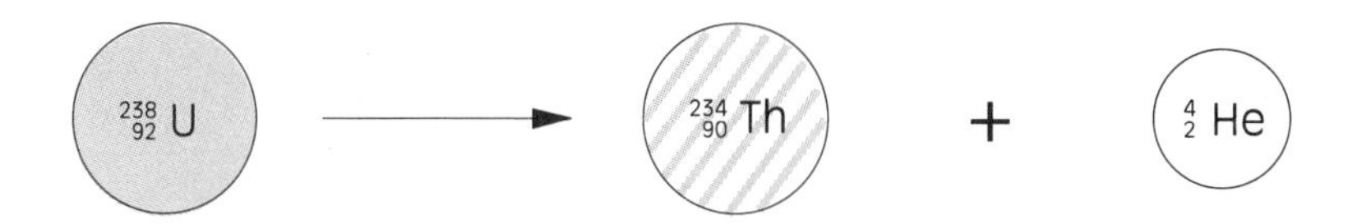

Figure 7.12
When a parent nucleus loses an alpha particle, the decay product has two fewer protons and two fewer neutrons.

The equation shows that the uranium 238 nucleus disintegrated, resulting in an alpha particle (the helium nucleus) and a nucleus with atomic number 90 (90 protons) and mass number 234 (234 neutrons + protons). The element with atomic number 90 is thorium, so the nucleus produced is thorium 234. Figure 7.12 may help you visualize what is happening.

The original uranium 238 nucleus is called the **parent nucleus**. The thorium 238 nucleus is called a **decay product**. In an alpha decay, the decay product always has a mass number that is four less than that of the parent, and its atomic number is two less than the parent's. Since the atomic number of the decay product can be easily calculated, finding its name is just a matter of looking up the element with that atomic number in the periodic table.

A new type of nucleus is always produced when a parent nucleus eliminates an alpha particle. A change in the type of nucleus is called a **nuclear transmutation**. Thus, alpha decay is one way in which nuclear transmutation is possible.

BETA DECAY

Some radioactive isotopes decay by emitting beta particles (high-energy electrons) from their nuclei. This may seem strange, since you know that electrons are not contained in the nucleus. However, protons and neutrons are not themselves the fundamental particles that they were once thought to be. In fact, under some circumstances a neutron can change into a proton and an electron. That is what happens in a beta decay. In this type of decay, the atomic mass of a nucleus does not change, because the total number of neutrons and protons remains the same. The number of neutrons decreases by one and the number of protons increases by one, however. As a result, in a beta decay, the atomic number of a parent nucleus increases by one and a new element is produced. Therefore, if a radioactive isotope decays by eliminating a beta particle, there will be a nuclear transmutation.

Nuclei of the isotope carbon 14 undergo beta decay. A nuclear equation for this change is

$$^{14}_{6}C \longrightarrow {}^{0}_{-1}e + {}^{14}_{7}N$$

The symbol $^{0}_{-1}e$ stands for the beta particle that is emitted. The parent nucleus had six protons, and a charge of -1 is lost from the nucleus. Thus, the nucleus of the decay product will have $6 - (-1) = 7$ protons. The element with this number of protons in the nucleus (and atomic number 7) is nitrogen. In this, and all other transmutations by beta decay, the decay product has an atomic number that is one more than that of the parent.

GAMMA DECAY

Unlike alpha and beta decay, gamma decay does not give off a particle. For this reason, there is no change in the type of nucleus in gamma decay. Instead, the gamma radiation that is given off carries excess energy away from the nucleus. Often this excess energy is present in a decay product that has been produced as a result of the alpha or beta decay of a parent nucleus. For example, cesium 137 undergoes nuclear transmutation by beta decay to produce barium 137. This barium 137 then loses energy by emitting gamma radiation. The result is a stable barium 137 nucleus. You can see this summarized in the nuclear equations

$$^{137}_{55}Cs \longrightarrow {}^{0}_{-1}e + {}^{137}_{56}Ba^{\star}$$
$$^{137}_{56}Ba^{\star} \longrightarrow {}^{137}_{56}Ba + \text{gamma radiation}$$

The * beside the symbol for barium signifies that this nucleus is unstable because it has excess energy. It is this energy that is given off as gamma radiation. Whereas a nuclear transmutation was responsible for the production of $^{137}_{56}Ba^{\star}$, there was no nuclear transmutation involved in its gamma decay.

RADIOACTIVITY CHANGES WITH TIME

The **activity** of a sample refers to the number of nuclei in the sample that undergo radioactive decay each second. Activity is measured in **becquerels (Bq)**, named after Henri Becquerel. In a very active sample, millions of nuclei may decay in each second. The activity of this sample would be millions of Bq. In a less active sample, 6000 nuclei may decay in a minute. This corresponds to an activity of 6000 decays per 60 seconds, or 100 Bq.

As radiation is emitted from a sample, fewer and fewer radioactive nuclei remain. What will happen to the activity of the sample as time goes on?

ACTIVITY 7C *How the Activity of a Radioactive Isotope Changes with Time*

Since you cannot see atoms or their nuclei, you may find it helpful to model their behaviour as they undergo radioactive decay. Using the model, you will examine how radioactivity decreases with time.

MATERIALS

100 popcorn kernels
plastic Petri dish
5 cm strip of masking tape
graph paper

PROCEDURE

1. Copy Table 7.3 into your notebook. (Add additional lines if necessary.)
2. Count out 100 kernels of popcorn and place them in the bottom half of a Petri dish (Figure 7.13). Imagine that these 100 kernels represent 100 nuclei of atoms of a particular radioactive isotope.
3. Fasten a 5 cm strip of masking tape along the outer edge of the Petri dish. Then cover the dish

Table 7.3 Sample Data Table for Activity 7C

Time (shakes)	Number of decayed nuclei	Number of undecayed nuclei remaining
0		
1		
2		
3		
4		
5		
6		
7		
8		
etc.		

SAMPLE ONLY

Figure 7.13
The popcorn kernels in the Petri dish represent radioactive parent nuclei. A piece of masking tape on the outside of the dish allows the student to determine if the "nuclei" have decayed.

Figure 7.14
Popcorn kernels that point at the masking tape have decayed. Count how many kernels point at the masking tape after each shake. Then remove this number from the Petri dish.

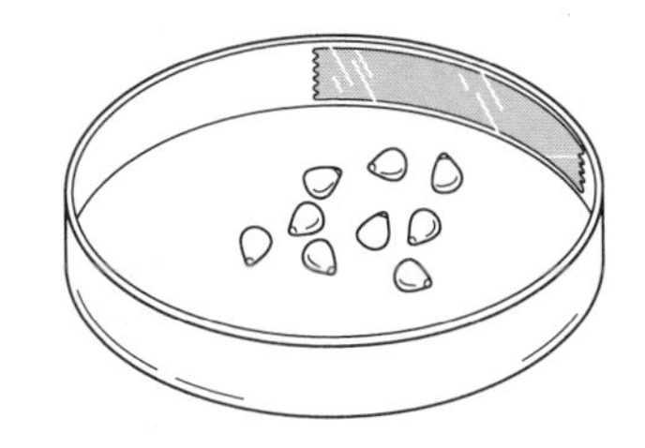

with its lid and shake it well. Place the dish on your desk and count how many of the kernels point towards the masking tape. These kernels represent nuclei that have undergone radioactive decay (Figure 7.14). Remove this number of kernels from the Petri dish. Record the number of kernels removed (decayed nuclei) and the number of kernels that remain (undecayed nuclei) in your table.

4. Re-cover the Petri dish and shake the remaining popcorn kernels again. Then remove the number of kernels facing the masking tape as before. Record the number of kernels remaining. Each shake represents the passage of the same amount of time. Even if no kernels are removed, a shake still counts as a time interval. Continue shaking the dish and recording the number of kernels removed and the number remaining until all the kernels are gone.
5. On a graph, plot the number of original nuclei remaining on the vertical axis and the time in shakes on the horizontal axis.
6. On a second graph, plot the number of nuclei decayed in each shake on the vertical axis and the time in shakes on the horizontal axis.

DISCUSSION

1. How many of the original radioactive nuclei were in the Petri dish when time equalled zero? What happened to the number of these nuclei as time went on (step 5)?
2. In general, did the number of nuclei that decayed in each shake increase, decrease, or remain the same as you proceeded with the experiment (step 6)?
3. Use your graph from step 5 to find
   (a) at what time the Petri dish contained 50, or half, of the original nuclei and
   (b) the time interval between when there were 50 of the original nuclei in the Petri dish and when there were 25.
   (c) How does the result you obtained in (a) compare with the result you found in (b)?
4. Would you get different results if you repeated the experiment? Would you get the same type of results? Explain your answer.
5. If the masking tape fastened to the edge were 1 cm long instead of 5 cm, what would happen to the rate of the radioactive decay? What would happen to the activity of the sample?

HALF-LIFE

The number of radioactive parent nuclei in a sample decreases as radioactive decay continues. Thus, the activity of a sample decreases as time proceeds (as long as the decay products are not radioactive too). This decrease in activity follows a regular pattern that can be described by the term **half-life**. A half-life is the time required for half the nuclei of a sample of a radioactive isotope to decay. The half-life is the same regardless of the starting amount of the radioactive isotope. This means that after one half-life, half the original amount will be left; after the second half-life, half of that, or one-quarter of the original amount, will be left; and so on.

Figure 7.15 shows the decay pattern for two of the many different isotopes of radon. Figure 7.15a shows the decay of radon 222. It has a half-life of 3.8 days, the longest of all radon isotopes. Figure 7.15b shows the decay of radon 216. Its half-life of 4.5×10^{-5} second is the shortest of all radon isotopes. Both these isotopes decay by emission of alpha particles.

Because the amount of all radioactive isotopes decreases by half in each half-life, graphs that show the amount of these isotopes at various times all look similar. Differences occur only in the time scale for the decrease.

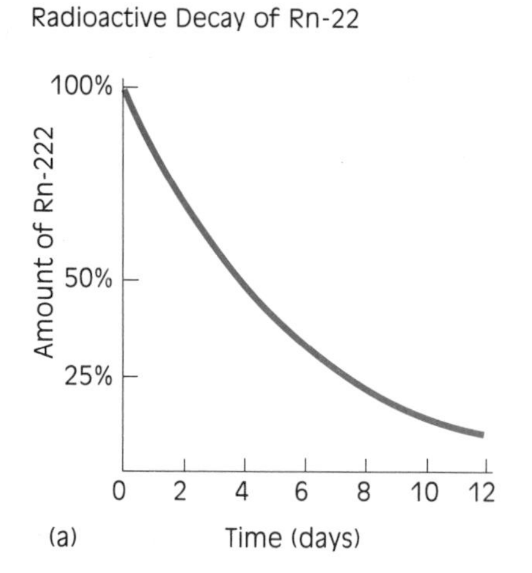

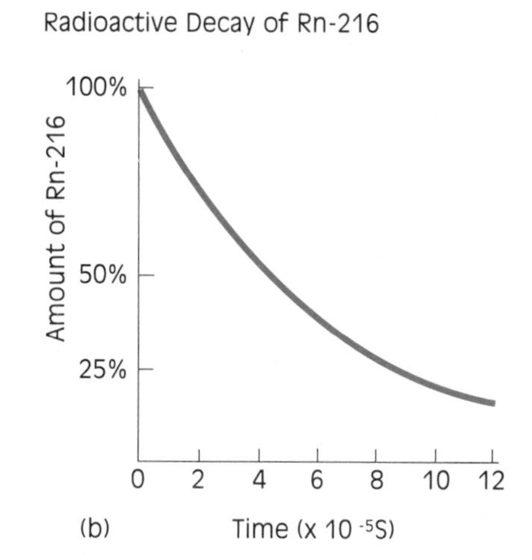

Figure 7.15
Even though the half-life of radon 222 (a) is much longer than that of radon 216 (b), graphs showing the decay are similar in shape. The time scale, however, is very different.

R E V I E W 7 . 2

1. (a) What force acts to hold nuclei together?
   (b) What force acts to break nuclei apart?
2. Using the symbol ○ for protons and • for neutrons, make a drawing of the nuclei of nitrogen isotopes $^{13}_{7}N$, $^{14}_{7}N$, and $^{15}_{7}N$.
3. Complete Table 7.4.

Table 7.4 Sample Data Table

Isotope	Symbol	Protons	Neutrons
Potassium 39			
	$^{37}_{17}Cl$		
	$^{35}_{17}Cl$		
		27	32
		88	134

SAMPLE ONLY

4. Complete Table 7.5 to summarize the properties of alpha, beta, and gamma radiation.

Table 7.5 Sample Data Table

Name of radiation	Electric charge	Description of radiation
Alpha		
Beta		
Gamma		

5. Explain why radioactive decay by the emission of an alpha or a beta particle produces a nuclear transmutation whereas radioactive decay by gamma radiation does not.

6. Some examples of radioactive decay are shown in the following list. What kind of radiation is emitted in each example?
   (a) $^{226}_{88}Ra \longrightarrow ^{222}_{86}Rn$
   (b) $^{60}_{27}Co^{*} \longrightarrow ^{60}_{27}Co$
   (c) $^{231}_{90}Th \longrightarrow ^{231}_{91}Pa$
   (d) $^{222}_{86}Rn \longrightarrow ^{218}_{84}Po$

7. Calculate the activity in becquerels of a radioactive substance that has
   (a) 125 emissions in 1 s,
   (b) 500 emissions in 10 s,
   (c) 2500 emissions in 20 s,
   (d) 4200 emissions in 1 min.

8. The activity of a sample of a radioactive isotope changes as shown in Figure 7.16.

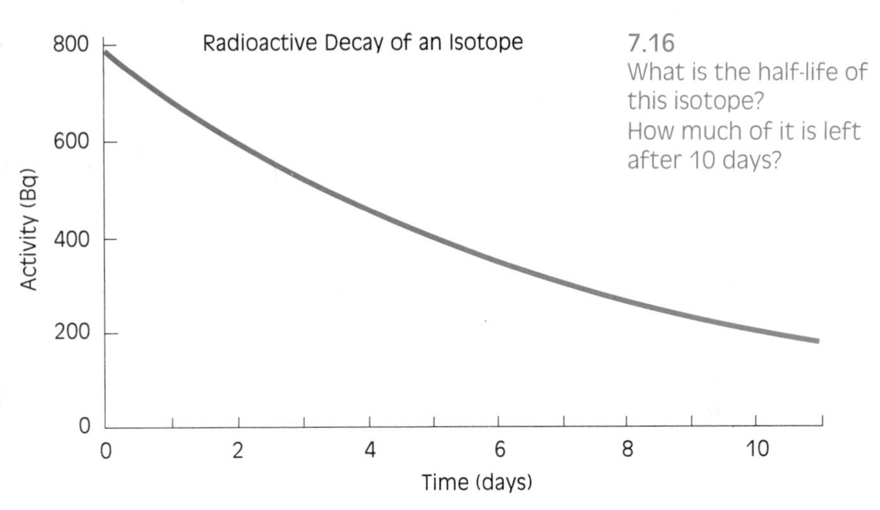

7.16
What is the half-life of this isotope?
How much of it is left after 10 days?

From the graph, calculate
   (a) the half-life of the isotope,
   (b) the amount of the isotope left after 10 days.

7.3 The Effects of Radiation on Living Organisms

Atoms of radioactive isotopes are found everywhere. For example, your body contains many radioactive atoms, and every minute about a quarter of a million of these atoms decay inside of you. Radioactive substances have always been present in the Earth's water and soil as well. Radiation is also formed in the Earth's upper atmosphere by cosmic rays, high-energy radiation from space. The total amount of radiation from all these sources is called **natural background radiation**. You cannot avoid being exposed to this radiation.

> **EXTENSION**
>
> It is easy to measure the background radiation in your classroom with a Geiger counter. To do this, count the number of clicks recorded by a Geiger counter during a one-minute period. Each click corresponds to an emission of radiation. Why should you count the number of emissions during one minute instead of for a second? Would you get the same result in a second trial?

HAZARDS OF RADIATION

Not all radiation is hazardous, but some types, such as the alpha, beta, and gamma radiation from radioactive isotopes, and short-wavelength ultraviolet and X-ray radiation, are very hazardous. These kinds of radiation have enough energy to knock electrons from the atoms of the material they hit. Atoms from which one or more electrons are removed are called **ions**. For this reason, these kinds of radiation are called **ionizing radiation**.

The loss of electrons can produce negative effects in cells. Cells that are dividing seem to be most vulnerable to ionizing radiation. In humans, blood-forming cells in the bone marrow and the cells lining the walls of the intestine divide more frequently than others. Large doses of radiation affecting these cells can cause "radiation sickness." Some symptoms are diarrhea, loss of appetite, changes in blood, and nausea.

Delayed effects of radiation can occur months or years later. For example, when the normal process of cell division goes out of control, the condition that occurs is called **cancer**. Radiation can initiate cancer in healthy cells. It can also cause changes in the hereditary material of people exposed to the radiation. These changes, called **mutations**, may be transmitted to future generations as well.

MEASURING DOSES OF RADIATION

As you know, the activity of a radioactive source can be measured in becquerels. But the number of becquerels from a sample tells nothing about the effect of the radiation on living or non-living materials. Nor does it apply to non-radioactive sources of ionizing radiation (such as X-ray sources, for example).

The effects of ionizing radiation doses can be measured by two units, the **gray (Gy)** and the **sievert (Sv)**. The gray measures the energy transferred to a material by a dose of ionizing radiation, whereas the sievert measures the effect of a dose of ionizing radiation on living things (Figure 7.17).

A gray is defined as the dose of ionizing radiation that delivers one joule of energy per kilogram of an absorbing material. As radiation passes through a material, some or all of the radiation is absorbed. The less penetrating the radiation, the more its energy will be absorbed by the material. In addition, radiation does not penetrate dense materials as easily as materials of low density, so dense materials absorb the energy of radiation more effectively. That is why lead makes a good shield from radiation.

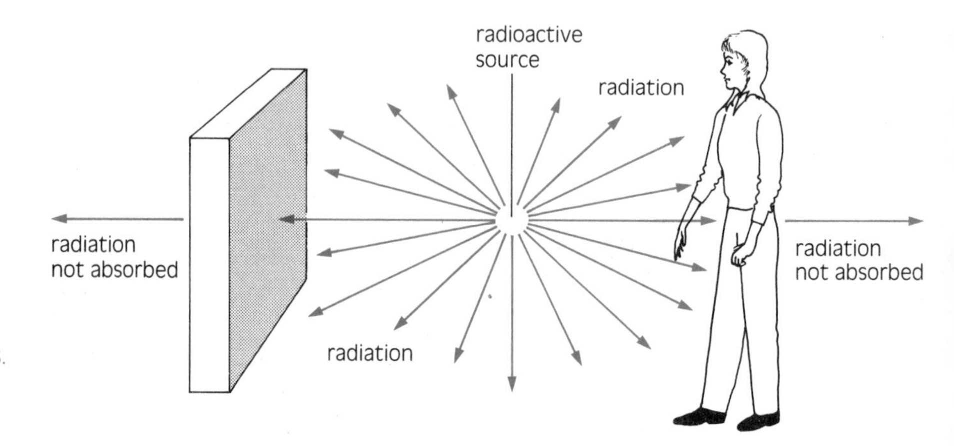

Figure 7.17
The activity of a source is measured in becquerels. A material receives a dose of one gray from the source when it absorbs one joule of energy per kilogram. The effect on living things of a dose of radiation from this source are measured in sieverts.

From this discussion, you can see that the energy delivered to a material depends on the type of radiation and the nature of the material. With some thought, you will also realize that the activity of the sample, the energy of its radiation, and the duration of exposure will also affect the dose of radiation, measured in grays, that a material receives.

The sievert is related to the gray, but it takes into account the biological effect of radiation. To compare radiation doses in grays and sieverts, you need to know the type of radiation that is involved. One gray of beta, gamma, or X-ray radiation is equivalent to 1 Sv in its effect on biological materials. One gray of alpha radiation, however, is equivalent to about 20 Sv.

Atomic bomb explosions and accidents have provided some information about the effects of different doses of radiation. Under 0.25 Sv, there are no observable effects. Above 1 Sv, there is damage to blood-forming tissue. If the dose is above 5 Sv, the person will die within a period of days or weeks.

For an individual living in Canada, the recommended maximum exposure is 0.005 Sv per year. This is approximately the dose of radiation that results from a series of X rays of the upper stomach. A typical dental X ray or chest X ray gives a much smaller dose, about 100 μSv (1 μSv = 1 micro sievert = 0.000 001 Sv).

DID YOU KNOW?

The becquerel, gray, and sievert are all part of an international system of units. Until recently, the activity of a radioactive sample was measured in curies, named after Madame Marie Curie (1867–1934). One curie = 3.7×10^{10} Bq. Instead of the gray, a unit called the rad was used. One rad is 1/100 of a gray. A unit called the rem (from rad equivalent man) was used instead of the sievert. One rem is 1/100 of a sievert.

PROTECTION FROM RADIATION

You can do a number of things to decrease your exposure to radiation. The first is shielding. You use shielding when you use sunscreen to protect your skin from the sun's harmful ultraviolet rays. People who routinely work with radioactive materials use lead shields to protect them from ionizing radiation (Figure 7.18).

As you move back from a source, its radiation spreads out over a larger volume and your dose becomes less. Thus, another way to protect yourself from radiation is to maintain distance between you and the source of radiation.

Duration of exposure to ionizing radiation affects the amount of energy delivered to your body. If you have ever had a sunburn, you know the effect of long exposure to ultraviolet radiation.

Time affects the dose of radiation in another way too. Since the activity of a radioactive source decreases with time, sources that begin as extremely active and hazardous ones will eventually reach levels of activity that can be handled more safely. If the half-life of the source is short, even a few minutes can be enough to decrease its activity significantly.

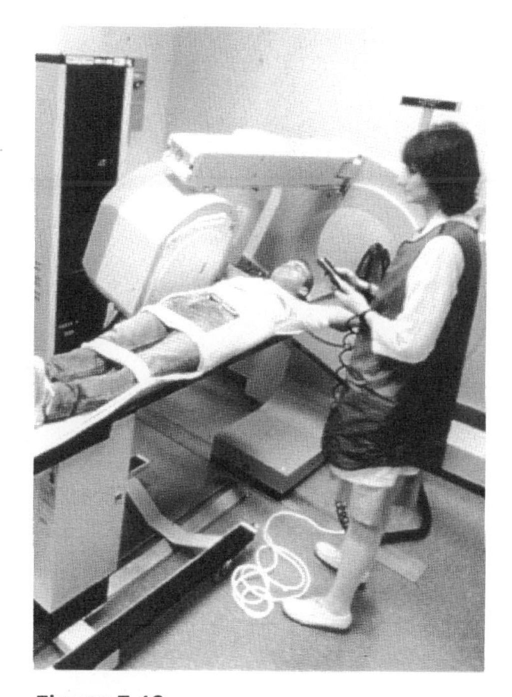

Figure 7.18
This technician works with radioactive isotopes in a hospital. A lead apron helps protect her from radiation.

Profile

MARIE CURIE

Marie Curie was one of the outstanding scientists in history. Near the end of the 19th century, scientists thought that atoms were tiny building blocks. They thought that atoms of a particular element were all the same, and that they never changed. Marie Curie's investigations of radioactivity and her discovery of new radioactive elements helped change these ideas.

Marie Curie was born Manya Sklodowska in Poland in 1867. After she finished secondary school, she worked for several years until she had earned enough money to leave Poland and attend the Sorbonne, a famous university in Paris. As a student, she lived in extreme poverty. Only by winning a scholarship was she able to complete her degree at the top of her class.

In 1895, Marie married Pierre Curie, a scientist at the Sorbonne. He had discovered piezoelectricity and investigated the loss of magnetism experienced by magnetic materials above a certain temperature (now called the Curie point).

Henri Becquerel's discovery that some unknown form of energy was emitted from uranium ores drew Marie Curie to study this exciting new field. She showed that radiation was emitted from the elements uranium and thorium in a process she named radioactivity. She also discovered that uranium ore contained elements even more radioactive than uranium. Pierre Curie gave up his work to become her assistant, and they began to search for these elements. By July of 1898, they had separated a small amount of an element that was many times more radioactive than uranium. They called this element polonium (atomic number 84), after Poland. Then they separated an even more radioactive element, which they called radium (atomic number 88).

Marie and Pierre Curie.

It took the Curies four years to obtain radium from tonnes of waste from a uranium mine. They did their work in a drafty, leaky shed that was very hot in summer and very cold in winter. In addition, the radiation to which they were exposed caused them fatigue, burns, and illness. It was only later that the dangers of radiation were understood.

For their pioneering work on radioactivity, the Curies and Becquerel were awarded the 1903 Nobel Prize in physics. In 1911, Marie Curie was awarded the Nobel Prize in chemistry for the discovery of polonium and radium.

The Curies could have made a lot of money from their early work if they had patented it, but they refused to do this. Instead, they chose to make the knowledge they gained freely available to all scientists.

Pierre Curie was killed by a horsedrawn cart in 1906. After his death, Marie Curie became the first woman professor at the Sorbonne. Because she was a woman, and because of her Polish background, she faced much discrimination, however. She was denied membership in the French Academy by one vote because she was a woman.

With the outbreak of World War I, Marie Curie raised money so that vehicles could be equipped with X-ray equipment. These vehicles were sent to field hospitals. There X-ray photographs produced with their portable equipment helped surgeons treat injured soldiers. Marie Curie herself drove one of those vehicles.

After the war, Marie Curie became the director of the Paris Radium Institute, an institute set up to study radiation and its effects on the body. She died of leukemia in 1934, almost certainly as a result of overexposure to radiation.

SOME APPLICATIONS OF RADIATION

Most people are familiar with the X-ray pictures used by doctors and dentists to show broken bones, the state of internal organs, or decay in teeth (Figure 7.19). Since they are produced by X radiation, these pictures are properly called radiographs rather than photographs.

Like all radiation, X rays are absorbed more effectively by dense materials than by materials of low density. Thus, since bones are denser than flesh, the bones absorb more of the X rays passing through the patient to the X-ray film. As a result, bones appear lighter in radiographs.

At a hospital, the radiography department produces X-ray radiographs. In many B.C. hospitals, there is a nuclear medicine department as well. Nuclear medicine covers a large range of applications of radiation to medicine. It is a rapidly changing field, and its medical value is steadily increasing.

In one application, the gamma radiation emitted by radioactive isotopes within a patient's body is used to produce pictures (Figures 7.20 and 7.21).

Nuclear medicine technologists can also use radioactive isotopes to analyze various hormones in blood and to test a patient's blood flow. To show blood flow, for example, a radioactive isotope, usually technetium 99, is injected into the patient's bloodstream. Then, as the blood circulates, the movement of the gamma-emitting technetium 99 can be shown on a gamma radiation camera.

Figure 7.19
Broken bones can be shown with X-ray radiographs.

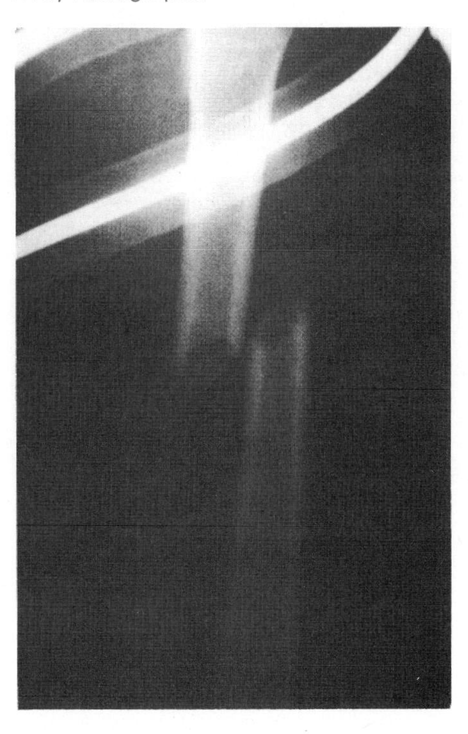

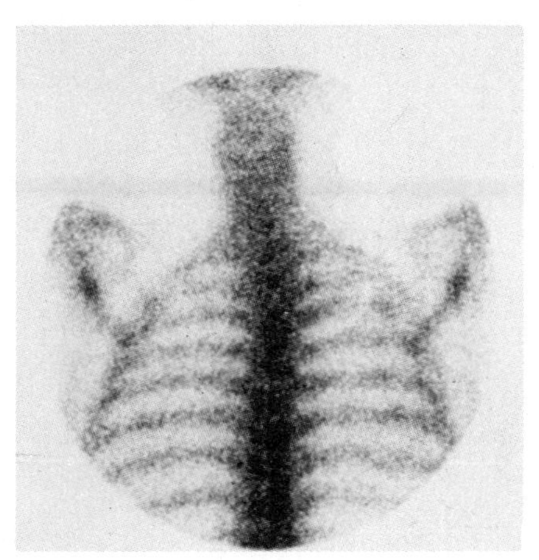

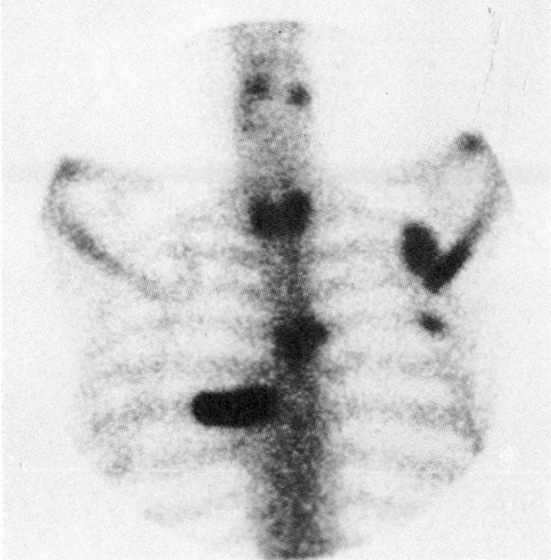

Figure 7.21
These nuclear medicine bone scan pictures were taken with gamma radiation. To do this, a radioactive isotope is attached to a phosphorus compound that is injected into the patient's bloodstream. As the blood circulates, the phosphorus compound, including the radioactive isotope, is absorbed by all bones, but it is absorbed most quickly at any locations where bone is growing rapidly. A normal bone scan is shown on the left. On the right, you can see dark areas where radioactive isotopes have been concentrated as a result of rapidly growing cancerous cells.

Figure 7.20
X-ray radiographs are made by passing X rays from an external source through a patient's body. The pictures produced with nuclear medicine use gamma radiation emitted by radioactive isotopes that are injected into a person's body.

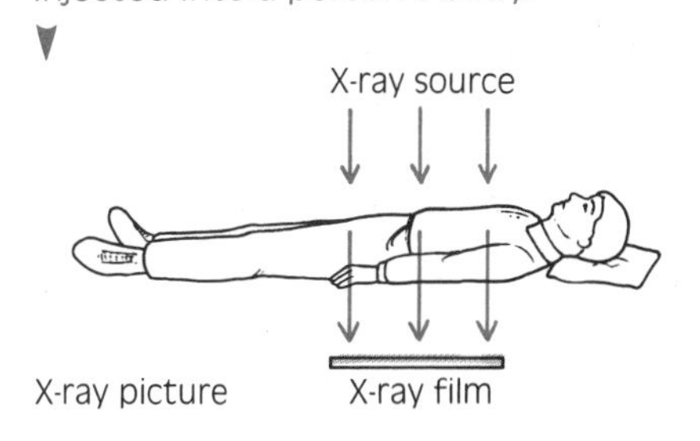

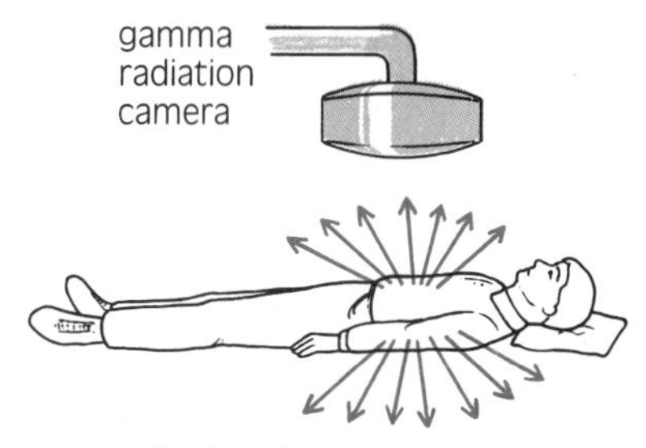

CAREER

NUCLEAR MEDICINE TECHNOLOGIST SUSAN WINTHER

Susan Winther works with radiation every day. She is a nuclear medicine technologist at Mills Memorial Hospital in Terrace, B.C. When she began, the nuclear medicine facility was just being added to the hospital. Under her direction, the nuclear medicine facility is now in operation and providing information for doctors.

What do you do in your work?
I use trace levels of radioactivity to perform diagnostic tests. For example, I can test for bone cancer and other bone diseases. I can test various body organs to see if they are functioning properly, and I can check circulation of blood through a patient's body. To do this, I use radioactive tracers—chemicals that have an attached radioactive isotope.

I begin the day by preparing the radioactive tracers. I do this in a "hot" lab. This is a lab that is equipped with lead shields and thick radiation-absorbing glass so that I can work safely with very active sources of radiation.

When I take a picture of a patient with the gamma radiation camera, the tracers show up because of the gamma radiation emitted. Where there are more tracer molecules, the image is darker. This image is recorded electronically in a computer. Then I analyse the computer image and transfer it to film. After I process the film, it goes to a nuclear physician for interpretation.

This is the gamma radiation camera Susan Winther uses. The patient lies beneath the camera while the camera collects gamma radiation emitted by the tracer inside the patient.

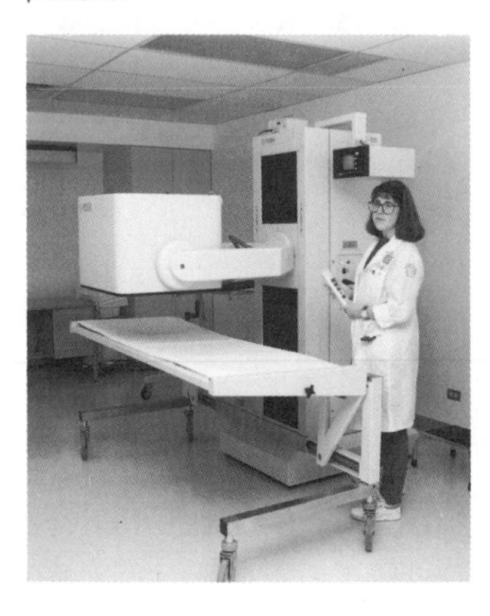

How did you decide to become a nuclear medicine technologist?
I was interested in a paramedical field, so I looked at becoming an X-ray technician, physiotherapist, lab technician, or ultrasound technician. When I leaned about nuclear medicine, I thought it looked like the most interesting job of all. What I liked was that it included lab work, as well as work with patients.

What training did you need for your job?
Of course, secondary school graduation with a good science background is required. Physics and chemistry are really important. Then two years of technical school. I went to the British Columbia Institute of Technology (BCIT) for this training. I also had two years of science at university.

What are the career possibilities for a nuclear medicine technologist?
There are many different jobs available to a person with my training. In British Columbia, there aren't too many jobs available in hospitals, although the number is growing. A nuclear medicine technologist can go into radiation protection for a private company or a government agency. Another area is for the licensing or controlling of radiation; I'm the radiation protection officer for the hospital, for example. I make sure that equipment and materials that involve radiation are used safely.

A nuclear medicine technologist can work part-time, and it's also a job that is easy to get back into after taking time off to have a family, for example. Nuclear medicine is a good stepping stone into other paramedical fields. Some nuclear medicine technologists go into medicine, too. Our training gives us some credit towards an MD degree.

What advice do you have for a student who would like to become a nuclear medicine technologist?
Make sure you know what you want. Spend some time in a nuclear medicine department to see what it's like. Be prepared to study and work hard.

I love my job. It's really interesting. Nuclear medicine technologists make up a small community in B.C. It's satisfying to be a part of that community.

Radiation can be used to treat some kinds of diseases as well. For example, you have a thyroid gland located near the base of your neck. While you are growing, this gland affects your body's physical and mental development. It also affects body activity when you are fully grown. In some people, the thyroid gland can become overactive. These people become extremely nervous and their heart rates increase. This condition can be treated by injecting iodine 131 into the bloodstream of the affected person. The thyroid gland concentrates iodine inside the gland. There beta emissions from the iodine 131 destroy some of the gland and decrease its activity.

Cancers can be treated with radiation too. Bone cancer, for example, can be slowed by injecting the radioactive isotope strontium 89 into the patient's bloodstream. The body uses strontium the same way it uses calcium, so the radioactive strontium soon enters the bones. Cancerous areas absorb the strontium more rapidly than normal bone, and the beta emissions from the strontium 89 help kill the cancerous cells. Another alternative is to treat the patient with X rays or gamma rays that are aimed at the cancerous growth within the patient's body. In both methods, cancerous cells are more affected by the radiation than normal cells, but patients treated with radiation still become very sick from the radiation.

REVIEW 7.3

1. What is natural background radiation? Give three examples of sources of this radiation.
2. What is ionizing radiation? What are its effects?
3. Describe the short-term and long-term effects of ionizing radiation on living cells.
4. What is a mutation? What can be the effect of mutation?
5. Infants are much more likely to be harmed by radiation than are adults. Why do you think this is so?
6. Explain the difference between a gray and a sievert. Why is the sievert the more important unit for living things?
7. What level of radiation
   (a) is the recommended maximum exposure for Canadians?
   (b) is the maximum that has no observable effects?
   (c) produces damage to the blood-forming tissue in the bone marrow?
   (d) will cause death within a short period of time?
8. Many people are concerned about overexposure to radioactivity and X rays. Medical X rays are known to be the largest source of radiation made by humans. When you have an X ray, you slightly increase your chances of developing leukemia, a form of cancer.
   (a) What kinds of questions might you ask your doctor or dentist if it is suggested that you need a series of X rays?
   (b) Would you consider the possible detection of disease in its early stages worth the risk associated with X rays?

7.4 Nuclear Energy

You already know that the nucleus of an atom is stable if the nuclear force acting within the nucleus is stronger than the repulsive electrical force from the nucleus's positively charged protons. But there is another way of understanding the stability or instability of atoms. It came from the work of American scientist Albert Einstein (1879–1955), shown in Figure 7.22.

Figure 7.22
Albert Einstein. His special theory of relativity explained the source of the energy of radioactivity.

MASS AND ENERGY

In 1905, Einstein published *The Special Theory of Relativity*. This theory makes many remarkable predictions, one of which is that mass and energy are different forms of the same thing. You can see this from the equation that relates them, $E = mc^2$. In this equation, E stands for energy, m represents mass, and c is the symbol for the speed of light in a vacuum. Since the speed of light in a vacuum is a constant, as mass increases, so does energy.

How does this equation explain radioactive decay? Careful measurements of mass show that the total mass of the decay products is slightly less than the mass of the original radioactive nucleus. The large amount of energy released in the decay corresponds to the change in mass according to the equation $E = mc^2$. In other words, an unstable radioactive nucleus has more energy than its decay products. By decaying, the nucleus loses some of its excess energy, just as a ball loses its potential energy as it rolls downhill.

NUCLEAR FISSION

In the 1930s, the Italian scientist Enrico Fermi (1901–1954) experimented with the nuclei of atoms by firing neutrons at them. He realized that since neutrons have no charge, they would not be repelled by the positive charge of the nucleus; thus, they would easily be able to penetrate it. He reasoned that if uranium nuclei were used as the target nuclei, it might be possible to produce new elements with atomic numbers greater than that of uranium.

Fermi succeeded in producing new elements. It was later found that neptunium (atomic number 93) and plutonium (atomic number 94) were produced. Fermi found something else too, but the experiments and their significance were not clearly understood until 1938, when they were explained by the Austrian scientists Lise Meitner (1878–1968) and Otto Frisch (1904–1979). They realized that some uranium nuclei had absorbed neutrons and then split into two roughly equal pieces. They called the process by which a large nucleus split into pieces nuclear **fission**.

Here is a typical example of nuclear fission represented by a nuclear equation:

uranium 235 + neutron ⟶ uranium 236 ⟶
strontium 90 + xenon 143 + 3 neutrons + energy

or

$$^{235}_{92}\text{U} + ^{1}_{0}\text{n} \longrightarrow ^{236}_{92}\text{U} \longrightarrow$$
$$^{90}_{38}\text{Sr} + ^{143}_{54}\text{Xe} + 3\,^{1}_{0}\text{n} + \text{energy}$$

The equation shows that after a neutron is added to a uranium 235 nucleus, the resulting uranium 236 nucleus undergoes nuclear fission. The products are strontium 90 and xenon 143. As well, three neutrons and heat energy are released.

In this nuclear fission, the total mass of the products is less than that of the neutron and uranium 235 nucleus with which the reaction began. The reaction releases about 10 times as much energy as the type of nuclear decay you studied earlier in this chapter, and in excess of a million times as much energy as is produced in even the most energetic chemical reactions. This energy comes from the conversion of mass to energy according to the equation $E = mc^2$.

The neutrons produced by the nuclear fission make a **chain reaction** possible. In a chain reaction, some of the products of the reaction go on to cause more reactions to occur. You can see how this occurs in the diagram shown in Figure 7.23.

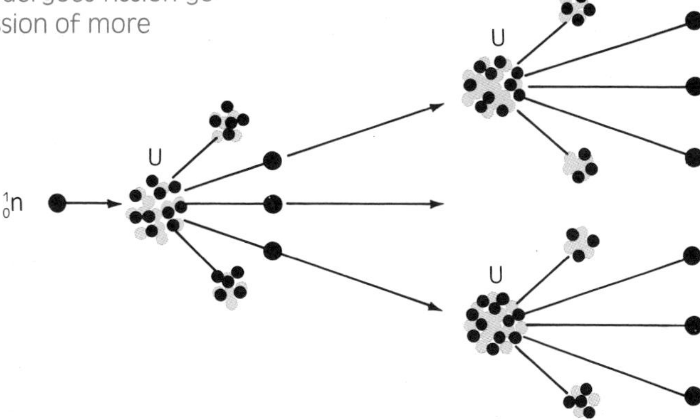

Figure 7.23
A chain reaction. Neutrons produced when a nucleus undergoes fission go on to cause the fission of more nuclei.

NUCLEAR REACTORS

A **nuclear reactor** is a structure designed for the controlled release of the heat energy from a sustained chain reaction. For a chain reaction to occur, neutrons produced in the reaction must be available to initiate new fission reactions. Neutrons may not be available to continue the process for a number of reasons, however.

Fast-moving neutrons will not continue the chain reaction because they move so quickly that few of them will be absorbed. For this reason, a substance called a **moderator** is used to slow the neutrons in a reactor. Ideally, a moderator contains atoms whose nuclei are of about the same mass as a neutron. Then, when fast neutrons strike the nuclei, the neutrons are slowed most effectively.

Two common moderators are heavy water and carbon in the form of graphite. In heavy water, or deuterium, the hydrogen isotope ^2_1H replaces normal hydrogen isotopes (^1_1H) in the water molecule. Ordinary water is not satisfactory as a moderator because normal hydrogen absorbs neutrons.

Besides the fission of uranium 235, reactions that absorb neutrons also occur in nuclear reactors. For example, naturally occurring uranium consists mainly of the isotope uranium 238. This isotope absorbs neutrons to make uranium 239, which does not undergo fission. For this reason, some types of reactors can operate only on uranium that contains more than the normal amount of uranium 235. Such uranium is said to be enriched.

A second factor affecting the number of available neutrons is the amount of fissionable material in a nuclear reactor. If it is too small, too many of the neutrons produced by the fission reaction will escape without continuing the reaction, and the chain reaction will cease (Figure 7.24). The smallest mass in which the chain reaction can be sustained is called the **critical mass**. For uranium, it is in the order of a few kilograms.

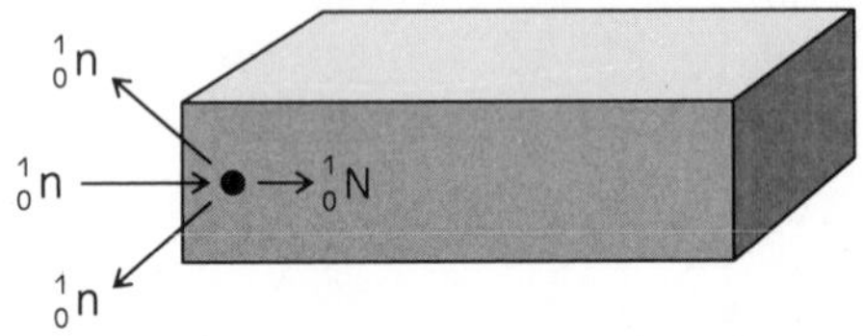

Figure 7.24
If the mass of fissionable material is too small, neutrons can escape from it and the chain reaction will cease.

Although a chain reaction will not proceed if too many neutrons escape from the fuel, you can see that too many slowly moving neutrons in the nuclear fuel could be disastrous. Under these circumstances, the reaction would quickly proceed out of control. The number of available neutrons would increase very rapidly, causing more fissions and huge amounts of energy in a very short time. This energy could cause a **reactor meltdown** that would destroy the structure of the reactor. Then radiation and radioactive material would be released into the environment. To keep the chain reaction under control, nuclear reactors use **control rods**, made of the elements cadmium or boron, to absorb neutrons in the reactor. The control rods can be moved in or out of the reactor to decrease or increase the number of neutrons.

Much of the energy released in a nuclear reactor is in the form of heat. This heat is used to produce steam, and the steam is used to turn a turbine that drives a generator (Figure 7.25).

PROBLEMS WITH NUCLEAR REACTORS

Other than the possibility of malfunctions, there are several serious unsolved problems in the use of nuclear reactors to produce energy for electricity. When nuclei undergo fission, many radioactive isotopes are produced. Some are products of the reaction, whereas others are produced in the interaction of neutrons with the reactor itself. These isotopes can leak from reactors.

Spent fissionable material has to be removed from a reactor. As yet, there is no really satisfactory way of disposing of this very hazardous material that contains high concentrations of radioactive isotopes, many of which have long half-lives.

Materials used to build nuclear reactors have to withstand high temperatures and pressures and the bombardment of radiation within the reactor. Because of these harsh conditions, reactors have a short lifetime of about 30 years during which they can be safely operated. Safe disposal of a reactor after its operating lifetime is over is a major problem.

Figure 7.25
This series of diagrams shows how a CANDU nuclear reactor (right) uses the fission of uranium to produce electricity. (a) Uranium dioxide power is pressed into fuel pellets. (b) The pellets are inserted into metal rods, which are then bundled. (c) The bundles are inserted into pressure tubes, each with a mass of about 300 kg. These are loaded into a structure called the calandria in the reactor core. (d) Cadmium control rods can be moved in or out of the calandria to regulate the number of neutrons striking the uranium 235. The heavy-water moderator surrounds the calandria to slow the neutrons from speeds of up to 42 000 km/s to about 3 km/s, so the neutrons can be absorbed by uranium 235. Heavy water acts both as a coolant to the reactor core and as a moderator. The coolant is heated by the fission reaction, and then it transfers its heat to ordinary water, which turns to steam. The steam drives turbines, which turn generators.

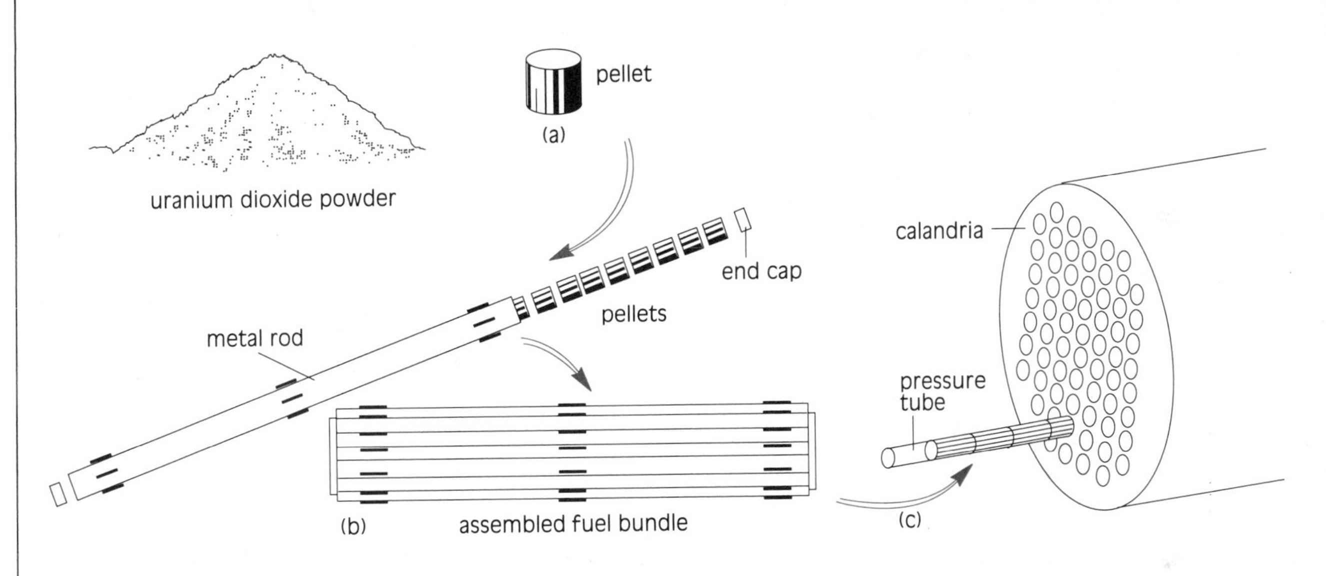

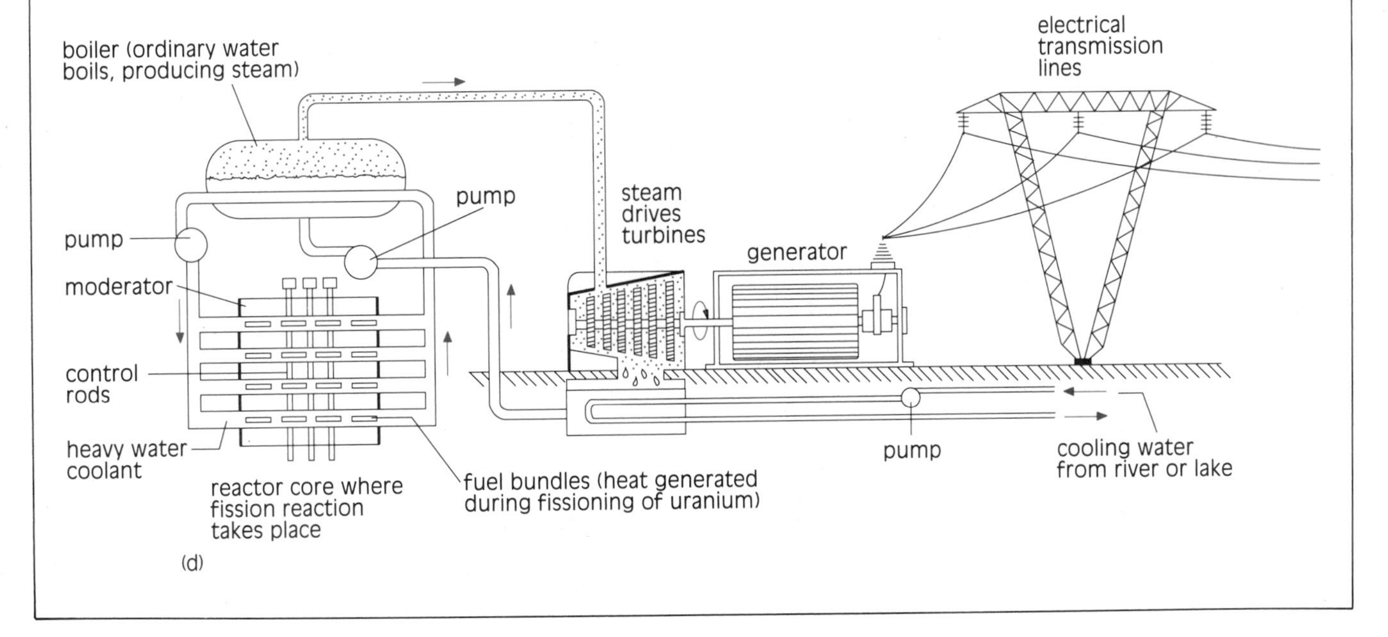

SCIENCE IN OUR WORLD

THE CHERNOBYL ACCIDENT

In the spring of 1986, a nuclear power reactor in Chernobyl, U.S.S.R., exploded. The reactor's hot core was exposed, and radioactivity was released into the atmosphere, raining onto millions of people. It was reported that the accident had caused the death of more than 30 people at or near the reactor site. Since then, it has been estimated that at least 5000 people have died and 50 000 have become ill.

What are the dangerous elements in radioactive fallout? How are they taken into the body, and how do they cause harm?

The most dangerous radioactive materials are the isotopes iodine 131, cesium 137, and to a lesser extent, strontium 90. Iodine 131, with a half-life of eight days, poses the most immediate danger. If a person inhales iodine 131 or ingests food or water contaminated with it, the isotope can become concentrated in the thyroid gland. The radioactivity can destroy thyroid function or cause cancer. Thyroid cancer is very rare among children. But in Byelorussia, the republic where Chernobyl is located, thyroid cancer was reported in 19 children in 1990. In the nearby Ukraine, 21 cases of thyroid cancer among children were reported.

Cesium 137 and strontium 90 have half-lives of about 30 years and can contaminate soils and food for a long time. After the Chernobyl explosion, cesium contaminated crops as far away as Italy, Wales, and Lapland, in northern Sweden. Cesium has the same outer electron structure as sodium and potassium, and it is taken up by the body as if it were sodium or potassium. Cesium 137 causes radiation damage throughout the body, increasing the chance of cancer and genetic defeats.

Strontium has the same outer electron structure as calcium and is incorporated into the body's bone structure along with calcium. The presence of strontium 90 in the bone can affect the bone marrow's production of red blood cells and cause blood cancer (leukemia). In 1990, leukemia rates in the regions around Chernobyl increased sharply. Scientists expect these rates to reach a peak between 1991 and 1993. Other types of cancer are expected to appear in another 5 to 10 years.

In the Soviet Union, a young girl stands beside a sign listing precautions against fallout from the Chernobyl power station. The precautions include warnings to limit children's playtime outside and to beware of dust on tree leaves.

NUCLEAR FUSION

When objects are fused, they are joined together. This is exactly what happens in nuclear **fusion**. Small nuclei are joined together to make a larger nucleus. In the process, a large amount of energy is released by the conversion of mass to energy. This is the process by which the sun and other stars produce energy.

A typical nuclear fusion reaction is

$$^{2}_{1}H + ^{2}_{1}H \longrightarrow ^{3}_{1}H + ^{1}_{1}H + \text{heat energy}$$

Since the starting material in this reaction is the hydrogen isotope deuterium, the term "hydrogen fusion" can be applied to it. Many other nuclear fusion reactions are possible that do not involve hydrogen.

For a given mass of fuel, a greater amount of energy is released in fusion reactions than in fission. Nuclei can fuse together only if they collide at very high speed; otherwise, the repulsive electrical forces between nuclei push the nuclei apart before they can get close enough to join. A collection of nuclei moves more quickly as its temperature is increased. But very high temperatures, about 100 000 000°C, are needed before nuclei move rapidly enough for fusion to occur. At this range of temperature, ordinary materials vaporize. For this reason, scientists have not yet succeeded in containing a fusion reaction. Unfortunately, some scientists have produced uncontrolled fusion reactions with hydrogen bombs. In these terrible devices, a fission bomb produces temperatures high enough to cause hydrogen fusion. The fusion causes the devastating effect of these bombs.

A DEFINITION OF RADIATION

From all you have learned in this chapter, you will know that there is no one definition for radiation. Part of the reason is that near the beginning of this century, scientists applied the term "radiation" to phenomena that were not fully understood. As a result, radiation can be electromagnetic or it can consist of alpha or beta particles. It can also consist of neutrons, protons, or other subatomic particles produced in nuclear reactors and at nuclear research facilities. Figure 7.26 summarizies these different types of radiation.

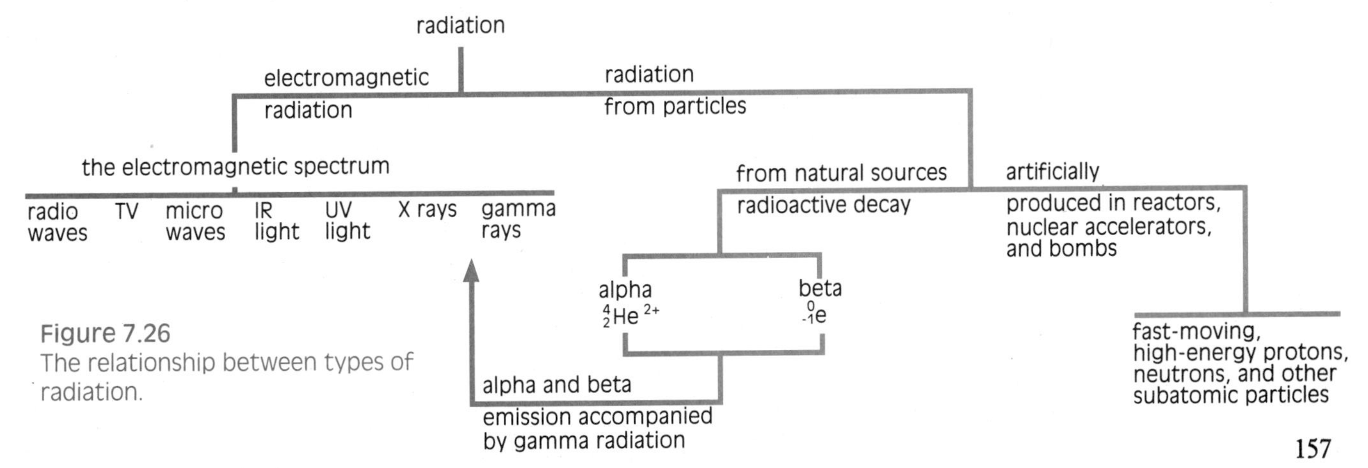

Figure 7.26
The relationship between types of radiation.

REVIEW 7.4

1. Give two alternate explanations for the instability of a radioactive nucleus.
2. The radioactive isotope polonium 218 decays by the emission of an alpha particle to produce lead 214. How does the starting mass of polonium 218 compare with the total mass of lead 214 and alpha particles produced in this transmutation? Explain your answer.
3. Explain why
   (a) neutrons can easily approach the nuclei of atoms, and
   (b) it is hard for protons to approach the nuclei of atoms.
4. Explain why it is easier to cause a nuclear fission reaction than a nuclear fusion reaction.
5. How does the amount of energy released in a nuclear fusion reaction compare with the amount of energy in a very energetic chemical reaction (like the explosion of TNT)?
6. (a) Explain the functions of a moderator and control rods in the operation of a nuclear reactor.
   (b) In what way is the function of a moderator the opposite of that of the control rods?
7. Some of the waste radioactive isotopes produced in nuclear reactors have short half-lives, whereas others have much longer half-lives. Explain why a radioactive isotope with a short half-life would generally be easier to dispose of.

CHAPTER • REVIEW

Key Ideas

- Radiation may be electromagnetic, or it may consist of particles.
- Electromagnetic radiation has an associated wavelength, frequency, and energy; the shorter the wavelength, the higher the frequency and the greater the energy.
- Isotopes of an element all have the same atomic number (same number of protons) but different atomic mass because they have different numbers of neutrons.
- Naturally occurring radioactive nuclei emit alpha, beta, or gamma radiation. Nuclear transmutations result from the emission of alpha and beta particles.
- The amount of a radioactive parent and the activity of a sample decrease with time.
- Any radiation that can remove electrons from atoms is called ionizing radiation. Ionizing radiation can cause cancer and mutations in living things.
- Radiation has harmful effects, but it also has many beneficial applications. For example, it can be used in nuclear medicine and for making X-ray images.
- Nuclear changes are accompanied by changes in energy as a result of changes in mass.
- Nuclei can undergo fission, in which nuclei break into two roughly equal smaller parts, or fusion, in which two small nuclei are joined to make a single larger one.

Vocabulary

electromagnetic radiation
visible spectrum
wavelength
electromagnetic spectrum
radioactive
nuclear force
Geiger counter
alpha radiation
beta radiation
gamma radiation
mass number
isotopes
radioactive decay
nuclear equation
parent nucleus
decay product
nuclear transmutation
activity
becquerel (Bq)
half-life
natural background radiation
ions
ionizing radiation
cancer
mutation
gray (Gy)
sievert (Sv)
fission
chain reaction
nuclear reactor
moderator
critical mass
reactor meltdown
control rod
fusion

1. Construct a concept map that connects the different types of radiation, their sources, and their effects.
2. Draw a labelled diagram showing the parts of a nuclear reactor.

Connections

1. In activities in this chapter, you may have used sealed radioactive sources. Some scientists believe that the hazard from these sources is very slight. Do you think, then, that a very strict set of rules might not be necessary for their use? Explain your answer.
2. In past centuries, one of the substances in artists' paints was lead. Today little or no lead is used in such paints. How could X rays be used to distinguish a genuine 18th or 19th century masterpiece from a modern fake?
3. In what way are all the isotopes of an element alike? In what way are the isotopes different?
4. Radioactive isotopes with high atomic numbers often decay by alpha emission. Radioactive isotopes with low atomic numbers decay by beta emission. Explain this difference.
6. Some nuclear reactors use ordinary water as a moderator. These reactors need fuel enriched in uranium 235 for their operation. Other nuclear reactors use heavy water and can operate on uranium that is not enriched in uranium 235. Explain why the reactors operate on different fuels. (The CANDU nuclear reactor developed by the scientists of the Atomic Energy Company of Canada is a heavy-water reactor.)
7. The sun and other stars release their energy by nuclear fusion. What does this fact tell you about stars?

Explorations

1. Use a reference book to find out what a breeder reactor is.
   (a) What are the advantages and disadvantages of this type of reactor?
   (b) Why is this type of reactor a potential source of bomb material?
2. Use reference material to find information about the Three Mile Island reactor meltdown of 1979 and the nuclear accident at Chernobyl in 1986. What were the causes of these accidents? What were their effects?
3. Prepare a poster that outlines the safe handling of radioactive isotopes.
4. By contacting your local hospital or the British Columbia Institute of Technology, find out what is involved in becoming an X-ray technologist or a technologist in nuclear medicine.

Reflections

1. (a) Go back to Activity 7A and complete the right-hand side of the page.
   (b) Briefly describe how your ideas of radiation and nuclear energy have changed.
2. People who are in favour of nuclear power suggest that use of this energy source to provide electricity will mean that our society will use less fossil fuel and so produce less carbon dioxide. As a result, they say, the greenhouse effect caused by the carbon dioxide will not be so pronounced in coming years. Critics of this position say that conserving energy could decrease carbon dioxide emissions far more, and no low- or high-level radioactive wastes would be produced. What do you think?

UNIT III

Investigating Matter

What do you see in the picture on the right? Probably you can identify geese, clouds, and the sun at daybreak. But do you see any elements of physics, biology, or chemistry? Scientists looking at this scene might see the wonders of the physics of flight or the biology of feathered creatures. They might also see the remarkable profusion of chemicals in nature.

Chemicals are everywhere. They are in the Earth, in water, in the clouds, and in the geese flying through the morning air. It is the reactions of chemicals that enable geese to fly and allow plants to use energy from the sun to make their own food. In addition, chemists have discovered many ways of using such chemical reactions in our daily lives. As a result, the products of industrial chemistry have an enormous impact on our society.

In this unit, you will first investigate the basic chemical elements and how they are classified. Then you will discover how elements combine to form compounds and how you can identify different types of compounds. Finally, you will find out about chemical reactions and how these reactions are used in making silverware, jewellery, cleaning agents, batteries, and other products of technology. All this information will give you a greater appreciation of the world around you.

CHAPTER

8 Elements and the Periodic Table

The students in this picture are surrounded by elements. The food they are eating, the air they are breathing, the chairs they are sitting on, and the clothes they are wearing are all made of elements. Everything is made of elements, including your own body, the Earth, and the entire universe. In fact, elements are often called the basic building blocks of matter. In this chapter, you will find out about elements and how they are classified.

ACTIVITY 8A What Do You Know about Elements?

List at least three elements you know in your learning journal, along with anything you know about them. (If you cannot think of an element, look at the list of elements in Appendix C.) Your teacher may want you to share this information with your lab partner. After you have compared what you have written with what your partner has written, you may wish to change some of the information in your learning journal.

As you study this chapter, update what you have written about elements. Finally, write any questions you have about elements in your learning journal as you go along.

8.1 The Early Search for Elements

The ancient Greeks were the first people to offer hypotheses about the nature of matter. Around 450 B.C., a Greek scholar named Empedocles suggested that all matter was made of four "elements": earth, air, water, and fire. About 50 years later, another Greek scholar, Democritus, proposed that matter was made up of tiny particles that could not be broken down further. He named these particles **atoms**, from the Greek word *atomos*, which means "indivisible" or "uncuttable." Aristotle, who was a well-respected scholar and teacher, opposed Democritus's ideas and supported the idea of four elements. Democritus's notion was in line with modern ideas, but it was not accepted for almost 2000 years. Aristotle's influence was too great, and the hypothesis of the four elements was too popular.

In the Middle Ages, scholars were searching for the philosopher's stone, a mythical rock that they believed could change cheap metals such as lead and copper into gold. Finding this stone was the goal of alchemy, a school of thought that blended magic, religion, philosophy, and science. In 1669, a Hamburg alchemist, Hennig Brandt, discovered a previously unknown substance with unusual characteristics. As part of his search for gold, he distilled urine. When he condensed the vapours underwater, a pasty white solid formed. He found that the solid glowed in the dark and burst into flames in warm air. Brandt called the substance phosphorus, from the Greek roots *phos*, "light," and *phoros*, "bringing." Brandt did not realize that he had discovered an element. Like most scholars of the time, he still accepted the Greek idea of four elements.

Robert Boyle (1627–1691), an English scientist, was convinced that there was no basis to the hypothesis of four elements (Figure 8.1). He was the first to believe that air is a mixture and not an element. Boyle was also certain that both gold and Brandt's white paste were elements. In 1661, he published a book titled *The Skeptical Chymist*, which included this definition: "I mean by element, certain simple unmingled bodies. . . . " Boyle's new definition became the basis for the modern definition of an **element**, a pure substance that cannot be chemically broken down into a simpler substance.

DID YOU KNOW?

The chemical symbol for mercury (Hg) comes from the Latin word *hydrargyrum*, meaning "liquid silver."

Figure 8.1
Robert Boyle challenged the beliefs of alchemists for years before he was able to prove the four-element idea of matter wrong.

ACTIVITY 8B The Nature of Air

In this activity, you will experiment with air to see whether it fits the definition of an element.

MATERIALS

safety goggles
steel wool (without oil coating)
wax or felt-tip marker
two small jars
dish filled with water
wooden splint

PROCEDURE

1. Put safety goggles on.
2. Moisten the steel wool and set up an apparatus like that shown in Figure 8.2. As soon as you invert the jar, mark the original water level with the marker. Leave the jar for a day.
3. After a day (at least 24 h), mark the water level again. Has the water level changed?

Figure 8.2
Apparatus to demonstrate the reaction of steel wool with air.

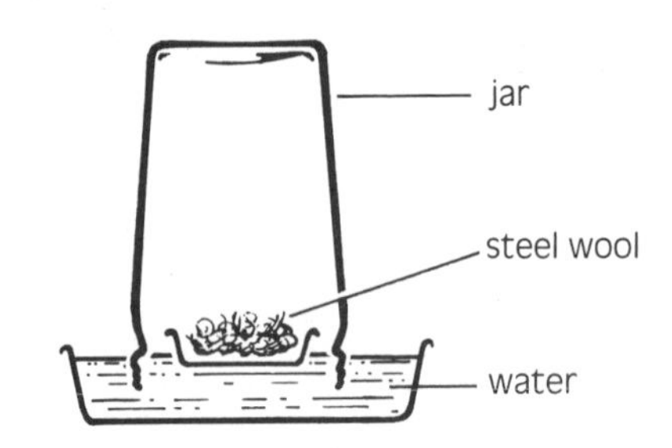

4. Lift the jar away from the water, holding it carefully in a vertical position. Test the air inside the jar by placing a lighted wooden splint inside the jar. For comparison, test the air inside the second jar, which has not been used in the experiment.

DISCUSSION

1. (a) What happened to the lighted wooden splint when you tested the air inside the first jar?
   (b) What happened to the lighted wooden splint when you tested the air inside the second jar?
   (c) Why was the air in the first jar different after the water level rose in the jar?
2. (a) Compare the original volume of air with the final volume of air. What fraction of the air was used up by the steel-wool reaction?
   (b) What do you think happened to the "missing" air?
   (c) Is air an element? Explain.
3. Does air fit Boyle's definition of an element? Explain.

Figure 8.3
Henry Cavendish spent his life doing scientific research. This rich English recluse is said to have cared nothing for money, people, or fame. He was the first to isolate hydrogen.

THE SCIENTIFIC REVOLUTION

A scientific revolution was taking place in the late 1700s, at the same time the Industrial Revolution was occurring. During this time, old ideas were challenged and new discoveries were made. The work of three scientists, Henry Cavendish, Joseph Priestley, and Antoine Lavoisier, contributed greatly to the concept of what an element is. These three men also discovered new elements.

In 1766, English scientist Henry Cavendish (1731–1810) mixed a metal with acid and prepared a flammable gas that was lighter than air (Figure 8.3). He did not know the gas he prepared was the gaseous element hydrogen.

The element oxygen was first prepared by Joseph Priestley (1733–1804), another English scientist (Figure 8.4). In 1774, Priestley made oxygen by focusing sunlight through a "burning lens" (magnifying glass) onto a mound of a mercury compound (mercury(II) oxide). The results were pure oxygen and mercury.

The first person to recognize oxygen as an element was the great French scientist Antoine Lavoisier (1743–1794), who is shown in Figure 8.5. Experimenting with Priestley's oxygen, Lavoisier concluded that air must be a mixture of at least two gases, one that supports burning and one that does not. (Activity 8B gives the same evidence.)

Figure 8.4
A non-conformist minister, Joseph Priestley was eventually forced to leave England because of his political views. Too poor to afford even a balance, he depended on friends to equip his kitchen laboratory. He was the first to prepare oxygen by heating mercury(II) oxide.

DID YOU KNOW?

The Earth's atmosphere is composed of about 78 per cent nitrogen and 20 per cent oxygen. At ordinary temperatures, these two elements do not react. However, when air is taken inside an automobile engine, compounds of nitrogen and oxygen, called nitrogen oxides, form significant sources of air pollution and smog.

Figure 8.5
Antoine Lavoisier was a prosperous businessman and had great intelligence, which often led him to be critical of others. During the French Revolution, he was executed by guillotine. He is often considered the father of modern chemistry.

Cavendish also heard about Priestley's new gas. He used Priestley's methods to prepare some oxygen and burned his own gas, hydrogen, in it. The two gases produced water (Figure 8.6). Up to that time, scholars still believed that water was an element. Cavendish published his results, reporting that water could be made from two gases, hydrogen and oxygen.

When Lavoisier read Cavendish's report, he understood what an element was. Lavoisier said that an element is the final form that can be reached by the chemical breakdown of matter. His explanation led to the discovery of more and more elements. Lavoisier made many more contributions to chemistry. His wife, Marie, was an active partner in his research, but her contribution is seldom mentioned. Lavoisier also published his ideas without mentioning the work of either Priestley or Cavendish. He even named the gases they discovered.

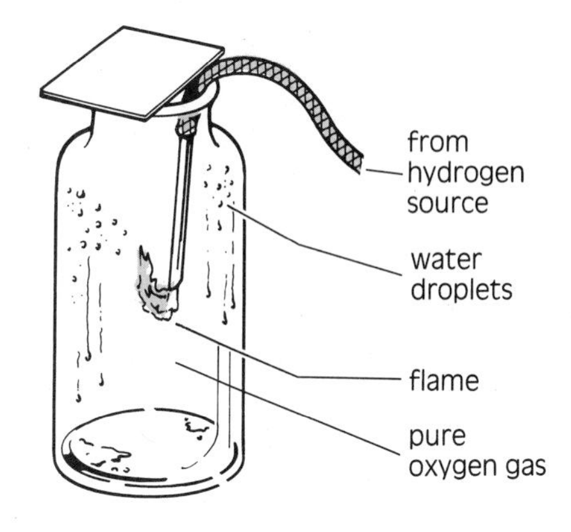

Figure 8.6
When Cavendish burned hydrogen in oxygen, water was produced.

ACTIVITY 8C Preparing the Element Hydrogen—Cavendish's Experiment (Teacher Demonstration)

In this activity, hydrogen gas will be prepared in much the same way that Cavendish prepared it in 1766.

MATERIALS

safety goggles
apron
apparatus similar to that in Figure 8.7:
- 500 mL Erlenmeyer flask
- ring stand
- burette clamp (to hold ring stand)
- test tube (free of cracks)
- tank of water or 1000 mL beaker
- 32 cm of rubber tubing
- stopper unit (one hole)
- glass tube to fill the stopper

200 mL of dilute hydrochloric acid
two medium-size pieces of mossy zinc
wooden splint

PROCEDURE

1. Put safety goggles and apron on.

> CAUTION!
> Hydrochloric acid may cause damage to the eyes or skin. If you get any in your eyes or on your skin, immediately rinse with water for 15 to 20 minutes and inform your teacher.

2. Your teacher will clamp the flask to the ring stand and pour 200 mL of hydrochloric acid into the flask (Figure 8.7a).
3. Next, about 800 mL of water is added to the tank or beaker.
4. Your teacher will fill the test tube with water, placing a finger over the open end and putting the test tube in the tank upside down, without losing any water (Figure 8.7b).

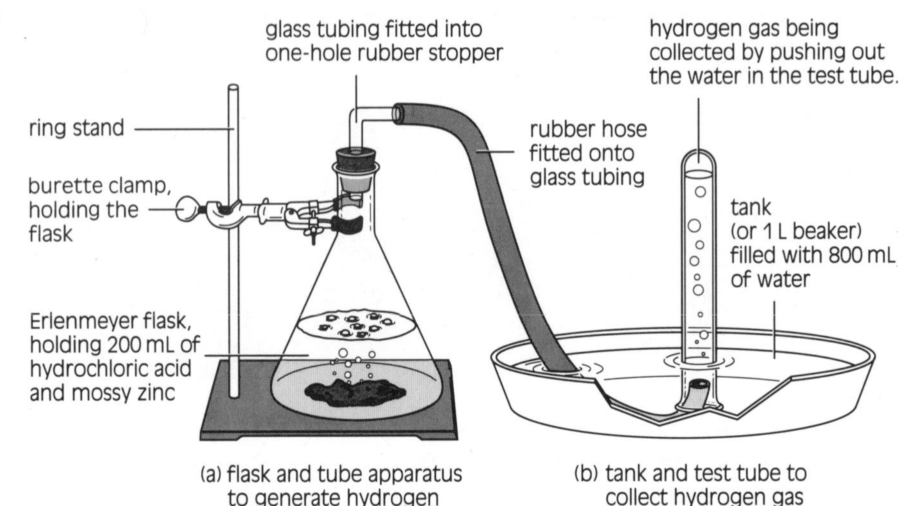

Figure 8.7
Apparatus for generating and collecting hydrogen.

5. The two medium-size pieces of mossy zinc are placed into the flask containing the hydrochloric acid. What happens in the flask? Record your observations in your notebook.
6. Your teacher will finish setting up apparatus as shown in Figure 8.7 (inserting the stopper with the attached rubber tubing). The initial gas generated is not to be collected. This gas will contain mostly air from the flask.
7. After about a minute or two, your teacher will start to collect the gas.

> CAUTION!
> Hydrogen gas is explosive. The test tube should not be pointed towards anyone when the hydrogen gas is tested with a flaming wooden splint.

8. When the test tube is full of hydrogen, a small quantity is removed and tested with a wooden splint (Figure 8.8). When a burning splint is held to the mouth of the test tube, the hydrogen reacts to make a "pop" sound. What was produced when the burning splint reacted with the hydrogen?
9. Your teacher will dispose of the hydrochloric acid.

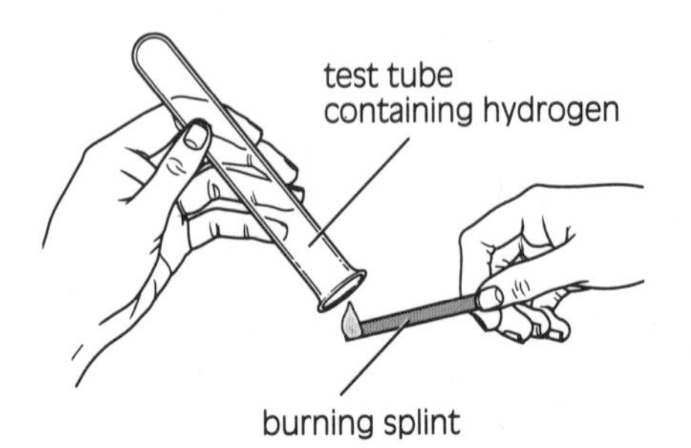

Figure 8.8
Testing hydrogen gas with a burning splint.

→

DISCUSSION

1. Where was the hydrogen produced?
2. (a) What happens when a burning splint is held to the mouth of the test tube full of hydrogen?
   (b) The burning splint caused hydrogen from the test tube to combine with oxygen to form water. Where did the oxygen come from?
   (c) Is water an element? Explain.
3. What did you observe about hydrogen when it was in the test tube?
4. The method of collecting the gas is downward displacement of water. What characteristic must a gas have if it is to be collected in this manner?
5. Why is hydrogen considered an element?
6. State some of the safety precautions observed in this lab.

PROPERTIES OF ELEMENTS

Early chemists could distinguish elements from other substances because elements would not break down chemically. Chemists could also distinguish one element from another because each element had its own set of **properties**. A property is any observable fact that can be used to distinguish one substance from another. For example, at room temperature, the element oxygen is a colourless and odourless gas that supports burning. Oxygen does not burn by itself. Rather, it causes fuels such as hydrogen to burn. Hydrogen is also a colourless and odourless gas, but it is less dense than oxygen and is explosive. Copper is a reddish solid that conducts electricity and heat well. These properties make copper an ideal material to use in electrical wires and on the bottom of pots and pans. Knowing their properties makes elements useful to us and easier to identify.

REVIEW 8.1

1. What was an element according to
   (a) Aristotle?
   (b) Boyle?
   (c) Lavoisier?
   (d) modern scientists?
2. What were alchemists looking for? Explain.
3. Describe how Priestley prepared oxygen.
4. Design an experiment that would demonstrate that fire and earth are not elements.
5. Why were elements like phosphorus not known in ancient times?
6. (a) What is a property?
   (b) List some examples of properties.
7. There are two substances with the same set of properties, except that one substance will conduct electricity and the other substance will not. Are they the same substance? Explain.
8. If you have learned more about the element(s) you wrote down in Activity 8A, add this information to your learning journal.

8.2 The Structure of the Atom

In 1781, Lavoisier explained how to tell elements from non-elements in the laboratory. Over the next three decades, many new elements were discovered. Then, in 1808, the English chemist John Dalton (1766–1844) published a theory to explain why elements differ from non-elements and from each other. This theory is called the **atomic theory,** and it states the following:

1. All matter is made of atoms, which are particles too small to see.
2. Each element has its own kind of atom. Atoms of the same element have the same mass. Atoms of different elements have different masses.
3. Compounds are created when atoms of different elements link to form compound atoms. In a compound, all the compound atoms are alike. (We now call Dalton's compound atoms **molecules.**)
4. Atoms cannot be created or destroyed.

The atomic theory caught on quickly because it explained the known facts so well. (We now know that atoms can be both created and destroyed, but in Dalton's time there was no evidence to suggest the splitting of atoms.)

But another 150 years passed before enough evidence was gathered to understand the structure of the atom. You may have already learned something about the structure of the atom in Chapter 3.

ACTIVITY 8D *What You Already Know about the Structure of the Atom*

Copy Table 8.1 in your learning journal. In the first column, write the answers to the following questions:

- What do you know about the structure of the atom?
- What do you think you know about the structure of the atom?
- What do you think you might learn about the structure of the atom? (Include any questions you have about the structure of the atom.)

Table 8.1 Sample Data Table for Activity 8D

What You Already Know	What You Learned

SAMPLE ONLY

In the second column, jot down new ideas, details, examples, or even diagrams about the structure of the atom that occur to you as you read the following subsection. After you have read it, put a check in the first column beside the ideas that your reading has confirmed and a question mark beside the ideas left unconfirmed by your reading.

THE MODERN THEORY OF ATOMIC STRUCTURE

The modern view of the atom is somewhat different from Dalton's version. Dalton's idea that atoms are indivisible particles is now known to be incorrect. Atoms can be split, by means of nuclear reactions, into still simpler particles called **subatomic particles** (Figure 8.9). Subatomic particles with a positive electrical charge are called **protons.** Those without a charge are called **neutrons.** Protons and neutrons are found in the nucleus at the centre of an atom. Particles with a negative charge are called **electrons.** Electrons fill the space around the nucleus.

Profile

JOHN DALTON

John Dalton continued to study and learn all his life.

In England during the 1700s, education was expensive. John Dalton's Quaker family could not afford to send him to school past the age of 11. But Dalton took every opportunity to educate himself. For example, when he was 11, he began to keep a journal to record changes in weather, amount of rainfall, hours of sunshine, and so on. He kept up this habit until the day he died. He also enjoyed classifying plants and collecting butterflies.

Dalton taught himself so well that he became a teacher while he was still in his teens. By 19, he was a headmaster. An elderly blind scholar taught him the most important lesson of all—how to be his own teacher. Dalton acquired a reputation as a fine scientist, and although he had no degree, he became a university professor.

When Dalton was in his twenties, during a walk with some friends, he discovered that the geranium flower he thought was blue was really pink. He then realized what he had suspected all along—that he was colour-blind. He began studying other people who were colour-blind and left careful instructions in his will to dissect his own eyeballs to try to find out what caused the condition.

Late in his life, Dalton attended a university ceremony held in his honour. He arrived wearing the traditional scarlet robe. His fellow Quakers (who never wore bright colours) were shocked, but Dalton announced that he could see no scarlet, only gray.

Dalton's early scientific work dealt with evaporation, heat, and the makeup of mixed gases. This work eventually led Dalton to formulate his atomic theory. Throughout his life, he continued teaching and pursuing his scientific investigations. He once said that his "head was too full of triangles, chemical processes and electrical experiments" to settle down and marry.

When Dalton died in 1844, he was given a funeral with full honours. His body lay in state for four days, and more than 40 000 people came to pay their respects to one of the most innovative thinkers of his time.

Figure 8.9
Protons and neutrons are found in the nucleus at the centre of an atom. In this drawing, the size of the nucleus is much exaggerated. Its actual diameter is about 1/100 000 the diameter of the atom.

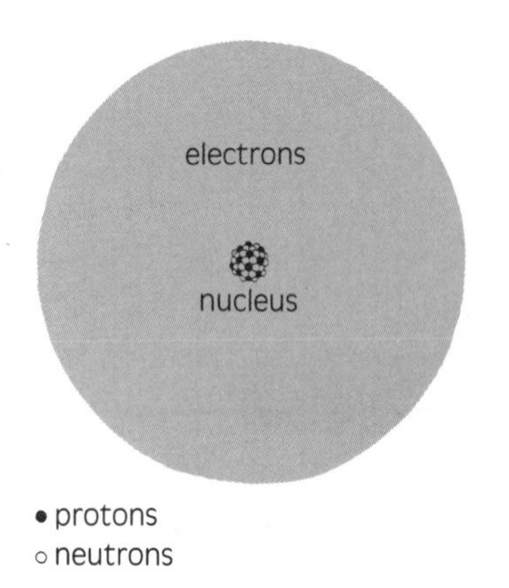

ATOMIC NUMBER

The **atomic number** of an element is the number of protons in the nucleus of an atom of that element. The atomic number for carbon is 6. This means that each carbon atom has six protons in its nucleus. The element lithium has an atomic number of 3, which means that it has three protons in its nucleus.

INSTANT PRACTICE

Copy Table 8.2 in your notebook and fill in the blank spaces. You may use Appendix C to help you find the symbols.

Table 8.2 Finding Atomic Numbers

Element	Symbol	Atomic number	Number of protons
Hydrogen		1	
Beryllium		4	
Carbon			6
Cobalt			27
Krypton		36	

MASS NUMBER AND ATOMIC MASS

The **mass number** of an element is the total number of protons and neutrons in the nucleus of the atom. As you may have learned in Chapter 7, atoms of the same element that have different mass numbers are called **isotopes** of the element.

INSTANT PRACTICE

Copy Table 8.3 in your notebook and fill in the blank spaces. You may use Appendix C to help you find the symbols.

Table 8.3 Finding Mass Number

Element	Symbol	Mass number	Number of protons	Number of neutrons
Hydrogen			1	0
Chromium			24	28
Beryllium			4	5
Carbon			6	6
Gold			79	118
Cobalt			27	32
Barium			56	81
Krypton		84	36	
Iron		56	26	

The **atomic mass** of an element is the average mass of an element's naturally occurring atoms, or isotopes, taking into account the percentage of each isotope that is present in an ordinary sample of the element.

For example, hydrogen has three naturally occurring isotopes. They are hydrogen 1 (one proton), hydrogen 2 (one proton and one neutron), and hydrogen 3 (one proton and two neutrons), as shown in Figure 8.10. Hydrogen 1 accounts for about 99.9 per cent of the naturally occurring hydrogen in the atmosphere. Hydrogen 2 accounts for about 0.015 per cent of the hydrogen. And a hydrogen 3 isotope occurs in the atmosphere once in every 10^{18} hydrogen atoms. If you were to take the average mass of all three hydrogen isotopes occurring in nature, it would be about 1.0008 — the atomic mass of hydrogen.

Figure 8.10
Isotopes of hydrogen. What do all the hydrogen isotopes have in common?

p = protons n = neutrons

1 p 0 n	hydrogen 1
1 p 1 n	hydrogen 2 (deuterium)
1 p 2 n	hydrogen 3 (tritium)

NUMBER OF ELECTRONS

In a neutral atom, the atomic number is also equal to the number of electrons:

atomic number = number of protons = number of electrons

Thus, the element lithium, with an atomic number of 3, has three electrons. The element carbon, with an atomic number of 6, has six electrons.

DID YOU KNOW?

On an average-sized globe, the map of Canada is 20 million times smaller than the actual country. If an atom were magnified up to that scale, its "map" could be drawn on a penny. The electrons would look like a blur, and you could not see the nucleus at all.

BOHR'S MODEL OF ELECTRON ARRANGEMENT

You have seen that electrons occupy the space outside the nucleus of an atom. When atoms approach one another, it is their electrons that interact. The first widely accepted explanation for the arrangement of these electrons was put forward by a Danish physicist, Niels Bohr (1885–1962), shown in Figure 8.11.

In 1913, Bohr proposed a model of the atom that helped explain the movement of the electrons around the nucleus of an atom. He suggested that the electrons move around the nucleus of an atom in **orbits.** Each orbit is a certain distance from the nucleus and contains a definite number of electrons. Orbits are sometimes called shells.

After each orbit is filled, the next orbit is then filled. Table 8.4 shows the maximum number of electrons that each of the first three orbits of any atom can hold.

Table 8.4 Maximum Number of Electrons in the First Three Orbits of an Element

	First orbit	Second orbit	Third orbit
Maximum number of electrons	2	8	8

Figure 8.11
Niels Bohr was awarded the 1922 Nobel Prize in physics for his work in determining atomic structure.

The element hydrogen, which has an atomic number of 1, has 1 electron in its first orbit (Figure 8.12a). The element nitrogen, which has an atomic number of 8, has 8 electrons. Two of the electrons are in the atom's first orbit, and the remaining 6 electrons are in its second orbit (Figure 8.12b). Phosphorus, which has an atomic number of 15, has 15 electrons. Two electrons are in the first orbit, 8 are in the second orbit, and 5 are in the third orbit (Figure 8.12c).

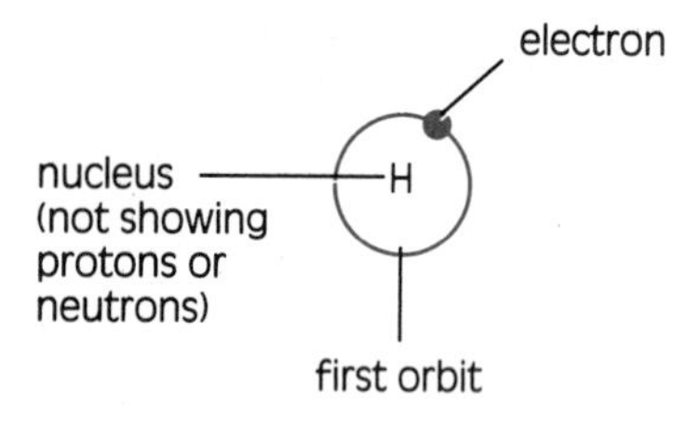

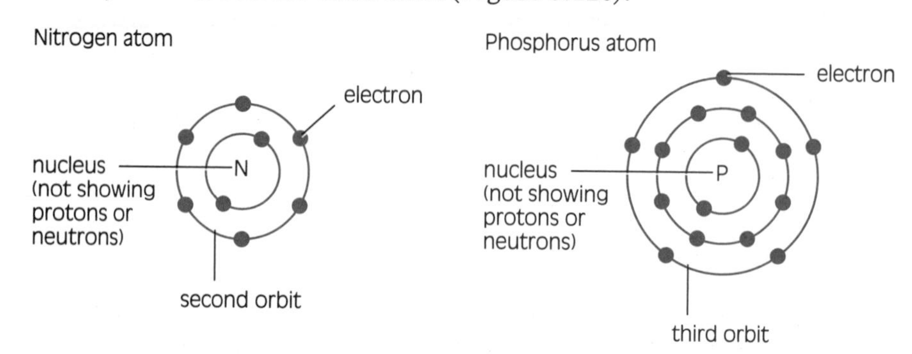

Figure 8.12
Bohr models of the atoms for hydrogen, nitrogen, and phosphorus.

INSTANT PRACTICE

Draw Bohr models of atoms, similar to those in Figure 8.12, for the following elements. The atomic numbers for these elements can be found in the periodic table in Appendix B or in Appendix C.

(a) hydrogen	(c) carbon	(e) sodium
(b) lithium	(d) helium	(f) argon

BOHR'S MODEL OF THE ATOM: AN EXPLANATION OF THE SPECTRUM

Earlier scientists had discovered that each element produces its own characteristic visible **spectrum** (plural: spectra), or series of coloured lines (Figure 8.13). They discovered this by looking at elements through a **spectroscope**, an instrument that works like a prism. Light that is passed through a spectroscope spreads out to form a spectrum.

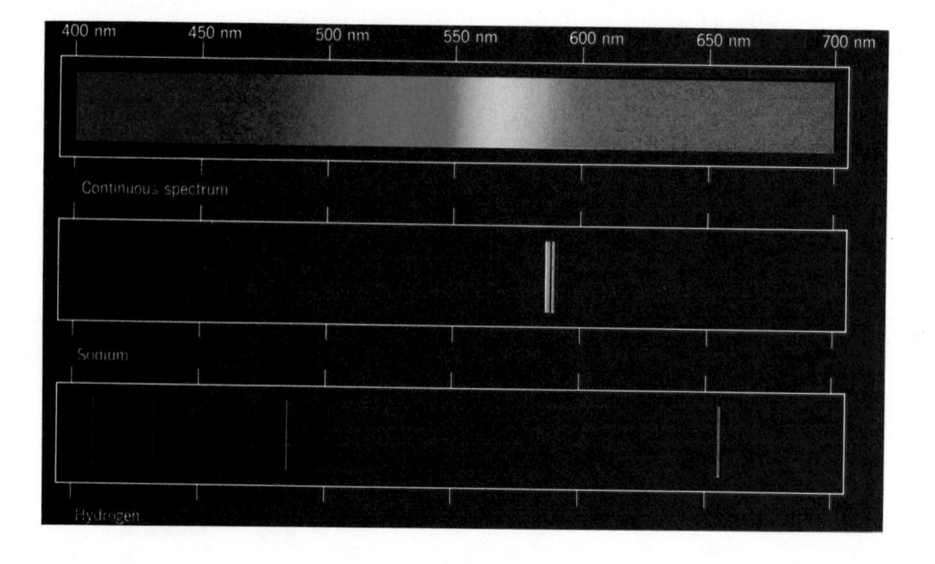

Figure 8.13
(*Top*) The spectrum produced by an incandescent light bulb. (*Centre*) The spectrum produced by sodium. (*Bottom*) The spectrum produced by hydrogen.

Bohr's model explained why each element has its own spectrum. In Bohr's model of the atom, electrons are arranged in orbits that surround the nucleus like layers of an onion. Space in the orbits is limited, and electrons are arranged in a definite pattern. Within the orbits, electrons move quickly. When electrons are energized by heat, electricity, or light, they use this extra energy to jump to a higher orbit. The electrons in the higher orbit are very unstable and tend to fall back into their normal, more stable, orbits. When the electrons drop back to their normal orbits, the extra energy is given off in the form of glowing light.

Figure 8.14 shows an atom that has been given extra energy (by heat or electricity). As a result, the electron is forced out of its regular location. As the electron returns to its usual place, it gives off light. Each element gives a different spectrum because each element has a different electron arrangement.

Figure 8.14
An electron "jumping" out of its regular orbit.

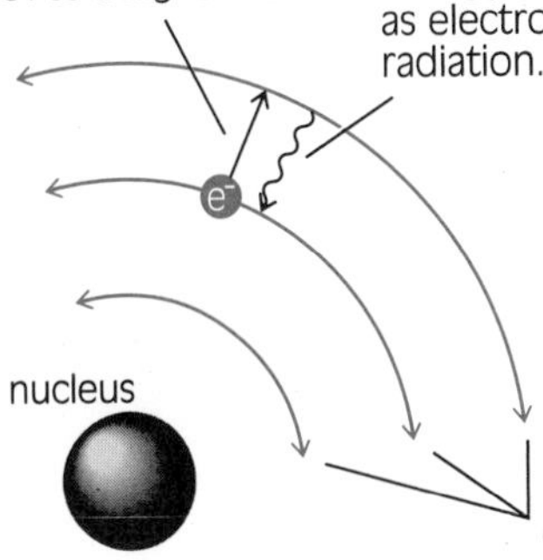

ACTIVITY 8E Spectra of Elements

In this activity, you will use a spectroscope to find the spectra of various sources of light.

MATERIALS

spectroscope
coloured pencils
assorted sources of light (such as candles and light bulbs)
gas discharge tubes containing a single element (if available)

PROCEDURE

1. Examine your spectroscope carefully. Your teacher will explain how to use it.
2. Use your spectroscope to observe light from the window, if possible, or the light from an incandescent bulb. Draw sketches in your notebook to record your observations. Include the colour and relative thickness of each spectrum you see.

CAUTION!
Do not look at the sun. You could lose your sight.

3. Observe the spectra of light from a variety of light sources. Possible sources are a fluorescent tube, a heat lamp, a candle, and a Bunsen burner.
4. If possible, observe the spectrum produced by the gas discharge tube containing a single element. Draw a sketch to record your observations. Observe the spectra of other gaseous elements, if available. You may wish to compare your observations of spectra with Figure 8.13.

DISCUSSION

1. How are the spectra of the light sources in step 3 of the procedure similar to the spectrum of sunlight (or that of an incandescent bulb)? How are they different?
2. How is the spectrum in step 4 of the procedure similar to the spectrum of sunlight (or that of an incandescent bulb)? How is it different?
3. (a) How do the various single-element spectra compare with each other?
   (b) Explain the differences using Bohr's model of the atom.
4. In step 4, a high current of electricity was passed through a gas discharge tube containing a single element, causing the electrons to "jump" to another orbit.
   (a) What causes the light given by each element?
   (b) Why do different elements have different spectra?
5. Could you use your notes from this activity to identify one of the elements in step 4 of the procedure if you saw it again a year from now? What would you need besides your notes?

EXTENSION

Find out how the spectroscope helps the forensic chemist—one who applies chemistry to the legal system—identify a hit-and-run car from paint chips on the victim's clothing.

REVIEW 8.2

1. Copy Table 8.5 in your notebook and fill it in. You may have to do some calculations.
2. How did you calculate
   (a) the number of protons?
   (b) the number of electrons?
3. (a) How are the atoms of different elements similar?
   (b) How are the atoms of different elements different?
4. (a) What is an isotope? Give at least one example.
   (b) Look at the periodic table in Appendix B and name at least two elements not mentioned in this chapter that have isotopes. How do you know that they are isotopes?
5. Draw Bohr models of the atoms for helium and chlorine.
6. Update the information from Activity 8A by adding the following information about your element:
   - a Bohr model of your element's atom
   - its atomic number
   - its atomic mass
   - the number of protons, electrons, and neutrons in its atom
   - any other information you can come up with

Table 8.5 Finding the Number of Protons, Electrons, and Neutrons in an Element (You may use Appendix C to find the symbols.)

Element	Symbol	Atomic number	Number of protons	Number of electrons
Hydrogen		1		
Helium		2		
Boron		5		
Neon		10		
Aluminum			13	
Calcium			20	
Bromine			35	
Xenon				54
Gold				79
Lead				82

DID YOU KNOW?

The chemical symbol for lead (Pb) is derived from the Latin word *plumbum* and is the root of the word "plumbing."

8.3 Organizing the Elements

Up to the mid-1800s, scientists were busy discovering elements. They also observed many properties of these elements and wanted to organize this vast amount of information.

At that time, the known elements were always listed alphabetically. Each time a new element was discovered, the lists had to be changed. Alphabetical lists were not useful for scientists because elements that had nothing in common were put side by side. For example, bromine, a gas, was beside boron, a solid. The element iodine has more in common with bromine than boron does, but iodine was farther down the list.

On what basis could all these elements be classified? Could they be classified by state and colour? No. Too many elements look alike. By taste? Definitely not. Too many elements are poisonous. Another objection was that state, colour, and taste could not be measured.

Atomic mass could be measured accurately. Each element has an atomic mass that is different from that of every other element. For this reason, several scientists set about arranging the known elements according to their atomic mass.

The best arrangement was produced by Dmitri Mendeleev (1834–1907), a Russian scientist (Figure 8.15). At the time, 64 elements were known. He wrote the name of each element on a separate card, along with its atomic mass and all other known properties: colour, solubility, density, flammability, and so on. Then he played "chemical solitaire," arranging the cards every way he could think of, to see if there were patterns that would help him make new discoveries of unknown elements.

Figure 8.15
Born in Siberia, Dmitri Mendeleev was the youngest of 17 children. After studying in Europe, he set up a graduate school of chemistry in Russia. His organization of the elements into a periodic table was a stroke of genius that made the study of chemistry manageable.

THE FIRST PERIODIC TABLE

Mendeleev pinned his cards on the wall and endlessly rearranged them. He sorted them both by atomic mass and by chemical properties. Often he found a place where an element seemed to belong because of its mass but did not fit because of its properties. When that happened, he let the properties rule. At last, he found an arrangement that made sense because it showed a regular pattern. The periodic law describes this pattern:

> If elements are arranged according to their atomic mass, a pattern can be seen in which similar properties occur regularly.

Mendeleev designed the **periodic table**, a summary of the structure and properties of the known elements, to illustrate his law (Table 8.6). Although he could not explain why the law worked, he proved its power to predict.

Table 8.6 Mendeleev's Periodic Table (Includes Symbols and Atomic Mass)

Group I	Group II	Group III	Group IV	Group V	Group VI	Group VII	Group VIII
H 1							
Li 7	Be 9.4	B 11	C 12	N 14	O 16	F 19	
Na 23	Mg 24	Al 27.3	Si 28	P 31	S 32	Cl 35.5	
K 39	Ca 40	–44	Ti 48	V 51	Cr 52	Mn 55	Fe 56, Co 59 Ni 59, Cu 63
(Cu 63)	Zn 65	–68	–72	As 75	Se 78	Br 80	
Rb 85	Sr 87	?Yt 88	Zr 90	Nb 94	Mo 96	–100	Ru 104, Rh 104 Pd 105, Ag 108
(Ag 108)	Cd 112	In 113	Sn 118	Sb 122	Te 128	I 127	
Cs 133	Ba 137	?Di 138	?Ce 140	—	—	—	—— ——
—	—	—	—	—	—	—	
—	—	?Er 178	?La 180	Ta 182	W 184	—	Os 195, Ir 197 Pt 198, Au 199
(Au 199)	Hg 200	Tl 204	Pb 207	Bi 208	—		
—	—	—	Th 231	—	U 240	—	——

While Mendeleev was arranging the table, he sometimes came to a place in which no known element would fit. When that happened, he used a blank card. By leaving those spaces blank, Mendeleev was predicting that elements would eventually be found to fill in the spaces. He not only predicted the elements but also predicted the properties the elements would have. He did this by examining the elements surrounding the blanks in his periodic table. Table 8.7 shows two of Mendeleev's predictions.

EXTENSION

Examine Mendeleev's first periodic table (Table 8.6). Compare it with the modern periodic table (Appendix B). What are the similarities? What are the differences? Compare the new and old values for atomic mass. Find out why some of the old values are different from the modern values.

Table 8.7 Two of Mendeleev's Predicted Elements

Property	Gallium		Germanium	
	Predicted 1871	Discovered 1875	Predicted 1871	Discovered 1886
Relative atomic mass	68	69.9	72	72.3
Density (g/cm^3)	5.9	5.94	5.5	5.47
Melting point	Low	30°C	High	2830°C
Solubility				
In acids	Will slowly dissolve	Slowly dissolves	Will be slightly soluble	Insoluble
In bases	Will slowly dissolve	Slowly dissolves	Will resist dissolving	Slowly dissolves

Soon more elements were discovered to fill in the blank spaces. These discoveries helped convince the scientific world that the periodic table worked. A century later, it is still used to summarize chemical facts.

ACTIVITY 8F Meet the Elements

In this activity, you will observe some of the properties of elements and record them on note cards.

MATERIALS

safety goggles
apron
a collection of elements
note cards
a conductivity apparatus
a list of compounds or jars of compounds from the supply room

PROCEDURE

1. Put on your safety goggles and apron.
2. Carefully observe each element in the collection of elements provided by your teacher. If the sample is sealed, you may observe it only visually. DO NOT OPEN SEALED SAMPLES.
3. For each element, record the following information (Figure 8.16):
   - name of the element (already on the sample)
   - symbol (from the list of elements in Appendix C)
   - colour (observe it visually)
   - state (observe it visually; is it a solid, liquid, or gas?)
   - appearance (observe it visually; is it shiny or dull?)
   - formulas of compounds that it forms (look at the list of compounds to get this information)
   - conductivity (see Figure 8.17; test only opened samples of elements).
4. Wash your hands thoroughly after you have completed this activity.

→

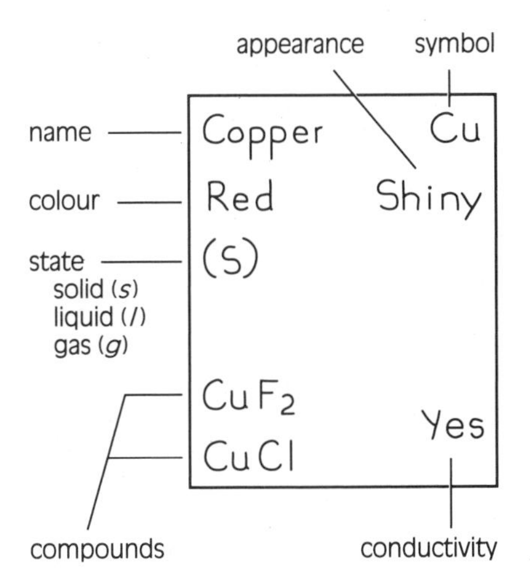

Figure 8.16
Sample note card, filled in for the element copper.

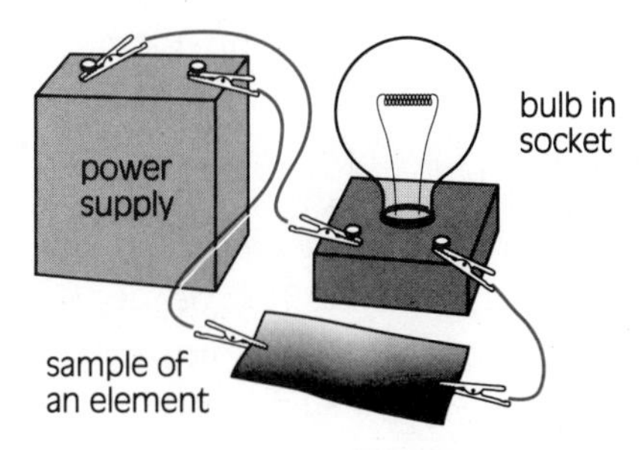

Figure 8.17
Apparatus for testing conductivity. If the bulb lights, the element conducts electricity.

DISCUSSION

1. (a) List the properties you observed in this activity.
   (b) List some properties you could have observed but did not because you did not have the instruments to measure these properties.
2. Sort your cards into two groups according to whether or not they conduct electricity.
   (a) What other properties do elements that conduct electricity have in common?
   (b) What other properties do elements that do not conduct electricity have in common?
3. Look at the periodic table in Appendix B.
   (a) On what side of the periodic table do you find those elements you observed that conducted electricity?
   (b) On what side of the periodic table do you find those elements you observed that did not conduct electricity?
   (c) If you had an unknown element that would not conduct electricity and was a gas, where on the periodic table would you expect to find it?
4. What evidence have you found that some elements are related, or have similar properties?

THE MODERN PERIODIC TABLE

Until the turn of this century, the periodic table was still arranged according to atomic mass. Then the element argon was discovered. Efforts to add argon to the periodic table revealed an inconsistency. The atomic mass of argon (39.95) is greater than that of potassium (39.10), suggesting that potassium should precede argon. But the physical and chemical properties of both elements suggested that their placement in the periodic table should be reversed. Elements in the present-day periodic table are arranged according to atomic number, and thus argon (18) precedes potassium (19). This arrangement has eliminated inconsistencies that previously existed.

REVIEW 8.3

1. (a) How were elements in Mendeleev's period table arranged?
   (b) What is Mendeleev's periodic law?
2. Explain why Mendeleev included blanks in his periodic table. How did these blank spaces strengthen the belief in his method of organizing the elements?
3. (a) Why were properties such as colour, state, and taste not used to arrange elements on the first periodic table?
   (b) Why do you think atomic mass worked and those properties mentioned in (a) above did not work when Mendeleev arranged his periodic table?
4. Why was Mendeleev able to predict the properties of the elements that belonged in the blanks in his table?

8.4 Chemical Families

Even before Mendeleev created the periodic table in 1869, chemists looked for ways to group elements based on their chemical and physical properties. In 1829, a German chemist, Johann Wolfgang Dobereiner (1780–1849), noted three groups of elements, each with similar chemical properties. But this left out a lot of other elements. When the elements are arranged in the periodic table in order of their atomic numbers, all the inconsistencies that occurred in Mendeleev's table disappear. Thus, the modern periodic table can be used to correlate chemical and physical properties. One such correlation occurs on the vertical columns on the modern periodic table. Vertical columns of elements are known as **families**. These families of elements have similar chemical and physical properties.

Table 8.8 lists the first 21 elements, their atomic numbers, a physical property (state), and two chemical properties. These properties are the types of compounds that each element forms with hydrogen and fluorine. Elements in the same family form compounds with hydrogen and fluorine that have the same ratio of atoms. For example, elements in the alkali family form compounds with hydrogen in a ratio of one atom of the alkali element to one atom of hydrogen.

Table 8.8 The First 21 Elements in the Periodic Table

Symbol	Atomic number	State at 25°C	Compound formed with hydrogen	Compound formed with fluorine
H	1	gas	H_2	HF
He	2	gas	—	—
Li	3	solid	LiH	LiF
Be	4	solid	BeH_2	BeF_2
B	5	solid	B_2H_6	BF_3
C	6	solid	CH_4	CF_4
N	7	gas	NH_3	NF_3
O	8	gas	H_2O	OF_2
F	9	gas	HF	F_2
Ne	10	gas	—	—
Na	11	solid	NaH	NaF
Mg	12	solid	MgH_2	MgF_2
Al	13	solid	Al_2H_6	AlF_3
Si	14	solid	SiH_4	SiF_4
P	15	solid	PH_3	PF_3
S	16	solid	H_2S	SF_2
Cl	17	gas	HCl	ClF
Ar	18	gas	—	—
K	19	solid	KH	KF
Ca	20	solid	CaH_2	CaF_2
Sc	21	solid	not known	ScF_3

NOBLE GASES

As you can see from Table 8.8, helium (He), neon (Ne), and argon (Ar) are the only elements that do not form compounds with hydrogen and fluorine. These three elements do not react under normal conditions and are part of the chemical family called **noble gases.** The term "noble" is used to suggest a very small degree of chemical reactivity. The noble gases are unlikely to react because their outer orbits are full (Figure 8.18). Chemists associate this lack of reactivity with the full outer orbits of the noble gas atoms. These gases were discovered after Mendeleev first proposed his table (Figure 8.19). No compounds of these elements have ever been found in nature.

Figure 8.18
The electron arrangement in noble gases.

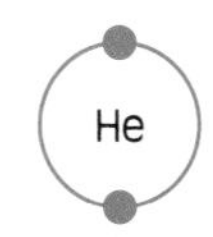

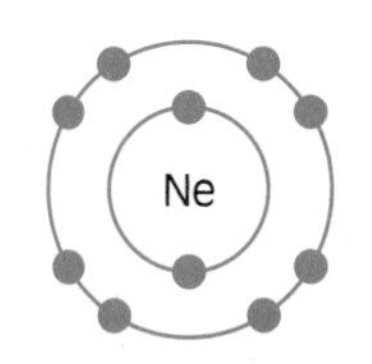

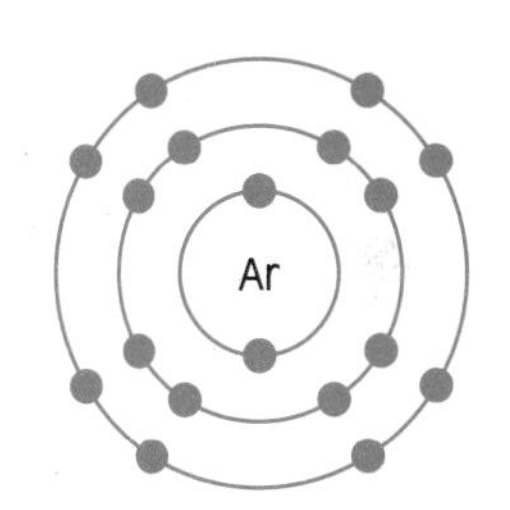

Figure 8.19
Location of noble gases on the periodic table.

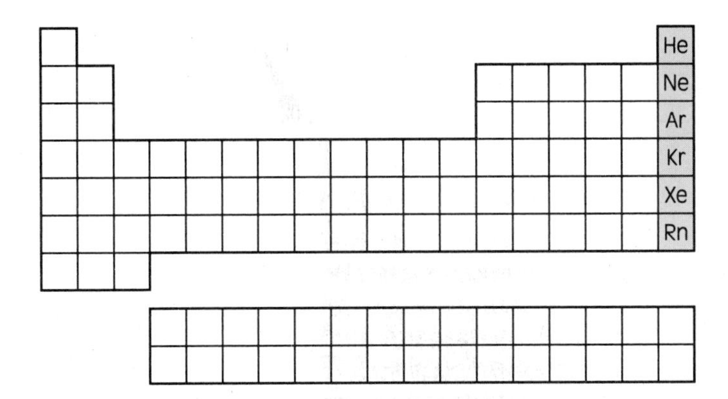

Argon, the most common noble gas, makes up 0.93 per cent of the air you breathe. Because argon is colourless, odourless, tasteless, and non-reactive, no one notices it.

Evidence of helium appeared as an unexplained spectral line in the sun's spectrum in 1868 ("helium" comes from the Greek word *helios*, meaning "the sun"). Later, helium was found to exist on Earth as well.

ALKALI METALS

In another group from Table 8.8, the solids lithium (Li), sodium (Na), and potassium (K) react with hydrogen and fluorine in the same way: one hydrogen atom reacts with one atom of each of these elements, and one fluorine atom reacts with one atom of each element (Table 8.9).

Table 8.9 Alkali Metals

Hydrogen compounds	Fluorine compounds
LiH	LiF
NaH	NaF
KH	KF

These three elements are part of the chemical family called **alkali metals** and are the most active metals on the periodic table (Figure 8.20). All the alkali metals are so reactive that they are never found as free elements (not combined with other substances) in nature. It is believed that the presence of one electron in their outer orbits is in some way associated with their reactivity. The violent reaction of sodium in water (Figure 8.21) is typical of alkali metals.

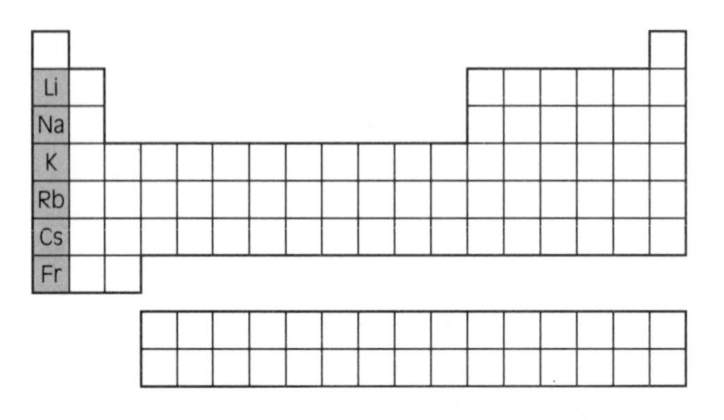

Figure 8.20
Location of alkali metals on the periodic table.

Figure 8.21
Violent reaction of an alkali metal, sodium, in water.

Alkali compounds are found everywhere on Earth. The most common are sodium compounds, which occur in plants, animals, soil, and sea water.

HALOGENS

Another family is made up of the elements fluorine (F) and chlorine (Cl). Both elements are gases and react with hydrogen in a similar way (Table 8.19). One hydrogen atom reacts with one fluorine atom (HF), and one hydrogen atom reacts with one chlorine atom (HCl). These two elements are part of a group of elements called **halogens** (Figure 8.22). The electrons in the outer orbits of all halogens need one electron to fill the orbits completely (Figure 8.23).

These elements are the most reactive non-metals. When fluorine was first purified, it was so reactive that it reacted with most of the containers it was put in. The most common halogen compounds are those of chlorine, which are found in living things, ocean water, and rocks.

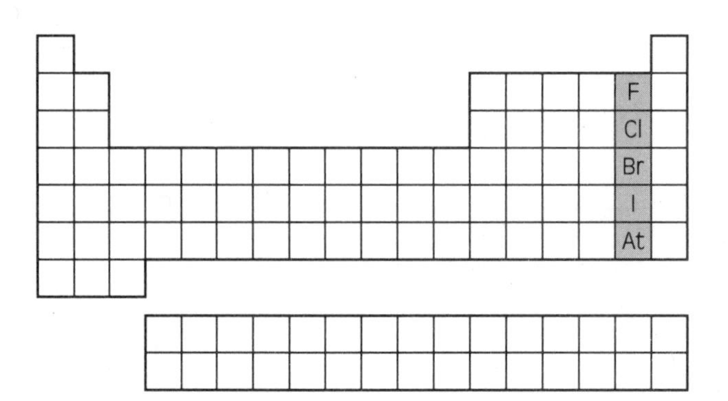

Figure 8.22
Location of halogens on the periodic table.

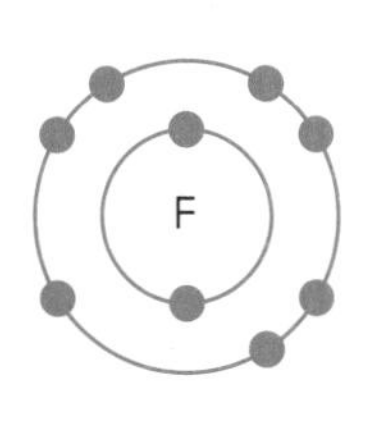

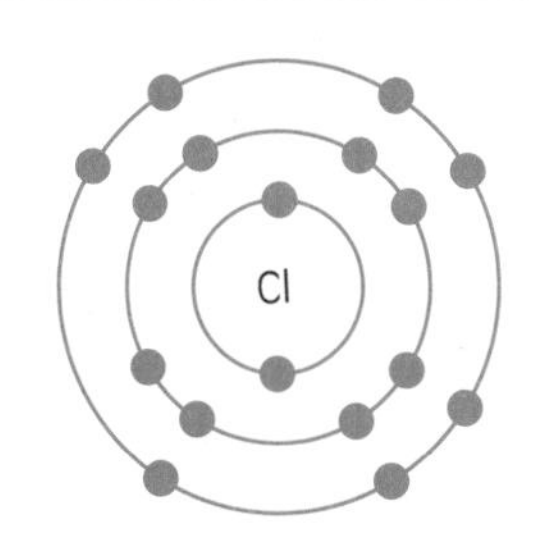

Figure 8.23
The electron arrangement in halogens. This arrangement requires an electron to fill its outer orbit.

A FAMILY OF ONE

Hydrogen is a family of one because in some reactions it acts like a metal and in other reactions it acts like a non-metal. Like alkalis, hydrogen has one electron in its outer orbit. Thus, hydrogen is both active and versatile. Virtually all the Earth's hydrogen exists in combination with other elements. Its reactivity is too great to allow it to be present as a free element in the atmosphere. Hydrogen is one of the main elements in all organic material, including living things, both animal and vegetable, as well as their fossils—petroleum, coal, and natural gas.

At room temperature, hydrogen is a gas and is the least dense of all gases; it will rise in air. This property makes hydrogen useful in balloons. It is extremely flammable, making it a good but very dangerous fuel supply (Figure 8.24).

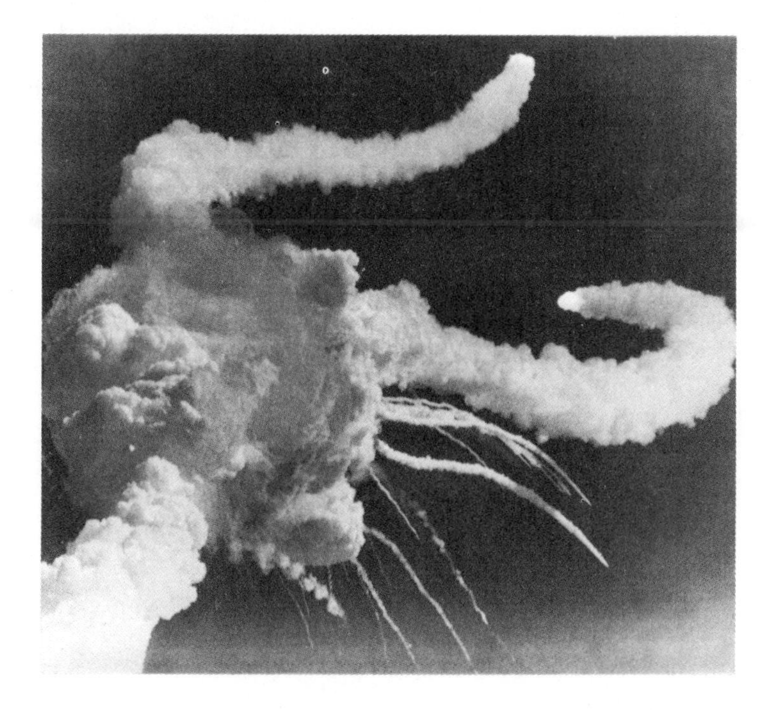

Figure 8.24
The U.S. space shuttle *Challenger* was destroyed during a launch on 28 January 1986, when a seal in its right booster rocket burned through and caused the rupture of the main liquid hydrogen fuel tank. *Challenger*'s seven crew members were killed in the explosive fireball formed by this hydrogen-oxygen reaction.

PERIODS: ROWS ON THE PERIODIC TABLE

The families of elements, which form the columns on the periodic table, have similarities. Some elements adjacent to each other are often not too different from one another. These adjacent elements form horizontal rows of elements on the periodic table called **periods**. The first row, or period, on the periodic table contains two elements, hydrogen and helium. The second row, or period, contains eight elements, starting with lithium and ending with neon. As you go from left to right within a row, the elements gradually change from metallic (lithium) to non-metallic (fluorine), then finally to the noble gases (neon) at the extreme right. Note also that the atomic number increases by one as you go from left to right. Along with changes in atomic number, do you notice any other changes as you go along a row or period?

REVIEW 8.4

1. Draw Bohr models for the first two (a) noble gases, (b) halogens, and (c) alkali metals.
2. What conclusion can you draw from looking at the number of electrons in the outer orbits of (a) noble gases, (b) halogens, and (c) alkali metals?
3. Why is hydrogen considered a family of one?
4. (a) List the elements in the third period of the periodic table.
   (b) What happens to the state, atomic number, number of protons, and number of electrons as you go from left to right along the third period of the periodic table?
5. (a) List the major differences between metals and non-metals.
   (b) Is hydrogen a metal or non-metal? Explain.
   (c) Is argon a metal or non-metal? Explain.
6. Update your learning journal from Activity 8A.

CHAPTER • REVIEW

Key Ideas

- All matter is made up of atoms. Atoms contain subatomic particles called protons, neutrons, and electrons.
- Atoms cannot be chemically created or destroyed.
- The atomic number of an element is the number of protons in an atom's nucleus. The atomic number also equals the number of electrons.
- Atomic mass is the average mass of the atoms of a given element as they occur naturally.
- Isotopes are atoms of the same element that have different mass numbers.
- The nucleus of an atom contains the protons and neutrons. The electrons of an atom are in orbits around the nucleus, with a limited number of electrons in each orbit.
- The modern periodic table is arranged according to atomic number, whereas Mendeleev's early periodic table was arranged according to atomic mass.
- The periodic table contains about 100 elements. It includes information about each element's name, symbol, atomic number, and atomic mass.
- The columns in the periodic table contain elements with similar properties. These columns of elements are called families.
- Alkali metals, halogens, and noble gases are three families on the periodic table.
- The horizontal rows, or periods, in the periodic table are arranged in order of increasing atomic number.

Vocabulary

atom
element
property
atomic theory
molecule
subatomic particle
proton
neutron
electron
atomic number
mass number
isotope
atomic mass

orbit
visible spectrum
spectroscope
periodic table
family
noble gas
alkali metal
halogen
period

1. Construct a concept map that includes the following words: periodic table, atom, electrons, protons, neutrons, atomic number, and atomic mass. You may include other words on the vocabulary list or any other words that you come up with.
2. Draw a diagram, with labels, that includes as many of the vocabulary words as possible.

Connections

1. (a) Where are the metals found on the periodic table?
   (b) Where are the non-metals found on the periodic table?
2. (a) If it were possible for a chlorine atom to gain a proton, how would that change the element?
   (b) If it were possible for a chlorine atom to lose a proton, how would that change the element?
   (c) If it were possible for a chlorine atom to gain an electron, how would that change the element?
   (d) If it were possible for a chlorine atom to lose an electron, how would that change the element?
3. A new element has been discovered, and all you know is that it is an alkali metal.
   (a) Predict the state at room temperature of this new alkali metal.
   (b) Predict the number of electrons in its outer orbit.
   (c) Predict its possible atomic number(s).
   (d) What do you base your predictions on?
4. A new element has been discovered, and all you know is that it is a halogen.
   (a) Predict the state at room temperature of this new halogen.
   (b) Predict the number of electrons in its outer orbit.
   (c) Predict its possible atomic number(s).
   (d) What do you base your predictions on?
5. A new element has been discovered, and all you know is that it is a noble gas.
   (a) Predict the state at room temperature of this new noble gas.
   (b) Predict the number of electrons in its outer orbit.
   (c) Predict its possible atomic number(s).
   (d) What do you base your predictions on?
6. Questions 4, 5, and 6 above suggested that a new alkali metal, halogen, and noble gas were discovered. Why do you think it is unlikely that any of these would be discovered?
7. Study the periodic table.
   (a) How many orbits do the elements in the third row have?
   (b) How many orbits do the elements in the fourth row have?
   (c) What conclusions can you draw about the relationship between the periods on the periodic table and the number of orbits the elements have in that period?

Explorations

1. Design a poster for one of the following families of elements: noble gases, alkali metals, halogens. Include as much information as you can about the family and its members.
2. Elements are the basic building blocks of all the substances in the world. Think of the structure of an atom. Which part of the atom do you think is involved in the chemical reactions that form these substances? Give reasons for your answer. Note: You have just made a prediction. You will see if it is correct in Chapter 9.
3. Use Table 8.10 to find a pattern for how the formulas are written. Then write the formulas for hydrogen and bromine, and barium and fluorine.

Reflections

1. Write down everything you know about the periodic table in two minutes. Spelling, grammar, and sentence structure are not important at this time. If you get stuck, write the title over and over until an idea pops into your mind. After you have finished, let your lab partner read what you have written. He or she is to put question marks beside ideas he or she is not sure of and add ideas you may have left out. Do the same for your partner.
2. What questions do you have about elements that have not been answered? Go back over Chapter 8 and look for questions that are still not answered for you. These questions may be answered in later chapters.

CHAPTER

9 Chemical Formulas and Compounds

In this picture, nitric acid pours from a damaged tank car. The acid is corroding metals, producing the yellowish-brown gas you can see rising from the scene.

In this example, an acid is having a destructive effect on the environment. Acids have many positive effects as well. Acids are found in foods and are used in many common cleaners. In this chapter, you will find out more about acids. You will also find out about two other types of chemical compounds, bases and salts.

ACTIVITY 9A Thinking about Substances

You already know from Chapter 8 that there are about 100 elements. Yet many more substances occur in nature, and still more substances can be made in the laboratory. How can there be so many different substances? Take a minute to write your thoughts in your learning journal. Share your ideas with your lab partner. After talking with your lab partner, you may wish to make modifications to your ideas.

9.1 Composition of Compounds

As you learned in Chapter 8, some substances are made of one element. Other substances, such as the acid shown at the beginning of the chapter, are made of two or more elements that are combined in a definite formula. These substances are called chemical **compounds**.

Chemical compounds are all around you. Your breakfast, lunch, and dinner contain compounds. So do your house, the furniture in it, and your clothing. Even your body contains chemical compounds. About 20 per cent of your body mass is composed of the element carbon. Yet you do not look like charcoal or diamonds, which are examples of pure carbon. The reason is that the carbon in your body is combined with other elements to form many compounds.

Every compound has its own chemical **formula**, which shows the proportion of elements present in that compound. For example, table sugar is a compound consisting of 12 parts of carbon, 22 parts of hydrogen, and 11 parts of oxygen. Thus, the formula for table sugar is $C_{12}H_{22}O_{11}$. In this formula, each element is identified by its symbol. The subscript at the lower right of each symbol represents the proportion of that element in the compound.

Another common compound is water. The formula for water is H_2O. The subscript 2 to the right of the H indicates that there are two parts of hydrogen. Because there is no subscript to the right of the O, we know that the compound contains one part of oxygen for every two parts of hydrogen.

The formula for table salt is NaCl. This formula tells you that table salt consists of one part of sodium for every part of chlorine.

Now let's see if you can identify the elements and their proportions in other chemical compounds.

EXAMPLE I

The formula for potassium dichromate is $K_2Cr_2O_7$. What elements are present in this compound? In what proportions are these elements present?

PREDICT THE ANSWER BEFORE YOU GO ON.

Solution:

First, look at the periodic table (Appendix B) to find the →

elements represented by the symbols K, Cr, and O. You will find that K = potassium, Cr = chromium, and O = oxygen.

Next, use the subscripts to determine the proportion of each element. You will find that the proportion is two parts potassium (K_2) to two parts chromium (Cr_2) to seven parts oxygen (O_7).

INSTANT PRACTICE

Following is a list of formulas for chemical compounds. What elements are present in each compound? In what proportions are these elements present?

1. NaCl
2. N_2O_3
3. $KHCO_3$
4. $FeSO_3$
5. SnO_2
6. SO_2
7. $C_6H_{12}O_6$
8. NaOH
9. KF
10. HCl

CLASSIFYING COMPOUNDS

By the 1500s, alchemists recognized a certain group of substances that had a sour taste. They called this group of substances **acids**, from the Latin word *acidus*, meaning "sour." Acids were known to react with some metals. By 1800, early chemists knew that the reaction of an acid with certain metals would produce hydrogen gas. Today chemists define an acid as any compound that

- turns blue litmus paper red
- tastes sour
- reacts with some metals to produce hydrogen
- conducts electricity when dissolved in water

Some common acids found around the home are shown in Figure 9.1.

Early scientists also recognized another group of compounds that they called alkalis. They made alkalis by dissolving such materials as wood ashes or seaweed in water and then evaporating the liquid. These compounds were used both as cleaning agents and to make other cleaning agents, such as soap. The alkalis are now called **bases**. A base is a compound that

- turns red litmus paper blue
- has a bitter taste
- feels slippery
- conducts electricity when dissolved in water

Some bases found around the home are shown in Figure 9.2.

When a base solution reacts with an acid solution, a third type of compound results. This compound often has a saltlike taste. Evaporating the liquid in such a solution yields a solid crystalline substance, which

DID YOU KNOW?

Sodium chloride is used to preserve food. Foods soaked in sodium chloride solutions, called brines, have large salt concentrations. Microorganisms are usually the cause of food spoilage. When a microorganism lands on salt-treated food, it takes in the salt. The excess salt causes water to flow into the microorganism until the cell walls of the microorganism burst. The salt thereby kills the microorganism, keeping it from multiplying on the food's surface, and prevents spoiling.

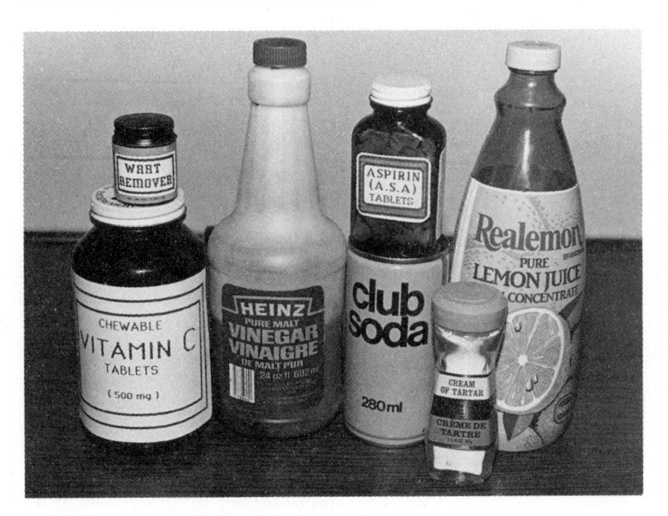

Figure 9.1
Some common household acids.

Figure 9.2
Some common household bases.

early chemists called "an earth." Modern chemists call this compound a **salt**. Table salt and bath salts are common examples of salts. Figure 9.3 shows some salts that might be found around the lab.

In ancient times, the most common salt, sodium chloride (NaCl), or table salt, was used to purchase slaves and to pay Roman soldiers. When there was no refrigeration, salt was used to preserve foods. Humans have used salt for at least 5000 years as a food preservative.

Not all compounds are acids, bases, or salts. Common table sugar may look like table salt, but it is not a salt and cannot be made by mixing an acid and a base. Nevertheless, this grouping of compounds (acids, bases, and salts) is still used today to classify many of the compounds in chemistry. Dyes such as litmus help classify some acids and bases. These dyes are called **indicators**. They use different colours to show whether an acid or a base is present in a solution.

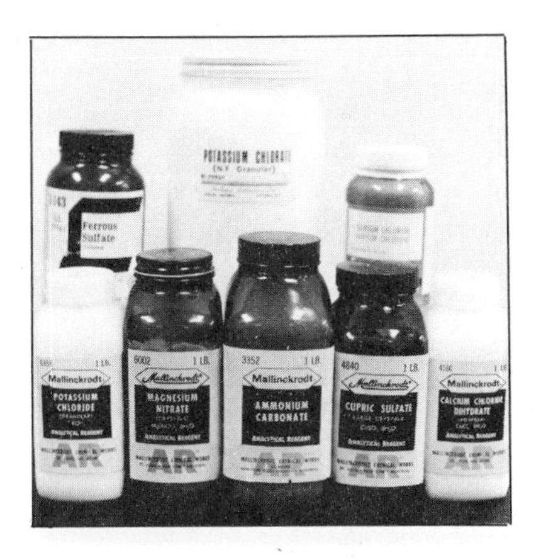

Figure 9.3
Some common salts.

ACTIVITY 9B *Testing for Acids and Bases*

Acids and bases can be distinguished from one another by the colour they turn certain indicators, such as litmus paper, bromothymol blue, and phenolphthalein.

In this activity, you will use several indicators to observe the behavior of an acid and a base and to classify six unknown solutions.

MATERIALS

PARTS I AND II

- safety goggles
- apron
- six test tubes
- six medicine droppers
- test tube rack
- phenolphthalein solution (in dropper bottle)
- bromothymol blue solution (in dropper bottle)
- red litmus paper
- blue litmus paper

PART I

- known acid solution
- known base solution

PART II

- sheet of clear plastic
- set of six unknown solutions (labelled A to F)

PART I

PROCEDURE

In this part, you will observe the behaviour of an acid and a base.

1. Copy Table 9.1 into your notebook.
2. Put on safety goggles and apron.

> **CAUTION!**
> With acids and bases, there is the risk of eye and skin irritation. If you get any acid or base in your eyes or on your skin, immediately rinse the area with water for 15 to 20 minutes, and tell your teacher.

3. Place three to four drops of phenolphthalein indicator in a test tube. Now add a drop of known acid solution. Record your observations in your table.
4. Place three to four drops of bromothymol blue indicator in a test tube. Now add a drop of known acid solution. Record your observations in your table.
5. Place a small piece of red litmus paper in a test tube. Now add a drop of known acid solution. Record your observations in your table.
6. Place a small piece of blue litmus paper in a test tube. Now add a drop of known acid solution. Record your observations in your table.
7. Repeat steps 3, 4, 5, and 6, using the known base instead of the known acid.

Table 9.1 Sample Data Table for Activity 9B (Part I)

Indicators	Colour	
	In acid	In base
Phenolphthalein		
Bromothymol blue	SAMPLE ONLY	
Red litmus		
Blue litmus		

DISCUSSION

1. (a) What does an acid do to phenolphthalein, bromothymol blue, blue litmus, and red litmus?
   (b) What does a base do to phenolphthalein, bromothymol blue, blue litmus, and red litmus?
2. If there is no change in an indicator, what conclusion can you draw?

PART II

PROCEDURE

In this part, you will classify six unknown solutions.

1. Copy Table 9.2 into your notebook, using a full page.
2. Put on safety goggles and apron.
3. Label your test tubes A to F, corresponding to the letters in Table 9.2, and obtain your unknown solutions. The test tubes should be about one-third full so that you can reach each sample with a medicine dropper. Place the test tubes in a test tube rack and put a clean medicine dropper in each.
4. Place the sheet of clear plastic on top of your table and put one or two drops of the phenolphthalein and bromothymol blue indicator in the appropriate squares of the grid. Tear the red and blue litmus paper into small pieces and place them in the appropriate squares.
5. Using the medicine dropper, add one or two drops of unknown solution A to the phenolphthalein square of the grid. Record your observations in your table.
6. Add one or two drops of unknown solution A to each of the bromothymol blue, blue litmus, and red litmus squares of the grid. Record your observations in your table.

Table 9.2 Sample Data Table for Activity 9B (Part II)

	Chemical Indicators			
Unknown solutions	Phenolphthalein	Bromothymol blue	Blue litmus	Red litmus
A				
B		SAMPLE ONLY		
C				
D				
E				
F				

7. Repeat steps 5 and 6 for the other unknown solutions (B, C, D, E, and F).

DISCUSSION

1. (a) Which solution are acids and which are bases?
   (b) Explain how you decided which solutions were acids and which were bases.
   (c) Write a definition for the terms "acid" and "base," using the test results as part of your definition.
2. (a) If you were given an unknown solution, what tests would you run to determine whether it is an acid or a base?
   (b) If all the indicators had no colour change, what conclusion could you draw?
3. If a piece of red litmus remained red and a piece of blue litmus remained blue in a solution, what is present in the solution?

EXTENSION

Place a fresh piece of blue litmus paper in a beaker of water. Using a straw, blow gently into the water for several minutes, until a change in the litmus paper occurs. What does the change show about exhaled breath when it dissolves in water?

PRACTICAL USES OF ACIDS AND BASES

Both industrial acids and acids that occur naturally affect us directly or indirectly (Table 9.3). Acidic compounds give the tangy taste you find in lemons, oranges, and apples. Hydrochloric acid is found in your stomach and helps in the digestion of proteins. Your stomach protects itself from its own hydrochloric acid with an acid-resistant lining. If the lining is damaged, the hydrochloric acid can cause ulcers to form (Figure 9.4).

Table 9.3 Common Acids

Where acid is found	Acid name	Acid formula
Lemons	Citric acid	$HOOCCH_2C(OH)(COOH)CH_2COOH$
Apples	Malic acid	$HOOCCH_2CH(OH)COOH$
Rhubarb	Oxalic acid	$HOOCCOOH$
Vinegar	Acetic acid	CH_3COOH
Sour milk	Lactic acid	$CH_3CH(OH)COOH$
Soft drinks	Carbonic acid	H_2CO_3
Automobile batteries	Sulphuric acid	H_2SO_4
Stomach	Hydrochloric acid	HCl

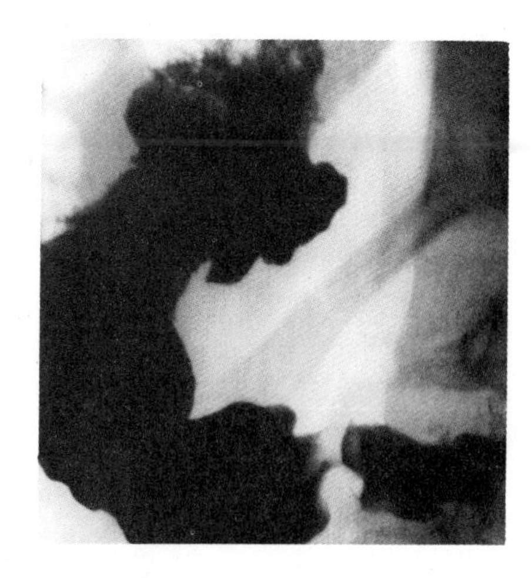

Figure 9.4
X ray of stomach ulcer.

One of the most important acids is sulphuric acid (Table 9.3). This versatile compound is used in such varied processes as manufacturing fertilizers and explosives, refining oil, and electroplating metals. When properly contained, sulphuric acid is a great help to people. But a related compound called sulphurous acid often "gets loose" during industrial processes. It contributes to acid rain.

Many bases are used as household cleaners (Table 9.4). They work well as cleaners because bases react with plant or animal matter to form compounds that dissolve in water. For example, kitchen waste clogging a drain can be softened by the action of drain cleaner, which contains a strong base, potassium hydroxide. The waste and the base react, producing a soluble compound that works like a soft soap. As this compound dissolves, the water can flow freely through the pipe again.

EXTENSION

Both drain cleaners and oven cleaners are strong bases and can be harmful to the environment. Find out about environmentally safe alternatives that could be used to unclog drains and clean ovens effectively. Include the components of these alternatives and why they are considered safe for the environment.

Table 9.4 Common Bases

Products that contain bases	Base name	Base formula
Oven cleaner	Sodium hydroxide	$NaOH$
Drain cleaner	Potassium hydroxide	KOH
Laxative	Magnesium hydroxide	$Mg(OH)_2$
Antacid	Aluminum hydroxide	$Al(OH)_3$

Oven cleaner also contains a strong base, sodium hydroxide. Baked-on grease will not come off with plain water or even with ordinary detergents. However, hardened grease reacts with oven cleaner, producing a soapy liquid that can be washed away with water.

NEUTRALIZING ACIDS

Milk of magnesia (Figure 9.5), commonly sold as a nonprescription stomach antacid, contains a base, magnesium hydroxide (Table 9.4). This base does not dissolve in water. Thus, it will not react with the tissue in your mouth or esophagus (the tube leading from the mouth to the stomach). But in your stomach, naturally occurring hydrochloric acid reacts with the milk of magnesia. As you have learned, when an acid reacts with a base, a salt solution is produced:

acid + base = water + salt

The reaction of an acid and a base to produce water and a salt is called **neutralization**. The base in milk of magnesia neutralizes the hydrochloric acid in the stomach to form water and salt.

Permanent waves, or perms, also use neutralization. The hairdresser pours neutralizer (acid) over hair treated with waving lotion (base) to form a salt on the customer's hair. When the customer's hair is rinsed, the salt dissolves in the water and washes down the drain.

Many common household products are acids or bases and are potentially dangerous. Acid spills can be neutralized with baking soda. Spills that involve bases can be neutralized with vinegar.

Figure 9.5
Milk of magnesia is a home remedy for acid indigestion.

ACTIVITY 9C Acids versus Bases

In this activity, you will neutralize an acid with a base.

MATERIALS

safety goggles
apron
evaporating dish
acid solution (dilute HCl)
base solution (NaOH)
conductivity apparatus
(for demonstration)
droppers
stirring rod
phenolphthalein

PROCEDURE

1. Put on safety goggles and apron.

CAUTION!
With acids and bases, there is the risk of irritating eyes and skin. If you get any acid or base in your eyes or on your skin, immediately rinse the area with water for 15 to 20 minutes, and tell your teacher.

2. Measure 20 drops of base (NaOH) into an evaporating dish. Add a drop of phenolphthalein. Record the colour change in your notebook.
3. Add hydrochloric acid (HCl), one drop at a time, stirring after each addition. Record the effect of the acid on the colour of the liquid in the dish. Continue to add acid, drop by drop, until the colour change is complete.
4. Allow the liquid in the dish to evaporate, using whatever method your teacher recommends. Examine the solid residue. Describe and sketch it in your notebook.
5. Dissolve the solid residue in water. Your teacher will test the solution for conductivity.
6. Wash your hands thoroughly with soap and water after you have completed this activity.

DISCUSSION

1. In view of the acid and base response to phenolphthalein, what kind of substance is phenolphthalein?
2. In this reaction, the hydrogen from the acid (HCl) and the hydroxide (OH) from the base (NaOH) react to form a new compound.
   (a) What is the probable formula for the new compound?
   (b) What is its name?
   (c) Is this substance a solid, a liquid, or a gas at room temperature?
   (d) There is a second new compound formed. What do you think it is? (Give name and formula.)
3. (a) In this activity, you reacted an acid with a base. What experimental evidence was there that a new compound (or compounds) was formed?
   (b) Can you devise an experiment to prove that a new compound (or compounds) was formed?
4. In step 5, the electrical conductivity of a substance was tested. What was that substance?

EXTENSION

Repeat Activity 9C, but this time use milk of magnesia with the hydrochloric acid. You will get the same reaction that occurs in your stomach when you take milk of magnesia for acid indigestion.

PRACTICAL USES OF SALTS

The type of salt produced by neutralization depends on the acids and bases you use. The reaction between waving lotion and neutralizer produces salt that dissolves in water. Not all salts dissolve in water. Those that do conduct electricity. Salts have very few common properties; as a result, the use for any given salt depends on the properties of the individual salt. Table 9.5 lists some of the more common salts and their formulas.

Table 9.5 Common Salts

Where found	Name	Formula
Table salt	Sodium chloride	$NaCl$
Baking soda	Sodium hydrogen carbonate	$NaHCO_3$
Match heads	Potassium chlorate	$KClO_3$
Fertilizer, explosives	Potassium nitrate	KNO_3

The most common salt is table salt (NaCl). It is found as a solid in underground mines and wells, in dried-up ocean beds, and dissolved in the ocean (Figure 9.6). Only 5 per cent of all the sodium chloride we use becomes salt for the table. The other 95 per cent is used for hundreds of other purposes, including leather processing, glass making, and the creation of other chemicals for industry.

(a)

(b)

Figure 9.6
Sodium chloride occurs in large deposits both below and above ground. (a) The interior of a salt mine in Texas. (b) The Bonneville Salt Flats in Utah.

REVIEW 9.1

1. Why are there so many more compounds than elements?
2. How do the elements in a compound differ from those of the pure element?
3. Solid black carbon and three colourless gases—hydrogen, nitrogen, and oxygen—combine to form a deadly poison called strychnine ($C_{21}H_{22}N_2O_2$). Why is there a difference in properties between strychnine and the four elements (carbon, hydrogen, nitrogen, and oxygen)?
4. Use the following formulas to state what elements are present in each compound and in what proportions.
   (a) H_2SO_3
   (b) HCl
   (c) $CuCO_3$
   (d) $NaOH$
   (e) CH_3COOH

5. (a) State some properties that all acids have in common.
   (b) State some properties that all bases have in common.
   (c) Name one thing all salts have in common.
6. Suppose the labels came off two identical bottles, one containing liquid drain cleaner; the other, vinegar. Without smelling them, how could you test and identify which bottle contained the vinegar? Explain why you should not smell the bottles.
7. (a) What is neutralization?
   (b) What is the word equation for neutralization?
8. Figure 9.6 shows one source of salt, mines. Describe another way to obtain solid salt (sodium chloride or table salt). HINT: Most of the world is covered by oceans (salt water).

9.2 Electron Transfer: The Ionic Compound

If shiny sodium metal is lowered into a jar of poisonous green chlorine gas, a non-metal, the result resembles fireworks (Figure 9.7)! When the sparks die down and the jar cools off, a new white substance can be seen. The compound sodium chloride has been formed.

Figure 9.7
Sodium burns rapidly in the chlorine atmosphere as the compound sodium chloride is formed. Note: This reaction should not be carried out in the classroom, since the reaction is very vigorous and numerous safety precautions must be taken.

HOW ELEMENTS FORM COMPOUNDS

Although the sodium-chlorine reaction is too dangerous for a school laboratory, it is an ideal way to explain how elements form compounds. Take a look at what happens when sodium meets chlorine. Figure 9.8 shows electron orbits of each atom just before they collide. Note that sodium has just one electron more than the stable noble gas neon, and chlorine has just one electron less than the stable noble gas argon. Many atoms tend to have a stable electron configuration for their outer orbit similar to the configurations of noble gases such as neon and argon.

In Figure 9.9, the sodium and chlorine atoms bump together. Look at their outer orbits where the two atoms are nearest each other. The outer orbits of both atoms are where the atoms will react with one another.

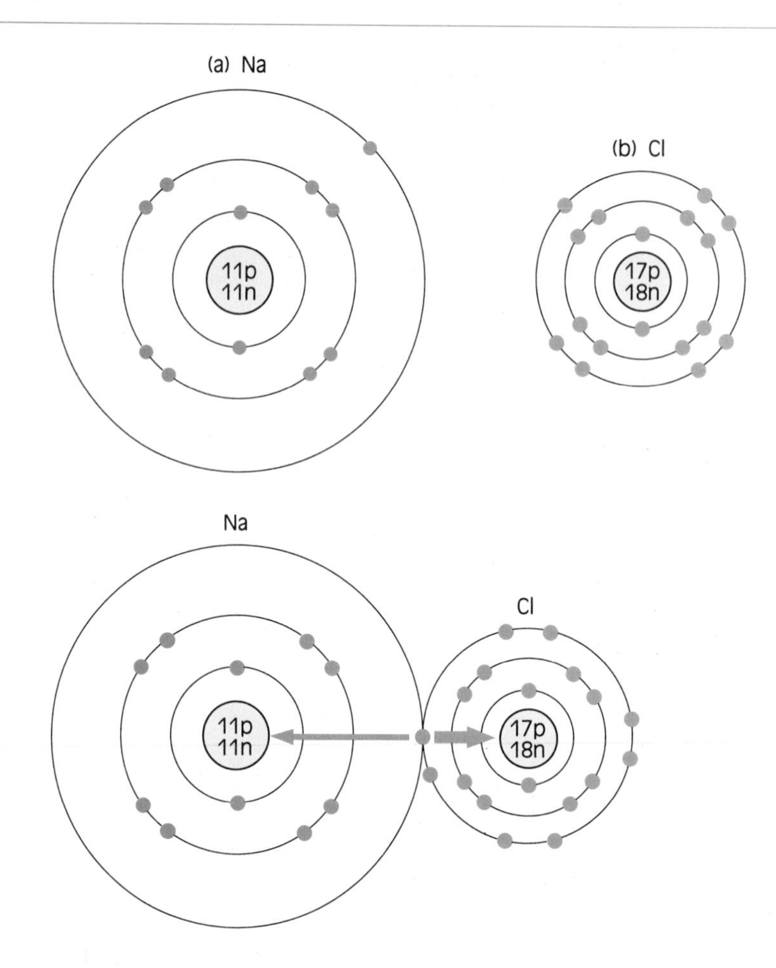

Figure 9.8
(a) Sodium atom (Na) and (b) chlorine atom (Cl). The chlorine atom has only half the diameter of the sodium atom. Its nucleus is much more powerful.

Figure 9.9
The only electron in the outer orbit of sodium is attracted towards the chlorine atom. Notice that the pull of the chlorine nucleus is stronger than the pull of the sodium nucleus.

Notice in Figure 9.9 that the sodium's outermost electron is closer to the chlorine atom's nucleus than it is to its own. The negative electrons and the positive protons obey the laws of electricity presented in Unit II. That is, their opposite charges cause them to attract each other. The positive electrical attraction from the more intensely positive chlorine nucleus is strong enough to pull this loosely held sodium electron into the chlorine's outer orbit (Figure 9.10).

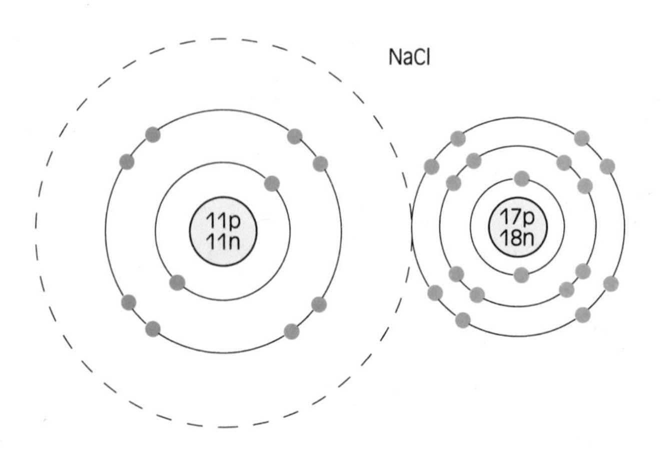

Figure 9.10
An ionic bond has been formed between sodium and chlorine.

The transfer of an electron from metallic sodium to non-metallic chlorine results in a new structure for both atoms. The sodium atom has given up an electron. Sodium now has a positive charge because it has 11 positive protons and only 10 negative electrons (Figure 9.11a). The electron arrangement is stable, similar to that of the noble gas neon. But the sodium no longer has the structure of a neutral atom.

The chlorine atom has accepted an electron. Chlorine now has an overall negative charge because it has 17 positive protons and 18 negative electrons (Figure 9.11b). The electron arrangement is stable, similar to that of the noble gas argon. But the chlorine no longer has the structure of a neutral atom.

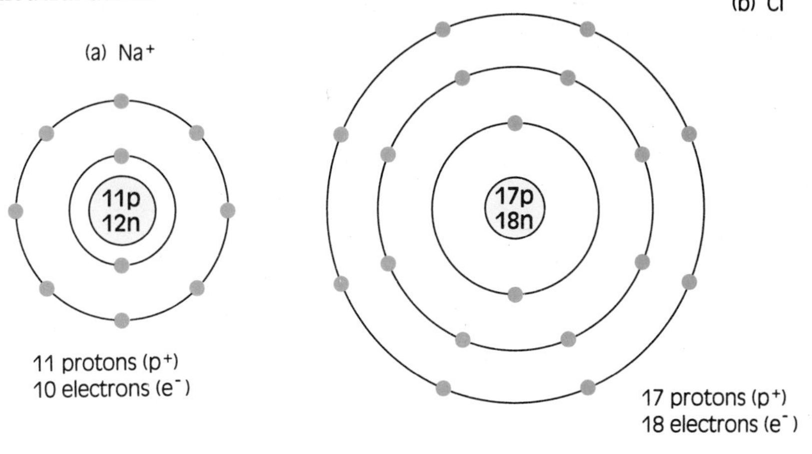

Figure 9.11
(a) Sodium ion (Na^{+1}) and (b) chloride ion (Cl^{-1}).

The movement of an electron from one atom to another is called electron transfer. It results in the formation of charged particles called **ions**. Both the sodium and the chlorine particles are ions because they have an electrical charge (Table 9.6). (Note that the ion formed by the chlorine atom is called a chloride ion.) Because the sodium ion and the chloride ion have opposite charges, they attract each other. The force of attraction between positive and negative ions is called an **ionic bond** (Figure 9.10). The attraction between positive sodium ions (Na^{+1}) and negative chloride ions (Cl^{-1}) is what holds the new compound together. Sodium chloride is therefore called an **ionic compound** because it is composed of ions and held together by ionic bonds.

EXTENSION

Examine crystals of sodium chloride with a magnifying lens or a microscope. What do you notice about the shape of the crystals? Try both regular table salt and pickling salt.

Table 9.6 Ionic Bonding in Sodium and Chlorine

Name of element	Atom (symbol)	Ion (symbol)	Name of ion
Sodium	Na	Na^{+1}	Sodium
Chlorine	Cl	Cl^{-1}	Chloride

IONIC COMPOUNDS AND COMBINING CAPACITY

Ionic compounds are formed by electron transfer between metals and non-metals. The term **combining capacity** refers to the number of electrons an atom must give up or gain to have a stable electron arrangement. Understanding combining capacity will enable you to predict the charges of ions formed by metals and non-metals. Figure 9.12 shows the electron arrangement of the first 18 elements.

You know that atoms have equal numbers of positively charged protons and negatively charged electrons and thus are neutral. When magnesium metal reacts with the non-metal fluorine, an ionic compound is formed. A neutral fluorine atom has 9 protons and 9 electrons, with 7 of the electrons in its outer orbit. Fluorine's outer orbit needs one more electron to have a stable electron arrangement similar to that of a noble gas. If the fluorine atom gains one electron, it will have 9 protons and 10 electrons. The fluoride ion that results will have a charge of -1 (Figure 9.13).

Figure 9.12
Electron arrangements of the first 18 elements.

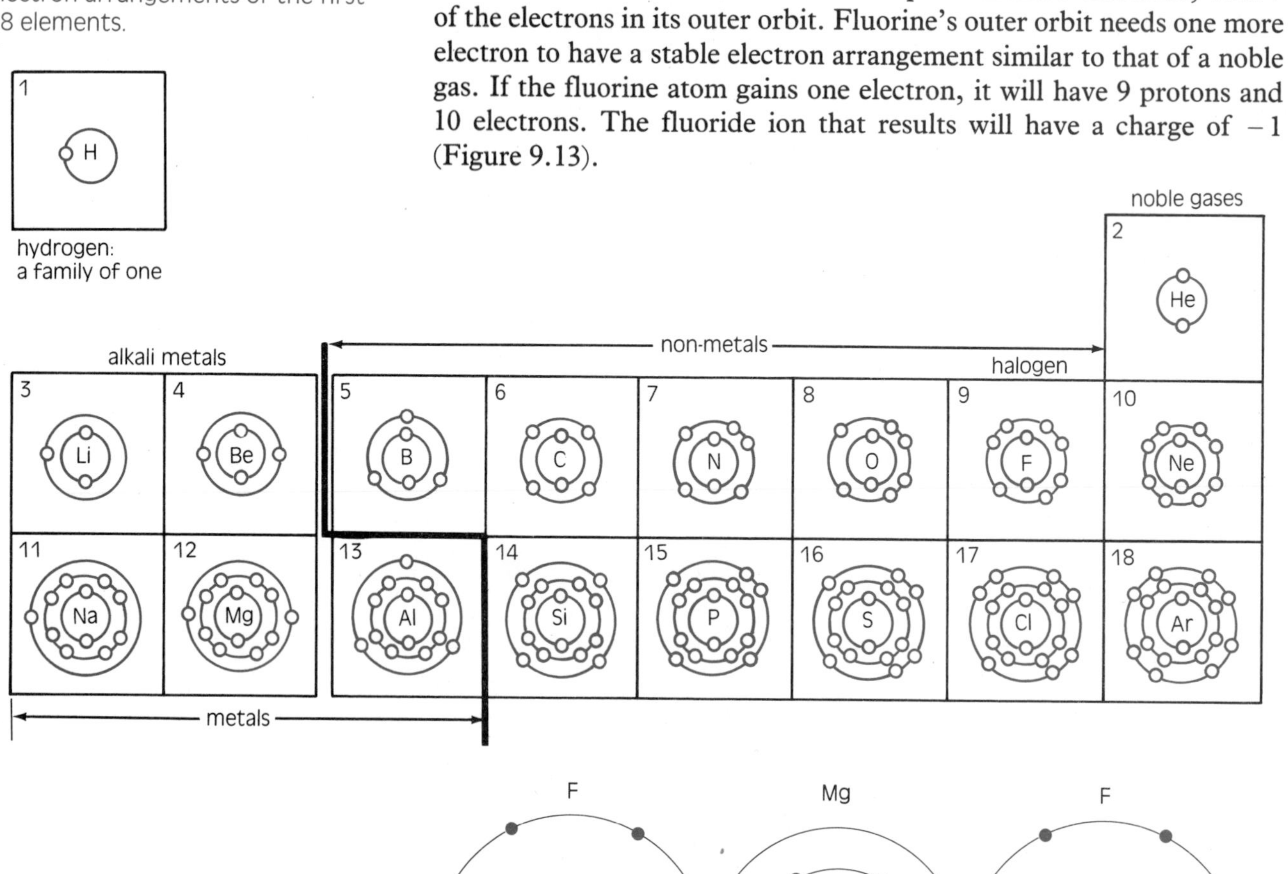

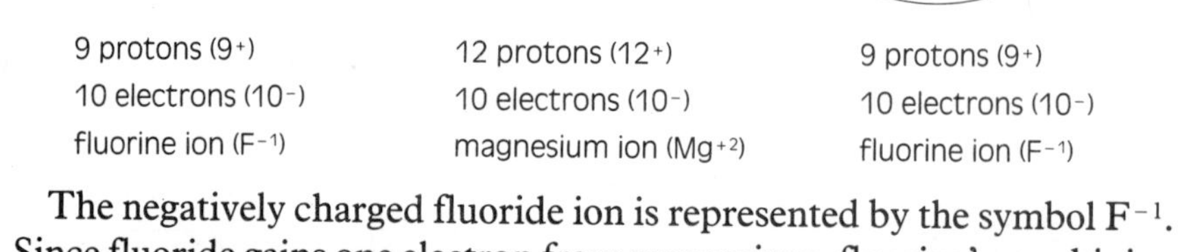

Figure 9.13
The ionic compound magnesium fluoride.

The negatively charged fluoride ion is represented by the symbol F^{-1}. Since fluoride gains one electron from magnesium, fluorine's combining capacity must be -1. This value agrees with the combining capacity of fluorine shown in Figure 9.14.

The ionic compound magnesium fluoride contains positively and negatively charged ions. A neutral atom of magnesium has 12 protons and 12 electrons. Its outer orbit has 2 electrons, which magnesium tends to lose when it forms compounds. When magnesium combines with fluorine, each magnesium atom loses two electrons, one to each of two fluorine atoms. As a result, the magnesium ion has 12 protons and 10 electrons (Figure 9.13). What is the charge for a magnesium ion? You are right if you said +2, represented by the symbol Mg^{+2}. What do you think the combining capacity for magnesium will be? Figure 9.14 tells you that magnesium has a combining capacity of +2. Thus, in the ionic compound magnesium fluoride, two fluoride ions (each an F^{-1}) are present for each magnesium ion (Mg^{+2}).

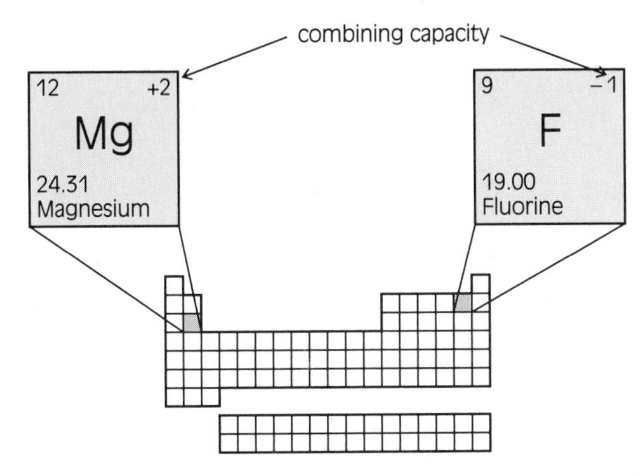

Figure 9.14
Information in the periodic table on magnesium and fluorine.

Sodium is an alkali metal. It has one electron in its outer orbit. When a sodium atom reacts with chlorine, it loses this electron and forms a positive sodium ion. The combining capacity for sodium is +1. When other members of the alkali metal family form ions, what do you think the combining capacity will be? Do you also notice any relationship between the charge on the alkali metal ions and their combining capacities?

ACTIVITY 9D *Ions, Combining Capacity, and Chemical Families*

You have learned a lot about elements, the structure of atoms, families of elements, ionic compounds, and combining capacity. You have also learned that tables are a good way to organize your knowledge and at the same time gain more knowledge by observing similarities and differences. This activity will help you summarize the information you will need to make your own chemical formulas in the next section.

PROCEDURE

Construct three tables using the periodic table and what you have already learned. The tables should include the following information.

Table 1: Alkali Metals

(a) Give the names of the alkali metals.
(b) Give the symbols of the alkali metals.
(c) Draw Bohr models for the first three alkali metals.
(d) Give the electrical charge for the alkali metal ions.
(e) Give the symbols for the alkali metal ions.
(f) Give the combining capacities of the alkali metals.
(g) Draw Bohr models to represent the ions formed by the three alkali metals.

→

Table 2: Halogens

(a) Give the names of the halogens.
(b) Give the symbols of the halogens.
(c) Draw Bohr models for the first three halogens.
(d) Give the electrical charge for the halogen ions.
(e) Give the symbols for the halogen ions.
(f) Give the combining capacities of the halogens.
(g) Draw Bohr models to represent the ions formed by the four halogen non-metals.

Table 3: Noble Gases (Inert Gases)

(a) Give the names of the inert gases.
(b) Give the symbols of the inert gases.
(c) Draw Bohr models for the first three inert gases.
(d) Give the electrical charge for the inert gas ions, if any.
(e) Give the symbols for the inert gas ions, if any.
(f) Give the combining capacities of the inert gases.
(g) Draw Bohr models to represent the ions formed by the inert gases, if any.

DISCUSSION

1. (a) For alkali metals, what connections can you draw from your data on the tables (names, symbols, charge on ion, symbol for ion, combining capacity, electrons in outer orbit)?
   (b) For halogens, what connections can you draw from your data on the tables (names, symbols, charge on ion, symbol for ion, combining capacity, electrons in outer orbit)?
   (c) For inert gases, what connections can you draw from your data on the tables (names, symbols, charge on ion, symbol for ion, combining capacity, electrons in outer orbit)?
2. For the metallic elements lithium, beryllium, boron:
   (a) Draw Bohr models for each element. How many electrons are in their outer orbits?
   (b) Do these metallic elements tend to gain or lose electrons? Give reasons for your answer.
   (c) What is the charge on each of the metal ions (include the ion symbol)?
   (d) What is the combining capacity for each of these metals?
3. For the non-metallic elements nitrogen, chlorine, fluorine:
   (a) Draw Bohr models for each element. How many electrons are in their outer orbits?
   (b) Do these metallic elements tend to gain or lose electrons?
   (c) What is the charge on each of the non-metal ions (include the ion symbol)?
   (d) What is the combining capacity for each of these non-metals?
4. Explain why it is difficult to discuss ions or combining capacity in connection with noble gases such as xenon.

REVIEW 9.2

1. (a) What charge results when a metal atom forms an ion? Explain.
   (b) What charge results when a non-metal atom forms an ion? Explain.
2. What kind of outer electron orbit does a stable ion have?
3. Write formulas for stable ions of the following elements: aluminum, iodine, barium, magnesium, and sulphur.
4. How are ionic compounds formed?
5. (a) What charge do members of the halogen family of elements have when their atoms form ions? Why?
   (b) What charge do members of the alkali metals family of elements have when their atoms form ions? Why?
   (c) What charge does the element magnesium have when its atoms form ions? (HINT: Look at the number of electrons in its outer orbit.)
6. Will a noble gas form an ion? Explain.
7. (a) How are ions different from atoms?
   (b) How are ions similar to atoms?
8. (a) What is combining capacity?
   (b) What are the combining capacities for the following elements: aluminum, iodine, barium, magnesium, and sulphur?

9.3 Formulas for Ionic Compounds

Hundreds of different ionic compounds are possible, resulting in hundreds of formulas. Fortunately, writing formulas for ionic compounds can be simplified by using combining capacities. The following example takes you through the formula for an ionic compound.

EXAMPLE II
What is the formula for the ionic compound formed by potassium and fluorine?

PREDICT THE FORMULA BEFORE YOU GO ON.

1. Symbols and order are

   K F

2. Combining capacities are

   K(+1) F(−1)

3. How can you achieve a balanced electrical charge? The charge on K (+1) is balanced by the charge on F (−1). This ion pair is neutral, so there will be one potassium ion for each fluorine ion (Figure 9.15). The formula is

   KF

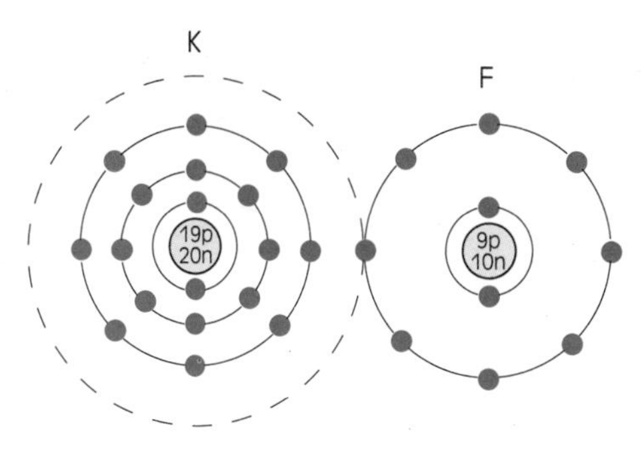

Figure 9.15
Ionic bonding of potassium and fluorine. One potassium ion bonds with one fluorine ion.

ACTIVITY 9E *Making Up Rules for Chemical Formulas*

In this activity, you will use calcium and chlorine to see how much you remember about writing chemical formulas. Use two-column notes, as shown in Table 9.7. In the first column, write your own rules. Do this by writing a set of instructions to a classmate on how to write the formula for the compound formed by calcium and chlorine. You may include questions to your classmates about ideas you are not sure of. Use the following vocabulary words: element, compound, periodic table, symbol, metal, non-metal, combining capacity, and any other key words you can come up with.

Table 9.7 Sample Data Table for Activity 9E

Your own rules	New learning after your readings and examples
	SAMPLE ONLY

After you have written your rules, compare them with the ones below. You may have combined some of the steps or rules, which is fine. Did you get the formula $CaCl_2$? Make your corrections and comments or ask questions in the right-hand column. Then ask a classmate to check them over or to help you with the answers to some of your questions.

RULES FOR WRITING CHEMICAL FORMULAS

Rule 1:
Write the symbol for the more metallic element first. Then write the symbol for the non-metallic element. The metallic elements are to the left in the periodic table (Appendix B). Calcium is more metallic than chlorine, so Ca comes first.

Rule 2:
Next, write the combining capacities for the elements. Combining capacities are found in the periodic table (Appendix B).

Ca^{+2} and Cl^{-1}

Rule 3:
The combining capacities should "balance" one another, giving a neutral compound (total positive combining capacity + total negative combining capacity = 0). To balance the combining capacity of one calcium ion (+2), you need two chloride ions (−1 each). The result is a total combining capacity of (−2), −1 plus −1.

Rule 4:
Subscripts are used to indicate the number of atoms each element has in the compound. If the element has only one atom, no subscript is needed for that element. The subscripts should be the smallest of whole numbers possible. The formula with one ion of calcium and two ions of chlorine is $CaCl_2$.

There is another simple way to obtain the formula of a compound. First, write the metal ion (positive ion) followed by the non-metal ion (negative ion). Don't write the charge on the ion. Second, criss-cross the combining capacities (numbers only), using them as subscripts.

Using calcium and chlorine,

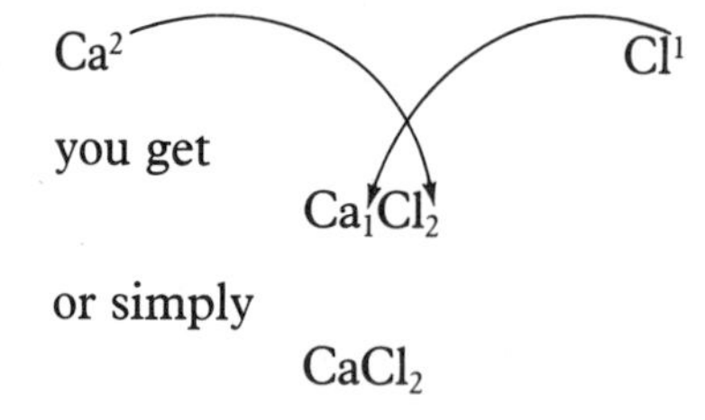

or simply

$CaCl_2$

(Remember, the subscript 1 is never written.)

> *EXAMPLE III*
> What is the formula for the compound formed by lithium and oxygen?
>
> PREDICT THE FORMULA BEFORE YOU GO ON.
>
> 1. Symbols and order are
>
> Li O

2. Combining capacities are

   Li(+1) O(−2)

   Using the criss-cross method, the result is

   Li^1 O^2

   Li_2O (Figure 9.16)

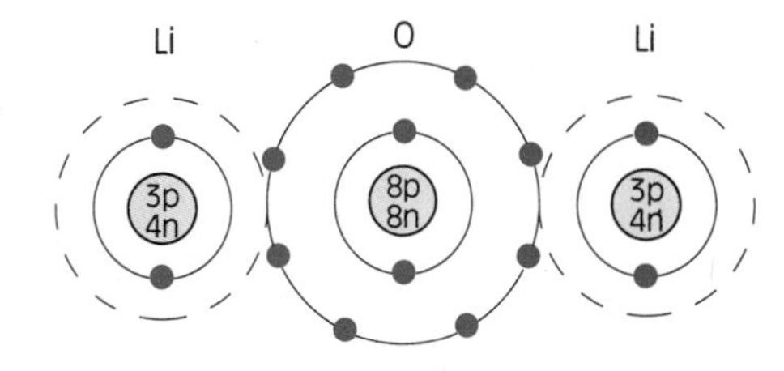

Figure 9.16
Ionic bonding of lithium and oxygen. Two lithium ions bond with one oxygen ion.

NAMING IONIC COMPOUNDS

In many cases, the name of an ionic compound is easily determined. When there are only two elements in an ionic bond, it is known as a binary compound and the name ends with "ide."

NaCl is sodium chloride.
MgF_2 is magnesium fluoride.
K_2S is potassium sulphide.

INSTANT PRACTICE

Use the periodic table (Appendix B) and what you have just learned to write formulas for compounds and to name the compounds.

1. Write formulas for compounds formed from the following sets of ions. Then write the names of these compounds.
   (a) Li^{+1} and Cl^{-1}
   (b) Ca^{+2} and O^{-2}
   (c) Na^{+1} and S^{-2}
   (d) Al^{+3} and I^{-1}
   (e) Ba^{+2} and F^{-1}

2. Write formulas for compounds formed from the following elements. Then write the names of these compounds.
   (a) sodium and bromine
   (b) potassium and oxygen
   (c) aluminum and sulphur
   (d) barium and chlorine
   (e) lithium and oxygen
   (f) silver and chlorine

3. Write formulas for the following compounds:
   (a) potassium chloride
   (b) sodium oxide
   (c) calcium bromide
   (d) magnesium oxide
   (e) aluminum fluoride

WRITING FORMULAS FOR ELEMENTS WITH MORE THAN ONE COMBINING CAPACITY

A few elements are able to form more than one kind of ion. The metallic element iron forms two completely different ionic compounds when it reacts with chlorine. One iron-chlorine compound is yellow; the other is green. Besides differing in colour, they also differ in solubility and many other properties. By analyzing these compounds, chemists have shown that their formulas are $FeCl_2$ and $FeCl_3$.

You know that the charge on a chloride ion is -1. So the iron ions in $FeCl_2$ must have a charge of $+2$. But the iron ions in $FeCl_3$ must have a charge of $+3$.

A system has been devised to tell one ion of the same metal from another. Roman numerals are included in the names of these ionic compounds to show the charge of the metal ions. For example, the formula $FeCl_2$ stands for iron(II) chloride; the formula $FeCl_3$ stands for iron(III) chloride.

Two other common metals that have more than one combining capacity are copper, which has combining capacities of $+1$ and $+2$, and lead, which has combining capacities of $+2$ and $+4$.

EXAMPLE IV

What is the formula for the compound lead(IV) iodide?

PREDICT THE FORMULA BEFORE YOU GO ON.

1. Symbols and order are

   Pb I

2. Combining capacities:

   Lead's combining capacity is in its name (IV).

   Pb(+4) I(−1)

   Using the criss-cross method, the result is

INSTANT PRACTICE

Use the periodic table (Appendix B) and what you have just learned to write formulas for compounds and then to write names of the compounds.

1. Write formulas for compounds formed from the following elements. Then write the names of these compounds.
   (a) iron(II) and oxygen
   (b) copper(I) and bromine
   (c) tin(II) and iodine
   (d) lead(IV) and chlorine
   (e) nickel(III) and oxygen
   (f) nickel(II) and oxygen

2. Write formulas for the following compounds:
   (a) tin(II) fluoride
   (b) iron(III) oxide
   (c) iron(II) oxide
   (d) copper(I) chloride
   (e) lead(IV) iodide
   (f) mercury(II) oxide

FORMULAS FOR CHARGED GROUPS OF ATOMS

When ionic compounds such as sodium chloride salt (NaCl) and copper(II) bromide salt ($CuBr_2$) are dissolved in water, the compounds form separate ions. NaCl forms Na^{+1} and Cl^{-1} ions; $CuBr_2$ forms Cu^{+2} and Br^{-1} ions. Experiments show that a compound such as sodium hydroxide (NaOH), a base, comes apart as ions of Na^{+1} and OH^{-1}. The OH^{-1} ion can be described as a **polyatomic ion**, a group of atoms that tend to stay together and carry an overall charge.

Table 9.8 lists some common polyatomic ions and their combining capacities. These are helpful in writing ionic formulas. Remember to treat the polyatomic ions as a single ion. If two or more of the polyatomic ions occur in the formula, the formula for the polyatomic ion is enclosed in parentheses. For example, the formula for calcium hydroxide, $Ca(OH)_2$, has one calcium ion and two polyatomic hydroxide ions.

Table 9.8 Combining Capacities of Some Common Polyatomic Ions

Name	Formula	Combining capacity
Ammonium	NH_4^{+1}	+1
Hydrogen carbonate	HCO_3^{-1}	−1
Chlorate	ClO_3^{-1}	−1
Hydroxide	OH^{-1}	−1
Nitrate	NO_3^{-1}	−1
Nitrite	NO_2^{-1}	−1
Carbonate	CO_3^{-2}	−2
Sulphate	SO_4^{-2}	−2
Sulphite	SO_3^{-2}	−2
Phosphate	PO_4^{-3}	−3
Phosphite	PO_3^{-3}	−3

EXAMPLE V

What is the formula for the compound calcium hydroxide?

PREDICT THE FORMULA BEFORE YOU GO ON.

1. Symbols and order are

   Ca OH

2. Combining capacities are

   Ca (+2) OH (−1)

Using the criss-cross method, the result is

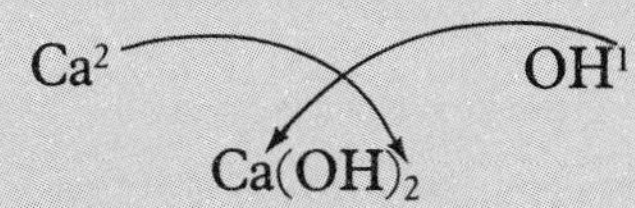

Note that parentheses ( ) are used when there is more than one polyatomic ion.

DID YOU KNOW?

Acid spills in the lab can be safely neutralized by sprinkling solid sodium bicarbonate ($NaHCO_3$) on them. The hydrogen ion (H^{+1}) from the acid combines with the polyatomic bicarbonate ion (HCO_3^{-1}) to form harmless carbon dioxide gas (CO_2) and water (H_2O). This is the same reaction that occurs in your stomach when you take an antacid tablet. Sodium bicarbonate combines with the hydrochloric (HCl) acid in your stomach to form carbon dioxide gas.

NAMING COMPOUNDS WITH POLYATOMIC IONS

When polyatomic ions are part of the formula, follow the same rules as when you name binary compounds. You will come across endings such as -ate, -ite, -ide, and -ium. They are all part of the name of the polyatomic ion, which you do not change.

$NaNO_3$ is sodium nitrate
$FeCO_3$ is iron(II) carbonate
NH_4Cl is ammonium chloride
$Ca(OH)_2$ is calcium hydroxide
$(NH_4)_2SO_4$ is ammonium sulphate

INSTANT PRACTICE

Use the periodic table (Appendix B), Table 9.8, and what you have just learned to write formulas for compounds formed from the following elements and polyatomic ions. Then write the names for the compounds.

1. potassium and bicarbonate
2. magnesium and nitrate
3. zinc and carbonate
4. aluminum and nitrite
5. calcium and carbonate
6. sodium and sulphate
7. lead(IV) and sulphate
8. tin(II) and nitrate
9. ammonium and chlorine
10. ammonium and fluorine

REVIEW 9.3

1. What is an ionic formula?
2. Write compound formulas and then name the compounds for the following:
   (a) sodium combining with bromine
   (b) magnesium combining with iodine
   (c) potassium combining with oxygen
   (d) potassium combining with carbonate
   (e) calcium combining with hydroxide
   (f) ammonium combining with phosphate
3. Name the formulas:
   (a) Al_2O_3
   (b) $FeCl_2$
   (c) $FeCl_3$
   (d) $NaHCO_3$
   (e) NH_4Cl
   (f) $AgNO_3$
4. Write formulas for the following ionic compounds:
   (a) lead(IV) oxide
   (b) iron(II) chloride
   (c) mercury(II) sulphide
   (d) chromium(III) oxide
   (e) copper(II) nitrate
   (f) iron(II) sulphate
5. Write names for the following ionic compound formulas:
   (a) Na_2O
   (b) $CuCl$
   (c) HgO
   (d) FeO
   (e) Al_2O_3
   (f) $CuClO_3$
   (g) $HgSO_4$
6. How would you recognize the formulas of an acid, a base, and a salt?
7. Give the name and formula for the only positive polyatomic ion.
8. Go back to Activity 9E. In the second column of the two-column notes, write to your lab partner telling him or her what you have learned and ask any questions you might have about formulas. If your teacher permits, share your ideas from the first column with one or more students. Compare your ideas or answer some of the questions you have from the first column. You may wish to add or change your ideas; do this in the second column.

9.4 Electron Sharing: The Covalent Bond

Most of the compounds you meet in your daily life are not ionic. Rather, they are neutral groups of atoms called molecules (Figure 9.17). Water consists of molecules made from two hydrogen atoms and one oxygen atom. The formula for one molecule of this compound is H_2O. Most substances consist of much larger molecules. For example, the formula for table sugar is $C_{12}H_{22}O_{11}$. This formula describes the composition of each individual sugar molecule.

You have seen that for ionic bonding to occur, the metal loses electrons and the non-metal gains the electrons. Both metal and non-metal acquire stable outer electron orbits similar to those of noble gases.

Frequently in nature, non-metals combine with other non-metals or even with each other. Let's look at what happens when two hydrogen atoms join to form the H_2 molecule.

If two hydrogen atoms come close enough, each electron is attracted towards the other's nucleus (Figure 9.18). Since both atoms are the same size, the pull is not enough to force either electron out of its own orbit. There will be no electron transfer.

The attraction does have an effect, however. Because of this pull, the two electrons spend most of their time between the two atoms, where the greatest attraction occurs (Figure 9.19). Each hydrogen atom in the pair shares the other's electron. The shared electrons provide the "glue" that holds the atoms together. The glue, or bond, is called a **covalent bond** and is a result of the atoms' sharing electrons. This creates a stable two-electron arrangement (like the outer orbit of helium) between the atoms (Figure 9.20).

Chlorine gas provides another example of electron sharing between two identical atoms. Each chlorine molecule consists of two chlorine atoms held together by a covalent bond. The bond consists of two shared electrons, one from each atom. The effect is like having an outer orbit filled with electrons, even though neither has all the electrons to itself (Figure 9.21).

Figure 9.17
(a) An ionic substance and (b) a molecular substance.

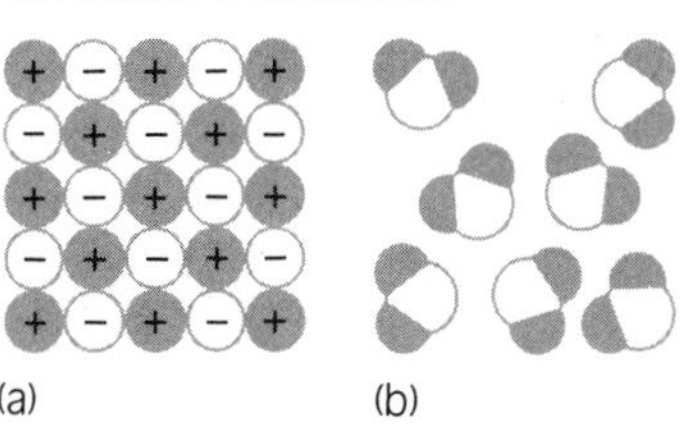

Figure 9.18
As two hydrogen atoms come close together, each of their single electrons is attracted to the positive nucleus of the other hydrogen atom.

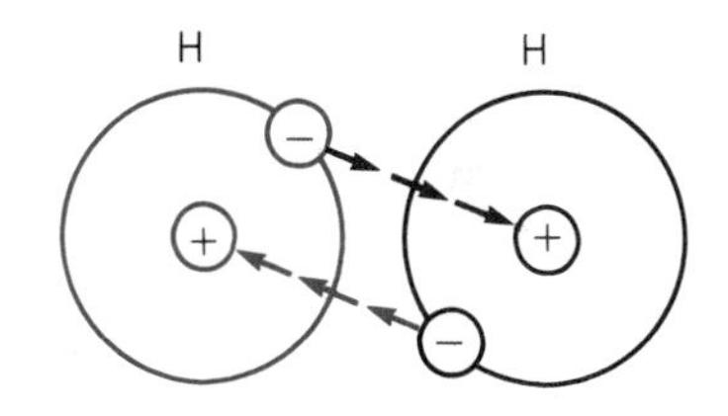

Figure 9.19
A hydrogen (H_2) molecule has been formed by the sharing of electrons.

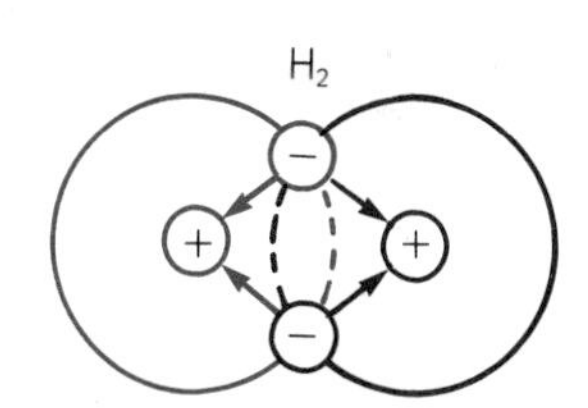

Figure 9.20
Two hydrogen atoms (a) will form a covalent bond to have a stable electron configuration similar to that of a helium atom (b).

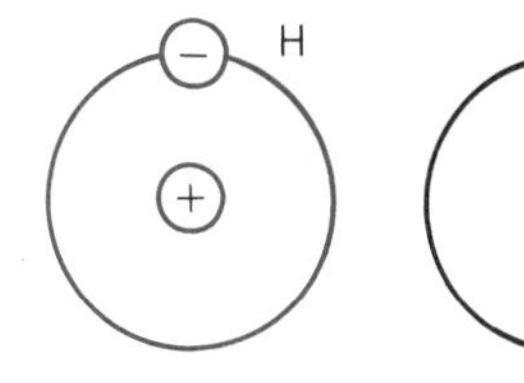
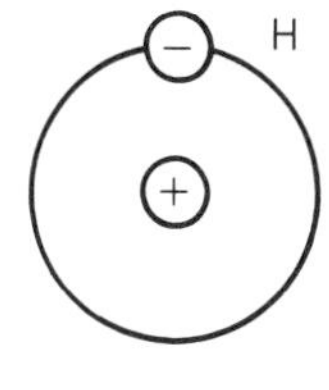

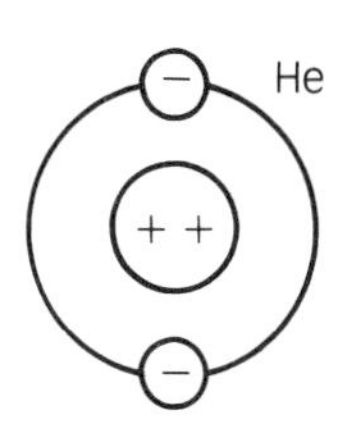

Table 9.9 lists elements that form diatomic molecules, or molecules with two atoms, with covalent bonds. This table includes only molecules made up of two of the same type of atoms. But atoms of different elements can form covalent bonds too.

Figure 9.21
Molecule of chlorine gas (Cl_2), with filled outer orbits of electrons.

Table 9.9 Elements That Form Diatomic Covalent Molecules

Name of element	Symbol for one atom of the element	Formula for one molecule of the element
hydrogen	H	H_2(gas)
nitrogen	N	N_2(gas)
oxygen	O	O_2(gas)
fluorine	F	F_2(gas)
chlorine	Cl	Cl_2(gas)
bromine	Br	Br_2(liquid)
iodine	I	I_2(solid)

COVALENT COMPOUNDS

A **covalent compound** is formed when there is electron sharing between non-metal atoms. Figure 9.22 shows some covalent compounds.

Figure 9.22
Covalent compounds.

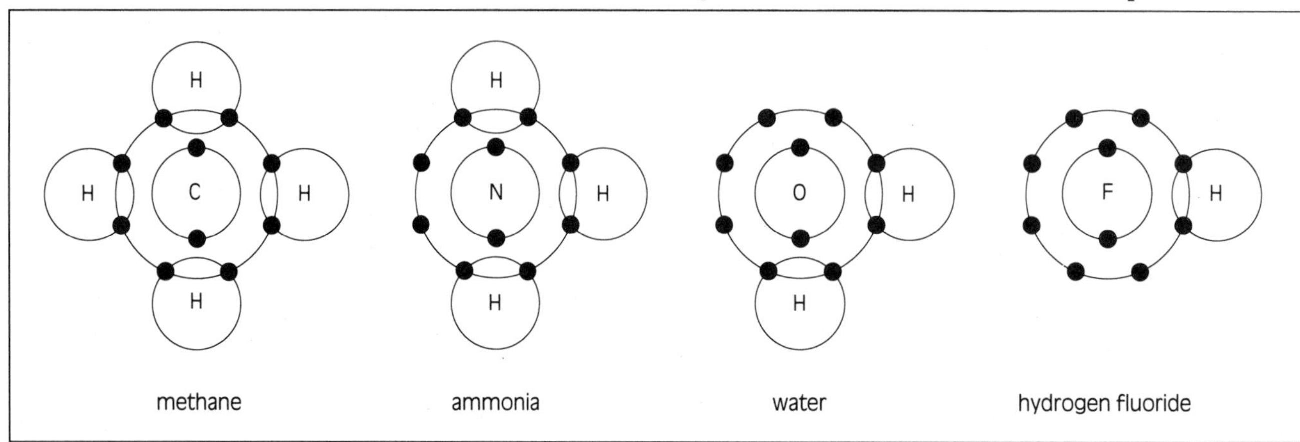

COMPUTER APPLICATION

USING COMPUTERS IN A CHEMICAL LABORATORY

Analytical chemists like Ram Sadana use computers to analyse samples, prepare graphs, and perform calculations.

Today's chemical laboratories are very different from those used by scientists 20 or 30 years ago. New instruments as well as new processes have changed every aspect of analysing—the search for "what" and "how much" has become very sophisticated. The laboratories of the next 5 or 10 years are expected to be even more incredibly advanced, in part as a result of the work of analytical chemists like Ram Sadana.

"Laboratory work contains many steps and many repetitions of steps, all of which must be extremely carefully done if the answers are to be trustworthy," notes Sadana. "In the past, technicians had to spend time ensuring that even the monotonous portions of their work were done to high standards. This was a difficult task and reduced the number of analyses that could be done each day."

One answer for Sadana's lab has been to use microcomputers to "smarten up" the instruments themselves. Older instruments would analyse a prepared sample, but a technician would have to take the reading from the instrument, prepare graphs and do statistics, then find the answer by comparing the results to known standards.

"Today's instruments," Sadana explains, "begin by doing the same task of analysing a prepared sample, but then their computerized portion does the graphs and calculations. The technician obtains a final answer at the push of a button."

EXTENSION

Using Table 9.9, write the formulas and draw covalent diagrams for the combination of oxygen with hydrogen and for the combination of fluorine with hydrogen. Check your formulas and diagrams with your lab partner.

In covalent bonding, a non-metal atom shares electrons with another non-metal until their outer orbits become stable (having an outer orbit similar to that of the nearest noble gas). This means that the earlier definition of combining capacity can be rewritten to include covalent compounds. The combining capacity of an atom tells how many electrons it must give up, gain, or share to have a stable outer orbit. The combining capacity of non-metal atoms can be determined from the periodic table (Appendix B) or Table 9.10.

Table 9.10 Combining Capacities of Non-Metal Atoms

4	3	2	1
C Si	N P As	O S Se Te	F Cl Br I

Figure 9.23
Models of covalent molecules.

Carbon has a combining capacity of 4. With hydrogen, it will form a covalent compound called methane (Figure 9.22). Each hydrogen atom shares its one electron with carbon. The four hydrogen atoms provide carbon with a full outer orbit of eight electrons. Carbon also shares its four outer electrons, giving each one of the hydrogen atoms a full outer orbit of two electrons.

The combining capacity of nitrogen is 3. Nitrogen and hydrogen form a covalent compound called ammonia (NH_3) (Figure 9.22). Figure 9.23 shows a model of covalent molecules.

WRITING FORMULAS FOR COVALENT COMPOUNDS

It is not necessary to draw all the electron orbits to predict formulas. The combining capacities may again be used in a short-cut method. Thus, for a compound of carbon and sulphur, the following method is used:

1. Write both symbols, left-hand element first, with combining capacities, as shown.

   C^4 S^2

2. Criss-cross.

   C^4 S^2

   C_2S_4

3. The first formula can be reduced:

   $C_2S_4 = C_1S_2$

4. The subscript "1" on the carbon is not needed. It can be dropped. The finished formula is

   CS_2

INSTANT PRACTICE

Write formulas for the covalent compounds formed by the following element pairs.

1. nitrogen-hydrogen
2. carbon-chlorine
3. nitrogen-bromine
4. phosphorus-sulphur
5. carbon-oxygen
6. hydrogen-sulphur

NAMING COVALENT COMPOUNDS

Many covalent compounds have simple formula names: H_2S is hydrogen sulphide (the gas given off as eggs rot). SeS_2 is selenium sulphide (found in some dandruff shampoos).

A number of covalent compounds have names that tell nothing about their formulas. Their common names were used for centuries before the formulas were known. For instance, H_2O is called water instead of hydrogen oxide. Other examples are NH_3, ammonia; and CH_4, natural gas or swamp gas.

The names of covalent compounds often include prefixes. Prefixes help us keep track when the same two elements form two (or more) entirely different compounds. For example, under some conditions, carbon and oxygen may combine to form the gas carbon monoxide, a deadly poison. In other cases, carbon and oxygen combine to form carbon dioxide, which is used by plants in photosynthesis.

The meanings of these and other common prefixes are shown below.

MON(O)- (1) as in carbon monoxide (CO)
DI- (2) as in carbon disulphide (CS_2)
TRI- (3) as in sulphur trioxide (SO_3)
TETRA- (4) as in carbon tetrachloride (CCl_4)
PENT(A)- (5) as in phosphorus pentoxide (P_2O_5)

Many of these formulas would have been hard to predict from their combining capacities. The names make it easy to write the formulas.

INSTANT PRACTICE

1. Write names for PBr_3, HF, CO_2, CF_4, and P_2O_3.
2. Write formulas for phosphorus pentasulphide, nitrogen monoxide, carbon tetraiodide, and nitrogen dioxide.

REVIEW 9.4

1. (a) Write formulas for the following compounds: carbon tetrafluoride, nitrogen monoxide, dinitrogen tetroxide, and silicon disulphide.
   (b) Write the names for the following compounds: H_2S, AS_2O_3, CBr_4, OI_2, and N_2O_4.
   (c) Why is it difficult to write correct formulas for the following compounds: chalk, gasoline, lye, alcohol, and rust?
2. (a) In what way are ionic and covalent bonds similar?
   (b) In what ways are they different?
3. What combining capacity would you expect for each of the following elements: bismuth, boron, silicon, tellurium, and astatine?
4. (a) Is Cs_3N the formula of a covalent compound? Is C_3N_4 the formula of a covalent compound? How do you know?
   (b) State a rule for predicting which element combinations will form covalent compounds.

CHAPTER • REVIEW

Key Ideas

- A chemical formula for a compound represents one molecule of a compound. It can be determined if you know the symbols and combining capacities of the elements that make up the compound.
- The name of a chemical compound can be determined from its formula, using the periodic table (Appendix C) and some simple rules for naming compounds.
- An atom has the same number of electrons as protons and is electrically neutral. An ion is a particle that has either gained or lost one or more electrons to achieve a stable outer orbit similar to that of a noble gas.
- If an atom gains electrons, it has a negative charge and tends to be a non-metallic element.
- If an atom loses electrons, it has a positive charge and tends to be a metallic element.
- Atoms are neutral, whereas ions are electrically charged.
- Acids have a hydrogen ion.
- Bases have a hydroxide ion.
- Salts do not have a hydrogen ion or a hydroxide ion.
- Ionic compounds occur because of electron transfer between the atoms. Covalent compounds occur because of electron sharing among the atoms.
- Ionic compounds are formed with negative ions (non-metals) and positive ions (metals). Covalent compounds are formed with non-metals.

Vocabulary

compound
formula
acid
base
salt
indicator
neutralization
ion
ionic bond
ionic compound
combining capacity
polyatomic ion
covalent bond
covalent compound

1. Construct a concept map using as many of the terms in the vocabulary list as possible. You may include other terms as well.
2. Study Figure 9.24. Then use words from the vocabulary list to write a caption for the diagram.

Connections

1. (a) What part of the atom is involved in a chemical bond?
   (b) What part of the atom is involved in an ionic bond?
   (c) What part of the atom is involved in a covalent bond?
   (d) How are ionic and covalent bonding different?
2. What is the relationship between ionic charge and combining capacity?
3. Describe how the ionic compound KI forms ions in solution.

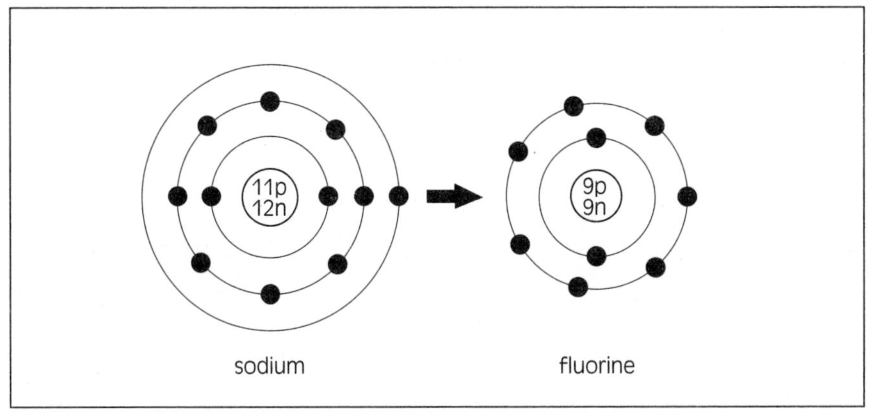

4. (a) How are the covalent compound CCl_4 and the ionic compound $CaCl_2$ different?
   (b) How are the covalent compound CCl_4 and the ionic compound $CaCl_2$ similar?

5. A new element has been discovered, element X. All we know is that it is an alkali metal. What is its formula when it combines with
   (a) iodine?
   (b) sulphur?
   (c) carbonate?

6. A new element has been discovered, element Z. All we know is that it is a halogen. What is its formula when it combines with
   (a) sodium?
   (b) calcium?
   (c) ammonium?

7. You are given an unknown substance. All you know is that it may be an acid, a base, or a salt. Design an experiment that will tell what the unknown substance is.

Explorations

1. Review your two-column notes from Activity 9E. Use these notes to make a flow chart that includes all the rules you can think of.

2. Write a story, with drawings, from the viewpoint of a chlorine atom/ion. Use the following as an outline:
   - In the acid HCl, chlorine is one part of the acid.
   - The acid HCl and the base NaOH neutralize one another and produce Na^{+1} ion and Cl^{-1} ion solution:

     $$HCl + NaOH \rightarrow NaCl + H_2O$$

   - Finally, the water evaporates and the ionic compound (solid) NaCl is left.

   In your story, describe how the various compounds are joined, bonding, and the electron transfer.

Reflections

Review your two-column notes in Activity 9E. What was most helpful in learning about formulas? What did you have the most problems with? What could you do next time when you have to learn something similar to writing formulas? Write your answers to these questions in your learning journal.

CHAPTER

10 Chemical Reactions

A dynamite explosion like the one shown in this picture is a dramatic example of a chemical reaction. But the quiet burning of a piece of paper, the slow rotting of a tree stump, and the simple frying of an egg are also chemical reactions. In fact, we rely on chemical reactions to survive. The clothing you are wearing, the food you had for breakfast, and the desk you are sitting at are the result of chemical reactions.

There are millions of chemical reactions going on around you and in your body. Green plants use energy from the sun to combine water and carbon dioxide, which react to form sugar and oxygen. When you eat these plants, the sugar and oxygen react in your cells to produce water, carbon dioxide, and energy. You need the energy from this reaction for your daily activities. The act of lifting this book, reading it, and storing the memory of it in your brain is a result of many chemical reactions going on in your body.

A chemical reaction can be expressed as a chemical equation. In this chapter, you will find out about different types of chemical equations. You will also learn how to balance equations, how to classify chemical reactions, and how to predict whether or not a chemical reaction will occur when substances are mixed together.

ACTIVITY 10A Thinking about Chemical Reactions

In your learning journal, divide two pages into four columns. Look at Figures 10.1 and 10.2. Chemical reactions are taking place in these photographs. In the first column in your learning journal, list as many things you can think of that are the result of chemical reactions. In the second column, describe the chemical reactions that are taking place in the photographs. In addition, describe where each chemical reaction takes place, materials involved in the chemical reaction, what the final result of the chemical reaction looks like, and anything else you can think of.

When directed by your teacher, get together with one or more students and compare what you have written in columns one and two. You may wish to make changes in these two columns after talking to other students. In column three, add any ideas or questions you or the other students have about chemical reactions. Leave the fourth column empty for now. You will fill it in after you have finished the chapter.

Figure 10.1
What things in this picture are the results of chemical reactions?

Figure 10.2
What chemical reactions are taking place in this picture?

10.1 Chemical Reactions and Equations

In any **chemical reaction**, one or more new substances are formed. A chemical reaction is also known as a **chemical change**. The new substances formed by a chemical reaction or chemical change have properties different from those of the starting materials.

Consider the substances shown in Figure 10.3. Sodium metal (Na) will react violently in water (Figure 10.3a). For this reason, contact of sodium with the skin (which is usually moist) must be avoided. Chlorine (Cl_2) is a pale yellow-green gas that is also very corrosive to animal tissue (Figure 10.3b). In chemical reactions, the starting materials, such as sodium and chlorine, are called the **reactants**. When sodium and chlorine are brought together, they undergo a violent chemical change. As the reaction proceeds, the sodium and chlorine are changed into a new substance with a new formula. The chemical name of this substance is sodium chloride (NaCl), better known as table salt (Figure 10.3). The substance produced by a chemical reaction is known as the **product**. In this case, the product is sodium chloride.

In this example, both sodium and chlorine can cause great bodily harm. But when these two substances react, their harmful properties disappear. They are replaced by the properties of the new substance, common table salt, a chemical our bodies must have to function. When sodium atoms and chlorine atoms react, the atoms are rearranged, forming the compound sodium chloride. It has a new chemical formula and a whole new set of chemical properties.

DID YOU KNOW?

A chemical reaction between powdered cement and water causes concrete to harden a few hours after it is mixed. For years, this reaction continues to make the concrete harder and harder.

(a)

(b)

(c)

Figure 10.3
(a) Sodium (Na) is a metal that is so soft it can be cut with a knife. The white crust was formed by the reaction of sodium with oxygen and moisture in the air. The freshly exposed surface of the metal will tarnish quickly because of this reaction. (b) Chlorine (Cl_2) is a pale yellow-green gas. (c) When sodium reacts with chlorine, sodium chloride crystals (table salt) are produced. How are the properties of sodium chloride different from those of sodium and chlorine? Note: This reaction should not be carried out in the classroom, since the reaction is very vigorous and numerous safety precautions must be taken.

EXTENSION

Matches are a fairly new invention. Matches help us to start fires, an example of a chemical reaction. Find out how matches were developed and how they came to be made in the form that we use today.

PHYSICAL CHANGE

A **physical change** does not form a new substance. Only the state, shape, or external appearance of matter is changed. Figure 10.4 shows the three states of water: gaseous water, liquid water, and solid water. All three of these have the same chemical formula, H_2O. The change from one state to another is a physical change because the substance has not changed and there is no new formula. For example, when ice melts, the substance produced is still water.

Figure 10.5 shows other examples of physical change. The Ping-Pong ball has been dented, but not punctured. When the ball is placed in boiling water for a short time, the heated air inside it expands, forcing the surface of the ball back to its original shape. This change in the ball is a physical change. Heating changes the volume of air in the ball without changing the substances that make up the air or the ball. When substances undergo a physical change, their atoms are not rearranged into new substances.

Examples of other physical changes are carving soapstone in art class, shaping a wooden leg for a coffee table in woodwork, and cutting paper into different geometric shapes in math class.

Figure 10.4
Water exists in three states: gas (steam coming from boiling water), liquid, and solid (ice).

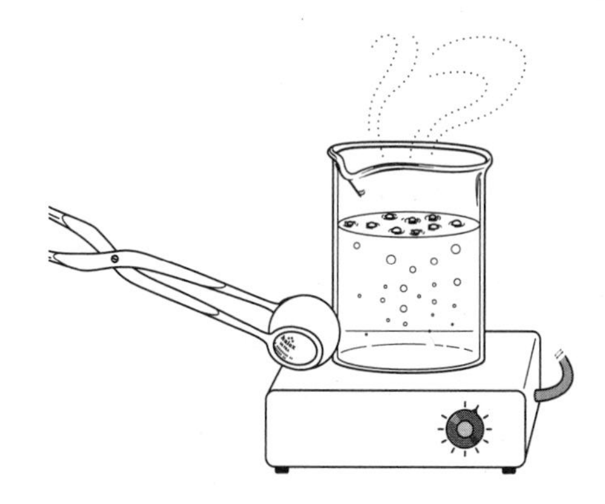

Figure 10.5
A Ping-Pong ball undergoes a physical change. What do you think would happen if the ball were placed in ice instead of boiling water? Would the result be a chemical or a physical change? Why?

ACTIVITY 10B *Chemical Reactions versus Physical Changes*

In this activity, you will observe a number of changes and classify them as chemical reactions or physical changes.

CAUTION!
You will be using a Bunsen burner in this activity because a very hot flame is required. Special care should be taken at all times while you are using the Bunsen burner.

MATERIALS

safety goggles
apron
copper wire (2 cm)
steel wool
tongs
Bunsen burner
wire solder
magnesium ribbon
nichrome wire
ceramic tile
two test tubes (20 mm × 150 mm)
test tube rack
two 100 mL beakers
wooden splint
dilute solution hydrochloric acid (HCl)
small piece of paper
3 per cent hydrogen peroxide (H_2O_2) solution
manganese dioxide (MnO_2)

PROCEDURE

1. Make a table to record your observations. Your table should include the starting materials, their names, and their properties, as well as any observations you can make.
2. Put on safety goggles and apron, to be worn at all times during the activity.
3. Polish the 2 cm piece of copper wire with a small piece of steel wool. Examine the wire carefully and record your observations. Hold the end of the wire with tongs and place the wire in the hottest part of the Bunsen burner flame. Remove the wire after 5 s and note any changes that you observe.
4. Repeat step 2 using pieces of wire solder, magnesium ribbon, and nichrome wire.

CAUTION!
Hold the heated wires in steps 3 and 4 over a ceramic tile. Parts of the wire might melt and burn the table. You should also place the hot wire on the tile to cool.

CAUTION!
Do not look directly at the piece of magnesium when it is in the flame. You may hurt your eyes.

5. Open up a small wad of fine steel wool so that there is space between the strands, and place it on a ceramic tile. Ignite it with a wooden splint. In your table, note any changes that occur.
6. Place a test tube in a test tube rack. Without spilling, carefully pour hydrochloric acid into the test tube to a height of approximately 2 cm. Add three or four small pieces of magnesium (or a 1 cm piece of magnesium ribbon) to the test tube. Note any

changes that occur. Immediately invert another test tube over the mouth of the first test tube (Figure 10.6) to collect the gas that is produced. When the reaction appears to slow down, test the collected gas by bringing a burning wooden splint close to the mouth of the test tube (Figure 10.7). Note that a "pop" or small explosion indicates the presence of hydrogen gas.

CAUTION!
Hydrochloric acid is corrosive to skin, eyes, and clothing. Wash spills and splashes off your skin and clothing immediately, using plenty of water. If you get acid in your eyes, immediately rinse for 15 to 20 minutes and inform your teacher.

CAUTION!
When testing the gas with a burning splint, do not point the test tube towards anyone.

7. Follow your teacher's instructions for discarding the contents of the test tube containing hydrochloric acid.
8. Place a small piece of paper in a beaker. Place a small piece of magnesium ribbon in another beaker. Add enough water to each beaker to cover its contents. Record your observations.
9. Place another test tube in a test tube rack. Pour some 3 per cent hydrogen peroxide solution into the test tube to a depth of about 3 cm. Add a small amount of manganese dioxide (enough to cover the tip of a wooden splint). Note any changes that occur. Put a glowing (not flaming) splint into the mouth of the test tube to test the gas produced. A glowing splint that bursts into flame indicates the presence of oxygen gas.
10. Wash your hands thoroughly after you have completed this activity.

Figure 10.6
An inverted test tube is used to collect gas from the test tube containing hydrochloric acid and magnesium.

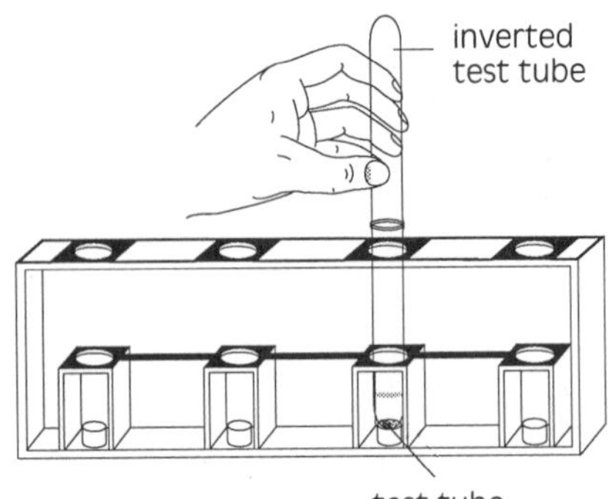

Figure 10.7
Testing the gas from magnesium and hydrochloric acid. Do not point the test tube towards anyone.

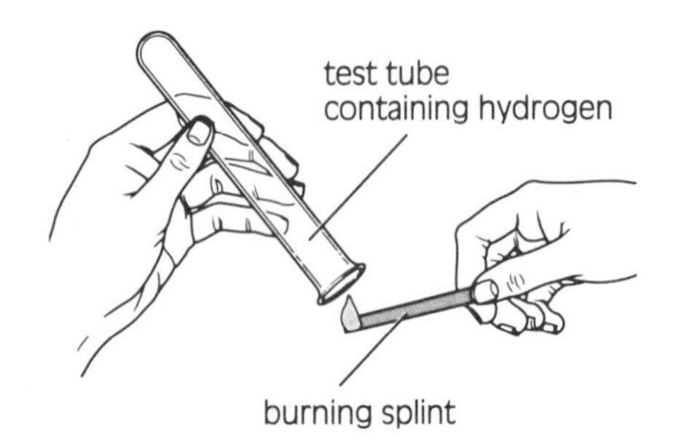

DISCUSSION

1. (a) Which of the "wires" from steps 3 and 4 experienced a chemical change when heated?
   (b) What was the evidence of a chemical change?
   (c) Which of the "wires" experienced a physical change when heated (steps 3 and 4)?
   (d) What was the evidence of a physical change?
2. Name the type of change that took place (chemical change or physical change) in each of steps 5 to 9. Give reasons for your answers.
3. (a) For steps 3 to 9, what substance or substances did you start out with? If you cannot name the substances, describe them.
   (b) For steps 3 to 9, what substance or substances did you end up with at the end of the change? If you cannot name the substances, describe them.
4. (a) Based on the observations in your data table, what is meant by a chemical change?
   (b) Based on the observations in your data table, what is meant by a physical change?
5. (a) Name three chemical reactions that occur in your home.
   (b) Select one of the reactions and explain how you know it is a chemical reaction.
   (c) Select one of the reactions and name or describe the substance or substances you started with and the substances you ended up with.
6. (a) Name three physical changes that occur in your home.
   (b) Select one of the changes and explain how you know it is a physical change.

CHEMICAL EQUATIONS

The easiest way to describe a chemical reaction is with an equation. In a sense, it provides a "before" and "after" picture of what happens. For example:

zinc + sulphur → zinc sulphide
(before)   (after)

The above statement is an example of a word equation. It describes the chemical reaction shown in Figure 10.8. In this reaction, zinc (Zn) combines with sulphur (S) to form zinc sulphide (ZnS), the substance used on the inner surface of TV screens. This word equation is read as follows: "zinc and sulphur react to produce zinc sulphide."

A word equation tells which substances react with each other and what new products are formed. It does not give any other information, however. For example, it does not give the amount of the reactants or the mass of the product. A **chemical equation** uses symbols and formulas to give all the essential information about a chemical reaction. Table 10.1 lists the symbols that may be used in chemical equations and what they mean.

Figure 10.8
Once ignited, a mixture of zinc (Zn) and sulphur (S) reacts rapidly to form zinc sulphide (ZnS). Note: This reaction should not be carried out in the classroom, since the reaction is very vigorous and numerous safety precautions must be taken.

Table 10.1 Meaning of Symbols

Symbol	Meaning
(*g*)	Means gas; the substance is in a gaseous state. For example: $O_2(g)$.
(*l*)	Means liquid; the substance is in a liquid state. For example: $H_2O(l)$.
(*s*)	Means solid; the substance is in a solid state. For example: $C(s)$.
(*aq*)	Means aqueous; the substance is dissolved in water. For example: $NaCl(aq)$.
$2H_2O$	The number used in front of the symbols and formulas helps balance equations (so that an equal number of each type of atom is on each side of the equation). This number is called a **coefficient**. The coefficient 2 used here means 2 molecules of water.
H_2O	The subscripts used just after the symbols indicate the number of atoms present for each element. There are 2 atoms of hydrogen and 1 atom of oxygen; the 1 is understood.
+	Means "plus" or "and." It is placed between the reactants. For example: $H_2(g) + O_2(g)$. It can also be placed between products that are formed.
→	Means "yields" or "produces." It separates the reactants (on the left) and products (on the right). For example: $H_2(g) + O_2(g) \rightarrow 2H_2O(l)$

REVIEW 10.1

1. For each of the following, tell whether it is a physical or a chemical change. Give reasons for your answer.
   (a) boiling water in a kettle
   (b) burning the asparagus while making supper
   (c) striking a match against sandpaper until it ignites
   (d) hanging wet towels on a rack until they dry
   (e) stroking a nail with a magnet until the nail becomes magnetic
   (f) dipping a silver spoon in egg until the spoon turns black
   (g) placing leaves in a compost box; by spring, a crumbly brown material replaces the leaves
2. Pure hydrogen gas can be burned as a fuel, and oxygen gas supports burning. Yet when hydrogen and oxygen are combined in a chemical reaction, water, which can put out fires, is produced. Explain the change in properties.
3. Identify the reactants and products in the following chemical reactions, using their formulas (or names).
   (a) $HgO(s) \rightarrow Hg(l) + O_2(g)$
   (b) $KClO_3(s) \rightarrow KCl(s) + O_2(g)$
   (c) $Pb(NO_3)_2(aq) + KI(aq) \rightarrow PbI_2(s) + KNO_3(aq)$
   (d) carbon + oxygen → carbon dioxide + water + energy
4. What is a chemical equation?

DID YOU KNOW?

The law of conservation of mass does not apply to the nuclear reactions that occur in the interior of the sun, in atomic bombs, and in nuclear reactors. In nuclear reactors, there is some loss of mass: the mass is changed into energy. The idea that mass and energy are interchangeable was first suggested by Albert Einstein.

Figure 10.9
Magnesium burning in air, with a very bright flame, to form magnesium oxide.

10.2 Balancing Chemical Equations

Bakers must know how much of each ingredient to put into a cake, and painters must know how much of each type of paint to mix to get a desired colour. In the same way, chemists must know the amounts of reactants and products in a chemical reaction. Whether a chemist is planning an experiment to be done in the lab or a chemical process to be used in industry, knowing the amounts of these substances will save time, effort, energy, and money. For chemical equations to be useful, they must contain information about the substances present and the amounts or proportions of these substances.

The **law of the conservation of mass** states that "the total mass of the reactants in a chemical reaction is equal to the total mass of the products." In a chemical reaction, atoms are neither created nor destroyed. They are only rearranged. Thus, there must be equal numbers of atoms on each side of an equation. An equation that has the same number of atoms of each type on each side of the arrow is a **balanced equation.**

Magnesium ribbon burns brightly in air (Figure 10.9). The oxygen (O_2) from the air combines with the magnesium (Mg) to produce the compound magnesium oxide (MgO), a white powder. This chemical reaction is represented by the following equation:

$$Mg(s) + O_2(g) \rightarrow MgO(s)$$

This equation is not balanced. The reactants include two oxygen atoms, but the product includes only one oxygen atom (Figure 10.10a).

To balance the oxygen atoms, you first place the coefficient 2 in front of the product, MgO. This number applies to the whole formula that follows it. The equation now reads

$$Mg(s) + O_2(g) \rightarrow 2MgO(s)$$

The oxygen atoms are balanced. But now there is one magnesium atom on the reactant side of the equation and two magnesium atoms on the product side of the equation. To balance the magnesium atoms, a coefficient of 2 must be placed in front of Mg:

$$2Mg(s) + O_2(g) \rightarrow 2MgO(s)$$

This equation is now balanced. There are two oxygen atoms and two magnesium atoms before and after the reaction has taken place (Figure 10.10b). Figure 10.11 shows another example of balancing an equation.

Balancing an equation is a trial-and-error process that requires patience and persistence. It can be viewed as a three-step process:

1. Write the symbols and formulas to represent reactants and products. After you have the correct symbols and formulas, the subscripts must not be changed when you balance the equation.
2. Place coefficients in front of the formulas as needed to make the number of atoms on each side of the equation equal.
3. Check each type of atom to see if there is the same number of that atom on each side of the equation.

EXAMPLE 1

Write and balance the equation for the reaction of hydrogen and oxygen to produce water.

TRY TO WRITE AND THEN BALANCE THE EQUATION BEFORE YOU GO ON.

1. Write the correct formulas for all reactants and products. Remember that hydrogen and oxygen form covalent bonds with themselves, so their formulas are H_2 and O_2.

   $$H_2(g) + O_2(g) \rightarrow H_2O(l)$$

2. Check each type of atom to see if there is an equal number of atoms on each side of the equation. In this example, oxygen atoms on each side of the equation are not equal. Place the coefficient 2 in front of water:

   $$H_2(g) + O_2(g) \rightarrow 2H_2O(l)$$

3. Again check each type of atom present to see if there is an equal number of atoms on each side of the equation. Hydrogen atoms are not equal on each side of the equation. Place the coefficient 2 in front of hydrogen.

   $$2H_2(g) + O_2(g) \rightarrow 2H_2O(l)$$

This equation is balanced; there are four hydrogen atoms and two oxygen atoms on each side of the equation.

Figure 10.10
Balancing equations.

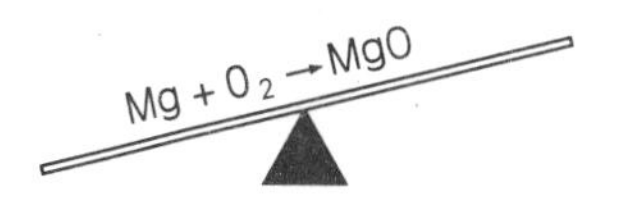

(a) Unbalanced Equation

Mass of the reactants is more than the mass of the products.

Reactants	Products
1 Mg atom =	1 Mg atom
2 O atoms ≠	1 O atom

2Mg + O_2 → 2MgO

(b) Balanced Equation

Mass of the reactants is equal to the mass of the products.

Reactants	Products
2 Mg atoms =	2 Mg atoms
2 O atoms =	2 O atoms

Figure 10.11
A coefficient in a balanced equation applies to the whole formula following it.

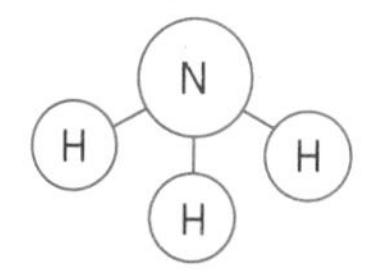

NH_3 means one molecule of ammonia consisting of 1 N atom and 3 H atoms.

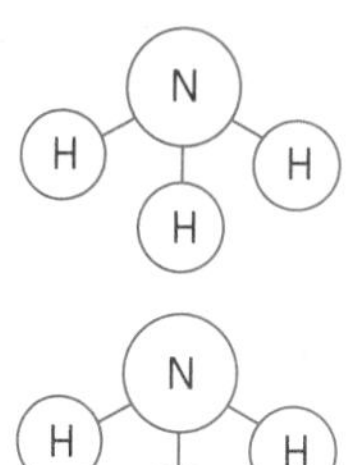

2 NH_3 means two molecules of ammonia consisting of 2 N atoms and 6 H atoms.

INSTANT PRACTICE

Copy the following equations into your notebook. Then balance the equations.

1. $Na(s) + Cl_2(g) \rightarrow NaCl(s)$
2. $K(s) + O_2(g) \rightarrow K_2O(s)$
3. $H_2(g) + Cl_2(g) \rightarrow HCl(l)$
4. $N_2(g) + H_2(g) \rightarrow NH_3(l)$
5. $CO(g) + O_2(g) \rightarrow CO_2(g)$
6. $Al(s) + Br_2(g) \rightarrow AlBr_3(s)$
7. $N_2H_4(g) + O_2(g) \rightarrow H_2O(l) + N_2(g)$
8. $CH_3OH(l) + O_2(g) \rightarrow CO_2(g) + H_2O(l)$
9. $CH_4(g) + O_2(g) \rightarrow CO_2(g) + H_2O(l)$

REVIEW 10.2

1. (a) What does balancing an equation mean?
   (b) What are the steps in balancing an equation?
2. (a) Why is the following equation not balanced?

   $$H_2(g) + O_2(g) \rightarrow H_2O(l)$$

   (b) Following is an attempt to balance the above equation. What is wrong with the way the equation is balanced?

   $$H_2(g) + O_2(g) \rightarrow H_2O_2(l)$$

3. Balance the following equations:
   (a) $K(s) + Cl_2(g) \rightarrow KCl(s)$
   (b) $K(s) + O_2(g) \rightarrow K_2O(s)$
   (c) $Na(s) + O_2(g) \rightarrow Na_2O(s)$
   (d) $Sr(s) + N_2(g) \rightarrow Sr_3N_2(s)$
   (e) $Na(s) + N_2(g) \rightarrow Na_3N(s)$
   (f) $Ca(s) + H_2O(l) \rightarrow Ca(OH)_2(s) + H_2(g)$
   (g) $CuSO_4(s) + Fe(s) \rightarrow Fe_2(SO_4)_3(s) + Cu(s)$
   (h) $Br_2(g) + KI(s) \rightarrow KBr(s) + I_2(g)$
4. For each of the following, write the correct formula equation and balance it.
   (a) copper(II) oxide + hydrogen → copper + water
   (b) lead(II) nitrate + potassium → lead + potassium nitrate
   (c) calcium + water → calcium hydroxide + hydrogen
   (d) lead(II) sulphide + oxygen → lead + sulphur dioxide
   (e) hydrogen sulphide → hydrogen + sulphur
5. (a) What is the law of conservation of mass?
   (b) Why is it not possible to start with 8 kg of reactant in a closed container and end up with 9 kg of product?
   (c) If you wish to make H_2O, you will need O_2 and H_2. Will you need the same amount of each? Explain.

10.3 Classifying Chemical Reactions

When a chemical reaction occurs, bonds are broken or formed or both broken and formed. The breaking and the formation of bonds involves energy. Thus, one method of classifying chemical reactions is to group them according to the energy they absorb or give off. A second method groups them according to how the atoms are rearranged.

CHEMICAL REACTIONS AND ENERGY

Some reactions produce energy. In these reactions, energy is usually given off in the form of heat or light. Such reactions are called **exo-**

thermic reactions. The prefix "exo" means outside, and "therm" means heat. The burning of a candle is an example of an exothermic chemical reaction. Once the candle is lighted, heat and light are given off. Usually an increase in temperature is noticed in exothermic reactions. The photograph at the beginning of this chapter provides a spectacular example of an exothermic reaction.

Exothermic reactions are important to our everyday life. For example, the burning of methane (CH_4), the major component of natural gas, heats many homes, helps to cook food, and provides hot water. The equation for the burning of methane is

$$CH_4(g) + 2O_2(g) \rightarrow CO_2(g) + H_2O(g) + \text{energy}$$

To show that this reaction is exothermic, the release of energy is given along with the products of the reaction in the equation. In fact, methane is a good fuel because it produces more heat per gram than other fuels. It also releases far less carbon dioxide per joule of heat than fuels such as gasoline, heating oil, or coal.

Some reactions require the constant input of energy. The energy required, usually heat or light, is absorbed from the surroundings. Such reactions are called **endothermic reactions.** The prefix "endo" means inside, and again, "therm" means heat. Cold packs used in first aid treatment (Figure 10.12) for injuries use an endothermic reaction to produce the cooling effect. The pack contains water and ammonium nitrate separated by a barrier. When the barrier is broken, the water and ammonium nitrate react, absorbing the heat energy from the injured body part.

Figure 10.12
An athletic cold pack is based on an endothermic reaction.

The breakdown of limestone ($CaCO_3$) is another example of an endothermic reaction. The heat for an endothermic reaction is written on the left side of the equation. This indicates that heat must be supplied with the reactants to produce the products. The equation for the breakdown of limestone is

$$CaCO_3(s) + \text{energy} \rightarrow CaO(s) + CO_2(g)$$

The heat is much like a reactant—without it, the reaction cannot take place. If the supply of heat is removed, the reaction stops. If an endothermic reaction takes place at room temperature, the reaction absorbs heat from the surroundings and the temperature of the surroundings drops.

DID YOU KNOW?

The bombardier beetle uses a "rocket gun" to protect itself. When threatened, the beetle mixes substances called hydroquinones and hydrogen peroxide in a reaction chamber within its body. The violent chemical reaction is an exothermic reaction, which produces smelly benzoquinones that are ejected towards the enemy at 100°C through a flexible rotating turret.

CLASSIFYING CHEMICAL REACTIONS ACCORDING TO ATOM REARRANGEMENT

Chemical reactions may also be classified according to the way the atoms in the reacting substances are rearranged. This section describes four types of chemical reactions. Knowing what type a reaction is enables you to predict the outcome of the reaction.

Figure 10.13
Synthesis of water.

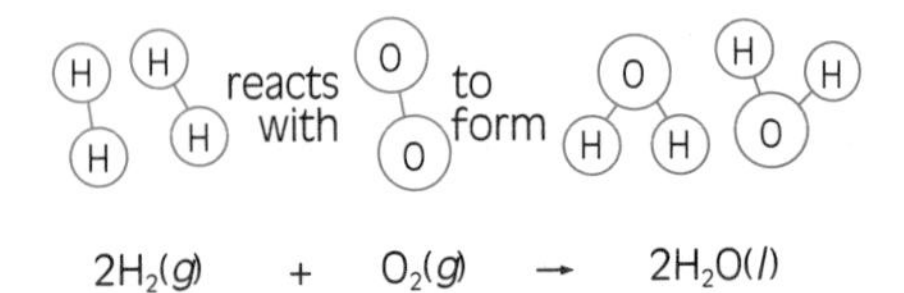

In **synthesis** reactions, the atoms of two or more substances are rearranged to form a compound. The formation of water is a synthesis reaction:

$$2H_2(g) + O_2(g) \rightarrow 2H_2O(l)$$

This reaction is shown in Figure 10.13. The gaseous elements hydrogen (H_2) and oxygen (O_2) combine to form liquid water.

The general form of a synthesis reaction is

$$A + B \rightarrow AB$$

The substances A and B may be elements or compounds, which combine to form the compound AB.

The synthesis reactions shown below take place during the formation of acid rain:

$S(s) + O_2(g) \rightarrow SO_2(g)$ two elements combine
$2SO_2(g) + O_2(g) \rightarrow 2SO_3(g)$ a compound and an element combine
$SO_3(g) + H_2O(l) \rightarrow H_2SO_4(l)$ two compounds combine

EXTENSION

Do a library project on hydrogen as a future source of fuel. Include the benefits to society, the cost, the disadvantages, and the technology needed to use hydrogen as a source of fuel.

Decomposition reactions occur when the atoms of one substance break down and are rearranged to form two or more simpler substances. Decomposition is the opposite of synthesis. The breakdown of water is a decomposition reaction:

$$2H_2O(l) \rightarrow 2H_2(g) + O_2(g)$$

This reaction is shown in Figure 10.14. By breaking down the compound water, you get the elements oxygen and hydrogen. This reaction is the opposite of the one in Figure 10.13.

Decomposition reactions have the general form

$$AB \rightarrow A + B$$

AB is the compound that is broken down, and the substances formed are A and B, which can be either elements or compounds.

Figure 10.14
Decomposition of water.

forms and

$2H_2O(l) \rightarrow 2H_2(g) + O_2(g)$

ACTIVITY 10C Synthesis and Decomposition Reactions

In this activity, you will observe both synthesis and decomposition reactions and classify them as endothermic or exothermic.

CAUTION!
You will be using a Bunsen burner in this activity because a very hot flame is required. Special care should be taken at all times while you are using the Bunsen burner.

MATERIALS

PART I

- safety goggles
- apron
- Bunsen burner
- iron wire (10 cm length)
- steel wool
- tongs
- wire gauze

PART II

- safety goggles
- apron
- magnesium carbonate ($MgCO_3$)
- lime water
- two test tubes (20 mm × 150 mm)
- ring stand
- one-hole rubber stopper
- glass tubing
- rubber tubing
- utility clamp

PROCEDURE

PART I

In this part, you will observe synthesis reactions.

1. Construct a table that includes the headings "Procedures," "Observations," "Reactants," "Products," and anything else you think should be recorded.
2. Put on safety goggles and apron, to be worn at all times during the activity.
3. Rub the wire vigorously with a piece of steel wool. Observe the wire carefully and record your observations in your table.
4. Hold the wire with tongs, and heat one end of the wire in the hottest part of the Bunsen burner flame for several minutes (Figure 10.15). Record your observations in your table.
5. Let the wire cool on the wire gauze pad. When it is cool, observe and compare both ends of the wire. Record your observations in your table.
6. Bend the wire at the end that was heated and record any changes in your table.

DISCUSSION

1. (a) What evidence was there that a chemical change took place?
   (b) What were the reactants and the products?
   (c) Write a balanced equation for this reaction, assuming that the iron atoms formed Fe^{+3} ions.
   (d) Why is this reaction a synthesis reaction?
2. Was the reaction endothermic or exothermic? Why?
3. (a) Name a product of a synthesis reaction, giving its chemical name.
   (b) What reactants are needed for this synthesis reaction?
4. Synthesis reactions happen in your body. What do you think some examples might be?

PART II

In this part, you will observe decomposition reactions.

1. Assemble the equipment as shown in Figure 10.16.
2. Add enough magnesium carbonate to the test tube to fill it about one-third full. Make certain that the magnesium carbonate lies along the side of the large test tube. This can be done by having the test tube at an angle (Figure 10.16). The magnesium carbonate should not remain at the bottom.
3. Gently heat the test tube by using the coolest part of the Bunsen burner flame (4 cm above the inner blue cone of the flame). Make sure to distribute the heat evenly underneath the magnesium carbonate by moving the flame along the sides of the test tube.
4. Continue heating for several minutes, constantly observing the contents of both test tubes. Record your observations in the table you made in Part I.
5. Remove the rubber tubing from the test tube containing the lime water. Then turn off the Bunsen burner. When the test tubes are cool, observe the contents of both tubes. Record your results in your table. →

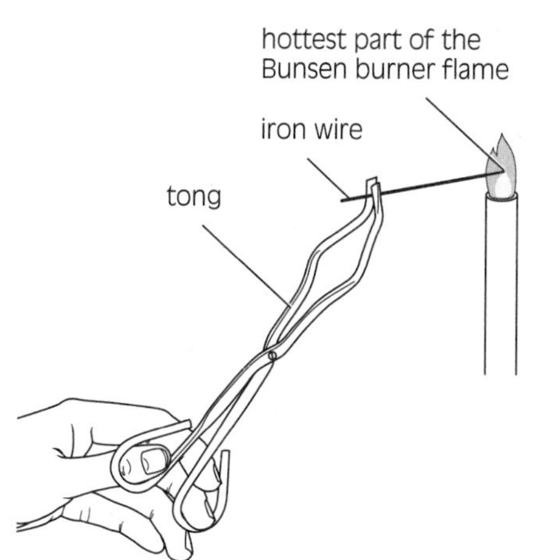

Figure 10.15
Heating iron wire in a Bunsen burner flame.

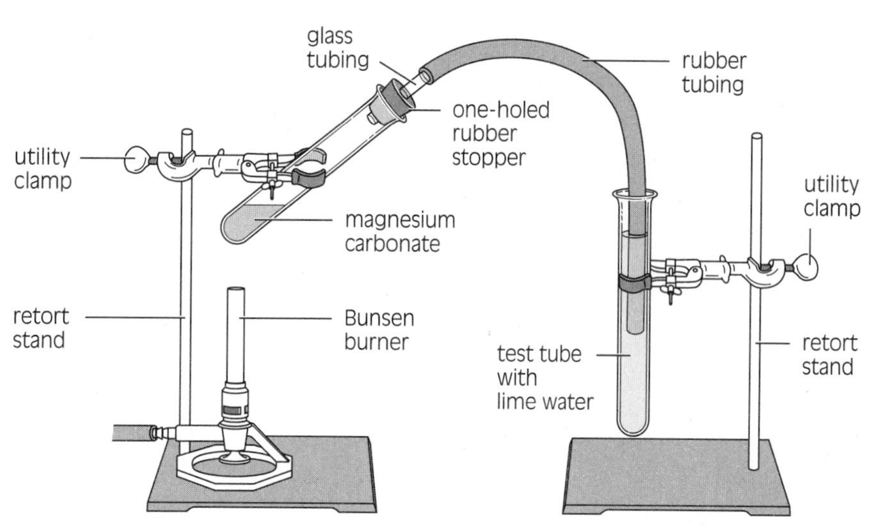

Figure 10.16
Equipment for the decomposition of magnesium carbonate.

DISCUSSION

1. (a) What evidence was there that magnesium carbonate took part in a chemical change?
   (b) What were the reactants and the products?
   (c) Write a balanced equation for this reaction.
   (d) How do you know this reaction is a decomposition reaction?
2. Was the reaction magnesium carbonate went through endothermic or exothermic? Why?
3. What colour did the lime water turn? Note: Lime water turns a milky colour in the presence of carbon dioxide. What conclusion can you draw from this?
4. (a) Name a reactant of a decomposition reaction, giving its chemical name.
   (b) What products do you think are produced for the decomposition reaction in 4(a)?
5. Decomposition reactions happen in your body. What do you think some examples might be?
6. Figure 10.17 shows the decomposition of water using a Hoffman apparatus.
   (a) Where is the reaction taking place?
   (b) What is the energy source for the reaction?
   (c) What is the reactant?
   (d) What are the products?

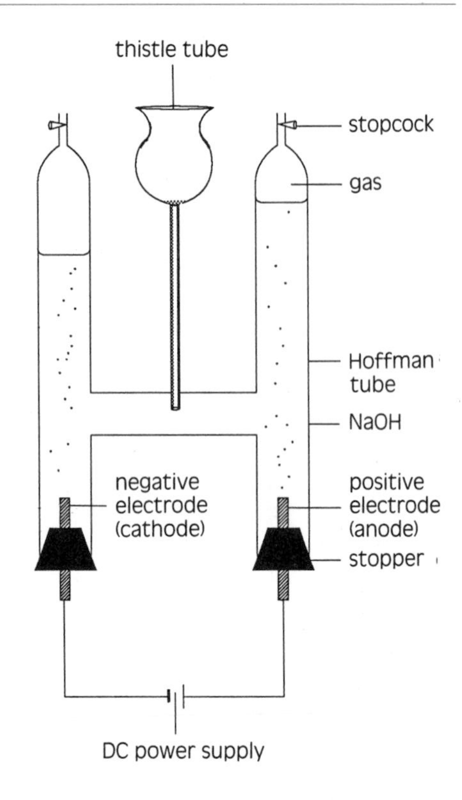

Figure 10.17
Decomposition of water using the Hoffman apparatus.

REPLACEMENT REACTIONS

In the next two kinds of chemical reactions, elements are replaced within compounds. For example, bromine (Br) replaces chlorine (Cl) in the compound NaCl to produce NaBr:

$$Br_2(g) + 2NaCl(aq) \rightarrow 2NaBr(aq) + Cl_2(g)$$

Single replacement reactions occur when a combined element in a compound is replaced by another, uncombined element. A single replacement reaction has the general form

$$Z + AB \rightarrow AZ + B$$

The uncombined element Z replaces the element B in the compound AB. The resulting compound is AZ. Iron (Fe) is the combined element in the compound iron(III) oxide (Fe_2O_3). Iron is replaced by the uncombined element aluminum (Al), as shown in Figure 10.18a.
The chemical equation for this reaction is

$$2Al(s) + Fe_2O_3(s) \rightarrow 2Fe(l) + Al_2O_3(s)$$

This is the most strongly exothermic of all chemical reactions. It is used by welders to melt metal parts that are to be fused together.

In Figure 10.18b, the replacement of iron by carbon in iron(III) oxide forms pure iron and carbon dioxide. This is the same reaction in which pure iron is prepared from iron ore—impure iron(III) oxide—by react-

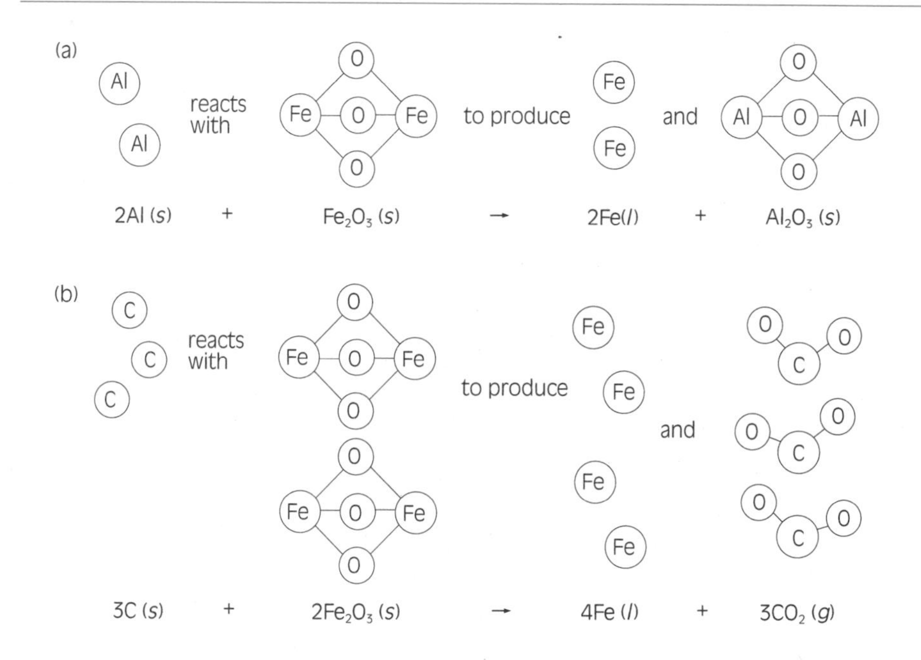

Figure 10.18
Single replacement. (a) Aluminum replacing iron. (b) Carbon replacing iron.

ing the ore with carbon in a blast furnace. The equation for the reaction is

$$3C(s) + 2Fe_2O_3(s) \rightarrow 4Fe(l) + 3CO_2(g)$$

Double replacement reactions occur when the elements in different compounds replace each other or exchange places. For example, when solutions of potassium iodide (KI) and lead(II) nitrate ($Pb(NO_3)_2$) are mixed, a solid compound, lead(II) iodide (PbI_2), forms in the solution (Figure 10.19).

The balanced equation for the reaction is

$$2KI(aq) + Pb(NO_3)_2(aq) \rightarrow PbI_2(s) + 2KNO_3(aq)$$

Figure 10.19
Potassium iodide (KI) solution and lead nitrate ($Pb(NO_3)_2$) solution react to form an insoluble yellow solid, lead iodide (PbI_2).

Both potassium iodide and lead nitrate are ionic solids. When potassium iodide is dissolved, it separates into K^{+1} ions and I^{-1} ions. When lead nitrate is dissolved, it separates into Pb^{+2} ions and NO_3^{-1} ions. When the two solutions are combined, Pb^{+2} ions from the lead nitrate solution and the I^{-1} ions from the potassium iodide solution form the solid lead iodide (PbI_2). Lead iodide will not dissolve in water, so it remains a solid. This solid can be separated from the solution by filtration. But the K^{+1} ions and NO_3^{-1} ions will remain in solution as long as water is present. That is because the compound potassium nitrate is highly soluble in water. But when the water is evaporated, solid potassium nitrate will remain.

The general form for a double replacement reaction is

$$AB + XY \rightarrow AY + XB$$

The elements *B* and *Y* exchange places, or we could also say that the elements *A* and *X* exchange places. Figure 10.20 shows another example of a double replacement reaction.

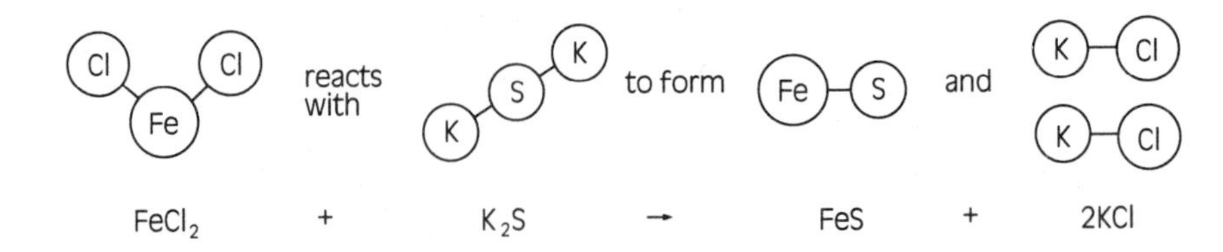

Figure 10.20
Double replacement. In the reaction of iron(II) chloride ($FeCl_2$) and potassium sulphide (K_2S), the elements chlorine (Cl) and sulphur (S) exchange places, or replace each other.

EXTENSION

Soft drinks contain a dilute solution of weak carbonic acid. Warm a small amount of a soft drink. What evidence is there of decomposition? What test could you perform to identify the product of decomposition? Write an equation to describe the decomposition of carbonic acid (see Chapter 9 for the formula for carbonic acids).

Neutralization is a type of double replacement reaction in which an acid and a base exchange ions. The result of the reaction is the formation of a salt and water. The general form for this reaction is

(positive ion) OH^{-1} + H^{+1} (negative ion) → HOH or (H_2O) + salt (positive ion joined to negative ion)

Here, the formula for water is written as HOH to show the hydrogen and hydroxide ions.

The reaction can also be expressed as follows:

$$AOH + HB \rightarrow HOH + AB$$

Can you identify the acid, the base, and the salt in this form? How?

The hydrogen ion from the acid combines with the hydroxide ion from the base to form water. The negative ion from the acid combines with the positive ion from the base to form a salt. For example, the reaction of the base sodium hydroxide (NaOH) with hydrochloric acid (HCl) involves an exchange of ions forming water (HOH) and the salt sodium chloride (NaCl):

$$NaOH(aq) + HCl(aq) \rightarrow HOH(l) + NaCl(aq)$$

ACTIVITY 10D Performing and Observing Replacement Reactions

This activity will provide you with the opportunity to observe two types of replacement reactions. Keep accurate observations because you will be asked to predict what the products of each reaction are.

MATERIALS

PART I

safety goggles
apron
dilute solution hydrochloric acid (HCl)
two test tubes (20 mm × 150 mm)
magnesium ribbon
wooden splint
evaporating dish
drying oven or incandescent lamp

PART II

safety goggles
apron
six test tubes (20 mm × 150 mm)
test tube rack
dilute solutions of:
- iron(III) chloride ($FeCl_3$)
- sodium hydroxide (NaOH)
- sodium chloride (NaCl)
- potassium nitrate(KNO_2)
- sodium carbonate ($NaCO_3$)
- calcium chloride ($CaCl_2$)

PROCEDURE

PART I

In this part, you will observe a single replacement reaction.

1. Copy Table 10.2 into your notebook, using half a page.
2. Put on safety goggles and apron, to be worn at all times during the activity.

CAUTION!
Hydrochloric acid is corrosive to skin, eyes, and clothing. Wash spills and splashes off your skin and clothing immediately, using plenty of water. If you get acid in your eyes, immediately rinse them for 15 to 20 minutes and inform your teacher.

Table 10.2 Sample Data Table for Activity 10D (Part I)

	Name	Formula	Properties
Reactant 1			
Reactant 2			
Products (If you do not know the name or formula, describe the properties.)			

SAMPLE ONLY

3. Add dilute hydrochloric acid to a test tube to a height of approximately 4 cm.
4. Drop two or three pieces of magnesium into the test tube.
5. After observing the reaction, test the gas produced with a burning splint and record your observations.
6. Allow the reaction to proceed until there are no noticeable changes. Empty the test tube into an evaporation dish.
7. Place the dish in a drying oven or under a lamp for 24 h. After all the liquid is gone, record your observations.

DISCUSSION

1. (a) What evidence was there that a chemical change took place?
   (b) What type of chemical change was it? Explain.
2. (a) What were the reactants?
   (b) What were the products?
3. Write a balanced chemical equation for the reaction.

PART II

In this part, you will observe double replacement reactions.

1. Label six test tubes 1 through 6 and place them in order in the test tube rack.
2. Add about 1 mL of each solution indicated below to the proper test tube.
   Test tube 1: iron(III) chloride
   Test tube 2: sodium hydroxide
   Test tube 3: sodium chloride
   Test tube 4: potassium nitrate
   Test tube 5: sodium carbonate
   Test tube 6: calcium chloride

CAUTION!
Most of these solutions are irritants or are poisonous or corrosive. Wash any spills and splashes immediately with plenty of cold water. If you get any solution in your eyes, immediately rinse them for 15 to 20 minutes and inform your teacher.

3. Copy Table 10.3 into your notebook, using half a page.

Table 10.3 Sample Data Table for Activity 10D (Part II)

	Tubes 1 and 2	Tubes 3 and 4	Tubes 5 and 6
Reactant 1			
Reactant 2			
Products (If you do not know the name or formula, describe the properties.)			

SAMPLE ONLY

4. Mix the contents of tubes 1 and 2 by carefully pouring the contents of one tube into another.
5. Mix tubes 3 and 4 together. Then mix tubes 5 and 6 together.
6. Observe each of the mixtures and record any evidence of a reaction in the products row.

DISCUSSION

1. (a) What evidence was there that a chemical change took place in each mixture?
2. (a) What were the reactants for each chemical change?
   (b) What were the products for each chemical change?
   (c) Write a balanced chemical equation for each reaction.

CAREER

CLINICAL PHARMACIST

When a doctor in a hospital prescribes a drug for a patient, the prescription is reviewed by a clinical pharmacist, an expert in the chemistry of medicine. Brenda Wang is a clinical pharmacist in a hospital that specializes in treating cancer patients.

What do you do in your work?
I look for potential problems with a prescription that the prescribing doctor may be unaware of. There could be dangerous interactions among different drugs given to one patient; a patient could have an allergy to a certain kind of drug; or there could be a problem with the dose or frequency of the use of the drug. It is very difficult for physicians to keep up with all the different aspects of pharmaceuticals as well as their own specialty. That is my role in the hospital—to provide them with the expertise.

What other tasks do you perform?
I review the medications given to a particular patient each week. I check that the proper dose has been given in the correct manner. Sometimes I need to suggest a different way of administering a drug so that the patient's body can accept and use the drug better.

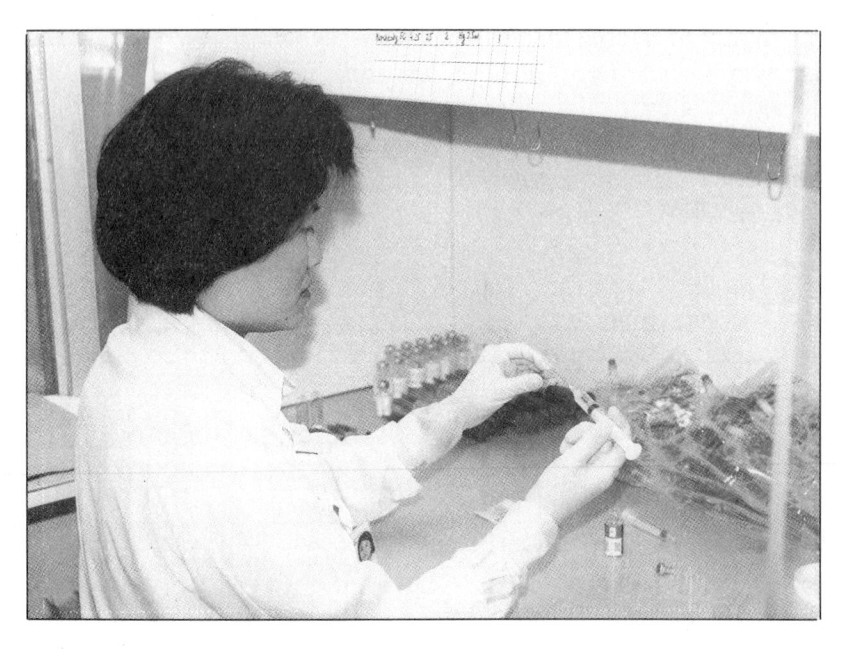

Brenda Wang prepares a dose of medication for a cancer patient.

How did you become a clinical pharmacist?
I studied at a university faculty of pharmacy for four years and then took a nine-month apprenticeship. I chose this field because I was interested in science and math, and I was always good at memorizing information. There is a lot of that in pharmacy courses.

How did you begin your career?
At first, I worked as a retail pharmacist, dispensing drugs over the counter in a drugstore. But I found being a retail pharmacist very monotonous. There was very little opportunity to follow a patient's progress after he or she left the counter.

What do you like best about your job?
I deal with patients over a long term, and I'm constantly in contact with very highly trained professional staff as well. It's a challenging and satisfying combination.

REVIEW 10.3

1. (a) What is an exothermic chemical reaction? Give an example.
   (b) What is an endothermic chemical reaction? Give an example.
   (c) How are endothermic and exothermic reactions similar? How are they different?

2. Classify the following chemical reactions as synthesis, decomposition, single replacement, double replacement, or neutralization. Give reasons for your classification of chemical reactions for 2(a), (b), and (e).
   (a) $2Fe(s) + O_2(g) \rightarrow 2FeO(s)$
   (b) $Fe(s) + CuSO_4(aq) \rightarrow FeSO_4(aq) + Cu(s)$
   (c) $4Cr(s) + 3SnCl_4(aq) \rightarrow 4CrCl_3(aq) + 3Sn(s)$
   (d) $2H_2O_2(l) \rightarrow 2H_2O(l) + O_2(g)$
   (e) $Ca(OH)_2(s) + 2HCl(l) \rightarrow CaCl_2(s) + 2HOH(l)$

3. Draw diagrams, showing the type and number of atoms present, for 2 (a) and (b). Your diagram should be similar to Figures 10.13, 10.14, 10.18, and 10.20.

4. (a) According to how atoms rearrange themselves, name and define the four types of chemical reactions.
   (b) Classify each of the following as one of the four major reaction types:
      (i) phosphorus + oxygen → phosphorus trioxide
      (ii) lithium chlorate → lithium chloride + oxygen
      (iii) silver nitrate + zinc chloride → silver chloride + zinc nitrate
   (c) Write the chemical equations for the above reactions.

5. Relationships between chemical families can sometimes be used to predict reactions.
   (a) If you know that sodium reacts with water to produce sodium hydroxide and hydrogen gas, what product do you think would result from the reaction of rubidium with water? Write a balanced equation for each reaction.
   (b) Use family relationships to predict the outcome of the following reactions and complete the equations:

      $H_2S(g) + \text{electricity} \rightarrow$

      $NaClO_3(s) + \text{heat} \rightarrow$

6. Copy the following equations into your notebook, then balance the equations where needed and tell what type of reaction they are.
   (a) $FeS(s) + HCl(l) \rightarrow FeCl_2(s) + H_2S(g)$
   (b) $NaOH(s) + H_2SO_4(l) \rightarrow H_2O(l) + Na_2SO_4(s)$
   (c) $Ag(s) + S_8(s) \rightarrow Ag_2S(s)$
   (d) $Fe_2O_3(s) \rightarrow Fe(s) + O_2(g)$
   (e) $Hg(l) + H_2SO_4(l) \rightarrow HgSO_4(s) + H_2(g)$

CHAPTER • REVIEW

Key Ideas

- A chemical reaction produces a new substance (or substances), called a product (or products), that has different properties from the starting materials, called the reactants.
- A physical change has the same substance before and after the change.
- A chemical reaction can be represented by a word equation or a chemical equation.
- The law of conservation of mass states that when a physical or chemical reaction takes place, the mass of the products is equal to the mass of the reactants.
- Chemical equations follow the law of conservation of mass and as a result can be balanced by the use of coefficients.
- Chemical reactions can be classified by the energy involved in the reaction or by the way the particles in the chemical reaction rearrange themselves to produce new substances.
- A chemical reaction can be classified as exothermic or endothermic, according to the energy displayed by the reaction.
- A chemical reaction can be classified as synthesis, decomposition, single replacement, double replacement, or neutralization, depending on how the atoms rearrange themselves during the reaction.

Vocabulary

chemical reaction
chemical change
reactant
product
physical change
chemical equation
coefficient
law of conservation of mass
balanced equation
exothermic reaction
endothermic reaction
synthesis
decomposition
single replacement
double replacement

1. Construct your own concept map for this chapter, including as many words as you can from the vocabulary list. You may include other terms as well.
2. Design a crossword puzzle using as many words from the vocabulary list as possible.

Connections

1. Based on the reactants only, classify each of the following reactions as synthesis, decomposition, single replacement, double replacement, or neutralization.
   (a) $Zn(s) + MgSO_4(s) \rightarrow$
   (b) $Ca(s) + O_2(g) \rightarrow$
   (c) $HgO(s) \rightarrow$
   (d) $HCl(l) + KOH(s) \rightarrow$
   (e) $ZnSO_4(s) + SrCl_2(s) \rightarrow$
   (f) $HNO_3(aq) + Sr(OH)_2(s) \rightarrow$
   (g) $Sr(s) + O_2(g) \rightarrow$
   (h) $AlCl_3(s) + Na_2CO_3(s) \rightarrow$
2. Complete the equations in question 1 by predicting the products.
3. Balance the equations in question 1.
4. (a) Describe how a chemical change and a physical change are the same.
   (b) Describe how a chemical change and a physical change are different.
5. (a) List all the things that a chemical equation tells you.
   (b) What does a chemical equation not tell you?
6. The following expression contains a major error. Explain what is wrong and rewrite the expression so that it becomes a correctly balanced equation.
   $Al(s) + H_3PO_4(l) \rightarrow H_3(g) + AlPO_4(s)$
7. Write balanced equations for
   (a) the reaction of iron with sulphur to produce iron(III) sulphide,
   (b) the reaction of calcium carbonate to produce calcium oxide and carbon dioxide,
   (c) the reaction of iron in copper(II) sulphate solution to produce iron(II) sulphate and metallic copper,
   (d) the reaction of sulphuric acid with sodium hydroxide to produce sodium sulphate and water.
8. Write generalized equations for exothermic and endothermic reactions.
9. Classify each of the following reactions. (Equations are correct.)
   (a) $2HgO(s) \rightarrow 2Hg(l) + O_2(g)$
   (b) $4As(s) + 3O_2(g) \rightarrow 2As_2O_3(s)$
   (c) $Na_2S(s) + 2AgNO_3(s) \rightarrow Ag_2S(s) + 2NaNO_3(s)$
   (d) $Pb(s) + 2AgNO_3(s) \rightarrow Pb(NO_3)_2(s) + 2Ag(s)$
10. Explain how an acid and a base neutralize each other.
11. What is the relationship between the law of conservation of mass and a balanced chemical equation?
12. Copy the following equations into your notebook and then balance them.
   (a) $C_4H_{10}(g) + O_2(g) \rightarrow CO_2(g) + H_2O(l)$
   (b) $N_2(g) + H_2(g) \rightarrow NH_3(g)$
   (c) $NH_3(g) + O_2(g) \rightarrow N_2(g) + H_2O(l)$
   (d) $H_2(g) + Fe_2O_3(s) \rightarrow Fe(s) + H_2O(l)$

Explorations

1. The energy involved in chemical reactions (energy that is absorbed or given off) can be used at home or in industry. Find out about some chemical reactions that give off a lot of energy and are inexpensive to carry out. State how the energy is used.
2. Chemical products and the trucks and railway cars carrying chemicals are marked with hazard symbols such as the ones shown in Figure 10.21. Find as many of these symbols as you can in your own home and then find out their exact meaning.

Figure 10.21
What do these symbols mean? Which of them can you find in your home?

3. Make a poster of one type of chemical reaction (synthesis, decomposition, single replacement, double replacement, or neutralization) you studied in this chapter. Show the atoms and molecules present and the energy involved. Your poster should include applications of the reactions in the "real world."

Reflections

1. Now that you have completed the questions under "Connections," go back to your concept map and add as many crosslinks as you can come up with.

2. Think of a non-living object that is around you. Now list the chemical reactions that you think took place to produce that object.

3. Refer to the four-column notes you made in Activity 10A. Are you able to answer some of the questions in the third column? Fill in the fourth column with as many answers as you can. Have you changed your mind about any of the ideas you listed in the third column as a result of doing this chapter? If so, what was it that changed your mind? Write your answer in the fourth column. Look at your descriptions of the chemical reactions you wrote down in column two. Add to or subtract from those descriptions. As often happens when scientists study a subject, they discover new questions. Have you thought of any new questions that you would like to know more about?

CHAPTER

Practical Applications of Electrochemistry

This photograph shows how scientific applications affect our daily lives. The detergent this student is using to wash the dishes, the glass he is washing, and the radio he is listening to are all products of technology, or the practical applications of science.

Have you ever wondered how scientists and inventors come up with applications like these? The answer to this question lies partly in the way scientists and inventors think. They are always looking for relationships between different concepts.

In this chapter, you will study the relationship between the concept of chemical reactions and the concept of electricity. This relationship is called electrochemistry. To understand electrochemistry, you must understand electrolysis, a chemical reaction that occurs when an electrical current is passed through an ionic solution. In this chapter, you will also learn about chemical reactions that are able to supply electricity.

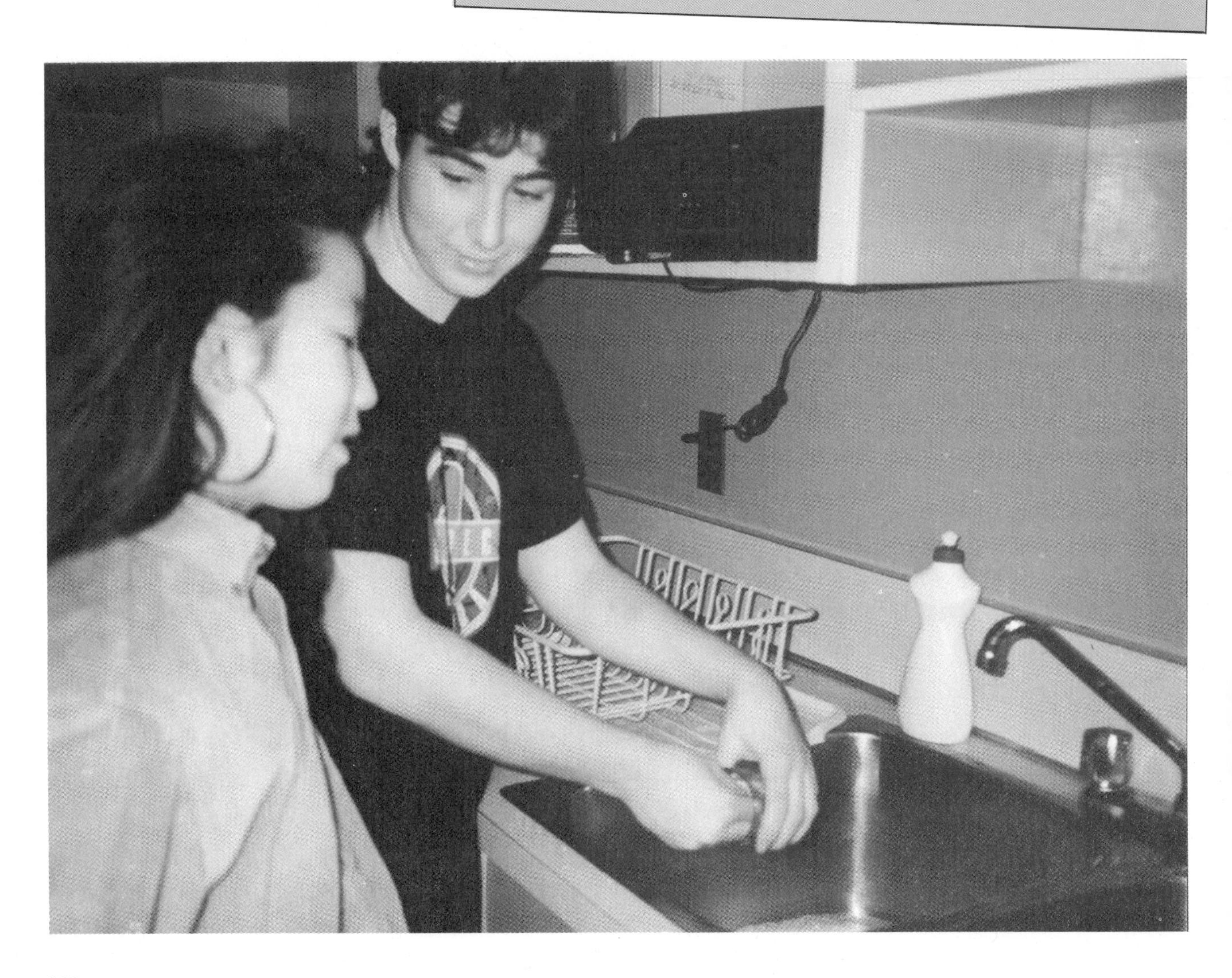

ACTIVITY 11A What You Already Know about Electrolysis

You already know some things about electrolysis; you have used the process in chemistry already. Let's find out what you know.

Divide a page in your learning journal into two columns. At the top of the first column, write the heading "Knowledge before Reading." At the top of the second column, write the heading "Knowledge after Reading." In the first column, write down all you know about electrolysis, using words, phrases, or even diagrams. Include examples of how electrolysis might affect your daily life, as well as any questions you have about electrolysis. Your teacher may want you to get together with one or more other students to compare ideas. Change your list as necessary after your discussion with other students. You may wish to refer to it as you read the following section to see if your ideas change.

After you finish section 11.1, or as you go through it, add any new ideas about electrolysis under the heading "Knowledge after Reading."

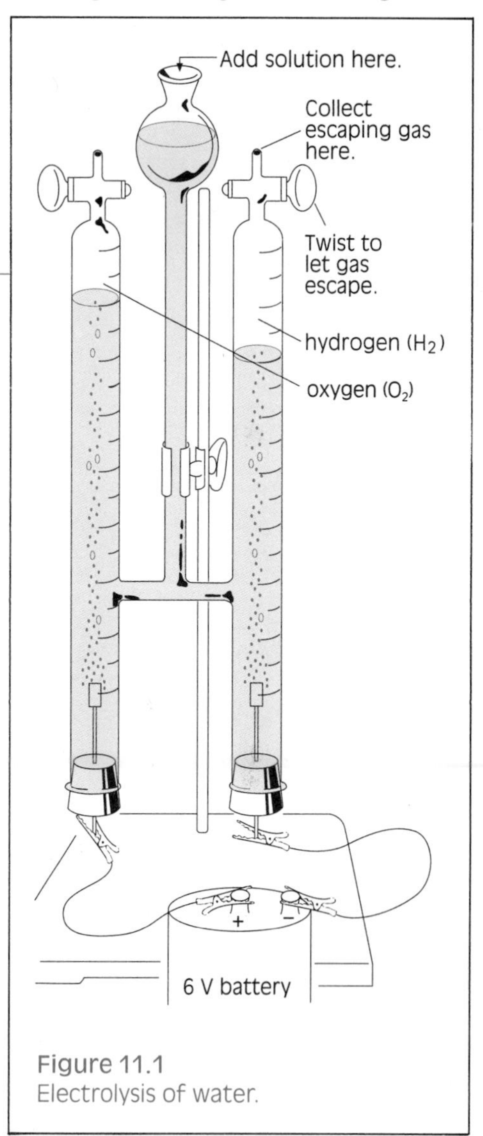

Figure 11.1
Electrolysis of water.

11.1 Electrolysis

In 1800, Alessandro Volta (1745–1827), an Italian scientist, made a device that used chemical reactions to supply a steady electrical current. This device came to be called a voltaic cell. It is an example of **electrochemistry**, the science that deals with the relationship of electricity to chemical reactions. A great many experiments were carried out using the voltaic cell. For example, water was decomposed by simply inserting wires from a voltaic cell into a container of water. It was found that hydrogen gas was produced at the end of one wire and oxygen gas at the end of the other. Using electrical current to decompose compounds such as water is known as **electrolysis** (Figure 11.1). Today electrolysis is used all over the world to produce hydrogen and oxygen for a variety of industrial uses.

There are many other commercial applications of electrolysis. For example, electrolysis is used to prepare certain chemicals and to refine metals. In British Columbia, such chemicals as chlorine, sodium chloride, and sodium chlorate are made by electrolysis, largely for use in the pulp and paper industry. Metals such as aluminum, lead, zinc, silver, and gold are refined by electrolysis. British Columbia is an ideal place for electrolytic industries because of its large supply of hydroelectric power.

ACTIVITY 11B Electrolysis (May Be a Teacher Demonstration)

In this activity, you will investigate the effect of a direct electrical current on a solution of water and sulphuric acid (H_2SO_4). The acid is needed because pure water will not conduct electricity. The sulphuric acid provides charged ions, which are free to move in the solution.

MATERIALS

safety goggles
apron
5 mL dilute sulphuric acid (H_2SO_4)
direct current (DC) power supply
100 mL distilled water
apparatus as shown in Figure 11.2

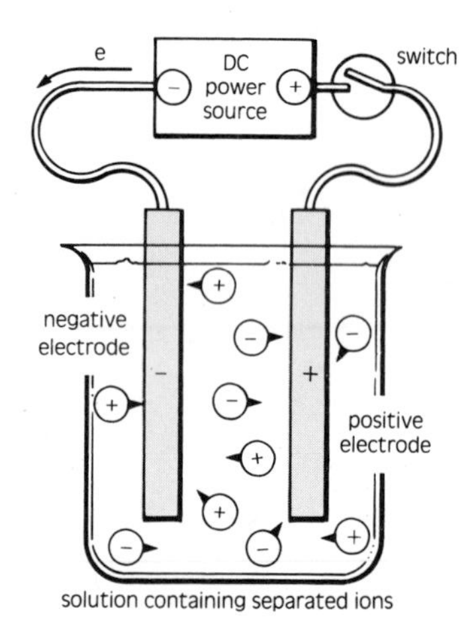

Figure 11.2
A typical electrolysis apparatus.

PROCEDURE

1. Put on safety goggles and apron.
2. Place distilled water in the beaker.
3. Add sulphuric acid to the beaker and stir until dissolved.
4. Attach a label marked " + " to one electrode, and a label marked " – " to the other. Lower the electrodes into the solution and connect them to the power source.

CAUTION!
Hydrochloric acid is corrosive to skin, eyes, and clothing. Wash spills and splashes off your skin and clothing immediately, using plenty of water. If you get acid in your eyes, immediately rinse them for 15 to 20 minutes and inform your teacher.

CAUTION!
Electrical shock hazard. Make sure the power is off before any part of the apparatus shown in Figure 11.2 is handled.

5. Turn on the power source and record in your notebook what you observe.
6. Disconnect the power source and record what you observe. Now lift the electrodes out of the solution to examine them more closely. Record your observations.

CAUTION!
Do not touch the electrodes!

7. Increase the voltage. Record your observations.
8. Wash your hands thoroughly with soap and water after you have completed the activity.

DISCUSSION

1. What evidence is there (a) that the solution contains charged particles, and (b) that it contains both positive and negative particles?
2. (a) How many reactions took place in the beaker?
   (b) Where did they occur?
3. Review your observations, then draw a diagram similar to Figure 11.2; include the movement of the positive and negative charged particles, the gases produced at each electrode, and anything else you think should be included.
4. (a) What is the energy source for this reaction?
   (b) What is the relationship between the amount of direct voltage and the speed of the reaction?
   (c) Is the electrolysis reaction endothermic or exothermic? Explain.
5. Compare electrolysis with the other decomposition reactions that you observed in Chapter 10.
   (a) What advantages does electrolysis have?
   (b) What disadvantages does it have?

ELECTROLYTIC CELLS

The container in which electrolysis takes place can be called an **electrolytic cell.** Two metal electrodes are connected by wires to a source of direct current. The negative electrode, the one connected to the negative terminal of the power supply, is called the **cathode**. The positive electrode, the one connected to the positive terminal of the power supply, is called the **anode**. The continuous flow of electrons (electrical charge) that makes up an electric current can only move around a complete circuit. But pure water does not conduct electricity. To make water a conductor and complete the circuit, the solution must contain charged particles. These are provided by **electrolytes**, substances such as acids, bases, and salts, which produce mobile ions when they dissolve in water (Figure 11.3). Just as electrons are responsible for the flow of electricity through a wire, ions provide the flow of electricity through a solution.

Figure 11.3
(a) Salts, (b) bases, and (c) acids are electrolytes because they form ions that are capable of moving in water, making it a conductor.

salt

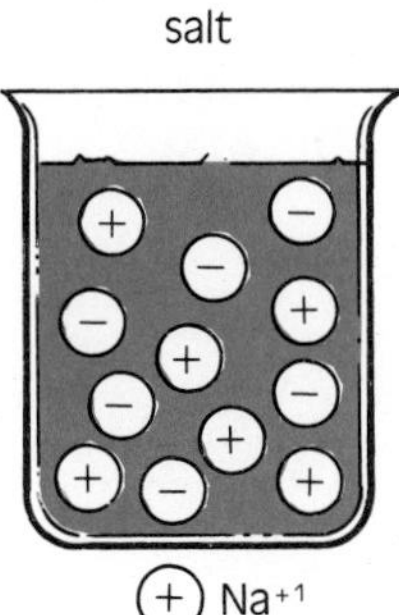

sodium chloride ($NaCl$)

Na^{+1}

Cl^{-1}

(a) When salt is dissolved in water, its positive and negative ions (Na^+ and Cl^-) break away from each other.

base

sodium hydroxide ($NaOH$)

Na^{+1}

$(OH)^{-1}$

(b) Here, sodium hydroxide has dissolved in water, breaking up into mobile Na^{+1} and $(OH)^{-1}$ ions.

acid

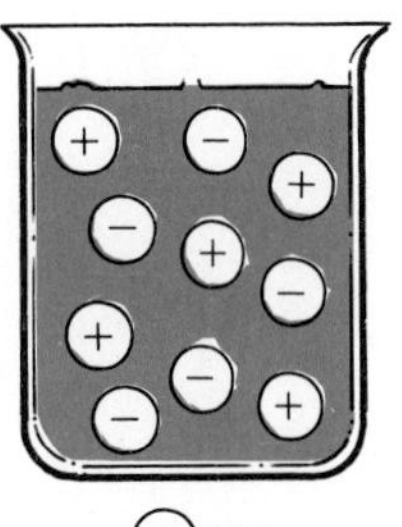

hydrochloric acid (HCl)

H^{+1}

Cl^{-1}

(c) Hydrogen chloride has dissolved in water to form mobile H^+ ions and Cl^- ions.

ACTIVITY 11C *Testing Solutions for Their Ability to Conduct Electricity (Teacher Demonstration)*

In this activity, you will test various materials for their ability to conduct electricity when pure and when dissolved in water.

MATERIALS

safety goggles
apron
testing apparatus similar to that shown in Figure 11.4

Group A: liquids
ethanol
methanol
oil
glycerol
distilled water
tap water

Group B: solutions
bleach
fabric softener
vinegar
lemon juice
soft drink
household ammonia

Group C: solids
laundry detergent
copper(II) chloride ($CuCl_2$)
boric acid
sugar
mothballs
antacid tablets

CAUTION!
Some of these solutions are irritants or are poisonous or corrosive. Always wear your safety goggles and apron when handling these solutions. Wash any spills immediately with plenty of water. If you get any in your eyes, immediately rinse for 15 to 20 minutes and inform your teacher.

Figure 11.4
Apparatus for testing the conductivity of liquids, solutions, and solids.

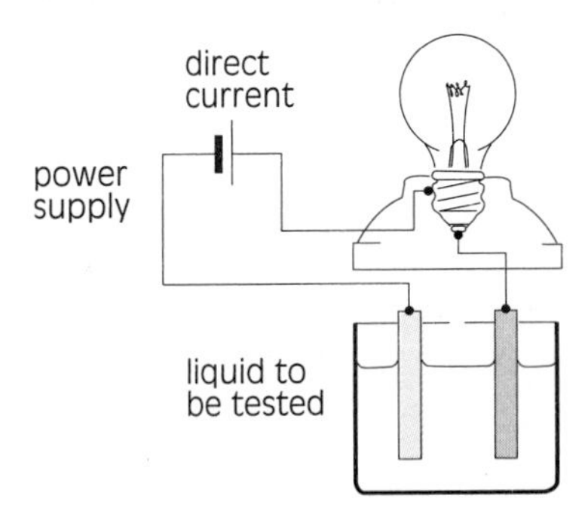

PROCEDURE

1. Before beginning this activity, devise a method for rating the conductivity of a liquid or a solid according to the brightness of the bulb in your testing apparatus.

Prepare a table to record your observations, listing the materials in one column and using a second column for the results of the test.

2. Put on safety goggles and apron.

CAUTION!
Electrical shock hazard. Make sure the power is off before any part of the apparatus shown in Figure 11.4 is handled.

3. Your teacher will test several Group A liquids for conductivity. Electrodes must be disconnected and cleaned between tests (rinsed in distilled water and dried). Record the results.

4. Your teacher will test Group B solutions for conductivity. Electrodes must be disconnected and cleaned between tests (rinsed in distilled water and dried). Record the results.

5. Your teacher will dilute each individual Group C solid with water and then test Group C solutions for conductivity. Electrodes must be disconnected and cleaned between tests (rinsed in distilled water and dried). Record the results.

DISCUSSION

1. (a) Which solutions were able to conduct electricity? Make a list starting with those solutions that conducted electricity poorly.
   (b) Why were some solutions able to conduct electricity better than others?
   (c) Why did some solutions not conduct electricity at all?
2. Why are those solutions that conducted electricity considered electrolytes?
3. Why are electrolytes considered part of an electrolysis set-up?
4. How could you make a solution that conducted electricity poorly a better conductor of electricity without increasing the current or changing the apparatus?

E X T E N S I O N

Plan an experiment to test your answer to question 4 in Activity 11C. Your teacher may allow you to conduct this experiment.

Figure 11.5
The electrolysis of copper(II) chloride ($CuCl_2$) solution.

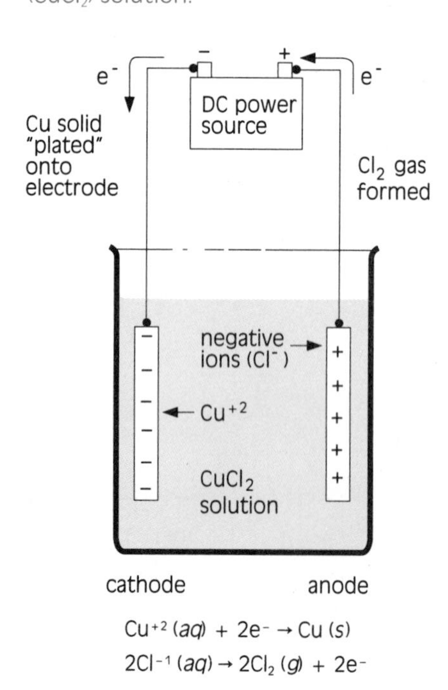

HOW ELECTROLYSIS WORKS

To bring about electrolysis, these requirements must be met:

1. The solution must contain an electrolyte, and the conduction solution must fill the gap between the two electrodes.
2. There must be a direct current.
3. The lead-in wires from the source of direct current must be attached to electrodes (such as carbon or platinum) that are dipped into the electrolyte solution.

In Figure 11.5, the electrolyte is copper(II) chloride ($CuCl_2$) solution. The negative chloride ions (Cl^{-1}) move to the anode, or positive electrode. There each chloride ion gives up an electron and forms a chlorine atom. These form gaseous chlorine (Cl_2) molecules that bubble off. The electrons given up by the chloride ions travel through the wire from the anode to the cathode. At the cathode, copper ions (Cu^{+2}) pick up (gain) the electrons lost by the chloride ions and become copper atoms. (These will form a thin coat of solid copper on the cathode). After the electrolyte is used up and there are no more ions in solution, electrolysis stops. The equations for this process are as follows:

$$CuCl_2(s) \rightarrow Cu^{+2}(aq) + 2Cl^{-1}(aq)$$

$$2Cl^{-1}(aq) \rightarrow 2Cl(g) + 2e^-$$

$$2Cl(g) \rightarrow Cl_2(g)$$

$$Cu^{+2}(aq) + 2e^- \rightarrow Cu(s)$$

REVIEW 11.1

1. (a) What is electrolysis?
   (b) What are electrolytic cells?
2. (a) What are electrolytes? Give several examples.
   (b) Which is more likely to carry an electric current, a pure liquid or a solution? Explain.
3. Is electrolysis a chemical change or a physical change? Give reasons for your answer.
4. (a) What ions will result when copper(II) bromide ($CuBr_2$) is dissolved in water?
   (b) Suppose the resulting solution is placed in an electrolytic cell. Which ions will move to the positive electrode? Why? What new chemical will be produced at the positive electrode?
   (c) Which ions will move to the negative electrode? Why? What new chemical will be produced at the negative electrode?
   (d) Explain what you would expect to happen when more copper(II) bromide is added to the solution. Use a diagram in your explanation.
5. A student has set up an electrolytic cell, but it does not conduct electricity. List all possible reasons why the electrolytic cell does not work.

11.2 Application of Electrolysis

Practical applications of electrolysis touch our daily lives in numerous ways. The knives and forks you eat with may have a thin metal coat of silver over them, which was placed there by electrolysis. This process is also used in the production of pure metallic elements such as aluminum and copper, pure non-metallic elements such as chlorine and hydrogen, and compounds such as sodium hydroxide. All of these materials influence your daily life in important ways.

Figure 11.6
Many consumer and industrial products, like these car bumpers, are electroplated for protection and improved appearance.

ELECTROPLATING

Electroplating uses an electric current to place a thin covering of one metal on the surface of another metal. This procedure is done to prevent rusting or to improve the appearance of metal objects. Steel car bumpers are coated with a thin layer of chrome to prevent rusting (Figure 11.6).

Inexpensive knives and forks, made of iron products, are coated with silver to make them look nicer and prevent rusting. Jewellery made with inexpensive metals is often electroplated with silver or gold to make it look expensive. Other metal objects can be electroplated with copper, chromium, tin, and nickel.

Figure 11.7 shows a simple set-up for electroplating silver onto a spoon. The cathode, or negative electrode, is the metal object to be electroplated, the spoon. The anode, or positive electrode, is a bar of pure silver, which will release silver ions into the solution. The electrolyte is a water solution of a silver salt, usually silver nitrate ($AgNO_3$). The power supply is a direct current.

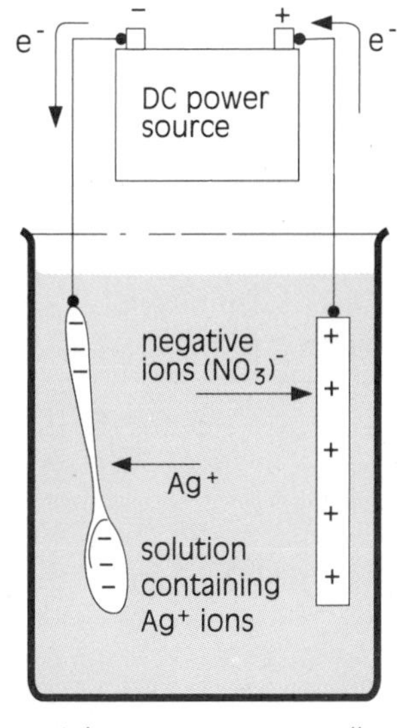

Figure 11.7
Electroplating of silver onto a spoon.

After the direct current has been switched on for several minutes, the metal being plated will have a coat of silver. How does this happen? The positive silver ions in the silver nitrate solution are pulled towards the negatively charged cathode. One at a time, the positive silver ions pick up negative electrons at the spoon and become silver atoms:

$$Ag^{+1}(aq) + e^{-} \rightarrow Ag(s)$$

These silver atoms are neutral and are no longer pulled by the electrical attraction or repulsion, so they begin to form a coat on the spoon.

During the electroplating, the silver ions are removed from the solution. At the same time, new silver ions are formed at the anode. The anode is a bar of silver and is thus made of silver atoms. As the current flows, each silver atom loses an electron. The resulting silver ions enter the solution and cross over to the cathode as a result of electrical attraction. This action continues until the current is switched off or until all of the silver atoms in the anode have been consumed. Similar methods can be used to plate inexpensive metals with gold or copper.

ACTIVITY 11D *Electroplating Metal with Copper*

In this activity, you will set up an electrolytic apparatus to plate a metal object with copper, from a solution of copper(II) sulphate and water. Review the terms "cathode," "anode," and "electrolyte." Where should the object to be plated be attached, at the cathode (negative electrode) or the anode (positive electrode)?

MATERIALS

safety goggles
apron
three 250 mL beakers
steel wool
electrode holder
bare copper wire
6 V battery
copper(II) sulphate ($CuSO_4$) solution
metal object to be plated
copper strip
two wire leads
alligator clips

PROCEDURE

1. Put on safety goggles and apron.
2. Pour about 225 mL of copper sulphate solution into a clean 250 mL beaker.

CAUTION!
Copper sulphate is a strong irritant and is toxic when ingested. Wash any spills immediately with plenty of water. If you get any in your eyes, immediately rinse for 15 to 20 minutes and inform your teacher.

3. Select an object to be plated (you may use a metal object of your own), such as a key, coin, nail, or metal strip. Clean the object of any dirt, coatings, or grease by polishing the object with steel wool. If the object is clean, the copper will adhere more effectively.
4. Place an electrode holder on a second empty beaker. Place a strip of copper metal into one side of the electrode holder (Figure 11.8).
5. On the other side of the electrode holder, suspend the metal object to be plated with a piece of bare copper wire, as shown in Figure 11.8. Hang the metal object below the 200 mL mark of the beaker.

CAUTION!
Electrical shock hazard. Make sure the power is off before you handle any part of the apparatus shown in Figure 11.8.

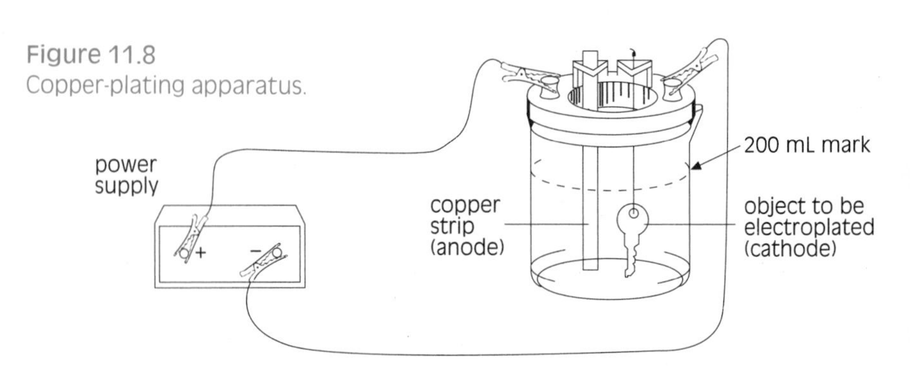

Figure 11.8
Copper-plating apparatus.

6. Connect the metal object to be plated to the negative terminal of the battery. Connect the copper strip to the positive terminal of the battery and turn on the power. Pour the 225 mL of copper sulphate solution into the beaker, holding the electrode holder. Turn on the power source.
7. Observe what happens to the solution, and at each electrode, for the next several minutes.
8. Turn off the power source first, then disconnect the wire from the electrodes.
9. Place an electrode holder on a third empty beaker. Remove both metal objects from the electrode holders. Return the copper strip. Rinse and dry the object you plated (you may want to polish the object by rubbing it with a fine abrasive such as chalk dust).
10. Pour the copper sulphate solution into a container designated by your teacher.
11. Wash your hands thoroughly after completing this activity.

DISCUSSION

1. What was plated onto the metal object? How do you know?
2. Where did the plating material come from?
3. Draw a diagram of the electroplating apparatus used in this experiment. Make it large enough for you to label the following:
   - copper strip
   - object to be electroplated
   - cathode
   - anode
   - electrolyte
   - power supply
   - movement of copper ions
4. When copper is removed from the copper sulphate solution, the solution becomes clear. The copper from the solution was plated onto the metal object. Why did the copper sulphate solution remain blue? (HINT: Where is there another source of copper ions?)
5. How was this experiment different from the description of how a spoon is electroplated?
6. (a) What is observed at the cathode when the object is electroplated? What happens to the electrons?
   (b) What is observed at the anode? What happens to the electrons?
   (c) What happens to the copper ions during the electroplating? How are electrons involved?
   (d) What happens to the copper rod during electroplating? Are electrons involved? If so, explain how.
7. Can you think of any applications for electroplating that have not been mentioned?

ELECTROLYSIS AND METAL REFINING

The impact of electrochemistry on modern society is found nearly everywhere. Important metals such as aluminum and copper are manufactured or refined exclusively by electrolytic reactions.

ACTIVITY 11E *What You Know about Refining Metals by Electrolysis*

In this activity, you will be given several statements about aluminum and copper to guide you in your reading about these elements in the next two sections. Copy the following statements onto a page in your learning journal:

- Pure aluminum is hard to produce because it has a high melting point.
- A college student discovered the process for the production of aluminum.
- To produce both aluminum and copper, a large amount of electricity is required.
- Copper that is 99 per cent pure is good enough for electrical wiring.
- Gold, silver, and platinum are "waste" products in copper refining.

Write "true" or "false" beside each statement and give a reason why you think each statement is true or false. There are no wrong reasons, just ideas at this point. After you have finished writing down your reasons, your teacher may want you to get together with one or more students so that you can compare your ideas. Add or subtract from your list of reasons based on your discussion with other students. Keep this information, since you may want to refer to it as you read and make further changes in your reasons.

THE PRODUCTION OF ALUMINUM

Aluminum is the third most common element (after oxygen and silicon) in the Earth's crust. In nature, however, aluminum is always combined with other elements to form compounds. Today objects made of pure aluminum are common. Yet 100 years ago, pure aluminum was found only on the chemist's shelf. In 1886, a student at Oberlin College in Ohio, Charles M. Hall (1863–1914), began a series of experiments to try to invent a cheap method of extracting the metal from its compounds. One of the problems Hall and other scientists of the time faced was that aluminum ore, or bauxite (Al_2O_3), has such a high melting point (about 2000°C) that no means could be found to melt it. Another problem was that aluminum is very reactive in the unrefined state. Thus, it is difficult to produce pure, uncombined aluminum by chemical reactions. During chemical reactions, aluminum reacted with the other materials, leaving it in a combined form. Efforts to produce aluminum by electrolysis also failed because the aluminum compounds were poor conductors of electricity.

DID YOU KNOW?

The aluminum ore bauxite gets its name from a town near Arles, France, called Les Baux. Aluminum ore was mined there for many years.

In 1886, Hall developed an electrolytic process that used a much lower temperature. He dissolved the aluminum ore (aluminum oxide—Al_2O_3) in a substance called cryolite (Na_3AlF_6). The resulting solution conducts electricity because it contains mobile aluminum ions. While in solution, the aluminum ions do not react. The dissolved ore is placed in an iron box that is lined with carbon (Figure 11.9). The box acts as the negative electrode (cathode). Large carbon rods, which constitute the positive

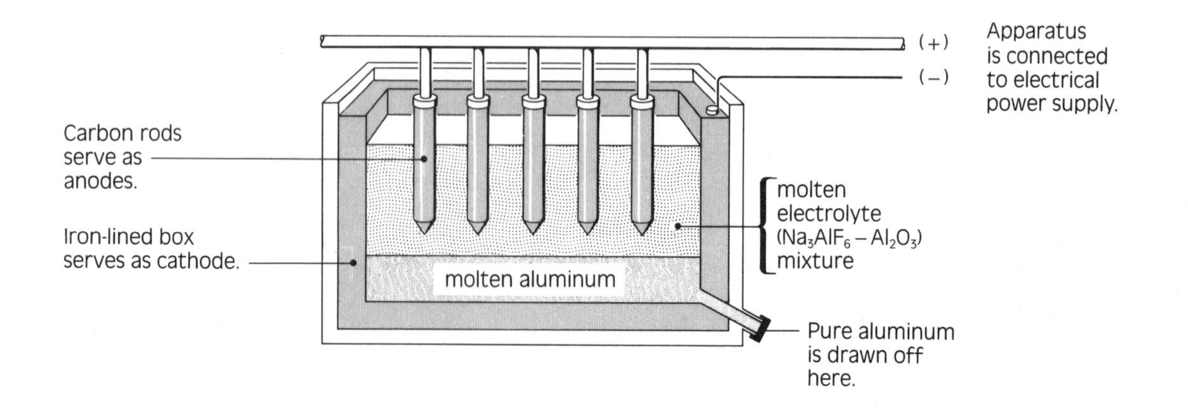

Figure 11.9
Diagram of the apparatus used to produce aluminum electrolytically by the Hall process.

electrode (anode), are lowered into the dissolved ore. When a direct current is passed through the box, the aluminum oxide breaks down to release positive aluminum ions. These are pulled towards the cathode. At the same time, negative oxide ions are released. These are pulled towards the anode. The aluminum ions pick up electrons from the cathode and form aluminum atoms. The pure metallic aluminum is then drawn off through a tube.

The overall equation for the reaction is

$$2Al_2O_3(s) + \text{electricity} \rightarrow 4Al(s) + 3O_2(g)$$

Large electrochemical plants such as those used to produce aluminum are nearly always built in mountainous regions where there are large supplies of rushing water to generate hydroelectric power. Being close to a plentiful and inexpensive source of hydroelectric power is often more important than being close to the source of the ore. The bauxite ore used at Kitimat, British Columbia, is shipped by freighter from mines in Jamaica and South America. Because Canada can provide relatively inexpensive electrical energy, it is one of the world's top producers of refined aluminum metal.

EXTENSION

Do a research project about the waste products from metal refining in British Columbia and how they are disposed of.

Figure 11.10
Purification of copper by electrolysis.

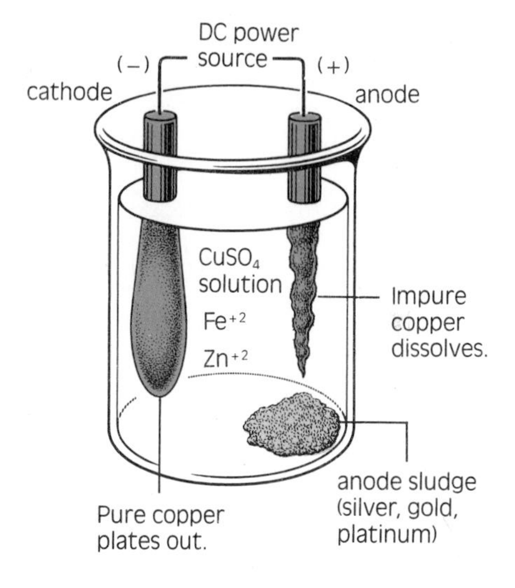

THE REFINING OF COPPER

Canada is the fifth largest producer of the metallic element copper in the world. Copper is mined in most provinces in Canada. Copper metal is refined by Cominco in Trail, British Columbia, for example.

After copper is removed from the ore, the copper metal is about 99 per cent pure — not pure enough to conduct electricity. The 1 per cent impurities consist of platinum, gold, silver, zinc, and iron. To accomplish further refinement, the impure copper is used as the anode in an electrolytic cell (Figure 11.10). The cathode is highly purified copper that was previously refined.

When the current is turned on, the impure copper at the anode forms ions of copper. The copper ions travel to the cathode, where they pick up electrons to form copper atoms. The copper then plates onto the negative cathode. The cathode is highly purified copper. The electrolyte is a solution of copper sulphate ($CuSO_4$).

When electrolysis is carried out in this way, copper forms ions that travel in the solution. The silver, gold, and platinum do not form ions, so they fall off the impure anode as it disintegrates and form a sludge at the bottom of the cell. Only the copper from the solution is deposited at the cathode.

Thus, the result of electrolysis is that copper is transferred from the anode to the cathode. The silver, gold, and platinum sludge is removed and sold for enough money to pay the cost of most of the electrolysis process. The resultant copper is about 99.96 per cent pure — pure enough to conduct electricity (Figure 11.11). This method of refining copper is one of the most important applications of electrolysis.

Figure 11.11
Copper cathodes, 99.96 per cent pure, are pulled from the electrolytic refining tanks. It takes about 28 days for the impure copper anodes to dissolve and deposit pure copper on the cathodes.

OTHER ELECTROCHEMICAL PRODUCTS

Each day you come into contact with chemicals produced from industrial electrolysis reactions. For example, the chlorine that may be used to kill bacteria in your drinking water is a product of industrial electrolysis. Chlorine is also used to make pesticides to protect crops and is vital for the manufacture of some plastics, like vinyl.

The CanadianOxy plant in Squamish produces valuable chemicals such as hydrogen (H_2), chlorine (Cl_2), and sodium hydroxide ($NaOH$) through an electrolytic process that begins with inexpensive table salt, sodium chloride ($NaCl$) (Figure 11.12). The electrodes are made of materials, such as carbon, that do not take part in the reaction, as they do in copper refining. The product at the anode is chlorine gas. The product at the cathode is hydrogen gas, a result of the breakdown of water. The water breaks down at the cathode, producing both hydrogen gas and hydroxide ions (OH^{-1}) in solution. These combine with the sodium ions (Na^{+1}) to form sodium hydroxide ($NaOH$). The newly formed sodium hydroxide is dissolved in the solution.

A disadvantage of electrolyzing salt as in Figure 11.12 is that it may be contaminated by the various chemicals present. One source of contamination is the unreacted salt, which will mix with the newly formed sodium hydroxide. To prevent contamination, separate cells are used for the reactions. Figure 11.13 shows an industrial method for the electrolysis of salt water. The CanadianOxy plant uses a similar process.

Figure 11.12
Electrolysis of aqueous sodium chloride (NaCl), or salt water, solution.

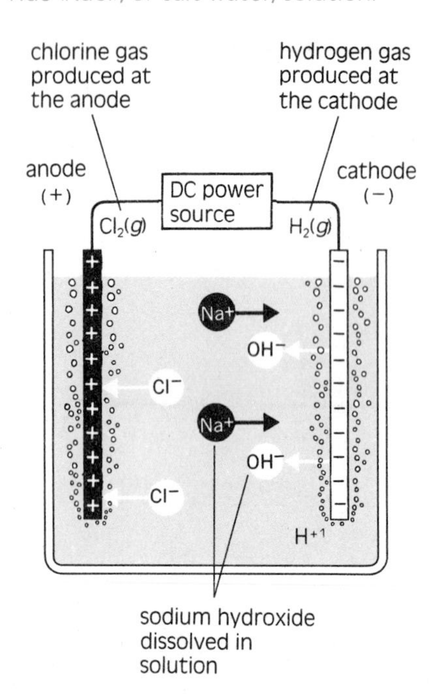

DID YOU KNOW?

Chlorine production for a year in North America amounts to over 10 million t. Chlorine ranks in the top 10 of all chemicals manufactured.

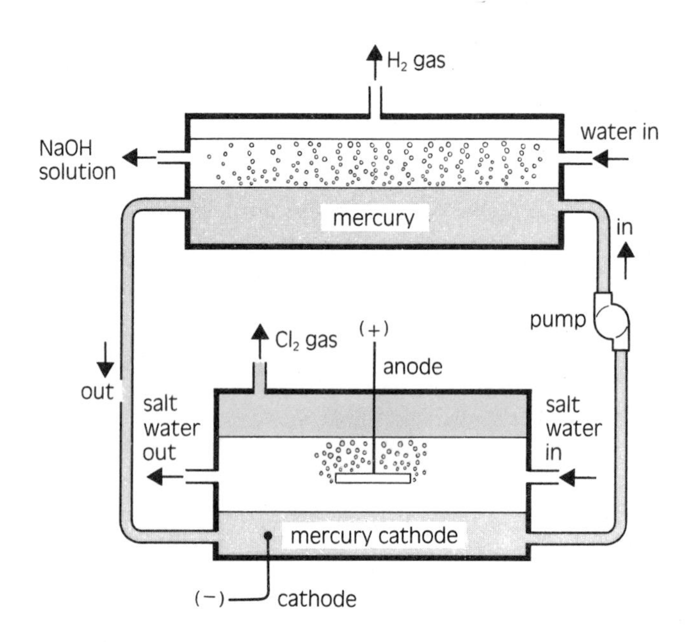

Figure 11.13
Electrolysis of aqueous sodium chloride (NaCl), or salt water, using a mercury cell. Chlorine gas (Cl_2) is produced at the anode. At the cathode, sodium ions (Na^{+1}) are formed into sodium atoms that dissolve in the mercury. The mercury is pumped to a tank, where it comes in contact with water. Sodium atoms react there with water to liberate hydrogen gas (H_2) and produce sodium hydroxide (NaOH).

In the lower cell, salt water is taken in and electric current is passed through it. The electric current breaks down the sodium chloride to chlorine gas and sodium metal:

$$2NaCl(s) \rightarrow Cl_2(g) + 2Na(s)$$

The chlorine is removed from the cell, and the sodium settles to the bottom of the cell and dissolves in mercury. The sodium-mercury mixture is then pumped to a second cell. There the mixture reacts with water to produce sodium hydroxide and hydrogen gas:

$$2Na(s) + 2H_2O(l) \rightarrow 2NaOH(s) + H_2(g)$$

Both the sodium hydroxide and hydrogen are removed. The mercury is returned to the upper cell to carry down more sodium.

The products of this process—chlorine gas, hydrogen gas, and sodium hydroxide — are used to make many other things used in the modern world. Besides being used to kill bacteria, chlorine is used in the paper-making process to bleach paper.

One of the oldest uses for sodium hydroxide is in the production of cleaning agents such as soap. Sodium hydroxide is also a base and is used to neutralize acids in various chemical processes. In addition, it is used to make oven cleaners. It is also used in the manufacture of textiles and the refining of petroleum. In British Columbia, we use a great deal of sodium hydroxide in the processing of pulp and paper. Hydrogen is used to manufacture hydrogenated vegetable oil and to make ammonia, which is used to make cleaning agents and fertilizers. In the future, hydrogen may be used as fuel.

EXTENSION

Do a library project on the use of chlorine and sodium hydroxide in the pulp and paper industry in British Columbia.

REVIEW 11.2

1. (a) List as many objects as you can think of that are electroplated.
   (b) What are the advantages of electroplating?
2. (a) If you were to electroplate copper onto an iron knife, what would you need to carry out this process?
   (b) Explain how the copper is electroplated onto the knife.
3. (a) List as many things as you can think of that are made of aluminum or have aluminum in them.
   (b) What are some of the uses for aluminum in a modern society?
4. (a) What were some of the problems early chemists had when they tried to purify aluminum?
   (b) How were these problems overcome?
   (c) Why is Kitimat, British Columbia, well suited for aluminum production?
5. (a) List as many things as you can think of that are made of copper or have copper in them.
   (b) What are some of the uses for copper in a modern society?
6. (a) What products are the result of the electrolysis of salt water?
   (b) What are some of the uses for the products that are the result of the electrolysis of salt water?
   (c) How does each of these products benefit a modern society?
7. (a) If chromium chloride ($CrCl_3$) is dissolved, what ions would be produced in the solution?
   (b) Which ions would tend to move to the negative electrode (cathode)?
   (c) Which ions would tend to move to the positive electrode (anode)?

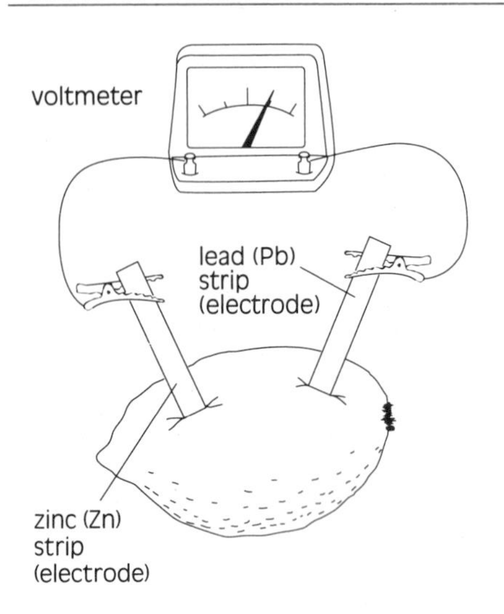

Figure 11.14
Making electricity from a lemon. (Note: Make sure the metal strips are not touching.)

11.3 Electrical Energy from Chemical Change

In the previous sections, you have studied how the flow of electricity can be used to cause and control chemical reactions. In this section, you will study how a chemical reaction that takes place in a cell can be used to generate a direct electrical current. Originally called a voltaic cell, this type of cell is now usually called a **chemical cell**.

CHEMICAL CELLS

A chemical cell consists of two different metals (the electrodes) and an electrolyte. Figure 11.14 shows a chemical cell made from a lemon. The electrolyte in the lemon is an acid, called citric acid, that citrus fruits contain. The citric acid ions are involved in chemical reactions at the metal electrodes. The citric acid ions also carry the electrical current through the lemon. If you squeeze the lemon and break some of its cell walls, the electrical current will be increased and more acid will be freed.

ACTIVITY 11F *Selecting the Best Metals for Electrodes*

In this activity, you will make several different chemical cells. Each cell will have the same type of solution (electrolyte) and the same size of electrodes, and the electrodes will be placed at the same distance apart and to the same depth in each cell. The only condition that will change is the electrodes, or kinds of metal strips. In this way, you will find out which combination of metal strips generates more electrical energy.

Figure 11.15
Apparatus for Activity 11F.

MATERIALS

safety goggles
apron
assorted metal strips
ammonium chloride (NH_4Cl) solution, electrolyte
apparatus as shown in Figure 11.15 (beaker with electrode holder, galvanometer, wires)

PROCEDURE

1. Prepare a data table similar to Table 11.1.

Table 11.1 Data Table for Activity 11F

Testing Different Metals in a Chemical Cell		
Metal A	**Metal B**	**Galvanometer deflection** (direction electrons flowed)
iron	iron	
iron	copper	
iron	zinc	SAMPLE ONLY
iron	lead	
iron	aluminum	
iron	magnesium	
(etc.)	(etc.)	

2. Put on safety goggles and apron.
3. Select a starting metal combination, then hook up the apparatus. Immerse the metal strips and record the voltage produced. (If the needle "wants" to go backwards, reverse the connections to the galvanometer.)
4. The electrode marked "–" shows where electrons move into the galvanometer from the wire. This tells you the direction in which the electrons are flowing. Which metal is losing electrons? Which metal is gaining electrons? Record your answers in your table.
5. Keep the same metal strip in one holder, but change the other and repeat the test. Try as many combinations as you can. For each combination, record which metal lost electrons, which metal gained electrons, and how much voltage was produced.
6. Wash your hands thoroughly with soap and water after you have completed the activity.

DISCUSSION

1. Which metal combination made the most energetic chemical cell (produced the greatest voltage)?
2. Name any combination that did not produce energy.
3. Can you suggest a reason why some metal combinations produce a large voltage and others very little or no voltage?

HOW CHEMICAL CELLS PRODUCE ELECTRICITY

Take some time to think about all that you have learned so far about chemical cells. Each cell has two electrodes that are made of two different metals. The electrodes are surrounded by an electrolyte solution that provides freely moving ions to complete the electrical circuit.

All chemical cells work because of a competition for electrons. The negative electrodes in a chemical cell readily give their electrons away. The zinc metal in Figure 11.16 is the negative electrode; it gives up its electrons relatively easily.

The positive electrodes in a chemical cell readily accept electrons. The silver metal in Figure 11.16 is the positive electrode. It does not give up its electrons as readily as zinc does. Thus, the electrons flow from one metal to another, through a wire, because of this competition for electrons. Ions in solution give up electrons to an electrode or gain electrons from an electrode because of this same competition for electrons.

EXTENSION

Design an experiment to see if the amount of current a chemical cell can produce is affected by varying the type of solution (electrolyte), the size of electrodes, the depth electrodes are immersed, and the distance between the electrodes. What combinations of the above factors would make the best chemical cell?

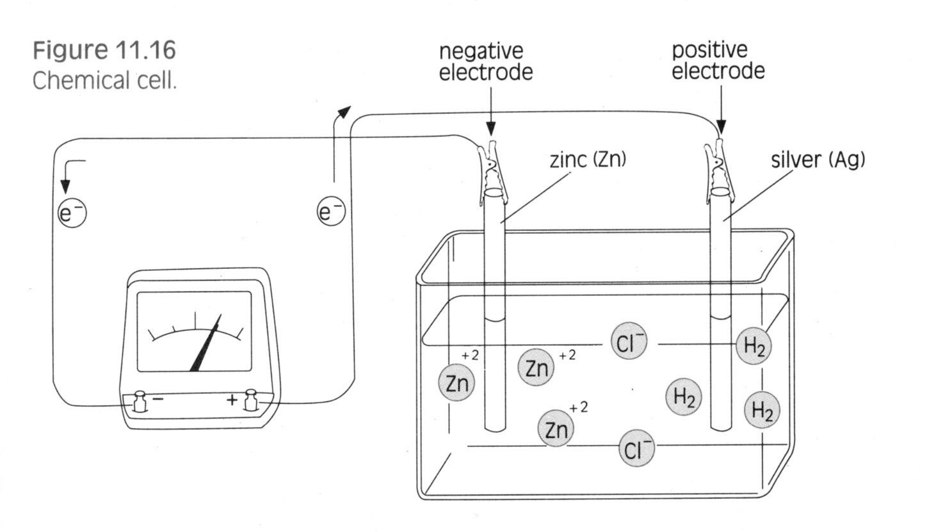

Figure 11.16
Chemical cell.

DID YOU KNOW?

Have you ever bitten down on a piece of aluminium foil and experienced an unpleasant sensation? That sensation was mild electrical shock. A battery is created when the saliva in your mouth, a metal filling in your tooth, and the aluminum foil all come into contact with one another, generating electricity.

BATTERIES AND CELLS

All the test cells in Activity 11F were wet cells because they contained a liquid electrolyte. Dry cells provide just as much voltage without the mess. They are not really dry, since they contain a damp paste that allows ions to move freely.

There are several types of dry cells and wet cells. **Primary cells** are chemical cells that use chemical energy to produce their own electrical energy. They can only be used once. When they run down, you must fill them up with more chemicals or buy new cells. A flashlight battery is a primary cell. **Secondary cells** can be recharged by feeding in a direct supply of electricity to rejuvenate the cell, giving it more electrical energy. Car batteries, which are wet cells, are made up of secondary cells.

A **battery** is a combination of two or more chemical cells. In everyday use, the word "battery" is often used in place of "cell." When you buy a flashlight battery, you are usually buying a single cell.

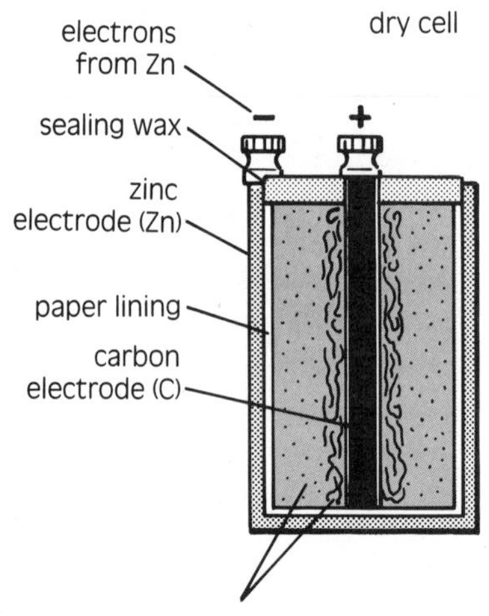

Figure 11.17
An ordinary dry cell is not really dry. Some moisture is needed so that ions can move towards the electrodes. If batteries are left in the sun, they will sometimes ooze this paste, rather like moisture.

CARBON-ZINC DRY CELL

This type of dry cell is commonly used in flashlights, toys, and portable radios. It has an exterior of either cardboard or metal, which acts as a container and insulation (Figure 11.17). The next layer is a zinc (Zn) cup that acts like an electrode. The zinc cup is filled with a moist paste composed of ammonium chloride (NH_4Cl), manganese dioxide (MnO_2), and particles of carbon. This paste is the electrolyte. The other electrode is a graphite rod—carbon—in the middle of the zinc cup. The chemistry of this type of cell is not completely understood. However, one of the principal reactions is the formation of zinc ions and the release of electrons at the zinc electrode:

$$Zn(s) \rightarrow Zn^{+2}(aq) + 2e^-$$

When the cell is connected in a circuit, the electrons flow from the negative zinc electrode through the wire to the positive carbon electrode. The chemical reaction at the carbon electrode takes away electrons, keeping the carbon electrode positive. The positive carbon electrode continues to pull electrons through the wire. The chemical reaction here is complex. One of the reactions appears to be:

$$2MnO_2(s) + 2NH_4^{+1}(aq) + 2e^- \rightarrow Mn_2O_3(s) + 2NH_3(aq) + H_2O(l)$$

The electrical circuit is completed by the movement of ions through the electrolyte paste.

Under heavy use, the carbon-zinc cell quickly appears to be "dead," generating no voltage. The cell can deliver an additional charge if it is not used but is left to "rest." That is because when the cell is in use, the products from the reactions gather around the electrodes. Then it becomes difficult for the reactions to occur at the electrodes, and current is decreased. After the battery is idle for a while, the products diffuse from the electrodes and the battery is able to function again.

The carbon-zinc dry cell cannot be recharged and has a shorter life span than a rechargeable battery. Through use, the voltage produced gradually decreases.

ALKALINE DRY CELL

The alkaline dry cell uses zinc and manganese dioxide as electrodes (Figure 11.18). Part of the electrolyte is potassium hydroxide (KOH), a base (alkaline). The zinc is porous, meaning that it has holes in it. Thus, it provides a larger area for the chemicals in the reaction to come in contact with each other. The chemical reaction between the zinc (Zn), which is the negative electrode, and the hydroxide ions (OH^-) from the base releases the electrons:

$$Zn(s) + 2OH^{-1}(aq) \rightarrow Zn(OH)_2(s) + 2e^-$$

The reaction at the positive electrode, between manganese dioxide (MnO_2) and water, requires electrons:

$$2MnO_2(s) + 2H_2O(l) + 2e^{-1} \rightarrow 2MnO(OH)(s) + 2OH^-(aq)$$

This reaction removes electrons from the positive electrode, leaving the manganese dioxide electrode positive. At the same time, the positive electrode draws electrons to it from the negative electrode.

These batteries stand up better to heavy use, have a longer life, and operate better in cold conditions than carbon-zinc cells. They are also more expensive than carbon-zinc cells.

Figure 11.18
An alkaline dry cell.

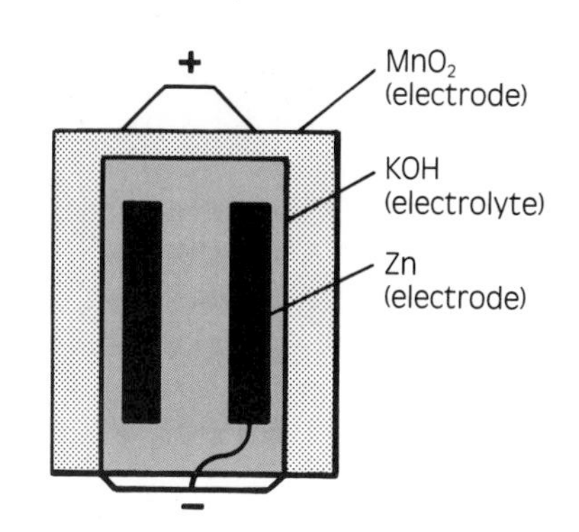

MERCURY DRY CELL

This was one of the first small cells produced for commercial use. It has a zinc negative electrode and a mercury(II) oxide (HgO) positive electrode. The electrolyte is a solution of potassium hydroxide (KOH). The potassium hydroxide is held in absorbent pads separating the electrodes (Figure 11.19). The mercury battery maintains a relatively constant voltage throughout its useable life.

Figure 11.19
Cross-section of a mercury battery.

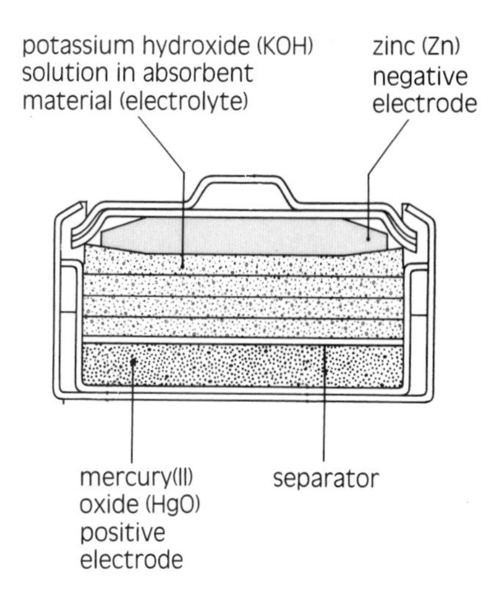

SILVER OXIDE DRY CELL

Silver oxide batteries are tiny but expensive. They are used as power sources in electronic watches and calculators. The positive electrode is silver oxide (Ag_2O), and the negative electrode is zinc. The electrodes react in a water solution (Figure 11.20).

LEAD STORAGE BATTERY

The common car battery is a lead storage battery that is made up of a number of cells connected in series (Figure 11.21). A single cell delivers 2 V. A 12 V lead storage battery used in most cars consists of six cells. The more cells, the greater the voltage.

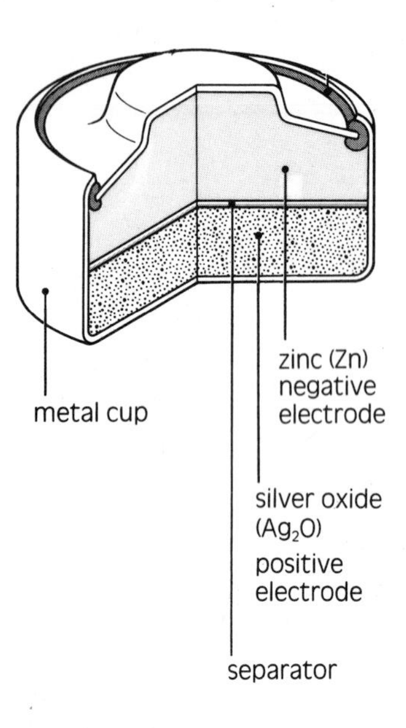

Figure 11.20
A silver oxide battery.

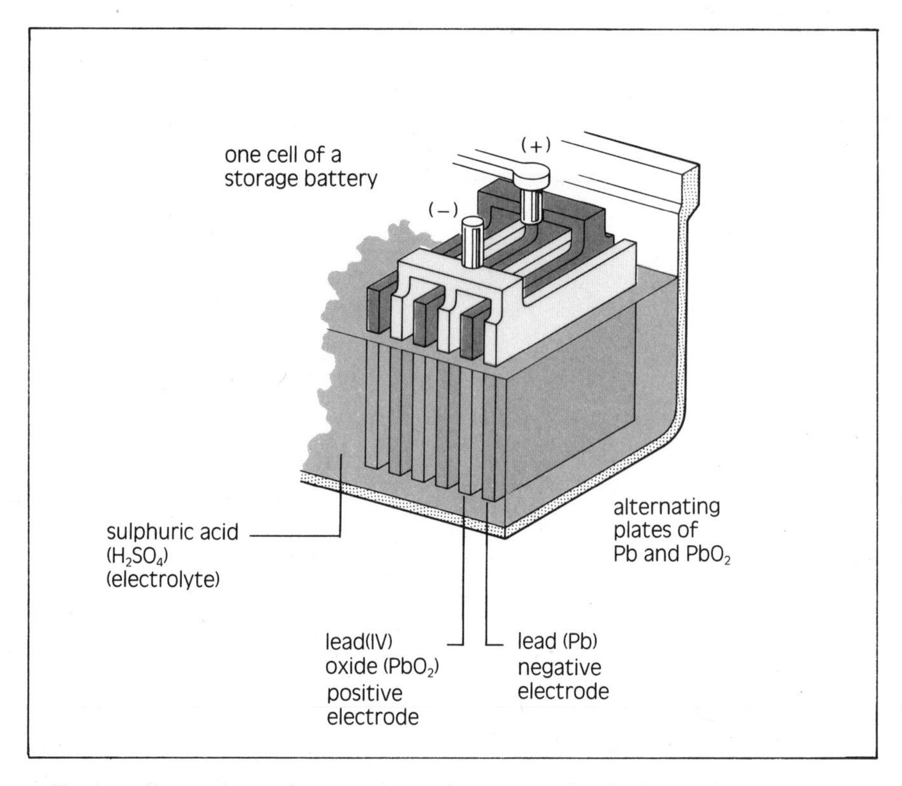

Figure 11.21
A 12 V lead storage battery, such as those used in most automobiles, consists of six cells, like the one illustrated here. Notice that the anode and cathode each consist of several plates connected together. This allows the cell to produce the large currents necessary to start a car.

Each cell consists of a number of negative lead electrodes connected together and a number of lead(IV) oxide (PbO_2) positive electrodes, also connected together. The electrodes are in an electrolyte of sulphuric acid (H_2SO_4). When the circuit is completed at the negative electrode, the lead reacts with the sulphate ion (SO_4^{-2}) from the sulphuric acid electrolyte to release electrons:

$$Pb(s) + SO_4^{-2}(aq) \rightarrow PbSO_4(s) + 2e^-$$

At the positive electrode, the lead(IV) oxide reacts with the hydrogen ion (H^{+1}) and the sulphate ion (SO_4^{-2}) from the sulphuric acid electrolyte to take up an electron:

$$PbO_2(s) + 4H^{+1}(aq) + SO_4^{-2}(aq) + 2e^- \rightarrow PbSO_4(s) + 2H_2O(l)$$

In lead storage batteries, the reaction can be reversed by connecting electrodes to a supply of direct current. The reversed reaction, known as recharging, is done by a generator or an alternator in the car, or by a battery charger if the battery is really run down. During recharging, the sulphuric acid that was used up in generating electricity is restored.

NICKEL-CADMIUM STORAGE CELL (BATTERY)

The nickel-cadmium storage battery is also rechargeable. The negative electrode is composed of cadmium (Cd), and the positive electrode is solid nickel(IV) oxide (NiO_2). The electrolyte is an alkaline (base). The nickel-cadmium battery has a longer life than the lead storage battery and is packed as a single unit like the common dry cell.

SCIENCE IN OUR WORLD

DISPOSING OF BATTERIES

Take a look around your home. How many appliances use batteries? Canadians use about 200 million disposable batteries a year, and the $3 billion world market is growing fast.

Unfortunately, there is a fast-growing problem too. Batteries are hard to dispose of safely. They contain toxic heavy metals such as cadmium, mercury, and lead. From a landfill, these chemicals can leach into soil and water. From an incinerator, fumes and ash can disperse into the air. Cadmium, mercury, or lead in the soil, water, or air can be absorbed by plants and animals either directly or indirectly. For example, people might absorb mercury indirectly by eating fish that have absorbed mercury. These toxic metals tend to build up in the bodies of animals rather than being flushed through. Over time, they can lead to severe and irreversible conditions such as nervous disorders, mental disturbances, or kidney failure.

So how do you decide which batteries to buy, and how do you dispose of them? As you have seen, batteries can be classified as primary cells (disposable) and secondary cells (rechargeable).

Most disposable batteries contain toxic materials, but some manufacturers are using fewer of these materials. Carbon-zinc batteries contain very small amounts of mercury and small amounts of cadmium. Mercuric oxide button cells, used in hearing aids and cameras, also contain small amounts of mercury. Eveready has brought out a zinc-chloride battery called Encore, which contains no mercury. Alkaline batteries, which are the most common small batteries used in Canada, contain about 90 per cent less mercury than they used to. Lithium batteries contain the least toxic materials, but they are expensive.

Rechargeable batteries may seem to be the answer to the problem, but it's not that simple. Nickel-cadmium cells are more expensive than disposable batteries, and though they can be recharged up to 500 times, their working life on a single charge is only about 70 per cent of that of an equivalent alkaline cell. They also must be used carefully if they are to perform well. But eventually they stop working, and the cadmium they contain is toxic.

You can take used lead-acid batteries to some service stations for recycling.

Lead-acid batteries like the one in your family car contain lead and sulphuric acid. When they will no longer hold a charge, most dealers and service stations will take them back for recycling.

Rechargeable lithium batteries have recently been developed and may be available in stores in two to three years. Scientists believe that these batteries will be as efficient as any battery on the market. They are also expected to be cheaper and safer when finally disposed of.

How can you reduce the problems associated with disposing of batteries? First, use plug-in power whenever you can. Manufacturing batteries takes 50 times more energy than they produce. Second, if you need to use batteries, use rechargeables. Find lithium rechargeables as soon as they are on the market.

When disposing of batteries, treat all batteries as hazardous waste. Cities and towns in some regions of British Columbia offer one or two "household hazardous waste weekends" a year. Seal and store your batteries in a cool, dry place and take them to the depot in your area on the appropriate day.

All these different types of cells and batteries provide our society with a convenient, portable source of direct electrical current. Cars, cordless screwdrivers, pacemakers—all depend on these important electrochemical devices. However, this useful technology also has a negative side. Many scientists are now voicing a growing concern over how we dispose of cells and batteries, since the chemicals they contain can cause significant environmental damage. A major challenge to modern chemists is the development of safe ways to recycle or reuse cells and batteries.

REVIEW 11.3

1. (a) What are the essential parts of a chemical cell?
   (b) With a chemical cell, what is the type of energy you start out with?
   (c) With a chemical cell, what is the type of energy you end up with?
2. Zinc metal gives up electrons, and carbon accepts electrons. Which is the positive electrode and which is the negative electrode in a chemical cell?
3. (a) What is a primary cell? Give an example.
   (b) What is a secondary cell? Give an example.
   (c) What is a battery?
4. (a) Describe how a carbon-zinc dry cell generates electricity.
   (b) Describe how a lead storage battery generates electricity.
5. Make a full-page table comparing the various types of dry cells and batteries. At the top of the table, include the following items as part of your comparison:
   - negative electrode
   - positive electrode
   - advantages of this dry cell or battery
   - where the dry cell or battery is best used
   - at least one other category that you come up with

   Along the side of the table, write the names of the dry cells and batteries you will be comparing: carbon-zinc dry cell, alkaline dry cell, mercury dry cell, silver oxide dry cell, lead storage battery, nickel-cadmium storage cell.

CHAPTER • REVIEW

Key Ideas

- Electrolysis is a chemical reaction that uses a direct electrical current passed through an ionic solution to decompose compounds.
- During electrolysis, the anode attracts negative ions from the solution and takes electrons away from these ions. The cathode attracts positive ions and gives electrons to these ions. The ions in solution conduct the electrical current.
- An electrolytic cell contains an ionic solution, both cathode and anode, and a wire to complete the circuit.
- Electrolysis has many practical applications. For example, it is used to refine copper, to produce aluminum, and to produce valuable chemicals such as sodium chloride and chlorine. It is also used in electroplating.
- Chemical cells generate a direct electrical current as a result of a chemical reaction. They include two different metals and a conductive solution.
- Primary cells use chemical energy to produce their own electrical energy and can only be used once.
- Secondary cells, such as car batteries, can be recharged and used again.
- A battery is a combination of two or more chemical cells.

Vocabulary Checklist

electrochemistry
electrolysis
electrolytic cell
cathode
anode
electrolyte
electroplating
chemical cell

primary cell
secondary cell
battery

1. Construct your own concept map using as many words as possible from the vocabulary list. You may also use words that are not included in the list.
2. Construct a word-search puzzle containing as many of the words in the vocabulary list as you can. Do not use the words as clues; your clues should be the definitions of the words.

Connections

1. (a) Describe what happens during electrolysis.
   (b) Describe what happens in a chemical cell.
   (c) In what ways are the process during electrolysis and the process in a chemical cell the same?
   (d) In what ways are the process during electrolysis and the process in a chemical cell different?
2. (a) In an electrolytic cell and in a chemical cell, what components are the same?
   (b) In an electrolytic cell and in a chemical cell, what components are different?
3. (a) List ways in which you could stop or slow down the process of electrolysis.
   (b) List ways in which you could stop or slow down the process in a chemical cell.
4. (a) Consider what you know about aluminum production, copper refining, and chemical production. List all the ways you can think of in which the countries of the world depend on one another for these activities.
   (b) If shipments of aluminum ore from South America were to stop today, how might Canada and the rest of the world be affected? List as many effects as you can.
5. How would your life be different if the only battery that existed was the carbon-zinc dry cell?
6. (a) What are the advantages of having so many types of batteries?
   (b) What are the disadvantages of having so many types of batteries?

Explorations

1. Choose some of the industrial processes covered in this chapter or other industrial processes you are familiar with. Then make a chart showing how the various industrial processes are interrelated.
2. Choose a process from the chapter. Make a flow chart showing the materials needed for that process and the products produced by that process. Include where these materials come from and where any waste products go. Be sure to include the energy source and the origin of the energy source.

Reflections

1. Go back to Activity 11A and read what you wrote in your learning journal. How have your ideas changed since you did that activity? What questions do you still have? Write your new ideas and your unanswered questions in the second column of your chart.
2. Go back to Activity 11E. What have you learned since you completed that activity? How have your ideas changed? Write the answers to these questions in your notebook.

UNIT IV

Examining the Earth

Have you ever thought about what would happen in an earthquake? What could happen in your community? Even an earthquake far away can sometimes affect us. For example, the earthquake in Anchorage, Alaska, in 1964 caused giant sea waves that flooded Port Alberni, over 1500 km away.

The photograph at the right shows the displacement of a creek bed along the San Andreas fault in California. The new creek bed is the dark line running from the top left of the picture to the bottom right. The old bed is on what is now higher ground, to the right of the new creek bed. This displacement is the result of thousands of earthquakes over millions of years. The idea that changes can take place extremely slowly, over millions of years, is important to Earth science. These changes have left clues about the Earth's past. By learning to recognize and understand these clues, we can find out more about the Earth's history.

In this unit, you will learn how to find out when a volcano last erupted, how old rocks are, and how long ago fossils were living organisms. You will also find out about earthquakes, about the kinds of waves earthquakes cause, and how these waves can help you discover what the Earth is made of. All of these discoveries will give you a better understanding of your home, the Earth.

CHAPTER

12 Geological Time

Evidence from the rocks indicates that the Earth has undergone many changes over its 4.5-billion-year history. Scientists conclude that parts of the Canadian landscape were once shaped by thick sheets of moving ice and at other times were occupied by steamy, tropical jungles. A variety of exotic prehistoric animals, including a type of North American rhinoceros shown in the illustration, are thought to have lived on the Earth.

How do scientists know that any of this ever actually happened? A simple answer is that they do not. You might be surprised to learn that scientists are not absolutely certain about anything. Scientists do make careful observations about nature, so that information may be collected and logically developed into explanations or theories. Yet these theories are nothing more than the best answers scientists have at the time. They are revised as new information is gathered.

In this chapter, you will learn about some of the clues and scientific logic that scientists who study the Earth, known as earth scientists or **geologists**, use to learn about the history of the Earth. In particular, you will study how the age of geological events can be determined.

ACTIVITY 12A Questions about the Past

Suppose you find the remains of a clam shell solidly embedded in a rock. Make a list of all the questions that you would like to ask about this object. In a group, discuss how you, as scientists, might go about finding answers to these questions if they cannot be found in the library.

12.1 Time and Change

First, let us consider how you find out the age of things in daily life. Sometimes it is not very difficult. For example, the dates on coins and newspapers are direct evidence of their age. Some other kinds of objects can be dated only by using indirect clues. We have only indirect clues for determining the age of most of the geological events in the Earth's past.

ACTIVITY 12B Estimating Age

In this activity, you will identify indirect clues you use in guessing the age of things. You will also find out different meanings of the word "age."

PROCEDURE

1. Examine the photographs in Figure 12.1. Estimate the age of each item shown and record it in your notebook.
2. For each item, describe the clues you used to estimate its age. How would the appearance of each item be different if it were older or younger than it is shown?
3. Study the pairs of photographs in Figures 12.1 and 12.2, and decide which item of each pair is older. List the indirect clues you considered in making this decision and give an explanation for your choice.
4. Describe what you understood the term "age" to mean for each of the items. For example, if you

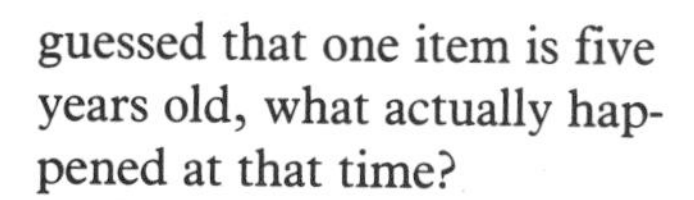
guessed that one item is five years old, what actually happened at that time?

(a)

(c)

(b)

(d)

Figure 12.1
Estimate the age of each item.

5. List your own examples of objects or events that have an unknown age. In small groups, discuss them and come to a decision about the following questions:
   (a) What indirect clues are useful for estimating the age of each item? What assumptions do you have to make?
   (b) What kinds of things are easiest to guess the ages of?
   (c) Why might it be easy or difficult to estimate the age of a person? a rock?

DISCUSSION

1. The water in an aquarium evaporates over time. Describe how you would estimate the amount of time elapsed since the water was last changed. What information would you need? What assumptions would you make?
2. Why might it be easier to estimate the length of time an apple has been left out in the sun than to estimate the time that water has been left in a sealed bottle?

(a)

(b)

(c)

(d)

Figure 12.2
Compare each photograph with the corresponding photograph in Figure 12.1. How can you tell which item in each pair is oldest?

HOW TIME IS RELATED TO CHANGE

Most of the indirect clues that we use to guess the age of something require us to observe how the thing has changed. Change may involve an increase in volume, as in the growth of trees, the growth of a crystal in solution, or the growth of a river delta. Change may involve a decrease in volume, as when salt is dissolved in water, a candle burns, or an old deer's teeth wear down. Change may also bring a transformation, like the rusting of a nail or the ageing of a person's face.

RATE OF CHANGE

You even mark the passage of time by measuring change. Think about this the next time you look at the changing face of a clock as the hours pass, or when you count the number of times the sun has risen and set (which is what you do when you use a calendar). But not all changes occur at a constant rate. It takes a very long time for mountains to wear down to little hills, or for deep canyons to be carved by rivers (Figure 12.3). However, a single catastrophic event can drastically change a landform in a matter of seconds. The eruption of Mount Saint Helens in 1980 and the Hope landslide of 1965 are two examples of rapid change (Figure 12.4). These rapid changes, though, are slight compared with some very slow but constant processes that act over an extremely long period of time, such as the erosion of the Grand Canyon.

One way you can estimate the age of something is by telling how much time has passed since an event. The statement "Karen is 15 years old" means that 15 years have passed since her birth. Geologists often do the same thing with geological events. For example, they believe that the dinosaurs disappeared about 65 million years ago and that the last great glacial advance in North America ended about 10 000 years ago.

DID YOU KNOW?

About one-third of the Earth's landmass was covered by glaciers as much as 2 km thick during the last ice age. Although the Earth is now in a relatively warm period and most of this ice has melted, Greenland and the continent of Antarctica are still almost entirely covered by ice.

Figure 12.3
Grand Canyon, Arizona, is about 1.5 km deep and averages about 15 km wide. Some scientists say it is about 10 million years old. How do you think they know that?

Figure 12.4
(a) The 1980 eruption of Mount Saint Helens. (b) The 1965 landslide near Hope, British Columbia.

(a)

(b)

ACTIVITY 12C Putting Things in Order

It is sometimes impossible to say how much time has passed since something has occurred. This activity is intended to help you to think about a different way in which the age of particular events may be described.

PROCEDURE

1. Think of any four events that have occurred in your life. Try to choose two important events and two unimportant ones. Write a three- or four-word description of each event, each one on a separate slip of paper.
2. Arrange the slips of paper in the order that the events occurred, with the most recent event at the top (Figure 12.5).
3. Share your list with a partner. Arrange your combined events in order of time as you did before. Keep a written record of the questions you asked yourselves as you figured out the correct order.

DISCUSSION

1. What kinds of things do you need to know before you can determine which of two events happened first?
2. Describe a situation in which you need to refer to a third event to decide which of two other events occurred first.
3. (a) Two volcanoes are located close to one another. Describe some of the ways that you might tell which has erupted more recently.
   (b) How might you solve this problem if the volcanoes were located on different continents?

Figure 12.5
Arrange the events with the most recent at the top.

RELATIVE AND ABSOLUTE AGE

It is easy to compare the age of events if you know the exact time that each occurred. For example, if Sharon fell off her skateboard in 1986 and Bindy passed her driving test in 1990, the **absolute age** — the time in years — of each event is known. Sharon's accident happened before Bindy's test. The order of these events is called their **relative age**.

However, if the absolute ages are unknown, the task becomes challenging. For example, how would you determine the relative age of these two events if you did not know the date of Sharon's accident?

Solving problems of geological time is like the above example. The absolute age of a rock or a geological event is rarely obvious, so geologists must assemble the pieces of the puzzle using logic and detective work.

REVIEW 12.1

1. In general, how can you tell if one object is older than another?
2. What kinds of changes do we commonly use to measure the passage of time?
3. Give two different meanings for the "age" of something. Illustrate your answer with an example.
4. Why is it fairly easy to estimate the absolute age of some objects but not of some others? Describe how these types of objects might be different.

12.2 Rocks and Structures

In geology, the relative and absolute age of events usually refers either to the formation of rocks or to actions that may later change the rocks. Before going on to apply some principles for finding the age of geological events, it will be necessary to discuss how rocks are formed and can change.

FAMILIES OF ROCKS

Rocks are classified into three groups, or families, depending on how they were formed.

Igneous rocks (from the Latin word *ignis* or "fire") were at one time in a hot, liquid state. Certain areas beneath the Earth's lower **crust** (the rocky surface layer) become hot enough to melt rock. **Magma** is the name given to this molten rock when it is underground. Since the weight of overlying rock creates great pressure, the magma is often squeezed up through weaknesses in the crust and thereby escapes to the surface. In some ways this pressure is like the pressure you would create standing with your full weight on a sealed tube of toothpaste. Your body weight creates the pressure that forces the toothpaste to leak out. Similarly, volcanic eruptions result when magma under pressure escapes to the surface.

If magma flows out on the Earth's surface as a liquid, it is known as **lava**. If it is blown out of the volcano in small particles, it is known as **ash**. Rocks formed of either lava or ash are known as **volcanic rocks**. You might know examples of these kinds of rocks such as basalt, obsidian, and pumice.

Magmas that cool and crystallize beneath the Earth's surface form **plutonic rock**. Granite (Figure 12.6a) is a common example of igneous rock formed in this way. As the ground above a solidified magma erodes away, the plutonic rock may eventually be revealed on the surface. The Coast Mountain Range of British Columbia consists of granites that are believed to have formed about 100 million years ago, at depths of 10 to 15 km.

Sedimentary rocks usually consist of rock fragments, such as mud, sand, or gravel, that have been squeezed or cemented together. These fragments originated from the **weathering**—physical or chemical breakdown—of rock exposed to water, wind, or ice on the Earth's surface. Typically these fragments are carried by rivers and eventually settle out on the sea floor in layers of **sediment**. (The word "sediment" comes from the Latin word meaning "to settle.") Layers of sedimentary rock can be seen in the cliffs of the Grand Canyon and in the Rocky Mountains, suggesting that these regions were once covered by ancient seas.

Common sedimentary rocks include mudstone, shale, sandstone, and conglomerate (Figure 12.6b), which originated from the sediments mud, clay, sand, and gravel, respectively. Plant or animal remains may be buried by sediment. If conditions are favourable, they may be preserved

Figure 12.6
(a) Igneous rock, granite; (b) sedimentary rock, conglomerate; (c) metamorphic rock, slate.

(a)

(b)

(c)

DID YOU KNOW?

A particular kind of sediment, called a chemical sediment, crystallizes out of sea water in response to a change in the temperature or chemistry of the water. Many kinds of limestone are formed in this way, as are gypsum (from which plaster is made), rock salt, and chert (a silica-rich rock). Flint and jasper are varieties of chert.

as **fossils**, which are often found within sedimentary rocks. Another common sedimentary rock, limestone, is occasionally composed of fossil shell fragments. Thick layers of limestone, such as those found near Cache Creek, British Columbia, might indicate that the region was once beneath an ocean with large reefs like Australia's Great Barrier Reef.

You will learn later how fossils can be useful in finding out the age of certain rocks.

Metamorphic rocks were once either igneous or sedimentary rocks that have been changed through the action of heat and pressure deep underground. The word "metamorphic" is derived from the Greek words meaning "change in form." You may recall that a similar word—metamorphosis—is used to describe the change in form of a caterpillar as it becomes a butterfly. Bread dough undergoes a kind of metamorphosis (or metamorphism, as it is called in geology) when it is baked.

A good example of metamorphism in geology is the change from mudstone or shale, which are sedimentary, to the metamorphic rock slate (Figure 12.6c). A deeply buried layer of shale is subjected to a great deal of pressure from overlying rock. Since the temperature rises with depth, the shale also becomes hot, but not hot enough to melt. As a result of this high temperature and pressure, sometimes crystals grow larger and new minerals start to form. The rock becomes hard, brittle, and shiny, and may develop flaky layers. Another example of metamorphism is the way limestone may be recrystallized to form marble.

As you read the preceding descriptions of rock families, it may have occurred to you that several things can happen to a rock that may change it into another kind of rock. In fact, rocks are never really created, they only get recycled. The rock recycling process, called the rock cycle, is illustrated in Figure 12.7.

Figure 12.7
The crust of the Earth is an active region. Look for the places where rocks are continually being changed or "recycled" from one form to another.

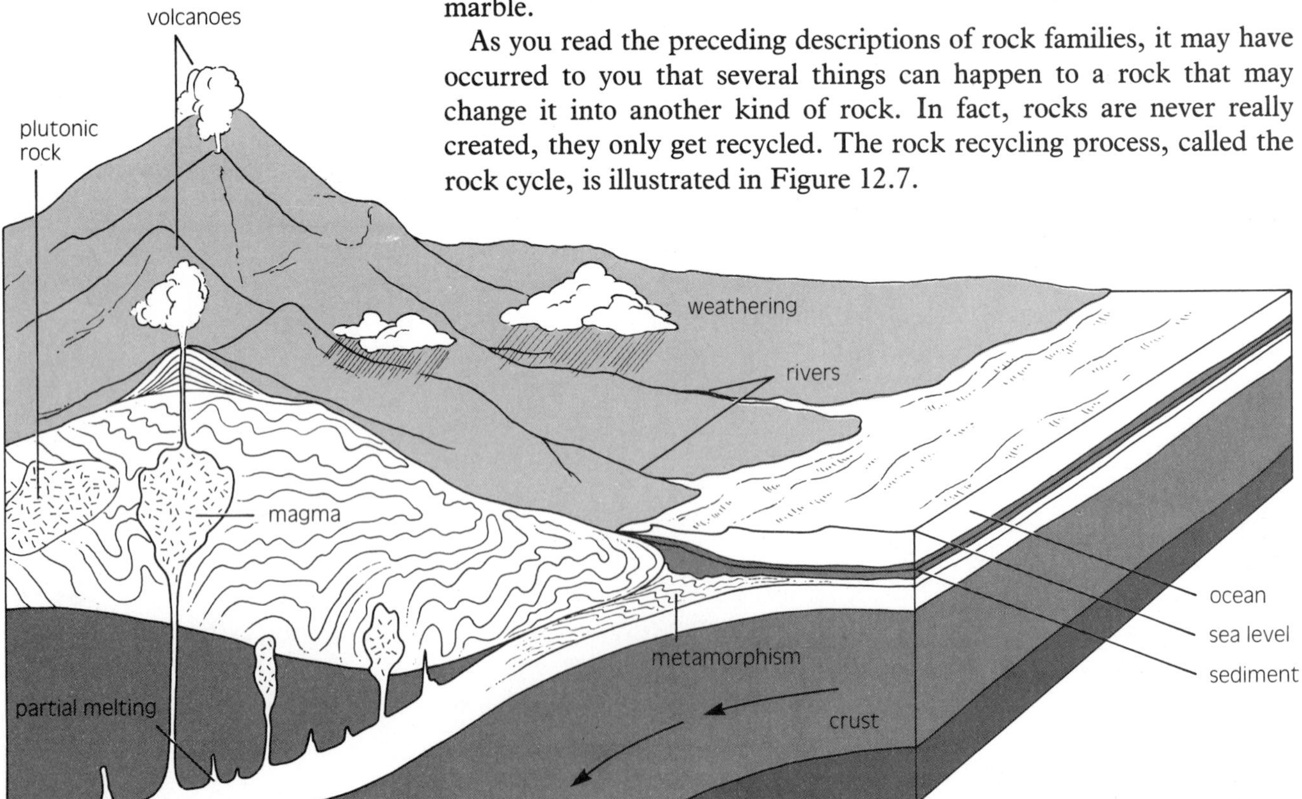

GEOLOGICAL STRUCTURES

Once a body of rock forms, forces may act to bend, break, or change it in some physical way, excluding metamorphism. These forces produce features in the rocks called geological **structures** (Figure 12.8). Several common ones are described below.

Fractures are cracks in a body of rock.

Faults are fractures along which there has been movement. For example, two slabs of concrete might shift along a crack in a sidewalk. This would be a fault on a small scale. Earthquakes generally result from the sudden movement of bodies of rocks on either side of a fault.

Dikes are igneous rocks that are formed in rock fractures when magma is squeezed upwards and later cools. Dikes often appear as a stripe of different-coloured rock cutting through another rock.

Folds occur when layers of sedimentary rock are squeezed and then buckle. You can demonstrate folding by compressing (pushing the ends together) a stack of paper or a piece of carpet.

Erosion surfaces occur wherever rock has been exposed in the past to weathering and erosion and is later reburied. Imagine a layer of snow that is eroded by rain, and then covered by fresh snow. The boundary between the two snow layers would appear rough and uneven. Buried erosion surfaces may have a similar appearance.

Strata (singular: stratum) means layers. Strata are common in sedimentary rocks. They result from the settling and depositing of different sizes and kinds of sediment in layers, one above another. If left undisturbed by other forces, they usually appear horizontal and parallel, like the layers of a cake.

Figure 12.8
Structures in rocks: (a) fracture, (b) dikes, (c) fold, (d) fault.

REVIEW 12.2

1. (a) Describe three basic ways in which rocks can form.
   (b) Name the rock family that corresponds to each of these origins.
2. Make a concept map to describe the appearance and formation of each of the families of rocks and to show how they are related to one another. Include concepts from the following list as well as any others that you wish to add.

sedimentary
metamorphic
plutonic
heat
erosion
compacted
settling
solid
magma
underground
igneous
volcanic
weathering
pressure
melting
cementing
deposition
liquid
lava
surface

3. Make a series of sketches to illustrate the steps involved in the formation of each of the geological structures described in this section.
4. Although the Earth is thought to be about 4.5 billion years old, no rock that old has ever been found. Refer to the rock cycle. Explain why it is unlikely that a rock this old will ever be found.

12.3 Finding Relative Age

Each rock and structure described previously was a result of a sequence, or order, of geological events. Recall that the relative age of rocks or geological structures refers to the order in which they occurred. In this section, you will discover and apply some of the rules that geologists use for finding out the relative age of geological events. Before you start the next activity, make some predictions of your own about the kinds of clues geologists might look for in establishing the relative age of rocks.

ACTIVITY 12D *Which Came First?*

Part I will help you discover some rules or geological principles. You will apply them in Part II, to solve some problems about the relative age of rocks.

PART I

PROCEDURE

1. Study the photographs of rocks and structures shown in Figure 12.8. For each photograph, do the following tasks.
   - Make a sketch of the feature illustrated.
   - Label as many rock types and structures as you can identify.
   - Write a brief description about how each was formed.
2. Number and label your diagrams to indicate the relative age of geological events that resulted in the formation of each feature.

DISCUSSION

1. In your own words, summarize the rules for finding out the relative age of sedimentary rocks and geological structures. (You will probably have to state at least two rules.)
2. How do these rules compare with clues you predicted at the start of this activity?

PART II

PROCEDURE

The illustrations in Figure 12.9 represent cross-sections—cutaway side views—of several geological structures. Each letter on the illustrations represents a particular event, such as the formation of a particular rock type or geological structure. For each cross-section, arrange the letters in order of their relative age. Put the letter of the most recent event at the top of the list and the letter of the oldest event at the bottom.

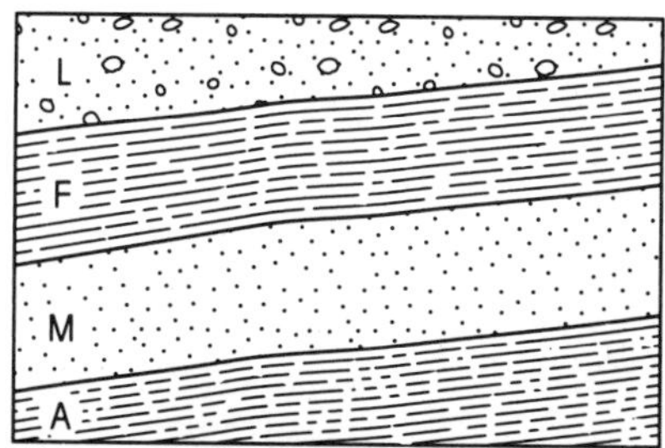

(a) Layered sedimentary rock (strata)

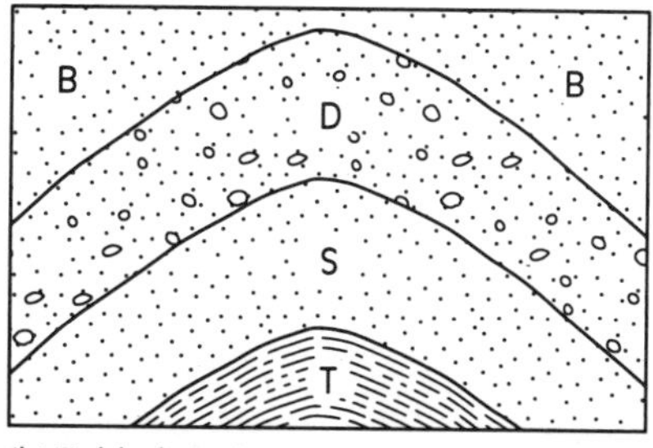

(b) Folded strata

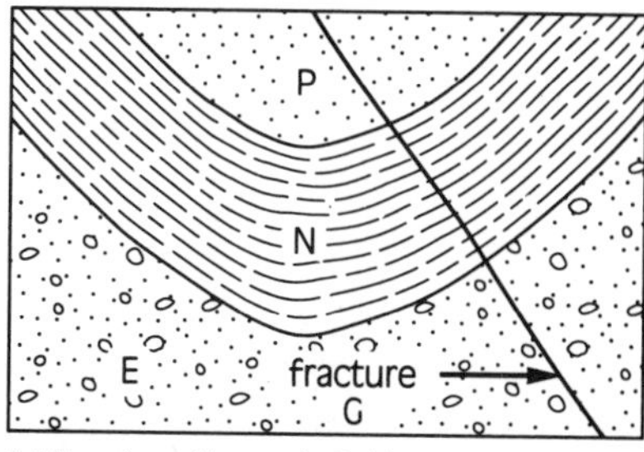

(c) Fracture through folded strata

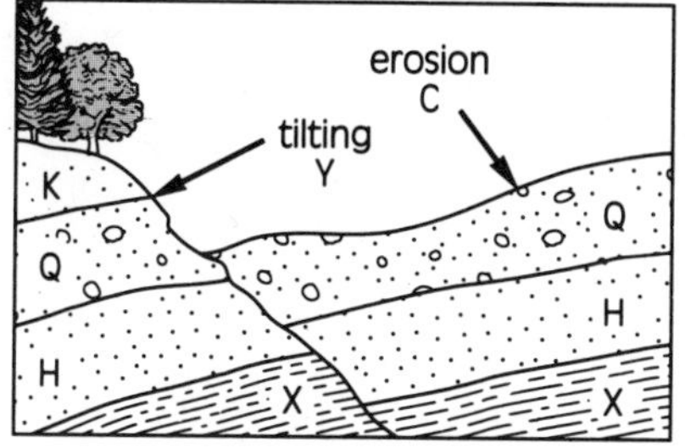

(d) Fault and sedimentary layers

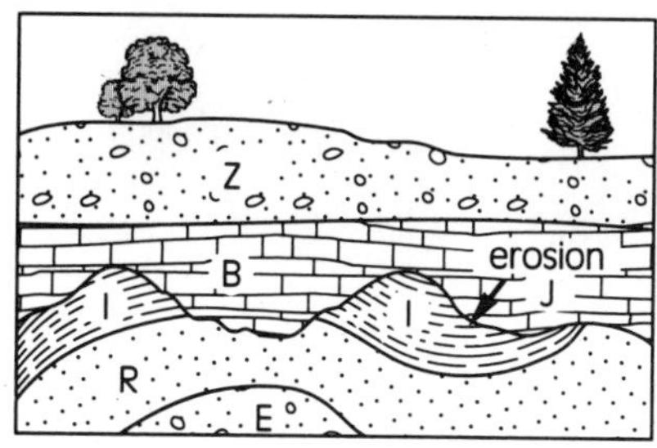

(e) An ancient erosion surface between layered structures

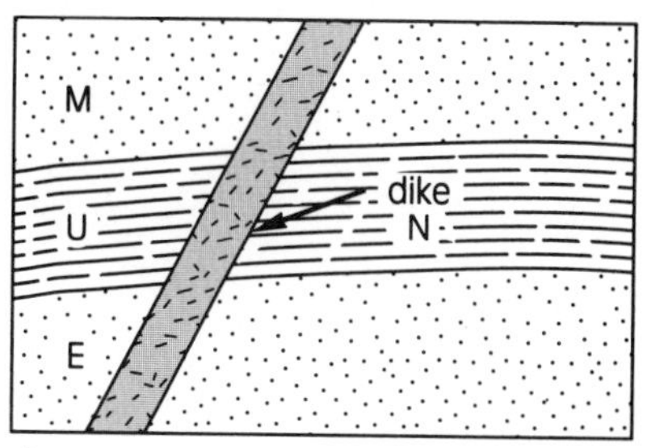

(f) Igneous dike and sedimentary strata

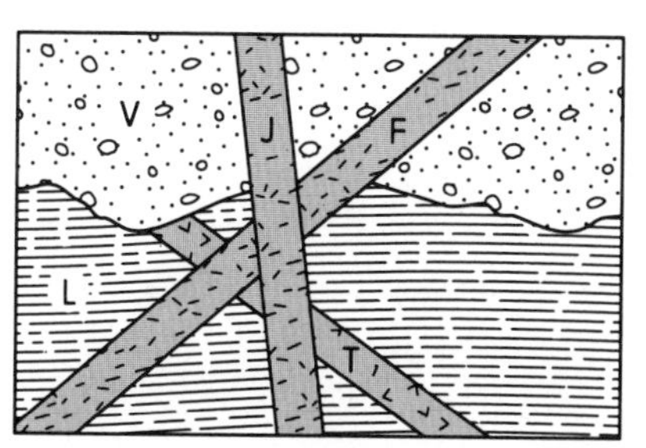

(g) Several dikes and sedimentary strata

conglomerate

shale, mudstone

sandstone

limestone

igneous rock

Figure 12.9
Cross-sections of geological structures.

DISCUSSION

1. In a sequence of layered rock strata, where would you find the oldest rock? The youngest?
2. Why do you suppose you were asked to list the relative ages with the oldest on the bottom and the youngest at the top of the list, instead of the reverse?
3. If a structure, such as a dike, an erosion surface, a fold, a fracture, or a fault, disturbs a rock body, which is younger—the structure or the rock body?

RULES FOR FINDING RELATIVE AGES

Consider the pile of junk on the floor of your bedroom closet (Figure 12.10). If you ever decided to clean it out, you would likely discover that the items at the bottom of the pile have been covered up for the longest time. The things you find on top have been put there more recently.

So it is with sedimentary rocks. Sand or mud settles out of water to deposit on the sea floor. The weight of water and overlying sediments eventually compresses the deeper sediments, to form sandstone and shale. Older layers are found towards the bottom in a series of layers. Younger layers are found towards the top (Figure 12.11a). This simple rule is called the **law of superposition**. (Superposition means "on top.")

EXTENSION

Make layers of sediment in an aquarium. Use several kinds of material such as soil, clay, sand, and pebbles. Stir them all up together vigorously. Allow them to settle. Which materials settle quickly? Where are they located?

Figure 12.10
In an undisturbed stratified deposit, the material at the bottom has been there the longest.

Now imagine that you finally manage to organize some of the mess in the closet. This event — reorganizing the pile of junk — is more recent than the accumulation of the pile itself. This is an application of a second rule, the **crosscutting rule**: Any event that disturbs a rock is always more recent (younger) than the rock itself. For example, after an earthquake, a new crack found in a building wall must be more recent than the wall itself. Likewise, if one dike cuts across another, the one that has been disturbed is the older (Figure 12.11b).

Figure 12.11
(a) Law of superposition: "In undisturbed sedimentary layers, the youngest layers are found towards the top, the oldest layers towards the bottom." (b) The crosscutting rule: "Any feature that disturbs a rock is always younger than the rock itself." Which dike is older, the dark one or the light one?

(a)

(b)

FINDING RELATIVE AGES INDIRECTLY: GEOLOGICAL DETECTIVE WORK

By now you might have the feeling that being a geologist is much like being a detective. You are right. Both use evidence and logic to make sense of observations. Consider the following: suppose that two trees fell in the forest. How would you find out which one had fallen first? Think about this for a few moments before you read on.

If someone had been there at the time these events happened, their absolute ages would be known. If not, we could still determine their relative ages through indirect methods. For example, if one tree had fallen across the other, the law of superposition would provide an answer. But suppose the trees fell in quite different locations (Figure 12.12). How then would you find out which had fallen first?

Figure 12.12
Which of these trees fell first? What assumptions are you making?

If a layer of snow had covered one tree, and the other tree was lying on top of the snow, you could relate the time when each tree fell to the time of a third event—the snowfall. One tree fell before the snowfall. The other tree fell after the snowfall, and was therefore the most recent event.

Now consider a geological example. Imagine that you are shown samples of sedimentary rocks from two different locations. One location is in the Rocky Mountains, the other is in the Queen Charlotte Islands. How are you to find out the relative age of these rocks?

Suppose that, thousands of years ago, a single volcano spread ash right across the continent. This layer of ash would have settled on sediments already deposited in both locations. Like the snowfall in the earlier example, this layer of ash could be used to determine the relative age of these rocks.

Examine Figure 12.13 and determine which would be older, a layer of conglomerate in the Rocky Mountains or a layer of shale in the Queen Charlotte Islands. Explain how you solved the problem.

Figure 12.13
Which rock layer would be older: the shale in the Queen Charlotte Islands or the conglomerate in the Rocky Mountains?

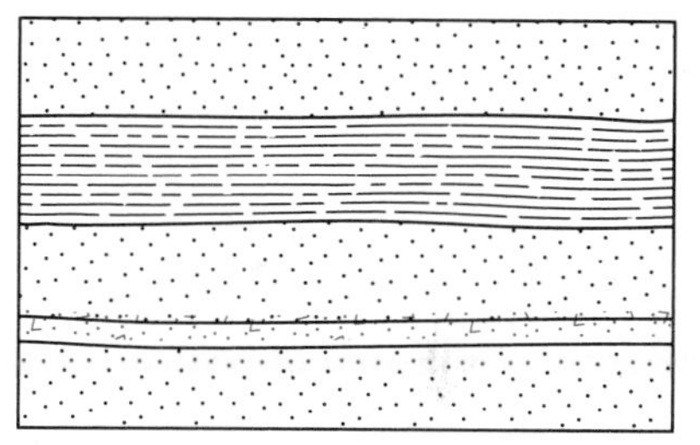

(a) Queen Charlotte Islands

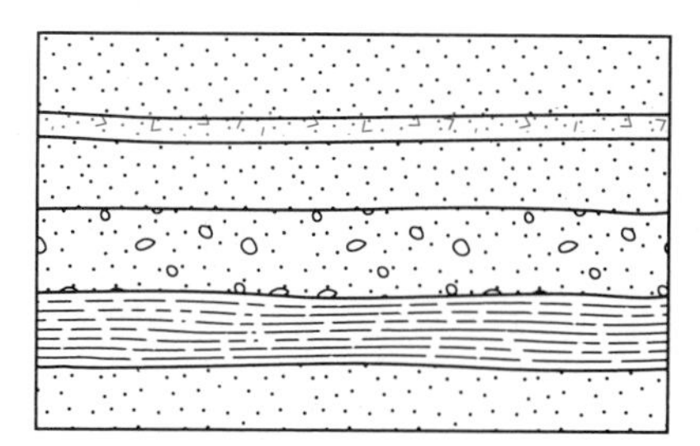

(b) Rocky Mountains

FOSSILS AS CLUES TO RELATIVE AGE

The world's oceans support an astounding variety of plants and animals. As these living things die, they settle to the sea floor. Most of them will decay or be eaten by scavengers, but some may become buried, and therefore protected, under new sediment. As the sediment hardens into sedimentary rock, these fossils—evidence of ancient life, in the form of shells, bones, and imprints—become encased within it. Although metamorphism may eventually change their composition, their impressions may be preserved for many millions of years. Undisturbed by weathering, a deeply buried layer of rock may become a kind of "time capsule" of fossils, representative of a particular interval of geological time.

Thick sedimentary strata, such as those found in the Grand Canyon in Arizona, reveal something interesting (Figure 12.3). Fossils found in deeper layers are quite unlike modern organisms. In fact, the deeper you go, the more unfamiliar the fossils appear. This evidence suggests that living things on Earth are continually changing—rather like styles, fashions, and car designs!

Every period in the Earth's history appears to have had its own characteristic collection of life forms. The fossils of these organisms provide important clues to finding the relative and absolute ages of rock layers. One of many such examples is the fossil known as *Olenellus*. *Olenellus* was a marine animal known as a trilobite, a many-legged relative of modern crabs, lobsters, and insects. *Olenellus* lived within a fairly narrow range of geologic time, first appearing in that form about 550 million years ago. It became **extinct**, or died out, about 450 million years ago. During that 100 million-year period, it was abundant and spread widely in many oceans. Like the modern crab, *Olenellus* made its living crawling on the sea floor scavenging for food. So *Olenellus* was likely to be buried by new sediment when it died. Its hard, armoured shell, which was shed as the animal grew, was more resistant to decay than soft body parts, and was more likely to be preserved as a fossil than for example, a jellyfish.

Olenellus is an example of an **index fossil**, a fossil that is useful for indicating the period of geological time when a sediment was deposited. A **paleontologist** is a scientist who studies ancient life forms. By knowing how to recognize index fossils, paleontologists have learned how to determine the relative age of rocks, and to estimate their approximate absolute age.

EXTENSION

If you had to choose, what kinds of things would you like to put in a time capsule that would be representative of the time you now live in? Which items would last the longest? Why?

ACTIVITY 12E *A Matching Puzzle*

Here is an opportunity to apply the same principles that paleontologists use in solving a relative age problem.

Sedimentary rocks containing fossils are located in many parts of the Earth. Unfortunately, there is no single place where all strata are found together, complete, and stacked neatly in order. To put all these layers in relative order we need to assemble many separate pieces. It is like sorting out a puzzle. Index fossils are valuable tools for allowing us to fit the pieces together.

The illustrations in Figure 12.14 show cross-sections of layered sedimentary rock from different locations. At least one index fossil was found in each layer. The key in Figure 12.15 gives the type of each index fossil as well as the name of the period of geological time during which it lived. The fossils in the key are arranged in random order.

PROCEDURE

1. You will be provided with a photocopy of the cross-sections from Figure 12.14. Cut out each of the six columns (do not separate the layers—it is important to know their relative ages).
2. Apply what you know about the law of superposition and index fossils. Arrange the columns side by side on your desk so that the fossils are in the order of their relative ages (see Figure 12.15).
3. Using additional information provided in the key, list the geologic periods in order. List the most recent period at the top and the oldest at the bottom.

DISCUSSION

1. (a) Describe in your own words how index fossils found in rock layers from two widely separated locations can be used to find out the relative age of the layers.
   (b) What information do you need to know beforehand?
   (c) What would you look for in other rock layers to find out which fossil is older?
2. In this activity, you may have noticed that some rock layers appeared to be missing from a sequence. What may have happened in nature to cause this? (HINT: What could have caused the rough and uneven surfaces where the missing layers should be?)

locality 1

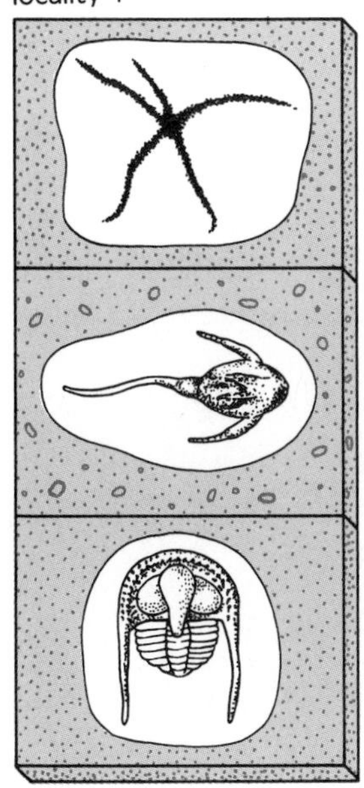

locality 2

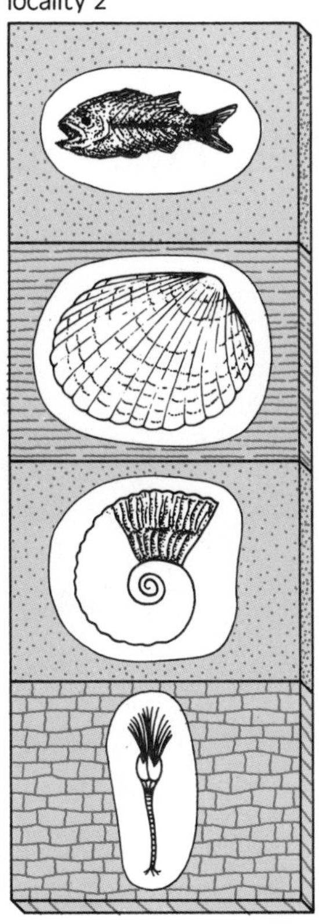

locality 3

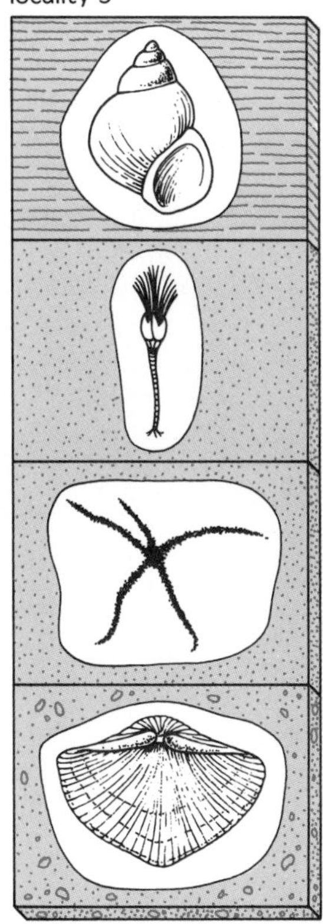

locality 4

locality 5

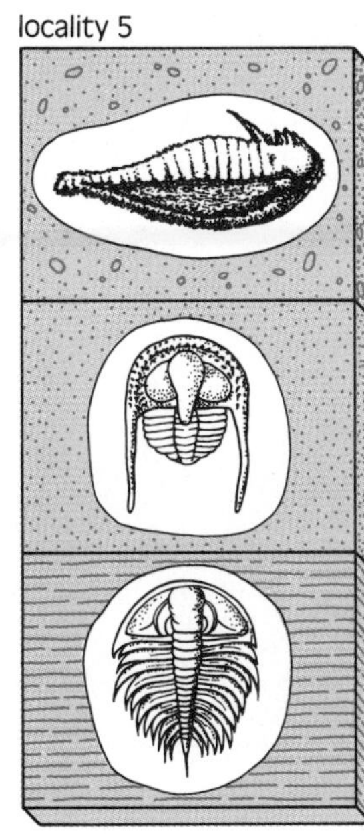

locality 6

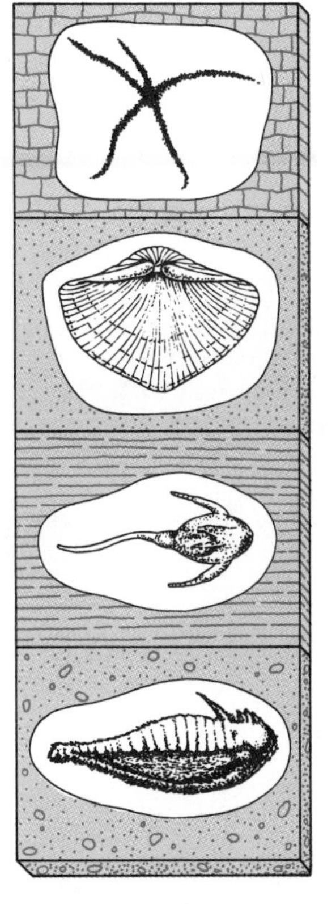

Figure 12.14
Geological cross-sections representing sedimentary rocks from six localities.

Key

Fossil	Type and period
	crinoid, Permian
	trilobite (*Olenellus*), Cambrian
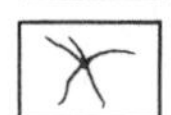	jawless fish, Devonian
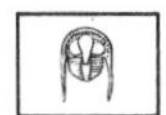	starfish, Pennsylvanian
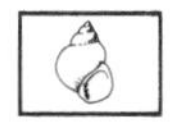	trilobite, Ordovician
	snail, Triassic
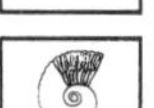	eurypterid, Silurian
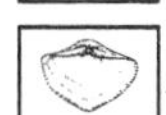	ammonite, Jurassic
	brachiopod, Mississippian
	bony fish, Cenozoic
	clam, Cretaceous

Figure 12.15
Align the columns in Figure 12.14 so that layers of similar age are matched.

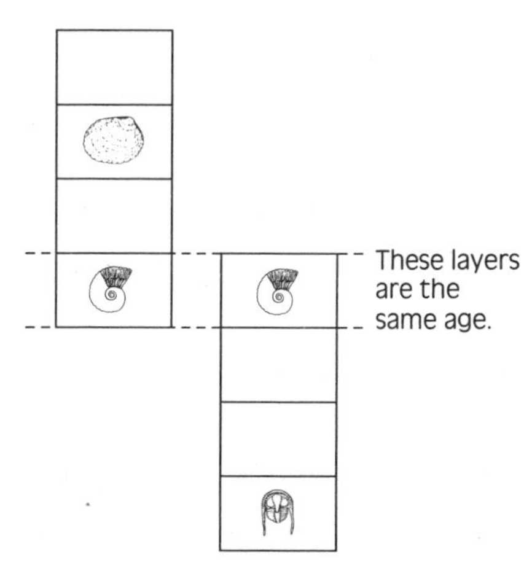

SCIENCE IN OUR WORLD

A WINDOW ON ANCIENT LIFE — THE BURGESS SHALE OF FIELD, B.C.

Most fossils show only the most durable parts of organisms — for example, a bone, a shell fragment, or a leaf impression. As a result, at most fossil sites, there is only a sketchy record of past forms of life. But, in 1909, American paleontologist Charles Walcott discovered the complete and extraordinarily preserved fossils of many animals in the Burgess Shale rock formation near Field, B.C. Walcott collected more than 80 000 fossils. He believed that they were ancestors of organisms living today.

In the 1960s, paleontologist Harry Whittington, of Cambridge University in England, visited the Burgess Shale at the invitation of the Geological Survey of Canada. Later, he and colleagues re-examined Walcott's fossils too. They used ultraviolet photography to make fossil images more visible. And they used dental drills to delicately remove shale from around the fossils. Microscopes helped them to draw detailed illustrations of each specimen. The drawings were used by experts at the University of Oslo, in Norway, to make models of what the animals probably looked like. The task can be compared to swatting several mosquitoes and dropping them in fresh cement, and then leaving instructions for someone, centuries from now, to chisel them out of the cement, draw them, and create a model of what mosquitoes once looked like.

Over the years, paleontologists reinterpreted many of Walcott's discoveries. They described unique invertebrates including many puzzling arthropods, and what appeared to be a forerunner of the vertebrates. Strange, soft-bodied creatures included *Opabinia*, which had five eyes and a trunk-like appendage. No other organisms anything like these fossils have ever been found. Thus, scientists concluded that the Burgess fossils represent many animals that are now extinct.

Other scientists studied the local environment at that time, over 500 million years ago — a muddy sea-bottom at the edge of a giant reef. They think that the soft organisms were enveloped in a slow-moving mudslide and moved to deep water, where scavengers and bacteria could not reach them. Later the mud was buried deeply and became the Burgess Shale. Through chance, the fossils were not destroyed by intense faulting or metamorphism in later geological time.

What became of *Opabinia* and all those unique life forms? The Burgess Shale shows us that life may change more frequently than we thought. In the past 500 million years, there may have been many other events similar to the extinction of dinosaurs 60 million years ago. If so, are we, too, only a temporary species on Earth, no more secure than any other form of multicellular life?

Opabinia. *Its "snout," tipped with grasping spines, is on the left. Three of five eyes are seen where the "snout" joins the body. Long flaps for movement hang down from the body, and fins are located on the right.*

THE STANDARD GEOLOGICAL COLUMN

A paleontologist's dream: to find a single location on Earth where all the sedimentary rocks that ever existed could be found together with all their index fossils. Such a "superstack" would be extremely useful. The relative age of any sedimentary rock layer containing an index fossil could be found by simply comparing it to the index fossils in the column. Unfortunately, such a column has never been found and is unlikely to exist.

However, geologists have been piecing together the puzzle, using a process similar to the one you used in Activity 12E. They have developed an imaginary "superstack" over the last 100 years or so. It contains all the sedimentary layers and fossils ever found on the Earth, organized in order of relative age (Figure 12.16). It is called the **Standard Geological Column**. Segments of the Standard Geological Column are named for places where sedimentary rocks and index fossils were first studied. Names like "Pennsylvanian" and "Mississippian" are taken from regions in the United States. "Ordovician" and "Silurian" are named after two Celtic tribes, the Ordovices and the Silures, that once lived in Wales. "Jurassic" comes from the Jura Mountains located between France and Switzerland.

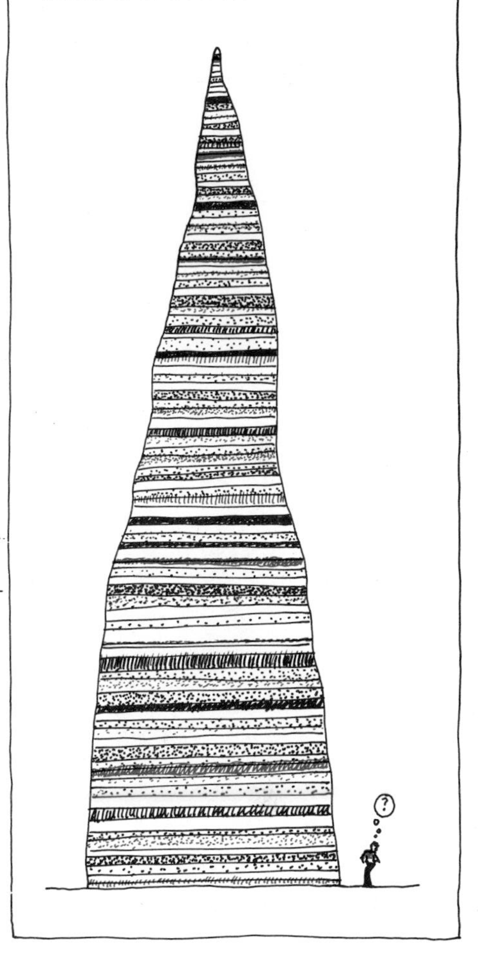

Figure 12.16
The Standard Geological Column is an imaginary collection of all the sedimentary rock layers and their fossils from everywhere in the world. In what ways is it useful?

REVIEW 12.3

1. While on a field trip to the Rocky Mountains, students discovered fossils in different strata. Jan and Paul got into an argument about which of the fossils they found was older. Suggest some ways that this argument might be resolved.
2. (a) What are some characteristics of index fossils?
   (b) Name an organism that you think could become an index fossil for the time we live in. Justify your answer.
3. Each cross-section of sedimentary rocks and "fossils" in Figure 12.17 was found in a different location. Study them carefully to determine the relative age of these "fossils." Arrange them in order of time, then answer the following questions:
   (a) Which "fossil" is the oldest?
   (b) Which "fossil" is the youngest?
   (c) Indicate where you think erosion surfaces are. (HINT: Erosion often removes entire layers before new sediments are deposited above them.)

Figure 12.17
Cross-sections representing sedimentary rock layers. The symbols represent index fossils.

μ
Ω
π
φ

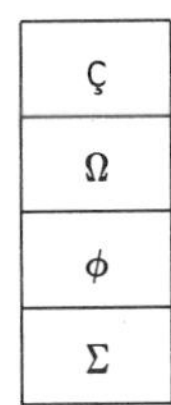

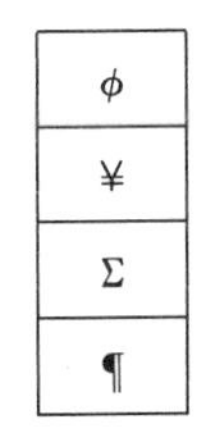

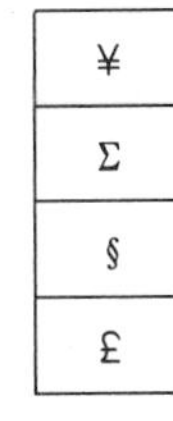

4. Examine the cross-section in Figure 12.18 and use it to do the following tasks.
   (a) Name the geological structures that you see and label them on a sketch.
   (b) List these structures, and the bodies of rock in which they are found, in the relative order in which they were formed.
   (c) Describe how you applied each of the two principles—the law of superposition and the crosscutting rule—in solving the problem in (b).
5. Draw a cross-section problem like the one above. Give it to another student to solve.
6. Suggest reasons why it is unlikely that there is a single location on Earth that shows a continuous record of sediments deposited throughout geological time.

Figure 12.18
Geological cross-section.

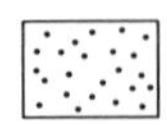 sandstone

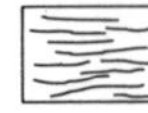 shale

 igneous rock

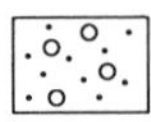 conglomerate

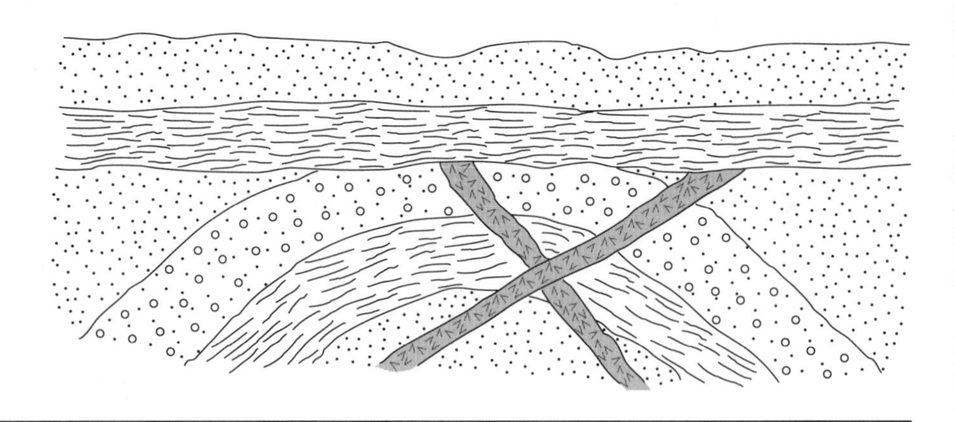

12.4 Finding Absolute Age

Geologists find it challenging to piece together the Earth's history. Sorting out the order in which many of the Earth's changes occurred is a great accomplishment, but it is not enough. The Standard Geological Column is useful, but it gives us only relative age. We would like to know more than which event came first—we want to know *when* each event took place. Since almost all the Earth's history occurred before there were written records, we must use clues found in the rocks to establish their absolute age.

You have discovered that the passage of time is marked by change. By knowing the *rate* at which something changes, you can calculate the amount of *time* that has passed by measuring the amount of *change* that has occurred. For example, imagine a candle that gets shorter by 2 cm each hour of burning (Figure 12.19). If you notice that it has become 6 cm shorter after burning for a while, you have enough information to calculate how long it has been burning.

burning rate: 2 cm/h
0 cm
2
4
6
3 hours

Figure 12.19
A candle burning at a steady rate can be used to mark the passage of time.

$$6 \text{ cm} \div 2 \text{ cm/h} = 3 \text{ h}$$

Similarly, if some kind of change can be measured in a rock and if the rate of change is known, the age of the rock may be estimated.

In this section, you will learn about certain kinds of measurable changes that occur in rocks and how to apply these measurements to estimate their absolute age.

RADIOACTIVITY

The atomic nuclei of some **radioactive** elements, such as uranium and plutonium, are unstable. Radioactive means that some of these nuclei are randomly and spontaneously changing, usually fragmenting into smaller pieces, as explained in Chapter 7. Much of the radiation emitted results from the escape of the smallest high-energy fragments. This kind of change is called **radioactive decay**. Many of the fragments become the nuclei of lighter elements, such as lead and helium.

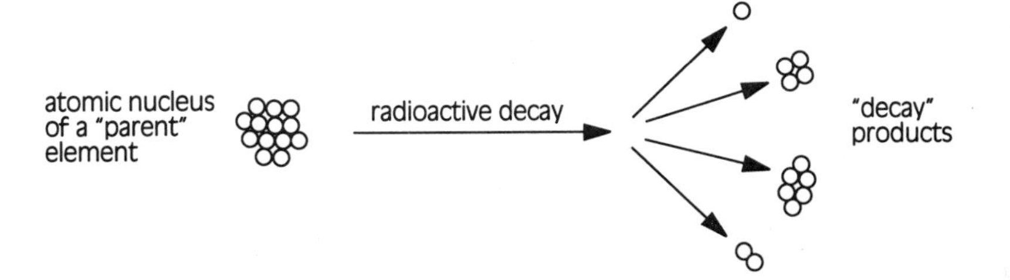

Figure 12.20
During radioactive decay, the atoms of a "parent" element break up into smaller atoms and atomic particles called decay products.

Some of the atoms found in mineral crystals are also radioactive isotopes (Figure 12.20). **Isotopes** are atoms of the same element that have different mass numbers, as you learned in Chapter 7. For example, uranium 238 (U-238) is a radioactive isotope of uranium that has a mass number of 238. U-238 decays in a number of steps to produce one atom of lead 206, eight atoms of helium 4, and six free electrons. The uranium is the **parent material**. The lead and helium are **decay products**.

When a rock begins to crystallize (as when a magma cools and solidifies to form granite), the decay products begin to accumulate in the rock as the radioactive parent material decays. Over time, then, the amount of the parent element decreases as the amounts of decay products increase.

Like the burning candle, the age of the sample may be calculated if we can measure the *amount* of change (the amounts of parent material and decay products), and if we know the *rate* at which the change is occurring. The decay rate of every radioactive element has been determined experimentally. The method of measuring and comparing the amounts of parent material and decay products in a mineral to determine its absolute age is called **radiometric dating**.

Figure 12.21
After each half-life (10 minutes), one-half of the remaining pie "decays."

Elapsed time (min)	Number of half-lives passed	% pie remaining
0	0	100
10	1	50
20	2	25
30	3	12.5

HALF-LIFE

To understand radiometric dating, you need to understand the concept of half-life. Imagine that a whole pie has been left unattended in your classroom (Figure 12.21). After 10 minutes, Blair cannot resist it any longer and eats half of it. Ten minutes later, Sandra eats half of what is left, leaving a quarter. Ten minutes later, Jasmine eats half of that. This process continues for some time.

If you plotted a graph comparing the amount of pie remaining with the passage of time, you would get something similar to Figure 12.22.

Although the amount of pie is decreasing with time, it does not "decay" at a constant rate. The first piece to disappear was quite large, whereas pieces taken later were very small. Radioactive materials decay in a similar way. How can the "decay rate" of the apple pie be described?

One feature of the apple pie model that remained constant was the amount of time required for one-half of it to "decay" — which in this example was 10 minutes. This is known as the **half-life**. Similarly, the decay rate of radioactive materials can be described using half-life. The half-life of a radioactive material can be found by using its decay graph to locate the *time* that corresponds to 50 per cent remaining. (In Figure 12.22, it is 10 minutes.)

Figure 12.22
If the "decay" process continues indefinitely, will the pie ever completely disappear?

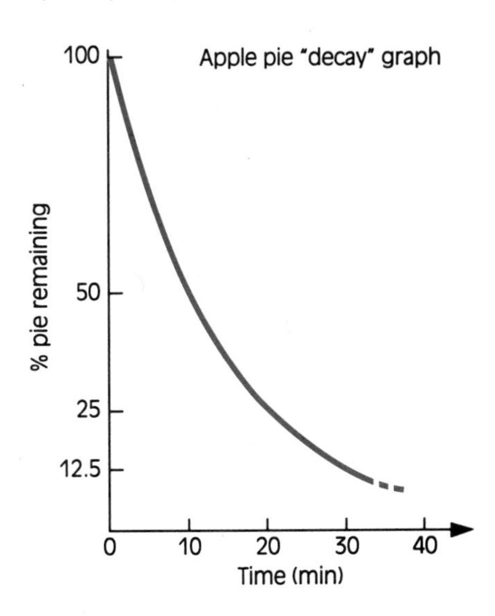

COMPARING METHODS

Many elements have radioactive isotopes. These isotopes have half-lives that range from fractions of a second to billions of years. Several radioactive isotopes are valuable for determining the absolute age of rocks. Each one has its special uses and limitations. Three such isotopes, together with their corresponding decay products and half-lives, are described in Table 12.1.

Table 12.1 Three Radioactive Isotopes

Isotope	Decay product	Half-life (years)
Uranium 235	Lead 207	713 million
Potassium 40	Argon 40	1.3 billion
Carbon 14	Nitrogen 14	5730

For an isotope to be suitable for determining the age of something, it must fulfill the following requirements.

1. Some of the isotope must have been present in the sample when it was first formed. For example, C-14 is usually found in **organic** (once living) material, such as plant or animal remains. Since it is not incorporated into magma, C-14 cannot be used for dating igneous rocks.

2. There must not be any contamination or loss of either the parent isotope or the decay products. The atoms must be completely sealed within the crystal, so that nothing can leak in or out. It is impossible to find the radiometric age of a rock that has been exposed to chemical weathering (such as oxidation or the dissolving effects of water) or heating. This is a particular problem when the isotope K-40 is used for dating, since its decay product, Ar-40, is a gas and is therefore easily lost from the sample.

3. Isotopes with short half-lives are useful only for dating fairly young rocks. Recall the apple-pie decay model, where 50 per cent of the pie disappeared every 10 minutes. In theory, the pie would never completely disappear, although there comes a point (usually after about 10 half-lives) when there is too little of the original material remaining to allow accurate measurement.

4. Isotopes with long half-lives are useful only for dating very ancient rocks. Suppose you wanted to use U-235 to check the absolute age of ash from Mount Saint Helens. If it takes 713 million (713 000 000) years for 50 per cent of the U-235 to be converted into Pb-207, how much product do you think would have formed since 1980? Not much!

DID YOU KNOW?

Crater Lake, Oregon, is the site of an explosive eruption of Mount Mazama 6640 years ago. The date of this eruption was determined by applying the carbon 14 method to a piece of burned wood found in the layer of ash.

CAREER

ISOTOPE LABORATORY TECHNICIAN

Have you ever wondered how we know the age of archaeological sites or how long ago a mammoth died? The University of Waterloo Radiocarbon Laboratory in Ontario determines the age of many kinds of natural materials. Jeanne Johnson is one of several technicians responsible for analyses that produce the dates vital to research in archaeology, glacial history, studies of sea-level and climate change, and even the search for well water.

What do you do in the laboratory?
I prepare and test samples to find out their age. The technique I use depends on the material to be analysed. We are asked to date materials such as bone, wood, peat, shells, coral, and even water.

What was one of your most interesting tests?
Mammoth remains were found in a farmer's field near Stratford, Ontario. Actually they found them in a former peat bog and wanted to know how long ago the mammoth was alive. Bone and tusk were brought in for testing. Radiocarbon dating was used since it would be reliable for material up to 50 000 years old. Carbon 14 atoms decay so that after 5700 years only half of them remain. By finding out the amount of carbon 14 remaining today, it would be possible to estimate how long ago death occurred. On my first try I used a bone and it appeared to be only 4000 years old. This was odd, since mammoths haven't been around for 10 000 years.

What was the problem?
The original bone sample was porous enough to have allowed recent atmospheric carbon to penetrate. I was then given a piece of the animal's tusk. This sample dated correctly, but the tusk, which was to be used in a display, lost 20 per cent of its length in the process.

What do you like best about working at the university?
It is a very relaxed, friendly atmosphere and we are working with interesting people doing research all over the world. The problems are varied and exciting. And I find I can schedule my work hours to match the demands of my growing family. It's an ideal arrangement.

How do people become isotope technicians?
Most of us are university graduates—I have a degree in general science, though not everyone does. You learn on the job, starting with the simpler analyses and going where your aptitude and interests lead you.

What does it take to be good at your work?
You do have to have an understanding of chemistry and enjoy laboratory work. You must understand the routine and why you are doing it.

What is most difficult about your job?
Maintaining your laboratory system at peak performance is essential. You also have to realize when there is a problem and correct it. A lot of care is needed since there is often only one sample—it may have cost thousands of dollars to obtain it in the first place.

REVIEW 12.4

1. How does an atom of a radioactive element change as it undergoes radioactive decay?
2. Examine the radioactive decay curve of Element X, shown in Figure 12.23.
   (a) Find its half-life.
   (b) Which element has this half-life? (HINT: See Table 12.1.)
   (c) Find the time required for half of the sample to decay again (from 50 per cent to 25 per cent). Compare this value with the half-life you calculated in (a) above. Is there any similarity?
   (d) A rock containing some Element X is found to contain only 40 per cent of its original amount. How old is the rock?
3. Can radiometric dating be used to find the absolute age of any material? List some characteristics that would make some materials unsuitable for this dating method.
4. Suggest a reason that carbon dating is not suitable for finding the age of ancient fossils that are millions of years old.

Figure 12.23
The radioactive decay curve for Element X.

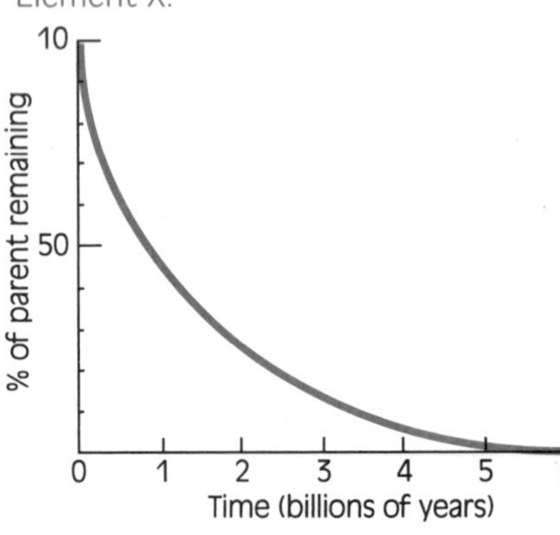

5. Which radioactive isotopes that you have studied in this chapter would not be suitable for radiometric dating when
   (a) testing materials that were once living?
   (b) testing inorganic materials (matter that has never been alive)?
   (c) dating rocks older than 1 billion years, such as meteorites?
   (d) dating materials younger than 1000 years old?
   (e) dating rock samples that had been heated at some time in the past? (HINT: This would drive off any gases trapped in the rock.)
6. Plot a decay curve for the radioactive isotope thorium 232, assuming its half-life is 14 billion years.

12.5 Finding Absolute Age of Sedimentary Rocks

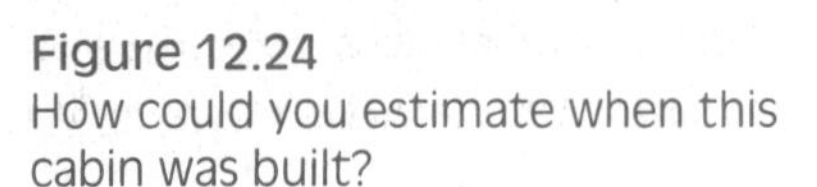

Figure 12.24
How could you estimate when this cabin was built?

Suppose you discover a deserted log cabin in the forest (Figure 12.24). Think of some ways that you might estimate how long ago it was built. What evidence would you look for? Stop and think about this problem and jot down some ideas, before reading further.

The style of the furniture and clothing worn in photographs, and perhaps scraps of newspapers found inside the cabin could all provide useful clues.

How about counting the tree rings on the logs from which the cabin was made? What would that be a measure of?

Now, suppose a granite pebble removed from a conglomerate is subjected to a radiometric test and shown to be about 125 million years old? What does that really mean? What actually happened to this pebble 125 million years ago?

Remember that radiometric age refers to the *time since a mineral grain crystallized*. The most common way that rock crystals form is by cooling from a melt, in the same way that igneous rocks form. Crystallization also occurs during metamorphism, when crystals change their size or composition in response to heat and pressure.

Thus, for both igneous and metamorphic rocks, the "atomic clock" starts running as soon as the mineral grains form. But sedimentary rocks are quite different. Rocks such as sandstone, shale, and conglomerate are made up of grains of sand, mud, or gravel derived from the weathering of older rocks. In the example above, radiometric dating of the granite pebble in the conglomerate reveals only that the granite crystallized from a magma 125 million years ago. It says nothing about when the pebble was weathered, when it was eroded, and when it was deposited to eventually become part of the sedimentary rock itself. Going back to the analogy of the log cabin, building a house out of 100-year-old logs does not make the house 100 years old. The age of the house (or sedimentary rock) means the time since the pieces were assembled, not the age of the pieces themselves.

So, how can the absolute age of sedimentary rock be found? To solve this problem, you will need to apply what you have learned about both absolute and relative age dating.

ACTIVITY 12F *Age Dating of Sedimentary Rocks*

Some rock structures are made up of both igneous rocks and fossil-bearing sedimentary rocks, allowing us to apply indirect methods to determine the absolute ages of sedimentary rock.

Apply what you have just learned, together with some logic of your own, to discover the absolute age of the sedimentary rock layers shown in the following cross-section.

1. Figure 12.25 shows layers of sedimentary rock (labelled A, B, C, D, and E) as well as some igneous rocks. The igneous rocks include a lava flow which is cut by one of three dikes, and a plutonic rock body. The absolute ages of these igneous rocks were determined through radiometric dating and are given on the diagram.

   Determine the range of the absolute age of each of the sedimentary layers. Construct your own table in which to record these ages.

2. In a sentence or two, describe how you solved this problem.

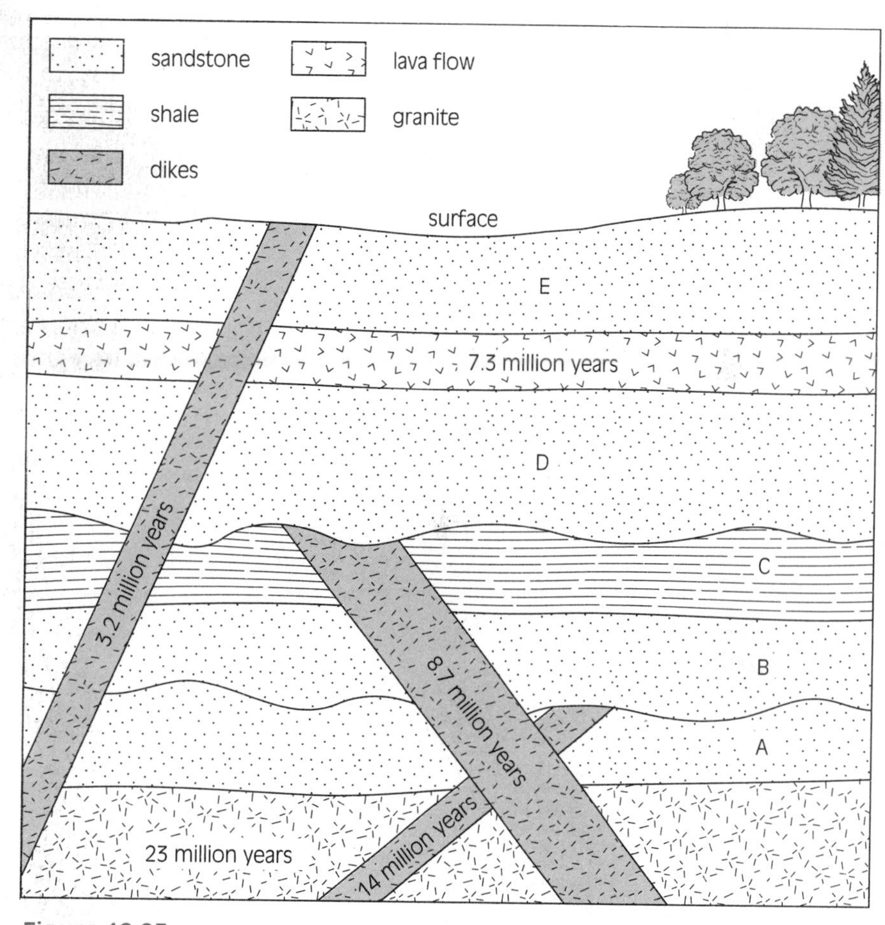

Figure 12.25
What is the maximum and minimum absolute age of each sedimentary layer in this cross-section?

3. To prepare for the next step, copy Table 12.2 into your notebook. ("Period" in the middle column refers to the segment of the Standard Geological Column in which each fossil was found.)
4. The cross-sections shown in Figure 12.26 are from five different localities. Each one shows igneous rocks and layers of fossil-bearing sedimentary rock. The absolute age of the igneous rocks (found by radiometric dating) is also shown. Determine the absolute age range of each fossil at each site, and record these ages on the table.
5. Once you know the absolute age of fossils, they can be used to find the absolute age of sedimentary rocks associated with them. Study the three cross-sections in Figure 12.27, and determine the absolute age of each sedimentary rock layer shown. Record these ages on a sketch you make of the cross-sections.
6. Describe in a sentence or two how you solved this problem.

DISCUSSION

1. What is the reason that radioactive isotopes cannot be used to directly measure the absolute age of a sedimentary rock?
2. Name and describe as many kinds of rocks and structures as you can that might be dated using radiometric methods.
3. Describe how fossils can be used to find
   (a) the relative age and
   (b) the absolute age of a sedimentary rock layer.

Table 12.2 Sample Data Table for Activity 12F

Fossil	Period	Absolute age
Clam	Tertiary	
Gastropod	Triassic	SAMPLE ONLY
Oyster	Cretaceous	
Ammonite	Jurassic	
Scallop	Mississippian	
Snail	Silurian	
Leaf	Recent	

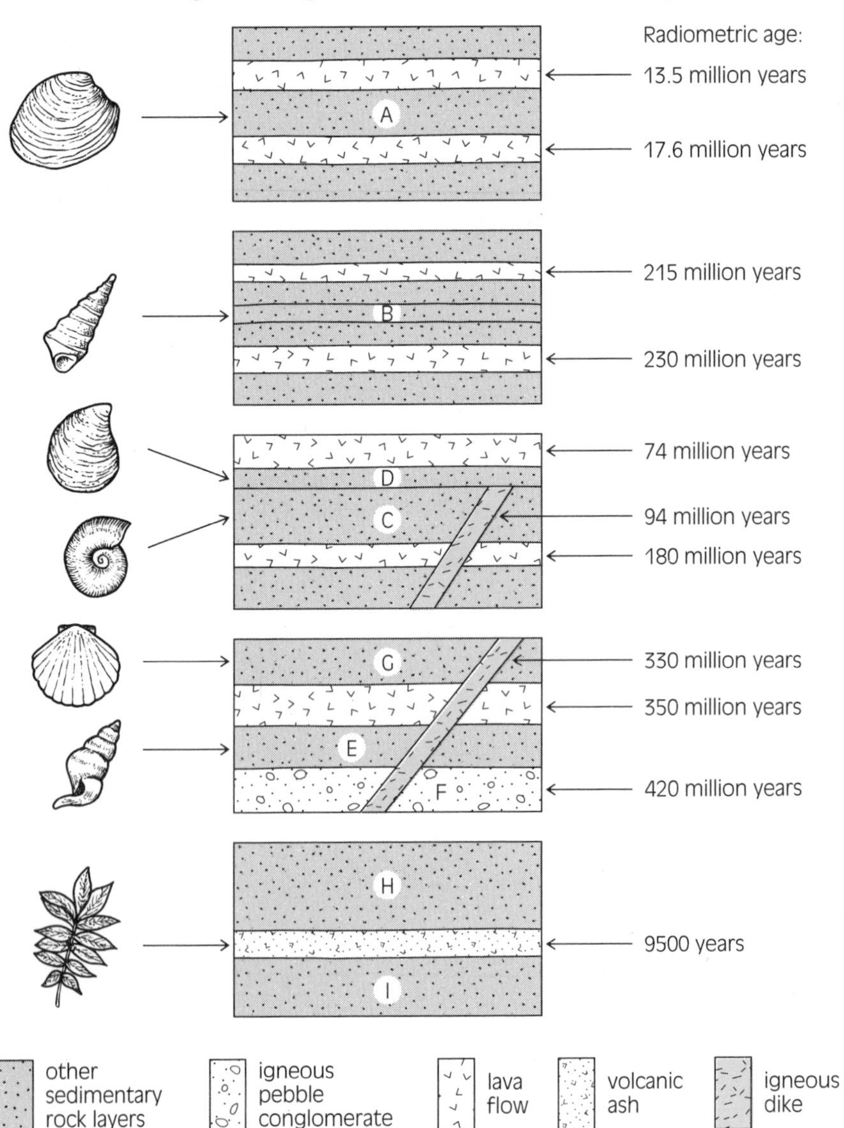

Figure 12.26
Use the absolute age of the igneous rocks to determine the age of each fossil.

Figure 12.27
From the information in Figure 12.26, find the absolute age of each sedimentary rock layer.

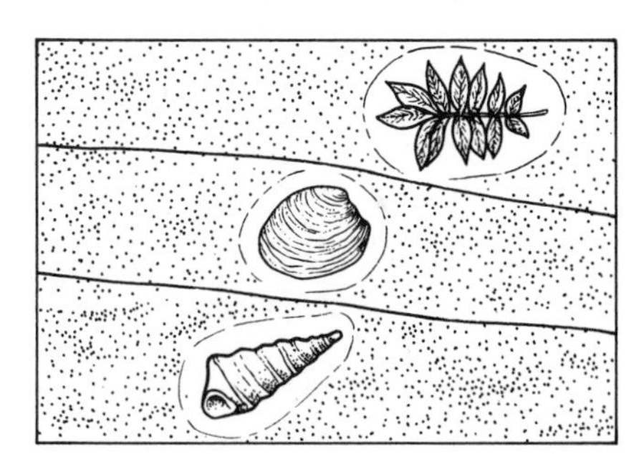

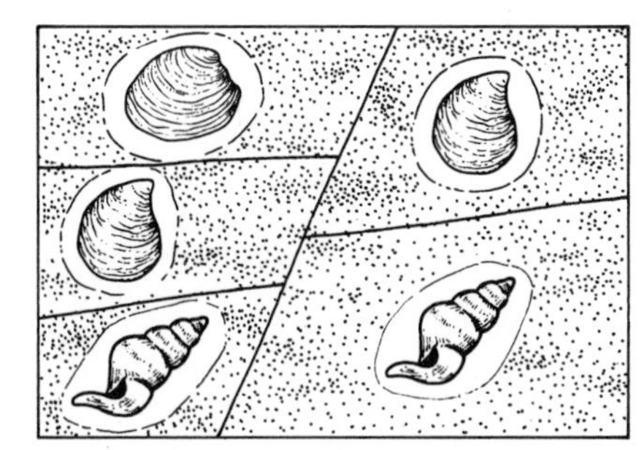

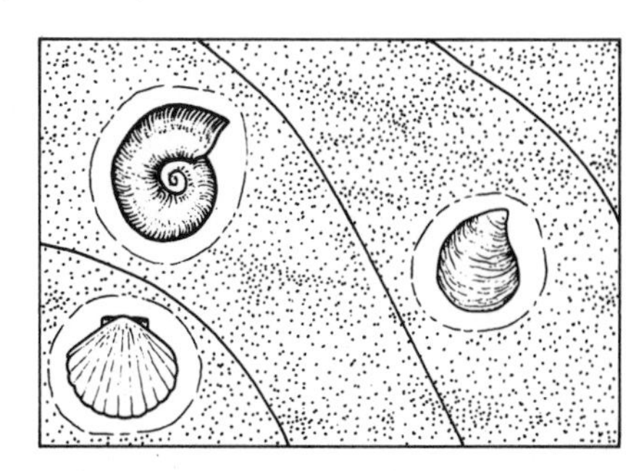

THE GEOLOGICAL TIME SCALE

Geologists have been able to assign absolute ages to each of the sections of the Standard Geological Column described earlier. What has emerged is an Earth calendar termed the **Geological Time Scale** (see Appendix D). It relates the time of geological events to an absolute time scale. You might describe the time of an event in the past year as occurring in February. In a similar way, a geologist might describe geological events (such as the formation of rocks, the folding of mountains, or the appearance of certain organisms) as occurring in the Devonian Period. If you are familiar with the divisions on the Geological Time Scale, the word "Devonian" is useful.

REVIEW 12.5

1. Explain why it is difficult to measure the absolute age of sedimentary rocks using radiometric methods.
2. List some similarities and differences between the Standard Geological Column and the Geological Time Scale.
3. List at least three ways in which fossils can be scientifically useful.

CHAPTER • REVIEW

Key Ideas

- The passage of time is marked by change.
- Geological time is measured in two ways. Relative age is the age of a geological event in relation to another event. Absolute age is the actual amount of time (expressed in years) that has passed since an event occurred.
- Relative age may be determined by applying the law of superposition and the crosscutting rule.
- The law of superposition states that the oldest rock layer in a sequence of undisturbed strata is found at the bottom.
- If some feature, such as a dike, cuts through a sedimentary rock, we know the interrupted rock is the oldest.
- Geological events include the formation of igneous, sedimentary, and metamorphic rocks; the formation of geological structures such as fractures, faults, folds, and erosion surfaces; and the deposition of sediments.
- Each period of geological time was represented by a particular assortment of living things; a few are preserved as fossils.

→

- The age of igneous or metamorphic rocks may be estimated directly by measuring the amount of radioactive decay that has occurred since a rock crystallized.
- The decay rate of a radioactive isotope is described in terms of its half-life—the time required for 50 per cent of the parent material to change into a decay product.
- The absolute age of sedimentary rocks may be estimated indirectly by examining the age of igneous rocks that were formed either before or after the sedimentary rocks.
- The Standard Geological Column and the Geological Time Scale describe all the Earth's rocks, geological events, and fossils in order of their occurrence.

Vocabulary

geologist	fold
absolute age	erosion surface
relative age	strata
igneous	law of superposition
crust	crosscutting rule
magma	extinct
lava	index fossil
ash	paleontologist
volcanic rock	Standard Geological Column
plutonic rock	radioactive
sedimentary	radioactive decay
weathering	isotope
sediment	parent material
fossil	decay product
metamorphic	radiometric dating
structure	half-life
fracture	organic
fault	Geological Time Scale
dike	

1. Create a concept map, using as many of the vocabulary words that refer to rocks and structures as you can.
2. Make a crossword puzzle, using as many vocabulary words as you can.

Connections

1. Project Apollo astronauts brought rock samples back from the surface of the moon. It was suspected that these rocks were very ancient—perhaps even as old as the solar system. Which parent isotopes would be the most useful for finding the age of very old rock—ones with relatively short half-lives, or ones with very long half-lives? Give reasons for your choice.
2. The geology of the Earth is sometimes compared with the pages of a book. In what ways are they similar? In what ways do they differ?
3. The age of a tree can be determined simply by counting the number of growth rings in its trunk. Can the age of the Earth be found in a similar way by counting the layers of sedimentary strata? Why or why not?
4. (a) Describe how the Standard Geological Column was compiled.
   (b) What further information was needed for the Standard Geological Column to become the Geological Time Scale?
   (c) How was this information collected?
5. Describe how fossils provide evidence for
   (a) the relative age of rocks,
   (b) the absolute age of rocks,
   (c) evolution.
6. Draw a diagram to illustrate how each of the following changes in the rock cycle might come about:
   (a) sedimentary to igneous,
   (b) metamorphic to sedimentary,
   (c) igneous to metamorphic.
7. You have been asked to find the relative age of two different sedimentary rock layers. They have not been found in the same sequence of strata.
   (a) Describe at least two ways to find out which layer is older.
   (b) Give two limitations to each method.
8. Read the following statements. Indicate whether they are examples of relative age or absolute age.
   (a) Mexico City is the oldest city in North America.
   (b) We moved to Vancouver two years before my sister was born.
   (c) The great Alaska earthquake occurred in 1964.
   (d) A fossilized leaf is found buried within Mount Mazama ash.
   (e) A layer of limestone is found beneath a layer of sandstone.
9. Not all organisms lived during only one period in the Earth's history. For example, the shelled animal *Lingula* (a brachiopod) is found in rocks as old as the Cambrian period, and is still found today off the shoreline of Japan.
   (a) Would the fossil, *Lingula*, be useful for finding the relative age of rocks? Why?
   (b) Could *Lingula*, the fossil, be classified as an index fossil? Explain.
10. What are some difficulties associated with using radiometric dating methods for finding the absolute age of
    (a) sedimentary rocks?
    (b) igneous rocks?
    (c) metamorphic rocks?

11. Describe how someone might go about finding the absolute age of
    (a) a layer of sandstone containing fossils,
    (b) lake sediments,
    (c) a layer of sedimentary rock sandwiched between two ancient lava flows,
    (d) a chunk of granite,
    (e) an igneous rock that has been heated and recrystallized through metamorphism,
    (f) a tree,
    (g) an old building.
12. Suggest some reasons why carbon 14 dating is particularly useful in archaeology—the study of ancient civilizations. Think of some practical examples of how it could be used.
13. Study the photographs of the Martian landscape (Figure 12.28). For each photograph,
    (a) identify the events that occurred,
    (b) arrange the events in order of relative age.

Explorations

1. Visit a local natural history museum and look for examples of rocks, rock structures, geological events and processes, and fossils that were outlined in this chapter. Use photographs, posters, or videotape to report to the class what you saw. Make sure you report what particularly interested you.
2. Choose one period of geological time from the Geological Time Scale. In small groups, do research to find out what kinds of life existed and what geological events occurred during the period you selected. Organize and report this information on a poster. As a class, assemble the posters from all the geological periods into a sequence and create an illustrated class display of the Geological Time Scale.
3. Invent a planet. Using your knowledge of geological time and processes, create a "biography" for this planet. Describe the sequence of events that shaped its past. Explain what clues to these past events were left that enable the present-day intelligent life forms to unravel this planet's past.

Reflections

1. Look at the answers you gave to the questions at the very beginning of this chapter. Have your ideas changed? How?
2. Write at least five questions about geological time that you would like more complete answers to.

Figure 12.28
Craters in the Martian landscape.

(a)

(b)

(c)

CHAPTER

13

Earthquakes and the Earth's Interior

Walking home along 4th Avenue, I hoped I wouldn't be late for dinner. As it was Good Friday, I was one of only about a dozen people on the street.

It came without warning. The ground shook so suddenly and violently it was as if it had been smacked with a giant hammer. The shaking continued—I wondered if it would ever stop. The sensation was like standing in a small rowboat in rough water. Parked cars bounced half a metre off the ground. Above the loud, low rumble I heard the buildings cracking and crashing as they rocked and shook. Debris was falling all around me. Crawling into the middle of the street, I was startled to see the movie theatre drop as if it had fallen into a hole. The marquee became level with the sidewalk, although the building was still intact and not a bulb was broken! The flower shop broke neatly in two. Many buildings on the north side of the street dropped three or four metres below it. After the shaking stopped, the buildings on the south side, apart from the odd broken window, looked fairly well undisturbed.

Those words are from the report of a survivor of the greatest earthquake to have struck North America this century. He was walking along the street in Anchorage, Alaska, shown in the photograph above.

Earth scientists have only recently begun to understand earthquakes. In this chapter, you will learn how, where, and why they occur. You will learn how the energy of earthquakes can be measured and how earthquakes can be located. You will also learn what special risks and hazards they pose. You will even learn how earthquakes can tell us what the interior of the Earth is made of.

ACTIVITY 13A Thinking about Earthquakes

You probably already know something about earthquakes, particularly since British Columbia is located along an active earthquake zone. In this activity, you will explain some of your ideas. It does not matter whether your ideas agree with what scientists or anyone else thinks. This activity is intended to help you identify what you already know and believe.

Make a list of statements about what you already believe or understand to be true about earthquakes. Consider such things as what causes them, where they occur, how often they occur, how long they last, and what some of their effects are. Select three of these statements, and record each one in brief form on a filing card. Share these ideas with other students within small groups. (Alternatively, your teacher may collect them and share them in a class discussion.) In your learning journal, write a brief summary of each idea contributed by your classmates. Beside each idea, indicate whether you agree or disagree with it. Make a note describing the reasons for your opinion. Keep this record so that you can review it at the end of the chapter.

13.1 On Shaky Ground

Centred about 200 km from the city of Anchorage, Alaska, the massive 1964 earthquake described in the opening to this chapter was felt as far as 1200 km away. The main shock is thought to have released about twice the energy of the earthquake that destroyed much of San Francisco in 1906, and about 30 times the energy of the more recent San Francisco quake of 1989.

Throughout history, people in many parts of the world have had reason to fear earthquakes. Records from China report about 9000 earthquakes over the last 3000 years. Tokyo suffers major destruction on the average of once each century. In 1923, it was nearly destroyed by a major earthquake that claimed 143 000 lives. The heaviest toll from an earthquake occurred in Shensi, China, in 1556. That earthquake caused 830 000 fatalities and was probably the world's greatest natural disaster. Perhaps one of the biggest earthquakes known to us is the one that reduced Lisbon, Portugal, to rubble in 1755. That quake shook buildings all over Europe.

Although earthquakes the size of the Alaskan quake only occur about once every year or two, hundreds of thousands that are strong enough to be felt without instruments occur every year.

Many cultures have proposed creative ideas about the Earth to explain these violent episodes. The early Tlingit of the northwest coast believed the Earth was supported on a giant post guarded by an old woman. When she was hungry, the post shook, causing earthquakes. Other cultures maintained that the Earth was supported on the backs of giant turtles, frogs, dogs, pigs, or gods with explosive tempers (Figure 13.1). Still others believed that the centre of the Earth held a raging furnace whose tremendous forces shook the crust from time to time. Ironically, this is not unlike what we believe today!

Figure 13.1
Early Tibetans believed that a giant frog carried the world on its back, and that earthquakes occurred when the frog twitched.

Figure 13.2
A cross-section illustrating the difference between the focus and epicentre of an earthquake.

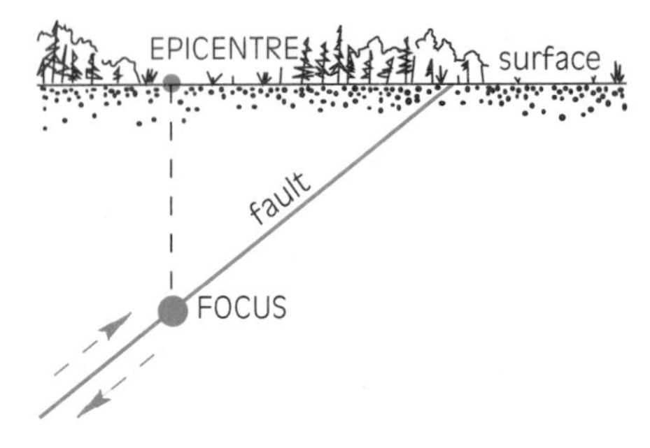

CAUSE AND EFFECTS

Earthquakes occur when rock masses in the outer layer of the Earth move suddenly along ruptures in the Earth's crust called **faults**. The rock movement may occur vertically, horizontally, or in combination.

The location on a fault where there is the greatest amount of movement is called the **focus** (Figure 13.2). The focus is usually some tens of kilometres below the surface. The **epicentre** is the point on the surface that is directly above the focus.

A great deal of energy is often released when rock masses move along a fault. This energy travels away from the focus in all directions as waves. Another example you may have observed of energy carried in this way is water waves generated by a pebble thrown in a pond. During some powerful earthquakes, ground waves up to 2 m high have moved on the land surface.

The severe shaking that causes most earthquake damage results from these earthquake waves passing over the Earth's surface. It may last from a few seconds to five minutes or longer. Buildings and bridges may collapse. Rock and soil may be loosened from steep slopes to fall as landslides. Both the Hope Slide of 1965 in British Columbia and the Madison Slide of 1959 in Montana are thought to have been triggered by earthquakes (Figure 13.3).

Certain types of soil will **liquefy** — turn to wet mud — when shaken. In other places, large cracks called fissures will open up in the ground. These effects may disrupt roads and power, sewer, and water lines. They also may damage the foundations of houses and other buildings.

Earthquakes also occur under the ocean. Sudden displacement along a fault on the sea floor often produces a **tsunami**, a Japanese word for "giant wave(s)." Unlike normal water waves, tsunami travel at speeds of up to 800 km/h in the open ocean. They remain small in height when the water is deep, and may pass unnoticed under a ship at sea. As the waves approach the shallow water near a shoreline, they slow down but

EXTENSION

To make a tsunami, submerge two bricks in a long, shallow tray of water. Lean one on the edge of the other and tie a string to the brick on top. Trigger an "earthquake" by tugging on the string, causing one block to slip suddenly past the other. Observe the motion of the water. What does it tell you about earthquake waves and their effects?

Figure 13.3
The 1959 earthquake near Yellowstone in Montana triggered the Madison Slide that blocked the Madison River and formed Earthquake Lake.

grow to surprising heights. Tsunami of almost 70 m in height have been reported. Coastal areas can suffer massive destruction and flooding. A huge tsunami generated by a powerful earthquake in Peru in 1960 destroyed some fishing villages on the coast of Japan. The tsunami from the Anchorage quake of 1964 flooded Port Alberni (Figure 13.4).

Even after the rumbling of a major shock has ceased, the danger has not passed. Damaged buildings and other structures are unstable and may later collapse. Ruptured gas lines pose the threat of fire, whereas broken water mains make firefighting difficult or impossible. For example, gas fires ignited by the 1989 San Francisco quake had to be doused with water pumped by fireboats from San Francisco Bay. Fresh water may become contaminated, and there is a major threat of disease.

Aftershocks are renewed tremors that follow a major earthquake. They may continue for months. Some may be large enough to cause further destruction, especially since many structures are already weakened from the main shock.

EXTENSION

Find some people who have experienced a strong earthquake. You might ask people who have lived in parts of the world that you may know to be geologically active. Interview them and make a written, audio, or video account of their stories. Bring the stories to class and describe some of the similarities and the differences among these experiences. What are some of the effects of earthquakes? In what ways are they dangerous? Have some of these accounts caused you to change your thinking about any of the ideas you listed at the beginning of the chapter?

Figure 13.4
A tsunami from the Alaska earthquake of 1964 caused considerable damage in Port Alberni, B.C.

A MODEL TO EXPLAIN EARTHQUAKES

Geologists have noticed that earthquakes keep occurring along the same faults. These earthquakes are separated by intervals ranging from a few years to a few hundred years. Energy is being "stored up" somehow, rather like when you stretch an elastic band or bend a stick. Eventually a limit is reached and the elastic band or stick breaks. At that point, the **potential energy**—energy that is stored—is suddenly transformed into energy of motion. Energy of motion is called **kinetic energy**.

Of course, we do not commonly think of rock as being able to bend. Under ordinary conditions, it is brittle and will tend to break rather than bend if large forces are applied. However, scientists believe that under great pressures, such as those present deep underground, rock becomes less brittle and more elastic.

Models are useful for understanding materials or processes that are otherwise hard to observe. A model has some characteristics that are similar to the thing it represents. For example, the idea of tiny moving particles called molecules is a useful model for describing a concept of matter. A model based on the rebounding ability of an elastic material helps geologists understand earthquakes.

ACTIVITY 13B Elastic Rebound

In this activity, you will use gelatin as a model of "elastic rock." By using this model, you can investigate how rock deep in the Earth may behave between and during earthquakes.

MATERIALS

one rectangle of stiff gelatin (between 5 and 10 cm on each side and 2 cm thick)
two stiff strips of wood, metal, or plastic (such as wood splints or rulers)
a felt-tip marking pen
sheet of wax paper

PROCEDURE

1. Place the gelatin brick on the wax paper. Use the marking pen to draw a horizontal line across the top surface of the gelatin. This will serve as a reference line.
2. Hold the strips tightly against each of the sides of the square. Slowly pull one of the strips towards you as you push the other away (see Figure 13.5).
3. Observe how the gelatin changes shape or breaks. Draw and describe any changes you notice. You may wish to draw "before" and "after" pictures.

DISCUSSION

1. What evidence is there that energy is being stored as force is applied to the block?
2. Account for the changes that occurred in your model. What do these changes represent in nature?
3. Make a sketch to show the relationship between the direction of the applied forces and the line along which a fault develops.
4. What happens to the energy in the deformed block at the instant the block breaks? What does this represent in nature?
5. What conditions are necessary for an elastic material such as gelatin to break? What information would help you to predict when the block will break?
6. Based on what you have discovered, propose an explanation that would account for the periodic release of energy along a fault.
7. Invent a model of your own that illustrates how energy is stored as something changes shape, and then is suddenly released. Using simple materials, demonstrate this model to others.

Figure 13.5
Experimenting with elastic rebound.

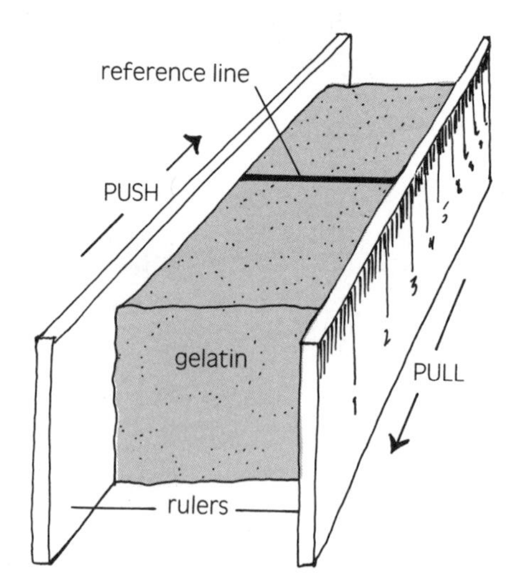

THE ELASTIC REBOUND THEORY

A **theory** is a proposed explanation of events, based on observation, testing, and reasoning. A theory that describes the storing and sudden release of stress in the Earth's crust is the **elastic rebound theory**. The elastic rebound theory is illustrated in Figure 13.6. Rock masses on each side of a fault are slowly pushed past one another, but the friction between the blocks also locks them in place. As a result, they bend in a way similar to the gelatin block you observed in Activity 13B (Figure 13.6a). When friction is eventually overcome, the energy that has accumulated over many years is released in a matter of seconds (Figure 13.6b). The process is repeated again and again, over periods of thousands of years.

(a)

(b)

(c)

Figure 13.6
Stages of elastic rebound.

REVIEW 13.1

1. Describe the difference between the epicentre and the focus of an earthquake.
2. List at least eight dangerous effects resulting from severe ground motion.
3. What are aftershocks? Why do they pose a special hazard?
4. Describe in your own words how the elastic rebound theory explains the periodic occurrence of earthquakes.

13.2 Measuring Earthquakes

The strength of earthquakes varies greatly. Although some are devastating, others may go unnoticed. It is important to be able to measure and compare the size of different quakes.

About 2000 years ago, the Chinese invented an instrument sensitive enough to detect even very faint ground motions. It consisted of an intricate arrangement of levers and wheels that released a ball in response to any kind of jarring movement (Figure 13.7). Although it was more sensitive than human senses, this device could not distinguish between large and small vibrations.

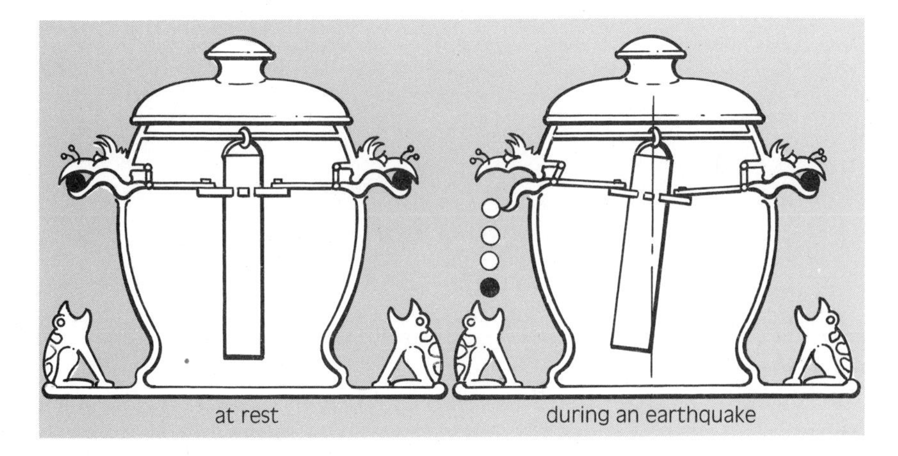

Figure 13.7
Chinese earthquake detector. Levers were attached to a pendulum suspended inside a bronze jar. When an earthquake shook the jar, the pendulum moved the levers, causing a ball to drop out of the mouth of the dragon into the mouth of the frog below. This device indicated the first ground motion from an earthquake.

EXTENSION

In 1971, the United States detonated an atomic bomb under the Aleutian Islands. Because these islands have experienced many earthquakes, some people feared that such a blast might jar the islands' faults and produce a new earthquake. In spite of this, when selecting a test site, the United States government chose one of the islands that had most recently experienced earthquake activity. Why do you think the government made this decision? Do you think the decision was a good one?

THE MERCALLI SCALE

Until this century, most scientists rated the strength of an earthquake shock using a scale based on judgement rather than measurement. One subjective scale, developed in 1902 by an Italian scientist, Giuseppi Mercalli, is still in use. Its formal name is the Modified Mercalli Scale of Earthquake Intensity, but it is generally called the **Mercalli scale.** (see Table 13.1). It is based on **intensity**, an assessment of damage or other observable effects at a particular location.

Table 13.1 The Modified Mercalli Scale of Earthquake Intensity

I	Earthquakes are not usually felt.
II	Earthquakes are felt only by persons at rest on the upper floors of buildings.
III	Vibrations felt are like those of light trucks passing.
IV	Vibrations felt are like those of heavy trucks passing. They are felt noticeably indoors. Dishes and windows rattle, houses creak.
V	Standing or unstable objects may fall; dishes may break; doors and pictures swing.
VI	Vibrations are felt by everyone. People often panic. Plaster or brick walls may crack slightly. School and church bells may ring from earthquake vibrations.
VII	Walking is difficult. Vibrations are felt by persons in moving vehicles. Poorly built structures and weak masonry are damaged.
VIII	There is a partial collapse of well-built structures. Chimneys, factory stacks, monuments, and towers fall. Tree branches may break off. Changes occur in underground springs and wells.
IX	General panic occurs. Buildings are shifted off their foundations. Damage occurs to dams and underground pipes. Sand and mud are ejected from cracks in the ground.
X	Water is thrown onto the banks of lakes and rivers. Bridges collapse. Landslides begin. Most structures are destroyed.
XI	Railroad tracks are bent severely. Broad cracks appear in the ground. All underground pipelines are destroyed.
XII	Damage is total. Rocks and other objects are thrown into the air as earthquake waves move across the ground surface.

THE SEISMOGRAPH

A major limitation of the Mercalli scale is that it is based on human judgement and is therefore unreliable. For example, a person's estimates of weight, size, or time cannot be as accurate as a measurement. An instrument that is used to measure the amplitude of ground motion resulting from earthquakes is known as a **seismograph**.

Trying to measure the distance the ground moves during a tremor presents a challenging problem. If you were an ant hanging onto a rope, you would certainly know you were moving from side to side, but you would be unable to measure the amplitude of that motion. To do that, you would need a stationary reference point, such as a point on the floor, to measure your motion against. There is a similar problem in trying to measure ground motion using an instrument that is fastened to the ground. To measure ground motion, there needs to be a reference point that remains stationary even when the ground is shaking.

You can demonstrate a principle that solves this problem in the following way. Place a smooth, cylindrical object (like a pencil or a piece of chalk) on top of a sheet of paper on your desktop. Holding the opposite edges of the paper, quickly shake it back and forth perpendicular to the pencil while keeping the paper flat against the desktop. How does the pencil move relative to the paper? How does the pencil move relative to a fixed point (such as the desktop)?

You have just demonstrated a principle known as **inertia**, defined as the tendency of a mass to stay at rest or continue moving unless the state of rest or motion is changed by external force. In this case, a stationary object (the pencil) tended to stay stationary even when another body (the surface it rests on) suddenly moves. We say that the pencil is showing inertia. The greater the mass, the greater is its inertia. You can experiment with this concept yourself using more massive objects, such as cylinders of metal of varying weight.

A seismograph contains a large mass of metal that is suspended so that it can move independently of its supports (Figure 13.8).

The rest of the instrument is anchored to a strong concrete foundation. During a tremor, the suspended mass tends to stay in one place as the ground vibrates under it. Because of its inertia, the mass becomes the stationary reference point. Sensors measure the difference in movement between the ground and the stationary mass. The amplitude of motion is electronically magnified up to 100 000 times and recorded by the vibrations of a pen over a roll of paper. The resulting pattern is called a **seismogram**.

DID YOU KNOW?

The ICBC buildings in North Vancouver were constructed over loose fill near a shoreline. Before construction began, a disk weighing several tonnes was repeatedly dropped from a crane over every square metre of ground surface. What was the purpose of doing this?

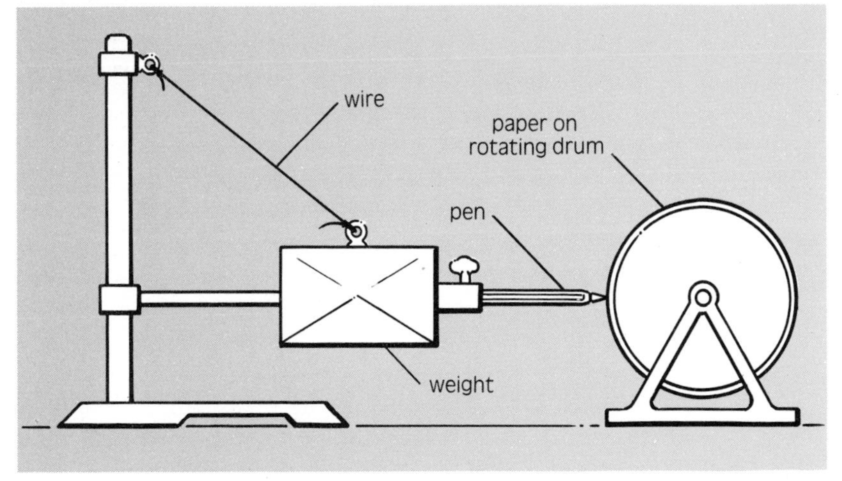

Figure 13.8
A simple seismograph.

The development of the seismograph meant that earthquake motion could be measured. A new earthquake scale — the **Richter scale** — was developed.

THE RICHTER SCALE

Suppose at this very moment the Earth begins to shake, causing windows and loose objects to rattle and walls to creak. These effects would be consistent with an intensity of about IV on the Mercalli scale. But how big was the earthquake itself?

Energy radiates in all directions from the epicentre, becoming weaker as it spreads out. However, when you experience a tremor you are unable to tell where the epicentre is or how far you are from it. What you experience could either be a minor tremor with the epicentre nearby, or a major earthquake centred hundreds of kilometres away.

The diagram in Figure 13.9 shows how the intensity of an earthquake varies with distance. The irregular pattern of intensity around the epicentre probably indicates a difference in the type of rock or surface soils.

The effect of varying surface materials was illustrated during a major quake in Mexico in 1985. Although the epicentre was near Acapulco on the Pacific coast, Acapulco itself received only minor damage. Mexico City, 300 km inland, suffered widespread damage. One explanation is that Mexico City is built on the clay and silt of an old lake bed, whereas Acapulco is built on much sturdier bedrock. The soft soil apparently sets up many ground vibrations. So the amount of damage to structures is not a reliable indication of the strength of an earthquake.

EXTENSION

Earthquake energy is carried away from the focus by waves. This can be demonstrated by touching the surface of a bowl of water or gelatin and watching how the energy travels. How does the strength of the waves change as they travel away from the focus? Can you think of another model that illustrates this effect?

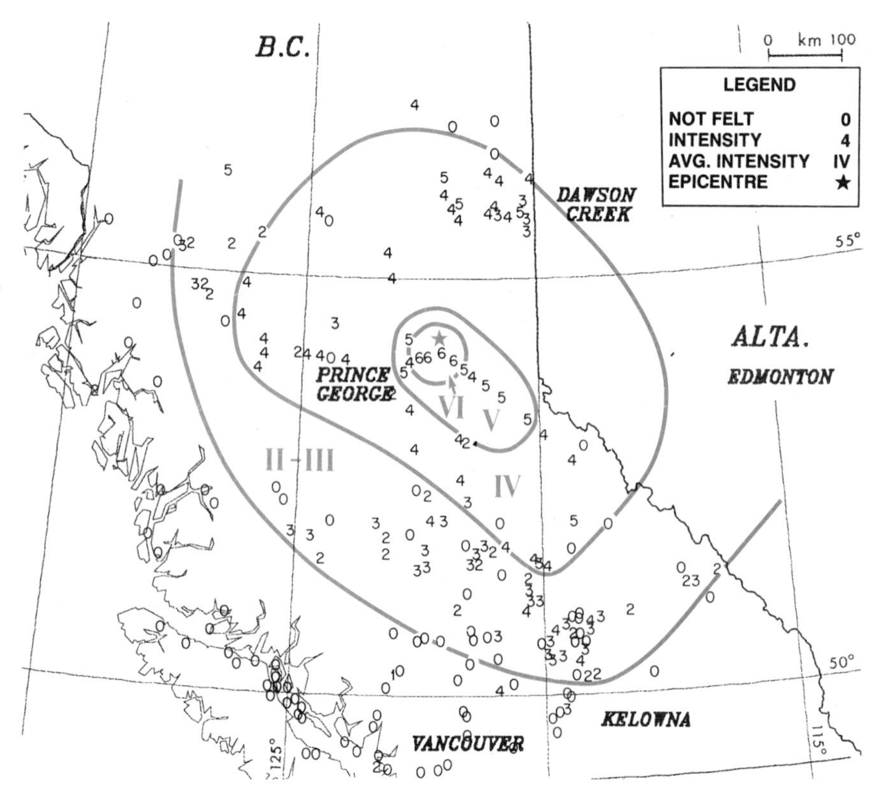

Figure 13.9
Distribution of Mercalli intensities felt from the earthquake of March 21, 1986, near Prince George.

In order to measure the strength reliably, Charles Richter developed a standardized scale of earthquake **magnitude** in 1935. Magnitude is based on the measurement of the amplitude of ground motion recorded on a seismograph. Unlike the Mercalli scale, which has an upper limit of XII, the Richter scale is open-ended — it has no maximum value. Magnitudes of any size are possible, although magnitudes greater than 10 are extremely unlikely.

Each point on the Richter scale represents ground vibrations 10 times (or tenfold) greater than the point below it. Each tenfold increase in vibration means about 30 times more energy is released. For example, an earthquake of magnitude 6 on the scale produces a ground motion 10 times greater than one with a magnitude of 5, and 100 times greater than one of magnitude 4. In addition, an earthquake of magnitude 6 releases about 30 times the energy of a magnitude 5 quake and 900 times the energy of a magnitude 4 quake.

A value of "0" on the Richter scale does not mean that there is *no* ground motion. On the contrary — the wind and ocean surf, as well as traffic and other human activities, cause a constant "hum" that seismographs easily detect. Only tremors that exceed these background vibrations can be identified. The value of "0" is set to match the amplitude of background vibrations.

DID YOU KNOW?

For high-energy earthquakes, there is a limit to the amplitude of ground waves. However, some earthquakes may last longer than others, and therefore release more energy. The Richter scale was recently revised to reflect these differences. As a result, new values have been assigned to some famous earthquakes.

Earthquake	Old scale	New scale
Alaska, 1964	8.5	9.2
San Francisco, 1906	8.3	7.9
Chile, 1960	8.6	9.5

EARTHQUAKE WAVES

A seismograph is used to measure earthquake waves. Examine the seismogram shown in Figure 13.10. Note the point at which the earthquake waves first arrive at the monitoring station. The seismogram shows waves with two distinctly different characteristics.

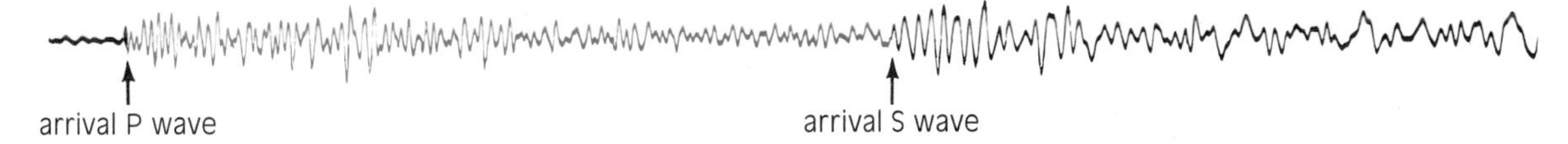

Figure 13.10
The seismogram illustrates both P wave (green) and S wave (black) arrivals at a recording station.

ACTIVITY 13C *Earthquake Wave Models*

In this activity, you will use a "Slinky," a loosely coiled spring, as a model for a sample of rock. By generating waves along the Slinky, you will be able to investigate and compare two different kinds of waves that may be produced by an earthquake.

MATERIALS

goggles
loosely coiled Slinky spring
masking tape

PROCEDURE

1. Attach one end of the Slinky to a rigid support. Stretch it over a table or across the floor.

CAUTION!
Be careful. Coil may spring back from extended position.

2. Place a small piece of masking tape on a single coil about halfway along the spring. The tape will serve as a reference point so that you will be able to see how a particular point on the spring moves as it vibrates.
3. Hold the free end of the spring. Experiment with different ways of producing vibrations that can be carried along as waves.

→

4. (a) Draw and describe the types of waves you are able to produce.
   (b) In each case, describe how the reference point (marked with the tape) moves as the spring vibrates. Draw an arrow to show the direction the wave moves along the spring. Draw another arrow to show how the reference point moves.
   (c) What happens when the wave reaches the fixed support at the end of the spring?

5. Two types of waves that you may have discovered are illustrated in Figure 13.11. Devise a test to compare the speeds of both wave types. Which travels faster?

6. Complete the following sentences in your learning journal:
   In this activity, I learned that . . .
   I was surprised that . . .
   I am still not sure about . . .
   Another example of this idea is . . .
   I wonder if . . .

Figure 13.11
(a) S wave. (b) P wave.

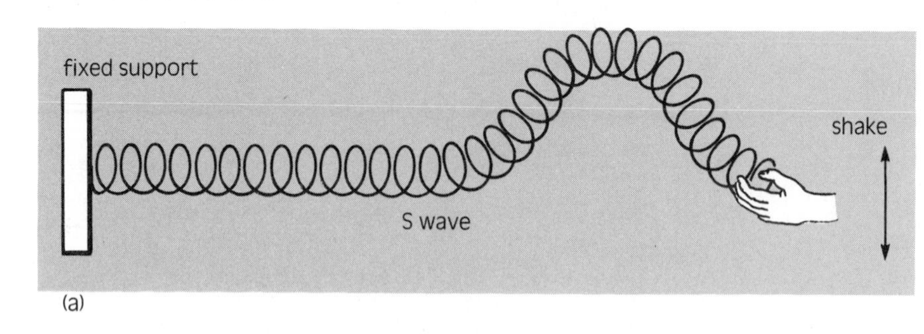

(a)

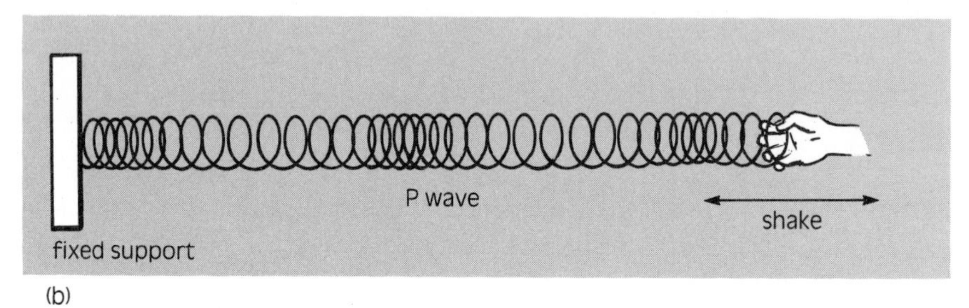

(b)

DISCUSSION

1. (a) In what ways are these Slinky wave models similar to earthquake waves?
   (b) In what ways are they different?

2. Examine the seismogram in Figure 13.10. The first wave on the recording is the **P wave** (coloured green), and the second is the **S wave** (coloured black). If both of these kinds of waves were produced at the same time at the earthquake focus, why do you think they arrive at the seismograph at different times?

3. Write some descriptive words that you think the "S" and the "P" might stand for.

4. Do you think sound waves are more like S or P waves? Why?

REVIEW 13.2

1. What advantages does the seismograph have over the early Chinese earthquake-detecting instruments?
2. Describe some of the limitations of the Mercalli scale.
3. Why is there a massive block of metal held within every seismograph?
4. Give as many reasons as you can why two cities may experience different intensities from the same earthquake.
5. What does the value of "0" represent on
   (a) the Mercalli scale?
   (b) the Richter scale?
6. Consider one earthquake of magnitude 7.5 on the Richter scale and another of magnitude 4.5. Compare them in terms of
   (a) total energy released,
   (b) amplitude of ground motion near the epicentre.
7. List all the similarities and differences you can think of between P waves and S waves.

13.3 Locating Earthquakes and Earthquake Zones

Whenever a major earthquake occurs, its location must be quickly identified. If the epicentre is found to be near a populated area, emergency aid can be dispatched to the victims quickly. How do **seismologists** (earthquake scientists) locate an earthquake's epicentre? Seismographs have been set up in thousands of locations worldwide so that vibrations anywhere on the Earth can be recorded and carefully studied. Differences among the seismograms from individual stations can be measured and interpreted to reveal epicentre locations.

EXTENSION

Construct a simple seismograph model by suspending a large mass on a moveable arm. Fasten a pen to the end of the arm. Slowly pull a paper tape under it while shaking the table on which your model rests. Compare the tracings with the seismograph record, or seismogram, in Figure 13.10.

ACTIVITY 13D *Locating an Earthquake Epicentre*

In this activity, you will learn how seismologists identify epicentre locations on a map.

MATERIALS

stopwatch or clock with a sweep second hand
graph paper
drawing compass
equal area projection map of North America

PROCEDURE

PART I

1. Your teacher will choose a particular spot at one end of your classroom to represent the epicentre of an earthquake. Two students will represent a P wave and an S wave. Several students equipped with watches or stopwatches will be positioned at 1.5 m intervals along the travel path of these two "waves."
2. When your teacher gives the signal that the earthquake has occurred, the "P wave" and "S wave" will begin to walk side by side across the room. Since P waves travel nearly twice as fast as S waves (in granite, P waves travel at a speed of about 6.0 km/s, and S waves at 3.5 km/s), one student will walk at a much brisker pace than the other.
3. Students positioned along the travel path represent seismographic stations. It will be their job to measure the *time interval* between the arrival of the waves at each "station." They are to start timing as the "P wave" passes in front of them, and stop timing with the arrival of the "S wave" (Figure 13.12). Report these measurements to the class and enter them in a data table.

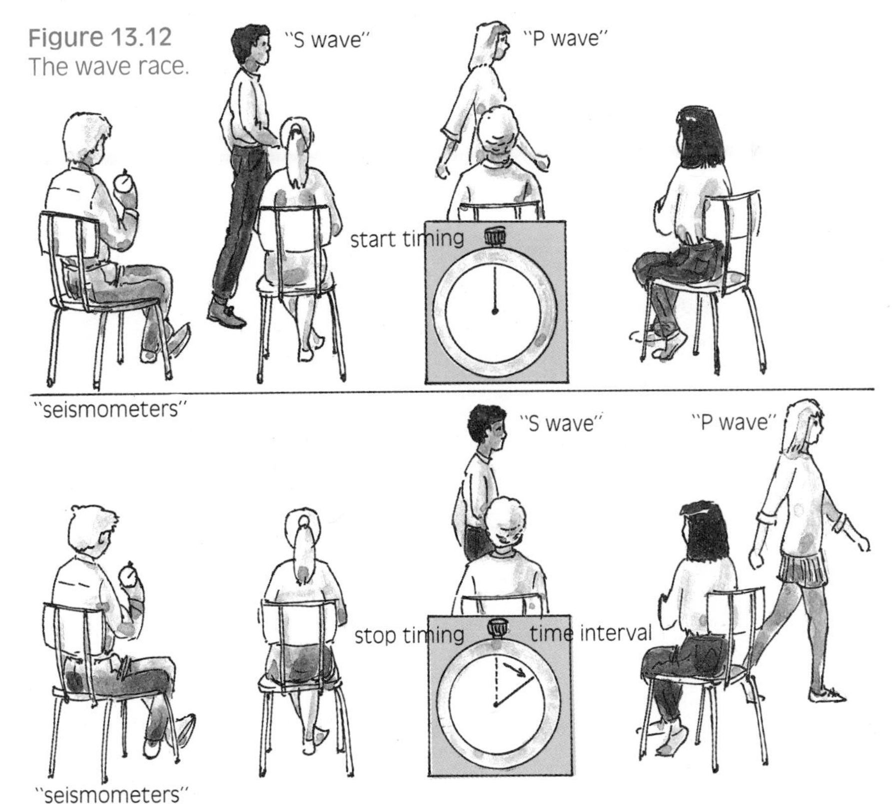

Figure 13.12
The wave race.

4. Plot a graph of these data. Label the x-axis "Distance from the epicentre (metres)" and label the y-axis "P and S wave time interval (seconds)."

5. What is the relationship between the distance travelled and the time interval? Describe how this principle might be applied to finding the distance between a seismographic station and the epicentre of an actual earthquake.

PART II

Table 13.2 shows the arrival times in hours (h), minutes (m), and seconds (s) of seismic waves at several North American seismograph stations. This earthquake occurred on June 23, 1946, and had a magnitude of 7.3 on the Richter scale.

The next steps will show you how this information can be interpreted to locate the epicentre on a map.

Table 13.2 Sample Data Table for Activity 13D

Seismographic station	P wave arrival time			S wave arrival time			P/S time interval			Epicentre distance (km)
	h	m	s	h	m	s	h	m	s	
Saskatoon	17	16	15	17	18	52				
Chicago	17	19	04	17	23	33				
Ottawa	17	19	57	17	25	19				
Halifax	17	21	06	17	27	21				
Salt Lake City	17	16	30	17	18	59				

SAMPLE ONLY

6. Copy Table 13.2 into your notebook. Calculate the time intervals between the arrivals of the P waves and the S waves at each station. Enter these time intervals in the data table.

7. The graph in Figure 13.13 shows the relationship between the P and S wave time intervals and epicentre distance. (It is similar to the graph you prepared in Part I.) Use it to estimate the distance to each of the five stations in Table 13.2. Enter the distances in your data table.

Figure 13.13
Earthquake wave time-travel graph.

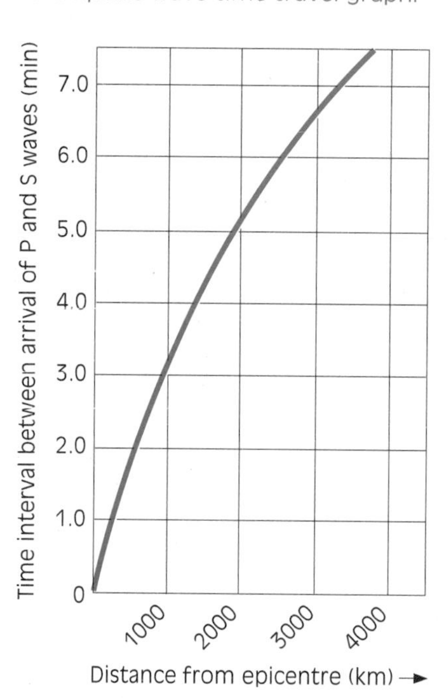

Figure 13.14
The epicentre of an earthquake 100 km from a seismograph in Victoria could be anywhere on the circle.

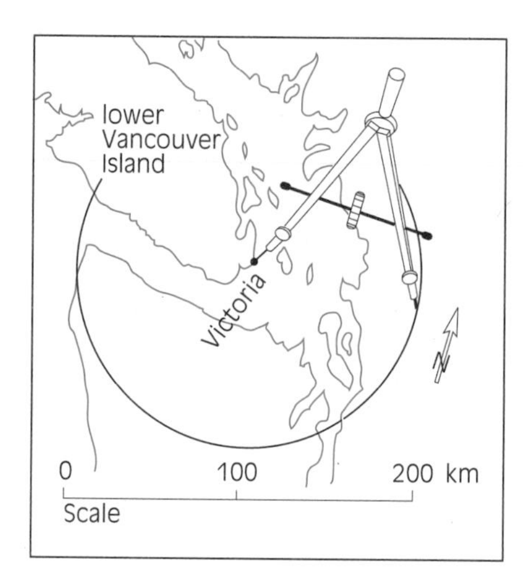

8. In each case, the distance to the epicentre is known, but the direction is not. The epicentre must therefore be somewhere along a circle drawn around the station. The radius of the circle corresponds to the distance between the epicentre and the station (Figure 13.14). Locate each of the seismograph stations on a map provided by your teacher.

   Locate each of the stations in Table 13.2 on a map of North America supplied by your teacher. Use a drawing compass to draw a complete circle around each station. Use the table you completed in step 7, as well as the scale provided on the map, to set the radius of the circle.

9. Describe the pattern that results from drawing all these circles on your map. Mark the suspected location of the epicentre with an X and label it "epicentre."

DISCUSSION

1. Examine Figure 13.15, which shows the seismograms of a small earthquake monitored at three different stations.
   (a) List as many similarities and differences among the seismograms as you can.
   (b) Which of the stations was the closest to the epicentre? Which was the farthest away? How do you know?

2. What single property of P and S waves makes it possible to measure the distance they have travelled?
3. What do the circles you drew on the maps represent?
4. (a) Why would a seismologist not be able to locate the epicentre of an earthquake using the data from one seismograph alone?
   (b) What is the usual minimum number of stations that must report before the epicentre can be located?
   (c) Can you think of a special case where fewer stations than this would suffice, or where a greater number than this would be necessary?
5. The "P" in the P wave actually stands for "primary," whereas the "S" in the S wave stands for "secondary." Why do you think these terms are used?

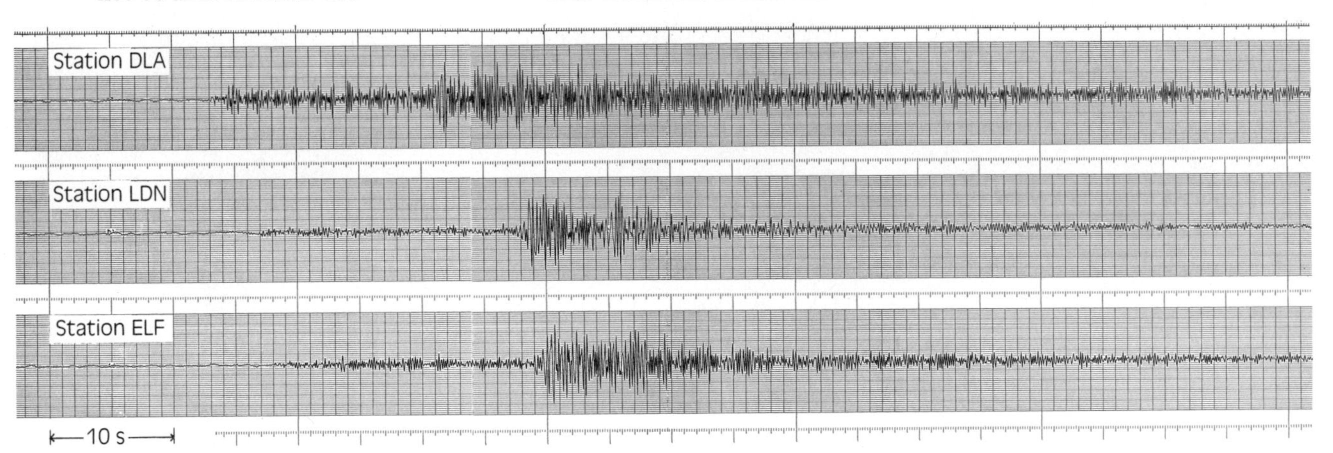

Figure 13.15
Three seismograms of an earthquake monitored at three different stations.

Earthquake in Lake Erie
February 2, 1976
Magnitude 3.4

Recorded by the University of Western Ontario Seismic Array
London, Ontario, Canada

EARTHQUAKE ZONES AND PLATE TECTONICS

By mapping the locations of earthquake epicentres around the world over many years, geologists have found that the epicentres tend to be clustered in certain areas. These areas can be identified as high-risk earthquake zones. They are regions where the likelihood of earthquakes is much higher than average.

These zones form a pattern, covering the Earth like seams on a baseball. After studying these zones and other data, scientists have concluded that there are regions where the Earth's outer layer is divided into separate, rigid slabs called **plates**. The plates drift around on the Earth at speeds of between 2 cm and 10 cm per year. The "seams," where earthquake activity occurs, represent places where the plates meet, called **plate boundaries** (Figure 13.16). These plate boundaries are marked by numerous active faults. There are three kinds of plate boundaries: colliding (moving together); rifting (tearing apart); and sliding sideways. Earthquakes are thought to occur as the plates alternately stick and slip against one another in the fault zones at plate boundaries.

The idea that the Earth's outer rigid shell is made up of moving plates is described in the **theory of plate tectonics**. Largely as a result of the efforts of a Canadian scientist, J. Tuzo Wilson, this theory has earned general acceptance among earth scientists.

DID YOU KNOW?

At the time of the 1964 Alaska earthquake there were some abrupt changes in land level. One coastal village was flooded as it sank below sea level, and in another area, a deep seaport was uplifted so that the bay became too shallow for ships to dock there anymore.

Profile

J. TUZO WILSON AND THE THEORY OF PLATE TECTONICS

The theory of plate tectonics was developed internationally, through the efforts of many scientists, in the mid-1960s. J. Tuzo Wilson was a major contributor along with Robert Dietz, Harry Hess, and Frederick Vine of the United States and Drummond Matthews of Great Britain. One of Dr. Wilson's contributions was to explain how large fracture zones, called transform faults, cut across plate boundaries and permitted plate movement. Another contribution was his concept of hot spots—magma flowing up from the mantle at fixed points beneath the Earth's surface to break through the migrating crust. Like many new ideas, Wilson's were not accepted with enthusiasm. The *Journal of Geophysical Research* rejected his article theorizing that the way the Hawaiian Islands are lined up on the Earth's surface indicates the direction the crust is moving. However, the *Canadian Journal of Physics* did publish the article and Wilson's explanation gained acceptance.

What combination of education and experience enabled Wilson to play a leading role in advancing this revolutionary theory? Wilson was born in Ottawa in 1908 and grew up there, attending private school. As a boy, he spent summer holidays gardening, feeding chickens, and picking berries, but at 15 he began mapping and prospecting in the bush north of Lake Superior. Later he attended the University of Toronto and enjoyed physics, despite the dull repetitiveness of the laboratory work. Summers of field work in geology inspired Wilson to become the first person in Canada to graduate in geophysics—the science that combines geology and physics. He pursued graduate work at Cambridge (England) and Princeton (U.S.A.).

Joining the Geological Survey of Canada gave Wilson the first of many opportunities to travel on every continent. During World War II, he served with the Royal Canadian Engineers. Afterwards, he became a professor of geophysics at the University of Toronto. During his career, he wrote over 200 journal articles and three books including *Continents Adrift and Continents Aground* (1976), *Unglazed China* (1973), and *International Geophysical Year: Year of the New Moons* (1961). Scientists have honoured Wilson with the Wollaston Medal of the Geophysical Society and the Maurice Ewing Medal of the American Geophysical Union.

Later in life, Wilson continued to consult with earthquake researchers throughout the world, especially in China. And at an age when most scientists retire, he became the first director of the Ontario Science Centre, a post he held for 11 years. When he began, the concept of science centres was new. Since then, it has been copied widely. He considered it to be "as challenging as any research project."

REVIEW 13.3

1. What information can be found in a seismogram of an earthquake that will allow a seismologist to find the distance to the epicentre?
2. Imagine that you are a seismologist monitoring a seismograph when suddenly the instrument begins to record the arrival of strong earthquake waves. Describe the steps that you would follow in locating the epicentre.
3. Describe at least two benefits of being able to locate the position of earthquake epicentres.
4. How do earth scientists explain why earthquakes tend to occur along particular zones, rather than randomly around the Earth?

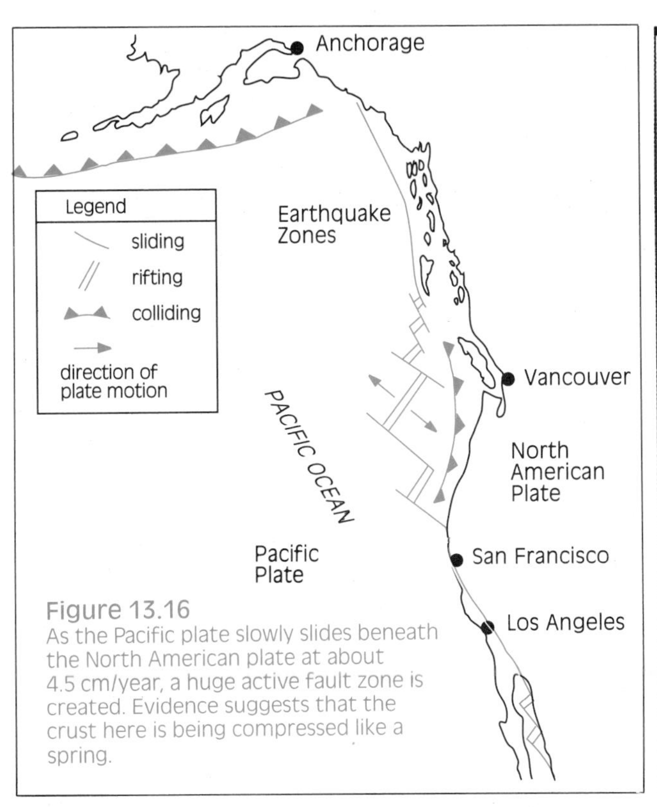

Figure 13.16
As the Pacific plate slowly slides beneath the North American plate at about 4.5 cm/year, a huge active fault zone is created. Evidence suggests that the crust here is being compressed like a spring.

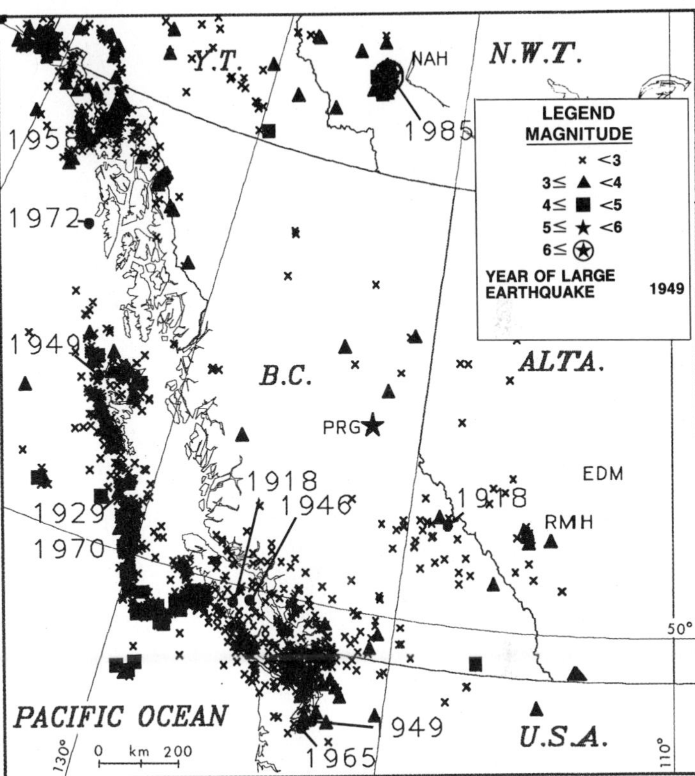

Figure 13.17
The symbols in the figure indicate earthquakes of British Columbia and vicinity, 1985-86. The additional dates indicate high-magnitude earthquakes in this century.

13.4 Earthquake Risk in British Columbia

High-risk earthquake zones usually have three characteristics:

1. They are located near a plate boundary.
2. There is a history of earthquakes in the region.
3. There are numerous active faults.

With this in mind, what is the earthquake risk for British Columbia?

The west coast of North America has a history of recent earthquake activity. Figure 13.17 shows the distribution of earthquakes recorded between the years 1985 and 1986 in British Columbia and vicinity. The clustering of earthquakes along the shoreline indicates the presence of an active plate boundary.

Figure 13.18
This aerial photograph of Vargas Island, on the west coast of Vancouver Island, shows beach terraces raised above sea level. What might have happened in the past to cause this?

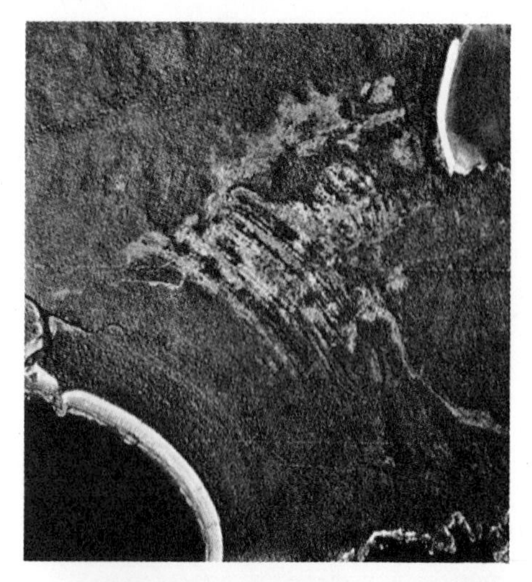

Obtaining a record of earlier British Columbia earthquakes is more difficult. Although a great many small- to moderate-sized quakes have occurred in British Columbia, there has not been a disastrous one in the past 300 years. One way to discover whether any major earthquakes occurred earlier is to apply principles of geological detective work, including absolute age dating (see Chapter 12).

What kind of geological evidence would you look for and expect to find if there had been a major prehistoric earthquake? Think about this and list some ideas of your own before you read further.

Following some major earthquakes this century, land has been permanently shifted relative to sea level. What kinds of evidence might indicate that the land had suddenly lifted up or dropped down in prehistoric times? If an area is near a coastline, you might expect that sea level (relative to the land) would have rapidly changed. If the crust was suddenly uplifted, you might expect to see flat, wave-cut beaches perched high above the shoreline. An example of this can be clearly seen on the west coast of Vancouver Island (Figure 13.18).

Similarly, if there were a sudden downdropping of the crust, coastlines would be flooded and nearshore vegetation would be buried by new ocean sediments. Geologists have found evidence for this in several places near the coasts of Vancouver Island, Washington, and Oregon (Figure 13.19). At least seven of these layers have been found. Carbon 14 dating indicates that they are less than 3500 years old. The youngest layer is about 300 years old.

Figure 13.19
A major earthquake could cause the crust to drop suddenly. Coastal vegetation could be flooded and buried. The thin dark layers of peat shown in this section from the Washington coast may have been formed in this way.

REVIEW 13.4

1. (a) What is meant by a "high-risk earthquake zone?"
   (b) Why is it useful to know the earthquake risk of an area?
   (c) How may high-risk earthquake zones be identified?
   (d) Name three areas in the world that are likely to be high-risk earthquake zones. Give reasons for your choices.
2. Summarize in your own words:
   (a) the changes that might occur in the crust on the west coast of British Columbia if an earthquake of magnitude 8 or greater on the Richter scale should occur,
   (b) the geological evidence that indicates that a major earthquake may have occurred in the past.
3. Do you believe scientists know for sure that we live in a high-risk earthquake zone? Give reasons for your answer.
4. Do scientists know how many major earthquakes have occurred on the west coast of Canada? Think carefully before answering.
5. Thin layers of sand have been found up to 3 km inland in certain low-lying areas on the Washington coast. Suggest what may have caused them. (HINT: Consider some of the effects that large earthquakes may have on coastal areas.)
6. How would you answer the question: "When is the next major earthquake going to occur on the southwest coast of British Columbia?"

13.5 Predicting Earthquakes

Twenty years ago, scientists were excited about the possibility of predicting earthquakes as readily (and at least as accurately!) as they can now predict the weather. Certain changes in the properties of rock (such as density, electrical conductivity, and gas content) were observed just before a nearby fault slipped. It appears that numerous tiny "microfractures" develop in rock that is strongly compressed, and that these microfractures fill with water just before the fault slips. It was hoped that by monitoring changes in these properties, the location, time, and magnitude of a future earthquake could be forecast.

Unfortunately, seismologists have not been able to accurately predict most earthquakes. Too many factors and conditions are involved. Every section along a particular fault is unique. For example, there are differences in the type of rocks, the depth and shape of the fault, and the amount of water present. Even the moon's tidal pull has an influence. It is not possible yet to estimate or calculate the effects of all these factors.

A typical weather forecast may be: "There is a 60 per cent probability of precipitation tomorrow," rather than a straightforward prediction like "it is going to rain." The uncertain forecast reflects the uncertainty of the conditions. Because of the many complex factors involved, earthquake predictions are much like weather forecasts (Figure 13.20).

EXTENSION

Faults are unstable structures that can shift at any time. You can make a model of an unstable structure by stacking blocks or sugar cubes into a column. Try to predict when the column will collapse. How does the likelihood of collapse change as the height of the column increases? In what ways is this model similar to the nature of earthquake prediction?

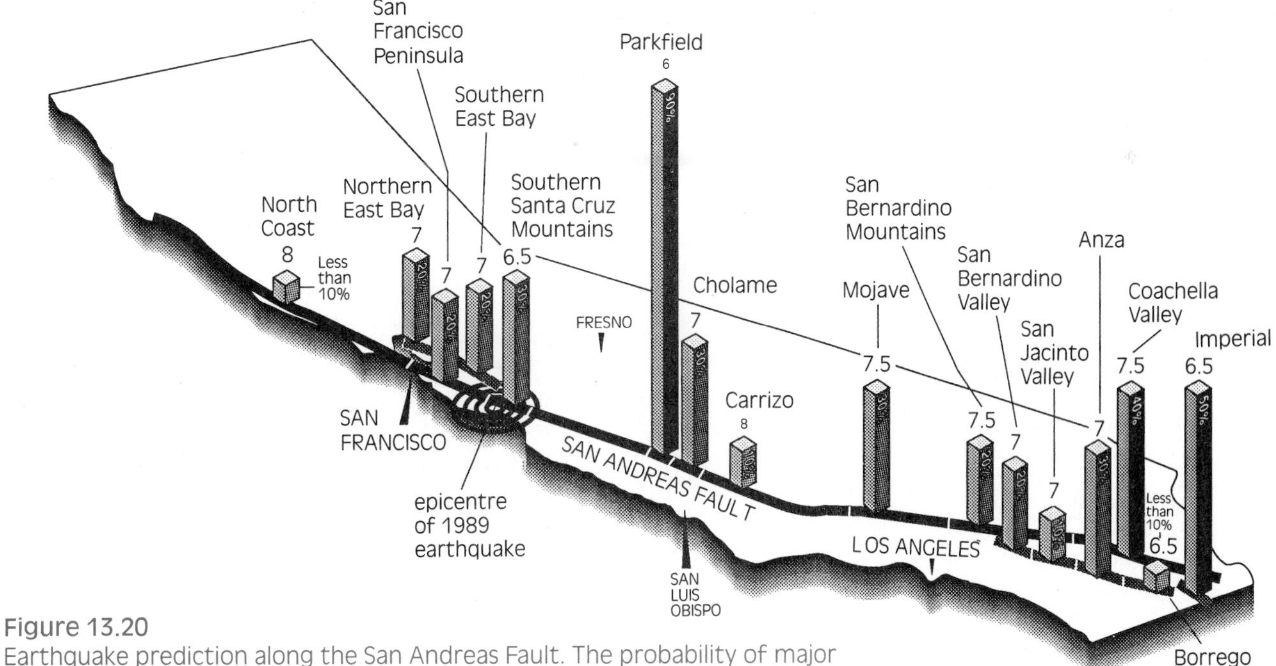

Figure 13.20
Earthquake prediction along the San Andreas Fault. The probability of major earthquakes occurring between 1988 and 2018 is shown for segments of some major faults in California. Calculations were made before the 1989 earthquake and are being re-evaluated. The expected Richter magnitude is given above each column. The higher the column, the greater the probability of an earthquake.

COMPUTER APPLICATION

THE YELLOWKNIFE SEISMOLOGICAL ARRAY AND THE NUCLEAR TEST BAN TREATY

Today, nuclear tests can occur in many parts of the world and at any time. Within an hour of a nuclear explosion, the federal government in Ottawa will know about it and will be able to pinpoint the test site to within 150 km. How is that possible? With the help of computers, a group of seismometers (called a seismological array) can rapidly detect nuclear explosions anywhere in the world.

A seismometer measures ground motion just like a seismograph. However, a seismometer produces numerical information rather than graphs. Not only does a seismometer record wave motion from nuclear tests and earthquakes, it also records motion from local disturbances like dynamite blasts and traffic. To lessen these local effects, seismometers are usually placed on bedrock in shelters and located far from human activity, in areas like the one near Yellowknife, Northwest Territories.

The Yellowknife Seismological Array has become one of the most up-to-date seismic detection systems in the world. There are 18 seismometers in the array, each one 2.5 km from the next. They are arranged like a giant cross. Waves travel from the focus of an earthquake through the Earth to arrive at each seismometer at slightly different times. By knowing the location of each seismometer in the array, seismologists can calculate the direction from which the waves arrived, their relative speed of travel, and the location of the focus.

Recently computers have enhanced the speed of transmission of the information. Computers at each seismometer convert the arriving wave motion into a code of numbers that is sent by radio to a control station west of Yellowknife. The computer at the control station takes the information from all the seismometers as it arrives and then transmits it to Ottawa via satellite. In Ottawa, the information is stored as a series of numbers on a computer and can be analysed at any time.

Computers have also enhanced the quality of the information received at each seismometer. The seismometers can be set at very sensitive levels. They routinely record small, distant earthquakes and low-level, distant nuclear tests, as well as explosions from local mines and noise from Great Slave Lake. Additional computer processing in Ottawa allows the seismologists to reject the "noise" sources and to separate the local from the distant seismic events.

The increased sensitivity to nuclear tests will increase Canada's ability to verify other nations' compliance with the Nuclear Test Ban Treaty. The knowledge gained from small, distant earthquakes will help to define earthquake zones in Canada and help us to understand the Earth's interior. In 1986, the array was the first ever to detect and track the shock wave, or boom, created by a meteorite as it passed into the atmosphere. Canada is also part of an international network of seismological stations and shares information and research projects with countries all over the world. In the coming years, new applications of the Yellowknife Seismological Array will continue to help us.

Control centre building in Yellowknife. Radio waves carry seismic information from each seismometer in the array to antennas on the tower (right). A cable takes the information to the computers in the control centre. Computers combine all the information and transmit it to Ottawa using the satellite dish (left). The all-terrain vehicle is needed to visit the seismometers in the array to perform maintenance.

REVIEW 13.5

1. What are some clues that might indicate that an earthquake could soon occur along a particular fault segment?
2. Suggest some reasons why it is so difficult to accurately predict the time and magnitude of future earthquakes.
3. What are some similarities and differences between predicting the weather and predicting earthquakes?
4. Describe some of the social problems that could result from a prediction of a possible earthquake near a city.

13.6 The Earth's Interior

Despite the problems associated with earthquakes, they do serve at least one useful purpose. Scientists have recognized that the energy from powerful earthquakes can penetrate and pass right through the Earth. Earthquake waves, like sound waves, travel through solid or liquid materials called **media** (singular: medium). These waves may change their **velocity** (speed and direction) if the properties of the medium, such as **density** (mass per unit volume), change. The energy waves can be used like a probe to learn about the Earth's interior.

The method of gathering information about materials that are some distance away by probing them with sound energy was borrowed from unlikely sources—bats and whales. These animals produce sounds with their voices that travel through the surrounding medium of air or water and reflect off nearby objects. The time it takes for the echo to return is a direct measure of the object's distance. Bats and whales can even distinguish size, shape, and texture from the information in the returning echo. Radar and depth-sounding are familiar human applications of this technique. It is also commonly used as a valuable prospecting tool.

Since wave speed varies with the density of the medium, it is possible to use radar and depth-sounding to find out the composition of different kinds of rock. Dynamite charges are used to produce enough sound energy to penetrate deeply buried rock. However, this technique is useful only to a few kilometres of depth. Energy from an earthquake is required to penetrate the Earth's core.

EXTENSION

Suppose a scientist predicts that a major earthquake will jolt southwestern British Columbia approximately one year from now. Arrange yourselves into small groups. Each person will role-play a character living in the area: the scientist, a mayor, the owner of a large business, a firefighter, another scientist, a hospital worker, and a homeowner with a young family. Create other characters too.

As one of these characters, prepare your point of view. Consider whether you would want the information to be released. What are your special concerns?

Present your views and debate the issue with the others in the group.

ACTIVITY 13E *Probing the Earth's Interior*

Whenever and wherever the Earth shudders, thousands of seismographs the world over record the arrival of the P and S waves as they travel through the planet. By comparing the arrival times and the distances travelled by these waves, we can gather information about the composition and structure of the inside of the Earth.

In Part I of this activity you will investigate how the velocity of energy waves (in this case, sound) is affected by the density of the medium through which they travel. Then, by examining some data on the velocity of earthquake waves in Part II, you will be able to determine how the density of the Earth's interior changes with depth.

MATERIALS

about 20 dominoes
metric ruler

PROCEDURE

PART I

1. Before you begin, make your own prediction about the following: Does sound energy travel faster through air or through steel? Explain the reasoning behind your answer.

2. This question may be partially answered by examining a model. In materials of low density, the particles are generally much more "spread out" than in more dense material. The difference between these two types of matter can be illustrated by arranging two lines of dominoes with different spacing distances as shown in Figure 13.21.

3. Predict which row would win the "race" if you pushed over the first domino in each line simultaneously. Test your prediction and observe the results.

4. Give an explanation for any difference you noticed.

5. Examine the prediction and explanation you made in procedure 1. If necessary, revise them. If you did revise them, explain why you changed your mind.

6. Design an experiment using real materials that would answer the question: "How does the density of a material affect the speed of wave energy passing through it?"

PART II

7. Figure 13.22 represents a cross-section of the Earth. Point X represents the focus of a particular earthquake. Points A, B, C, D, and E represent seismographic stations at five positions on the Earth's surface. The lines connecting X to each of the other points represent the paths taken by earthquake waves as they pass through the body of the Earth. (We will assume for now that they are relatively straight.) Trace or copy this diagram in your notebook.

Figure 13.21
Which row of dominoes will win the "race"?

Figure 13.22
Paths taken by earthquake waves through the Earth's interior.

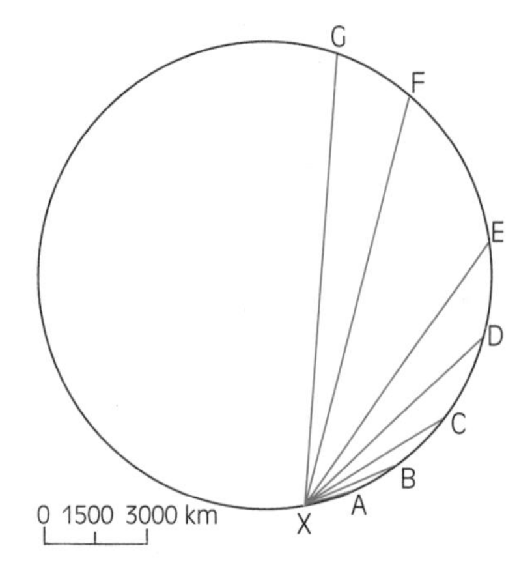

8. To calculate the average velocity of earthquake waves travelling along each path, you need to know both the distance the waves have travelled and the time it took to cover that distance. Remember that

   velocity (km/s) = distance (km)/time (s)

   (a) Distance
   With a metric rule, measure the distances between the focus of the earthquake and each of the seismographic stations. Use the scale of 1 cm:1500 km to convert to real distance (km), and enter these values in a reproduction of Table 13.3.

   (b) Time
   Travel times are the times taken for seismic waves to reach a seismograph. Travel times for the P and S waves are also given in Table 13.3.

   (c) Velocity
   Calculate the average velocity of each wave along each travel path by applying the above formula. Enter results in your data table.

Table 13.3 Sample Data Table for Activity 13E

Travel times (s)			Real distance (km)	Average velocity (km/s)	
Path	P wave	S wave		P	S
XA	220	420			
XB	370	670			
XC	530	960			
XD	690	1270			
XE	830	1500			
XF	1180	No waves			
XG	1380	No waves			

SAMPLE ONLY

9. Study the data you have compiled. What would account for a change in the velocities of waves as they pass closer to the Earth's centre?

DISCUSSION

1. (a) What property of a medium has the greatest effect on the velocity of a wave passing through it?
   (b) According to your data, how do you think this property changes towards the Earth's centre?
2. (a) The mass of the Earth has been estimated at about 5.98×10^{27} g, and its volume at about 1.08×10^{27} cm³. Calculate the average density of the Earth using the formula

   density (g/cm³) = mass (g)/volume (cm³)

   (b) The average density of rock near the Earth's surface is between 2.6 and 3.0 g/cm³. Compare this value with the value for the Earth's average density that you calculated above. Account for any difference.
3. Use Table 13.4 to answer the following questions.
   (a) Generally, what kinds of materials block S waves?
   (b) Suggest reasons why S waves are blocked along certain paths while P waves are not. (There are two possible answers to this question.)
   (c) Estimate the size of the zone that blocks these waves.
4. (a) People drilling wells or digging mines have observed that the temperature rises (on average) 1°C per 30 m of depth. It is assumed that the temperature at the Earth's centre is probably quite high. Keeping in mind how the density of rock changes towards the centre, in what state (solid, liquid, or gas) is the material at the Earth's core likely to be? Explain.
5. Earth's magnetic field is believed to result from currents of fluid matter circulating within the core.
   (a) Name a dense, magnetic substance that might be contained within the core to account for the Earth's magnetic field. (HINT: What two commonly occurring elements are magnetic?)
   (b) Suggest what might be causing the fluid core to circulate.

Table 13.4 Earthquake Wave Velocity within Different Media

Earth materials	P wave (km/s)	S wave (km/s)
Gravel	2.0	1.1
Sandstone	3.5	2.2
Limestone	5.2	3.5
Granite	5.0	3.0
Ice	3.5	1.7
Water	1.5	No waves
Air	0.33	No waves

LAYERS OF THE EARTH

As waves of energy enter a medium of a different density, they may be reflected or refracted. You notice this effect every time you hear an echo or bend light through a lens. Scientists have noticed similar effects as earthquake waves pass through the Earth. They have concluded that the interior of our planet is structured somewhat like the layers of an onion (Figure 13.23). Each layer is of different thickness and composition. Deeper layers are under greater pressure, compressing materials to greater density.

These layers are described as follows:

1. **Atmosphere** (about 100 km thick): the outermost envelope of gases, consisting mostly of nitrogen (78 per cent) and oxygen (21 per cent).
2. **Hydrosphere** (up to 11 km thick): a discontinuous layer of water found in oceans, lakes, rivers, and glaciers, covering about 70 per cent of the surface.

DID YOU KNOW?

The Earth's diameter varies from 12 719 km at the poles to 12 762 km at the equator. The deepest oil well ever drilled was about 10 km deep. Using this scale, if the world were the size of an apple, we would not yet have penetrated the skin! As a result, we cannot learn about the inside of the Earth by drilling for samples.

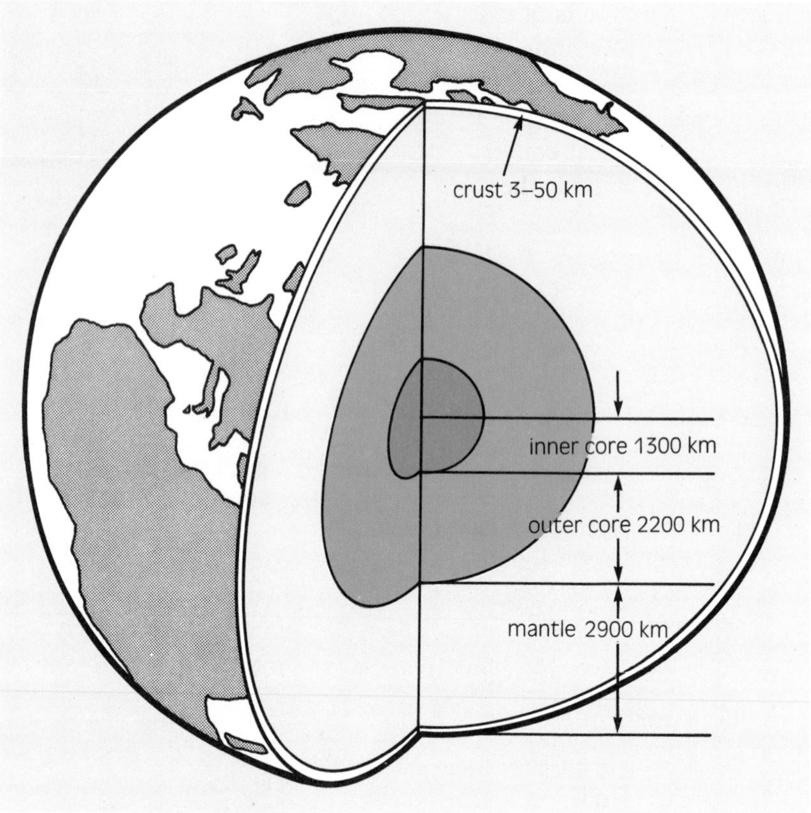

Figure 13.23
Layers of the Earth.

DID YOU KNOW?

Scientists currently believe that in an early stage of its formation the Earth was entirely molten. Minerals of different densities became separated: the denser elements such as iron and nickel sank to the core, whereas the lighter elements such as silicon, oxygen, and aluminum rose like light cream to the surface, forming the crust.

3. **Crust**: comparatively thin, rocky layer. It is thinnest under the oceans (3-8 km) and thickest over the continents (20-50 km). It is broken up into about a dozen large plates. The plates extend down into the next layer, the mantle.
4. **Mantle** (2900 km thick): The mantle is a complex layer, partly solid and partly liquid. The top 100 km or so is rigid and is known as the **lithosphere**. It, combined with the crust, forms the rigid layers of plates that make up the Earth's surface. The rest of the mantle is under very high pressure and is very hot. It tends to behave like an extremely thick fluid, slowly flowing without breaking. Hot material rises from below, then cools and sinks in slowly circulating **convection currents** (Figure 13.24). These currents are thought to be the driving force behind plate tectonics. The plates of the crust and lithosphere float very slowly on the moving mantle, like rafts. Whole continents are carried on these plates as they drift across the Earth's surface.

The collision of plates causes crustal rocks to be crumpled and folded into mountain ranges. Intense friction and increased temperatures at descending plate boundaries cause rock to melt, later to be squeezed to the surface as volcanic eruptions.

Convection currents are suspected to be the source of the energy released when compressed masses of rock snap along faults during deep earthquakes.

5. The centre region is called the **core**, subdivided into the liquid **outer core** and the solid **inner core**. They have a combined radius of about 3500 km. The core is thought to be composed largely of iron and nickel, which are both dense magnetic elements. Currents generated in the core by heat convection and the Earth's rotation are believed to be responsible for the Earth's magnetic field.

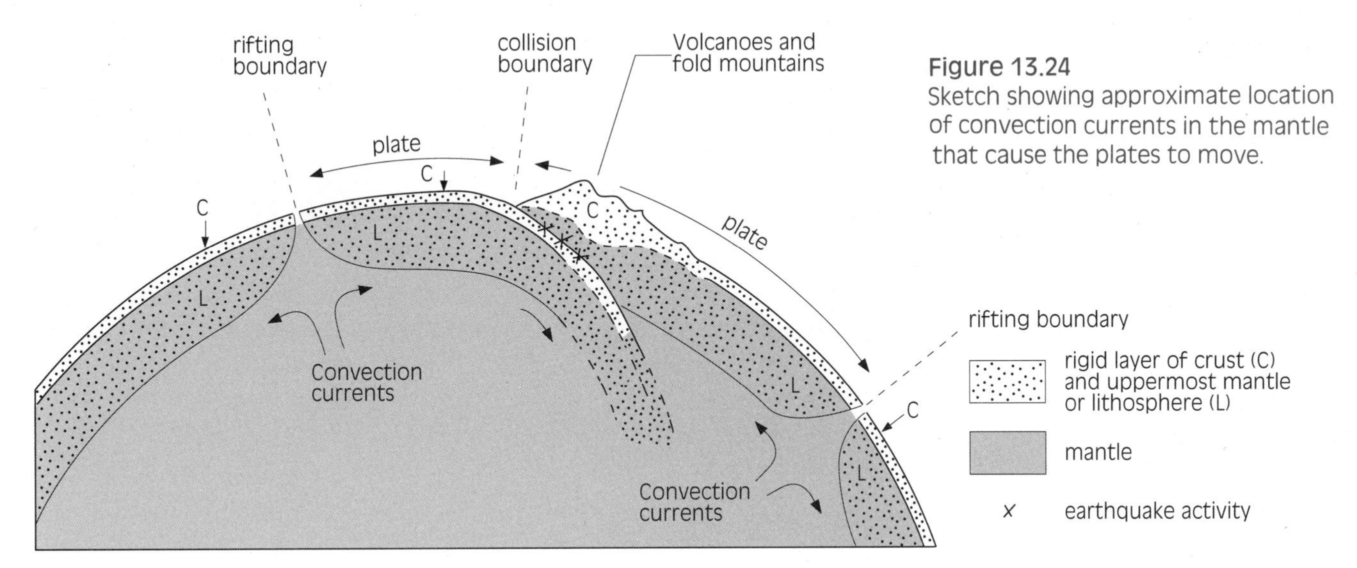

Figure 13.24
Sketch showing approximate location of convection currents in the mantle that cause the plates to move.

REVIEW 13.6

1. List each of the layers of the Earth in order of increasing density. Indicate which of these are solids, liquids, and/or gases.
2. Which layer do you think is affected the most by its interactions with other layers?
3. How do the properties of each layer change with distance from the Earth's centre?
4. Describe the evidence that leads scientists to believe that
   (a) the density of the Earth increases towards the centre,
   (b) the core of the Earth is liquid,
   (c) the interior of the Earth is layered.
5. What is the current explanation for the forces that cause the plates of the crust and upper mantle to move about on the lower mantle?

CHAPTER • REVIEW

Key Ideas

- Earthquakes result from the sudden release of energy when rock bodies slip on a fault.
- The elastic rebound theory accounts for the storing and periodic release of earthquake energy in terms of the friction on faults between crustal plates.
- The crust and upper mantle of the Earth are made up of interlocking plates that are always moving against one another.
- Earthquake intensity depends on the energy of the earthquake, the distance from the epicentre, and the type of ground material present.

- The Mercalli scale is a subjective scale of intensity that is based on an assessment of damage or other observable effects.
- The Richter scale is a measure of the total release of earthquake energy or magnitude as monitored on a seismograph.
- Every unit increase on the Richter scale represents a tenfold increase in the ground motion and about a thirtyfold increase in the energy released.
- The seismograph records the amount of ground motion during an earthquake.
- There are two basic kinds of earthquake waves that can penetrate the Earth: P waves and S waves.
- Earthquake zones are the regions of active crust at or near plate boundaries.
- The idea that the Earth's outer shell (crust and upper mantle) is made up of several moving and interlocking plates is described by the theory of plate tectonics.
- The arrival times of P and S waves can be used to determine the distance to and location of an earthquake's epicentre.
- The interior of the earth is layered, and increases in density towards the centre.

Vocabulary

fault
focus
epicentre
liquefy
tsunami
aftershock
potential energy
kinetic energy
models
theory
elastic rebound theory
Mercalli scale
intensity
amplitude
seismograph
inertia
seismogram
Richter scale
magnitude
S wave
P wave
seismologist
plate
plate boundary
theory of plate tectonics
medium
velocity
density
atmosphere
hydrosphere
crust
mantle
lithosphere
convection currents
core
outer core
inner core

1. Summarize what you have learned about how earthquakes occur by constructing your own concept map. Try to include as many of the concepts in the vocabulary list as you can, and give appropriate labels to the connecting links.

Connections

1. Bodies of rock may slip either vertically or horizontally on a fault (see Figure 13.2).
   (a) Which of these types of faults do you think is most likely to produce a tsunami? Explain your answer.
   (b) The San Andreas Fault is a horizontal slip fault. Is there much threat of tsunami if and when a major shock occurs in California?
2. (a) Describe the origin of the forces that cause plates to move.
   (b) What geological features and effects are produced as a result of plate movement?
3. The town of Hollister, California, is situated along a section of the San Andreas Fault that is slowly and gradually "creeping" about 2 cm/year. Mendocino is a town situated along a section of the fault that has not moved in several decades. Which town is at greater risk from an earthquake? Explain your reasoning.
4. Outline the physical principles on which the seismograph operates.
5. List as many similarities and differences as you can between the Mercalli and Richter scales.
6. Using the Mercalli scale, what intensity of earthquake would you consider to be life-threatening if you were in
   (a) an apartment building?
   (b) a house in the suburbs?
   (c) an open field?
   (d) a stone hut?
7. Indicate which of the following descriptions applies to a P wave or an S wave or both:
   (a) travels fast
   (b) travels like a sound wave
   (c) will travel through solids, liquids, or gases
   (d) velocity increases through denser materials
   (e) can be reflected
   (f) wave vibrations are perpendicular to the wave direction
8. Do you think the speed of an earthquake wave is always constant as it travels through the Earth? Explain why or why not.

9. How does the density of the medium affect the velocity at which the wave energy travels through it?
10. (a) List in order the six layers that make up the Earth.
    (b) Name the layer(s) possessing the following characteristics:
    - solid
    - liquid
    - hottest
    - convection currents
    - densest
    - thinnest solid layer
11. The moon does not have a magnetic field. What does that indicate about its interior?
12. Using what you know about increasing temperature and pressure, give your own explanation for why the outer core of the Earth is liquid and the inner core is solid.
13. Photographs of the surfaces of Mars, Mercury, and the moon do not show evidence of the plate tectonic activity that is found on Earth. On the basis of this observation, what can you conclude about the internal structure of these bodies?

Explorations

1. What can a community do to prepare itself for a damaging earthquake? Although no one likes to think about this possibility, it is important to plan ahead.
   (a) Form small planning groups to discuss the following problems. You may either record the ideas of your group on a sheet of poster paper for the class discussion or compile them and present them as an organized group report.
   What preparations can your family make?
   What plans should be in place for your community, city or town, and province?
   What should you personally do if a tremor strikes while you are at home? In school? Downtown?
   What special hazards would you face in the aftermath? How can these hazards be minimized?
   Identify some other problems.
   (b) Study the plan prepared by another group. Evaluate their plan and offer your own suggestions for improvement or additions.
2. Design (and possibly build) an instrument that will monitor ground vibrations. Be prepared to describe the basic principles of its operation.
3. Keep a record of earthquakes that occur during the year. Collect newspaper reports of each earthquake, and mark their locations on a world map.

Reflections

1. (a) Examine the ideas that you had about earthquakes at the beginning of this chapter. Which of those ideas have you changed your mind about? Describe how and why you might have changed your thinking.
   (b) Pose a number of new questions concerning what you wonder about, what you would like to know more about, and things about which you are still uncertain.
2. To the concept map in the vocabulary list exercise, add additional concepts of your own. Indicate how these concepts relate to each other, by drawing and labelling the connectors between them.

UNIT V

Exploring the Basis of Life

Have you ever been told you look like someone else in your family? Perhaps you and a parent, brother, or sister share features such as the shape of your mouth or face or the colour of your eyes or hair. In the photograph, the two sisters and their mother show a family resemblance in their cheekbones and foreheads, for example. How do these physical characteristics get passed on from parents to their children?

How we inherit such features from our parents is one of the most amazing processes in biology. More is being learned all the time, and newspaper headlines regularly announce recent advances in this field, called genetics.

Today's discoveries in genetics are possible only because of earlier research about the way living things reproduce and pass on characteristics from parents to offspring. And those studies depended on one of the most important scientific discoveries of all—that all living things are made up of tiny structures called cells.

Even before cells were discovered, however, the first step in this scientific journey was the invention of the microscope. Without it, scientists could not have found out anything about cells. Today, much improved versions of the microscope help to make the genetic discoveries behind the headlines possible.

The first chapter of this unit is about the microscope. Once you learn how to use this invaluable scientific tool, a whole new world awaits you.

CHAPTER

14 The Microscope

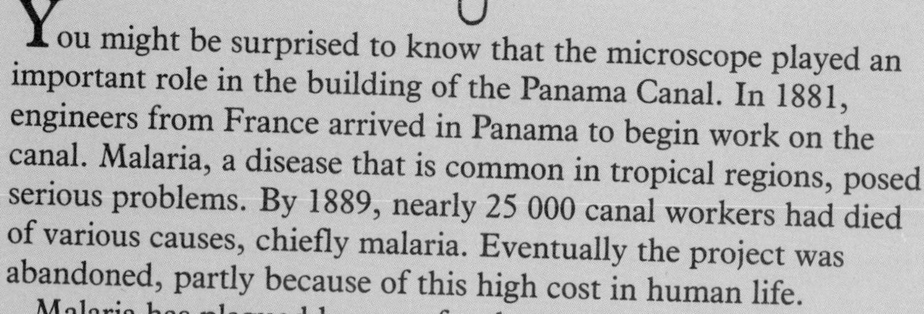

You might be surprised to know that the microscope played an important role in the building of the Panama Canal. In 1881, engineers from France arrived in Panama to begin work on the canal. Malaria, a disease that is common in tropical regions, posed serious problems. By 1889, nearly 25 000 canal workers had died of various causes, chiefly malaria. Eventually the project was abandoned, partly because of this high cost in human life.

Malaria has plagued humans for thousands of years. For most of that time, its cause was unknown, so people had little idea of how to treat or prevent it. Most of the organisms that cause disease cannot be seen without a microscope. Before this instrument was invented, people did not know that these tiny forms of life existed.

Plasmodium, the organism that causes malaria, was not discovered until 1880, when a French doctor named Alphonse Laveran saw it with a microscope (photo on left). In 1889, Ronald Ross, a British doctor, discovered that mosquitoes can transfer *Plasmodium* from the blood of malaria patients to the blood of healthy people.

The United States began a new attempt to build the Panama Canal in 1904. As part of a campaign to improve workers' health, as many mosquitoes as possible were eliminated from the surrounding area. As a result, many fewer workers died, and the project was completed in 1914.

The invention of the microscope was thus a vital step in controlling malaria in Panama—and in enabling the Panama Canal to be completed. The microscope has been called the most important of all tools for progress in biology and medicine. In this chapter, you will learn not only about the microscope and how it works, but also how to use it to make your own discoveries.

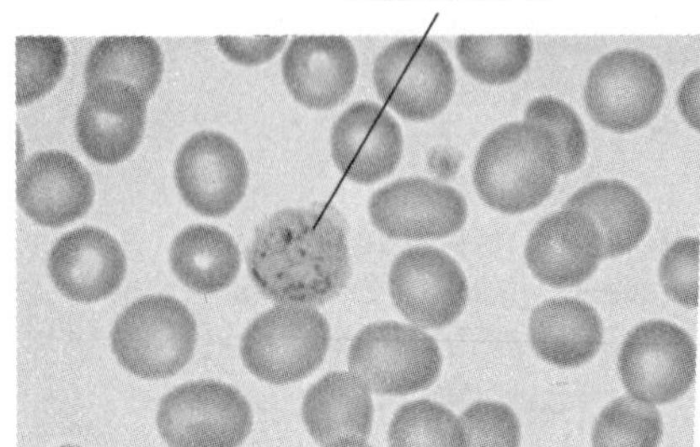

ACTIVITY 14A Microscope Knowledge

Microscopes are a common tool of modern science. You have probably seen them in use on television or read about their use in books or magazines. You may have used one yourself. This activity will help you find out what you know about the microscope.

PROCEDURE

1. In your learning journal, make a table with three columns. The first column should contain the numbers from 1 to 10, which correspond to the numbers shown in Figure 14.1. You will need more space in the second and third columns than in the first.
2. In the second column, list what you think would be a good name for each of the numbered parts. Even if you don't know what the part does, you can still give it a name—based on its location or appearance, perhaps.
3. In the third column, give a brief description (a few words long) of what you think this part of the microscope does. If you have no idea of the function of a numbered part, leave the appropriate space in the column blank.

DISCUSSION

1. What do you think is the function of a microscope?
2. (a) Can you think of any other instruments that perform a similar function?
   (b) How are they different from a microscope?
3. Suggest some possible uses for microscopes.

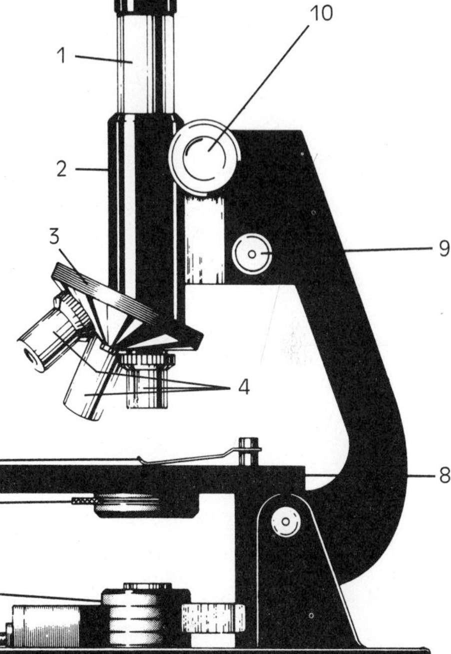

Figure 14.1
What are the names of these numbered parts? What are their functions?

14.1 A History of the Microscope

While swimming or wading, have you ever looked down and noticed that your legs looked shorter than you know they are — or that they appear to be bent in the wrong place? This optical illusion occurs because you are not looking at your feet through air or water alone, but through both. When light passes through only one substance, it travels in a straight path. When it crosses a barrier between two substances, the light bends. When you are looking at an object in the water, light reflected off the object is bent by the water's surface (Figure 14.2). Yet the part of your brain responsible for vision still assumes that light travels in a straight line from the object to your eye. As a result, you misjudge the distance from your eyes to the underwater object, making it seem closer to the surface than it really is.

EXTENSION

Find out more about malaria. What are its symptoms, and how many people are affected by it? What is the life cycle of *Plasmodium*—where does it live and what does it do inside mosquitoes and inside humans? How is malaria treated? How is it controlled today?

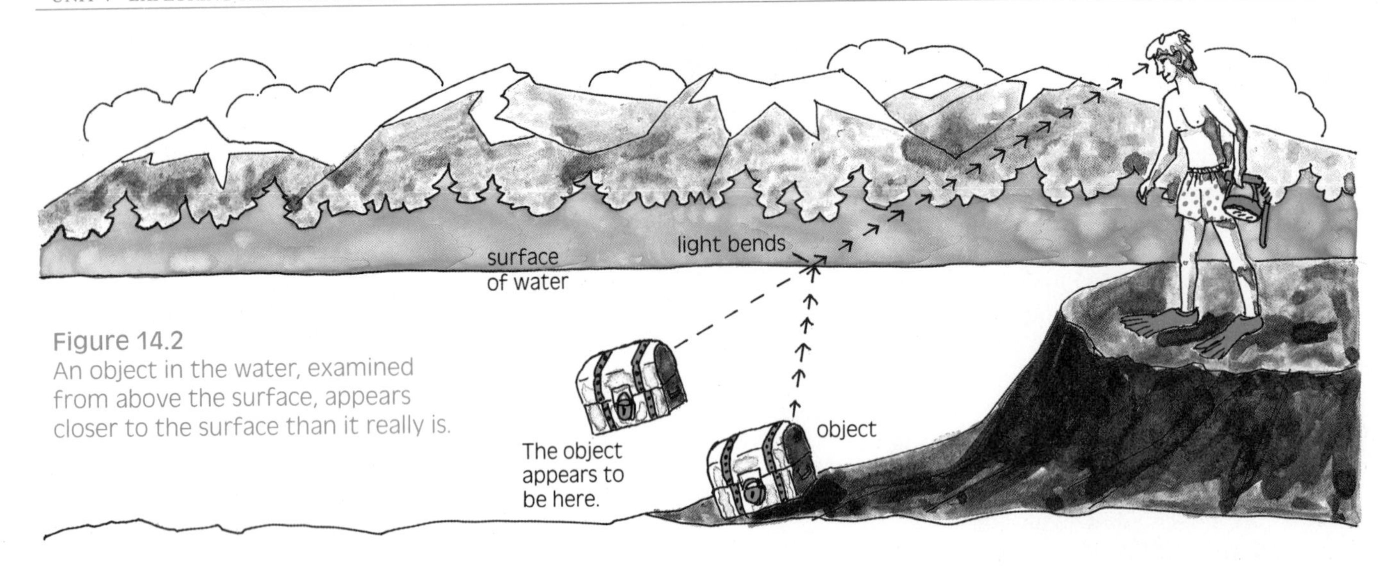

Figure 14.2
An object in the water, examined from above the surface, appears closer to the surface than it really is.

A flat barrier, such as the water surface in Figure 14.2, bends all the light beams in one direction. Curved surfaces bend light beams in different directions, depending on where the beams hit the surfaces. When an object is viewed through a surface that curves outward, the light beams are bent in such a way that the object appears to be larger than it really is (Figure 14.3). You are seeing an enlarged **image** (a representation or picture) of the object: the object itself has not changed in size. In other words, the curved surface produces an enlarged, or **magnified**, image of the object.

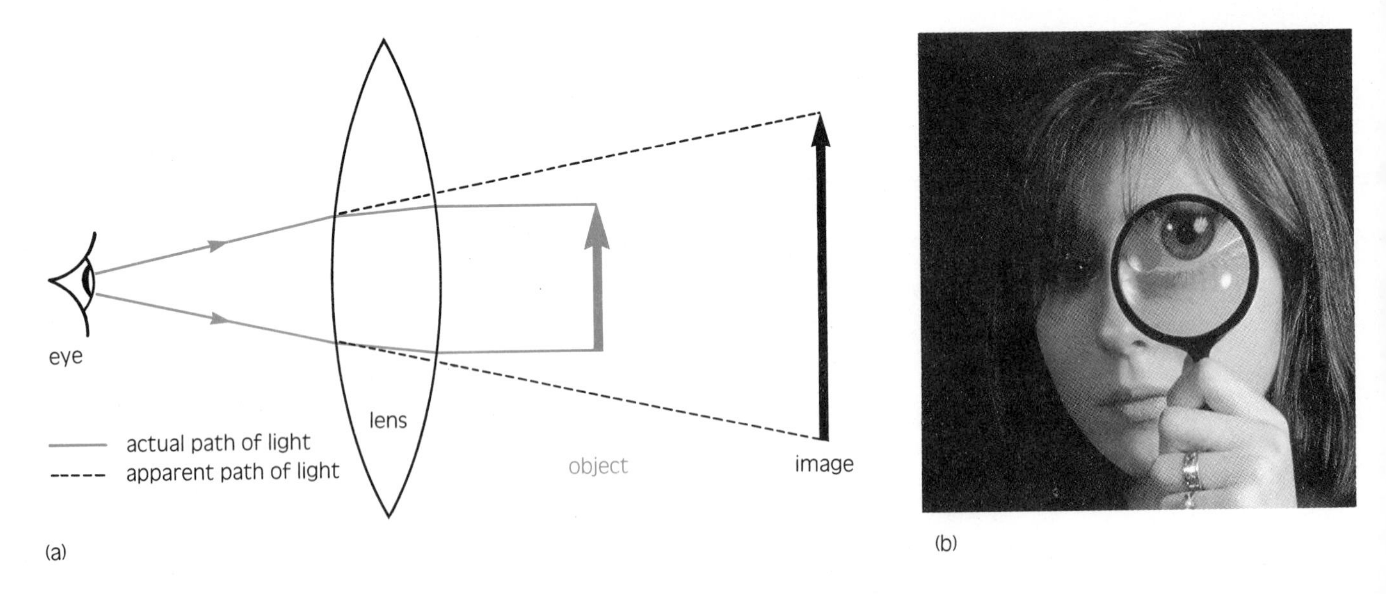

Figure 14.3
(a) Most lenses have two curved surfaces. A magnifying glass, for example, has a lens that looks much like this one. When you look at an object with a magnifying glass, the lens bends light so that the image you see is enlarged. (b) Someone looking at you through a magnifying glass would also see an enlarged image!

Probably even before people began to write down their experiences, they knew that the outwardly curved surface of a drop of water could produce magnified images (Figure 14.4). It is difficult to build a permanent or portable magnifying device with water, however. Glass is a much longer-lasting material. After people discovered how to manufacture glass, various kinds of magnifying devices were constructed and widely used.

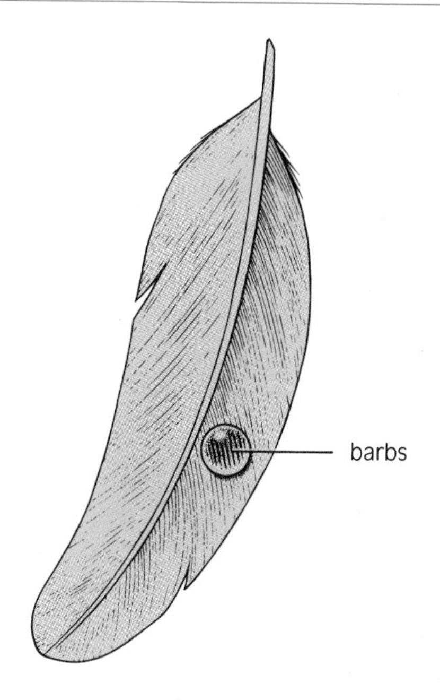

Figure 14.4
Have you ever noticed that the part of an object lying just below a drop of water appears to be magnified? Below the drop of water on this feather, you can see the individual barbs that together form the feather's vane.

THE DEVELOPMENT OF LENSES

People had discovered how to make glass by 2000 B.C. Yet historical records do not suggest that they appreciated the ability of curved glass to enlarge images until much later. Records indicate that by A.D. 30, people in various parts of the world had discovered that blobs of clear glass could bend the sun's beams so that they became concentrated at a single point. The concentrated sunlight caused this point to become very hot. Roman doctors called these glass blobs "burning glasses" and used them in various treatments. We now call these objects **lenses** (singular: lens). A lens is a device with at least one curved surface, which changes the path of light travelling through it.

One of the first people to realize that glass lenses could be used to produce magnified images (or at least one of the first to write this discovery down!) was Ibn al Haitham. This Egyptian astronomer lived from 965 to 1038. He did not suggest any practical use for his ideas, however, and it was not until 200 years later that the first spectacles (glasses) were invented. These helped farsighted people, who cannot see things that are very close to them. Over the next 300 years, glass-makers refined their skills, and eventually they were able to produce lenses in different shapes and sizes, which could correct many eyesight problems. Such lenses were also suitable for use in a variety of magnifying devices.

THE FIRST MICROSCOPES

The Italian scientist Galileo Galilei (1564–1642) has sometimes been given credit for inventing the microscope. However, the first records of a microscope suggest that it was built about 1590 by two Dutch spectacle-makers, Hans and Zacharias Janssen. Their microscope consisted of a tube about 2.5 cm in diameter and 50 cm long. It contained two lenses, one mounted at each end. The Janssens did not seem to realize the importance of their invention and did little with it. About the same time, an instrument-maker named Cornelius Drebbel, who was also Dutch, constructed a slightly better microscope and brought it to the attention of scientists. Soon Galileo learned about it and built his own microscope. However, he became more interested in a similar invention (also a tube containing lenses), the telescope (Figure 14.5). He left further development of the microscope to others.

Not all scholars of the day were willing to accept these new inventions and what could be seen with them. Many refused to look through telescopes or microscopes, believing that they were "unnatural" devices,

DID YOU KNOW?

Galileo commented that his microscope made flies look as big as lambs. At that time, personal hygiene such as bathing was less commonly practised than it is today, and insects such as flies and fleas were found everywhere. Indeed, fleas were so handy that they were a popular object of study. A common name for the microscope at that time was the *vitrum pulicare*, Latin for "flea glass."

Figure 14.5
With his telescopes, Galileo made some startling discoveries about the relationship between the Earth and the sun. Before turning to the telescope, Galileo made many detailed observations with a microscope. His descriptions of what he saw were among the first recorded biological observations made with this instrument.

which would mislead the human senses. Occasionally, this conviction posed real danger for people who used these instruments. In the early 1600s, for example, a Swedish scholar named George Stiernhielm persuaded a clergyman to examine a flea, its image magnified by the use of a lens. Frightened by the amazing size of what he saw, the clergyman denounced Stiernhielm, calling him a sorcerer. If Queen Christina of Sweden, who was a keen student of science, had not come to his aid, Stiernhielm might have been put to death for being a witch!

FURTHER DEVELOPMENTS

Some of the best early microscopes were made by Anton van Leeuwenhoek (1632–1723). Like the Janssen brothers and Drebbel, he was Dutch. Completely self-educated, Leeuwenhoek was one of the most important early microscopists (people who conduct studies using microscopes) (Figure 14.6).

Figure 14.6
During his lifetime, Anton van Leeuwenhoek held a variety of jobs, including those of bookkeeper, tax collector, cashier, sheriff, and owner of a dry goods store. He spent most of his time, however, constructing microscopes and using them to observe the microscopic world. He eventually made over 200 such instruments, some of which could magnify images 200 times.

The first microscopists faced many problems because it was difficult to produce lenses that were not flawed in some way. These flaws tended to distort the image: that is, they altered its appearance so that the true nature of the specimen (the object being examined) could not be seen clearly. Most microscope-makers attempted to correct such flaws by using more than one lens per microscope. Leeuwenhoek, in contrast, believed that the use of multiple lenses only produced *more* distortion, and so he took a different approach. Choosing only very high-quality glass and shaping it with great care, patience, and skill, he concentrated on producing microscopes with very small single lenses. An instrument with a single lens, like the one in Figure 14.7, is called a **simple microscope**.

With his simple microscopes, Leeuwenhoek made observations of a wide variety of materials, including his own saliva, drops of water, soil, and parts of plants, animals, minerals, and crystals. As careful as he was

Figure 14.7
The photo (a) shows a replica of one of Leeuwenhoek's microscopes. As you can see from the diagram (b), this device was actually quite small. Although they produced very clear images, instruments like this could also cause eyestrain.

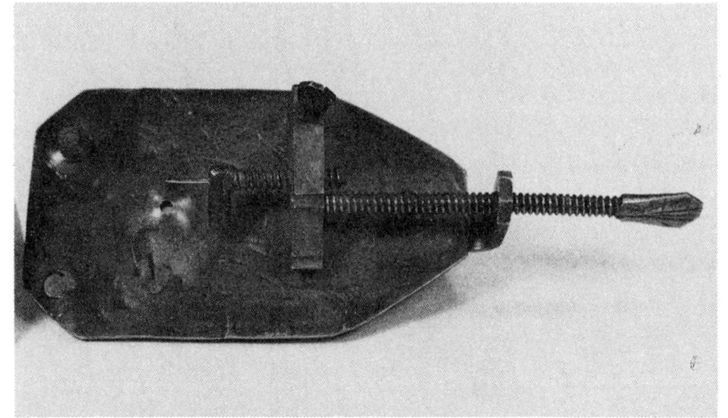

(a)

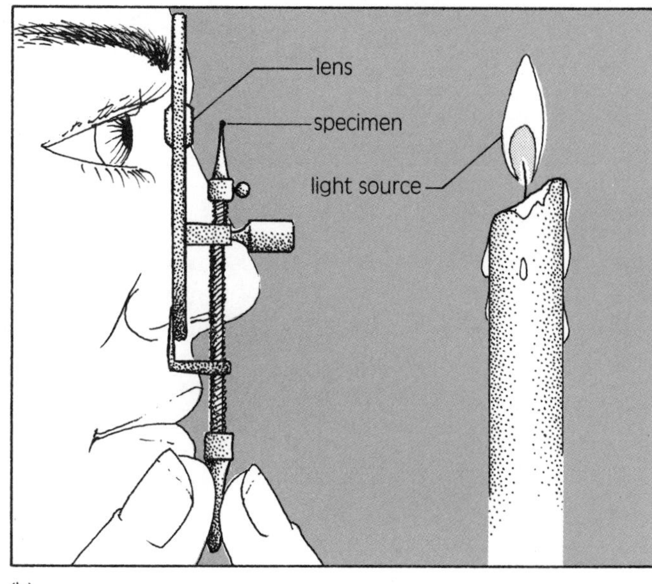

(b)

curious, Leeuwenhoek measured the objects he observed and wrote long and precise descriptions of what he saw, often accompanied by drawings. He shared these notes with his contemporaries, describing his discoveries in many letters to friends and colleagues. Over 200 of these letters were translated into English and Latin (Leeuwenhoek wrote in Dutch) and published by the Royal Society, a scientific organization in England. Although the things he described seem very familiar today, Leeuwenhoek was the first to see many of them. He discovered, for example, many types of micro-organisms — organisms so small they can be seen only with a microscope.

Another important microscopist working about the same time was an English scientist, Robert Hooke (1635–1703). In contrast to the simple microscopes of Leeuwenhoek, Hooke's microscope had two lenses, mounted at either end of a long tube (Figure 14.8). A microscope with more than one lens is called a **compound microscope**. Compound microscopes were more common at that time than were the type of instruments used by Leeuwenhoek. Although the simple microscopes he constructed were very fine for their day, Leeuwenhoek was mistaken in thinking that using more than one lens could not work well. As time would show, compound microscopes held far more potential for producing highly magnified images than did single-lens instruments.

Although Hooke used a different type of microscope, his procedure was similar to Leeuwenhoek's. He observed a wide range of everyday as well as unusual objects, recording his findings carefully in notes and drawings. Looking at a thin slice of cork, for example, Hooke observed that it looked much like a honeycomb (Figure 14.9). It seemed to be made up of many subunits, which resembled small rooms in the cork "building." He called each of these subunits a cell, from the Latin word *cellula*, which means "a small room." Today we apply the term **cell** to the basic units that make up all living things.

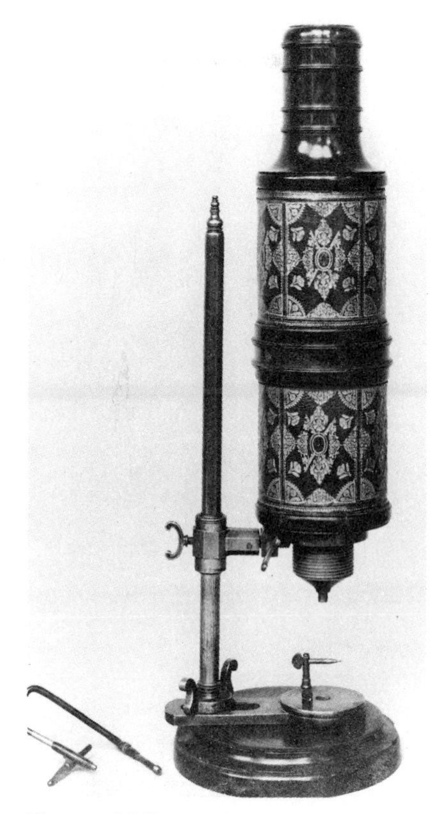

Figure 14.8
Robert Hooke had an expert instrument-maker construct several microscopes like this one for him. These compound microscopes could produce images that were magnified as much as 200 times.

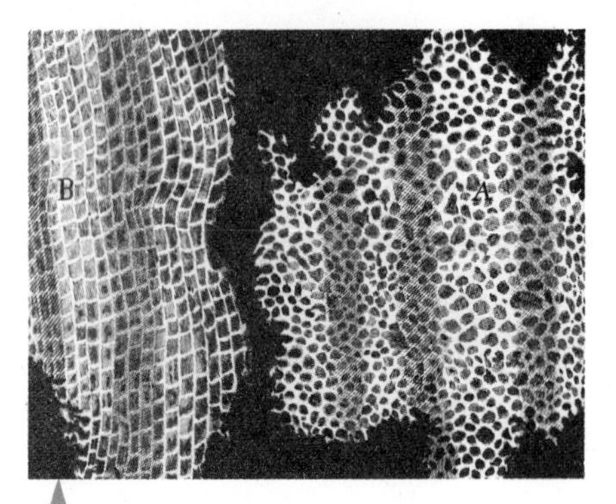

Figure 14.9
These drawings by Robert Hooke show the cork subunits that he termed "cells." The drawing on the left shows a piece of cork sliced lengthwise. The one on the right shows a cross-section of a piece of cork.

The work of Robert Hooke and Anton van Leeuwenhoek was unusual for its care and extensiveness. It was, in fact, among the best microscopic work done for the next 200 years. It was not until the early 1800s that some of the most serious problems with constructing microscopes were solved. By that time, microscope-makers had learned to make lenses that caused less distortion and to mount them properly. New methods for observing microscopic details are still being invented, and adjustments to the design of microscopes continue. In many ways, however, the type of microscopes most people use—and the type you will use—is similar to that employed 100 years ago (Figure 14.10).

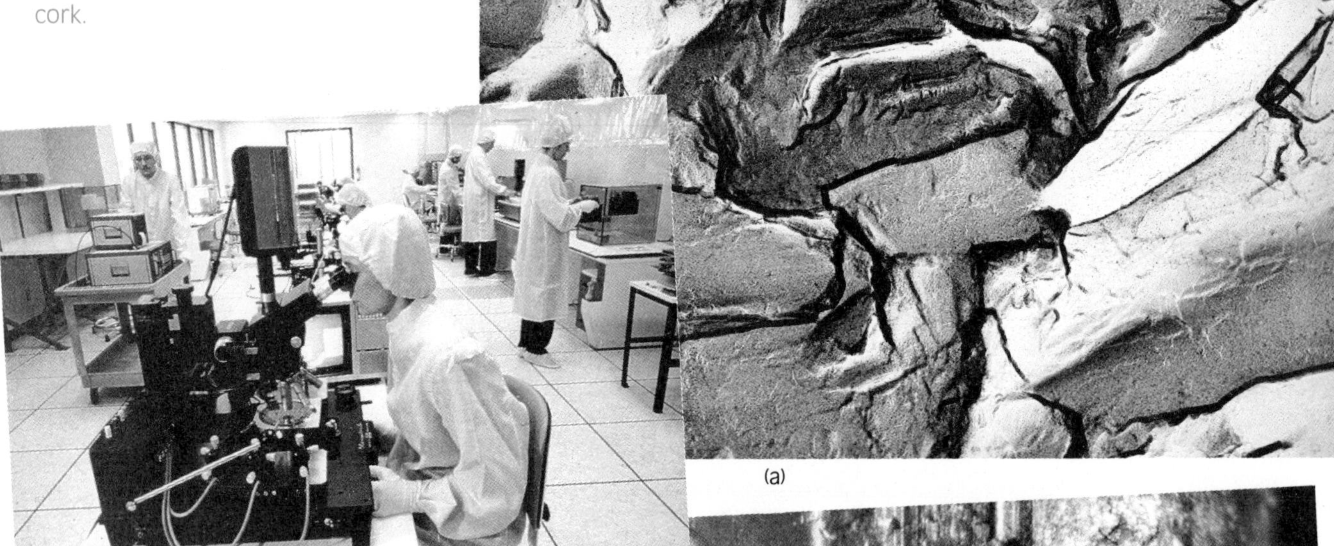

(a)

(b)

Figure 14.10
As well as being vital for medicine and for research in biology, microscopes are used in industry. For example, in (a), a zinc test rod is examined under a microscope so that a metallurgical engineer can see how a fracture occurred. In (b), workers use microscopes to assemble tiny computer chips. (c) A microscopic view of a bullet can help a forensic scientist to identify the weapon it was fired from.

(c)

REVIEW 14.1

1. What do spectacles, microscopes, and telescopes have in common?
2. Does a microscope, magnifying glass, or other instrument that contains a lens (or lenses) change the actual size of the object being examined? If not, what does it do?
3. Why was the discovery of how to make glass important for progress in science?
4. How did Anton van Leeuwenhoek's microscopes differ from those used by most other microscopists at that time?
5. (a) Who was the first person to apply the word "cell" to the many similar tiny subunits of a specimen that can be seen with a microscope?
   (b) How do we use this term today?
6. (a) In what way is the enlarged version of an object seen with a magnifying glass similar to a photograph?
   (b) In what way are they different?
7. There are no historical records to show who discovered that glass lenses could be used to assist human eyesight, or how this discovery was made. Invent your own history; make up a story about someone who discovered spectacles. What sort of person was it? What led him or her to make this discovery?

14.2 Using a Modern Light Microscope

The microscopes used most commonly today—as in the past—are called **light microscopes** because they make use of light to form images. In this type of microscope, beams of light pass through a specimen and are bent by the microscope's lenses. The bending of the light produces a magnified image. Most of the instruments used today are improved versions of Hooke's compound light microscope.

ACTIVITY 14B *Identifying the Parts of the Microscope*

To use a microscope safely and effectively, you must first have some knowledge of how the various parts of the instrument work and what relationship they have to each other. This activity will give you the opportunity to explore the parts and their relationships and to summarize what you learn for later use.

CAUTION!
Do not turn the knobs or move any of the microscope parts at this time.

MATERIALS

compound light microscope

PROCEDURE

1. In your notebook, construct a three-column table with the headings "Microscope part," "Location," and "Function." You will need more space in the second and third columns than in the first.
2. Read the next section, "The Compound Light Microscope." As you do so, examine your own microscope and identify the parts described. Also, fill in the table you made in step 1. All the microscope parts whose names are written in bold type should be included in your table. Your descriptions of the location and function of each part should be as brief as possible.

THE COMPOUND LIGHT MICROSCOPE

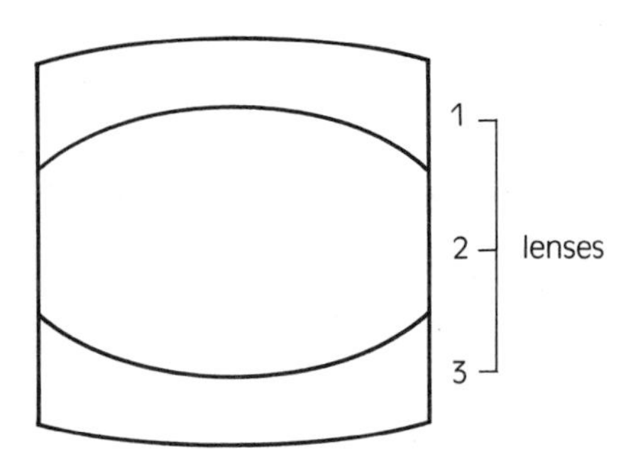

Figure 14.11
The three lenses that make up this lens system are designed to function together, as if they were a single lens. Their different shapes and characteristics make it possible for them to compensate for each other's flaws. This system eliminates much of the distortion produced by a single lens.

Compound light microscopes (often called simply "compound microscopes") contain at least two lens systems. A lens system may consist of a single lens, or it may include two or three lenses placed close together and designed to function as a single lens (Figure 14.11). Multiple lenses are used because blurring often occurs at the edge of a lens; the lenses in a multiple-lens system are designed to compensate for each other's blurring. The **body tube** holds the lenses in place at the proper distance from each other (Figure 14.12). This "proper distance" allows the particular combination of lenses to produce the clearest possible image.

The lens (or lens system) at the bottom of the body tube is closest to the object being examined. It is called the **objective lens,** or more simply, the objective. Most modern compound microscopes have more than one objective lens. Two, three, or four such lenses are usually mounted on a structure called the **nosepiece.** The nosepiece revolves so that different objectives can be moved into position at the end of the body tube. Each objective magnifies the image by a different amount.

The microscope shown in Figure 14.12 has three objective lenses. The low power objective has a magnifying power of 4×. This lens can produce an image four times larger than the object itself. The medium power objective on this microscope has a magnifying power of 10×, and the high power objective has a power of 40×. Four-, 10-, and 40-power objectives are a common combination on today's microscopes; another is 10×, 40×, and 100×. What is the magnifying power of each of the objectives on your microscope?

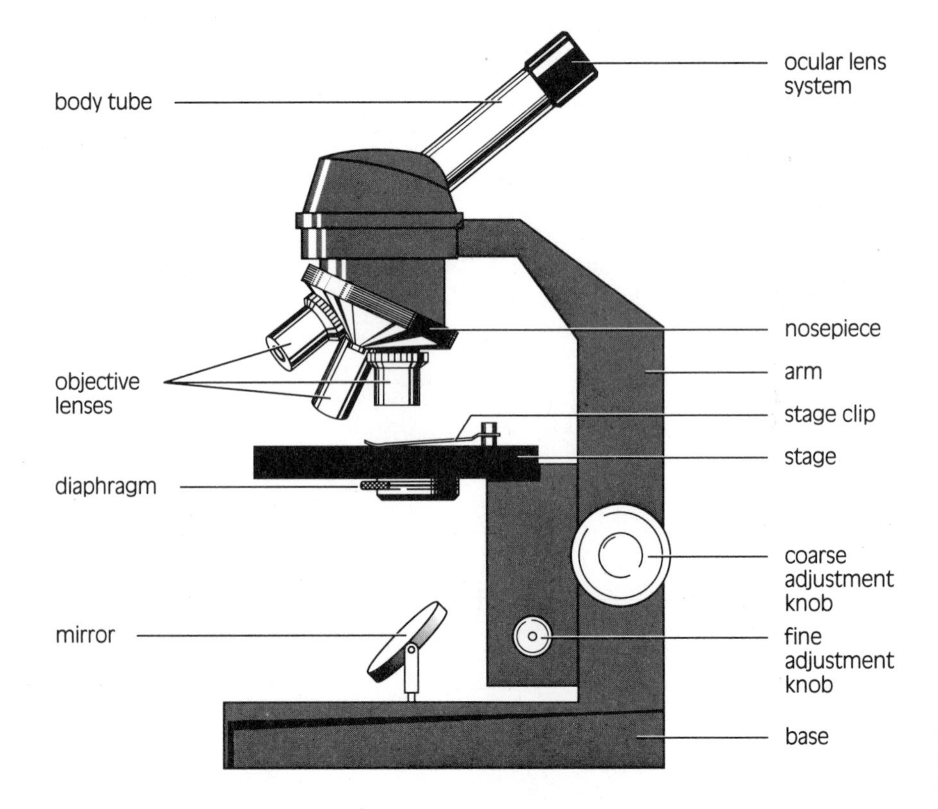

Figure 14.12
This microscope, unlike that shown in Figure 14.1, has an inclined body tube. Instead of a lamp under the stage, it has a mirror. This mirror catches the light from a separate lamp and reflects it up through the hole in the stage.

The lens system near the top of the body tube is called the **ocular lens**, or simply the ocular. (The Latin word *oculus* means "eye.") It is also sometimes called the eyepiece. You look into this part of the microscope. The ocular magnifies the image produced by the objective, so the image that you see is enlarged even more. The most commonly used ocular lenses have magnifying powers of 5× or 10×. What is the magnifying power of the ocular on your microscope?

To find out the total magnification of an image, you need to consider the magnifying power of both lens systems. For example, if a 40× objective lens was used with a 10× ocular, the image seen would be $10 \times 40 = 400$ times larger than the specimen being examined. What is the maximum magnifying power of your microscope?

There are many models of compound microscopes in use today. Some, like the one shown in Figure 14.1, have body tubes that are more or less vertical: they run straight up and down. Others, like that shown in Figure 14.12, have inclined (slanted) body tubes. The slanting of the body tube permits comfortable viewing. Microscopes with vertical body tubes usually have a hinge near the base of the instrument so that they too can be tilted.

Another feature that varies among microscopes is the **stage** (Figure 14.13), which is the platform located immediately below the objective lenses. It supports the object being viewed. On many microscopes, there is a pair of stage clips, designed to hold a rectangular piece of glass — the **slide** — in place. The specimen to be examined is usually placed on a slide. Other microscopes have a mechanical stage, with a jaw-like device to hold the slide. In the centre of the stage there is a hole, through which light shines from below to illuminate the specimen. Most microscopes have a built-in lamp that shines light directly upwards through this hole. Other microscopes have a mirror, which reflects light through the hole from the surrounding room.

EXTENSION

Some compound light microscopes have an additional lens system called the condenser lens. Do research in the library to discover where on a microscope this lens is located and what function it serves. Look at your microscope — handle it with care, and do not remove any parts. Does it have a condenser lens?

(a)

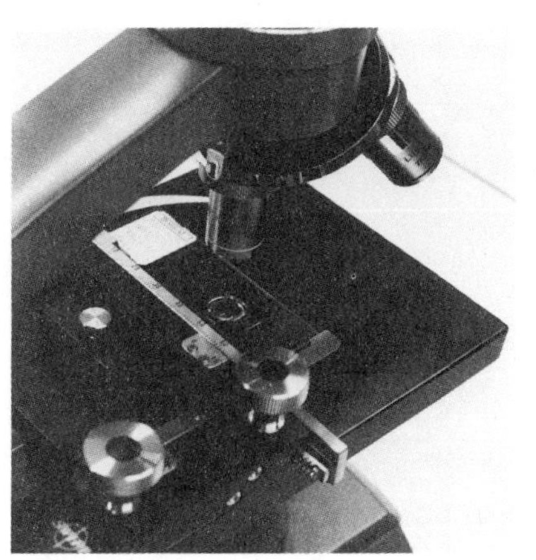

(b)

Figure 14.13
Some microscope stages, like that shown in (a), have stage clips to hold the slide in place. Other microscopes are equipped with mechanical stages (b). On such models, movable jaws are used to grip the slide.

Located between the stage and the lamp or mirror is the **diaphragm**. There are two basic types of diaphragm (Figure 14.14). Both are used to control the amount of light shining up into the body tube.

Figure 14.14
A disk diaphragm (a) consists of a ring in which there are openings of different sizes. This ring can be rotated until an opening of the desired size is in place just below the hole in the stage. An iris diaphragm (b) surrounds a hole that opens and closes like the pupil that is surrounded by the iris of your eye. It is adjusted with a small lever, found just under the stage.

(a)

(b)

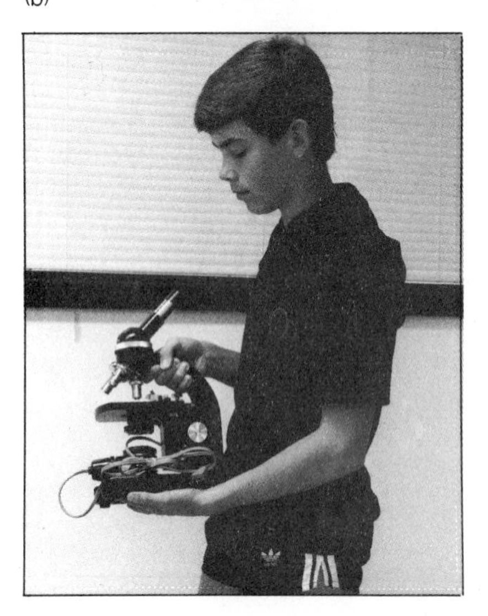

Figure 14.15
The correct way to carry a microscope. Keep the microscope upright, with one hand under the base and the other holding the arm.

You focus a magnifying glass by moving it closer to or farther away from the specimen until the part you want to see is in sharp focus. Compound microscopes are focused in much the same way; however, they have two knobs for adjusting the distance between the lens and the specimen. These are called the **coarse** and **fine adjustment knobs**. The coarse adjustment knob is always the larger of the two. The distance between the objective and the specimen, which is altered by adjusting these knobs, is referred to as the **working distance**.

The working distance is changed — and a focused image obtained — in different ways for the two major types of microscopes illustrated in Figures 14.1 and 14.12. In microscopes with inclined body tubes, the stage (with the specimen) is raised or lowered by knobs below the stage. In microscopes with vertical body tubes, the tube itself moves to raise or lower the objective lenses; the focusing knobs are located nearer the top of the microscope. In both types of microscopes, the larger coarse adjustment knob produces large changes in the working distance, whereas the smaller fine adjustment knob produces very small changes.

CARE OF THE MICROSCOPE

The best way to become familiar with a microscope is to use one. Before starting out, you will need to keep a few safety tips in mind. Modern light microscopes are powerful, expensive instruments and should be treated with care.

When carrying your microscope, use both hands: one below the microscope, supporting the base, and the other holding onto the arm (the curved structure connecting the body tube to the base) (Figure 14.15). Always carry your microscope in an upright position; in many types of microscopes, the ocular is removable, and it could fall out if the microscope were tipped over. If your microscope is stored in a box, always use a hand to support the bottom of the container when carrying it — the handles on top may not be secure. Before putting your microscope away, always rotate the low power objective into position, to reduce the chance of damage if the microscope is knocked. If there is a lamp, roll up its electrical cord. Then replace the instrument in its box or under its plastic cover and return it to its storage place.

Microscope lenses can be damaged by the oils and acids present in human skin, so the lenses should not be touched with your fingers. Even very soft material, such as tissues or handkerchiefs, can scratch the lenses. As you may know if you use a camera, only special lens paper should be used to clean lenses. Your teacher will have a supply of lens paper on hand. If the adjustable parts of your microscope do not move smoothly and easily, advise your teacher; this may be an indication of serious problems.

Probably the greatest risk to lenses is posed by glass, broken while using the microscope. When focusing, if you bring the objective too close to the glass slide on the stage, the slide may break. The resulting broken glass can easily scratch the objective lens. (See Safety Rules at the front of the book for safe disposal of broken glass.) If you always follow the proper procedures when using the instrument (as outlined in Activity 14C), broken glass should not be a problem. However, if an accident does happen, call your teacher for assistance before going any further. Various types of chemical crystals and sand can also scratch lenses. These materials are best examined with a type of light microscope in which the objective lenses are sealed inside a protective case (Figure 14.16).

If your microscope has an electrical cord for the lamp, review the instructions for safe handling of electrical appliances in Safety Rules at the front of this book.

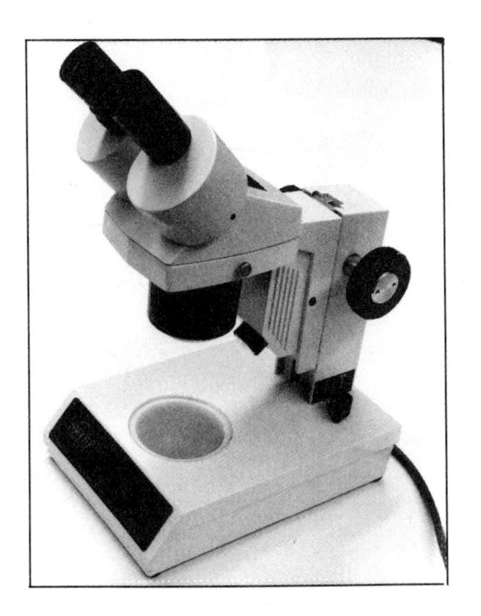

Figure 14.16
Some stereo microscopes, like this one, have the objective lenses in a protective case. They are thus useful for examining crystals and other specimens that might damage ordinary objectives. Notice that stereo microscopes have two ocular lenses—one for each eye. Because they also have an objective lens for each eye, stereo microscopes produce images with more depth. They are particularly good for examining very three-dimensional objects, such as plants or animals. Why do you think they are called "stereo" microscopes?

ACTIVITY 14C *Using the Microscope*

The microscope provides a "window" to a whole new world—one we would not even know existed if not for these instruments. In this two-part activity, you will learn how to prepare a wet mount. This type of mounting is particularly important for looking at living organisms or cells. In addition, you will learn how to focus the microscope in order to examine the specimen you have mounted. When you focus an optical instrument—whether you are using binoculars, a camera, a telescope, or a microscope—you obtain a clear, sharply defined image of the object being viewed.

MATERIALS

Parts I and II
compound microscope

Part I
slide, coverslip, dropper, forceps (tweezers), scissors, bond paper with writing on it, water, toothpick or needle

Part II
wet mount from Part I

CAUTION!
Slides and coverslips are delicate pieces of glass or plastic that can be easily broken. Handle them carefully so that you do not get cut or scratched.

PART I

PROCEDURE

1. Make sure that your slide and coverslip are clean and dry (Figure 14.17).

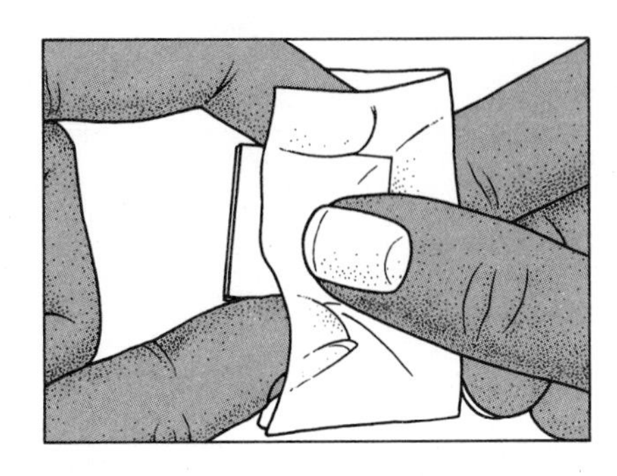

Figure 14.17
Coverslips are thin and can break easily. To dry a coverslip without breaking it, carefully wipe both sides at once.

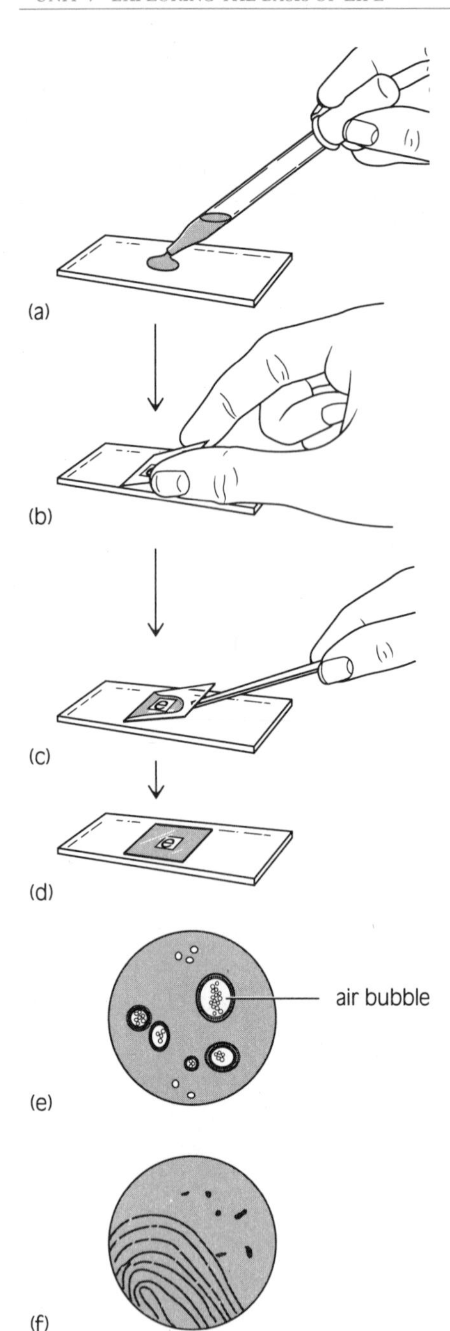

Figure 14.18
The preparation of a wet mount (a–d). Part (e) shows what you may see through the microscope if you do not successfully eliminate air bubbles from under the coverslip. Part (f) shows the view through the microscope if you do not clean the slide and coverslip properly before using them.

2. From the printed paper, cut out a small letter *a*, *e*, *f*, *h*, or *r*.
3. Put one drop of water on the slide with the dropper (Figure 14.18a). Using the forceps, place your letter on top of the water.
4. Hold the coverslip with your thumb and forefinger along opposite edges. Allow one of the free edges of the coverslip to touch the drop of water as shown in Figure 14.18b.
5. When the water has spread evenly along the edge of the coverslip, gradually lower the coverslip to cover the specimen. The goal here is to avoid trapping air bubbles beneath the coverslip. You may find this easier to do if you rest the edge of the coverslip on a toothpick or a needle and lower it with this tool, rather than with your fingers (Figure 14.18c, d).
6. Gently wipe any excess water from the top and bottom of the slide, and be sure that the top of the coverslip is dry. Leave a little bit of water around the edges of the coverslip. This will help to prevent the evaporation of the water from under the coverslip.

DISCUSSION

1. Why do you think it is important to have your slide and coverslip clean and dry before you start?
2. Why do you think it is important to avoid getting air bubbles under the coverslip?
3. Why do you try to prevent all the water from evaporating from under the covership? Is this important when you are looking at a letter? When might it be important?

PART II

PROCEDURE

1. Make sure that the low power objective is directly above the stage before proceeding. (If you wear glasses, remove them before looking into the ocular. You can usually focus your microscope to compensate for vision problems, and you will avoid scratching your glasses.)
2. Adjust the light source and diaphragm every time you use the microscope. To do this, turn on the lamp or adjust the mirror so that light shines through the hole in the stage and up the body tube. Look into the ocular and adjust the diaphragm so that the light is bright, but still comfortable to look at. (Glare can quickly cause eyestrain.) If your microscope has a mirror, do not use direct sunlight as a light source. Direct sunlight can damage the delicate tissue of the eye.
3. Place your slide from Part I on the stage. Be sure that the coverslip is on top of the slide (so that you will be looking through the coverslip when you examine the specimen). Centre your slide on the stage so the letter is in the path of light coming from the lamp or mirror. If your microscope has stage clips, it is often better to use only one of them. One clip will hold the slide in place while you adjust its position. If your microscope has a mechanical stage, the slide should be placed between the jaws, not under them. In this case, the slide is moved by rotating the appropriate knobs.

CAUTION!
Use the jaws or the clips carefully so that they don't "snap," breaking the slide.

4. Looking from the side of the microscope—so you can see the parts moving—use the coarse adjustment knob to bring the low power objective lens and the specimen as close together as possible (Figure 14.19). Note which way you turn the knob to do this: towards you or away from you. Make sure the lens and the coverslip do not touch.
5. Complete the focusing in one of two ways, depending on your microscope. (a) If a blurred outline of the specimen is now in view, additional focusing (to sharpen the image) should be done using *only* the fine adjustment knob. (b) If you cannot see a blurred image, you will have to use the coarse adjustment again. While looking into the ocular, *slowly* turn the coarse adjustment knob in the direction *opposite* to that of step 4, so that the distance between the objective and the specimen *increases*. Stop when a blurred image appears and then use the fine adjustment knob to sharpen the image. If the object does not seem to come into view, repeat steps 3 and 4.
6. Look through the microscope at the illuminated circle containing the specimen. This is called the **field of view**, or more simply, the field. The field is the area of the slide that you can see (or view) through the microscope. The field may include all of the specimen or, if the specimen is very large, only a portion of it.
7. Move the slide towards and away from you, and from side to side. Notice which way the letter seems to move when you do this.

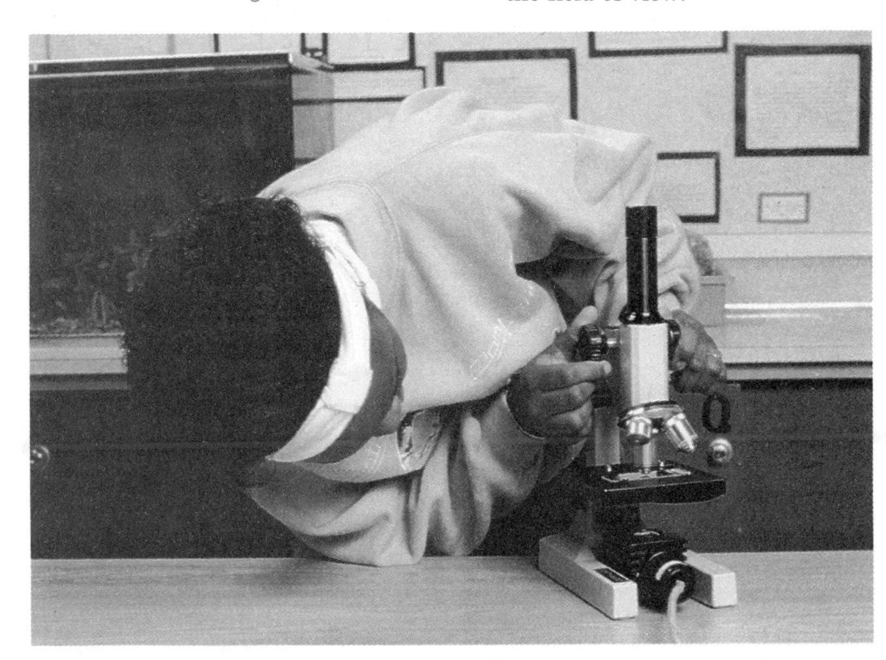

Figure 14.19
Do not let the objective lens touch the slide. The best way to avoid this is to watch from the side as you use the coarse adjustment knob to decrease the working distance.

DISCUSSION

1. Which way do you turn the coarse adjustment knob in order to increase the distance between the objective and the specimen: towards you or away from you?
2. When you look into the microscope, how does the movement of the image (towards you and away from you, or right and left) compare with the actual motion of the slide?
3. If your letter is right side up when you place your wet mount on the stage, how is it oriented in the field of view?

ACTIVITY 14D *Higher Magnifications*

Each of the objectives on a microscope provides a different view of the specimen being examined. Although there are many advantages to using the lenses with higher magnifying power, there may also be certain disadvantages. Do this activity to find out why.

MATERIALS

Parts I and II
compound microscope, scissors

Part I
coverslip, slide, toothpick or needle, piece of bond paper with printing on it, forceps, dropper, water, strand of hair

Part II
feather (or piece of feather)

PART I

PROCEDURE

1. Following the steps described in Activity 14C, prepare a wet mount of a letter. Focus on the

letter using the low power objective. Be sure your letter is exactly in the centre of the field of view and that the diaphragm is open as much as possible (without causing discomfort to your eyes).

2. Draw the letter as you now see it, in relation to the size of the field of view. To do this, draw a circle to represent the field, and then add the letter, making it fill only as much of the circle as the image fills of the field. Label this drawing "low power."
3. Rotate the nosepiece until the medium power objective clicks into place.

CAUTION!
Watch from the side as you do this to ensure that the lens does not hit the stage.

With most microscopes, the image will remain approximately in focus even when the nosepiece is rotated and the other objectives are used. You may want to adjust the focus slightly, but this should be done with the fine adjustment knob *only*.

4. Draw the letter as you see it now in relation to the size of the field of view. Remember to enclose it in a circle representing the field. Label this drawing "medium power."
5. Repeat steps 3 and 4, using the highest power objective. Label this third drawing "high power."
6. Return to low power and remove the slide from the stage. Wash and dry the slide and coverslip. Check that the objectives are dry. If not, lens paper will be needed to wipe them.
7. Prepare a wet mount of a strand of your own hair (approximately 2 cm long). Repeat steps 1 to 5.
8. Return to low power and remove the slide. Trade slides with three of your classmates and examine their hair specimens. In your notebook, record any similarities or differences between the different strands.

CAUTION!
The low power objective must always be in place before you remove the slide from the stage.

DISCUSSION

1. (a) What happened to the area around the letter as you changed from low to high magnification?
   (b) If you were examining a larger specimen (for example, one of the letters in the title to this chapter), what effect would changing the magnification have on the amount of that specimen that you could see?
2. (a) Did the field (and thus the specimen) seem to be as brightly illuminated when the higher power objectives were used?
   (b) How might (or did) you correct for this change in illumination?
3. How did the image of the hair change as you moved to higher powers of magnification?
4. Was the hair from other individuals similar or different in appearance to your own? Explain what you have learned about human hair by using the microscope.

PART II

PROCEDURE

1. As illustrated in Figure 14.20, cut a small piece (about 10 mm long by 5 mm wide) from the vane of a feather. Place this on a dry slide without a coverslip.

Figure 14.20
Cut a small piece from the vane of a feather.

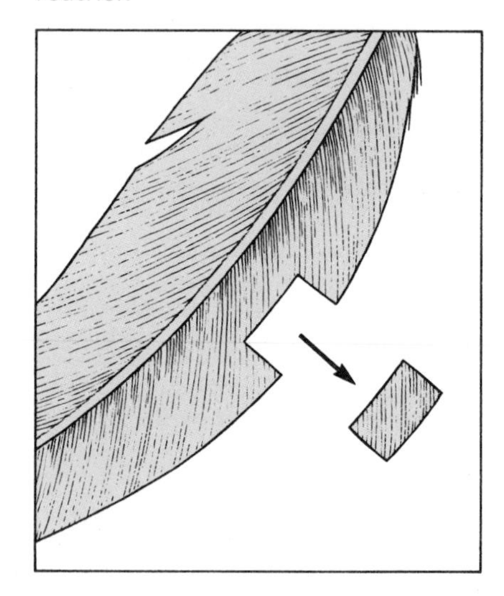

2. Use the low power objective to focus on this feather specimen.
3. Focus on the specimen using the medium and high power objectives. You will probably notice that it is not possible to get the whole feather in focus at the same time. This is because the field of view is really three-dimensional: as well as a diameter, it has a depth (Figure 14.21).
4. Using the fine adjustment knob, bring the top surface of the feather into focus. Then bring the bottom surface into focus. Finally, focus on a part of the feather that is between top and bottom. The amount (thickness) of the image that is in sharp focus at one time is referred to as the **depth of field** (Figure 14.22).

Figure 14.21
(a) The field of view is not flat, as might appear at first glance, but is slightly three-dimensional. (b) The green area indicates the portion of the specimen that is in focus at any one time.

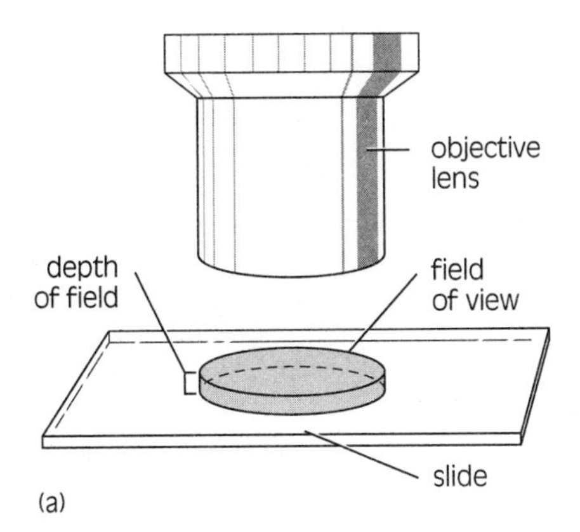

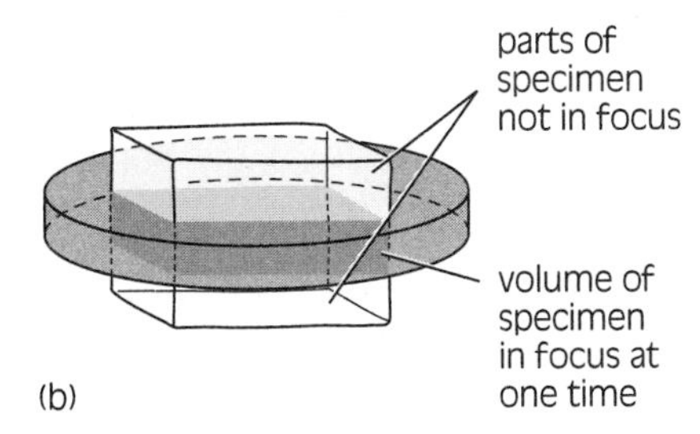

Figure 14.22
The depth of the field must be considered when using a camera lens as well as a microscope lens. Different portions of a picture will be in focus, depending on how the lens is adjusted.

5. Move the low power objective back into place and remove the slide from the stage. Carefully spread the barbs of the feather apart with a toothpick or needle.
6. Return the slide to the stage to see if you can discover what holds the barbs together. View the feather with the objective you feel is most useful for this purpose. Make a drawing of what you see. Label your drawing with the power used.

DISCUSSION

1. What happens to the depth of field as you change from low to medium to high power?
2. What type of specimens would have their images most affected by the depth of a particular field? For example, compare the effect of depth of field on the image of a letter with its effect on the image of a feather.
3. What are the advantages and disadvantages of higher magnifications? Consider the depth and the size of the field of view as well as light intensity.

EXTENSION

Look at a prepared slide of a flea with your microscope. Use whatever magnification you feel is appropriate. Draw what you see. What did you learn about the flea that you would not have known if you had not looked at it with a microscope?

REVIEW 14.2

1. Compare the location and function of the objective and ocular lenses.
2. (a) How do you change from one objective lens to another?
   (b) From what position should you look while doing this?
   (c) Which objective lens should be in place when you put the microscope away?
3. Describe the proper procedure for carrying a microscope.
4. Why should you not touch microscope lenses with your fingers or wipe them with anything but lens paper?
5. What is the "working distance" of a microscope? How is it related to producing a clear (focused) image of the specimen?
6. (a) If you were preparing a wet mount of the letter *f*, how would you position it on the slide so that you could read it right side up when viewing it through the microscope?
   (b) Why wasn't *x* one of the letters suggested for Activity 14C, Part I?
7. As you move from low to medium to high power, how does each of the following change:
   (a) the size of the field of view?
   (b) the illumination?
   (c) the depth of field?
8. (a) Why might you choose to switch from the low power objective to the high power lens to look at an object?
   (b) For what sort of specimen might you find that high power does not provide as useful a view of the specimen as medium or low power?

14.3 Beyond the Light Microscope

Specimens that are microscopic in size are made up of even smaller parts. When an image is magnified, all the parts that make up the specimen are also magnified. As well, all the distances *between* these parts are magnified. Using our eyes alone, these distances are often too small to see, and the parts all appear as a single mass. With magnification, however, the increased size of the distances makes it possible to see the parts as separate. Recall the letter you observed in Activity 14C. Was the letter solid, or could you see ragged edges and spaces in it that were not visible without magnification?

An optical instrument's ability to show objects or parts of objects as separate is referred to as its **resolving power**, or its resolution. The sharpness of the image that you see after focusing depends on the resolving power of the microscope. Magnification and resolution are equally important when viewing microscopic objects and organisms. High magnification alone is not enough to ensure that the images are useful. It doesn't help to make things larger if you can't see the enlarged detail clearly; small blurred parts then simply become larger blurred parts.

There is, therefore, a limit to the amount of *useful* magnification that a microscope can produce, and this limit is determined by the instrument's resolving power. The resolving power of a compound light microscope is limited by the length of the light waves: objects closer together than one-half of this wavelength cannot be seen as separate, using a light microscope. No improvements to lenses (to produce higher magnification) or lighting systems (to provide brighter illumination) will overcome this limitation.

The best light microscopes today can usefully magnify objects about 1500 times. In addition, they permit us to distinguish between objects that are only 0.0002 mm apart. With the unaided eye, most people cannot see objects as separate if they are closer together than 0.1 mm. The light microscope obviously allows us to see much more than we could with our eyes alone. With it, we can see things as small as individual cells—and even some of the subunits that make up these cells. Light microscopes do not, however, allow us to see some of the smaller—and very important—parts of cells. The discovery of these had to wait for the invention of a different type of magnifying device, the electron microscope.

DID YOU KNOW?

The arthroscope is an invention that combines fibre optics with micro-electronics. It has changed the way joint surgery is done. (*Arthro* is the Greek word for "joint.") The objective lens of this instrument is at the end of a long, thin probe (less than 5 mm thick), which can be inserted into a small incision in a joint. Glass fibres in the probe transmit images to a miniature video camera, allowing a surgeon to view the inside of the joint on an external monitor. Using tiny surgical instruments inserted into other incisions, he or she can even perform an operation by watching the monitor.

THE ELECTRON MICROSCOPE

Electron microscopes use beams of electrons instead of light. Because of the short wavelength of the electron beam, far greater resolving power is possible than with light. As a result, much higher powers of magnification are useful (Figure 14.23).

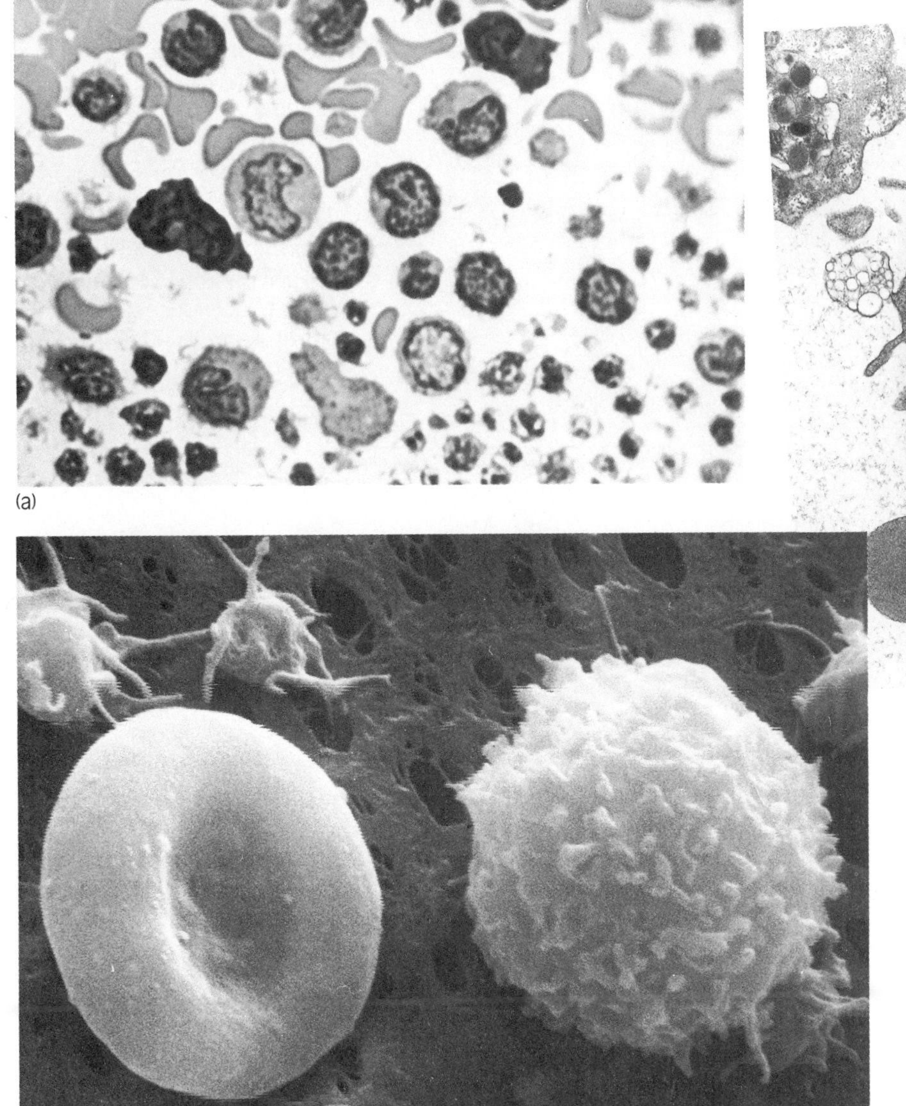

(a)

(b)

(c)

Figure 14.23
Photographs of red and white blood cells. These pictures were produced by three different types of microscopes: (a) a light microscope, (b) a transmission electron microscope, and (c) a scanning electron microscope.

Figure 14.24
This is what the surface of a leaf looks like when viewed with a scanning electron microscope.

The magnification powers of an electron microscope are amazing. Think of a human hair enlarged to the diameter of a railway tunnel; modern electron microscopes can give that power of magnification. They can magnify objects over 250 000 times, and can resolve (show as separate) objects that are as close together as 0.000 000 7 mm. With these instruments we can see the internal structure of cells and other specimens that are *very* small (Figure 14.24). We can even see some individual molecules.

The invention of the light microscope, and later the electron microscope, paved the way for the discoveries of many of the things you will read about in the other chapters in this unit. As research continues, our knowledge about organisms and about the processes of life grows. Microscopes have truly opened up a whole new world for people to explore!

DID YOU KNOW?

For most microscopic measurements, units much smaller than the millimetre are used. Scientists often use a unit called the micrometre, which is represented by a letter of the Greek alphabet. The abbreviation for micrometre is μm. One micrometre equals 1/1000 of a millimetre; that is, $1 \mu m = 0.001 mm$.

REVIEW 14.3

1. (a) What is resolving power?
   (b) What is the difference between the resolving power of our eyes and that of a microscope?
2. How are resolving power and magnification related to each other?
3. What sets the limit on the amount of useful magnification that a microscope can produce? Explain your answer in full.
4. What is the main difference between light and electron microscopes?
5. Robert Desowitz, a microbiologist who studies tropical diseases, has written, "Each time I turn on the microscope lamp, I still feel, as did Leeuwenhoek, like a voyager embarking on a journey to a distant and exotic land." Why do you suppose he feels this way?

CHAPTER • REVIEW

Key Ideas

- The microscope uses glass lenses to produce magnified images of very small objects.
- The invention of the microscope caused a revolution in science—and other fields—because it enabled people to see things that they had not previously imagined could even exist.
- Modern microscopes are sophisticated instruments, which must be used with care.
- The image of a specimen is focused by altering the working distance—the distance between the objective lens and the specimen.
- The magnification with which you view a specimen affects both the depth and the size of the field of view. It also affects how brightly the specimen is illuminated.
- The amount of useful magnification a light microscope can produce is limited by its resolving power.
- Electron microscopes permit us to see things that are much smaller than can be viewed with a light microscope. This is because an electron beam permits much greater resolution—and thus higher useful magnification—than does a beam of light.

Vocabulary

image
magnify
lens
simple microscope
compound microscope
cell
light microscope
body tube
objective lens
nosepiece
ocular lens
stage
slide
diaphragm
coarse adjustment knob
fine adjustment knob
working distance
field of view
depth of field
resolving power

1. Create a concept map to show what you have learned about the microscope.
2. Briefly describe the functions and locations of any five of these parts of your microscope: objective lens, ocular lens, nosepiece, stage, diaphragm, coarse adjustment knob, fine adjustment knob.

Connections

1. Providing examples from this chapter, discuss why it is important for individuals to communicate with others about their discoveries or studies.
2. What type of mount is most appropriate for looking at live specimens with a light microscope?
3. Give two reasons why specimens need to be relatively thin to be viewed under a light microscope.
4. Name two features of the field of view that decrease as magnification increases.

5. When viewing most specimens, particularly under high power, a good microscopist frequently changes the fine adjustment. Why is this so?
6. Compare the light and electron microscopes with respect to (a) the useful magnification they can produce and (b) their resolving power.

Explorations

1. Early Roman engravers used water-filled vessels as magnifiers. You can use a Florence flask to demonstrate how they might have done this. Design an experiment to show the various ways in which this flask can be used to produce images that are smaller and larger than the objects being examined. Things to consider are how much water you should place in the flask to produce each type of image and where the specimen should be placed in relation to your eye. See what happens if you place a candle on one side of a filled flask and a piece of paper on the other side (Figure 14.25). How would you focus the image of the candle seen on the paper?

CAUTION!
Use safety goggles. Get approval from your teacher before trying out your experiment.

2. Design a poster that advises microscope users on one aspect of the proper care or use of their instrument. Your poster should show how to correctly use the microscope (or a part of it) as well as the dangers of improper use.
3. Imagine that you are trying to invent a device that will allow you to investigate the deepest parts of the ocean from a position on land. (You don't have to design or explain the device, just imagine that you have invented it.) What might you set out to discover about this part of the planet? What will your instrument have to be able to do in order to allow you to make these discoveries? For example, does your device need to be able to "see" only, or must it also be able to sense other aspects of the environment? How will you tell others about your discoveries? Do you think that you might have trouble getting people to believe that what you discovered does, in fact, exist? Why or why not? If so, how will you go about convincing them?

Reflections

1. Look back over your learning journal and what you learned in this chapter, and consider how your ideas about microscopes have changed. Before reading this chapter, why did you think microscopes were used? Did you think microscopes were difficult or easy to use? What are your answers to these questions now? If they have changed, what parts of the chapter (for example, the activities or the text) were responsible for altering your ideas?
2. Do you think that the invention of the microscope was important for progress in science and medicine? Explain the reasons for your answer and, if possible, provide examples of the benefits provided.

Figure 14.25
A Florence flask acts on light from a candle.

CHAPTER

15 The Cell

The microfossils shown below may not look very impressive. Yet the discovery of these preserved remains of micro-organisms caused great excitement. The reconstructions beside the photographs show these tiny, ancient fossils more clearly. Until recently, the oldest known fossils—the first signs of life on Earth—were little more than 600 million years old. (The world is estimated to be about 4.5 billion years old.) Within the last 20 years, however, scientists have developed techniques that enable them to examine very thin slices of rock with an electron microscope. Their discoveries suggest that life may have existed as early as 3.5 billion years ago and almost certainly existed 2 billion years ago. These microfossils are, therefore, the remains of some of the oldest inhabitants of the Earth. A number of these organisms joined together in a string, but each had a body consisting of only a single cell. Yet, from these small beginnings, all life on Earth—from fleas to whales and from dandelions to Douglas fir trees—eventually developed. The history of life on this planet began with single cells.

Your individual history—indeed, the history of every living thing—also began with a single cell. Each person is the result of one cell that divided, giving rise to two, which then divided into four. This process continued until the trillions of cells that make up a human body were produced. Cells are important for anyone interested in science—or in how their body works—to know about!

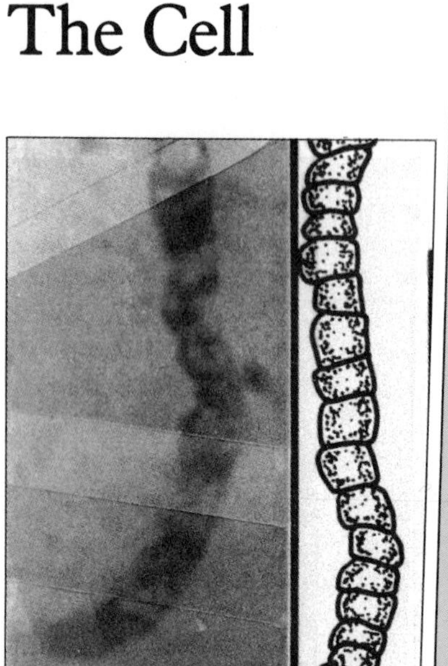

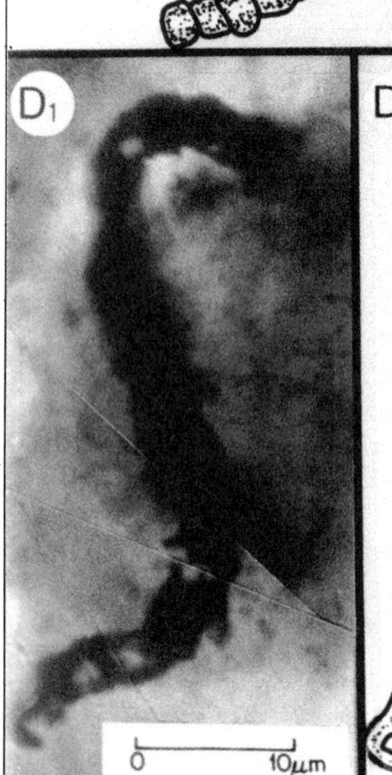

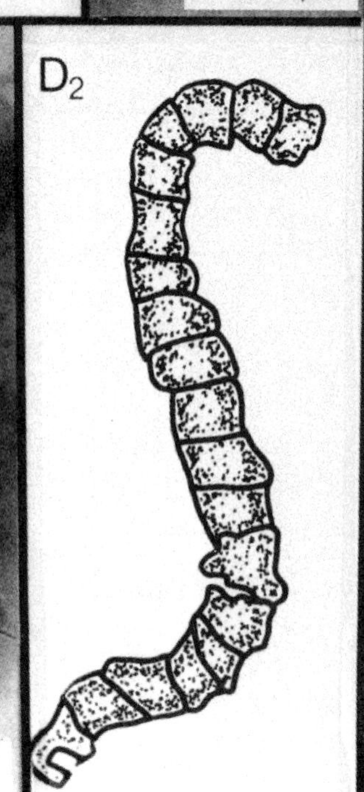

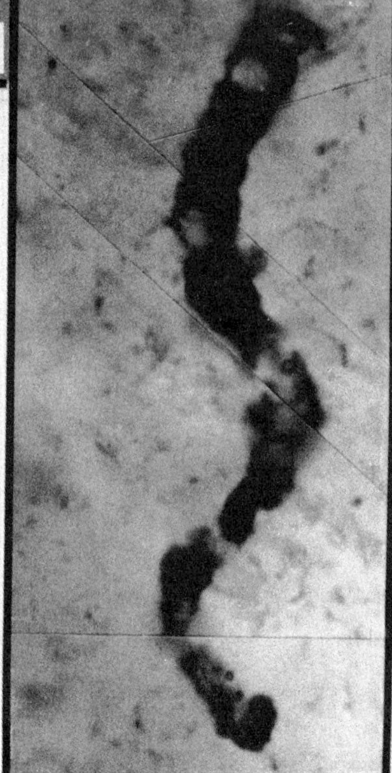

ACTIVITY 15A Exploring Cells

Recall from Chapter 14 that Hooke, Leeuwenhoek, and other early microscopists looked at a variety of materials and sometimes even saw individual cells. However, there were no books available to help them understand what they were seeing. Everything was a new discovery that they had to try to explain for themselves. At first, using the microscope often produced more questions than answers. It took many years of conducting experiments and examining specimens before scientists could understand the structure and function of what they saw.

In this activity, you will be in the same situation as the early microscopists. You may find yourself asking many of the same questions they did.

MATERIALS

compound microscope
leaf of an *Elodea* plant
prepared slide of animal cells

PROCEDURE

1. Prepare a wet mount of the tip of an *Elodea* leaf.
2. Using the low power objective, position the slide so that you are looking at the edge of the leaf's tip. Look for a single layer of green, oblong cells.
3. Switch to a higher power and examine these cells more closely. Draw a sketch of several cells, including what you see inside them. For later reference, number (on the sketch) what you feel are important parts of the cell.
4. Divide a sheet of paper into three columns. In the left column, list 5 to 10 questions about what you are seeing. (Your questions might start with "What are" or "What is the function of.")
5. In the middle column, answer as many of these questions as you can. Guessing is OK! If you don't have any ideas about the name or function of a particular cell part, leave the space blank.
6. Repeat the process with your specimen of animal cells. First look under low power to locate and identify single cells. Then repeat steps 3 to 5.
7. The third column of your chart can be completed at the end of the chapter.

DISCUSSION

1. Were all the individual cells in your *Elodea* leaf identical in size, shape, or internal appearance? Explain your answer.
2. (a) Were all the animal cells identical? If not, how were they different?
   (b) In what ways were all the animal cells similar to each other?
3. (a) What features did the plant and animal cells seem to have in common?
   (b) Were there any features that only plant or only animal cells seem to have?

15.1 The Cell—The Basic Building Block

Although Hooke had seen and described cells, the microscopes of the 1600s were not powerful enough to enable him to see the many smaller structures and the amazing activities inside cells. By the early 1800s, improvements in microscopes had allowed scientists to see many more details. Microscopists began to carefully investigate the structure of organisms. Eventually, scientists recognized that the cell was the basic building block in the bodies of both plants and animals.

The ideas of a number of researchers working during the 1800s contributed to what we now call the **cell theory**. This theory has three main parts, which can be summarized as follows:

- All living things are made up of one or more cells.
- The cell is the functional unit of life.
- All living cells come from pre-existing cells.

The development of the cell theory, one of the most important theories in all of science, revolutionized the study of life. Before the cell theory was accepted, many people thought that living things could come from non-living material. They believed that dew drops on plants could turn into insects, for example, or that rotting meat sometimes produced (rather than attracted) flies. The third part of the cell theory rejected this idea.

Development of the first two parts of the cell theory was perhaps even more important. Now researchers realized that all living things had a similar basic structure and thus shared many of the same life processes. They could begin to look for general explanations of these life processes, rather than having to find a different explanation for every type of organism. They began, therefore, to seek an understanding of the structure and function of cells and how they affect the survival of organisms — the subject of this chapter. The third part of the cell theory is vital to an understanding of the process of reproduction and the mechanism of heredity, which are the subjects of the next two chapters.

EXTENSION

Do research into the history of the cell theory. Who developed the various parts of the theory? How did they arrive at their conclusions? How long did it take for the theory to be developed and then to be accepted? How did the cell theory change people's views?

CELL STRUCTURE

In many ways, it is useful to compare an individual cell to an individual organism—a human, for example. In order to survive, you must provide your body with nutrients. Some of these nutrients supply energy, which allows you to move about and permits your body's organs to carry out their vital functions. Other nutrients provide the raw materials, such as proteins, vitamins, and minerals, that you need to build and repair body tissues.

An individual cell needs to take in nutrients for exactly the same reasons. Like your body, it needs raw materials to grow and repair itself. Energy is required for many of the chemical reactions that go on inside a living cell. The nutrients needed by you and by any of your body's cells are the same. In fact, it is the food *you* eat that supplies nutrients for your cells.

Before food can be used by the cells, it must be broken down into pieces small enough to enter individual cells. This is done by the digestive organs of your body, each of which is specialized to carry out a particular task.

Your body also has a variety of other organs and structures, each performing a necessary life function, such as collecting body wastes or breathing. This is another similarity between organisms and cells. Cells too possess a variety of structures, most of which are specialized to carry out one or more functions vital to the life of the cell. Before examining these structures in detail and learning about their functions, let us look briefly at some of the most important cell parts. You may have noticed some of these structures while doing Activity 15A.

All cells are surrounded by a thin, flexible covering called the **cell membrane** (Figure 15.1). With a few exceptions, every plant and animal cell also possesses a **nucleus** (plural: nuclei). It is one of the most easily

Figure 15.1
Photograph of an animal cell (from a human pancreas) as seen with an electron microscope. When an animal cell is viewed with a *light* microscope, however, the nucleus usually appears darker than the cytoplasm.

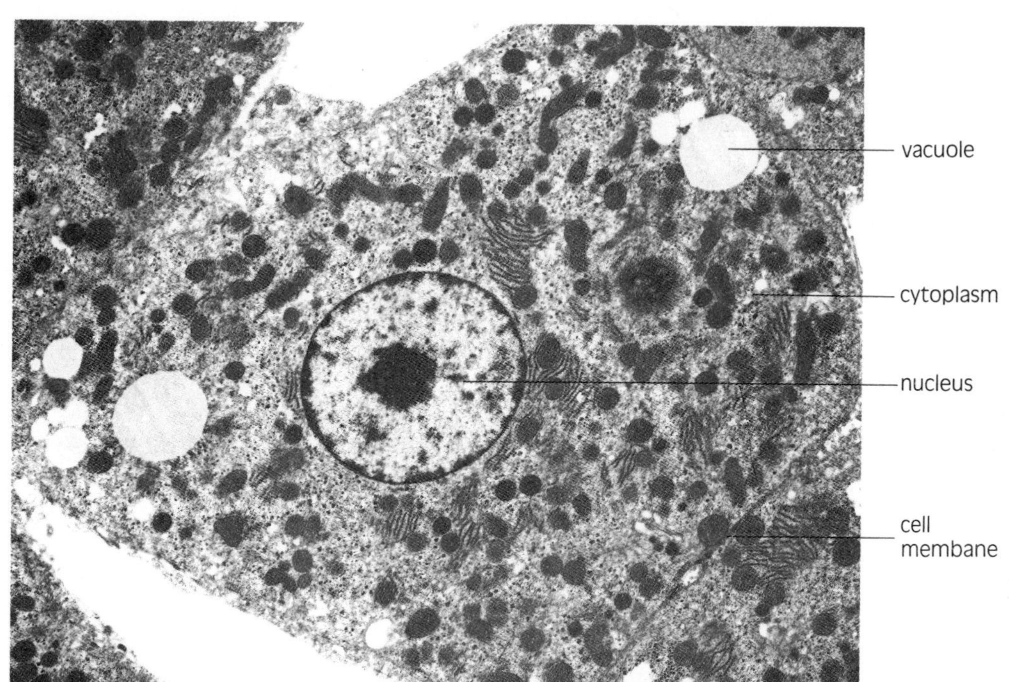

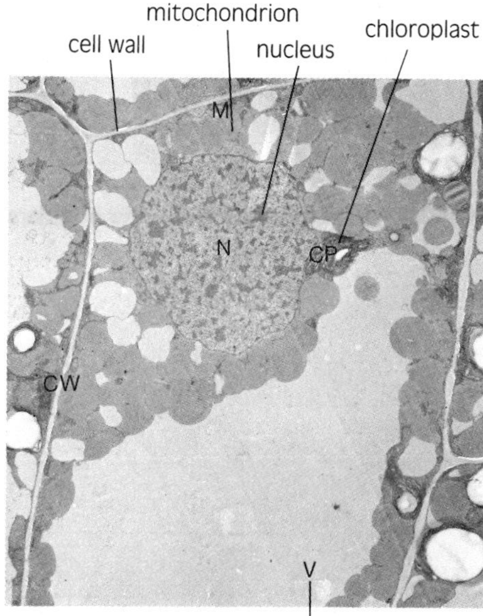

Figure 15.2
This photograph shows a plant cell (Eastern white pine) as seen with an electron microscope. Note that the large vacuole has pushed the other cell structures towards the sides of the cell. You will learn more later about the structure labelled "mitochondrion."

seen of the cell's internal structures, usually appearing as a large mass, slightly darker than its surroundings, sometimes near the cell's centre. The rest of the cell material, which is inside the cell membrane but outside the nucleus, is called the **cytoplasm**. Scattered throughout the cytoplasm (and considered part of it) are a variety of structures known as **organelles**. Many of these organelles are bounded by membranes. One of the organelles you can see in Figure 15.1 is a **vacuole**. Like the organs of your body, different types of organelles have differing structures and functions.

Plant cells have several structures not possessed by animal cells. One of the most obvious is the **cell wall**, a non-living layer that lies outside the cell membrane. In addition, only plants have **chloroplasts**, organelles that are often easily identified by their green colour (Figure 15.2). Although both plant and animal cells have vacuoles, those in plant cells are commonly much larger than those in animal cells. In some plant cells, most of the cytoplasm may appear to be taken up by one of these vacuoles.

In the next sections, you will learn more about some of these structures, particularly those that are responsible for (1) controlling the passage of materials in and out of a cell, (2) supplying the cell's energy requirements, and (3) controlling the overall activities of a cell.

ACTIVITY 15B The Structure of Cells

This activity will give you the opportunity to identify some of the parts of a cell and to compare the cells of plants and animals. Although most of the structures and activities of plant and animal cells are the same, there are differences between them. These differences relate to the different lifestyles of plants and animals. You will also use a technique that microscopists call staining. Stains make certain structures easier to see.

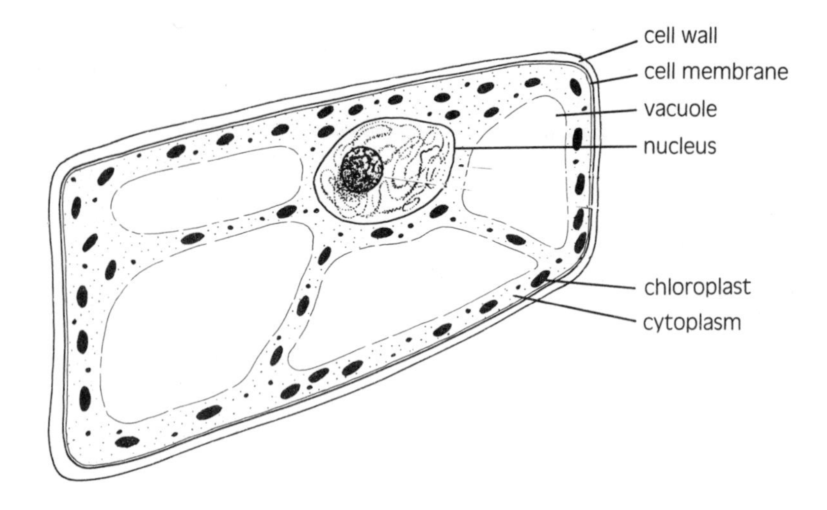

Figure 15.3
This sketch shows a plant cell as you might see it with a compound light microscope. Many of the cell's organelles are too small to enable any detail to be seen. The vacuoles shown here are quite large, and they have pushed the cellular structures to the outer portions of the cell. The vacuoles are usually smaller in younger cells than in older cells.

MATERIALS

Parts I and II
compound microscope

Part I
small section of white onion with the dry outer skin removed, microscope slide, coverslip, water, forceps, dropper, IKI (iodine solution), apron, safety goggles

Part II
prepared slides of animal cells

PART I

The cell walls of plants often appear as a dark, distinct boundary when viewed with a microscope (Figure 15.3). At other times, the cell wall appears as a clear area between cells. Just because this space appears empty, however, doesn't mean there is nothing there! Another feature that varies among cells is the size and number of vacuoles. In many onion cells, there is a single vacuole that is so large that it pushes all the cytoplasm to the outer edge of the cell. The appearance of this "edge" will vary, depending on your view of the cell.

Remember that cells are three-dimensional, not flat, as some drawings or photographs might suggest (Figure 15.4). Like a human body (which also has only two dimensions in a photograph), a cell has depth, as well as length and width. You will probably have to move the fine adjustment knob back and forth to bring different parts of the cell into clear focus.

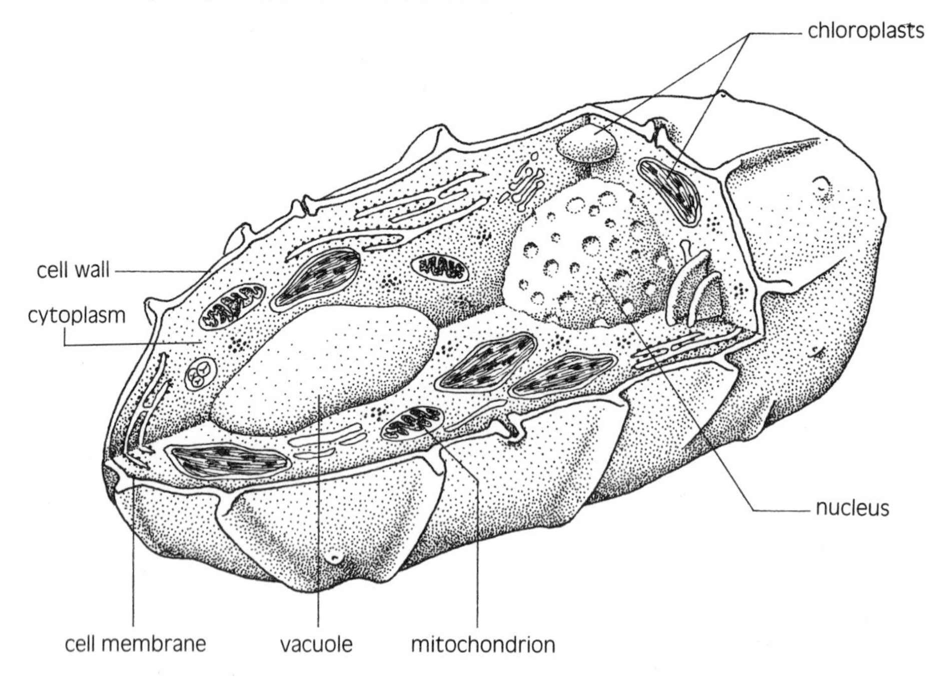

Figure 15.4
A three-dimensional drawing of a typical plant cell as it might be seen with an electron microscope. Note that many more organelles are now visible.

There are many organelles in the cytoplasm that are too small to be seen clearly with a light microscope. Some of these may appear as dots of various sizes in the cell you are looking at. Remember that just as people are individuals, who may look quite different from each other, so are cells. No two cells in your piece of plant tissue are identical, and none will be exactly the same as the one shown in Figure 15.3.

PROCEDURE

1. Put on your apron.
2. Strip off a small section of the onion membrane (Figure 15.5a).
3. Make a wet mount of this piece of onion.
4. Using low power, locate several onion cells.
5. If you think it would be useful, switch to a higher power and examine the cells in more detail.
6. Draw two of the onion cells and label the structures you can identify. Underneath your drawing write down what you think the functions of these structures might be.
7. Remove the slide and carefully draw off the water (Figure 15.5b).
8. With the dropper, place a drop of iodine solution at the edge of the coverslip (Figure 15.5c). The iodine will spread underneath the coverslip.

CAUTION!
Iodine is highly toxic if inhaled or swallowed. It is a strong irritant to eyes and skin. It will also stain the skin, clothing, desktops, and floors. Use safety goggles and an apron when handling iodine, and wipe up any spills.

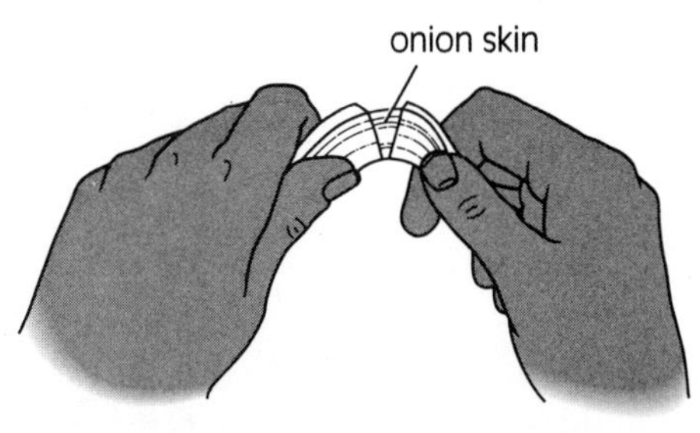

(a) Fold back the section of onion so that the outermost layer breaks in half. This will allow you to see the portion of the skin that you need.

(b) Place a piece of paper towel at one edge of the coverslip to draw off the water.

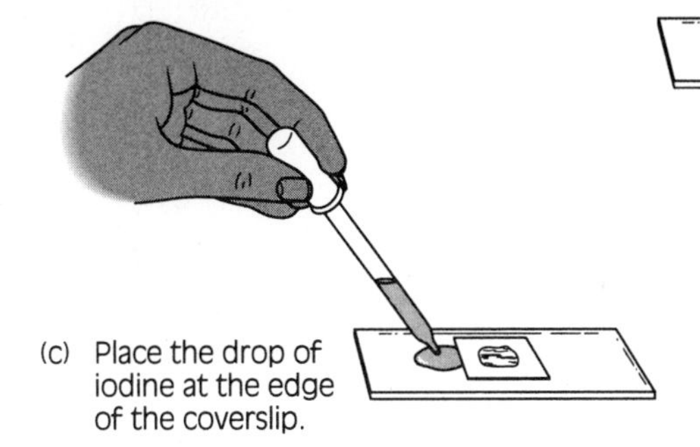

(c) Place the drop of iodine at the edge of the coverslip.

Figure 15.5
(a) How to obtain the onion specimen. Parts (b) and (c) show how to add iodine as a stain.

9. Repeat steps 4 to 6 with your stained slide.

DISCUSSION

1. What structures did you see
   (a) without the iodine and
   (b) with the iodine?
2. Did both onion cells have the same number of vacuoles?
3. (a) Where was the nucleus located in each of the two onion cells (near the centre or along the side of the cell)?
   (b) Why do you think the nuclei were in these positions?
4. (a) Refer to the drawings you made in Activity 15A. Were there any structures that you found in the *Elodea* cells that were not present in the onion cells?
   (b) Were there any structures in the onion cells that you could not find in the *Elodea* cells?

PART II

PROCEDURE

1. Under low power, focus on several of the cells on a prepared slide of animal cells.
2. Switch to high power and examine these cells, or a single cell, more closely. You should be able to identify the parts labelled in Figure 15.6, although you will not see as much detail with your microscope.
3. Draw two of the cells. Use shading to indicate the parts that you see as darker than the surrounding material. Underneath your drawing, write down what you think the functions of these structures might be.
4. If more than one type of prepared animal cell is available for examination, look at one of these other types as well, repeating steps 1 to 3.

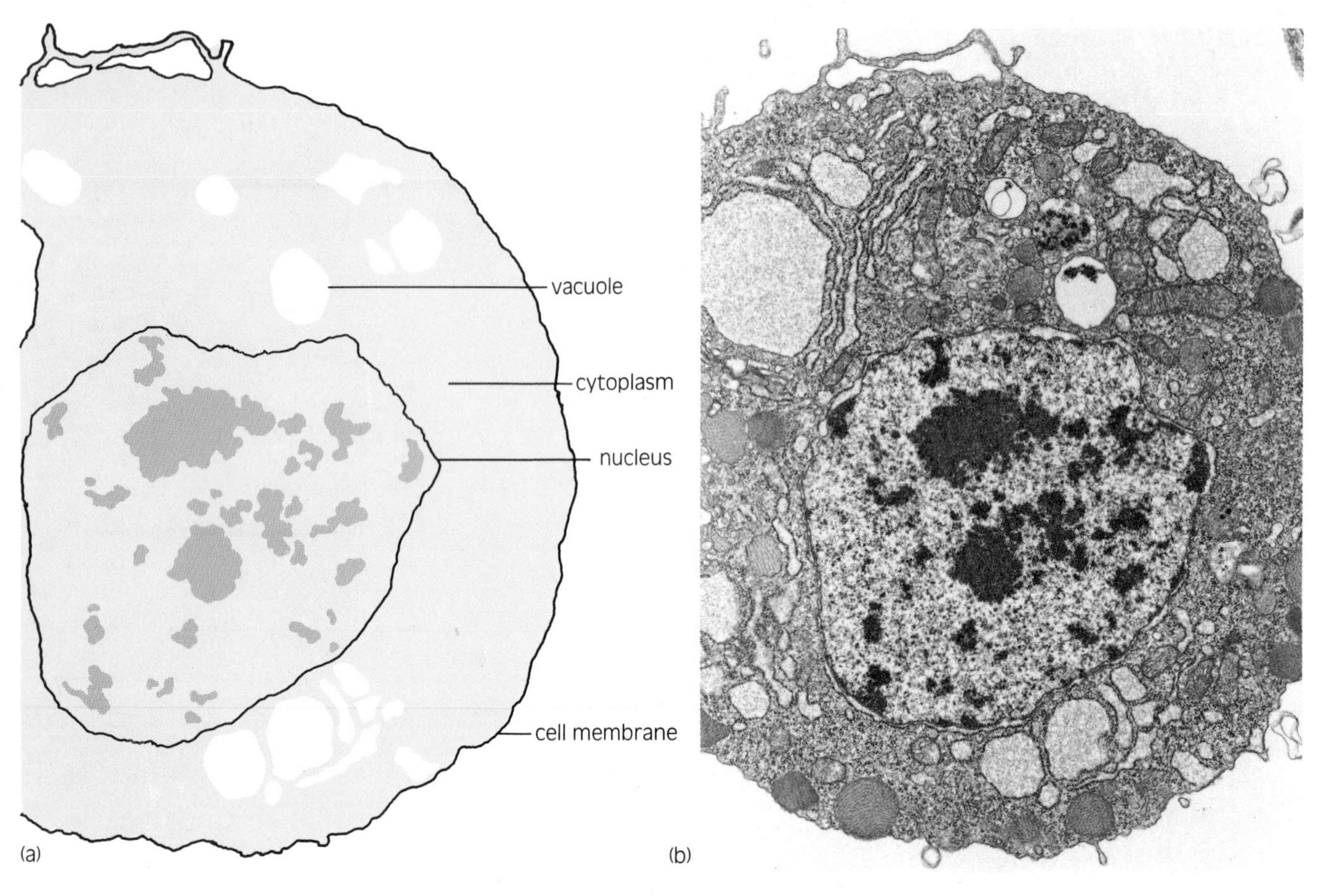

Figure 15.6
(a) A drawing of the main structures of a typical animal cell. (b) The same cell as seen with a powerful microscope.

DISCUSSION

1. Where was the nucleus located in the cells you examined?
2. Were these cells flat or three-dimensional? How did you know?
3. (a) If you examined more than one type of animal cell, were there any differences among the various types? If so, what were they?
   (b) In what ways were the different types of cells similar?
4. Were these cells different from the animal cells you examined in Activity 15A? Explain your answer.

REVIEW 15.1

1. (a) What are the three parts of the cell theory?
   (b) Why was development of the cell theory important for progress in science?
2. List three features that animal and plant cells have in common.
3. What are two differences between animal and plant cells?
4. (a) What are organelles?
   (b) How are cell organelles similar to human organs?
5. What general features do an individual organism and an individual cell have in common? (Describe at least two features.)

CYTOTECHNOLOGIST CAREER

Have you or has anyone in your family ever had a test to check for cancer, such as a Pap smear? You worry and wonder what the result will be. While you are wondering, the test sample may be in the laboratory at Victoria General Hospital. Jo Mason is the cytotechnologist there, and her job is to examine cells in test samples.

What do you do in your work?
I study cells in the specimens to determine whether they are abnormal or not. Then I identify the abnormality. Some of the conditions we diagnose in cytology are cancer, infections, and diseases caused by viruses, such as AIDS and herpes. A fair amount of my work comes from patients with lung cancer.

Cytotechnologists examine cells that have been washed, scraped, or brushed from the body, or that have been extracted through a very thin needle. We do not work with blood samples.

How do you study the cells?
The cells I want to examine are spread on a slide and then stained. That makes it easier to differentiate the cytoplasm from the nucleus.

I use a light microscope to study the cells. To first look at them, I use a magnification of $100\times$. For closer investigation, I use $400\times$. Depending on the type of specimen, I can spend from 1 to 15 minutes looking at one slide. You have to look at a field of cells and see that there's one in there that's different—and maybe just slightly different. Then you have another look at it on high power.

What do you find to be the best part of your work?
I like the challenge of finding out what the next patient's specimen will contain. And I like working directly with others in the hospital to help diagnose patients. They know I am interested in the patients' progress. When we get more information about a patient later on, we often look back at our original cytology diagnosis. This helps us improve our skills, and it's a great learning experience for me.

What does it take to be a good cytotechnologist?
You have to be able to sit quietly and concentrate for long periods and pay attention to detail. You need a very critical eye. You also need to be able to work on your own, compile information, and make decisions.

How did you get into this work?
I had planned to go into nursing but a friend suggested that I might be good at cytology. So I became a student at the Cancer Control Agency of B.C. in the early years of their now-famous central diagnostic lab.

Have there been any changes in cytology over the years?
Yes, and I have to read and attend workshops, meetings, and tutorials in order to keep up with the advances in the field. This is not just for my own learning and job security, but for the patients' well-being. We're here for the patients.

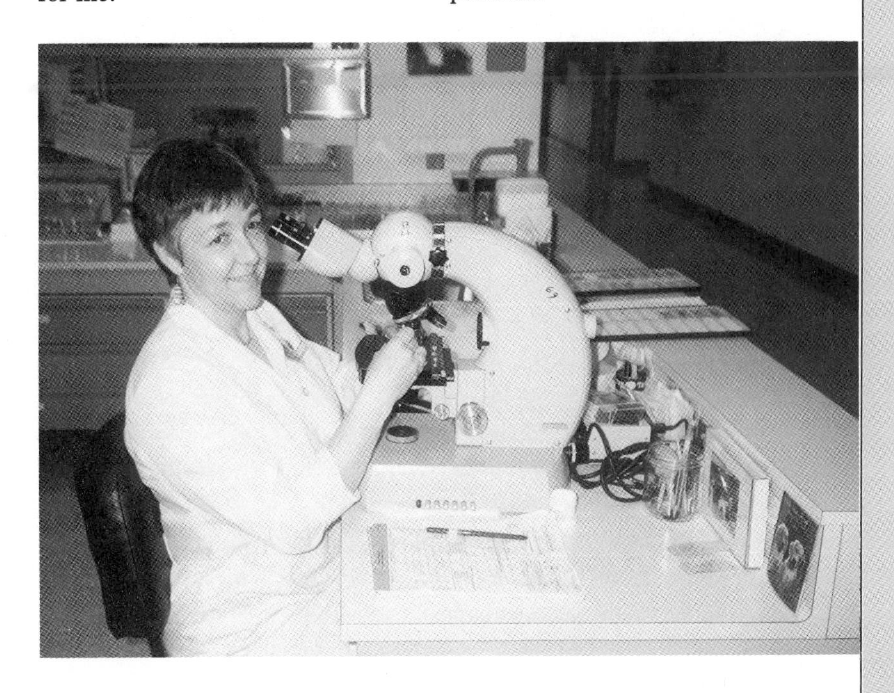

15.2 Moving Materials In and Out of the Cell

Like an organism, a cell and its organelles need a regular supply of nutrients, to provide both raw materials and energy. There must, therefore, be some way of getting these nutrients into the cell. Also, many of the chemical reactions within the cell produce waste products that must somehow leave the cell. In other words, every chemical reaction requires reactants and produces end products, and all these substances — both useful and harmful — must cross the barriers that surround cells. In animal cells, this barrier is the cell membrane. In plant cells, both the membrane and the cell wall must be crossed.

THE CELL BOUNDARIES

The wall that surrounds plant cells is composed primarily of a type of non-living substance called cellulose. This and other materials make the cell wall fairly rigid, providing support for the cell and the plant. Although the wall is quite thick relative to the cell membrane, it contains many pores (openings). These pores are large enough to permit most types of molecules to pass freely through the wall.

The cell membrane in both plant and animal cells is a much more complex structure than the cell wall. It is made up of a variety of materials, mainly proteins and fats. The membrane is generally thinner and more flexible than the wall, yet it is a much better "gatekeeper." A healthy cell membrane allows useful substances to enter the cell and ensures that important and needed materials stay inside. In addition, such a membrane keeps most harmful substances out of the cell.

A membrane like this is said to be "selectively permeable." This means that it lets some materials through, but not others. Unlike the pores in the cell wall, those in the membrane are very small. Small molecules, such as water, carbon dioxide, or oxygen, can pass through these pores easily. Other molecules, such as glucose, cannot pass through the pores. Instead, they are transported across the membrane by special carrier molecules embedded in the membrane. Still other molecules, such as starch, cannot pass through at all, because of their size or other characteristics (Figure 15.7).

DID YOU KNOW?

The cellulose of plant cell walls is used in a variety of industrial processes. It forms the major part of paper, cotton, flax, rayon, celluloid, wood, and hemp (used for making rope). The cell walls of some plants also contain lignin. Lignin is used in the manufacture of synthetic rubber, adhesives, and other products.

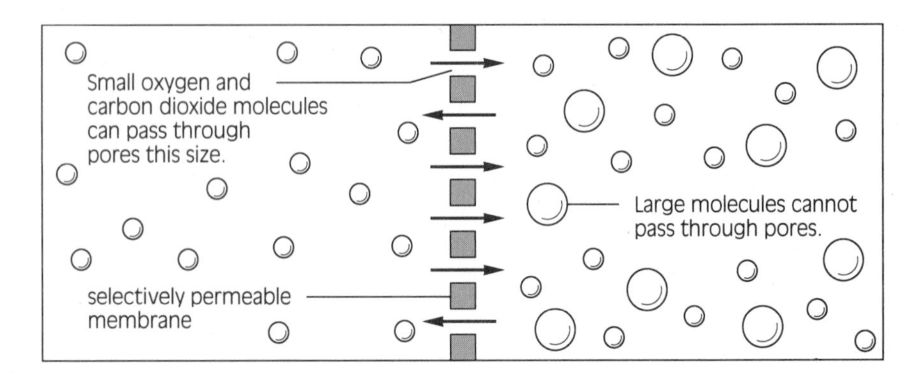

Figure 15.7
A model of a selectively permeable membrane.

DIFFUSION

Substances cross the cell membrane in several different ways. The most important method, which is used by the largest number of molecules, is **diffusion**. Diffusion is the movement of molecules from an area of high concentration (where there are more of those molecules) to an area of low concentration (where there are fewer). These molecules move through the pores in the cell membrane.

The kinetic molecular theory of matter states that all the molecules that make up matter are in constant motion. In gases and liquids, these moving molecules tend to spread out, distributing themselves evenly throughout the available space. This constant molecular motion results in diffusion.

You can demonstrate diffusion by opening a jar of strong-smelling perfume at one end of a room. The molecules of perfume will tend to move out of where they were highly concentrated (in the bottle), and spread themselves into areas of lower concentration (throughout the room). The same is true for molecules of a solute dissolved in a liquid (solvent). If you add food colouring to water, the coloured molecules will eventually spread out until they are evenly distributed and the solution has become evenly dyed.

Both these examples show how diffusing materials move from an area of high concentration to areas where they are less concentrated. Eventually the concentration of these molecules will be equal throughout the space (Figure 15.8).

This movement of molecules is essentially what happens when substances cross the cell membrane by diffusion. Substances that a cell needs (for example, water or nutrients) will usually be found in low concentrations within the cell—this is why it needs them. When these substances are more concentrated outside the cell, the needed molecules will diffuse across the cell membrane until the concentration is equal on both sides. Similarly, when a cell's waste products have become more concentrated inside than outside the membrane, they will leave the cell by diffusion. In this way, the cell is protected from the accumulation of its harmful waste products (Figure 15.9).

DID YOU KNOW?

Cell membranes cannot actually *control* which molecules cross by diffusion. Molecules small enough to pass easily through the membrane's pores will move towards an area of lower concentration, regardless of whether the cell needs them or not. Suppose poisonous molecules such as carbon monoxide or cyanide get into the body. They too will diffuse into some cells where their concentration is low. Since both these poisons can kill cells, they can easily kill the entire organism.

Figure 15.8
The large open circles in this diagram represent water molecules. The smaller, closed circles represent another substance—perhaps food colouring. Note that in (a) and (b) more of the coloured molecules move away from the area of high concentration than towards it.
(c) Eventually, the coloured substance is spread evenly throughout the beaker. Are the coloured molecules still moving at this point? Do the water molecules also move?

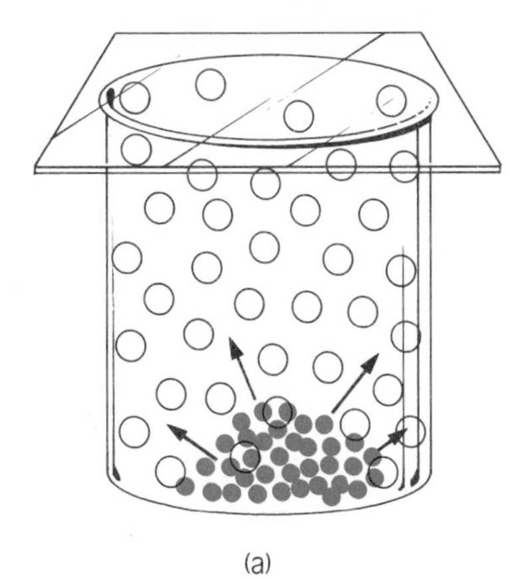
(a)

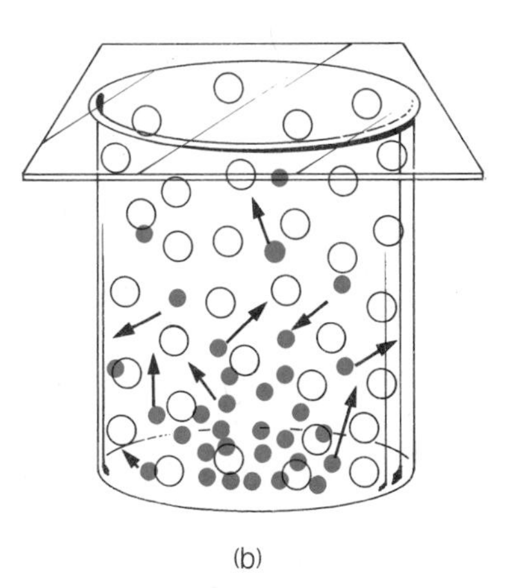
(b)

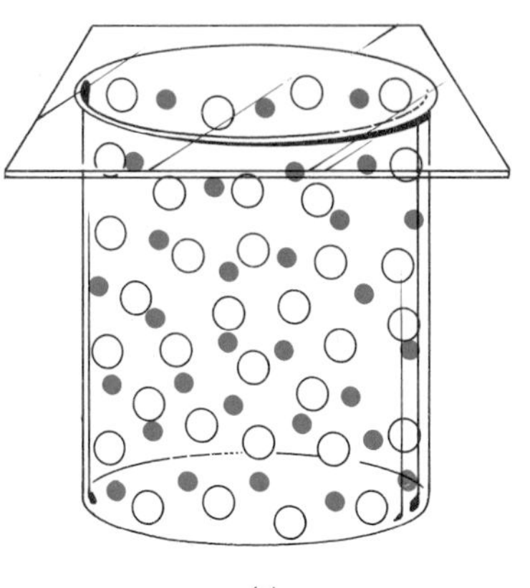
(c)

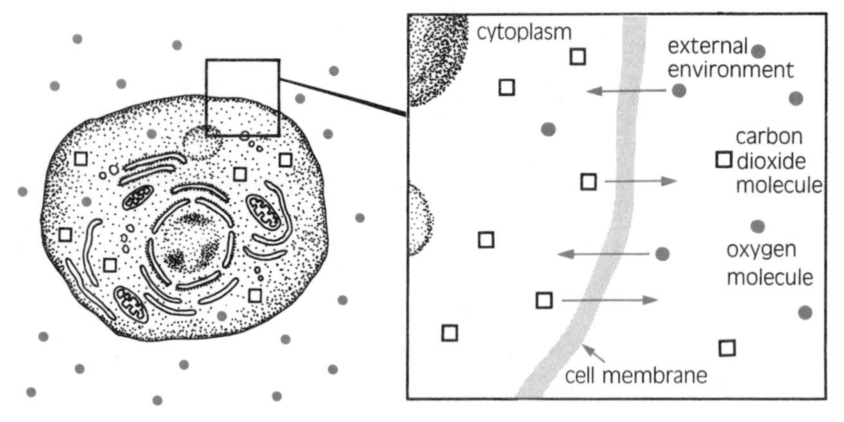

Figure 15.9
This drawing shows oxygen molecules diffusing into the cell and carbon dioxide molecules moving out. Why is this movement taking place?

OSMOSIS

Cells are made up largely of water (about 70 per cent). In addition, they contain a wide variety of other molecules dissolved in this water. Water is also the main solvent found outside the cell. Water molecules can pass through the cell membrane quite easily. The process by which water moves across the membrane is referred to as **osmosis**. Osmosis is actually a special case of diffusion, but the use of two different terms allows us to specify easily what kind of materials are diffusing. We use "osmosis" to refer to the movement of water (the solvent) and "diffusion" to refer to the movement of materials dissolved in the water (the solutes).

Osmosis is the process by which cells take in water or lose water. It helps them maintain the proper balance of water inside and outside the cell. The direction of water movement depends on the relative concentration of water on either side of the cell membrane.

ACTIVITY 15C *Osmosis and a Selectively Permeable Membrane*

This activity demonstrates osmosis.

MATERIALS

two lengths of dialysis tubing (each 6–10 cm long)
concentrated sugar solution (e.g., pure corn syrup or molasses)
two pieces of glass tubing (each 20–50 cm long), each with a rubber stopper on one end
two twist ties or rubber bands
two beakers of distilled water
two stands and 2 clamps
two small pieces of masking tape
ruler
apron

PROCEDURE

1. Put on your apron.
2. Soak the lengths of dialysis tubing in water for a few minutes. Rub them between your thumb and forefinger to open them (in the same way you might open a plastic lunch bag).
3. Tie a knot in one end of each length of tubing about 1 cm from the end.
4. Fill one of these dialysis-tubing "bags" with distilled water. Fill the other with the concentrated sugar solution.
5. Fit the open end of each of these bags over the rubber stopper on the end of a glass tube. Use a twist tie or rubber band to close the top of the bag around the glass tube above the stopper.
6. Rinse the outside of the dialysis bags with water.
7. Lower each of the dialysis bags into a beaker of distilled water (filled to just above the top of the bag) and clamp the glass tubes to the stands as shown in Figure 15.10.

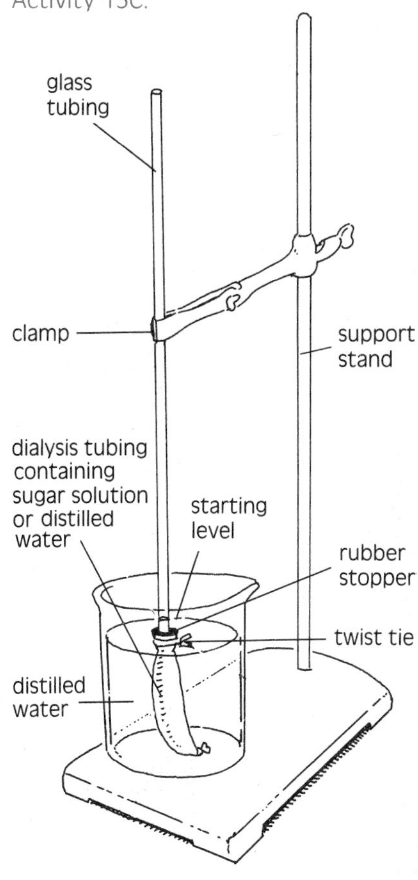

Figure 15.10
How to set up the equipment for Activity 15C.

8. Mark the level of the liquid in each of the glass tubes with a piece of masking tape. The top of the tape should be even with the top of the liquid. This will be your "starting level."
9. Predict what you think will happen to the level of the liquid in each of the glass tubes. Do you think that it will rise, fall, or stay in the same place? Why do you think that?
10. After 5 min, measure the difference (if any) between the starting level of the liquid inside the glass tube and the level presently observed. Record these values in a table of your own design.
11. Repeat step 10 after 10 and 15 min.

DISCUSSION

1. (a) What happened to the level of the liquid in the glass tube attached to the bag containing the sugar solution?
   (b) Was this what you predicted would happen?
2. (a) What happened to the level of the liquid in the glass tube attached to the bag containing distilled water?
   (b) Was this what you predicted would happen?
   (c) What did this particular set-up represent? In other words, what was the value of including it in your experiment?
3. Did anything move into the dialysis bags in either of these two set-ups? What evidence do you have for your answer?
4. Is water able to move into the dialysis bag? Is sugar able to move out? Explain your answers.
5. How does the dialysis tubing compare to a cell membrane?
6. What do you predict would happen if a dialysis bag containing distilled water were placed in a beaker containing a sugar solution? Support your answer.

MAINTAINING WATER BALANCE

Maintaining the correct proportion of water and solutes inside the cell is essential to a cell's survival. However, the cell membrane itself can't actually *control* osmosis. The direction of water movement depends entirely on the relative concentration of water inside and outside the cell.

Imagine an organism living in a pool of salt water (Figure 15.11). (Salt water contains more dissolved solutes than does fresh water and, therefore, has a lower concentration of water). Many such organisms maintain a concentration of solutes inside their cells that is more or less equal to that in the surrounding salt water.

If a rain shower suddenly adds a large amount of fresh water to this pool, the organism is faced with a problem. Suddenly it has a lower concentration of water inside than outside its cells. Water starts to move into the cells by osmosis, causing them to swell. Some cells might even burst and die. In contrast, if a lot of water evaporates from the pool, the water concentration becomes higher inside the organism's cells than outside, and the cells start to lose valuable water to the environment.

Figure 15.11
Along the edge of the ocean, rock formations create pools—called tide pools—that are underwater at high tide, but exposed and isolated from the ocean at low tide. The organisms in tide pools can tolerate many changes in the relative concentration of solutes to water in their environment.

EXTENSION

Place a piece of skinned (epidermis removed) rhubarb in a Petri dish. Add sugar (dry solid sucrose) to the exposed tissue. Observe and record what happens over 10 minutes. What do you think caused any changes you observed?

Animals solve these important water balance problems in a wide variety of ways, some of which you will learn about in future studies. The loss or gain of too much water has serious consequences for plants as well. In plants, the vacuole and the cell wall play important roles in maintaining water balance. Vacuoles are fluid-filled sacs, which store wastes until the cell can dispose of them. They also store nutrients and water until the cell needs them. This water storage is useful because the vacuoles continue to hold some water even when the cell loses water to an external environment with lower water concentration. Although the rigid cell wall usually prevents the cells from collapsing completely, plant cells can shrink somewhat if they lose too much water. This is what causes wilting (Figure 15.12).

When the concentration of water in the cells is lower than in the surrounding environment, water enters the cells by osmosis, causing them to swell. In such situations, the cell wall usually prevents the plant cell from bursting. Once the vacuole holding the excess water has swollen so much that it pushes against this rigid wall, it can expand no further. The internal pressure of the vacuole against the cell wall slows the movement of water into the cell. As a result, the cell's proper water balance is maintained.

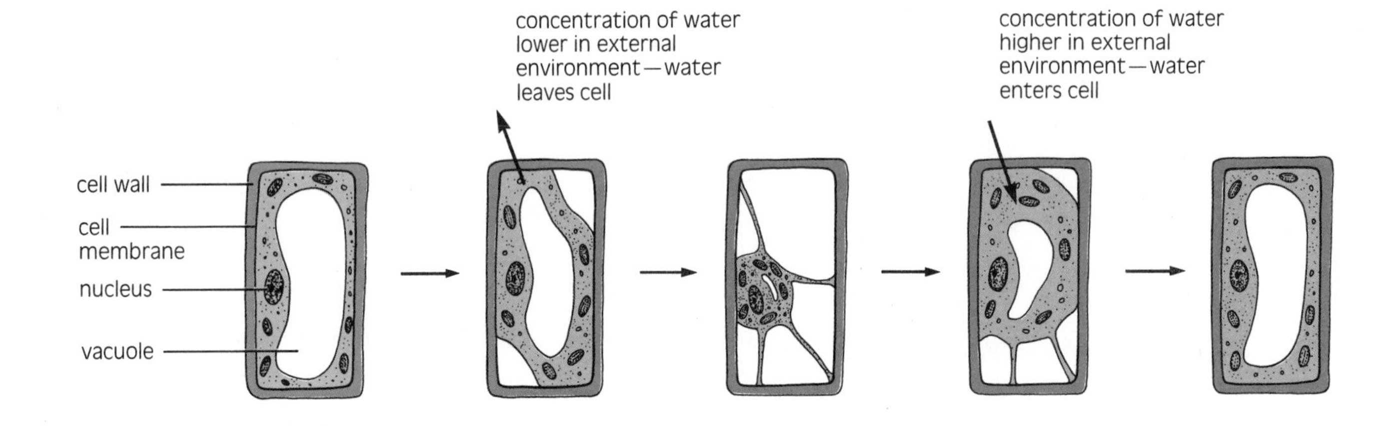

Figure 15.12
If the plant cell loses too much water to its environment, it will start to shrink. If water is then added to the plant's environment, the cells will take in water again.

REVIEW 15.2

1. What is a selectively permeable membrane?
2. (a) Explain the process of diffusion. (Use a labelled diagram in your answer if you think it would be helpful.)
   (b) What is the difference between osmosis and diffusion?
   (c) In what way are these two processes similar?
3. What determines the direction of water movement in or out of cells?
4. What are the functions of plant vacuoles?
5. How does the cell wall aid plants in maintaining proper water balance?
6. What is the difference between the cell wall and the cell membrane?
7. Why are the processes of diffusion and osmosis important for our survival?

15.3 Energy for Cells at Work

All the chemical processes occurring within cells require energy. Both plant and animal cells obtain their energy from "energy-rich" compounds, such as glucose. The major difference between plants and animals lies in how they obtain these compounds.

PHOTOSYNTHESIS

Plants manufacture their own energy-rich compounds by the process of **photosynthesis**. Photosynthesis occurs in the chloroplasts, organelles that contain a substance called chlorophyll, as well as other materials needed for photosynthesis (Figure 15.13). Because chlorophyll is green, the chloroplasts are also green, and so are the parts of the plant containing them. Chlorophyll absorbs light from the sun and converts it into chemical energy. This energy is then used to drive the complex series of chemical reactions that are part of photosynthesis. The overall process can be described by the following word equation:

$$\text{carbon dioxide} + \text{water} \xrightarrow[\text{chlorophyll}]{\text{light energy}} \text{glucose} + \text{oxygen}$$

or by the balanced equation:

$$6CO_2 + 6H_2O \xrightarrow[\text{chlorophyll}]{\text{light energy}} C_6H_{12}O_6 + 6O_2$$

Plants use the glucose they produce in various ways. Some is used right away to supply energy for the cell's life functions. Other glucose molecules may be joined together to form starch. Because starch molecules are too large to escape through the cell membrane, energy can be stored in this form. When the cell needs it, the starch is broken down again into individual energy-rich glucose molecules.

Animals are not capable of photosynthesis. Many obtain their energy-rich compounds by eating plants. During digestion, plant material is broken into individual glucose molecules and other substances used by the animal's cells. Many types of animals do not eat plants. Instead they obtain energy-rich molecules (and other nutrients) by consuming other animals. Still others eat both plants and animals. All animals must obtain the energy they need from the bodies of other organisms—they cannot manufacture their own.

For animals as well as plants, however, photosynthesis is the *original* source of energy-rich glucose molecules. How do the cells of both plants and animals extract the energy from these molecules? This is accomplished by another important process that occurs within cells, **cellular respiration**.

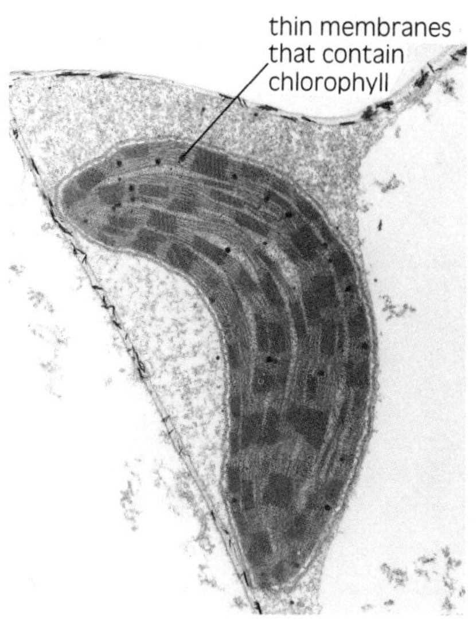

Figure 15.13
This photo shows a chloroplast as seen with an electron microscope. The membrane that surrounds this organelle can be seen, as can its internal structures. The chlorophyll that makes the chloroplast green is embedded in the many thin membranes that run throughout the chloroplast. The number of chloroplasts varies from one type of plant cell to another, ranging from only 1 to approximately 200 per cell.

DID YOU KNOW?

The leaves of many trees (for example, maples, oaks, and poplars) change colour in the fall. This is because the chlorophyll they contain in summer—which makes them appear green—breaks down and is not replaced. Other substances in the cell, which have a red, yellow, or brown colour, are then revealed. Although these other substances are always present in the vacuoles, you are not aware of them during spring and summer because they are concealed by the abundant chlorophyll.

ACTIVITY 15D *Comparing Photosynthesis and Cellular Respiration*

Use a table to help you become familiar with the relationship between photosynthesis and cellular respiration. Design a table that includes the following categories:

- materials required
- whether energy is produced or needed
- organisms in which it occurs
- where it occurs in the cell

Write a brief description in each category for photosynthesis. As you read the next section, complete the table for cellular respiration.

CELLULAR RESPIRATION

Respiration in cells, or cellular respiration, is quite different from what is sometimes called respiration in animals. The latter is simply breathing — inhaling and exhaling air.

Cellular respiration, like photosynthesis, takes place within cells. It consists of a series of complex chemical reactions, requiring the presence of many substances. During cellular respiration, oxygen is used and glucose is broken down. Energy is released for the cell's chemical reactions.

The overall process can be summarized in simple word and balanced equations:

glucose + oxygen ⟶ carbon dioxide + water + energy

$$C_6H_{12}O_6 + 6O_2 \longrightarrow 6CO_2 + 6H_2O + \text{energy}$$

Figure 15.14
A mitochondrion from a bat pancreas cell, as seen with an electron microscope.

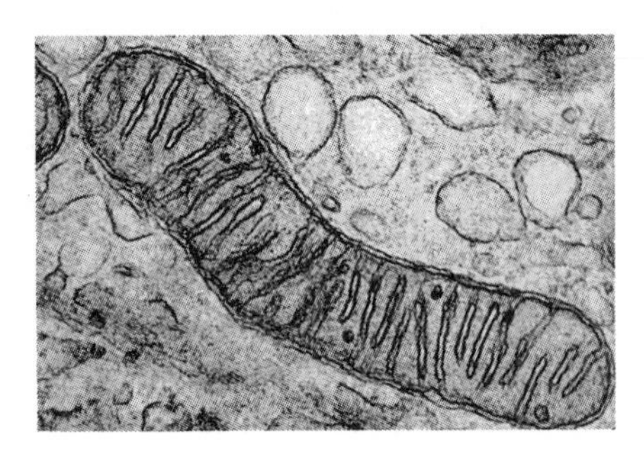

Look back at the equations describing photosynthesis and see how they compare to these.

Most of the reactions that are part of cellular respiration take place in organelles called **mitochondria** (singular: mitochondrion) (Figures 15.14 and 15.4). Mitochondria are found in both plant and animal cells, since both types of organisms need to extract energy from glucose. Once this energy has been extracted, it is stored in specialized molecules. These move out of the mitochondria and through the cytoplasm to locations in the cell where energy is needed.

As the equation for cellular respiration shows, to provide their cells with energy, organisms must have a supply of oxygen. They obtain this oxygen from the atmosphere — the air around them. Photosynthesis is a major producer of the oxygen in the atmosphere. Photosynthesis, in turn, requires carbon dioxide — an end product of cellular respiration — as a reactant. Thus, all living things are dependent on each other. They are all parts of a cycle connecting photosynthesis and cellular respiration (Figures 15.15 and 15.16).

As you might imagine, disrupting this cycle would have serious effects. Some human activities do interfere with parts of the cycle. For example, the many trees and other plants in tropical rain forests produce large amounts of oxygen. Destroying these forests reduces the supply of this vital gas.

DID YOU KNOW?

The number of mitochondria a cell contains is related to its energy needs. A normal liver cell, for example, contains 1000 to 1600 mitochondria. A human egg cell requires greater amounts of energy for the many reactions that are part of growth and development. It may contain as many as 300 000 mitochondria.

Other human activities affect cellular respiration. For example, manufacturing paint and tanning leather release arsenic into the environment. This chemical interferes with some of the reactions that are part of cellular respiration. The result is that less energy is available to affected organisms, and an imbalance is created in the overall cycle.

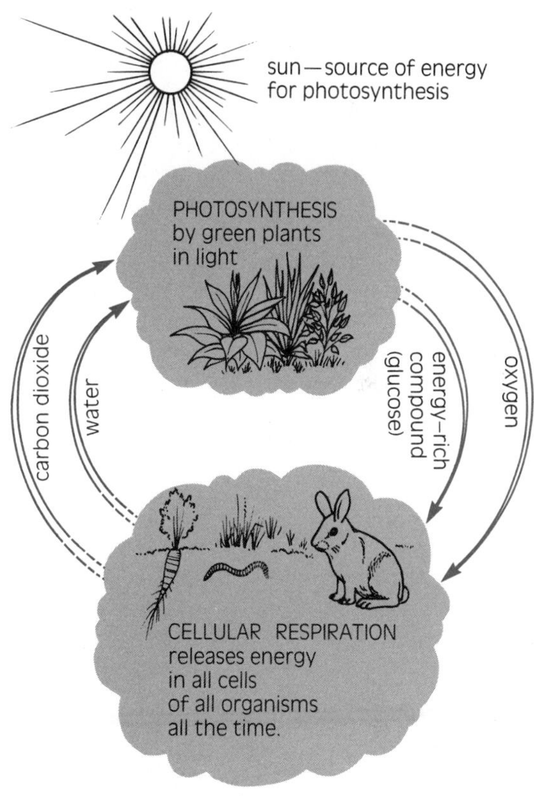

Figure 15.15
The cycle showing the relationship between photosynthesis and cellular respiration. Notice how some types of matter flow among living things. Energy, however, does not cycle between photosynthesis and cellular respiration. The sun must constantly provide a fresh supply.

◄ Figure 15.16
Humans fit into the cycle of photosynthesis and cellular respiration. What parts of the cycle are occurring here?

REVIEW 15.3

1. Why doesn't photosynthesis occur in animal cells?
2. What are mitochondria and what is their function?
3. (a) What are the word equations for photosynthesis and cellular respiration?
   (b) How are these two processes related to each other?
4. (a) Where does photosynthesis occur in plant cells?
   (b) Where does respiration occur in plant and animal cells?
5. Why must respiration occur in both plant and animal cells?
6. How are breathing (respiration by an animal) and cellular respiration similar?

15.4 Controlling the Cell's Activities

DID YOU KNOW?

Some types of animal cells do not have nuclei. For example, when human red blood cells mature, their nuclei break down so that the cells will have more space to carry oxygen, which is their primary function. Other types of cells may seem to have more than one nucleus. For example, the large cells of human muscle fibres are formed by the fusion of several smaller cells, each of which retains its own nucleus. However, all animal cells begin their lives with a single nucleus.

All living things are able to respond to the environment. You respond largely by using your brain and other parts of your nervous system. These complex structures ensure that your activities are appropriate for your environment. Cells are alive, and they also respond to their environments. Although cells lack your extensive nervous system, they too have a very efficient "controller"—the nucleus. The nucleus regulates a cell's activities so that it can survive in its environment.

Most of a cell's activities consist of chemical reactions. Some of these reactions result in growth. Others bring about the repair of damaged organelles. Still others (such as cellular respiration) supply the energy needed for the rest of the reactions to take place. In general, a cell's reactions depend on three things: (1) the necessary raw materials (the reactants), (2) energy, and (3) the appropriate catalysts to speed up the reactions.

In a cell, these catalysts are called enzymes. Enzymes are proteins that speed up chemical reactions. They are essential for most of a cell's reactions; without them the reactions would proceed too slowly to maintain life. Enzymes are very specific; that is, a particular enzyme will speed up only one type of chemical reaction. An amazing number of different enzymes exist, each playing a role in the proper functioning of the cell. The lack of even one enzyme can have serious consequences for any organism (Figure 15.17).

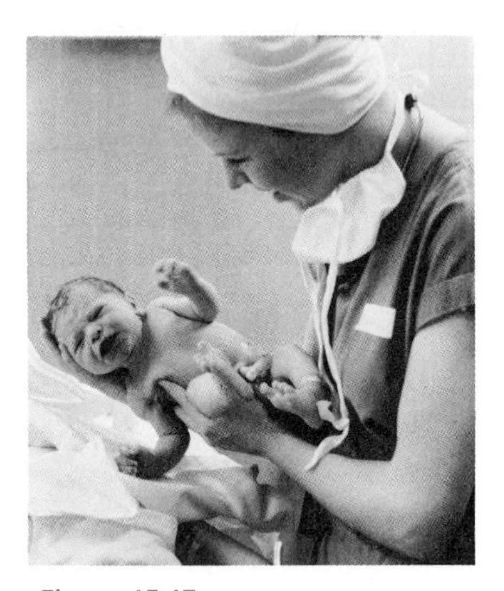

Figure 15.17
Today, all babies born in Canada (and many other countries) are tested for phenylketonuria (PKU) within two weeks of birth. PKU is a disease caused by the absence of an enzyme. Without this enzyme, a harmful substance builds up that can cause mental retardation. This can be prevented if the child is placed on a special diet that is very low in the harmful substance.

Controlling enzyme production is a very effective way to control the chemical reactions within a cell. This is exactly what the nucleus does. It determines which enzymes will be produced and when, and these then regulate the rates of chemical reactions. In this way, the nucleus can control almost all the cell's activities.

THE CONTENTS OF THE NUCLEUS

With a few exceptions, all plant and animal cells contain a single nucleus (Figure 15.18). The boundary of this structure is a membrane called the **nuclear membrane**. The nuclear membrane controls the passage of materials (for example, energy-storage molecules) between the nucleus and the cytoplasm.

Inside the nucleus lies the key to controlling enzyme production: **DNA** (*d*eoxyribo*n*ucleic *a*cid). There are a number of DNA molecules within the nucleus. The exact structure of these DNA molecules differs from one kind of plant or animal to another and, to a lesser extent, from one individual to another. In fact, it is the structure of the DNA that gives each kind of organism—and each individual—many of its own specific characteristics.

Each of an organism's DNA molecules is part of a different **chromosome**. Chromosomes are long, thread-like structures found only in the nucleus. Each chromosome is made up of one DNA molecule and a

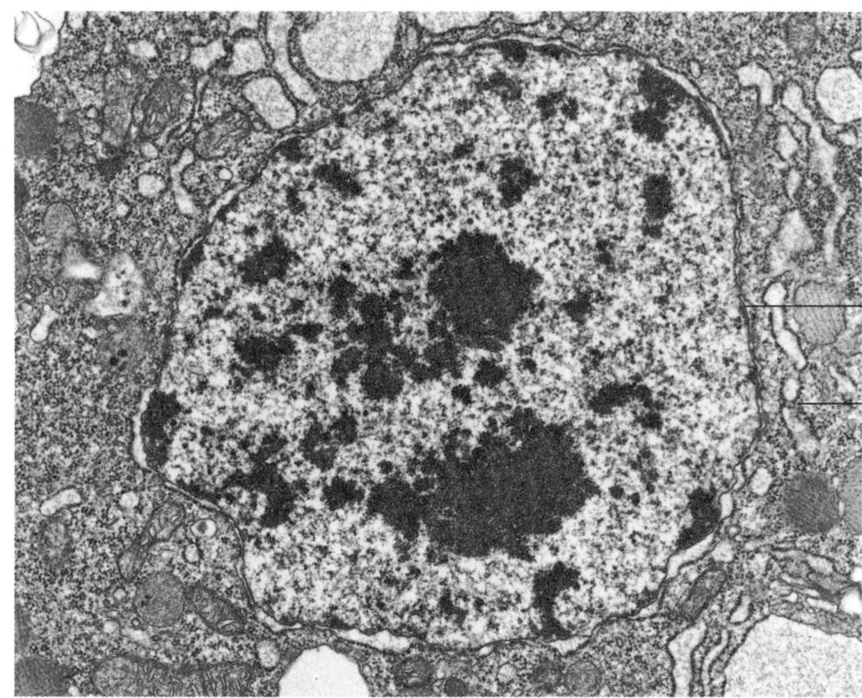

Figure 15.18
A nucleus as seen with an electron microscope.

variety of proteins. Different kinds of organisms have different numbers of chromosomes, but this number is constant for all members of the same species. For instance, cells of a human normally contain 46 chromosomes; mosquito cells, 6; dog cells, 78; garden pea cells, 14; and sugar cane cells, 80.

In 1953, American scientist James Watson (1928–) and British scientist Francis Crick (1916–) made an important announcement. They had discovered the details of the structure of DNA. Other scientists had already determined that DNA consisted of a number of smaller molecules, called nucleotides, which are arranged in a chain (Figure 15.19). There are four different types of nucleotides, distinguished from each other by a part of the molecule called the base. The four different bases are referred to as A,T,C, and G. Each of these four types of nucleotides occurs many times in the DNA chain.

British scientist Rosalind Franklin (1920–1958) produced X-ray pictures of the DNA molecule in the early 1950s. These suggested the chain had a coiled shape, known as a helix. Using this and other evidence, Watson and Crick proposed that the DNA molecule was a *double* helix. Later research showed their model was correct (Figure 15.20).

A series of the nucleotides along one side of the DNA ladder makes up a **gene** (Figure 15.21). Each gene contains several hundred nucleotides, which make up a coded "message." The four different bases (A,T,C,G) are the code symbols or "letters." The number and sequence of these bases form what is called the genetic code.

Many genes are code messages that call for the production of specific enzymes. When a cell needs a particular enzyme (to speed up a chemical

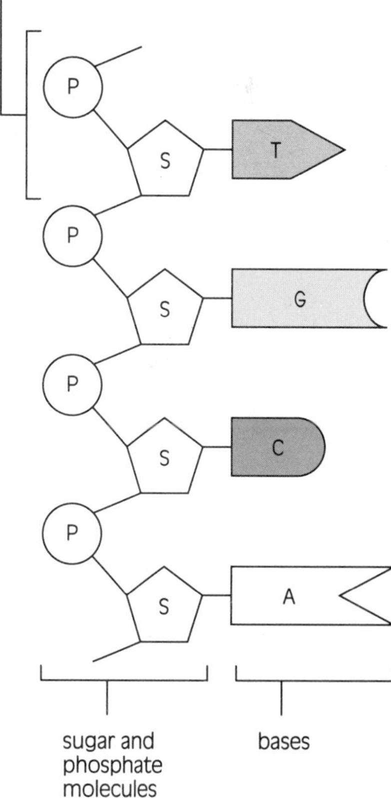

Figure 15.19
Each nucleotide consists of three units: a sugar molecule (S), one or more phosphate molecules (P), and one of four different types of bases (A,T,C,G). Chains are created when chemical bonds form between the phosphate unit of one nucleotide and the sugar unit of another.

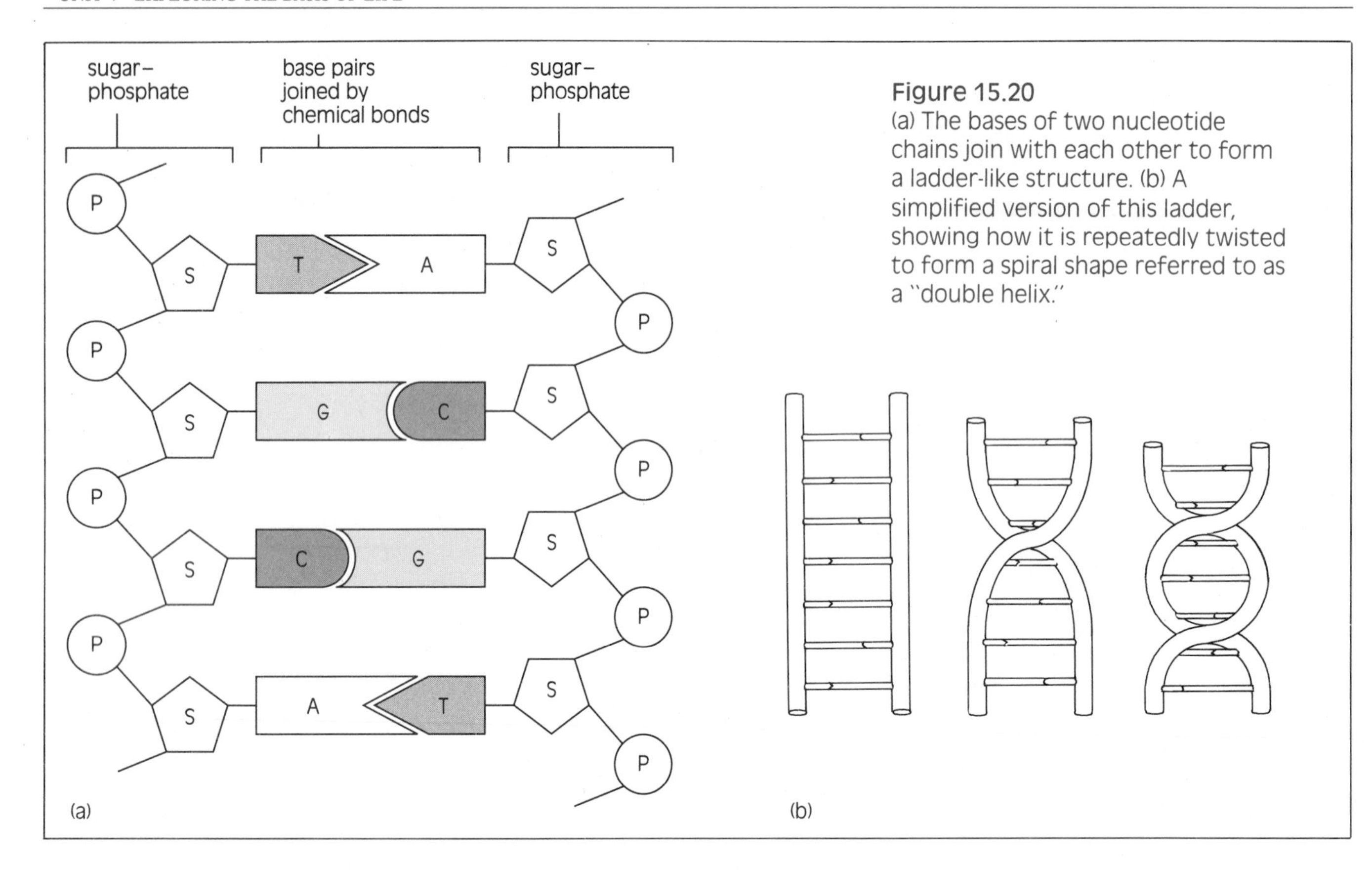

Figure 15.20
(a) The bases of two nucleotide chains join with each other to form a ladder-like structure. (b) A simplified version of this ladder, showing how it is repeatedly twisted to form a spiral shape referred to as a "double helix."

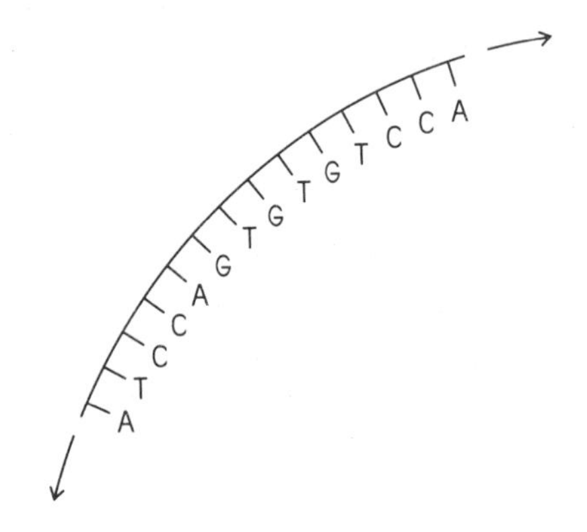

Figure 15.21
A gene consists of a long chain of nucleotides.

reaction), it sends a message to the nucleus. The gene that codes for that enzyme is then "turned on." It responds by providing the cell with the instructions ("written" in the genetic code) to produce the enzyme. Genes provide the instructions for all the enzymes and other proteins needed by the cell.

Think of a chromosome as a book with thousands of sentences in it. (Most chromosomes contain thousands of genes.) Each "sentence" (gene) is hundreds of nucleotide "letters" long, although the DNA "alphabet" consists of only four different letters. The nucleus, then, is like a library. Its "books" contain all the information necessary to produce the enzymes needed by a cell. Each type of enzyme speeds up a specific chemical reaction. Whether this reaction is part of photosynthesis, cellular respiration, or some other cellular process, the products of the reaction are needed by the cell for proper functioning.

REVIEW 15.4

1. Describe how the nucleus controls a cell's activities.
2. Why are enzymes important to a cell?
3. What are chromosomes?
4. Describe the structure and function of a gene.
5. Using words, a diagram, or a concept map, explain the relationships between the following: nucleus, nuclear membrane, DNA, chromosome, gene.

15.5 Single-celled Organisms

The organisms discussed so far in this chapter — plants and animals — have bodies that are made up of more than one cell. Although some of these cells could survive on their own, most depend to some extent on all the other cells in the body. Muscle cells, for example, rely on blood cells to bring oxygen to them. Each cell in your body started life as a basic, unspecialized cell. It soon began to develop features that made it into a skin cell, a muscle cell, or some other type of specialized cell. Such features enable these cells to carry out their own tasks very efficiently. However, as a result of specialization, many of them cannot carry out all the tasks necessary for independent survival.

There are many other types of organisms whose "bodies" consist of only a single cell. Such cells must be able to perform all life functions in order to survive. These single-celled organisms are microscopic. Despite their tiny size, they are a vital part of life on Earth.

Figure 15.22
The protists shown in (a) are seaweeds. As you might expect, their bodies are made up of many cells. The amoeba shown in (b) is one of the most familiar single-celled protists.

(a)

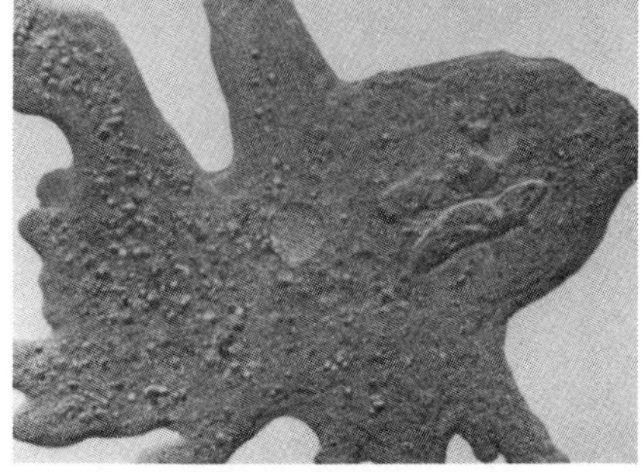

(b)

THE PROTISTS

Many different types of single-celled organisms exist. The ones called **protists** all share certain features; for example, they all have a nucleus and various organelles. Protists come in many different shapes and sizes. Although some species can grow quite large and have many-celled bodies, most protists have bodies that consist of a single cell (Figure 15.22).

As well as a nucleus, protists have many of the same organelles possessed by plant and animal cells. Most protists, for example, have mitochondria, where cellular respiration occurs. Some protists contain chloroplasts and can thus manufacture their own food. Others survive by eating a variety of organisms, including other protists.

Most protists live in water. They are vitally important parts of the food webs in oceans, lakes, rivers, and ponds. Those that photosynthesize contribute significantly to the world's supply of oxygen. Water pollution that kills these useful protists can have quite serious consequences.

Not all protists are beneficial. Some can cause disease in humans and other animals. One kind of amoeba, for example, can live in the human digestive tract, where it causes amoebic dysentery. This disease is most often picked up by drinking water or eating food contaminated with these amoebas. Dysentery is both serious and widespread: at any one time, 2 to 3 million people in North America may show symptoms of dysentery. In some tropical countries, more than half the population is affected. The symptoms range from mild stomach discomfort to persistent serious diarrhea, which causes the body to lose many valuable nutrients. Many thousands of people die from the disease each year. *Plasmodium*, the organism that causes malaria, is also a type of protist. (see first page of chapter 14).

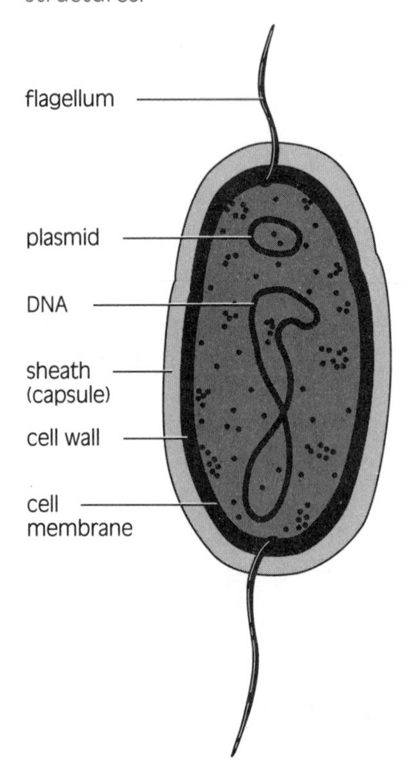

Figure 15.23
A representative bacterium, as it might look if seen with an electron microscope. The tiny, tail-like structures that help the bacterium move are called flagella. Some species of bacteria have only one flagellum. Other species may be completely covered with flagella, or may lack them entirely. Although most types of bacteria have a cell wall and some have a capsule, other types have neither of these structures.

BACTERIA

Bacteria (singular: bacterium) are the smallest and simplest forms of life on Earth. All bacteria have bodies consisting of a single cell, most of which are less than 0.002 mm long. Bacteria do not have most of the organelles found in animal, plant, or protist cells, nor do they have nuclei and nuclear membranes. They do not possess chromosomes like those found in plants and animals. The genes controlling their life functions are found on a single large molecule of DNA in the cytoplasm. In addition, bacteria may contain one or more plasmids, small circular DNA molecules that contain non-essential genes (Figure 15.23). The microfossils shown in the picture that opens this chapter resemble modern bacteria very closely. It is thought, therefore, that bacteria were the first living organisms on Earth.

Despite their small size and simple structure, bacteria are an essential form of life. They are found almost everywhere and are often very abundant. A spoonful of garden soil, for example, may contain as many as 10^{10} bacteria. They are part of most food webs, providing food for a variety of organisms.

Bacteria are used in various industrial processes, such as tanning leather and making yogurt. They are also champion recyclers! The decomposing action of bacteria breaks down dead plant and animal material into its basic subunits, which can then be used by other living organisms. Some bacteria can photosynthesize; these supply needed oxygen to the atmosphere. In fact, without bacteria we would have run out of food, raw materials, and oxygen long ago.

BACTERIAL DISEASE

The reputation of bacteria as harmful organisms is not completely deserved. Only some types cause disease. Because bacteria are so abundant, however, harmful types are often difficult to avoid. These cause disease by invading the bodies of other organisms and interfering with normal cell functions. Bacteria, protists, and other agents that cause disease are called pathogens.

There are a number of effective methods of preventing and treating bacterial diseases. One of the best methods of prevention is sanitation, which kills bacteria before they can enter our bodies in great numbers. Using soap and water before eating (to kill bacteria on our hands, dishes, and food) is a simple sanitary procedure. Sewage is treated so that the abundant bacteria it contains do not contaminate the water supply. Sanitation also prevents many diseases caused by protists.

Once bacteria have caused a disease, antibiotics may be used to kill them or slow their growth. You may have applied antibiotic cream or ointment to an infected cut — the infection occurs after bacteria have entered the wound. You might also have taken antibiotic pills. These treat internal bacterial infections, such as strep throat.

Your body has its own built-in defence system. Many pathogens have special molecules, called antigens, on their outer surfaces. Once inside

EXTENSION

Discover more ways in which bacteria are beneficial. Select one contribution made by bacteria or one use for them. Write a brief report explaining how this benefit is achieved. Alternatively, do research to expand on one of the examples in the text.

your body, these antigens stimulate the production of molecules called antibodies. The antibodies bind to the antigens (Figure 15.24).

In some cases, the antibodies kill the disease-causing agent. In other cases, they slow down its movement. Then other parts of the body's defence system can catch and destroy it. In either case, once the antibody-producing cells have been stimulated, some of them remain in the body and can reproduce rapidly if the body is infected again by the same pathogen.

Vaccines are another means of treating many bacterial diseases. They make use of the body's own defence system. Drug companies manufacture vaccines from weak or dead pathogens. The antigens from these weakened pathogens produce only very mild disease symptoms when injected into the body. A healthy body can easily fight off the mild illness that might be caused by a vaccine. It does this by producing antibodies that bind to the antigens. The benefit of a vaccine is that some of these antibodies remain in the body, ready to begin fighting immediately if a vigorous version of the same pathogen invades the body.

Several diseases can now be prevented by vaccines. Diphtheria, for example, affects the respiratory tract, heart, nervous system, and other parts of the body. In the past, children often died from this bacterial disease. In countries such as Canada, where diphtheria vaccine is routinely given to infants, the disease is now much less common. Vaccines have protected millions of people around the world from this and other serious bacterial illnesses. Researchers are working on vaccines to protect against other diseases, such as malaria, that are caused by protists.

Figure 15.24
The body produces a different antibody for the antigen from each different type of pathogen. Each antibody is shaped to "fit" that antigen, the way a lock fits a key.

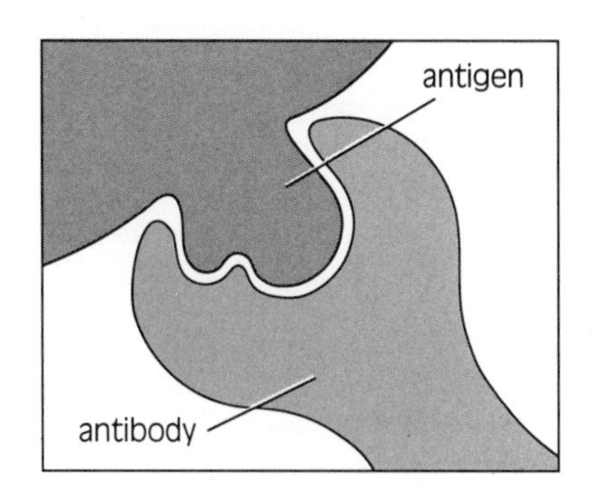

VIRUSES

Although they are included in this section, **viruses** lack most of the characteristics of living cells. Indeed, most scientists do not regard viruses as living things. They do not move about by themselves, they do not use energy, and they never grow. They do, however, share one important characteristic with living things—they reproduce. However, viruses cannot produce more viruses unless they get inside a living cell. An organism or cell that has been invaded by a virus is called the virus's host. Once inside the cell, the virus uses the host's enzymes and DNA to reproduce itself.

Viruses are so small they can be seen only with an electron microscope. Because they range from 0.000 02 mm to 0.0003 mm in diameter, viruses are 10 to 100 times smaller than the smallest living organisms, bacteria. A virus consists simply of a piece of genetic material covered by a protective coat made mainly of protein (Figure 15.25). The genetic material is either DNA or a similar molecule called RNA (*r*ibo*n*ucleic *a*cid).

Many viruses reproduce by injecting their genetic material into a host cell. The viruses' genetic instructions, "produce a virus," will be carried out over and over again. After many viruses accumulate within the organism, the cells do not function normally, and disease results. As far

Figure 15.25
Some viruses that affect humans.

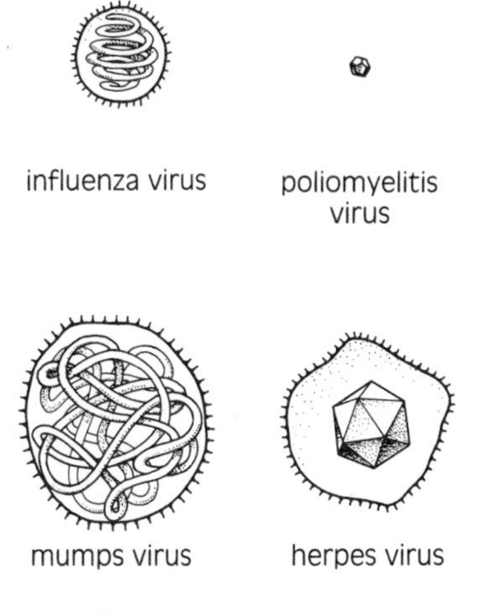

as is known, all species of organisms, including bacteria, have at least one virus that affects them. Many different viruses cause disease in humans. Each one affects a particular type of body cell or tissue. Table 15.1 gives some examples of viral diseases of humans, their effects on the body, and their methods of transmission. Organisms, such as insects, that spread diseases when they bite are called vectors.

The best way to prevent viral disease is vaccination. A number of viral vaccines have been developed, and researchers hope to develop vaccines for other serious diseases, such as acquired immune deficiency syndrome (AIDS). Vaccination has eliminated smallpox, a serious and once-common disease (Figure 15.26).

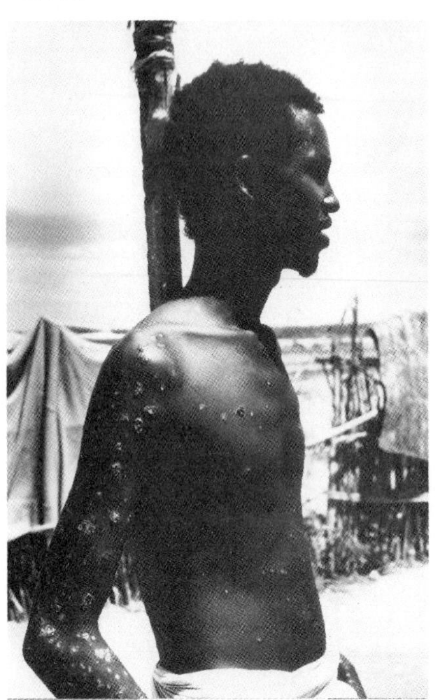

Figure 15.26
Smallpox once killed many thousands of people every year. When Europeans first brought the disease to Canada, smallpox killed more aboriginal people than did war. In the 1960s, a worldwide vaccination program was started. Its success has been one of the triumphs of modern medicine. The photo shows Ali Maow Maalin, of Somalia, the last known person to get the disease (in 1977).

Table 15.1 Some Viral Diseases of Humans

Viral disease	Part of the body affected	Common method of transmission	Vaccine available?
Common cold	Respiratory tract	Water-borne, airborne	No
Infectious hepatitis	Chiefly liver	Airborne, water-borne	No
Encephalitis	Brain	Insect vector	No
Warts	Skin	Contact	No
Rabies	Brain and spinal cord	Mammal vector	Yes
Mumps	Chiefly salivary glands	Water-borne (e.g., saliva)	Yes
Measles	Many body parts	Airborne, water-borne	Yes
Acquired immune deficiency syndrome (AIDS)	Immune system (T4 cells)	Body fluid contact	No
Mononucleosis	Immune system (B cells)	Water-borne (e.g., saliva)	No

REVIEW 15.5

1. What is the difference between the bodies of most plants or animals and those of bacteria and most protists?
2. How are bacterial cells different from those of other organisms?
3. Why are viruses not considered living things?
4. Are all bacteria harmful to living things? Explain your answer.
5. Describe several ways in which bacteria and viruses can be transmitted from infected organisms to healthy ones.
6. (a) What is a vaccine?
   (b) How does it work?

SCIENCE IN OUR WORLD

SEXUALLY TRANSMITTED DISEASES

There is one kind of disease that many people are reluctant to discuss even with their doctor—a sexually transmitted disease (STD). An STD is an infection that is usually passed from one person to another during sexual activity, sometimes through close genital, oral, or anal contact as well as through sexual intercourse.

Like other diseases, STDs are caused by pathogens such as bacteria and viruses. More than 25 STDs have been identified. There are cures for some. For others, there is no cure, and infected people need to learn ways to control the symptoms and avoid spreading the disease. There is no immunity to the common STDs and there are no vaccines to prevent them, so it is possible to get them more than once. It is also possible to get more than one at a time.

Symptoms vary with the STD. Sores, blisters, genital discharge, itching, or painful urination may be signs that a person is infected. There may, however, be no symptoms at all. Sometimes the symptoms disappear after a few weeks, but the pathogen remains in the body. The STD may come back later in a more advanced stage. Without treatment, some STDs result in severe complications, such as heart disease, brain damage, other infections, sterility (inability to have children), and death.

Some STDs can be identified only by a lab test. People with any doubt about whether or not they have an STD should go to a doctor or community clinic for testing. STDs are too serious to ignore.

There is much less risk of getting an STD if proper precautions are taken. Here are some ways to protect yourself:

- Abstain from sexual activity. That is the only absolutely safe way.
- If you are sexually active, a condom is *always* necessary. Since some STDs show no symptoms, you have no way of knowing whether or not your partner is infected.
- If you are sexually active, use a condom properly to reduce your risk. Follow the directions on the package and do not use petroleum-based lubricants, which can cause the condom to fail.
- Do not allow semen to enter the vagina or anus.
- Drug users are at special risk and must never share a needle or syringe, as these can pass blood from one person to another.
- Remember that drugs and alcohol make people less cautious so that they may not take the necessary precautions.

Some Common Sexually Transmitted Diseases

Disease	Method of Transmission	Cause and Treatment
Chlamydia	Direct contact with infected genital area	*Chlamydia trachomatis* (bacterium) Cured with antibiotics
Gonorrhea	Direct contact of infected mucous membrane with genitals or anus	*Neisseria gonorrhoeae* (bacterium) Cured with antibiotics
Syphilis	Direct contact with infectious sores, rashes, or mucous patches in genital area or mouth	*Treponema pallidum* (bacterium) Cured by antibiotics in early stages only
Herpes	Direct contact with blisters or open sores	Herpes simplex (virus) No cure, but treatment for some symptoms
Genital warts and probably cancer of the cervix	Direct contact with warts or infected mucous membranes	Human papilloma virus Removal of warts (but this may not eliminate virus) or removal of cancerous cells
Acquired immune deficiency syndrome (AIDS)	Sexual contact or blood-to-blood contact with infected individual Carried in blood, semen, and vaginal fluids	Human immunodeficiency virus (HIV) No cure to repair the body's immune system, so body cannot fight infections and diseases

15.6 Cell Size

All cells are small. This is true of a cell that makes up the entire body of an organism such as a protist or bacterium. It is also true of a cell that is only part of a plant or an animal. Although there is some variety in the size of cells, most are less than 0.2 mm in diameter. Why are cells so small?

Imagine what would happen if you failed to stop growing in your late teens, and you just continued to get bigger throughout your life. What problems would your body have? Your cells would have similar problems if they continued to grow past a certain size. You might have some ideas about what problems cells might have. If so, write them down in your learning journal before you do the next activity. The findings of this investigation might fit nicely with your ideas or might cause you to change them.

ACTIVITY 15E *Size and Surface Area*

In this activity, you will use imaginary cubes to investigate the problems of increasing cell size.

PROCEDURE

1. Imagine that you have three cubes, each representing a cell. They increase in size from cube A to cube C, as illustrated in Figure 15.27.

Figure 15.27
A model of cell size and surface area for Activity 15E.

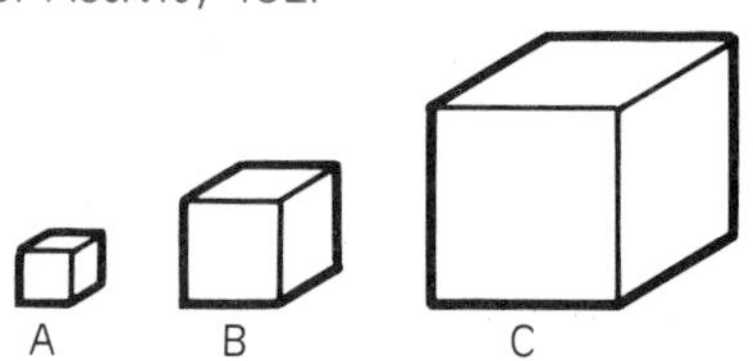

2. Calculate the surface area, the volume, and the surface-to-volume ratio for each of the cubes and enter them in a table similar to Table 15.2. The length of a side of each cube has been filled in for you. The surface-to-volume ratio provides a way to think about the surface area in relation to the volume of the cells. The bigger the ratio, the more surface area per unit of volume a cell has.

Table 15.2 Sample Data Table for Activity 15E

Cube	Length (cm) (L)	Surface area (cm^2) ($6 \times L^2$)	Volume (cm^3) (L^3)	Ratio: $\frac{\text{surface area}}{\text{volume}}$
A	2			
B	4	SAMPLE ONLY		
C	8			

DISCUSSION

1. How much greater is cube B's length than cube A's? How much bigger is cube B's surface area than cube A's? What is the difference between the volumes of the two cubes? Is the ratio of surface to volume also bigger for cube B?
2. Compare cubes C and B in the same way you did cubes A and B. Is the surface-to-volume ratio getting smaller or larger as the cube gets larger?
3. Make a general statement about the effects on surface-to-volume ratio of increasing the length of a cube's side.
4. How would you now answer the question: why are cells so small? Have your ideas changed? Why?

LIMITS ON CELL SIZE

Recall that cells obtain many of the necessary energy-rich molecules, raw materials, and other substances by diffusion and osmosis. They get rid of many wastes in the same way. Both these processes take place across the cell membrane—the surface area of the cell. Both processes can occur at only a certain rate: the rates cannot get faster and faster indefinitely. The bigger a cell becomes, the more materials it needs and the more wastes it produces. Yet, at the same time, its surface area gets smaller relative to its volume. So to function properly, cells can reach only a certain size—which is pretty small!

This is why "size control" is as important for cells as it is for you. What happens when cells reach a certain size is one of the topics discussed in the next chapter, on reproduction.

EXTENSION

After doing Activity 15E, imagine that a cell is spherical instead of cubic. Would the relationship between the size of the cell and its surface-to-volume ratio be the same as the one you described for cubes? You can test this by doing some calculations. The formulas for the area and volume of a sphere are as follows: $A = 4\pi r^2$; $V = 4/3\pi r^3$.

REVIEW 15.6

1. What effect does increasing the size of a cell have on its surface-to-volume ratio?
2. Why is it important for a cell to have a relatively large surface-to-volume ratio?
3. In general, how small must cells be?

CHAPTER • REVIEW

Key Ideas

- The cell theory states that (1) cells are the basic structural units of all living things, (2) cells are the basic functional units of all living things, and (3) all cells come from pre-existing cells.
- Cells possess a variety of internal structures. Each type performs a different function related to the cell's survival.
- Although all plant and animal cells possess a nucleus, a cell membrane, and various organelles, there are differences between these two types of cells. For example, only plant cells have cell walls and chloroplasts, and the vacuoles of plant cells are larger than those of animal cells.
- Many materials move across the cell membrane, which is selectively permeable, by the processes of diffusion and osmosis.
- Photosynthesis, which occurs in the chloroplasts of plants, uses the energy provided by the sun to produce energy-rich compounds such as glucose.
- Cellular respiration, which occurs in the mitochondria of both plant and animal cells, breaks this glucose down and releases energy.
- The DNA molecules within the nucleus control the cell's activities, mainly by controlling the production of enzymes. Each enzyme speeds up a particular cellular reaction.
- The bodies of plants and animals are made up of many specialized cells, which are dependent upon each other for survival. Other types of organisms, such as most protists and all bacteria, have bodies that consist of a single unspecialized cell.
- Viruses, which are not living organisms, reproduce by using their host's cellular processes and structures. Viruses are known primarily for causing disease.
- Sanitation, vaccines, and antibiotics are used to prevent or treat diseases.

Vocabulary

cell theory
cell membrane
nucleus
cytoplasm
organelle

vacuole
cell wall
chloroplast
diffusion
osmosis
photosynthesis
cellular respiration
mitochondria
nuclear membrane
DNA
chromosome
gene
protist
bacteria
virus

1. Make a concept map using at least 10 of the words in the vocabulary list.
2. "Cell," "nucleus," and "virus" also have non-biological meanings. What are they? How do these meanings relate to the scientific use of these terms?

Connections

1. Mitochondria are often called the "powerhouses" of the cell. Why is this a good name for them?
2. (a) What are some of the main differences between plant and animal cells?
   (b) How are these differences related to the ways in which plants and animals live?
3. (a) What makes bacteria different from all other living organisms?
   (b) Why do you think the cells of more complex organisms (e.g., animals and plants) developed certain structures that never appeared in bacteria?
4. At some point in their lives, the nuclei of red blood cells break down. After this occurs, these cells lose the ability to grow, change, and reproduce. Why do you think this is so?
5. Even if your body needs water badly, drinking salt water from the ocean is not wise. It will disrupt the proper functioning of your body and make you ill. Why is this so?
6. Why do you think that so many diseases are caused by very small organisms such as protists and bacteria, or by viruses, which are even smaller?

Explorations

1. Photosynthesis is vital to the survival of all organisms on Earth. It is the original source of energy for living things. It also provides much of the oxygen used by humans and all other air-breathing organisms. Three major types of organisms photosynthesize—plants, some protists, and some bacteria. Find out how much each type contributes to the world's supply of oxygen.
   With the class, discuss the relationship between humans and these oxygen-producing organisms. What might happen if one of these groups disappeared from the Earth or was greatly reduced in numbers? What human activities have an effect on these organisms?
2. Do research on the disease AIDS. The library, your local health unit, and community groups are good sources of information. Different members of your class may want to focus on different topics, such as the following:
   (a) What causes AIDS?
   (b) How is AIDS transmitted?
   (c) What are the symptoms?
   (d) How can AIDS be prevented?
   (e) How is AIDS treated?
   Share your research with other members of your class—for example, by preparing a pamphlet or a poster.
3. Find out what a dialysis machine is (Figure 15.28). Prepare a report that describes who needs this machine and how it works. Compare the dialysis machine to a cell membrane.

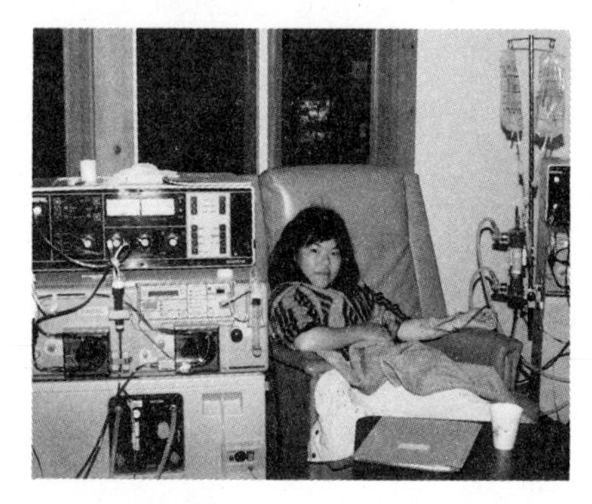

Figure 15.28
A dialysis machine helps this girl.

Reflections

1. Look back at the three-column charts you made in Activity 15A. Can you now answer all the questions there? In the third columns, fill in any answers you have learned. In the second columns of these charts, you may have listed your ideas about the names or functions of parts of the cell. Have you since changed any of these ideas? What caused you to do this?
2. The discovery that all living organisms are made up of cells is often described as the most important discovery in the history of biology. Why do you think this might be so?

CHAPTER

16 Reproduction

The large circular cell in the middle of this picture is a human egg cell. Clustered around the outside of the egg are many much-smaller sperm cells. If just one of these sperm can penetrate and enter the egg, the result may be a new human being. Where do the egg and sperm originate? How do these two types of cells interact to produce a new organism? These topics are part of the story of reproduction—the subject of this chapter. Like organisms, cells themselves reproduce. So another part of this story explains where the trillions of cells that make up the body of an adult human come from. As you will see, reproduction is a process that takes place at many different levels and in many different ways.

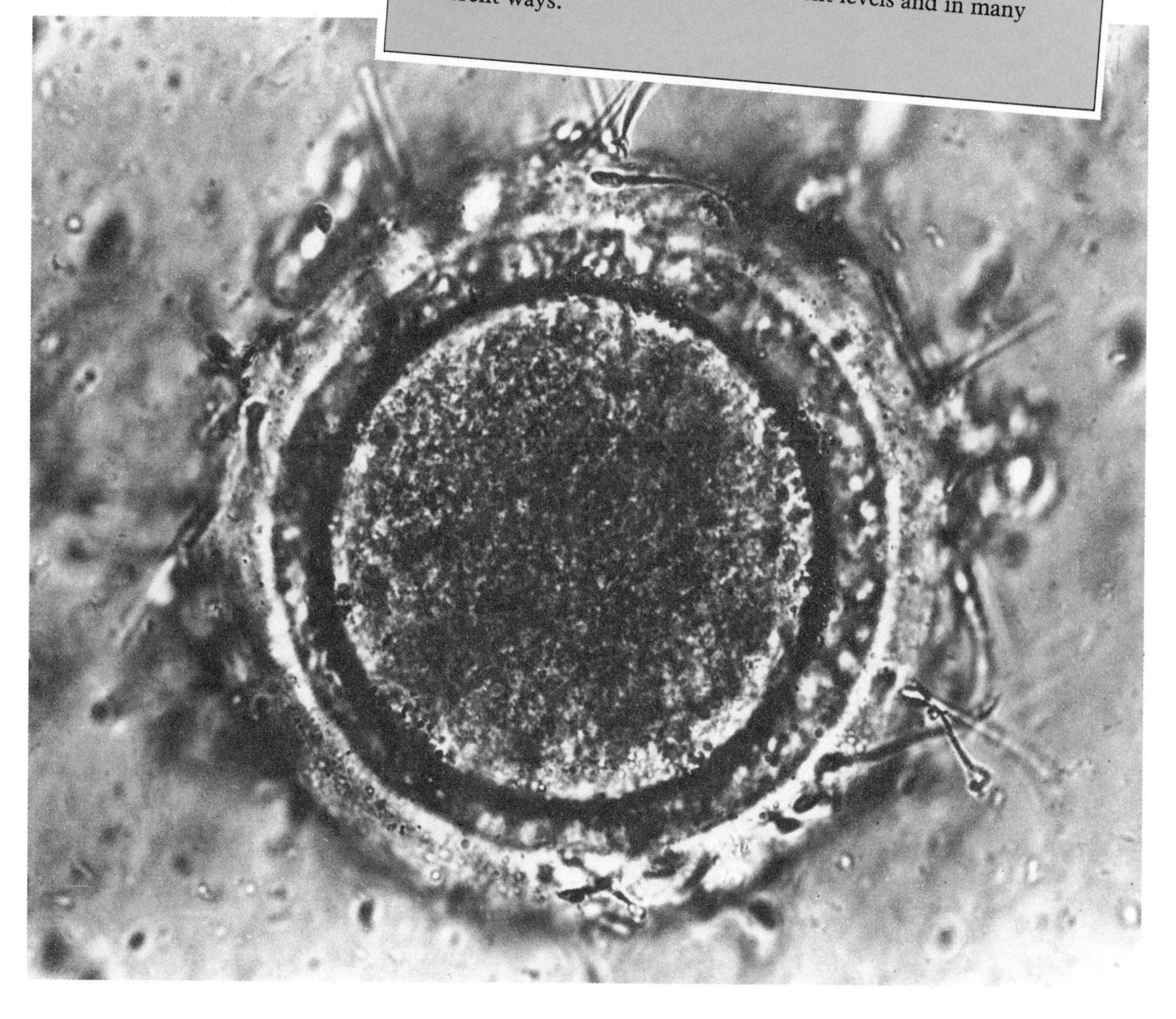

ACTIVITY 16A What Is Reproduction?

In your learning journal, divide two or more pages into three columns. In the first column, list some questions you have about reproduction. You might also include questions that you think other people might have. Have you considered all aspects of the subject, such as how cells, or organisms other than humans, reproduce? Perhaps you know, or can predict, the answers to some of these questions. If so, write your answers in the second column. Leave the third column blank for now. You will fill it in at the end of the chapter. Maybe you will find that your ideas have changed after reading this chapter. You may have discovered the answers to questions you did not have ideas about. As often happens to scientists when they are studying a subject, you may think of new questions.

16.1 Cell Reproduction

Figure 16.1
The body of the baby in this photo contains many fewer cells than does the body of the father. However, the cells in both parent and offspring are roughly the same size.

According to the cell theory, new cells arise from already existing cells. The new cells appear when existing cells divide in two. This division process, called **cell reproduction**, enables organisms to grow. As you saw in the last chapter, an individual cell can reach only a limited size before its volume becomes too large for its surface area. An organism, therefore, grows because the number of cells it contains increases, rather than because the cells grow larger (Figure 16.1).

Cell reproduction also enables organisms to replace cells that die. Every second, millions of your body cells are injured or die. If the remaining cells did not reproduce, your body would gradually decrease in size and eventually die.

Throughout Chapter 15, an individual cell was compared with an individual organism. The discussion of cell reproduction in the two paragraphs above may have suggested some new comparisons to you. Think of the main stages, or phases, in a person's life: birth, growth, reproduction, and death. Because reproduction results in a new human, whose life will then follow the same course, these stages make up a cycle, by which human life continues. Cells too have a life cycle, the phases of which make up what is called the **cell cycle**.

THE CELL CYCLE

Like the human life cycle, the cell cycle includes a phase of growth and a phase of reproduction. Most of a typical cell's life is spent growing, before it divides. This growth occurs during a phase called **interphase**. The Latin word *inter* means "between," so scientists used the term "interphase" to describe the period between cell divisions.

During interphase, proteins and other raw materials are manufactured or taken in by the cell. With these materials, additional organelles are formed and the cell membrane is enlarged to permit cell growth. This continues until the cell becomes too big to function properly and must divide (reproduce).

Another important event also occurs during interphase. Before a cell can divide, a new set of chromosomes must be made by copying the set already present in the nucleus. Recall that the genes that make up these chromosomes control all the cell's activities. What would happen if some of these genes were missing from a cell—or if a cell had several extra copies of a gene? A cell is like a complex machine, which would break down if some of its gears were missing or if there were extra parts confusing the works. The plans for making such a machine must be correct—and precisely the same—each time the machine is built.

During interphase, while the chromosomes are uncoiled and invisible, each chromosome is carefully duplicated so that there are exactly two copies of every gene. The duplicate chromosomes are attached to each other at a region called the **centromere** (Figure 16.2). The joined duplicates are called paired **chromatids**. As long as the two chromatids are attached at the centromere, the structure is still one chromosome.

Figure 16.2
A single duplicated chromosome is made up of two paired chromatids joined at a region called the centromere.

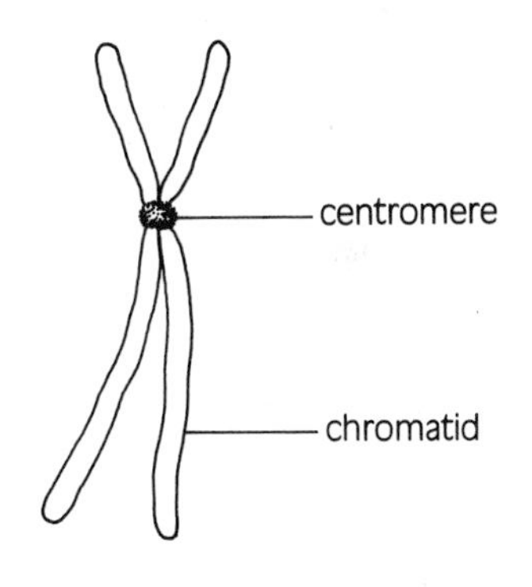

The other phase of the cell cycle is **cell division**, or cell reproduction. During cell division, the chromosomes become tightly coiled, and thus thickened (Figure 16.3). They can now easily be seen with a microscope.

Cell division begins with **mitosis**. This is the process by which paired chromatids in the nucleus separate and move to opposite ends of the cell. For cell reproduction to be completed, the organelles in the cytoplasm must also divide into two roughly equal halves. This is accomplished by the process of **cytokinesis**. Together, mitosis and cytokinesis result in cell division, or the production of two new cells. The two new cells are often referred to as daughter cells.

cell division (cell reproduction) = mitosis + cytokinesis.

Figure 16.3
In (a), the chromosomes in the cell's nucleus are not visible. In (b), however, the thread-like chromosomes in the cell's nucleus can be seen. This tells us that the cell is beginning to divide.

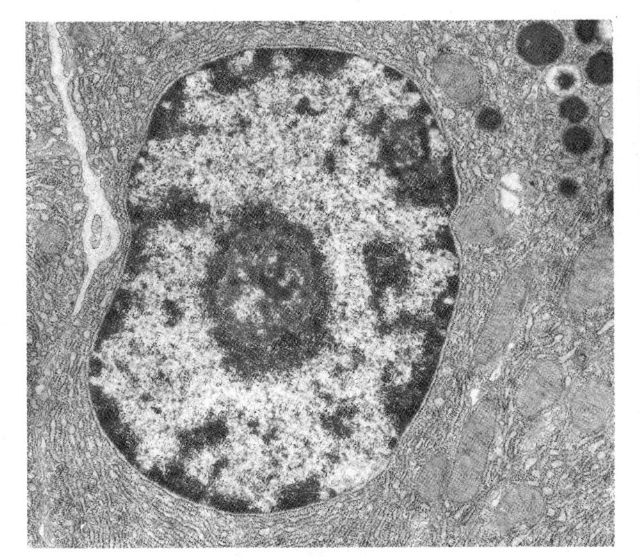
(a)

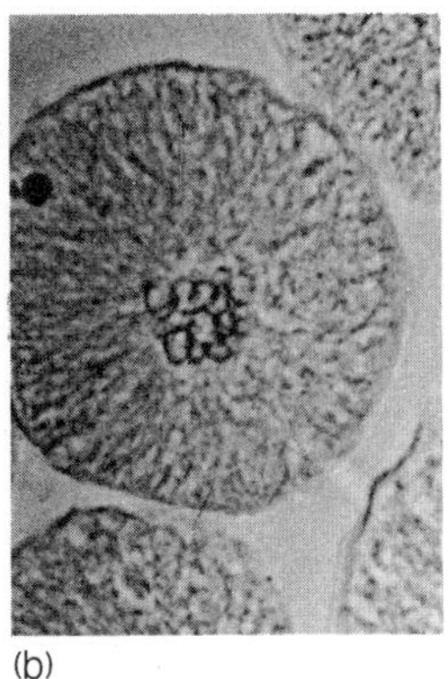
(b)

MITOSIS

DID YOU KNOW?

During interphase, some cells begin to become specialized—as skin cells, or muscle cells, for instance. These specialized cells usually do not divide again, but function for a while and then die. They are then replaced by the offspring of nearby unspecialized cells, which are still dividing.

Mitosis is a complex process. It has been described as resembling the intricate movements of soldiers marching on a parade ground—or Royal Canadian Mounted Police performing the Musical Ride. Mitosis results in the equal division of the material in the cell's nucleus and the formation of two new nuclei—one at each end of the cell. For convenience, biologists have divided the process of mitosis into four stages (Figure 16.4).

In each stage, several different events occur. Diagrams showing cells undergoing mitosis may make you think that the four stages are distinct from each other and that only one thing happens in each stage. Actually, the events of mitosis occur in a continuous sequence—like a movie. The events of one stage flow directly into the events of the next. Follow the description in the text by referring to the four components of Figure 16.4(a–d).

Figure 16.4
The stages of mitosis. In this case, cytokinesis is shown occurring during telophase of mitosis: the cytoplasm of the cell is pinching into two parts.

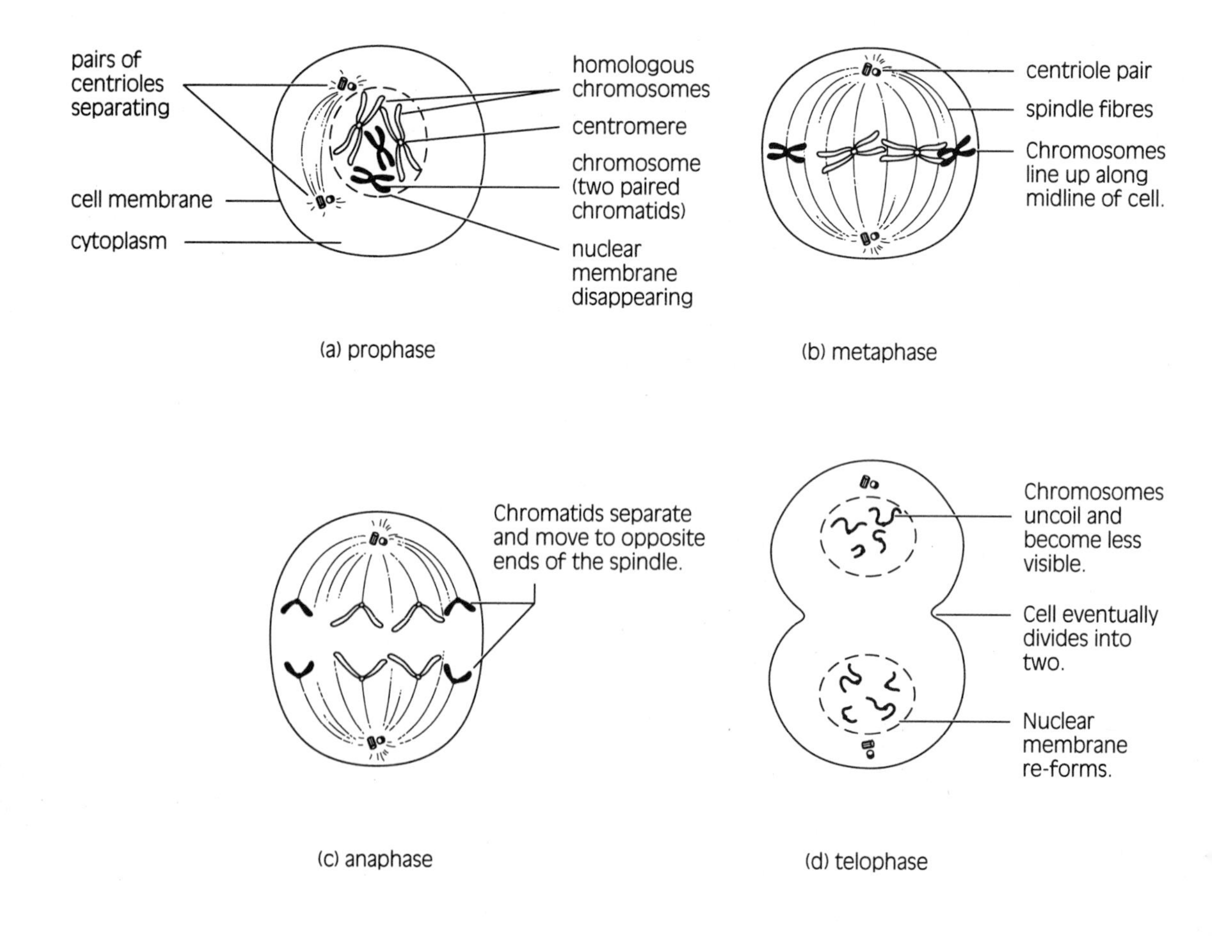

During **prophase**, the first stage of mitosis, the previously invisible chromosomes begin to coil and condense into visible threads, which become increasingly more visible. Eventually, it can be seen that each chromosome consists of two chromatids joined together at the centromere. (Recall that each chromosome was duplicated during interphase.) During prophase, the paired chromatids are distributed more or less evenly throughout the nucleus.

Other events also occur during prophase. In most animal cells, there is a pair of structures called centrioles. Like the chromosomes, these are duplicated during interphase. At the beginning of prophase, two pairs of centrioles are present in the cell, located near the nucleus. As prophase progresses, the pairs begin to move apart, one pair moving towards each end of the cell. The spindle — a system of tiny fibres — forms between the centriole pairs. Although plant cells do not have centrioles, they do form a spindle during prophase. Finally, the nuclear membrane breaks down and disappears during prophase.

Metaphase begins when the paired chromatids begin to move towards the centre of the cell. There, they line up along the middle of the spindle.

Anaphase starts when each pair of chromatids splits at the centromere. Each chromatid then becomes an independent chromosome. Now there are twice the usual number of chromosomes within the cell membrane. As anaphase continues, the two new chromosomes from each former pair of chromatids move towards opposite ends of the cell. Their movement is aided by the spindle fibres. By the end of anaphase, the cell contains two distinct groups of chromosomes, which are genetically identical (contain the same genes).

The events that occur during **telophase**, the final stage of mitosis, are in many ways a reverse of those in prophase. Nuclear membranes form around the two new sets of chromosomes; the spindle disappears; then the chromosomes begin to uncoil and gradually become less visible. Telophase ends when the chromosomes have returned to their interphase condition.

Mitosis is now complete, but cell reproduction is not. Two new nuclei have been formed, but these are still contained within a single cell membrane. To produce two daughter cells, the contents of the parent cell's cytoplasm must also divide.

DID YOU KNOW?

Healthy skin cells give off a chemical that inhibits, or discourages, mitosis. Damaged cells around a wound stop making this chemical. As a result, skin cells in the damaged area begin reproducing quickly and, in doing so, heal the wound. Scientists have discovered several chemicals that inhibit cell division, which are effective in the treatment of cancer by chemotherapy.

CYTOKINESIS

Cytokinesis is the name given to the division of the material outside the nucleus — the organelles and other substances in the cytoplasm — into roughly equal halves. This division usually begins during the later stages of mitosis. Cytokinesis is different in animal and plant cells because plants have a relatively rigid cell wall (Figure 16.5).

After cytokinesis, each daughter cell starts out with only half as many mitochondria and other organelles as its parent had. As the cell grows during interphase, additional cytoplasmic material will be made. The daughter cell will eventually reach the size of its parent (Figure 16.6).

Figure 16.5
Cytokinesis in (a) plant and (b) animal cells. A new cell wall will form along the cell plate in (a). What kind of cell was shown dividing in Figure 16.4? In what state of mitosis are the cells shown in (a) and (b)?

Cell plate (a layer of membranes) enlarges until it reaches outer surface of cell, splitting cell in half.

spindle

chromosomes

Furrow forms in cell membrane and deepens until cell is pinched in half.

centriole

(a) plant cell

(b) animal cell

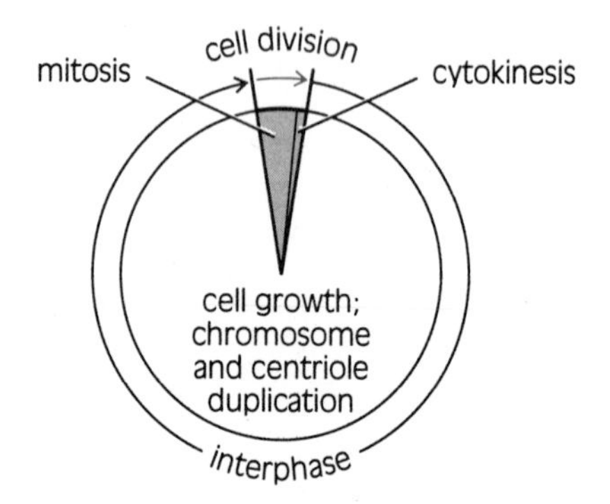

Figure 16.6
This circle represents the entire life of a cell. The cell cycle consists of two major phases: interphase, and the cell division phase, which takes up only about one-twentieth of the cell's life. The whole cell cycle usually takes 10 to 30 hours in plant cells and 18 to 24 hours in animal cells. However, in some cells, it may take weeks to complete. In contrast, in the cells of some embryos, the cell cycle may be as short as 20 minutes. Such cells grow very little during interphase, and spend relatively more time in mitosis.

ACTIVITY 16B *Observing Cell Division*

This activity will allow you to observe the stages of the cell cycle in both plants and animals.

MATERIALS

microscope
prepared slides of onion root cells and animal embryo cells

PROCEDURE

1. Place the prepared slide of onion root cells on the stage and examine the cells under low power. Estimate the proportion of cells that are actively reproducing. (HINT: Cell division is most active near the root tip.)
2. Switch to the medium or high power lens and locate cells that are reproducing. You will probably have to move the slide about to locate cells that are in different stages of mitosis or that are undergoing cytokinesis.
3. Sketch and label at least one cell in each of the different stages of mitosis. If you can, also sketch and label a cell undergoing cytokinesis and one in interphase.
4. Repeat steps 1 to 3 with the prepared slide of animal embryo cells.

DISCUSSION

1. Approximately what fraction or percentage of the onion root cells that you observed were actively reproducing (undergoing mitosis or cytokinesis)?
2. Why do you suppose that so few onion cells were dividing? (HINT: Remember that what you are looking at is like a snapshot—only a moment in the lives of these cells.)
3. Were any of the animal or plant cells undergoing cytokinesis? If so, what stage of mitosis were they in at that time?

CHANGES IN THE RATE OF CELL REPRODUCTION

As an organism gets older, its rate of cell reproduction slows. Cells are not replaced as rapidly as when the individual is young and growing. Eventually, the reproduction rate becomes so slow that the organism cannot replace cells as fast as they die. This slower rate of cell reproduction is part of what causes ageing — the gradual deterioration of an organism's body.

Sometimes, however, the cells reproduce too quickly because the DNA is damaged, and the nucleus of a cell loses control of division. Such damaged cells form clumps, or growths, called tumours. Some tumours are benign: they stop growing before causing damage. Other tumours, however, are malignant. They continue to grow, often spreading throughout the organism. In this case, an organism is said to have a disease called **cancer**. Eventually, malignant tumours get so large that they disrupt normal functioning, by crowding and invading the body's organs.

The type of damage to the DNA that produces cancer is caused primarily by **mutagens**. Mutagens are substances or agents that cause mutations (changes) in the genetic code. They do this by changing the order of the nucleotide bases that make up a gene. This then makes the instructions for the production of proteins (including enzymes) incorrect, and defective ones are produced. Mutagens come in many forms. For example, ultraviolet light from the sun is a mutagen. It can cause cancer in skin cells that are exposed to it for too long. The tar found in tobacco is a very active mutagen that can cause cancer in cells of the throat and lungs.

Various human activities release polluting chemicals into the environment, some of which may be mutagens. For example, even a very small amount of dioxin is thought to be carcinogenic (cancer-causing). Dioxin is produced by various industries, including chemical manufacturing and pulp and paper processing. Many of the substances used to kill insect pests and weeds are also believed to be mutagens, and may thus cause cancer.

Although much research is being done, there is no known cure for cancer. Treatment involves killing cancer cells or slowing their growth. This is usually done by surgically removing tumours, by treating them with chemicals (chemotherapy), or by exposing them to radiation (radiotherapy). For most types of cancer, treatment is more successful if the tumour is found early. Regular physical examinations, including tests for specific types of cancer, can help detect cancer early.

Currently, statistics show that one in three Canadians will get some form of cancer during their lives. Preventing this disease is the best approach to dealing with it. One way to do this is to avoid exposure to as many mutagens as possible. Stopping — or deciding not to start — smoking is an easy way to avoid one of the most serious mutagens (Figure 16.7). Reducing your exposure to the sun is another way to avoid mutagens (Figure 16.8).

E X T E N S I O N

The late Terry Fox focused interest on cancer and its victims through his Marathon of Hope. Annual Terry Fox Runs are held in his memory throughout Canada and around the world to raise money for research into the treatment and cure of cancer. Fox's home province, British Columbia, has made a contribution by founding the Terry Fox Research Laboratory. Find out about the work of cancer researchers in British Columbia through the Canadian Cancer Society, British Columbia and Yukon Division.

Figure 16.7
Many public places, including cafeterias, do not allow smoking. It has been shown that breathing the "second-hand smoke" exhaled by smokers and produced by burning cigarettes can be as harmful to people's health as smoking itself.

Figure 16.8
Lotions that act as sunscreens can help protect you from ultraviolet rays. Lotions that are labelled "broad spectrum" and that have a sun-blocking factor of 20 or over are believed to provide the best protection. However, you must still be careful not to spend too much time in the sun.

REVIEW 16.1

1. What important events occur in the cell cycle during interphase?
2. (a) What role does mitosis play in the overall process of cell division?
   (b) What would happen if mitosis did not occur correctly?
3. List the four stages of mitosis and, in point form, describe what happens in each.
4. What is the purpose of cytokinesis?
5. (a) How is mitosis different in plant and animal cells?
   (b) How is cytokinesis different in plant and animal cells?
   (c) How are mitosis and cytokinesis similar in plant and animal cells? (List at least five ways; these can be general or very specific.)
6. (a) What are two results of a change in the rate of cell reproduction?
   (b) What can cause a change in the rate of cell division? Give several examples.
7. In Greek, the word *pro* means "before" or "in front of." *Meta* means "between," and *ana* means "throughout." *Telo* means "completion" or "end." How do these names relate to the events of each stage of mitosis?

Profile

DR. JULIA LEVY

Are you stubborn? Full of energy? Do you love asking questions? These are just the qualities Dr. Julia Levy finds useful as a cancer researcher.

Levy supervises a team of scientists researching a new way of treating cancer. Photodynamic therapy is a technique of destroying tumours with light. A patient with cancer is injected with a photosensitizer—a chemical that is sensitive to light. After 24 to 48 hours, the photosensitizer will have accumulated in the patient's tumours. A fibre optic probe is then inserted into each tumour, and laser light is shone through. As the photosensitizer interacts with the light, it undergoes a chemical change and produces substances that destroy the tumour cells.

Surprisingly, when Levy was in high school, she didn't like science. In university, she became interested in microbiology and then took a course in biochemistry. She says, "I suddenly started understanding, at the molecular level, how living forms operate, and it was like a miracle! I wanted to ask more questions." She decided on a scientific research career.

When you get to the edge of what's known in a scientific field, then the whole world opens up, she says. "You're never going to run out of things to do in science. The more you know about an area, the more there is to know."

Levy started working with photosensitizers in the 1970s. She soon realized her research had important possibilities for cancer treatment. She also realized that developing the drugs for commercial use would be the best way to fund the research. So in 1980, she and some colleagues at the University of British Columbia founded Quadra Logic Technologies to develop, test, and market photosensitizers.

Quadra Logic was stepping into the business world of the giant pharmaceutical companies. It costs about $50 million to patent and test a drug and get government approval to sell it. Clinical trials must be done on about 1000 patients. To raise money, Quadra Logic sold shares through the Vancouver Stock Exchange.

Quadra Logic was in the right place at the right time. It caught the attention of a large pharmaceutical company that was looking for a small company to do research and development on photosensitizers. Quadra Logic and American Cyanamid made a business agreement. Quadra Logic would develop and test photosensitizers, and American Cyanamid would market and distribute them around the world. Quadra Logic is now the world leader in photodynamic therapy research and development. Levy is Vice President of Discovery, and the research scientists on her team work at UBC and Quadra Logic.

Levy says her advice to high school students is especially directed to girls. "Don't cut out your options for the future. By the year 2000, science courses will be even more relevant, because if we're going to survive as a nation we have to become a technology-based society. That's where the jobs are going to be."

And what else helps in her research? "Inspiration. You've got to love what you do."

16.2 Asexual Reproduction

EXTENSION

Your teacher will provide you with prepared slides of organisms reproducing by binary fission. Observe how (in what direction) each of these organisms splits. Using a reference book, find out what sort of environments these organisms live in, what they eat, and other features of their lifestyle.

Most of the cells in an organism reproduce as part of the cell cycle. Organisms, too, usually reproduce as part of their life cycle. Reproduction can occur in various ways. Cells, as well as some types of organisms, reproduce by different methods of **asexual reproduction**—all of which share three features. First, only one parent is involved—only one individual reproduces. This, as you saw earlier, is true of cells: one cell reproduces by dividing into two daughter cells.

Second, asexually produced offspring are genetically identical to the parent and to each other: they have an exact copy of the parent's DNA. This, too, is true for cell reproduction. The parent cell's chromosomes double during interphase. During mitosis, each daughter cell receives a set of chromosomes with identical DNA.

Third, in asexual reproduction, the cell or cells that produce the offspring are not usually specialized for reproduction. This is particularly true for each of the body cells in a many-celled organism, and for single-celled organisms. Such cells have other functions to perform, and cannot be specialized solely for reproduction.

All the ordinary body cells in many-celled organisms — including humans — reproduce asexually. So do single-celled organisms, such as bacteria and protists. In addition, some types of many-celled organisms often reproduce asexually. This is true, for example, of some of the larger types of protists as well as some plants and animals.

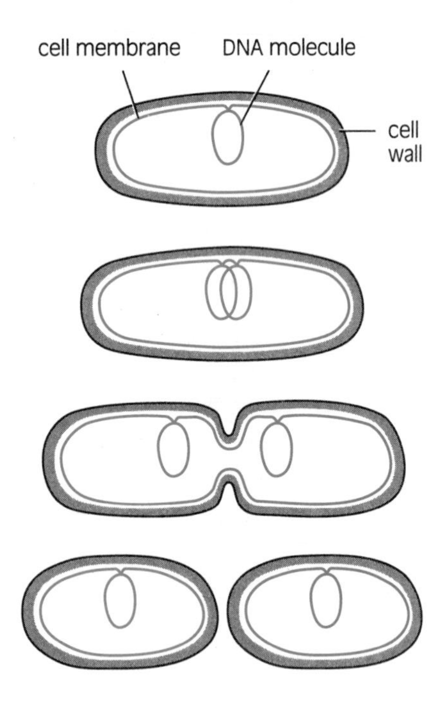

Figure 16.9
During binary fission, the bacterial cell begins to lengthen and its single molecule of DNA is duplicated. Each of the two molecules becomes attached at a separate point on the cell membrane. The cell is then pinched in two between these two points, so that each of the new cells receives a copy of the DNA.

METHODS OF ASEXUAL REPRODUCTION

There are five major types of asexual reproduction used by organisms. With each method, the parent's DNA is first duplicated and then divided equally between the offspring.

In **binary fission**, the organism splits into two equal-sized offspring. Binary fission occurs primarily in single-celled organisms (and in a few many-celled types). It is the usual method of reproduction for bacteria. Because a bacterial cell has only a single molecule of DNA, the division of its genetic material does not require the complicated process of mitosis. The overall process of cell division is much the same, however (Figure 16.9).

Binary fission is also the usual method of reproduction for protists. Protist cells have a complete nucleus, and more complex chromosomes than bacteria. In these organisms, therefore, mitosis occurs, followed by the division of the parent's cytoplasm into two roughly equal parts (Figure 16.10). Binary fission in protists is much like cell division.

Budding is another type of asexual reproduction. The offspring begins as a growth, called a bud, on the parent (Figure 16.11). When it has developed enough to survive on its own, the small bud detaches. Unlike binary fission, budding results in the reproducing organism's being divided unequally. It occurs both in single-celled organisms (e.g., yeast) and in some many-celled organisms (e.g., some plants and animals).

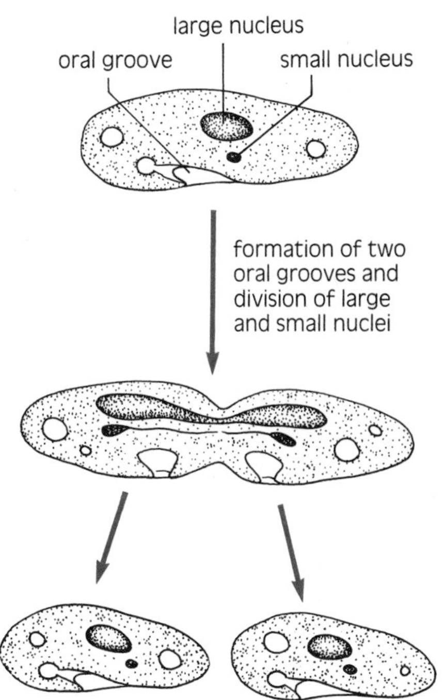

Figure 16.10
Binary fission in paramecium, a type of single-celled protist. A paramecium contains a large and a small nucleus, both of which must be duplicated before the cell divides. The oral groove is the paramecium's mouth, and it too must be duplicated so that each new cell can take in food.

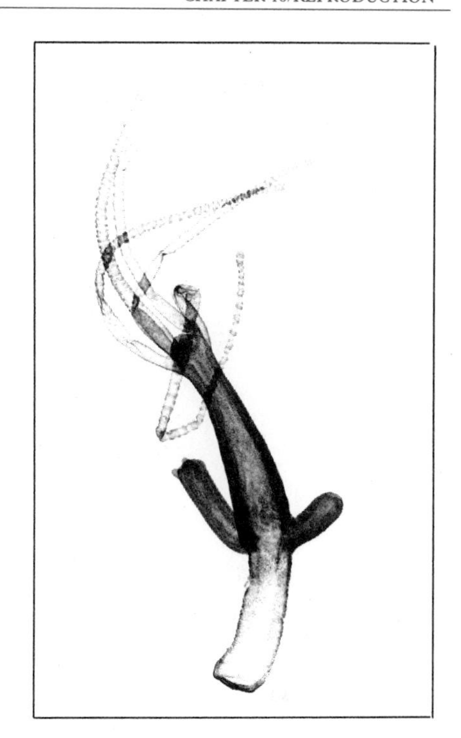

Figure 16.11
A hydra is a very small animal that lives in water. It obtains food by using tentacles, which surround its mouth. The small bulges on either side of this hydra are buds. Eventually they will develop tentacles and come to resemble a small version of the adult organism. At this point, they will detach and become independent.

Spore formation is slightly different from the other types of asexual reproduction. The organism produces cells, called spores, that are somewhat specialized for reproduction. Each spore contains a nucleus and a small amount of cytoplasm. Many types of spores also have a protective covering, which helps them avoid becoming dried out after they are released from the parent. If a spore lands in a favourable environment, it will develop into a new organism. For example, *Penicillium* reproduces by spore formation (Figure 16.12).

During reproduction by **fragmentation**, fragments (pieces) of the parent's body break off, and some of them grow into whole new organisms. Sometimes fragmentation (and thus reproduction) occurs by accident. Often, however, an organism breaks off a piece of its body deliberately, so that it can reproduce in this way. In some cases, these fragments are fairly small; in others, they amount to almost half the organism. In this fashion, one organism can produce several offspring with genes identical to its own. Starfish and flatworms are two animals that reproduce by fragmentation (Figure 16.13).

When plants break off pieces that grow into new plants, they exhibit one form of **vegetative reproduction**, a type of asexual reproduction that occurs only in plants. Most forms of vegetative reproduction do not require that the plants be broken into fragments first. Many plants send out special stems from which offspring emerge at intervals. These offspring grow their own roots and become independent plants. Some plants, such as strawberries and spider plants, send out stems called runners, which usually travel along the surface of the ground (Figure 16.14). Other plants, such as grasses, lilacs, and many ferns, send out underground stems called rhizomes.

Figure 16.12
Penicillium reproduces by spores.

Some types of woody shrubs reproduce asexually simply by using their ordinary stems. The branches of such plants may take root wherever they touch the ground, and new plants may then develop. Currants, willows, and forsythias, among others, can reproduce in this way.

Figure 16.13
A starfish (also called a sea star) can easily grow a new arm if one is lost. In addition, if a detached arm contains a portion of the starfish's central disk, it can grow a whole new body! By fragmentation, a single starfish can give rise to several new individuals.

Figure 16.14
Spider plants send out runners. Along these runners, a new plant may develop. The runners with new plants can be seen in the lower left.

THE OUTCOME OF ASEXUAL REPRODUCTION

There is a special name for the identical offspring produced by one parent through asexual reproduction. As a group, they are called a **clone**. (The individual members of such a group are also sometimes called clones.) All the members of a clone have the same genetic characteristics. If these characteristics help the organisms to function in their environment, the clone will be successful. More identical individuals will be produced, and they will succeed in the same environment.

Many types of asexually reproducing organisms are very successful. This is partly because asexual reproduction can occur extremely rapidly. Some kinds of bacteria, for example, divide every 20 to 30 minutes. At this rate of offspring production, a population living in an unchanging environment could grow quite large very quickly!

However, what would happen to the clone if the environment changed in some way? Perhaps the land on which a particular clone of plants was growing suddenly becomes flooded. If the clone's characteristics were suited only to dry conditions, none of the plants would survive. Or perhaps a new virus appears in an area and infects one of the organisms in a clone. Because all members of the clone are genetically the same, the virus can easily infect the others too. In this way, a deadly virus could rapidly wipe out all of a parent's offspring.

What would happen, do you think, if a parent had offspring that were *not* all identical, but that varied, so that some had different characteristics from others? Some of these offspring might not be as perfectly suited to the original environment as the parent itself. If the environment changed, however, at least some of these differing individuals might turn out to be *better* off than the parent. For example, say that a type of plant produces varied offspring. If there were a sudden drought, the parent and some of its offspring might be unable to survive the dry conditions. Other offspring might have different characteristics—ones that would be suited to a dry environment.

This is exactly the situation faced by many types of organisms, including humans. The conditions they live in are not always predictably the same. So producing offspring identical to themselves is not the best strategy. In these circumstances, organisms that produce differing offspring are often more successful. Then, if the environment changes, at least some of the offspring may have features that could help them survive.

Asexual reproduction is not the way to produce varied offspring. Reproduction that does result in differing offspring is the subject of the next section.

REVIEW 16.2

1. List three characteristics of asexual reproduction.
2. List five different methods of asexual reproduction and briefly describe each one.
3. (a) What is the difference between binary fission in bacteria and the same process in single-celled protists?
   (b) Why is there a difference?
4. (a) What is the difference between binary fission and budding?
   (b) How are these two processes similar?
5. (a) In what way is spore formation different from other types of asexual reproduction?
   (b) In what ways is it the same?
6. Describe some methods of vegetative reproduction.
7. (a) What are two advantages of asexual reproduction?
   (b) What is a potential disadvantage?
8. Oysters are a major part of the diet of starfish. Because of this, people who catch oysters for a living used to try to kill starfish by cutting them up and throwing them back into the sea. Why might this practice not reduce the numbers of starfish?

16.3 Sexual Reproduction

Unlike asexual methods, **sexual reproduction** requires two parents, and the offspring produced are not genetically identical to the parents or—usually—to each other (Figure 16.15). Another difference is that sexual reproduction always requires the formation of specialized cells, which function *only* in reproduction (Figure 16.16). To understand why specialized cells are required, you need to take another look at chromosomes.

Scientists often examine chromosomes by preparing a karyotype: a picture of the set of chromosomes possessed by a particular individual. The karyotype of a human female is shown in Figure 16.17.

The normal number of chromosomes in a human body cell is 46. Notice the way in which these are arranged in Figure 16.17. Scientists have discovered that the 46 chromosomes form 23 pairs. The chromosomes of all sexually reproducing organisms form pairs, just like those of humans.

The members of such a pair of chromosomes are said to be **homologous**, which means that they are similar in overall appearance and are made up of the same kind of genes. For example, both the chromosomes in one pair may contain a gene for hair colour, among many other genes. Cells containing pairs of homologous chromosomes are said to be **diploid**. (This comes from the Greek words *diploos*, for "double," and *eidos*, for "form.") A diploid cell thus contains two of each type, or form, of chromosome. All normal body cells are diploid. A human diploid cell has a total of 46 chromosomes. Other species have different numbers of chromosomes.

Figure 16.15
Why are these puppies different from each other? Why do some of them look quite different from their mother?

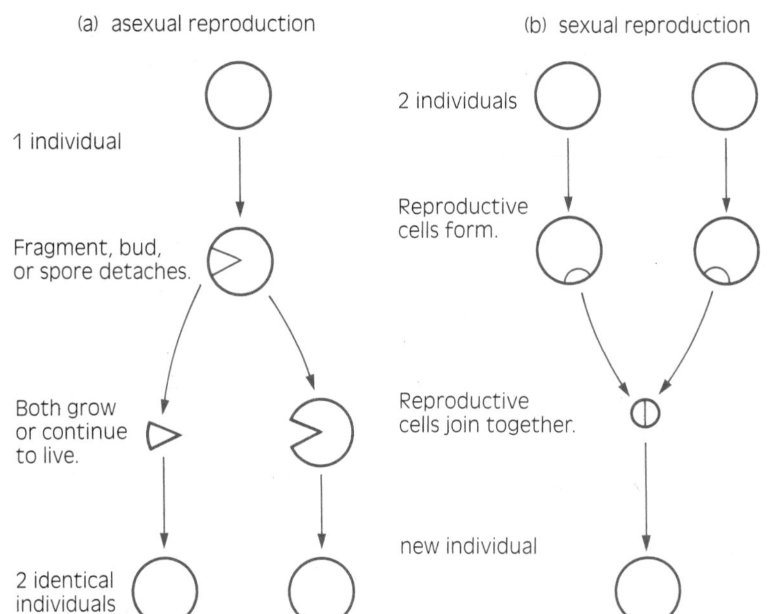

Figure 16.16
The difference between asexual and sexual reproduction. Although only certain methods of asexual reproduction are represented by this diagram, all the methods share these general characteristics. Often, more than two offspring are produced by an asexually reproducing parent. Similarly, many types of sexually reproducing parents can produce more than one offspring.

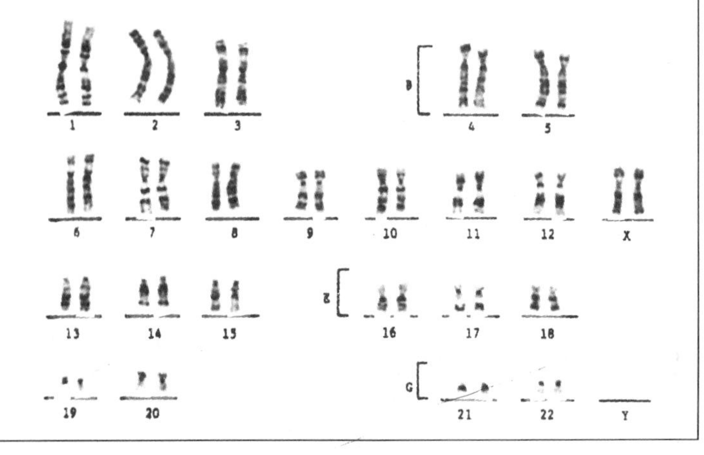

Figure 16.17
The karyotype of a human female. Note that the chromosomes are arranged in pairs. Twenty-two of the pairs are numbered. The members of the 23d pair, both labelled "X," are special, and identify this individual as a female. (In a male, this pair of chromosomes is different.) It is difficult to see the separate chromatids in each chromosome because they are very close together.

ACTIVITY 16C Why Are the Chromosomes in Diploid Cells Paired?

Where did the two chromosomes in a homologous pair come from? See if you can discover the answer in the following discussion. Perhaps you have already thought of a possible response. If so, write your answer in your learning journal. Check it against the answer you would give after reading the remainder of this section.

TWO PARENTS NEEDED

In sexual reproduction, both parents provide chromosomes for the offspring. Two cells, one from each parent, unite to form the first cell of the new organism. (See the photo that opens this chapter.) If these two cells each contained a full set of chromosomes, the cell resulting from their fusion would have two full sets of chromosomes. The complex machinery of the cell could not function properly with these extra chromosomes.

This problem is avoided because these are specialized reproductive cells, each containing only half the number of chromosomes in the parents' body cells. When two reproductive cells—one from each parent—join together, the first cell of the offspring will have exactly the same number of chromosomes as did each of its parents.

These specialized reproductive cells are called **gametes**. Each gamete contains one member of each homologous pair of the parent's chromosomes. Human gametes contain 23 different chromosomes. Gametes, and other cells that contain only *one* of each type of chromosome (rather than a pair), are said to be **haploid** (Figure 16.18). (In Greek, the word *haplo* means "single.")

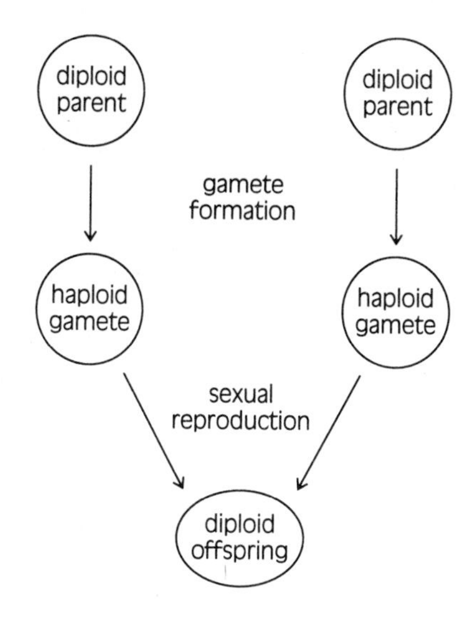

Figure 16.18
Each haploid gamete contains one member from each homologous pair of the parent's chromosomes. When the two gametes are joined during sexual reproduction, therefore, the offspring will have pairs of homologous chromosomes—one member from each parent. The offspring is thus diploid, like its parents.

MEIOSIS

The production of haploid gametes requires a special type of cell division. First, the material in the nucleus divides by a process called **meiosis**, which reduces the number of chromosomes by half. Then the cytoplasm divides by cytokinesis—the same process you studied earlier in the chapter.

Meiosis actually consists of two parts, meiosis I and meiosis II. The events that occur in both meiosis I and II are divided into four stages: prophase, metaphase, anaphase, and telophase. These stages are given the same names as in mitosis because they are somewhat similar.

Before meiosis I starts, during interphase, the chromosomes duplicate; each one forms a pair of joined chromatids. Between meiosis I and II, many cells go through a type of interphase again, but the chromosomes are not duplicated during this phase. As you read about the two parts of meiosis in Figure 16.19, notice how meiosis I reduces the number of chromosomes and meiosis II separates pairs of chromatids.

The process of cell division is complete after the cytoplasm divides. Four cells have been produced from the one original parent cell. Each of these four cells is a haploid gamete—the reproductive cell of a sexually reproducing organism.

Figure 16.19
The process of meiosis. At the end of meiosis I, the pairs of homologous chromosomes have been separated. A second division process is necessary to separate the pairs of joined chromatids. The cells of some organisms skip telophase I and cytokinesis, and go directly from anaphase I into meiosis II. What do you think the diagrams for such cells would look like?

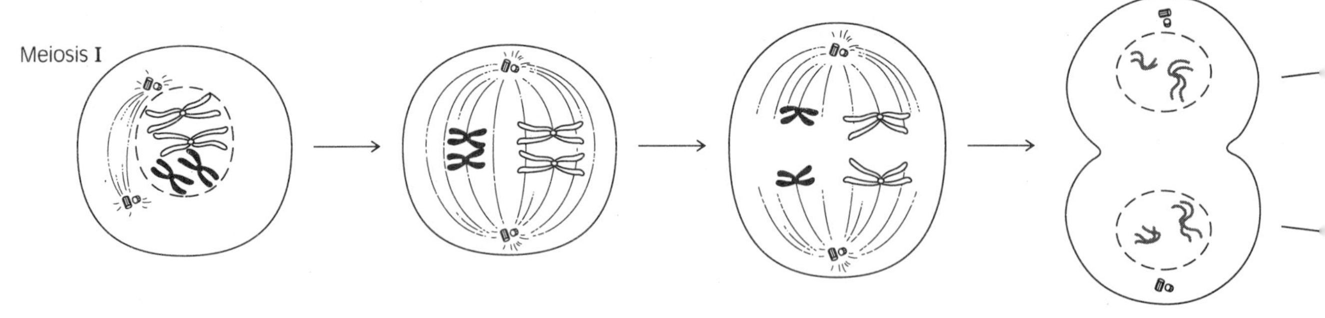

(a) prophase I
- Chromosomes coil and become visible.
- Nuclear membrane disappears.
- Spindle forms.
- Homologous chromosomes move towards each other.

(b) metaphase I
- Pairs of homologous chromosomes move to midline of cell.

(c) anaphase I
- Homologous chromosomes move apart, one of each pair going to each end of cell.
- Each cluster of chromosomes at end of cell has one of each type of chromosome—half the number the parent cell had.
- Paired chromatids remain attached.

(d) telophase I
- Paired chromatids uncoil, become less visible.
- Nuclear membrane forms around each chromosome cluster.
- Spindle disappears.
- Cytokinesis occurs.

ACTIVITY 16D *Modelling Mitosis and Meiosis*

This activity will give you a chance to create a model of meiosis and compare the stages in this process to those of mitosis. Meiosis produces haploid offspring cells, whereas mitosis produces offspring cells that are diploid. While doing this activity, see if you can discover how this difference between the two processes comes about.

MATERIALS

paper, pencil, eraser, scissors, modelling clay in two different colours, and other materials of your choice, such as toothpicks, pipe cleaners, or yarn

PROCEDURE

1. First make a model of meiosis. Make four single chromosomes out of clay, two about 4 cm long and two about 2 cm long. One long and one short chromosome should be one colour (colour a), and the other long and short chromosomes should be the other colour (colour b).
2. Duplicate the clay chromosomes (creating joined chromatids) as part of the interphase that precedes meiosis.
3. With these clay chromosomes, and using any of the suggested materials, model the stages and events of meiosis I and meiosis II. In your notebook, describe the appearance and number of each of the different items in your model at each step. What does each item represent?
4. Using clay or other materials, design a model or a display that compares meiosis and mitosis. Be sure that all the stages of both processes are illustrated or modelled. Indicate the differences as well as the similarities between them. You might want to work on this project with another student or in a small group. Your goal should be to produce something that would help you explain mitosis and meiosis—and the differences between them—to someone who knows nothing about these processes.

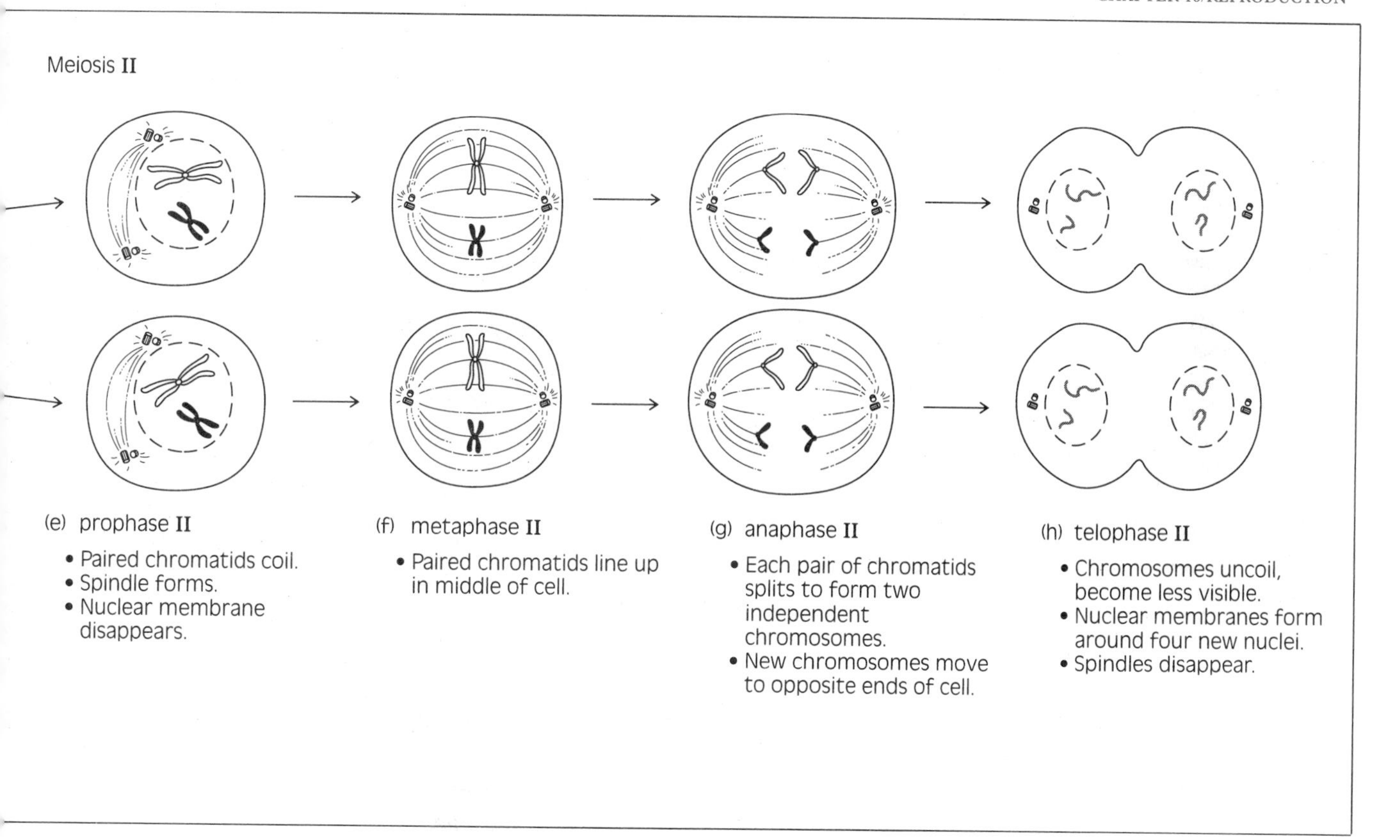

METHODS OF SEXUAL REPRODUCTION

You have seen how gametes are formed after a cell divides by meiosis. Because each gamete has only half the number of chromosomes of a normal body cell, two gametes must join together to form a new organism. In sexual reproduction, the two gametes usually come from two different parents. Most kinds of organisms produce two types of gametes, male and female, which are quite different from each other. The mature male gamete is called a **spermatozoon** (plural: spermatozoa), or **sperm**. It is usually small and able to move about so that it can reach the larger female gamete, the egg cell, or **ovum** (plural: ova).

If these two types of gametes meet successfully, a process called **fertilization** occurs. This process involves two steps. First, a single sperm penetrates the outer covering of the ovum. Then, the nuclei of the sperm and ovum fuse so that a single diploid cell is formed. This cell, called a **zygote**, is the first cell of a new organism—the offspring.

The zygote cell then begins to divide by mitosis, producing two new cells, which will themselves later divide. Once a zygote begins to divide in this way, it is called an **embryo**. The embryo continues to increase in size (by cell division) and develops body structures. Eventually, it grows and develops enough to become an independent organism (though it may still require some parental care), and it is born or hatched.

The basic steps of gamete production, fertilization, and embryo development are followed by all sexually reproducing organisms. However,

DID YOU KNOW?

Some types of organisms can reproduce both asexually and sexually. For example, although starfish and planarians can reproduce by fragmentation, they more commonly reproduce sexually. Strawberries also produce and exchange male and female gametes at certain times, as do the spider plants seen in Figure 16.14. Do you know what role insects (especially bees) play in assisting sexual reproduction in plants?

Figure 16.20
A ponderosa pine tree has both male and female cones. The male cones are much smaller than the female cones.

many aspects of the process vary widely. For example, in some types of organisms, individuals can produce both male and female gametes. In many types of plants, for instance, a single plant can produce both female and male gametes in separate parts of the plant (Figure 16.20). Although some of these plants can fertilize themselves, most exchange gametes with other plants. This is also true of the few types of animals in which one individual can produce both sperm and ova. Earthworms are hermaphrodites, meaning that each individual worm produces both male and female gametes. Earthworms cannot fertilize their own gametes, however, but must mate with another individual. They exchange sperm, and both worms produce offspring.

In most familiar types of animals, and in many plants, individuals can produce only one type of gamete. The eggs are produced by females, and the sperm are produced by males. Even so, there are many variations in the reproductive process — in fertilization, for instance. In animals that live in water, such as fish, fertilization is usually external: it occurs outside the mother's body. Ova and sperm are released into water and must find each other for fertilization to occur (Fig. 16.21).

Because gametes dry out very easily, external fertilization is possible only for animals that reproduce in water. Among land animals, the male inserts sperm into the female's body, which provides a protected internal environment for fertilization.

Figure 16.21
The eggs of salmon are fertilized externally, outside the mother's body. (a) The female deposits her eggs in a nest she has made in the bottom of a streambed. (b) The male deposits sperm over these eggs. When the two types of gametes meet, fertilization can occur. (c) The female then covers the fertilized eggs with sand to protect them.

Embryo development also occurs differently in different species. In some types of animals, including humans, the embryo develops within the mother's body. It grows there until its internal organs have developed enough that it can survive more or less on its own, at which point it is born.

In other types of animals, such as birds and many reptiles, the embryo develops outside the mother's body, inside a shell. This shell protects the embryo inside from damage and from drying out. In it, the embryo grows until it is ready to hatch.

In many ways, a plant's seed is like an egg. It contains a developing embryo surrounded by tissue that provides essential nutrients (much like an egg's yolk). The outermost layer of many seeds is specialized to protect the embryo inside.

There are many other differences among species with respect to sexual reproduction. Each type of organism has special structures and behaviours to aid the reproductive process. The subject of reproduction is truly a broad one!

ACTIVITY 16E Researching Sexual Reproduction

In this activity, you will research reproduction in an animal of your choice. Focus on discovering how the reproductive structures or behaviours of this type of organism are suited to the environment in which it lives. Your report could be in written form, or it could consist of a series of photographs or sketches, with captions. You should include information on the following:

- where fertilization and development take place
- any special structures of the reproductive system
- mating behaviours demonstrated by the animal
- the length of time the embryo takes to develop before hatching or birth
- the kind of care provided to the young organism by its parents
- any other factors that might affect reproductive success in this organism—that is, that might increase or decrease the number of offspring it can produce

REVIEW 16.3

1. List three ways in which sexual reproduction differs from asexual reproduction.
2. (a) In what ways are the two chromosomes in a homologous pair similar to each other?
   (b) How many pairs of homologous chromosomes are there in a human body cell?
3. (a) What is the difference between a diploid cell and a haploid cell?
   (b) Which type of cell results from mitosis and which type results from meiosis?
4. (a) How many chromosomes does every human body cell contain?
   (b) Are these cells haploid or diploid?
   (c) How many chromosomes does a human reproductive cell contain?
   (d) Are these cells haploid or diploid?
5. Why is meiosis necessary for sexual reproduction?
6. List the stages of meiosis and briefly describe what happens in each.
7. What is another name for the reproductive cells of a sexually reproducing organism?
8. What are the three basic steps in sexual reproduction? Write a sentence or two about the different ways each of these can occur (in different types of organisms).

16.4 Human Reproduction

Like many organisms, humans have specialized and complex reproductive structures, which are quite different in males and females. Although these structures are present in human babies at birth, they are not functional. Between the ages of about 12 to 17 in boys and 10 to 14 in girls, the reproductive structures gradually mature and begin to function. This period in an individual's life, during which he or she becomes sexually mature (and thus able to reproduce), is called **puberty**. Puberty is started by hormones produced by the pituitary gland, which is located near the brain. Hormones are substances secreted by glands and other body structures. They regulate the activity of various body organs and cells.

One of the most important functions of the reproductive system of both sexes is to produce gametes. In the female, the gametes are produced in two organs called **ovaries**. The male also has two gamete-producing organs, called **testes** (singular: testis). Recall that gametes are produced when diploid cells undergo meiosis to produce haploid cells. Much of this process occurs in the ovaries (which produce ova) and the testes (which produce sperm). In addition to these organs, males and females have other structures that play roles in gamete production, fertilization, or (in females) development of the embryo. These structures — the various parts of the reproductive systems of males and females—are called **primary sexual characteristics**.

Males and females also have other visible differences, called **secondary sexual characteristics**. These features do not play a direct role in gamete production, fertilization, or embryo development. Like the primary sexual characteristics, these features appear during puberty as a result of the action of hormones. The primary and secondary sexual characteristics are summarized in Table 16.1.

EXTENSION

Some athletes take drugs to improve their performance. Anabolic steroids, which are some of the most commonly used drugs, are also among the most dangerous. These steroids are derived from male sex hormones, such as testosterone. Find out what effect these drugs have on the bodies of athletes. Why are they used, and why are they dangerous?

Table 16.1 Primary and Secondary Sexual Characteristics

	Male	Female
Primary sexual characteristics	• Two testes, which produce sperm and hormones	• Two ovaries, which produce ova and hormones
	• Various glands and storage sacs	• Two oviducts (Fallopian tubes), which carry the ova to the uterus
	• Various tubes, which carry semen (made up of the sperm and a lubricating, protective fluid)	• Vagina, which receives penis during intercourse and serves as birth canal
	• Penis, which transfers sperm to the female	• Uterus (womb), where offspring develops
Secondary sexual characteristics	• More muscular body than that of females	• Rounded figure resulting from deposits of fatty tissue, especially in breasts, buttocks, and upper legs • Wider pelvis than in males
	• Conspicuous hair in pubic area and armpits, on face and other parts of the body	• Conspicuous hair in pubic area and armpits

THE MALE REPRODUCTIVE SYSTEM

The male reproductive organs are shown in Figure 16.22. The testes are enclosed in a sac called the scrotum. Each testis contains thousands of thread-like seminiferous tubules, coiled and tightly packed inside it.

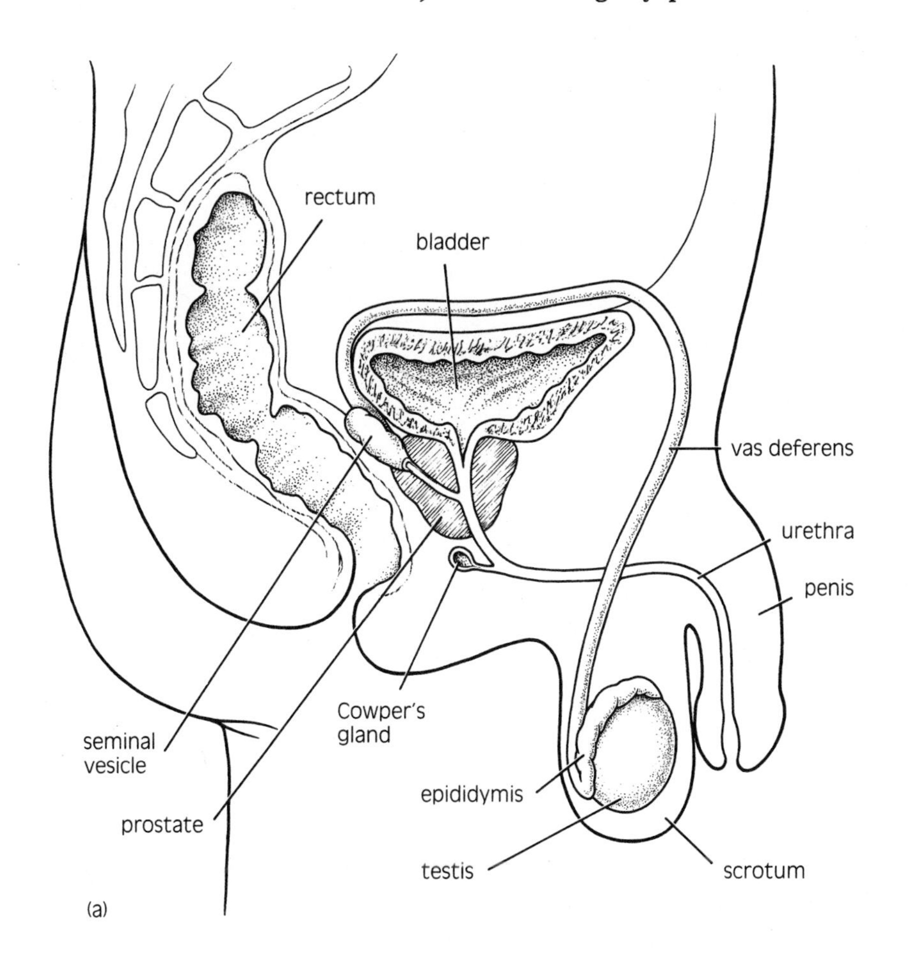

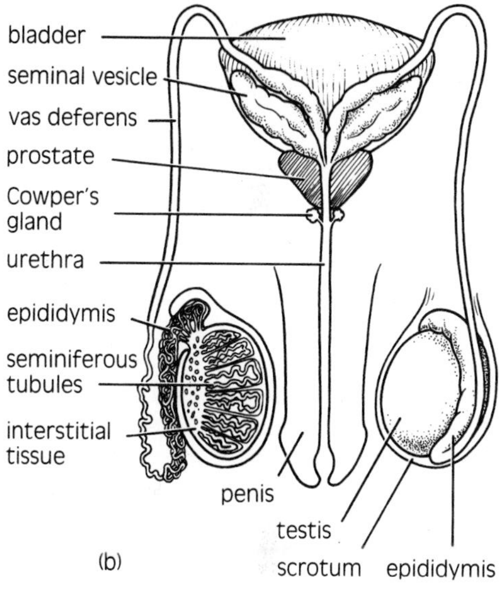

Figure 16.22
The male reproductive system, shown from the side (a) and the front (b). In (b), one of the testes has been drawn as if it were dissected, to show the seminiferous tubules inside. The internal coiling of the epididymis is also shown.

Inside the tubules, sperm are produced and stored. Each tubule is surrounded by a region of tissue in which are scattered the interstitial cells that produce the male hormone, testosterone. This hormone, acting with hormones from the pituitary gland, causes the seminiferous tubules to produce sperm.

Each sperm is composed mostly of a head (which contains the nucleus) and a tail-like flagellum (Figure 16.23). Many steps are involved in the production of mature sperm and their movement through the male reproductive tract. These are shown in Figure 16.24.

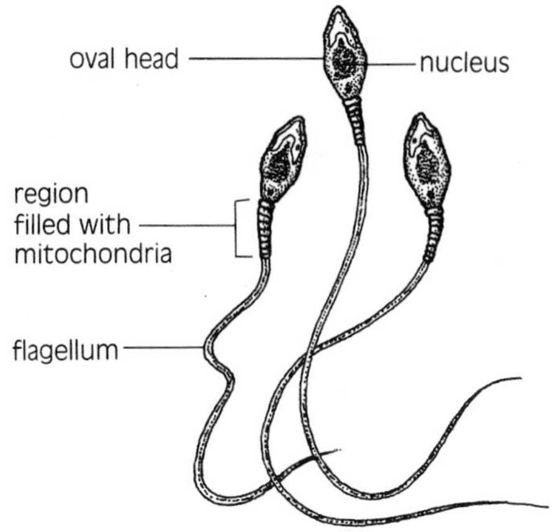

Figure 16.23
Spermatozoa. The flagellum helps the sperm to swim on its way to meet an ovum.

Figure 16.24
The development of mature sperm and the path they take through the male reproductive system.

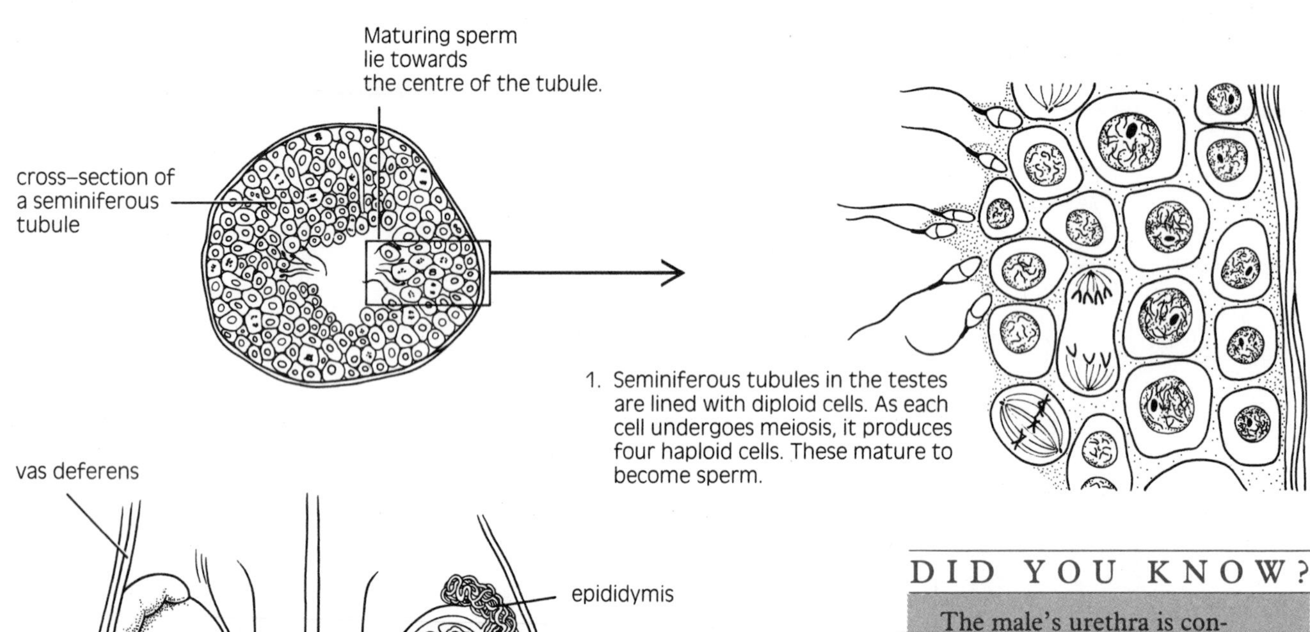

1. Seminiferous tubules in the testes are lined with diploid cells. As each cell undergoes meiosis, it produces four haploid cells. These mature to become sperm.
2. Mature sperm move into epididymis.
3. Secretions from epididymis enable sperm to move.
4. Sperm move into vas deferens.

DID YOU KNOW?

The male's urethra is connected to the bladder as well as to the vas deferens. It thus transports both urine and sperm. These two fluids cannot travel through the urethra at the same time, however. During intercourse, a ring of muscles contracts to close off the tube leading from the bladder to the urethra.

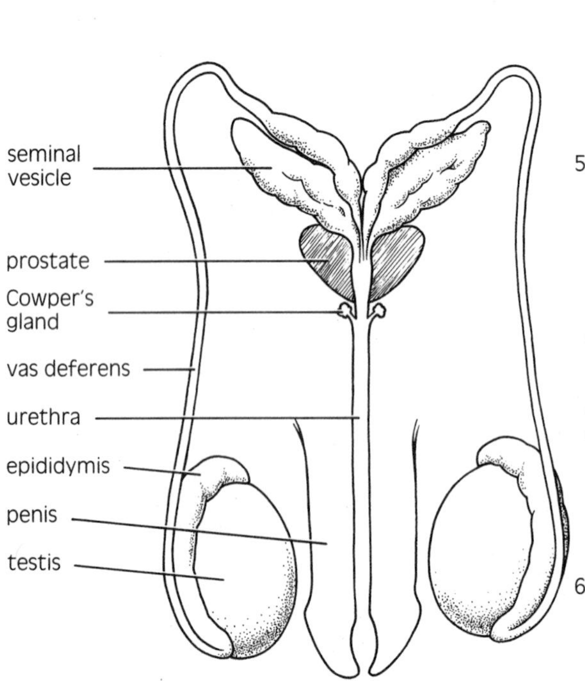

5. Secretions from seminal vesicles, prostate gland, and Cowper's glands are added as sperm move along vas deferens and into urethra. These secretions form the seminal fluid.

   Seminal fluid:
   - protects sperm from acidic environment of female reproductive tract
   - provides lubrication
   - provides energy-rich molecules to sperm

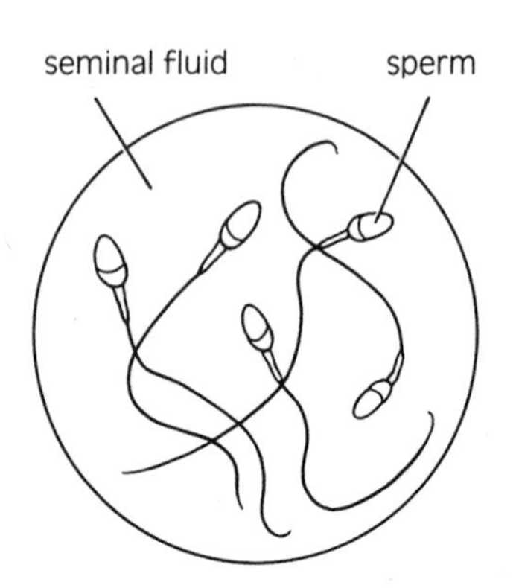

6. The semen (seminal fluid and sperm) is ejaculated.

THE FEMALE REPRODUCTIVE SYSTEM

A female's ovaries are located in the abdominal cavity near the pelvis (Figure 16.25). Each ovary is about 3 cm long and 2 cm wide. The ovaries not only produce ova, but also release the female sex hormones, estrogen and progesterone. At birth, the ovaries of an average female contain roughly 1 million immature ova. Each ovum is surrounded by a jacket of cells called a follicle. During a woman's life, only about 400 of these ova will actually develop to maturity. An ovum and its follicle develop together, one approximately every month from the end of puberty until the woman enters menopause (usually sometime in her fifties). By the end of menopause, no more ova are maturing, and the woman can no longer reproduce.

EXTENSION

What is a Pap test? Do research to discover what this procedure is designed to test for. How has it helped women to get early treatment for cancer of the cervix? Find out when women should begin having Pap tests and how frequently the tests should be done.

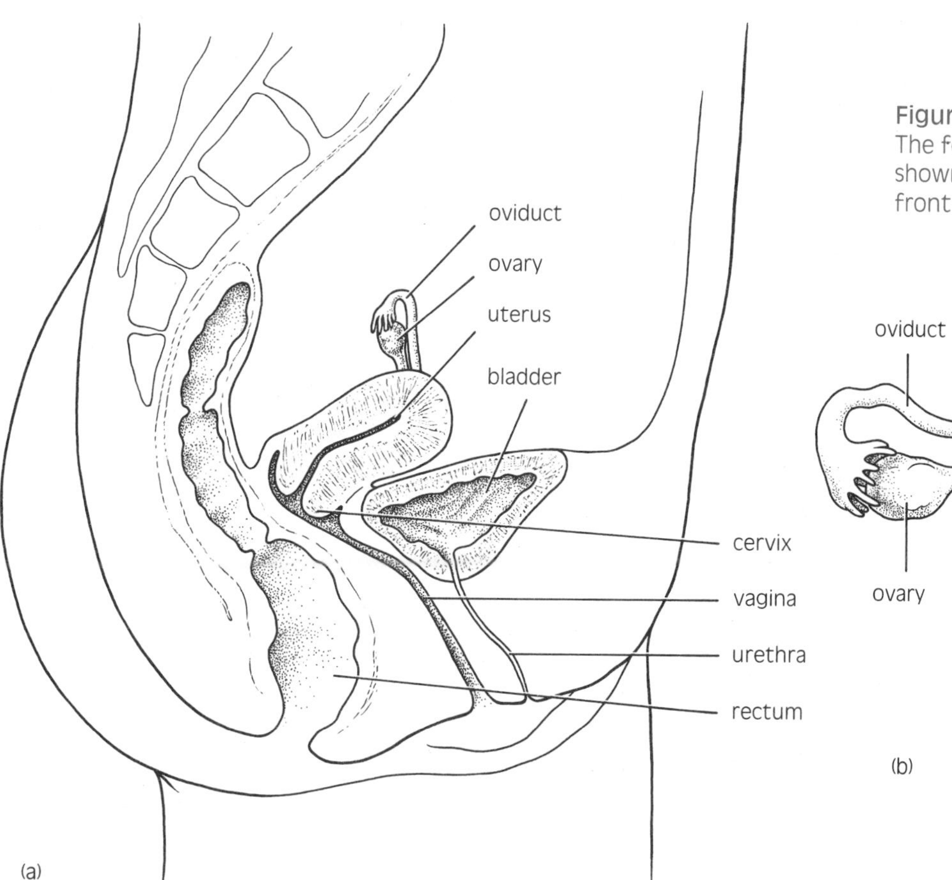

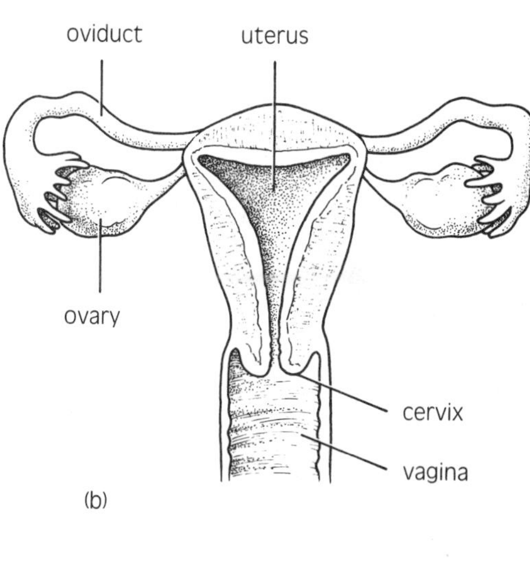

Figure 16.25
The female reproductive system shown from the side (a) and the front (b).

The development of an ovum involves a series of events triggered by changing hormone levels. These events occur during what is called the **menstrual cycle**. The menstrual cycle lasts an average of 28 days. However, its length can vary widely from person to person and even from cycle to cycle within the same individual. The most externally obvious event of this cycle is a woman's menstrual period — four to five days when blood flows from the vagina. Figure 16.26 describes the events that take place during the menstrual cycle.

Figure 16.26
The menstrual cycle.

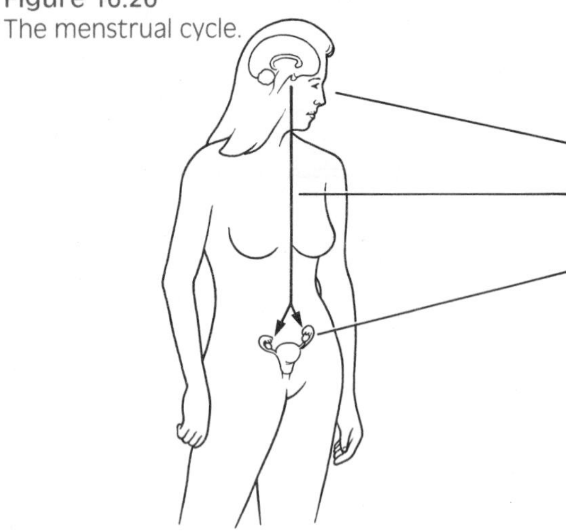

1. The pituitary gland under the brain sends out hormones.
2. Blood carries the hormones throughout the body. They affect certain organs and cells.
3. One of these hormones causes several immature ova and their follicles to grow. The growing follicles secrete the hormone estrogen.

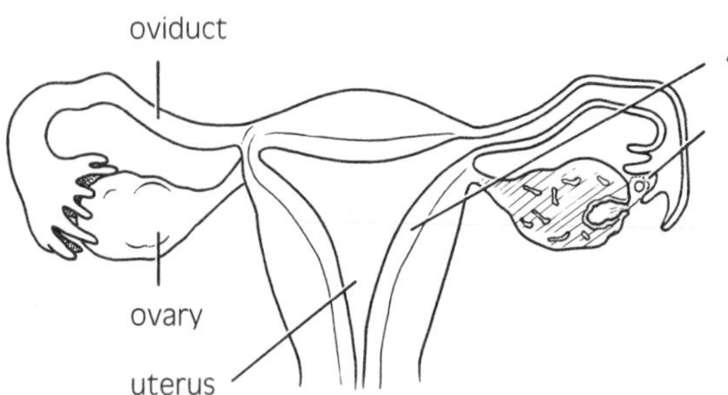

4. Estrogen causes the lining of the uterus to thicken.
5. One ovum and its follicle develop more than the others. This ovum is released from the follicle and ovary in an event called ovulation.

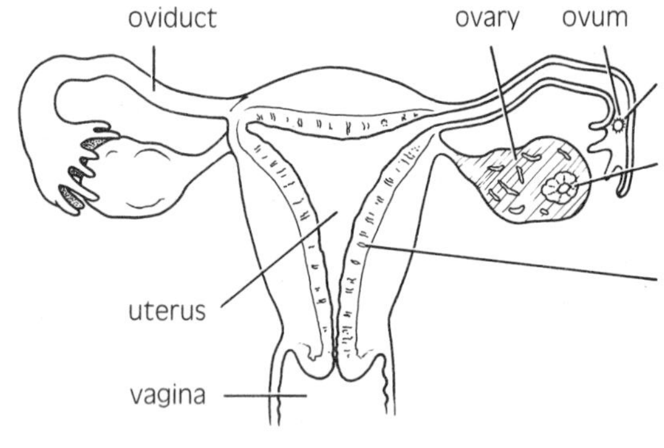

6. The released ovum enters the abdominal cavity and is drawn into the oviduct.
7. The empty follicle left in the ovary forms a mass of cells called the corpus luteum, which secretes the hormones estrogen and progesterone.
8. Progesterone causes the thickened walls of the uterus to prepare for an embryo.

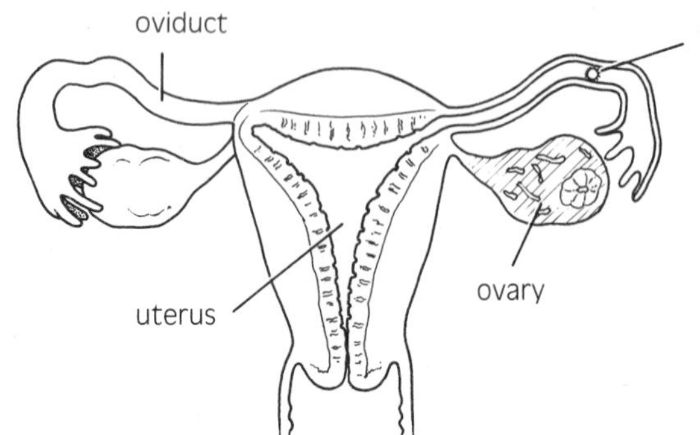

9. The ovum travels along the oviduct. One of two things happens:

(a)
- Ovum is not fertilized by a sperm and dies 24–48 hours after its release from ovary.
- Corpus luteum breaks down, so secretion of estrogen and progesterone decreases.
- Lining of uterus is partly absorbed, and the rest is released during the menstrual period.
- Menstrual cycle repeats.

OR

(b)
- The ovum is fertilized by a sperm in upper oviduct.
- Pregnancy results.
- Menstrual cycle stops during pregnancy.

PREGNANCY

During intercourse, about 3-4 mL of semen is delivered through the urethra from the penis into the vagina. Approximately 150 million to 450 million sperm are contained in this amount of semen. They begin to travel through the female reproductive system in an attempt to fertilize an ovum. Because sperm can survive for 25 to 72 hours, many of them die along the way. In some cases sperm can survive up to 5 days.

If fertilization does take place, it usually occurs in the upper part of the oviduct, where the surviving sperm meet the ovum. There one sperm penetrates and enters the ovum. Although only one sperm is needed to fertilize the ovum, millions are produced. Many sperm "lose their way" in the female reproductive system. Some die as a result of acidic conditions inside the female's body. Others are killed by the woman's white blood cells, which perceive sperm as a foreign substance. Usually only a few thousand make it to the oviduct. There, many sperm are needed to produce an enzyme that helps break down the covering of cells around the ovum. Once one sperm has penetrated the covering, a chemical reaction prevents the entrance of others.

The fertilized ovum, or zygote, moves down the oviduct towards the uterus. By the time it reaches the womb, about five days later, the zygote has divided a number of times and is an embryo consisting of about 100 cells. If all goes well, the embryo becomes implanted in the lining of the uterus, which nourishes and protects it for a short period of time.

Soon a large structure called the placenta begins to develop between the embryo and its mother. The placenta is composed partly of membranes that surround the embryo and partly of tissue from the wall of the uterus. In the placenta, the blood vessels of the embryo and those of the mother lie very close together (but not touching) (Figure 16.27).

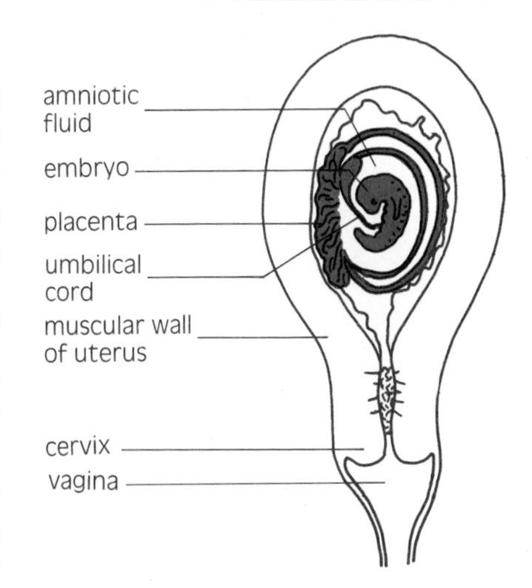

Figure 16.27
In the placenta, the blood vessels of the mother and the embryo lie side by side. Nutrients absorbed into the embryo's vessels are carried to the embryo by the umbilical cord. The embryo is surrounded by a cushioning layer of fluid called the amniotic fluid.

Gases and nutrients are exchanged by diffusion through the placenta. The mother's blood supplies the embryo with materials such as oxygen, energy-rich molecules, and the raw materials for building cells. Wastes produced by chemical reactions in the cells of the embryo diffuse into the mother's blood and are dealt with by her body. Undesirable substances, including mutagens such as drugs and alcohol, can also cross the placenta from mother to embryo. The mother's diet and lifestyle, therefore, have substantial effects on the health of the embryo.

The placenta secretes a hormone that acts on the corpus luteum, ensuring that it will continue to exist and secrete progesterone. Progesterone maintains the lining of the uterus. The menstrual cycle thus stops during pregnancy. Ovulation does not occur again until after the baby's birth.

The placenta is vital to the continued growth of the embryo. After 8 to 10 weeks, many changes have occurred (Figure 16.28). From this point on, the developing offspring is referred to as a **fetus**. Growth and development of the fetus continue within the uterus for approximately seven more months. From fertilization to birth, the development of a human offspring takes about nine months.

DID YOU KNOW?

An ovum can survive in the oviduct for 24 to 48 hours after ovulation. Sperm can live for up to 5 days. Therefore, fertilization (and thus pregnancy) can result from intercourse that occurs up to three days before and two days after ovulation. The exact timing of ovulation is difficult to predict. There are some clues, such as a slight increase in body temperature.

Figure 16.28
Some of the many events that occur during the first 8 to 10 weeks in the life of the developing embryo.

time (weeks)	size (mm)	appearance (3 × actual size)
0	0.2	single-celled zygote
1	0.2	ball of 100 cells implants itself in the uterus
2	1.5	ball folds to form inner layer of tissue; umbilical cord has formed
3	2.5	heart is beating; segments form along back; gill pouches appear; a fold is forming the spinal cord and brain
5	5.0	embryo is not yet recognizably human
8–10	30.0	now the embryo is called a fetus; human features are appearing; internal organs are beginning to function; first bone cells are laid down

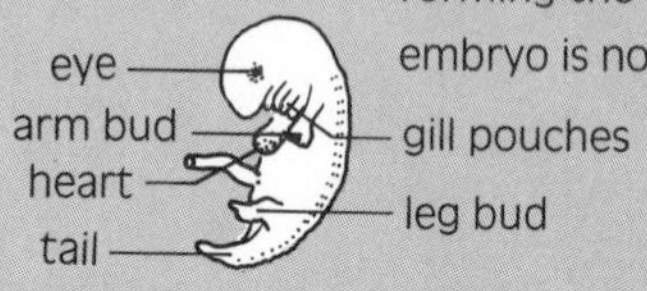

eye
arm bud
heart
tail
gill pouches
leg bud
artery 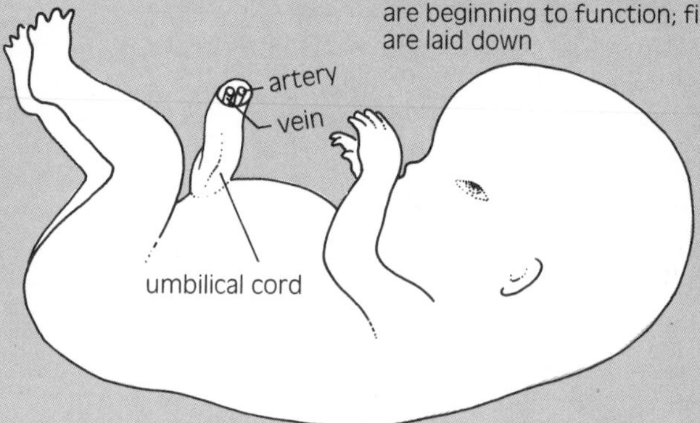
vein
umbilical cord

BIRTH

As you have seen, much of the reproductive process depends on the action of hormones. So does birth. Different hormones secreted by the mother, the fetus, and the placenta trigger and then assist in the process of birth. These hormones:

- loosen the connections between the bones of the pelvis
- help to open the cervix, a ring of muscular tissue that surrounds the opening to the uterus
- cause the muscular walls of the uterus to contract

When the cervix is opened sufficiently, strong rhythmic contractions of the uterus force the baby through the vagina, and eventually out of the mother's body. The vagina, a muscular tube, is very elastic. It can expand to receive a penis during intercourse, as well as to permit the passage of a baby during birth.

After birth, it is often apparent—to some at least—that the baby has certain features, such as hair colour or nose shape, that resemble those of its father, and others that are more like its mother's. How it acquires, or inherits, these characteristics is the subject of the next chapter.

ACTIVITY 16F *Investigating Reproduction*

Prepare a report that can be shared in some way—such as verbally or in the form of a poster—with your class. Select one of the following topics:

- drugs and the development of the embryo and fetus
- the dangers of childbirth for the mother
- the effect of alcohol on the developing embryo and fetus
- causes of miscarriage
- birth by Caesarean section
- the effects of smoking on the developing embryo and fetus
- causes of infertility
- "natural" childbirth
- the mother's diet and its effect on the developing embryo and fetus
- the effect of the mother's age on the baby's health

REVIEW 16.4

1. How do primary sexual characteristics differ from secondary sexual characteristics?
2. (a) Define the term "puberty."
   (b) When does puberty occur in male and in female humans?
3. (a) What is seminal fluid?
   (b) What is its function?
4. (a) In what ways are the human male and female reproductive systems different? List at least three differences.
   (b) In what ways are they similar? List at least three similarities.
5. (a) What events lead up to a menstrual period?
   (b) What event prevents the occurrence of a menstrual period? How does it do this?
6. (a) What is the placenta?
   (b) Why is it so important to the developing embryo and fetus?
7. Where does fertilization occur in humans?

CHAPTER • REVIEW

Key Ideas

- Cells reproduce by dividing, which involves both mitosis and cytokinesis. Cell division enables organisms to increase in size as well as maintain and repair body tissues.
- Mitosis is part of the process of cell division. It divides the chromosomes into two equal and identical halves and ensures that the genetic material in each half is identical to that of the parent cell. The process is divided into four stages: prophase, metaphase, anaphase, and telophase.
- Asexual reproduction involves only one parent, produces offspring that are identical to the parent, and usually doesn't involve specialized reproductive cells. Binary fission, budding, fragmentation, spore formation, and vegetative reproduction are five methods of asexual reproduction.
- Sexual reproduction requires two parents, each of which contributes chromosomes to the offspring. Sexually produced offspring therefore differ from their parents and from each other.
- Meiosis is the process that produces cells with half the number of chromosomes of the parent cell. This process is used to form reproductive cells called gametes.
- In sexual reproduction, male and female haploid gametes are formed. These fuse to form a diploid zygote.
- There are many different methods of sexual reproduction. Each type of organism has its own specialized reproductive structures and behaviours, which are suited to the environment in which it lives.
- Human males and females have different primary and secondary sexual characteristics. These characteristics mature or appear at puberty, as a result of the action of hormones.
- The human male reproductive system is made up of various structures that produce and transport sperm and seminal fluid, and deposit them in the body of the female during intercourse.

- The human female reproductive system produces and transports ova. Some female structures provide protection for a developing embryo.
- If fertilization occurs, the human offspring (first called an embryo, then a fetus) develops in the female's uterus for approximately nine months before birth. The placenta and other structures are vital to the offspring's proper development.

Vocabulary

cell reproduction
cell cycle
interphase
centromere
chromatid
cell division
mitosis
cytokinesis
prophase
metaphase
anaphase
telophase
cancer
mutagen
asexual reproduction
binary fission
budding
spore formation
fragmentation
vegetative reproduction
clone
sexual reproduction
homologous
diploid
gamete
haploid
meiosis
spermatozoon (sperm)
ovum
fertilization
zygote
embryo
puberty
ovaries
testes
primary sexual characteristics
secondary sexual characteristics
menstrual cycle
fetus

1. Construct two concept maps, one for the words related to sexual reproduction and one for the words related to asexual reproduction. After you have done this, see if you can link the two into one concept map.
2. Make up a crossword puzzle using at least 12 words on this list.

Connections

1. What is one of the major differences between an asexually produced spore and a gamete?
2. Describe how mitosis and other processes ensure that each daughter cell receives a copy of its parent's genes.
3. There is a saying, "Don't put all your eggs in one basket." What does it mean? How does it relate to the difference between asexual and sexual reproduction?
4. Which of the two parts of meiosis—meiosis I or meiosis II—is most like mitosis? Why?
5. (a) What role does meiosis play in human reproduction?
   (b) At what stage in the process of human reproduction is mitosis most important?
6. (a) What is the difference between the events that occur during anaphase I and anaphase II of meiosis?
   (b) What is the difference between the events that occur during anaphase of mitosis and anaphase I of meiosis?
   (c) How do the differences referred to in (b) affect the number of chromosomes present in cells at the end of mitosis or meiosis?
7. Compare two imaginary kinds of sexually reproducing organisms. One lives on land and rarely meets other organisms of its own kind, with whom it might reproduce. The other lives in lakes and often meets other organisms with whom it can mate. Which of these two kinds of organisms is more likely to use an internal method of fertilization, and which is likely to employ external fertilization? For which type of organism might it be useful for an individual to produce both male and female gametes? For which is this ability not particularly advantageous? Explain your answers to these questions.

Explorations

1. Imagine that you are part of an expedition to a planet in another solar system that has been found to be different from Earth, but inhabitable by organisms from Earth. Your job is to determine which of Earth's organisms would be most successful on this planet. Which organisms would you choose—those that reproduce sexually or asexually? Those that have internal or external fertilization? What factors will determine your choices? Besides thinking about the organisms' methods of reproduction, what other factors will you have to consider when deciding what organisms to take? Draw up a plan that describes what factors will be important to your decisions. You might want to include some specific examples of organisms you will take and

give your reasons for taking them.

2. Do research to discover how twins occur. There are actually two types of twins—identical and non-identical (fraternal) twins (Figure 16.29). What is the difference in origin between the two types? What effect does this difference have on the extent of the similarity between the twins? Identical twins are often referred to as monozygotic twins, whereas fraternal twins are called dizygotic twins. Why are these names appropriate?

3. For each sexually reproducing organism to replace itself, each pair of adults should have exactly two offspring in their lifetime. Individuals of most kinds of organisms actually produce many more offspring than this over their lives, yet populations of these organisms do not usually increase in size. What happens to the surplus offspring?

4. Why do you think that males produce so many sperm while females produce only one mature ovum per month—and only about 400 over their entire lives?

5. There are several methods available for preventing pregnancy—that is, methods of birth control. Select one of the methods listed below and do research to determine (a) what is involved in using the method, (b) why it works, and (c) its effectiveness. Your school nurse, doctor, local public health unit, and library will all have information. Present your findings to the class in the form of a poster or a pamphlet, or in some other way. Some commonly used methods of birth control are contraceptive pills, condoms, spermicides, intrauterine devices (IUDs), the rhythm method, and diaphragms.

Reflections

1. Refer to the three-column notes you made in Activity 16A. Are you now able to answer some of the questions for which you previously left the second column blank? Fill in the third column with as many answers as you can. Have you changed your mind about any of the answers you wrote in the second column as a result of reading this chapter? If so, what changed your mind? Have you thought of any new questions that you would like answered?

2. "Which came first, the chicken or the egg?" What do you think? Why?

(a)

(b)

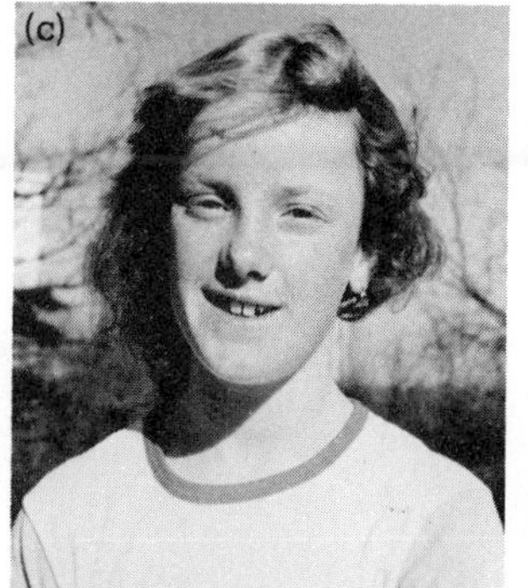
(c)

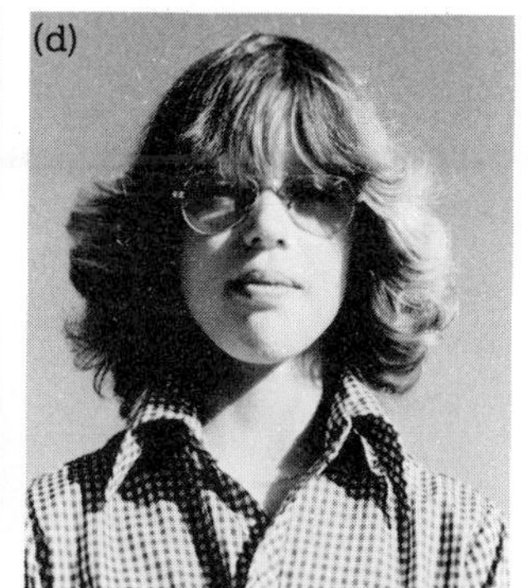
(d)

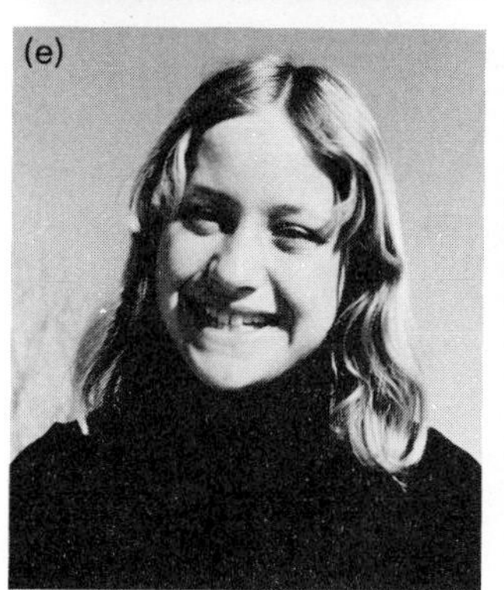
(e)

(f)

(g)

(h)

Figure 16.29
Can you tell which of these sets of twins are identical and which are fraternal?

CHAPTER

17

Heredity

The three women in this photograph resemble each other in several obvious ways—much as sisters or brothers share certain characteristics. Yet, until shortly before this picture was taken, these women had never met. How can their similarities be explained?

In fact, these three women *are* sisters. Given up for adoption shortly after birth, each of them was raised by a separate set of parents and grew up in a different family environment.

Sue Miller (left) did not know she had any natural sisters until her sister-in-law (the wife of her adoptive brother) saw a sales clerk who looked remarkably like Miller. That sales clerk was Miller's older sister Leanne Hamm (centre), who had been actively searching for her birth family. Their story in the *Vancouver Sun* brought them in touch with the third sister, Leilani Ozzard (right). Their shared features—among them big smiles, high cheekbones, and sparse eyelashes—convinced them of their relationship. They also all knew that a form of the inherited disease muscular dystrophy ran in their birth family.

How do characteristics such as eye colour or nose shape get passed on from parents to offspring? What determines whether a child will have curly or straight hair or will be male or female? These and other related questions are among those asked by individuals who study heredity—the subject of this chapter.

ACTIVITY 17A An Inherited Trait

Many people have hair on the middle joint of one or more of their fingers (Figure 17.1). Others lack hair on this joint. The presence or absence of hair in this location is an inherited characteristic, or **trait**. In other words, it is a trait that is passed from parents to offspring. This activity will give you an opportunity to collect data on this particular inherited characteristic.

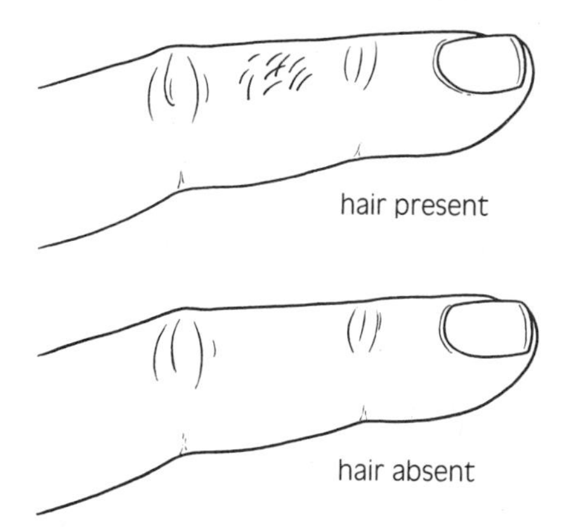

Figure 17.1
Many individuals have hair on the middle joint of one or more of their fingers. Other people lack this hair.

PART I

PROCEDURE

1. The "finger-hair" trait occurs in two forms: present or absent. Look to see which form of the trait you possess.
2. Make a class table on the blackboard or a large piece of paper. Record the form of this trait possessed by each member of the class.
3. Calculate the fraction or percentage of the class that has hair on the middle joint of their fingers and the fraction that does not.

DISCUSSION

1. What do you think the presence of hair on the fingers involves? In other words, what do you think you inherited?
2. Within your class, which is more common: the presence or absence of hair on the middle finger joint?

PART II

PROCEDURE

1. In a table of your own design, record whether you have or lack hair on the middle joint of one or more of your fingers. The table should also allow you to record data on the form of this trait possessed by any other members of your family that you are able to ask. In addition, it should allow you to record the sex and relationship to you of each family member. Alternatively, you might decide to study some other family that you know: a neighbour's family, for example, or that of a friend in another class or school. In this case, record the sex, relationships, and trait forms of all the members of this family.
2. For all the individuals in the family whom you can visit or reach easily by phone, record whether or not they have hair on the middle joint of one or more of their fingers. Although it is best to include many individuals in your table, even a small table with only two or three family members in it can provide useful information. When collecting this information, be sure that you are testing the same characteristic in every person. Each individual should examine all of his or her fingers for the presence or absence of hair on the middle joints. Even if only one finger has hair there, that person has the "hair-present" form of this trait.

DISCUSSION

1. Can you find any pattern in the forms of this trait inherited by the family members you studied? If so, describe the pattern.

PART III

PROCEDURE

1. Look at Figure 17.2. This diagram shows family relationships in a standard way.

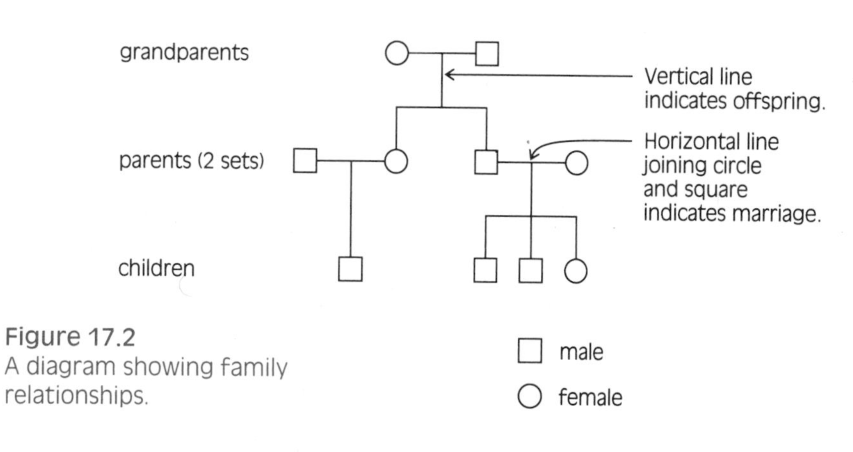

Figure 17.2
A diagram showing family relationships.

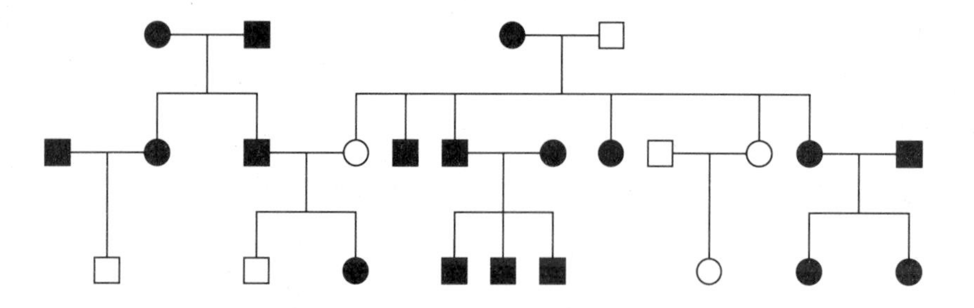

Figure 17.3
One family's pedigree for the finger-hair trait. The filled-in shapes represent individuals who have hair on the middle joint of one or more of their fingers. The open shapes represent those who do not.

2. In Figure 17.3, more individuals are included in the diagram, and it has been completed to show the form of the finger-hair trait possessed by all members of a particular family. Such a diagram is called a pedigree. Pedigrees show which members of a family have one form of a particular trait, and which members possess an alternative form.
3. Use the data in the table you made in Part II to draw a pedigree of the finger-hair trait for the family you studied.
4. Make the family pedigrees from step 3 available, so that everyone can examine them. This will provide you with a large amount of information, which makes it easier to form conclusions. Compare your family pedigree with those drawn by other students and with that shown in Figure 17.3. See if you can detect any general patterns in the inheritance of the finger-hair trait.

DISCUSSION

1. Based on the pedigrees made by members of your class,
   (a) Do both parents need to have hair on the middle joints of their fingers for any of their offspring to have the hair-present form of the trait?
   (b) If neither parent has hair on his or her middle joints, is it possible for any of their children to have it?
2. Is it possible for some, but not all, of the children in a family to have hair present on the middle joints of their fingers?
3. If both parents have the hair-present form of the trait, does this mean that *all* their children must have hair present? Support your answer by referring to specific pedigrees you have examined.
4. Does the hair-present form of the trait appear to be more common in one sex than in the other?

PART IV

The results of this activity may have raised a number of ideas and questions in your mind. Some of them probably relate to how characteristics are actually inherited. In your learning journal, write down your ideas about inheritance. Make a list of any questions you have. You might want to predict the answers to some of these questions.

17.1 The Beginnings of Genetics

Characteristics such as finger hair, hair colour, and eye colour, along with many others, are said to be inherited. This means that they are the result of **heredity** — the passing on of characteristics from parents to offspring. The field of science that deals with heredity is called **genetics**.

Throughout history, people have recognized that offspring and their parents often resemble each other in many ways. In other words, the influence of heredity has long been recognized. People used this knowledge, long before they understood how heredity worked, to produce preferred types of plants and domestic animals by allowing only certain of them to reproduce. This practice is termed **selective breeding**. From a group of cows, for example, farmers would select for breeding only those that gave the most milk. This practice greatly increased the percentage of offspring that were also good milk-producers, but it did not guarantee this outcome. Sometimes poor milk-producers were born.

How characteristics were passed from one generation to the next, however, was unknown. Scientists and agricultural workers suggested various hypotheses, but none seemed to explain all the observations made about heredity. In addition, few of these researchers actually tested their hypotheses by collecting data about inherited characteristics. They did not always carefully measure the characteristics they were studying, nor did they carefully record the numbers of offspring that did or did not display a parental feature.

Such record keeping turned out to be vital for understanding the mechanism of heredity. The first person to conduct carefully planned experiments on heredity, and to accurately record his results, was Gregor Mendel. His precision, and his insight in interpreting the results of his experiments, allowed him to take a major step forward in explaining how heredity works. For this, he is often called "the father of genetics."

DID YOU KNOW?

Before heredity was understood, people had some unusual ideas about the origin of certain species. The giraffe was thought to have resulted from the mating of a camel and a leopard. (The camel, of course, was supposed to have provided the long neck, and the leopard, the spots.) The ostrich was thought to have originated from the mating of a camel with a sparrow!

MENDEL'S EXPERIMENTS

Gregor Mendel (1822–1884) spent most of his adult life as a monk in an Austrian monastery (Figure 17.4). During this time, he did many experiments, exploring a variety of biological problems. His research on heredity took him over 10 years to complete.

Mendel's writings suggest that before starting his research, he had already thought about how heredity might work. His ideas included the following hypotheses:

- There are elements that are passed on by a parent to its offspring.
- Different elements control different traits.
- These elements remain unchanged during the life of any organism. They are passed on to new offspring in the same form in which they have been inherited from parents.
- The elements are passed on in the gametes (the reproductive cells) of organisms.

After formulating these hypotheses, Mendel set out to carefully test his ideas by doing experiments with pea plants. Garden peas were an excellent choice for Mendel, for several reasons (Table 17.1). One useful feature of peas is that they possess a number of inherited traits that come in only two forms. For example, for the trait "seed colour," Mendel's peas showed only a green form and a yellow form. For his experiments, Mendel used peas with two contrasting forms of seven different traits:

- seed shape (smooth, wrinkled)
- seed colour (yellow, green)
- seed coat colour (grey, white)
- pod shape (inflated, constricted)
- pod colour (green, yellow)
- flower position (at sides, at tips)
- stem height (tall, short)

Figure 17.4
Gregor Mendel, an Austrian monk, who is often called "the father of genetics."

Table 17.1 Some Features of Garden Peas and Their Usefulness in Mendel's Research

Features of peas	Usefulness to Mendel
Peas produce many offspring in a short period of time.	He could study many generations of peas.
Individual pea plants produce both male gametes (pollen) and female gametes (ova), and usually fertilize themselves.	He could use self-fertilizing plants to develop pure lines for various traits. He could later prevent self-fertilization, and breed two pure lines with each other to see what the offspring would inherit.
Peas have several traits for which there are only two forms.	He could study simple traits with only two forms, rather than more complex traits with many forms.

Figure 17.5
In any one group (population) of humans you will find a range of heights, rather than the two heights for peas that Mendel studied. In humans, height is a continuous trait.

Many traits in people and other organisms come in more than two forms. Characteristics with many more than two forms are called continuous traits. They are called "continuous" because, in a large group of organisms, a whole range of these forms can be observed. Hair colour and eye colour in humans, for example, are continuous traits (Figure 17.5).

Before beginning his experiments, Mendel allowed plants showing each of the trait forms to fertilize themselves for several generations. For example, he let plants with green seeds fertilize themselves until he had a group of plants that *always* produced offspring with green seeds. (Such a group of plants is called a pure line for that form of the characteristic.) Similarly, he let plants with yellow seeds self-fertilize until he had a pure line for that form of the seed-colour trait. He continued doing this until he had obtained pure lines for each form of the seven traits he was interested in. This process took him several years to complete.

Then, Mendel began breeding (crossing) the different pure-breeding pea plants with each other. For example, he crossed plants from his pure line of green-seeded peas with pure yellow-seeded plants. In breeding experiments like Mendel's, the term **hybridization** is used to describe the crossing of two individuals, each with a different form of the same trait. Mendel's plants were all members of the same species (type) of organism — they were all garden peas. However, they made up several different varieties of this species (each pure line for a particular form of a trait is called a variety). Crossing, or hybridizing, these different varieties produces offspring that look different from one or both parent plants. These offspring are called hybrids.

Mendel's first experiments involved breeding varieties that differed in one characteristic only. Such crossing experiments produced plants that were hybrid with respect to one trait only. Let us examine the results of one of these experiments in more detail.

DID YOU KNOW?

Mendel used the word "hybridization" to refer to the crossing of peas that were of different varieties, but were members of the same species (type) of pea. Today, scientists more commonly use "hybridization" to refer to the mating of individuals in two different species. For example, dogs and wolves are different species. Usually, dogs mate only with other dogs, and wolves breed only with wolves. However, members of these two species have been known to breed with each other — to hybridize. The resulting hybrids are not quite the same as either of the two parent species, but they show some of the characteristics of each.

HEIGHT IN PEA PLANTS

One of the traits Mendel examined was height of the pea plant stem. He crossed pure-breeding short plants with pure-breeding tall plants to produce hybrid offspring. The first generation of offspring in a crossing experiment — in this case, the hybrids — is called the first filial or F_1 generation. Mendel found that these F_1 plants were all tall; the short form of the trait seemed to have disappeared entirely.

Mendel then allowed the tall F_1 hybrid plants to fertilize themselves. He found that unlike the tall pure-bred parents, the tall *hybrid* F_1 plants did not produce only tall offspring. In other words, the offspring of the next (F_2) generation were not all exactly like their parents (the F_1). Although most of the F_2 plants were tall, some were short. There were approximately three tall plants for each short plant in the F_2 generation.

ACTIVITY 17B *Drawing Conclusions*

On the next page, Table 17.2 gives you the results of Mendel's experiments with pea plants. Look at the results for tall and short stems (trait 7). Imagine that you are Mendel's colleague. In your learning journal, write down any conclusions you might draw from these results and any questions you might have. For example, consider:

- Why does the short form disappear in the F_1?
- Why does it reappear in the F_2?
- There are approximately three tall plants for each short plant in the F_2. Does this 3:1 ratio have any significance?

MENDEL'S CONCLUSIONS

Mendel did experiments with plants showing contrasting forms of all seven traits. In each case, the results were similar (Table 17.2). The F_1 plants all exhibited only one form of the trait under study. Among the F_2 plants, however, only about three-quarters of the offspring showed this form, and the remaining offspring showed the alternative form. These results led Mendel to make a number of important conclusions.

1. His hypothesis about the existence of elements was correct. Passed on by parents, these elements remained unchanged in the F_1, even though some had no visible effect on the plants. This was the only way, he reasoned, that such traits could reappear in the F_2. For example, the element that produced short pea plants must have been passed on from the short parent to the tall F_1 offspring. It must have remained unchanged in these hybrid F_1 plants and been passed on to the F_2 offspring, because the trait controlled by this element was again visible in some of them.

2. The male and female gametes produced by parents with contrasting forms of the trait must each contain different forms of the element. For plant height, one form of the element produces tall plants; the other produces short ones. The F_1 peas would have received one of each of these elements from their parents. Each individual offspring must, therefore, contain a pair of these elements (Figure 17.6).

Table 17.2 The Results of Mendel's Experiments for Seven Traits

Trait no.	Parental lines	F_1 appearance	F_2 appearance	F_2 ratio
1	Round × wrinkled seeds	All round	5474 round; 1850 wrinkled	2.96:1
2	Yellow × green seeds	All yellow	6022 yellow; 2001 green	3.01:1
3	Grey × white seed coat	All grey	705 grey; 224 white	3.15:1
4	Inflated × constricted pods	All inflated	882 inflated; 299 constricted	2.95:1
5	Green × yellow pods	All green	428 green; 152 yellow	2.82:1
6	Flowers at sides × flowers at tips	All at sides	651 at sides; 207 at tips	3.14:1
7	Tall × short stems	All tall	787 tall; 277 short	2.84:1

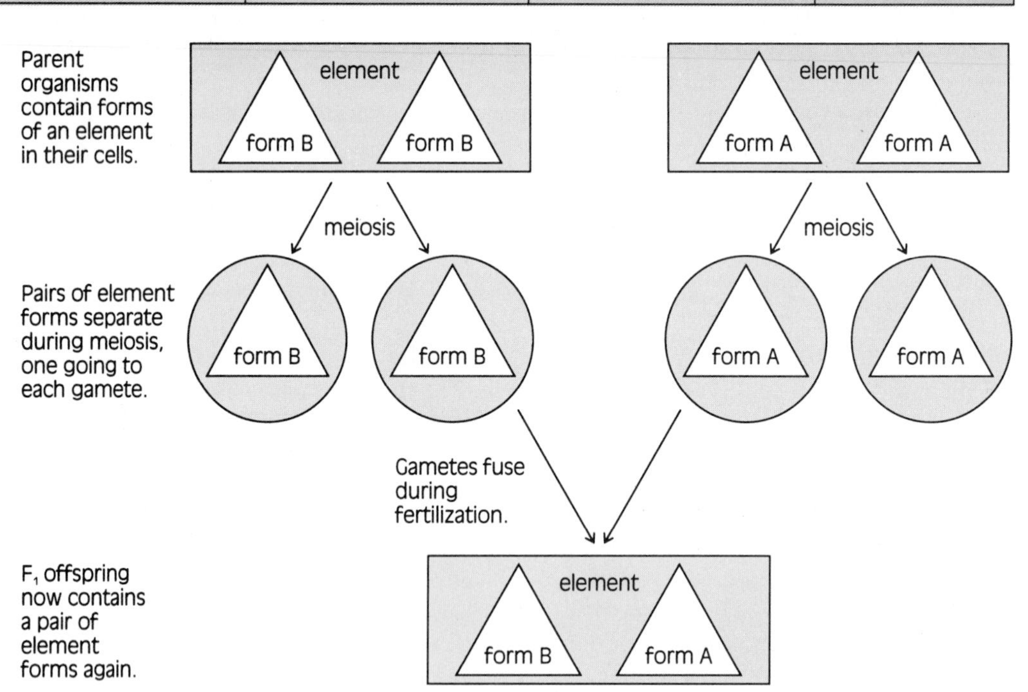

Figure 17.6
A model of Mendel's conclusions about how forms of elements are passed from parents to their offspring. Could you draw a similar diagram for two F_1 offspring being crossed to produce an F_2 offspring? How many different combinations of F_2 offspring could you have?

3. One form of each element is **dominant**, and the other is **recessive**. In peas, the tall form of the element controlling height is dominant. This is shown by the fact that all the F_1 offspring in the height experiment were tall. If both forms of the element for height exist in a single individual, the tall form dominates the short form. The recessive short form remains hidden, and the plant is tall. If a plant is short (as some of Mendel's parent plants were), both the elements it contains must be of the short form. If an element of the tall form were present, the plant would be tall.

In 1866, Mendel published his results. Unfortunately, relatively few people read his report, and, at the time, no one seemed to understand its significance. It was not until 34 years later (16 years after Mendel's death) that the report was rediscovered by geneticists. Other discoveries had enabled these later researchers to better understand Mendel's work and its importance. When the value of Mendel's research was finally recognized, his findings were widely publicized. In fact, a whole branch of genetics came to be named after him. This branch, called Mendelian genetics, studies simple traits that are usually controlled by single "elements."

EXTENSION

Find out why Mendel's results were virtually ignored for such a long time. They could have been of great interest to other scientists of his day. For example, Charles Darwin, who proposed the theory of evolution by natural selection, could have found Mendel's work useful.

ACTIVITY 17C Some Inherited Traits in Humans

Does heredity in other organisms follow the same general pattern as Mendel discovered in peas? This activity will give you the opportunity to investigate this question by studying two more human traits. Like the finger-hair trait, and like the characteristics Mendel examined in peas, each of these traits comes in two different forms.

MATERIALS

PTC (*p*henyl*t*hio*c*arbamide) paper

PROCEDURE

1. Prepare a table to record data on earlobe shape and the ability to taste a chemical called PTC. For each trait, include columns to record your form of the trait, the class total for each form, and the percentage of the class with each form.
2. What is the shape of your earlobes? They may be *free*, and hang down below the point where the bottom of your earlobe joins your head. Alternatively, they may be *attached* (Figure 17.7). Record your answer in your table.
3. Can you taste a chemical called PTC or not? Test this with paper that contains a small amount of this chemical. Tear off about 1 cm^2 of PTC paper. Place it on your tongue for only a few seconds—no longer. Record whether or not you tasted anything.

CAUTION!
PTC is toxic if ingested. Do not chew or swallow the paper. Dispose of it properly in the waste container.

Figure 17.7
Are your earlobes free or attached?

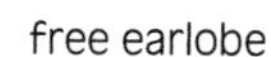

free earlobe attached earlobe

4. Collect results from the rest of the class and calculate the percentage of your class that (a) has free earlobes, (b) has attached earlobes, (c) can taste PTC, and (d) cannot taste PTC.
5. To your table, add the data you collected in Activity 17A on the presence or absence of hair on the fingers.

DISCUSSION

1. (a) In your class, which is the more common form of earlobe shape?
   (b) Which is more common—the ability or inability to taste PTC?
   (c) Are you able to conclude which form of these traits is dominant? Explain your answer.
2. Do you think that the presence of hair on the middle joint of one or more fingers is dominant or recessive? How might using the family pedigree you made in Activity 17A help you answer this question?

REVIEW 17.1

1. (a) In Mendel's height experiment, why were there no short pea plants in the F_1 generation?
   (b) Why did this form of the height trait reappear in some of the F_2 generation?
2. Why did Mendel conclude that the elements he had hypothesized did, in fact, exist?
3. What is the difference between the dominant and recessive forms of elements?
4. Look at Table 17.2. For each trait, which form is dominant and which is recessive? How can you tell?
5. Think about Mendel's "elements." What do you think they are?

17.2 Mendelian Genetics

Mendel's conclusions were surprising because he did not know anything about chromosomes or genes. These structures and their importance to heredity were not discovered until much later. Without knowing that genes existed, Mendel was able to correctly predict many aspects of how they worked. The "elements" that he described were what we now call genes. There is a gene that determines height in peas, just as there is a gene that controls whether or not you will have hair on the middle joints of your fingers.

ALLELES

The different forms of genes—which produce either short or tall peas, for example—are called **alleles**. Each pea plant possesses a pair of alleles for height, one on each of its homologous chromosomes (Figure 17.8). Similarly, every human possesses a pair of alleles for hair on the middle finger joints and another pair for PTC-tasting ability—as well as many additional pairs controlling other inherited traits. The genes controlling some traits, such as coat colour in rabbits, exist in many different forms—they have many different alleles.

Some of these alleles are dominant, while others are recessive. In pea plants, for example, the allele for tallness is always dominant and the allele for shortness is always recessive. If an individual pea plant has both an allele for tallness and an allele for shortness, the dominant (tall) allele will determine the height of the plant. In other words, the allele for tallness is "expressed," while the allele for shortness remains hidden. Similarly, in humans, the allele for being able to taste PTC is dominant, and the allele for not being able to taste this chemical is recessive. The alleles for having hair on the middle finger joints and for having free earlobes are also dominant.

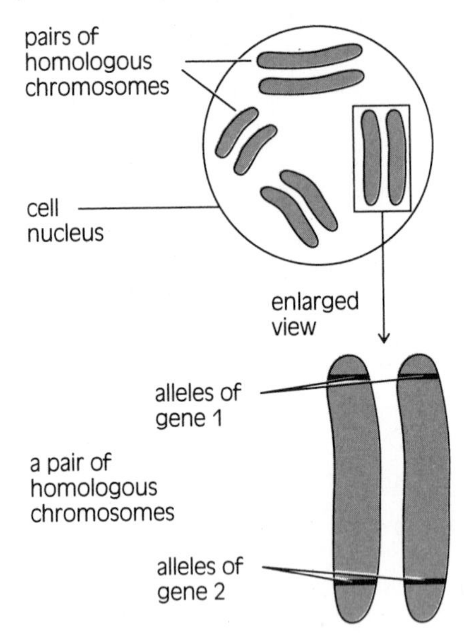

Figure 17.8
In human body cells, almost all genes come in pairs. The members of the pairs are called alleles. For every gene, there are two alleles, one on each homologous chromosome. Although this drawing shows only two genes, thousands of others will also be located on these chromosomes.

It is customary to represent specific genes by a single letter. The gene controlling height in peas, for example, is commonly labelled "t." The dominant (tall) allele of this gene is indicated by the capital letter (T). A lower-case letter (t) is used to indicate the recessive (short) allele. Individuals in which both alleles are of the same type (such as TT or tt) are said to be **homozygous** for that gene. Individuals that have two different alleles of a particular gene (such as Tt) are said to be **heterozygous** for that gene. (The capital letter is first—Tt rather than tT.)

Figure 17.9 shows the two alleles for height possessed by the peas in Mendel's height experiments. During meiosis, the members of each pair of alleles become separated and move into different gametes. Each gamete produced by Mendel's pure tall plants contained a T allele. Each gamete produced by short plants contained a t allele. When these homozygous plants (TT and tt) were crossed, therefore, all the F_1 offspring were heterozygous for height (Tt).

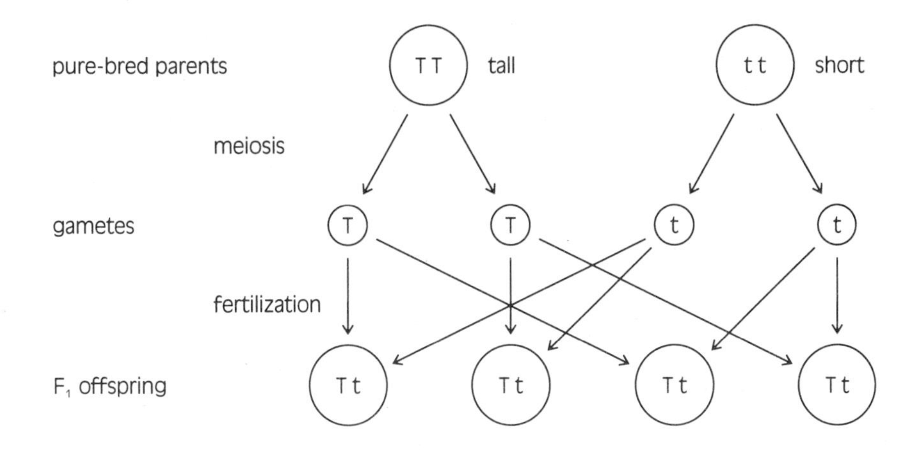

Figure 17.9
Mendel crossed individuals that were pure-bred (homozygous) for different forms of the pea height trait. From these crosses he obtained hybrid (heterozygous) F_1 offspring.

Figure 17.10 shows the results of breeding among peas in the F_1 generation. When the alleles of a heterozygous F_1 plant became separated during meiosis, two different types of gametes were produced. Half of the gametes contained the T (tall) allele, while the other half had the t (short) allele. At fertilization, these gametes then produced F_2 individuals with three different combinations of alleles: TT, tt, and Tt. The particular combination of alleles an organism possesses is called its **genotype**. TT, Tt, and tt are three different genotypes.

DID YOU KNOW?

In Greek, the word *homo* means "alike" and the word *hetero* means "different." Another Greek word, *zygo*, means "a yoke," which is something that joins two things (frequently domestic animals) into a pair so that they can work as a single unit. Thus, the word "homozygous" refers to alleles that are alike and found in a pair. Similarly, "heterozygous" applies to paired alleles that are different.

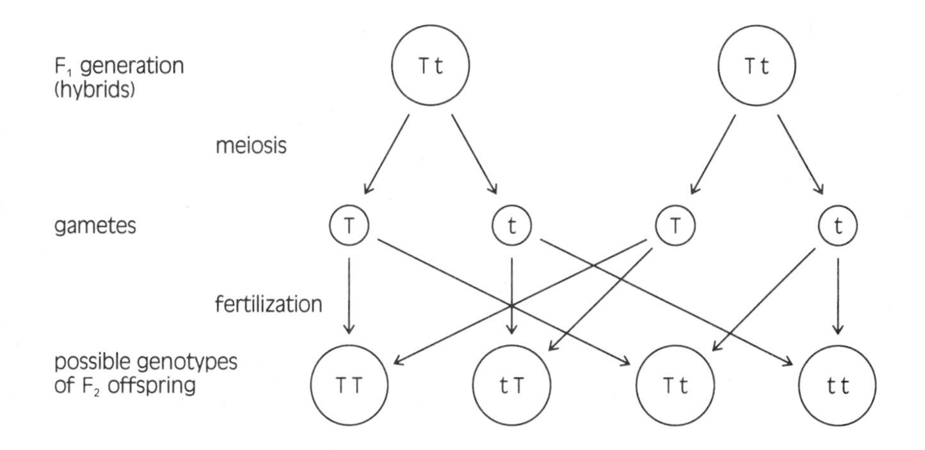

Figure 17.10
The alleles for the height gene present in the F_1 and F_2 offspring of Mendel's experiments. Which F_2 offspring are homozygous and which are heterozygous?

Pea plants with either the TT or Tt genotype turn out to be tall, whereas the tt genotype produces short plants. The outward (physical) appearance of an organism is called its **phenotype**. As was shown by

DID YOU KNOW?

Dominant alleles are not necessarily better, or more common, than recessive alleles. In some cases the possession of the dominant allele is definitely less desirable. For example, in humans, dominant alleles can cause such conditions as extra fingers or toes, a type of dwarfism, and several rare but serious diseases.

Mendel's tall F_1 peas, the phenotype does not necessarily tell you which alleles an organism has for a particular trait. For example, a person might have hair on the middle joint of one or more fingers—that is his or her phenotype for that trait. This person's genotype might be either HH (two hair-present alleles) or Hh (one hair-present allele and one hair-absent allele).

Heredity does not work in isolation from the environment. It is possible, therefore, for organisms that have identical genotypes to exhibit different phenotypes. A pea plant, for example, may have alleles that would ordinarily make it tall. However, if that plant does not receive sufficient light or nourishment when it is growing, it may be short. The environment, as well as heredity, plays a role in the development of all organisms—including people. Genotypes provide the *potential* for particular traits to develop; how a trait *actually* develops often depends on the environment. As a result, there is often much more variation among the phenotypes of organisms than among their genotypes.

REVIEW 17.2

1. (a) What do we now call Mendel's elements?
   (b) What are the different forms of these elements now called?
2. (a) How many alleles of a particular gene are there on a single chromosome?
   (b) How many alleles of a particular gene does a diploid cell possess?
   (c) What type of cells possess only one allele of a particular gene?
3. Where are the members of a pair of alleles located?
4. Define the terms "homozygous" and "heterozygous."
5. (a) What determines an individual's genotype?
   (b) What determines this same individual's phenotype?
6. In peas, a gene called P controls flower colour. The dominant allele (P) produces purple flowers. The recessive allele (p) produces white flowers.
   (a) What is the genotype of a white pea plant?
   (b) What is the phenotype of a plant with the genotype Pp?

17.3 Genetics and Probability

One reason Mendel's work was so successful is that he understood the importance of some basic mathematical principles. He used the laws of chance, or **probability**, in his research. These laws provide rules that predict the likelihood of a particular event occurring.

What happens when you toss a coin? It can land in only one of two ways—with either the head side or the tail side up. There is an equal probability (chance) that either a head or tail will appear. In mathematical terms, the probability of a coin's landing head up is one out of two (written 1/2) or 50 per cent—the same as the probability of its landing tail up. You cannot be certain what the outcome of a particular toss will be, but you can predict the probability of its being either a head or tail.

What if you had tossed a coin 20 times, turning up a head 7 times and a tail 13 times? What is the probability that the next toss will turn up a head? You might be tempted to say that heads need to "catch up," so the probability that a head will appear will be greater this time. However, the probability that the coin will land head up is still 50 per cent. This is because each toss of the coin is independent — what happens on any one toss is not affected by the result of any other toss.

Now consider what happens if you toss two coins at once. Both tosses are independent of one another — how one coin lands does not affect how the other one lands. The probability that the first coin will land head up is 1/2. The probability that the second coin will land this way is also 1/2. You can calculate the probability that both coins will land head side up by multiplying the two probabilities together: the chance that two heads will appear together is $1/2 \times 1/2 = 1/4$.

It is important to remember that this calculation does not tell you that two heads are *certain* to turn up one out of every four times you toss the two coins. It just tells you that such an event is *likely* to occur this often. Although the rules of probability allow you to predict the chance of an event's occurring, there is no certainty that the event will occur this often — or even that it will occur at all.

PROBABILITY AND THE GENOTYPES OF OFFSPRING

When paired chromosomes are pulled apart during meiosis, chance determines which member of the homologous pair will go into each of the gametes. This means that the laws of probability that apply to coin tossing can also be used to determine the likelihood that an offspring will have a particular genotype. For example, in Mendel's height experiment, heterozygous F_1 plants (Tt) were allowed to fertilize themselves. Figure 17.11 shows the probability for each different genotype in the F_2 offspring.

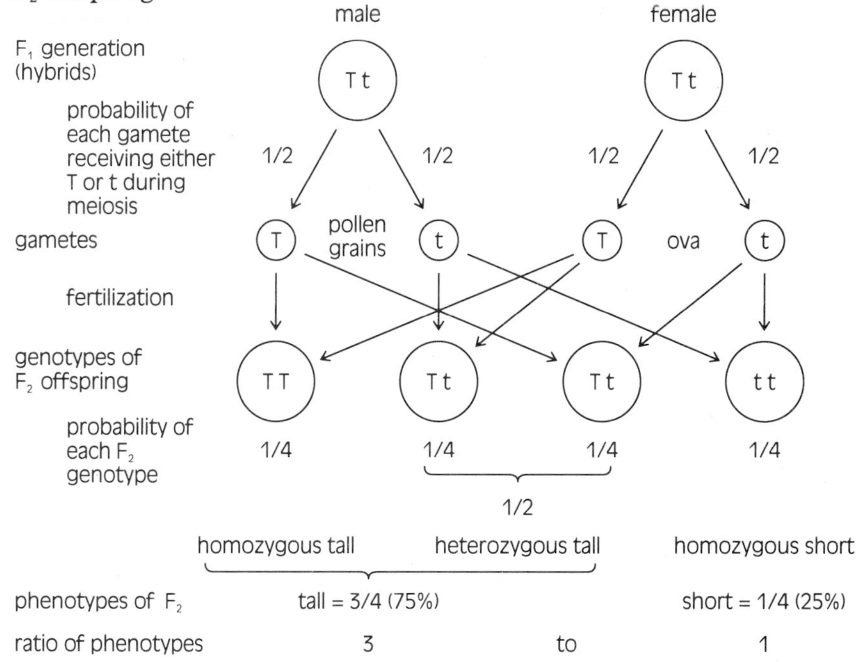

Figure 17.11
The probability of obtaining different genotypes in the F_2 offspring.

Figure 17.12
(a) The Punnett square shows possible offspring genotypes for height in peas. The parents are both heterozygous (Tt). The information in the square is interpreted in (b). The ratio of tall to short among the offspring is predicted to be 3:1.

possible gametes from female parent

possible gametes from male parent

	T	t
T	TT	Tt
t	Tt	tt

(a)

Probability	
TT (tall)—1/4	3/4
Tt (tall)—1/4 x 2 = 1/2	
tt (short)—1/4	

(b)

What is the chance of an offspring's receiving the combination TT? For this to happen, two independent events must occur together: both the pollen grain and the ovum that fuse must contain T alleles. During meiosis, the homologous chromosomes of the F_1 plants separate. Each gamete can receive either a T or a t allele. The probability that an individual pollen grain will receive T (the first independent event) is 1/2. The probability that the ovum it meets contains T (the second independent event) is also 1/2. The chance of a T gamete's meeting a T gamete (for a TT offspring) is thus $1/2 \times 1/2 = 1/4$.

One way to predict the chance of various possible genotypes occurring is to use the **Punnett square**. It shows all the possible combinations of alleles that parents might give to their offspring, as well as the probability of each combination's occurring. A very useful tool for geneticists, the Punnett square was designed by the British biologist R. C. Punnett (1875–1967). Figure 17.12 shows the use of a Punnett square to determine the probability of various genotypes in the offspring obtained by crossing heterozygous (Tt) pea plants with each other.

The shaded boxes show the different genotypes. The chance of an offspring's being TT is one out of four, or 1/4. Two of the four boxes contain the Tt combination, so the probability that an offspring will be Tt is $2 \times 1/4 = 2/4 = 1/2$. The total probability of an offspring's having a tall phenotype is the sum of the probabilities for TT and Tt. It is, therefore, $1/4 + 1/2 = 1/4 + 2/4 = 3/4$ (or 75 per cent). The probability that an offspring will have the tt genotype (and thus the short phenotype) is 1/4 (or 25 per cent).

INSTANT PRACTICE

1. In pea plants, the allele for green pods is dominant (G) and that for yellow pods is recessive (g). A pea plant with green pods that is homozygous for the pod colour trait is crossed with a plant having yellow pods.
   (a) What are the possible gametes produced by the parent plants?
   (b) Draw Punnett squares to find out the following:
   - the possible genotypes and phenotypes of the F_1 offspring
   - the possible genotypes and phenotypes in the F_2 generation

   (c) What is the probability of having each possible genotype and phenotype in the F_2 generation?
2. In humans, the allele that codes for an ability to taste PTC is dominant (T), and the allele that codes for an inability to taste this chemical is recessive (t). A male who is heterozygous for this trait marries a female who cannot taste PTC.
   (a) What are the genotypes of the male and female?
   (b) Draw a Punnett square to show the possible genotypes of their offspring.
   (c) What is the predicted percentage of their offspring that will be able to taste PTC?
   (d) What percentage is predicted to be unable to taste PTC?

SCIENCE IN OUR WORLD

HUNTINGTON'S DISEASE

Dr. Nancy Wexler calls it "the most fascinating detective story in the entire world." She is trying to solve a mystery—what is the location of the allele that is responsible for Huntington's disease?

Huntington's affects the nervous system. Its symptoms include loss of motor control, loss of memory, personality changes, and depression.

Huntington's is a genetic disease caused by a dominant allele. Each child of a parent with the disease has a 50 per cent chance of inheriting it. Symptoms do not appear until a person is 35 to 45 years old, so someone with Huntington's may have had children and unknowingly have passed on the allele.

Wexler, now in her mid-forties, is herself at risk: her mother was diagnosed with Huntington's disease in 1968. That year, her father formed the Hereditary Disease Foundation in the United States. Its goal is to find cures for hereditary illnesses. His daughter is now president.

Wexler's father invited a team of biologists, biochemists, and geneticists to a series of workshops and fired their enthusiasm for seeking a cure for Huntington's disease. The first task was to locate the allele. The scientists had to know which chromosome the allele was on, and what section of the chromosome contained it. Once the allele is

Dr. Nancy Wexler

found, there is a better chance of finding a treatment or cure.

To assist the scientists in their search, Wexler and others collected blood samples from families that showed the disease throughout several generations.

Meanwhile, the scientists had no information about which chromosome contained the Huntington's allele. They did not know its structure or how it works to selectively kill brain cells. Two researchers in Massachusetts, Jim Gusella and David Housman, decided to see if a newly discovered kind of DNA "marker" would lead them to the allele.

Wexler explains: "A marker is a way of holding your place in the chromosome. It's a tiny variant in the DNA itself that has a specific home in a chromosome and is inherited from generation to generation, exactly like a gene." The idea was to analyse the DNA of affected families to see if the disease-causing allele was inherited along with any known marker. If the two show up together, the allele must be located on the same chromosome near the marker.

In 1983, Gusella was lucky with this new genetic mapping technique. On one of his first experiments, he identified a section of chromosome 4 as the likely location. His findings reduced the range of possible locations from 3 billion to about 4 million!

Shortly afterwards, Wexler asked Gusella to join with other scientists, sharing their ideas and materials instead of working independently. Scientists at six labs in the United States and England are now part of a collaborative research group.

However, the allele's location remains elusive. In 1989, after several years of looking at the tip of chromosome 4, new evidence suggested that the allele might be closer to the middle—several million nucleotide bases farther away. Researchers continue to search in both locations on the chromosome.

In the long search for the allele, scientists in the collaborative group have become friends as well as colleagues. As president of the Hereditary Disease Foundation, Wexler has helped them work together through both progress and disappointment.

REVIEW 17.3

1. (a) What does it mean to say that two events are "independent"?
   (b) How can you figure out the probability of two independent events' occurring together?
2. Suppose you are walking through the school and there is an equal probability that each person you meet will be a male or a female. You have passed 30 people, 12 of them male and 18 of them female. What is the probability that the next person you meet will be male?
3. (a) What are the genotypes of a female who has hair on the middle joints of her fingers and is heterozygous for this trait, and a male who lacks hair in this location? (Use H and h.)
   (b) Draw a Punnett square to show all the possible genotypes of their children.
   (c) What is the predicted percentage of these offspring that will not have hair on the middle joints of their fingers?
4. Two black guinea pigs were mated, and over several years produced 38 offspring. Of these, 29 were black and 9 were white. Give the possible genotypes of the parents and offspring.
5. Imagine that a group of sasquatches have been discovered in the mountains of British Columbia. The scientists studying them observe three pairs, each with several offspring. Phenotypes of the pairs and their offspring are given in Table 17.3 for one easily observable trait: the presence or absence of hair on the toes. What do you predict are the genotypes of the parents in each of these three pairs?

Table 17.3 Phenotypes of Three Sasquatch Families

Pair number	Male parent	Female parent	Offspring
1	Hairy toes	Hairy toes	3/4 hairy toes 1/4 smooth toes
2	Hairy toes	Hairy toes	All hairy toes
3	Smooth toes	Hairy toes	1/2 hairy toes 1/2 smooth toes

17.4 More Complicated Patterns of Inheritance

Mendel chose his experimental traits very carefully—or was very lucky in his choice. Only one gene controlled each of the traits he studied. For each of these genes, there were only two alleles—one was completely dominant and the other completely recessive. Some human characteristics, including PTC tasting, are also inherited in this way. Most traits, however, are controlled in more complicated ways: by many genes acting together, or by genes with several different alleles. In addition, the different alleles of some genes have more complex relationships with each other than those Mendel studied.

ACTIVITY 17D *A More Complicated Problem*

Imagine that you are a researcher working on genetic problems at the same time as Mendel. You are studying an imaginary plant called the stardrop. You have done an experiment in which you crossed pure-bred (homozygous) red stardrops with a pure-bred yellow variety of the plant. Instead of getting results like those Mendel obtained for pea plants, you obtain offspring that are all orange. When you cross these orange individuals with each other, you find that the F_2 offspring exhibit three different phenotypes (orange, red, and yellow), instead of the two Mendel found in his experiments.

In a second experiment, you find that when you cross short stardrops with tall ones, you obtain offspring with a whole range of heights: short, tall, and everything in between. In other words, your plants' offspring exhibit many different phenotypes for height, rather than just tall and short. Finally, having studied stardrops for many years, you know that their seeds come in three different colours, rather than the two for pea plants.

You are wondering how to explain the difference between Mendel's observations and your own. Perhaps you can predict some possible explanations. If so, after reading the rest of this section, compare your predictions with the explanations of geneticists. When you have finished the section, write a letter to Mendel proposing some additions to his hypotheses about heredity.

INCOMPLETE DOMINANCE

When members of a pure line of white carnations are crossed with pure red carnations, the resulting hybrid offspring are pink (Figure 17.13a). This situation is a case of **incomplete dominance**—individuals that are heterozygous for flower colour are different from either homozygous parent. Because neither the allele for red nor the allele for white is completely dominant, the hybrid offspring are pink—a phenotype that is intermediate between the phenotypes of the red and white homozygous parents. The red allele is slightly dominant over the white allele, as demonstrated by the fact that the carnations are bright pink, rather than pale pink. Therefore, we use the capital letter (R) for the red allele and the lower case letter (r) for the white one.

When these intermediate individuals (the pink F_1 offspring) are crossed with each other, the flower colours of the F_2 offspring do not fall into the typical 3:1 pattern found by Mendel. Instead of two possible phenotypes, there are three. A ratio of 1 red to 2 pink to 1 white results (Figure 17.13b). These results confirm that the alleles have been sorted into different gametes during meiosis. However, the incomplete dominance produces a different pattern of F_2 phenotypes than the one Mendel found in peas.

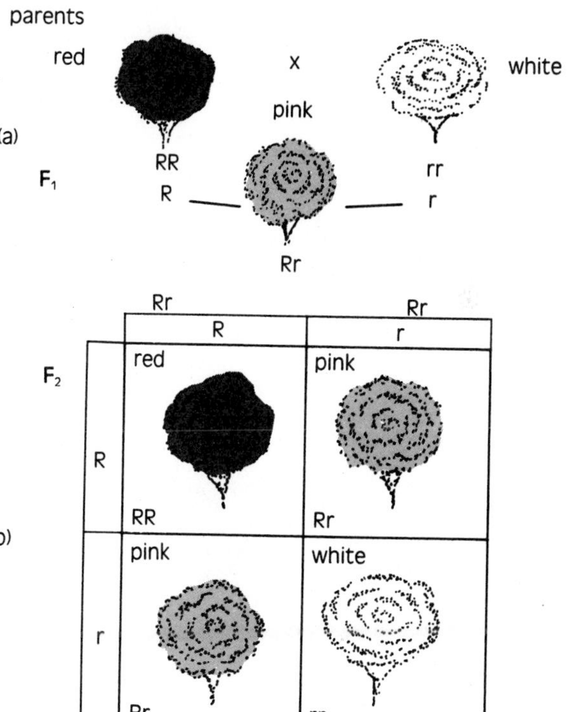

Figure 17.13
The alleles for the trait of flower colour in carnations demonstrate incomplete dominance. (a) The heterozygous F_1 offspring are a different colour from either homozygous parent. (b) A Punnett square shows the F_2 offspring, the result of crossing two pink plants. The phenotypes as well as the genotypes are shown in this square.

MULTIPLE ALLELES

For each of the traits Mendel examined, there were only two different alleles in existence. For example, there are only two alleles for height in pea plants, just as there are only two alleles controlling PTC-tasting ability. The genes that control many other traits, however, have several or many different allele forms. A gene of this type—with more than two alleles—is said to have **multiple alleles**. An *individual* can possess only two of these alleles—one on each chromosome of a homologous pair. Within a *group* of individuals (a population), however, several or many alleles of the same gene may be expressed.

Human blood types provide an example of multiple alleles. Every individual's blood is one of four types: A, B, AB, or O. Each of these is a different phenotype. The type of blood you have is determined by which antigens are found on the surface of your red blood cells:

- type A has antigen A
- type B has antigen B
- type AB has both antigens A and B
- type O has neither antigen

The different blood phenotypes are controlled by a gene that geneticists call I. There are three different alleles of the I gene, termed I^A, I^B, and i. The i allele is recessive to either of the I forms of this gene. The phenotypes produced by all possible combinations of these alleles are shown in Table 17.4.

EXTENSION

The ABO blood groups are very important. Perhaps you know your own blood type, or can figure it out if your parents know theirs. Before donating blood or receiving a blood transfusion, people must have their blood type determined. Write a report describing why this is so. In your report, include answers to the following questions: (1) Why are people with type O blood often referred to as "universal donors"? (2) Which people are "universal recipients"? (3) Is there the same proportion of people in each blood group?

Table 17.4 Genotypes and Phenotypes for the ABO Blood Groups

Genotype	A antigen	B antigen	Phenotype (blood group)
I^AI^A	+	–	A
I^Ai	+	–	A
I^BI^B	–	+	B
I^Bi	–	+	B
I^AI^B	+	+	AB
ii	–	–	O

The I^A and I^B alleles show a type of dominance relationship that has not yet been discussed: **codominance**. Neither the A nor the B allele is dominant over the other; both contribute equally to the phenotype. In incomplete dominance, heterozygous individuals have an intermediate phenotype. In the case of codominance, however, heterozygous individuals show *both* phenotypes. People who have an I^A and an I^B allele, therefore, have both A and B antigens on their red blood cells. They do not have an antigen that is some mixture of the two. As a result, people with the heterozygous genotype I^AI^B have the phenotype (the blood type) AB.

For the blood-type trait, there are only three different alleles. For many genes, however, a larger number of alleles exist in the population. For example, coat colour in rabbits is controlled by a gene called C, which has four different alleles. More than 80 alleles have been discovered for the gene that controls production of a single type of enzyme (called glucose-6-phosphate dehydrogenase) in humans! Another complicating factor is that the multiple alleles for these genes can have all sorts of dominance relationships. This can sometimes make determining the exact genotypes of parents and offspring very challenging!

GENE INTERACTION

Most of the traits discussed so far are completely or primarily controlled by one gene. Often, however, a single trait is controlled by two or more genes acting together. For example, the coat colour of Labrador retrievers (a type of dog) is determined by the action of two genes, referred to as B and E. The B gene has two alleles: the dominant form B, which produces the most commonly seen black coat colour, and the recessive form b. A Labrador retriever with the genotype bb is brown. The second gene, E, also has two alleles. The E allele is dominant over the e allele, as well as over both the alleles of the B gene. E produces a golden coat colour. The e allele is recessive to E, B, and b. The phenotypes associated with all possible combinations of the alleles of these two genes are shown in Table 17.5.

Table 17.5 Coat Colour in Labrador Retrievers

Phenotype	Possible genotypes
Golden coat	Eebb or EeBB or EeBb or EEbb or EEBB or EEBb
Black coat	eeBb or eeBB
Brown coat	eebb

Interaction can also involve many more genes and be far more complex. To some extent, almost all the genes in an organism's body interact with other genes. Just as all cells and organs are dependent on the functioning of other cells and organs, similarly, genes rarely act on their own. They are usually dependent on the functioning of other genes in the body, as well as on the environment. For example, the gene or genes that determine flower colour depend on the genes that control the actual production of flowers. (The flower-colour genes will have nothing to control if these other genes do not produce a flower!)

Traits that show a continuous range of phenotypes, rather than two or three distinct forms, are among those controlled by many interacting genes. Height in humans is an example of such a trait. A person's height results from the activity of many genes acting upon each other and on different parts of the body. The environment too plays an important role. For example, genes that would ordinarily cause cells to reproduce and an organism to grow cannot be effective if the cells are not provided with suitable nutrients.

Gene interaction is a complex subject, which you will learn more about if you study biology in future years. The important point to remember is that such interactions do exist, and they affect the phenotypes of organisms. Various types of dominance relationships and genes with multiple alleles also complicate inheritance. Although Mendel's research provided a breakthrough in the study of genetics, and a good starting point for understanding heredity, the inheritance of some traits is often far more complex.

EXTENSION

Consider the inheritance of two traits at the same time. In peas, yellow seed colour (Y) is dominant over green seed colour (y). Round seed shape (R) is dominant over wrinkled seed shape (r). Suppose that a pea plant that is homozygous for yellow seeds (YY) and round shape (RR) is crossed with a plant that is homozygous for green seeds (yy) and wrinkled shape (rr). The genotypes of the parents, then, are YYRR and yyrr. What are the phenotypes of the F_1 offspring? Remember that these traits are inherited independently (inheritance of one form does not affect inheritance of the other).

Next, cross two of the offspring in this F_1 generation. What is the ratio of phenotypes in the second generation (the F_2)? Take your time with this problem—it's a challenge!

REVIEW 17.4

1. (a) What is the difference between complete and incomplete dominance?
   (b) What is codominance?

2. Imagine that you have two pure lines for each of three species of plants. For each species, one pure line produces only blue flowers and the other pure line produces only red flowers. You do a series of crossing experiments on each species. For species #1, you find that the flowers of the F_1 offspring are all blue. The flowers of the F_1 offspring of species #2 have a mottled red and blue appearance. For species #3, the F_1 offspring all have purple flowers.
   (a) In which of these species are the two alleles for flower colour codominant?
   (b) In which species is one flower-colour allele completely dominant over the other?
   (c) Which species displays incomplete dominance for the flower-colour alleles?

3. (a) Can an individual ever possess more than two alleles for a particular gene? Explain your answer.
   (b) Can there be more than two alleles for a particular gene present in a population? Explain your answer.

4. Give three examples of traits that are controlled by, or depend on the action of, more than one gene.

17.5 Sex Determination in Humans

Refer to Figure 16.17 in the previous chapter, which shows the karyotype of a human female. Notice that one pair of chromosomes is not numbered, unlike all the others. Instead, the chromosomes in this pair are both labelled "X." These unnumbered chromosomes are the sex chromosomes. In Figure 16.17, each is an **X chromosome**. Figure 17.14 shows the karyotype of a human male. Here, the sex chromosomes in the "pair" are different from each other. In addition to one X, there is another, smaller chromosome. It is called the **Y chromosome**. The sex chromosomes are primarily what determine the sex of an individual. Human females have two X chromosomes, whereas males have an X and a Y.

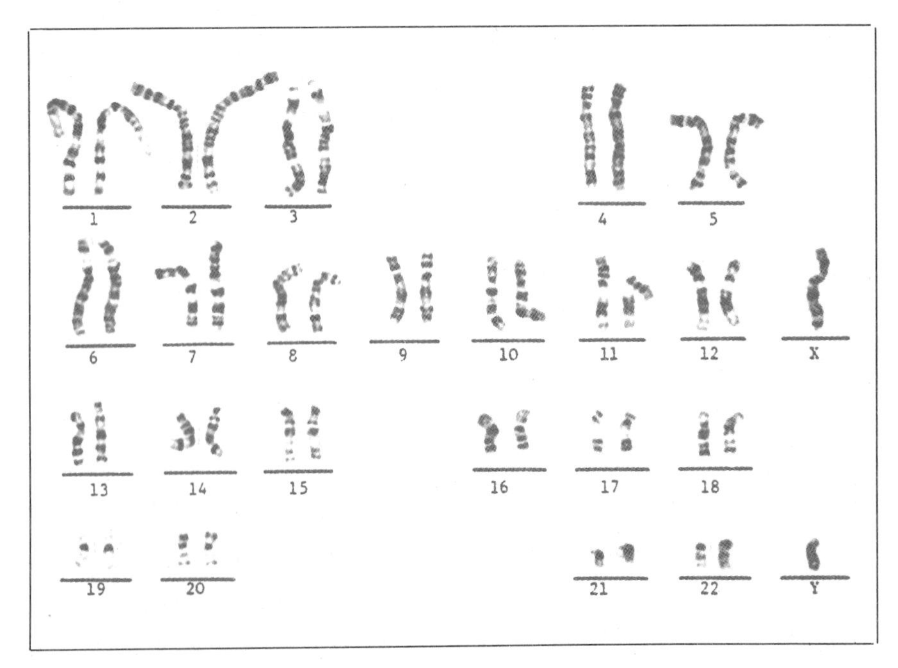

Figure 17.14
The karyotype of a human male.

Like the other chromosomes in the cell, the sex chromosomes move into different gametes during meiosis. All gametes produced by the female (the ova) will contain an X chromosome. As a result, all offspring receive an X chromosome from their mother.

Only half the male's gametes (sperm) will contain an X chromosome; the other half will contain a Y chromosome. The sex of the child thus depends on which sex chromosome it receives from its father. There is a 50 per cent chance that the fertilized egg will receive a second X chromosome from the father. It will then have the genotype XX and will develop into a female. Similarly, there is a 50 per cent chance that the fertilized egg will receive a Y chromosome from the father. If it does, its genotype will be XY and it will develop into a male (Figure 17.15).

Figure 17.15
A Punnett square showing the probability of a child being male or female.

	male	female
parents	XY	XX
gametes	X and Y	X and X

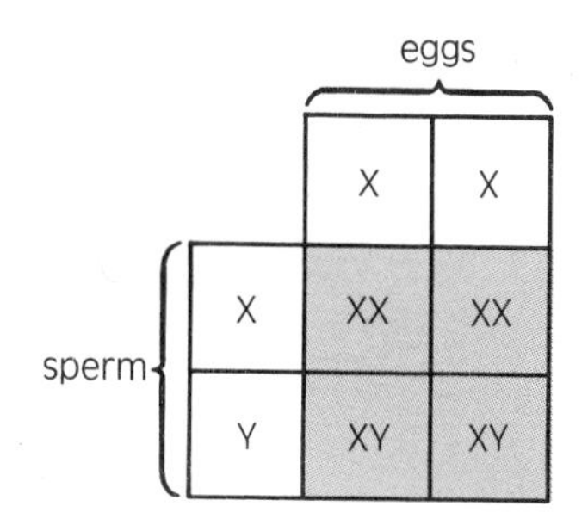

The predicted ratio of offspring is 1/2 male, 1/2 female.

SEX LINKAGE

As well as the genes that determine sex, the X (and possibly also the Y) chromosomes of humans carry genes controlling other traits. Although they behave as a pair during meiosis, the X and Y chromosomes are not completely homologous. The X chromosome is larger and contains many more genes than the Y. As a result, the X and Y chromosomes do not contain pairs of alleles for the same genes.

One gene that exhibits this lack of pairing controls the ability to distinguish the colours red and green. This gene is found only on the X chromosome — there is no matching allele on the Y chromosome. The dominant allele of this gene produces normal colour vision. Individuals who have only the recessive allele are colour-blind for red and green. (To a greater or lesser extent, they cannot distinguish between these two colours.)

Females must inherit two recessive alleles — one on each of their X chromosomes — to exhibit colour-blindness. Although a relatively large number of females may "carry" the recessive allele, few of them will be affected by it. The allele for normal colour vision on the other X chromosome will be dominant. Males, in contrast, have only one X chromosome. All males that receive the recessive allele will therefore, exhibit colour-blindness to some extent. This is because the Y chromosome does not have a paired dominant allele to hide the effect of the recessive allele on the X chromosome (Figure 17.16). As a result, red-green colour-blindness is 16 times more common in males than in females.

Characteristics such as colour-blindness, which are controlled by genes found on the sex chromosomes, are called **sex-linked traits**. All the sex-linked genes that have been carefully studied by geneticists are found on the X chromosome. As a result, sex-linked traits are usually exhibited by males rather than females. Females are more commonly simply carriers of these traits.

Several sex-linked genes control serious diseases. With hemophilia, the blood does not clot properly. Although no cure is available, several forms of successful treatment have been developed. However, if injured, individuals with hemophilia bleed profusely and can sometimes die.

DID YOU KNOW?

Are there genes on the Y chromosome that code for traits other than sex — as there are on the X chromosome? Although no examples of such genes have been confirmed in humans, several have been suggested, including one that produces hairy ear rims. The existence of genes on the Y chromosome controlling traits other than sex has been shown for other types of organisms, however. In a species of fish, for example, a gene on the Y chromosome determines the development of a coloured spot at the base of the tail fin. Could females ever display this spot?

Various other inherited diseases affect humans. Whereas some, like hemophilia, are sex-linked, others are not, and are therefore equally common in both sexes. Often these diseases are quite serious; some are fatal. Treatment can help in come cases, but another, more recent, approach is to actually try to alter heredity. Some of the exciting advances in genetics that make these alterations possible are discussed in the next chapter.

EXTENSION

Find out what progress has been made in the treatment of hemophilia in recent years. How is the disease treated today? What chance does a hemophiliac have of living a normal life span? How has AIDS been transmitted to some hemophiliacs?

F = allele for full colour vision (dominant)
f = allele for colour-blindness (recessive)

Because this gene is carried on the X chromosome, the alleles are written X^F (full colour vision), X^f (colour-blindness).

Possible genotypes

X^FX^F female with full colour vision

X^FX^f "carrier" female—has full colour vision, but carries allele for colour-blindness

ova from carrier female

sperm from male with normal vision	X^F	X^f
X^F	X^FX^F	X^FX^f
Y	X^FY	X^fY

Figure 17.16
The inheritance of colour-blindness. The Punnett square shows the possible genotypes of offspring resulting from a male with full colour vision and a female who carries the recessive gene for colour-blindness. It is probable that half the female offspring will carry the allele for colour-blindness, but will have full colour vision. Half the male offspring are predicted to be colour-blind. Why will none of the female offspring be colour-blind themselves?

REVIEW 17.5

1. How is sex determined in humans?
2. Give two examples of sex-linked traits in humans.
3. Why do more males than females exhibit these sex-linked characteristics?
4. Can a male be a carrier of hemophilia, but not have the disease himself? Explain your answer.
5. In what circumstances would a female have the disease hemophilia? Explain your answer.

CHAPTER • REVIEW

Key Ideas

- Gregor Mendel is often called "the father of genetics," because his research demonstrated several important principles of heredity and revolutionized how genetics was studied.
- Inherited traits are controlled and passed on by genes.
- The two or more forms of a gene are called alleles.
- An individual has two alleles for every gene, one on each member of a pair of homologous chromosomes (with the exception of genes on the sex chromosomes).
- During meiosis, the two alleles of a gene become separated and move into separate gametes. Each parent, therefore, contributes one allele to the pair possessed by the offspring.
- The alleles possessed by an individual make up its genotype. The genotype and, to some extent, the environment determine the physical appearance, or phenotype, of the organism.
- In many cases, one allele is completely dominant over any other alleles, which are then referred to as the recessive alleles. Recessive alleles for a particular trait are not expressed if a dominant allele for the same trait is part of the genotype.
- The offspring's genotype, and thus its phenotype, depends on which alleles it inherited from its parents.
- The laws of probability can be used to predict what the genotypes and phenotypes of the offspring of particular parents are likely to be. The Punnett square is a useful aid in this process.
- The inheritance of traits is often more complicated than Mendel proposed, involving genes with multiple alleles and with dominance relationships such as codominance or incomplete dominance. The interaction of genes with each other and with the environment also plays a role in many traits.
- In humans, sex is determined by which sex chromosomes an individual receives from his or her parents. Female offspring have two X chromosomes; males have one X and one Y chromosome.

Vocabulary

trait
heredity
genetics
selective breeding
hybridization
dominant
recessive
alleles
homozygous
heterozygous
genotype
phenotype
probability
Punnett square
incomplete dominance
multiple alleles
codominance
X chromosome
Y chromosome
sex-linked trait

1. Construct a concept map to show relationships between the following words: genotype, phenotype, trait, alleles, dominant, recessive, heterozygous, homozygous. You might want to include other words in your concept map, such as gene.
2. Select the word from the vocabulary list that matches each of the following descriptions:
   (a) females have two of them; males have only one
   (b) an aid for calculating genotype probabilities
   (c) a kind of trait more common in males than females
   (d) the passing on of traits from parents to offspring
   (e) a situation in which both alleles are equal in dominance
   (f) a field of science

Connections

1. Do Mendel's conclusions explain everything about heredity? Why or why not?
2. Distinguish clearly between the two terms in each of the following pairs:
   (a) gene and allele
   (b) homozygous and heterozygous
   (c) genotype and phenotype
   (d) the probability of certain experimental results and the actual experimental results
3. (a) Refer to Figure 17.16. Using similar Punnett squares, determine the genotypes of the offspring that would be produced by the following crosses of individuals carrying different allele combinations of the gene for red-green colour-blindness:
   - a homozygous female and a male, both with full colour vision
   - a carrier female and a male with full colour vision
   - a colour-blind female and a colour-blind male

- a colour-blind female and a male with full colour vision
- a carrier female and a colour-blind male

(b) In each case, what percentage of the male offspring are likely to be colour-blind?

(c) In each case, what percentage of the female offspring are likely to be colour-blind?

4. Galactosemia is an inherited disease that is due to the lack of a specific enzyme. Without treatment, mental retardation results. Figure 17.17 shows a pedigree for a family; the shaded shapes indicate members with galactosemia. The gene that determines the presence or absence of galactosemia is not sex-linked and has only two alleles. One of the two alleles is completely dominant over the other.
   (a) Is the allele that produces galactosemia dominant or recessive? Explain how you can tell.
   (b) Give the genotypes of as many of the individuals in this pedigree as you can.

Figure 17.17
A family pedigree for galactosemia.

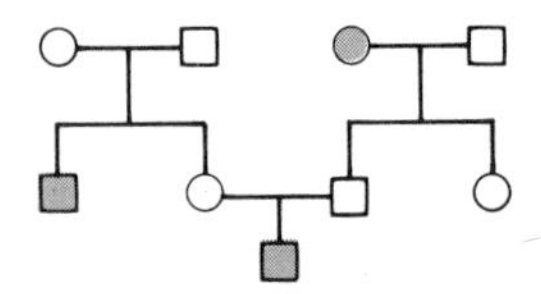

5. A man with blood type A had a father whose blood was type O. This man marries a woman whose blood type is AB.
   (a) What are the genotypes of the man, his father, and his wife?
   (b) What are the possible blood types this couple's offspring might have?

6. Women whose first babies are larger than average at birth tend to always have babies who are larger than average. (That is, their later children are also large.) Similarly, if a woman's first baby is smaller than average, her other children will also tend to be smaller. Do you think that this pattern can be explained solely on the basis of genes, or might some other factors be involved? Explain your answer in full.

Explorations

1. The laws of probability predict that 50 per cent of the time, a zygote (a newly fertilized egg) will have two X chromosomes and will develop into a female. Similarly, 50 per cent of all zygotes will develop into males. However, it has been found that in many large human populations, there are approximately 105 males born for every 100 females.
   (a) Why do you think this difference might occur?
   (b) Although the proportions of the two sexes are slightly different at birth, they seem to have become equal by the time people are 25 years of age. (Among a group of 25-year-olds, there are equal numbers of males and females.) How might you explain this change?

2. What is it that actually causes maleness in humans: is it the lack of a second X chromosome or the presence of a Y chromosome? Do research to help you answer this question.

3. Occasionally, errors occur during meiosis, and individuals end up with only one sex chromosome, or with an extra one. Klinefelter's syndrome, Turner's syndrome, and the XYY (or "super male") syndrome are all caused by such errors. Find out what you can about one of these conditions, and write a report that describes your findings. Your report should explain what causes the condition, what effects it has, and how it is treated (if treatment is necessary).

Reflections

1. Return to your learning journal where you did Activity 17A, Part IV. Have any of your ideas been confirmed by this chapter? What ideas have changed? Look at your list of questions. What answers can you now give to these questions? Are there any questions that you still do not know the answer to? Have you thought of any new questions while reading the chapter? Pick one of these unanswered or new questions and do some extra "digging" (perhaps in the library) to see if you can discover the answer.

2. People have been breeding domestic animals selectively for thousands of years. Selective breeding is how the many different varieties of cows, dogs, cats, and other animals have been produced. During the last 100 years, various individuals have suggested that selective breeding of humans might also be a good idea. What reasons might people have for wanting to do this? What information would they need before they could do so successfully? Do you think selective breeding of humans is a good idea? Why or why not?

CHAPTER

18 Recent Advances in Genetics

Engineers produce blueprints (plans) for bridges and buildings. The giant mouse on the left in this photograph is also a product of engineering: genetic engineering. Its DNA "blueprint" has been altered. A gene that controls the hormone responsible for growth in rats has been added. As a result, the mouse is "rat-size," rather than "mouse-size." Experiments such as this one show that it is possible, in at least some cases, to alter the inherited characteristics of organisms by manipulating their genes.

Such *direct* manipulation of inherited traits is a new tool of genetics, which is still being developed. Various *indirect* methods of influencing heredity have a much longer history. For thousands of years, for example, farmers have attempted to improve their herds or crops by allowing only individuals with desirable traits to reproduce. This selective breeding has produced varieties of crops resistant to plant diseases or to cold weather, and types of cattle that produce leaner meat or more milk. Today, this process is much more efficient and precise than it was in earlier days.

Our new-found abilities in genetics have far-reaching effects, particularly as they relate to humans. What are the moral and social implications of influencing human heredity? Although some uses of genetic technology seem very desirable, others hold potential dangers. This chapter explores some of the recent developments in genetics that have allowed us to influence heredity. Heredity may be influenced either indirectly, by controlling reproduction, or directly, by manipulating the genes themselves. As you understand more about genetics, you will become better equipped to decide which techniques can be usefully applied to our own species—and which ones might be dangerous.

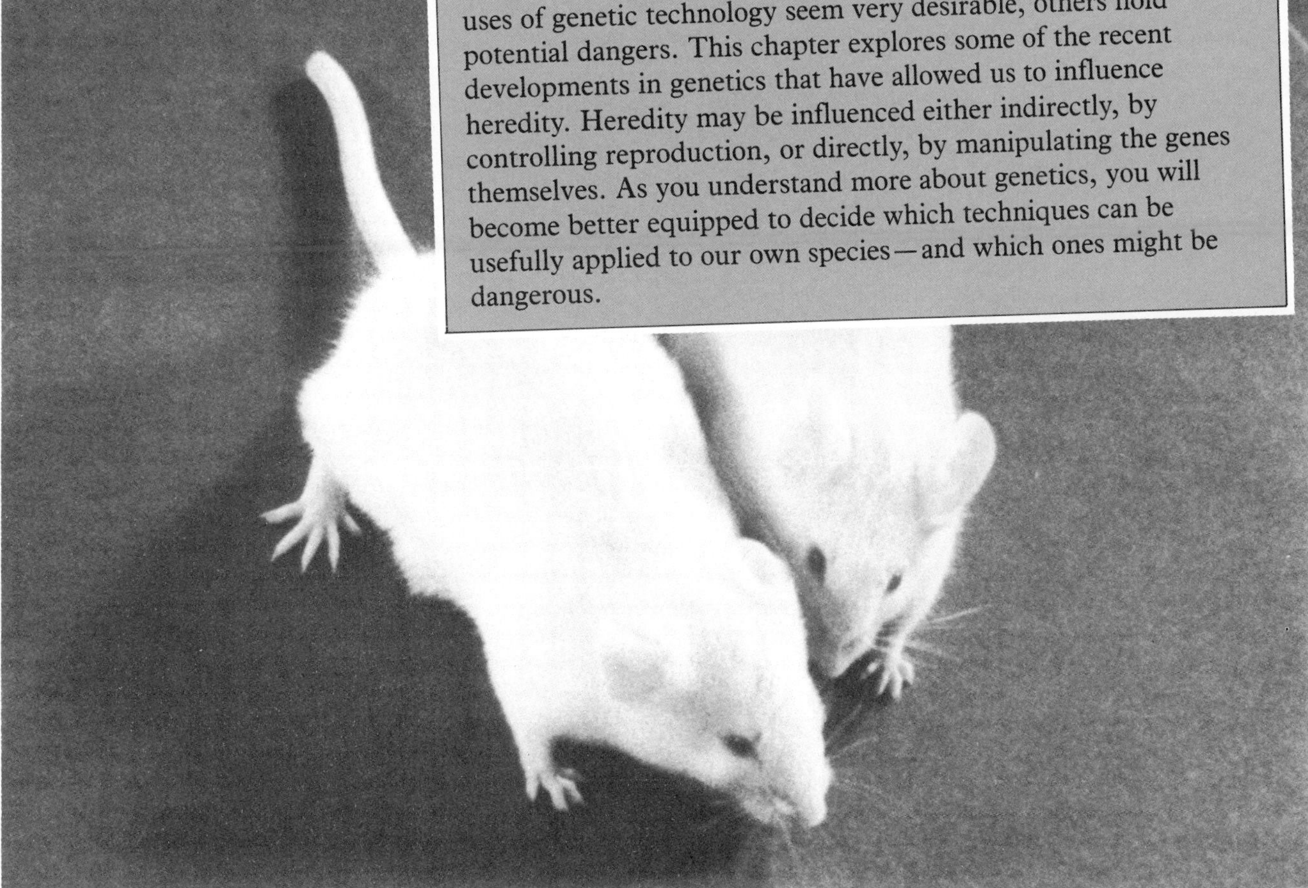

ACTIVITY 18A Issues in Genetics

What do you think might be some advantages of being able to influence human heredity? What disadvantages or hazards do you imagine might result from this ability? Think about the following situations in terms of these potential advantages or disadvantages. In your learning journal, answer the questions if you can. Briefly explain your reasons for each answer. The following questions raise issues, and there are no right answers. These issues represent real choices that people are making today or will be making in the near future.

1. Should people be allowed to selectively breed animals for whatever traits they consider desirable, regardless of how it affects the animals themselves?
2. Should couples be able to decide the sex of their children in advance?
3. Imagine that one woman agrees to have a child for another woman, who cannot do so. After the first woman has given birth, she changes her mind and decides to keep the child. A judge rules that she can do this. Is this is a fair decision?
4. A couple analyse their family pedigree. They discover that any children they might produce have a high probability of inheriting a serious genetic disease. Should they go ahead and have children anyway?
5. It is now possible to test for some disease-causing genes in embryos. A couple find that their developing embryo has inherited a genetic disease. The child will become ill soon after birth and die before reaching two years of age. Should the couple continue the pregnancy and have the child?
6. Someday it may be possible to produce one or more clones of an individual human. Is this a good idea?
7. Imagine that a company has developed a drug that can correct the effects of genetic disease that causes muscle weakness. When taken by individuals *without* the genetic disorder, this drug produces unusually strong muscles. It thus improves athletic ability. However, when these individuals get old, they experience serious side effects, which require expensive treatment. Should this drug be sold to any people who can buy it, regardless of whether they have the disease?
8. Some people think that human traits such as musical ability or intelligence are inherited—that these traits depend on your genes and not on your environment. Do you agree?
9. It is possible to obtain human sperm from sperm "banks." Using a syringe, a doctor puts the sperm into a woman's reproductive tract so that fertilization can occur. Should these women be allowed to select specific types of sperm—ones from very intelligent or very athletic men, for example?

18.1 Manipulating Reproduction

Selective breeding is one of the most widely used methods of influencing the inheritance of genetic traits. The goal of selective breeding is to produce offspring with desirable traits. This goal can be achieved in a variety of ways.

In the past, male and female animals with desirable traits were simply put together at appropriate times and allowed to mate. The present-day merino sheep is the result of matings between males and females of two different varieties of sheep. One variety had a coat of long coarse fibres, and the other had a coat of short fine wool. Many of the offspring of such pairs had long fine wool, which was preferred. These offspring were then bred among themselves to produce a line that bred true for this type of wool (Figure 18.1).

Varieties of plants have also been interbred to produce new types of offspring. The breeders used crossing methods (often termed cross-

◄ Figure 18.1
The present-day merino sheep (left) was produced by crossing two other varieties of sheep, one of which is shown on the right.

pollination) similar to Mendel's. A type of bread wheat in use today, for example, is the result of a cross between emmer wheat (which has large kernels) and goat grass (which is more resistant to diseases and cold weather).

Microscopes and other equipment have made possible a dramatic increase in our knowledge of reproduction and our ability to control it. For example, seeds can now be stored at very low temperatures for a long time. As a result, the genes of certain plants will be available whenever they are needed — perhaps at some time in the future when environmental conditions change. In addition, in modern greenhouses and laboratories, cross-pollination of useful plant varieties can be carried out rapidly and on a much larger scale than Mendel could ever have imagined.

Grafting is another way to manipulate reproduction to get plants with desirable traits. After planting an apple seed, you might have to wait for 10 to 20 years until the resulting tree has grown enough to produce a good crop. However, the waiting period can be reduced considerably by grafting. In this process, buds or twigs from a tree producing a particularly desirable type of apple can be grafted, or attached, to the trunk of an established tree of another variety (Figure 18.2).

Grafting is often used to produce fruit trees, grapevines, and roses in a shorter time than it would take to grow these plants from seed. Remember that the grafted twig or bud contains the same genes as the plant it came from. Grafting, therefore, is a means of rapidly producing multiple

Figure 18.2
Grafting a twig from a Spy apple tree onto a trunk of a Spartan apple tree. The well-established roots and trunk of the Spartan tree will provide water and nutrients for the grafted Spy twig.

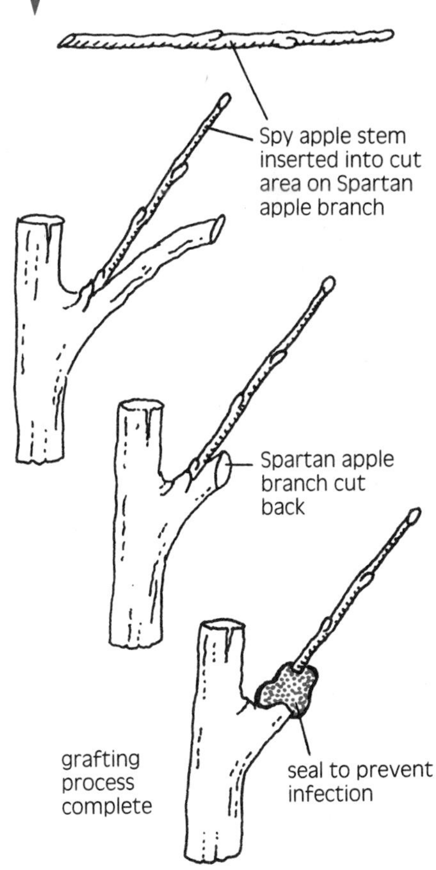

copies of this desirable plant. Most of them will produce seeds and new plants in their turn.

Modern techniques that perform the same functions as seed storage, cross-pollination, and grafting also exist for various types of animals. Some have even been applied to humans. These procedures make it possible to unite the eggs and sperm of females and males with desirable traits, even if the animals themselves are not present. In some cases, they allow animals (including humans) who might not otherwise be able to have offspring to do so. Several of these techniques are described in the rest of this section.

ARTIFICIAL INSEMINATION

In the **artificial insemination** process, sperm are placed in a female's reproductive tract by artificial means (such as a syringe), rather than during sexual intercourse. This technique is widely used on farms and ranches to breed domestic animals such as cows, sheep, and horses. Sperm collected from a male with desirable genetic traits are often frozen for storage and transportation. The sperm can later be inserted through the vagina of a similarly desirable female. The procedure is done just after the female has ovulated, when there is a higher chance of the sperm's fertilizing an ovum.

Artificial insemination is also used frequently in zoos, particularly for species of animals that are endangered (few in number, and in danger of becoming extinct). In zoos, artificial insemination is used more to increase the number of individuals in these endangered species than it is to select for desirable traits. For example, a zoo may have females of a certain species, but no males, or no males capable of producing sperm. The females in such a zoo can be artificially inseminated with sperm from males elsewhere. They can then produce offspring, thus increasing the number of individuals of this species.

Figure 18.3
Sperm from a scimitar-horned oryx living in the Metro Toronto Zoo were sent to a zoo in New Zealand. There, they were used for the artificial insemination of a female oryx, who later produced offspring. This process increased the genetic diversity of the oryx population in the New Zealand zoo.

Zoos also use artificial insemination to increase the genetic diversity of endangered species (Figure 18.3). To understand genetic diversity, remember that the alleles carried by two humans represent only *some* of the alleles that exist in the entire human population. The same is true of the single pair of leopards, for example, that one zoo might have. Other individuals of these species may have different beneficial alleles. It is important to retain as many different allele combinations as possible — this increases the genetic diversity of the species.

The best way to increase genetic diversity is to make sure that offspring are produced by many different parents, rather than by the same pairs all the time. Using different pairs often requires mating animals housed at different zoos. Because many animals may experience shock and die if captured or transported, moving animals from one zoo to another for breeding is often impossible. However, males can be tranquillized, and their sperm collected and quickly frozen. The sperm can then be transported to zoos around the world so that a number of females can be artificially inseminated. This technique also allows zoos to obtain sperm from wild animals, increasing genetic diversity still further.

Artificial insemination is also useful for human couples when the male is infertile. Male gametes are obtained from a sperm bank, where sperm donated by anonymous individuals are collected and stored. The female charts her menstrual cycle to determine the timing of ovulation. Sperm are then injected through the vagina with a syringe. If fertilization occurs, a normal pregnancy can take place.

In Chapter 17, you learned how the sex of the offspring is determined at fertilization. To review, the ova of humans and many other animals carry an X chromosome. If a sperm carrying a Y chromosome fertilizes the ovum, a male will result. Sperm carrying an X chromosome will produce a female. In many animals, including humans, sperm with an X chromosome are different (e.g., in size) from those with a Y. It may be possible, therefore, to sort and separate the two types of sperm.

Theoretically, therefore, the sex of an offspring could be preselected. Through artificial insemination, the female could receive sperm carrying either X chromosomes or Y chromosomes. However, some scientists are not convinced that accurate sorting is currently possible. Experiments on humans and other animals have not been successful at sorting 100 per cent of the sperm. In addition, the sorting techniques require special equipment and skills. As a result, sperm used for artificial insemination are not now usually sorted before use, and selecting an offspring's sex is not a common occurrence.

EXTENSION

"Outbreeding" is a technique that helps to keep the alleles in a population of animals or plants well mixed. This means that individuals breed with others outside their own family or local group (for example, outside their own herd or flock). Outbreeding maintains a diversity of allele combinations. The opposite of outbreeding is inbreeding. Do research to investigate the advantages and disadvantages of both outbreeding and inbreeding. How are these processes used by farmers and ranchers?

IN VITRO *FERTILIZATION*

The Latin words *in vitro* mean "in glass." Quite literally, **in vitro fertilization** occurs in a glass Petri dish, outside a female's body (Figure 18.4). The female is first treated with hormones so that she "superovulates," or releases several ova at one time. (Normally, only one is produced during ovulation.) These ova are then removed and placed in the dish with suitable sperm. If fertilization occurs, the embryos are allowed to develop for several days before one or more of them are placed in the female's uterus, using a syringe inserted through the vagina. There, the embryo or embryos may implant successfully and a normal pregnancy may follow. Although offspring produced by *in vitro* fertilization are often referred to as "test tube babies," most of their development actually occurs in the mother's uterus. In fact, if the embryos are not either frozen or placed into the uterus, they will soon die.

In vitro fertilization can be used to solve infertility problems in humans. In some couples, both the male's sperm and the female's ova are capable of producing offspring. For some reason, however (usually because the woman's oviducts are blocked), successful fertilization cannot be achieved. *In vitro* fertilization can enable some of these couples to have children. Sperm and ova are obtained from the couple and mixed in a Petri dish. If fertilization occurs, the resulting embryo(s) are placed into the mother's uterus.

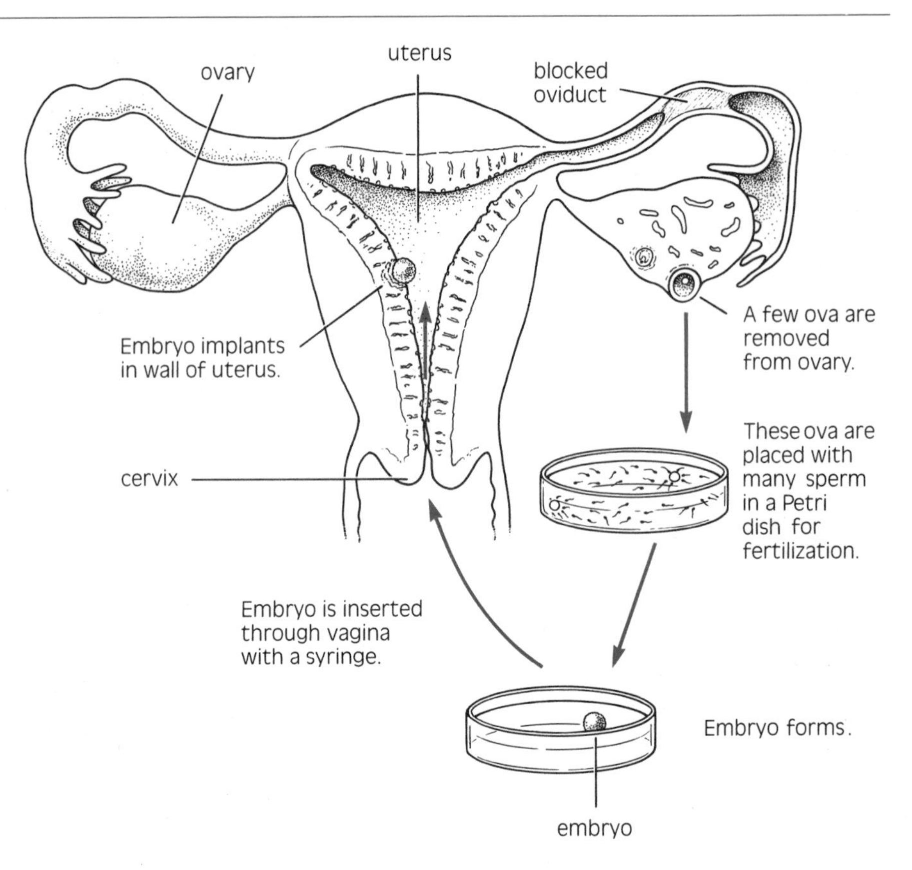

Figure 18.4
The process of *in vitro* fertilization. More than one ovum may be fertilized, and one or more embryos may be put in the uterus.

The first "test tube baby" was born in England in 1978. By 1990, over 3000 infants had been produced in this way. Sometimes more than one of the developing embryos implants in the wall of the uterus. For example, *in vitro* fertilization produced quintuplets (five babies born at once) in Toronto in 1988.

In vitro fertilization is the only way for some couples to have their own children. However, it is very expensive and has a low success rate. (Relatively few of the embryos develop successfully.) In places where the government health care system does not pay for the procedure, many infertile couples cannot afford it. Where tax money is used, people question whether the high expense for a procedure with such a low success rate is justified, particularly when other types of health care need more funds.

Sometimes, some of the human embryos produced in the Petri dish are frozen and stored, to be implanted at a later date. This practice increases the chances of at least one successful pregnancy and, in some cases, enables a couple to have children of different ages. However, it also raises questions. What should be done with stored embryos if the parents die, get divorced, or decide not to attempt fertilization again? In Australia, two embryos were in storage when their parents were both killed in a plane crash. While moral questions arose around what should happen to the embryos, legal experts were dealing with the problem of distributing a large estate left by the couple to family members—which may include these potential offspring.

SURROGATE MOTHERHOOD

In many species, single females can successfully carry (during pregnancy) and give birth to only one or two offspring a year, regardless of how many ova they release at ovulation. However, by means of *in vitro* fertilization, several ova can be removed and fertilized. The resulting embryos can then be injected into one or more **surrogate mothers**. A surrogate mother is a "substitute" female, whose uterus is used to carry the developing embryo until it is born. Zoos and animal breeders use this technique to increase the number of offspring of an individual.

Surrogate mothers can be other (perhaps healthier) members of the same species, or even members of a different but closely related species. For example, zoos have used domestic cows as surrogate mothers for gaur embryos (Figure 18.5).

Figure 18.5
A gaur is a species of ox, native to India, which is threatened with extinction. In zoos, domestic cows are used as surrogate mothers for gaurs. This has allowed the captive population of this species to increase more rapidly than it might have otherwise.

In vitro fertilization and surrogate mothers are also used by zoos to increase the breeding success of certain species. Some species will not mate unless they are in specific types of habitats. Members of some endangered cat species, for example, require a large, isolated habitat before they consider it safe enough to mate and produce offspring. In zoos, this is difficult to provide. *In vitro* fertilization and surrogate mothers help to ensure reproduction in these species (Figure 18.6).

Surrogate mothers can also be used when the ova are fertilized by artificial insemination. A female with desirable traits is given hormones to make her superovulate, and is then artificially inseminated. The resulting embryos are removed from the mother and inserted into other females. This procedure, called **embryo transfer**, is commonly used on many ranches and farms today. Embryo transfer can increase the number of offspring from a genetically desirable female.

Some human couples overcome their fertility problems by employing a surrogate mother. This is a woman who has agreed to carry and give birth to a child for another couple. Most commonly, the surrogate mother has her own ovum artificially inseminated with sperm from the male of the couple. She usually receives a fee or expenses for her services. After the birth, she is expected to allow the couple to adopt the child.

It is possible for a surrogate mother to carry an embryo produced by *in vitro* fertilization. The couple wanting a child would have their own sperm and ova mixed in a Petri dish. The embryo would then develop

Figure 18.6
An Indian desert cat kitten with its surrogate mother. In the Cincinnati Zoo, an ovum removed from an Indian desert cat was fertilized *in vitro* and then placed in the uterus of a domestic cat. The desert cat embryo developed successfully, adding a new member to this endangered cat species.

DID YOU KNOW?

Embryos that have been removed from a female can be frozen and moved long distances. Using embryo transfer, farmers and ranchers can "share" the desirable genetic traits of female animals, just as artificial insemination allows them to share the desirable traits of males.

in the uterus of the surrogate mother. The expense and technical difficulty involved in using *in vitro* fertilization with surrogate motherhood result in its being less commonly used.

For humans, surrogate motherhood can pose legal and moral difficulties. In some actual cases, the surrogate mother has decided that she does not want to give up the infant. She has gone to court for the right to break the agreement and keep the child. The couple that made the agreement with her have defended their right to adopt the baby. Who should be allowed to raise the child? Such questions will continue to arise as new technologies for manipulating reproduction are developed. Some of them may be very difficult for society to deal with. What do you think about these issues?

ACTIVITY 18B *The Effects of Manipulating Reproduction*

Imagine that you read the following article in your local newspaper. (It is not a true account, but it could happen in the future.)

"Researchers at University Hospital recently announced that they have improved the procedures for separating sperm carrying X and Y chromosomes. They conclude that their technique is almost 100 per cent accurate. Fifty couples have used the procedure in experiments, and 48 have had children of the selected sex. The researchers have also simplified the sorting procedure and considerably reduced its cost. They have asked that the provincial health care plan pay the full costs of the procedure for couples seeking to choose the sex of their children. Government officials are considering this proposal. A task force is travelling the province seeking the advice of experts as well as the opinions of citizens on this issue."

PROCEDURE

1. Form groups of four or five students. One group will represent the task force. Each of the other groups will represent the opinions of one expert or citizen.
2. Each group playing the role of an expert or citizen should select a person or group to represent. Consider possibilities such as a woman who is a carrier of a sex-linked genetic disease, a couple who have three boys and would like to have a girl, a representative of a women's rights organization, a taxpayer, and a government official who predicts future trends in the labour force.
3. Imagine that the task force is holding a public meeting in your community. Prepare a list of points that the person you are representing might make at this meeting. Include reasons for the person's opinions. You might do research to come up with information (for example, statistics from relevant studies) to support your adopted point of view.
4. Select a member of your group to present these points at the meeting. Think about how your representative might answer questions or respond to others with differing points of view.
5. Meanwhile, the task force group should prepare a list of questions to ask the group representatives.
6. The group representatives present their adopted points of view to the class. Task force members ask questions. Other class members act as the audience. After the presentations, the audience may wish to ask questions, raise additional points, or disagree with the presenters.

DISCUSSION

1. Did you feel that all the participants in this public meeting argued on the basis of facts alone, or were personal values and opinions sometimes involved? Give examples.
2. (a) Did participants present information that supported only their own point of view?
   (b) If presenters used facts or statistics, did these always agree? Or is it possible to find different studies on the same topic that have conflicting findings—which can then be used to support different points of view?
3. What were some of the issues raised during the hearing?

REVIEW 18.1

1. What is selective breeding? Give at least two examples of how selective breeding has been used in agriculture.
2. (a) Describe the process of grafting.
   (b) What advantage does it provide?
3. (a) What is artificial insemination?
   (b) Describe two situations in which artificial insemination is used.
4. (a) What is *in vitro* fertilization?
   (b) In what situations is *in vitro* fertilization used?
   (c) What are some advantages of *in vitro* fertilization?
   (d) What are some disadvantages of using this process in human reproduction?
5. (a) What is a surrogate mother?
   (b) What advantage do surrogate mothers provide for the breeding of endangered species and domestic animals?
   (c) What is one of the problems caused by the use of human surrogate mothers?
6. (a) List three ways human couples with fertility problems can still have children.
   (b) What infertility problems might lead couples to use each method?
7. Which of the methods of manipulating animal reproduction discussed in this section is most like grafting in plants? Which is most like seed storage? Which is similar to cross-pollination?

18.2 Genetic Counselling

As you learned in the last chapter, a variety of diseases — for example, hemophilia and muscular dystrophy — are inherited. They are caused either by mutations, by alleles passed on by the parents, or by errors occurring during meiosis (Figure 18.7). These errors may mean that the affected individual is missing a chromosome or pieces of a chromosome, or instead has an extra chromosome or pieces of a chromosome.

In domestic animals and plants, farmers try to eliminate such genetic disorders by selective breeding. They attempt to prevent individuals carrying disease-causing alleles from reproducing — and passing the problem on to future generations. By manipulating reproduction in this way, they are attempting to influence the inheritance of genetic diseases.

Figure 18.7
This girl is swimming in a Swim-a-Rama for Cystic Fibrosis. Like the other swimmers, she has cystic fibrosis, an inherited disease that makes breathing difficult. Most cases of cystic fibrosis are caused by a pair of recessive alleles.

To some extent, humans also employ a type of selective breeding for this same purpose. Some people are at greater risk than others for passing on genetic diseases to their children. A couple may be considered to be at risk in these situations if:

- They already have a child with some kind of disease that is known or suspected to be genetic.
- Several members of their extended family display a certain genetic disease.
- The prospective mother is over 35 years of age or the father is over 50. (Older parents are more likely to have chromosomal errors — caused by mutations or other events — in their gametes.)

Couples in these situations often receive **genetic counselling**. During this process a couple obtain information to help them decide whether or not to have a child. The usual steps in genetic counselling are outlined in Figure 18.8.

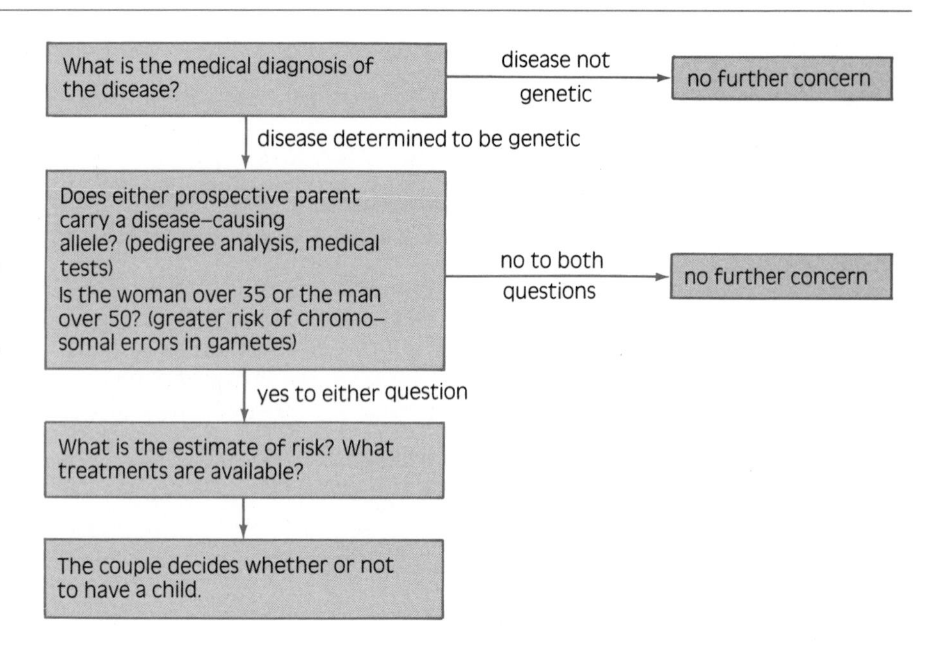

Figure 18.8
A flow chart illustrating the process of genetic counselling. Tests can detect the presence of disease-causing alleles for various genetic diseases. Such tests, along with other information, can help a couple decide whether or not to have a child.

EXTENSION

Sickle-cell anemia is a genetic disease that occurs among people of African descent. Another disease, malaria, is also common in Africa. Malaria is caused by a protist that is carried by mosquitoes and injected into human blood when these insects bite. Research the relationship between these two diseases. What does this relationship tell you about how a disease-causing allele might offer some advantages to individuals in certain populations?

Some genetic diseases are controlled by a single gene. In those cases, an estimate of the risk that a child will inherit the disease can be shown with a Punnett square (see Chapter 17). For example, if both prospective parents carry a recessive allele for a certain type of cystic fibrosis, there is a 1/4 chance that their child will inherit the disease (Figure 18.9). Often, however, the situation is not as clear-cut as this. Test results may be inclusive or incomplete, making the calculation of risk difficult.

Some diseases show some tendency to be inherited, but the exact genetic mechanism of inheritance is not understood. Diabetes, for example, occurs more often in some families than others, but how it is inherited is uncertain. Several genes may be involved, and environmental factors may also play a very important role. There are many diseases that show complicated patterns of inheritance. For these, it is more difficult, although still possible, to estimate the risk involved.

A couple who risk passing on a genetic disease have many questions to consider. Should they have children? How high must the risk be before they decide not to have a family? How serious is the disease? How effective is the treatment, if any is available? How much financial and emotional support can they expect from their families or community? As you might imagine, these decisions can be very stressful for a couple. The genetic counsellor can provide support at this time.

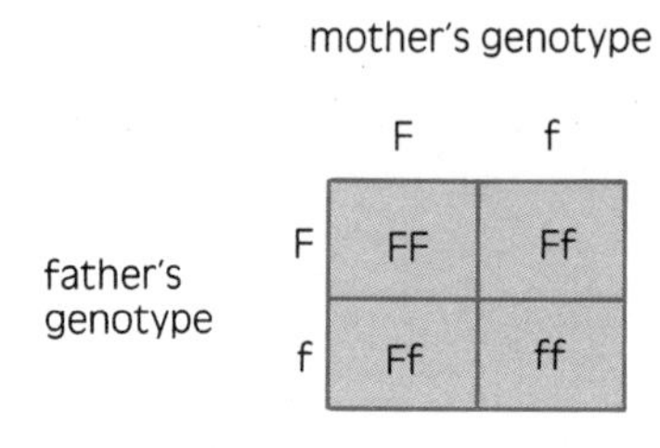

F = dominant allele
f = recessive allele
FF—not affected
Ff—carrier of cystic fibrosis
ff—affected with cystic fibrosis

Figure 18.9
A Punnett square showing the probability that a child of two carriers will inherit two recessive alleles, and thus have cystic fibrosis.

PRENATAL DIAGNOSIS

So far, we have discussed genetic counselling only for couples who have not yet made the decision to have a child. In some cases, however, the couple is at risk and the woman is already pregnant. The couple may then be seeking a **prenatal diagnosis**—information about the health of the fetus. If the genetic counsellor thinks that there is substantial risk of the child's inheriting a genetic disease, tests can be done while the offspring is still in the mother's uterus.

An **ultrasound examination** is one technique that can provide some information about the fetus. In this procedure, high-frequency sound waves are projected through the mother's abdomen. These waves bounce off the structures inside and can be displayed on a television screen to give a "picture" of the fetus. Most doctors consider this procedure to be safe.

An ultrasound image can reveal various types of external physical malformations. It can also show several types of heart disorders and kidney abnormalities. An ultrasound examination is often used in normal pregnancies to determine the size—and thus the age—of the fetus, so that the date of birth can be more reliably predicted.

Another common technique used for prenatal diagnosis is **amniocentesis**. This involves removing a sample of the fluid (called the amniotic fluid) that surrounds and protects the fetus in the uterus (Figure 18.10). The procedure is done by inserting a needle into the mother's abdomen and through the wall of the uterus, and withdrawing 5–20 mL of amniotic fluid. To avoid injuring the fetus, the doctor must determine its exact location before the needle is inserted. This is usually done by ultrasound examination.

The amniotic fluid contains cells that have been cast off by the fetus during development (much as you cast off skin and hair cells during your life). First, the cells are cultured (placed in a nutrient medium so that they grow and divide, thereby increasing in number). Then, tests are performed for various types of diseases. A karyotype can be produced, for example, which will reveal the existence of missing or extra chromosomes (Figure 18.11). Biochemical tests can be performed to test for various enzymes, the presence or absence of which indicates inheritance of certain genetic diseases. Analysis of the DNA can detect the presence of specific alleles that are known to cause disease.

Amniocentesis (and sometimes an ultrasound examination) can also reveal the sex of the fetus. This information can be useful when a fetus is at risk for a serious sex-linked disease. Finding out the sex of the fetus has raised social issues. Some couples strongly prefer to have sons rather than daughters. In some places, a female fetus is sometimes aborted, even if it is healthy. What do you think of this practice?

Amniocentesis has some medical limitations. There is some risk to the fetus and the mother (by infection from the needle, for example). The procedure cannot be performed until there is sufficient amniotic fluid present, usually between the 14th and 18th weeks of pregnancy.

Figure 18.10
(a) Amniocentesis involves removing a sample of amniotic fluid for analysis. (b) A doctor obtains a sample of amniotic fluid from the uterus.

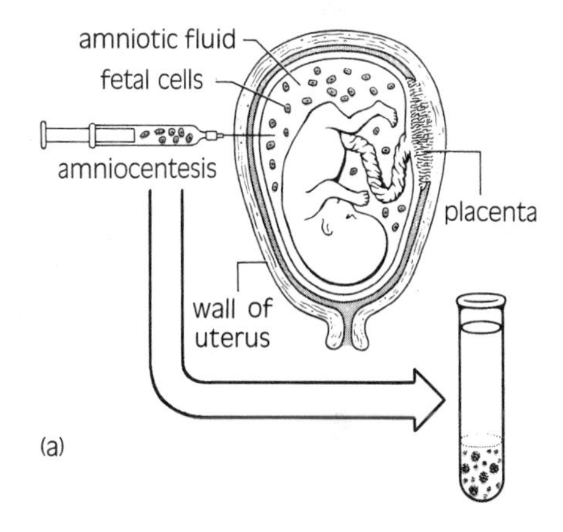

(a)

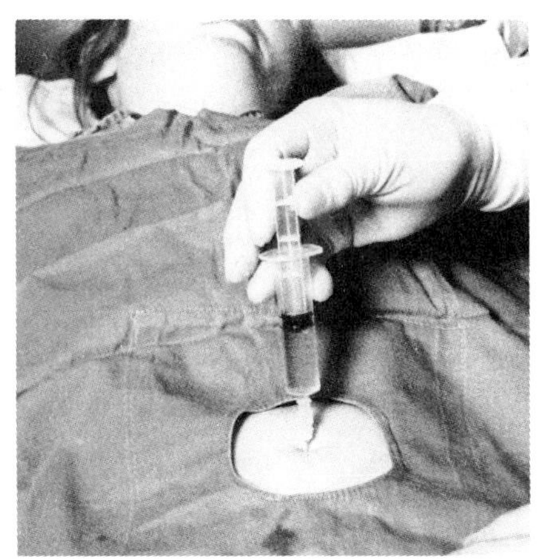

(b)

(a)

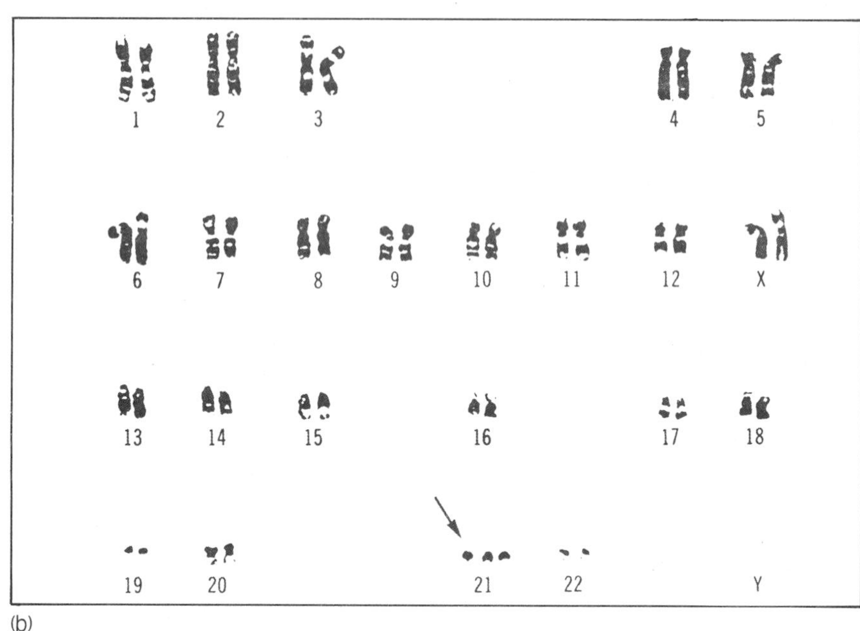

(b)

Figure 18.11
The teenager shown in (a) has Down's syndrome. Down's syndrome results from an error occurring during meiosis. Affected individuals receive an extra copy of chromosome #21, shown by the arrow in (b). Is this the karyotype of a male or female?

chorion
fetal portion of placenta
maternal portion of placenta
umbilical cord
uterus
chorionic villus
sampling tube

Figure 18.12
In chorionic villus sampling, a sample of cells is removed from a chorionic villus by inserting a special tube through the vagina. The chorionic villi extend into the placenta and become part of it.

Because it takes time to culture the cells and perform the tests, the results may not be obtained until a month after the fluid has been removed. Therefore, information about the health of the fetus may not be available until it is five months old.

Another technique for obtaining fetal cells is termed **chorionic villus sampling**. In this procedure, cells are removed from a chorionic villus (plural: villi). These villi are extensions of the chorion, the outermost of the membranes surrounding the fetus (Figure 18.12). This test can be performed as early as eight weeks into the pregnancy. Because the chorionic villi contain many more fetal cells than the amniotic fluid, culturing does not take as long. Test results are usually available a few days after the sample has been taken.

Amniocentesis and chorionic villus sampling are very useful in prenatal diagnosis. However, they enable doctors to detect only about 250 of the more than 3000 genetic diseases known to exist. Because both techniques involve some risk to the mother or fetus, they are used primarily when there is known to be a danger of genetic disease.

Not all prenatal tests are 100 per cent conclusive. In some cases, the results can provide only an estimate of the probability that the fetus has a certain disease. Sometimes tests indicate a genetic disease, but the child may turn out to be healthy or may be only minimally affected by the disease.

In over 90 per cent of the cases in which prenatal tests are carried out, the parents are reassured that the fetus is healthy. In the remainder of cases, a genetic disease is diagnosed or the test results are inconclusive. These parents face a difficult decision. In some cases, the parents may choose to have a therapeutic abortion (an operation in which the embryo or fetus is removed from the uterus and the pregnancy thereby terminated). In other cases, the parents may decide to continue the pregnancy and prepare for the birth of a child with a genetic disease.

Some genetic diseases can be treated successfully after the child is born. Prenatal diagnosis gives parents the time to learn about these treatments. A very few genetic diseases can be treated before birth by operating on the fetus. For example, hydrocephalus, a disease that causes fluid to collect around the brain, is sometimes treated by surgery on the fetus to drain the fluid. These operations are risky, however, and not always completely successful.

Prenatal diagnosis is an important tool of genetic counselling. Both before and during pregnancy, counselling gives parents information to help them make decisions about having children.

ACTIVITY 18C *Genetic Diseases*

In this activity, you will investigate a genetic disease.

PROCEDURE

1. Form into groups of two to four people. Each group should choose a genetic disease to research. You might choose one of the following: Down's syndrome, phenylketonuria (PKU), Duchenne muscular dystrophy, sickle-cell anemia, galactosemia, thalassemia, anencephaly, spina bifida, cystic fibrosis, Tay-Sachs disease.

2. Decide how to split the work among group members. Each member of the group might choose a different source to find out information. For example, possible sources might include the library, community health services, a hospital, or organizations such as the Muscular Dystrophy Association of Canada.

   Each person should try to answer the questions below. Alternatively, you might decide to have each person answer one of these questions, using whatever sources he or she can find.

   (a) If it is known, what is the genetic origin of this disease? That is, is it the result of a single dominant allele, for example? Or must an affected person inherit a recessive allele from each parent? Or is it the result of an error during meiosis?
   (b) What problems are caused by the genetic abnormality—how do the alleles involved affect the body? How is the individual's health affected?
   (c) Are some people more seriously affected than others?
   (d) How often does this disease occur? (This statistic is often given as the number of cases per 1000 births.) Is the frequency of the disease increasing, decreasing, or staying about the same? Why?
   (e) How is the disease diagnosed? Can it be detected with prenatal tests?
   (f) What treatment—if any—is available, and how effective is it?

3. Pool your information with that of other members of your group. What sources of information were most helpful?

4. Share the group's findings with the class, with a display or a written report, for instance.

SCIENCE IN OUR WORLD

A CASE STUDY IN HUMAN GENETICS

Jennifer and Sam have one child, Andrew. When Andrew was five months old, he was diagnosed as having hydrocephalus. With this condition, fluid builds up in the skull and there is pressure on the brain. As a result, Andrew's head was enlarged and he was mentally handicapped.

Sam and Jennifer wanted another child, but they were worried that the next child might have the same disease, especially since Jennifer's sister also had a son with hydrocephalus.

On the advice of their doctor, the couple went for genetic counselling. A doctor specializing in genetics knew that this form of hydrocephalus was inherited and was sex-linked. A genetic counsellor took a family medical history for the disease. She constructed a pedigree, which showed that Jennifer and her sister were carriers of a recessive allele for the disease. There was a 50 per cent chance of a male offspring having the same disease.

Jennifer and Sam were faced with a difficult decision. If they had a girl, there would be no risk of the condition, although the girl might be a carrier. If they had another boy, there was a 50 per cent chance that he too would have hydrocephalus.

The counsellor suggested that if the couple went ahead with the pregnancy, there were several options:

1. The sex of the fetus could be determined by chorionic villus sampling or amniocentesis. If the fetus was male, Sam and Jennifer would discuss the implications with their doctor. For example, they could consider monitoring the fetus closely with ultrasound, which might detect early signs of hydrocephalus, but which could not guarantee that the disease would not develop later. They could also consider therapeutic abortion. Surgery on the fetus while in the uterus might be a possibility.

2. No tests for sex would be performed. Jennifer and Sam would take the chance that they might have a boy with the genetic defect. Ultrasound tests would monitor the size of the fetus's head in order to prepare for a safe birth.

3. If the child was born with hydrocephalus, then a new surgical procedure could be carried out. A shunt would be inserted into the baby's skull to drain the fluid and thus relieve the pressure on the brain. This procedure might reduce or even eliminate the likelihood of mental retardation.

After much thought, Sam and Jennifer chose the second suggestion, and they were also willing to consider the third option. A baby boy was born, and fortunately he did not have the genetic defect.

You can see that the genetic counsellor helped Jennifer and Sam understand the risks and their options. During counselling, the family's way of life and resources were also considered. The final decision was the couple's. Think about what you would have done if you were faced with the same decision.

REVIEW 18.2

1. (a) What is genetic counselling?
   (b) As a result of genetic counselling, parents often face difficult decisions. What factors make these decisions so difficult?
2. (a) List three methods of prenatal diagnosis.
   (b) Briefly describe each method.
   (c) What are the advantages and disadvantages of each method?
3. If there is a risk that parents will pass on a genetic disease, what types of information can help them decide whether to have children?
4. (a) What decisions might parents make if they find out, through prenatal diagnosis, that the fetus has a genetic disease?
   (b) What factors might affect their decision?

18.3 Manipulating Genes

Genetic counselling identifies the risk that certain couples *may* pass on a genetic disease. Which alleles the offspring actually inherits, however, depends on chance. There is no guarantee that a particular allele or combination of alleles (whether harmful or beneficial) will end up in a new individual. Because of this chance factor, manipulating reproduction by the techniques discussed so far may not be the most efficient means of altering heredity. The techniques you have been learning about cannot change the genetic material. Another way to influence heredity would be to actually alter the specific genes that have been inherited.

Recent advances in technology have made it possible to change the genetic material itself. Researchers can now alter the genes or chromosomes of individual organisms—and thus alter the traits controlled by these genes. The technologies involved are constantly changing as new improvements appear. In this section, you will learn about some of these revolutionary techniques. The many new discoveries and developments in genetics make it an exciting field of science to investigate.

DID YOU KNOW?

Chance plays a major role in determining which alleles will be passed on to the next generation. The vast majority of sperm and ova never end up as part of a zygote. The particular combinations of alleles they carry are thus not preserved.

CLONING

In Chapter 16, you learned that a group of genetically identical offspring produced by a single parent is called a clone. **Cloning** is the production of genetically identical offspring. In nature, some types of cloning, such as asexual reproduction, have been around for millions of years. Recently, however, scientists have discovered new methods of cloning—using genetic technology. For example, clones can now be produced from single plant cells (Figure 18.13). In nature, asexual reproduction by these plants usually requires many cells, such as those that make up rhizomes or runners.

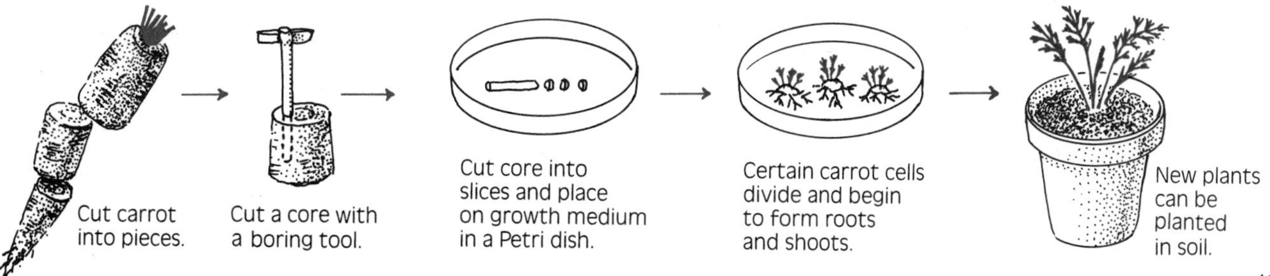

Figure 18.13 The steps in cloning a carrot—producing an entire new plant from a single cell.

In addition, scientists are now able to clone animals that never reproduce asexually in nature. The first successful clones of this kind were made by a British geneticist, J. B. Gurdon (1933–), in the early 1960s. Figure 18.14 shows how he cloned a toad. Later experiments showed that the tadpole's diploid cell did not need to come from the intestine. Many other cells worked equally well.

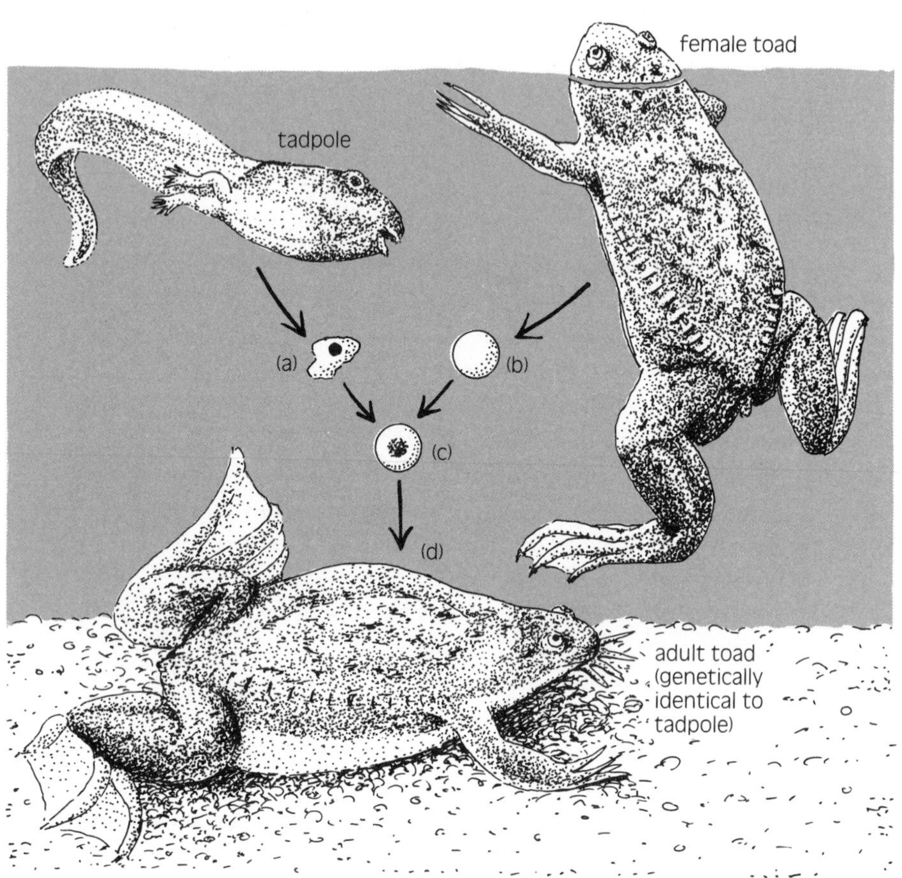

Figure 18.14
(a) Gurdon removed an intestinal cell from a tadpole.
(b) He took a mature, unfertilized (and therefore haploid) ovum from an adult toad of the same species and destroyed its nucleus.
(c) He implanted the diploid nucleus from the tadpole cell into the ovum.
(d) The ovum began to divide just as if it had been fertilized. Put in a suitable environment, it developed into a toad with the same traits as the tadpole that donated the nucleus.

In this way, Gurdon could choose a specific set of genes (that is, the genotype of a particular tadpole) that he wanted to reproduce. The various alleles did not have to get separated into different gametes during meiosis and did not have to be mixed up with those of another individual during sexual reproduction. By cloning, he could produce many copies of exactly the same genotype. The same is true for other organisms that are cloned—the carrot in Figure 18.13, for example.

Very recently, more complex types of organisms, including mammals, have also been cloned by transferring nuclei between cells. Although this is much more difficult to achieve, mice and cows, among other animals, have been cloned. In most cases, the resulting embryos develop inside surrogate mothers. Although it is not yet possible, someday humans may also be cloned. What issues do you think this practice might raise?

The great value of cloning is that it can produce many identical copies of organisms in a relatively short period of time. For example, one could produce a number of cows with a single, superior genotype in much less time than it would take to breed them selectively or to produce them by artificial insemination (since each cow can give birth to only one or two calves a year).

Cloning technology, however, is not used just to copy whole organisms. It is employed more often today for copying single genes, or portions of chromosomes. The techniques used to clone specific genes are fairly new. The first cloning of a human gene, for example, took place in 1980. To clone genes, scientists must be able to locate and remove them from the chromosomes on which they are located. They can then be added to the chromosome of another organism. The giant mouse in the photo that opens this chapter had its genes manipulated in this way. Changing or manipulating the genes in an individual organism is referred to as **genetic engineering**. Cloning is just one of the techniques used in genetic engineering.

Scientists in the genetic engineering field directly alter the hereditary material, the DNA. They may remove pieces of it, add to it, or exchange it between organisms. Because they actually alter the genes, they also alter the traits controlled by these genes — they change what organisms have inherited from their parents. Some researchers do this to learn more about heredity — to discover what genes control what traits, and how various types of genes work. This information is used by other researchers, who use genetic engineering techniques to prevent or correct inherited defects.

DID YOU KNOW?

Identical copies of animals can also be made in a different fashion from Gurdon's cloning method. When an embryo has fewer than 100 cells, it can be split into two or more segments, and in some cases into individual cells. Each of these can then be placed in a different surrogate mother for development. Because the cells in each segment all contain the same alleles, the offspring are genetically identical to each other. Genetically identical sets of horses, cows, and other domestic animals have been produced in this way.

GENE SPLICING

One technique for manipulating genes is to splice together pieces of DNA from different organisms. This process *recombines* specific pieces, to produce what is termed **recombinant DNA**. Another name often used for this process is **gene splicing**. The production of recombinant DNA is perhaps the major focus of genetic engineering today. Various methods are used. They all allow scientists to remove or insert specific genes into the hereditary material.

The most commonly used procedures for producing recombinant DNA involve the basic steps shown in Figure 18.15. By this process, bacteria can be used as miniature factories, producing clones of specific genes quickly and efficiently.

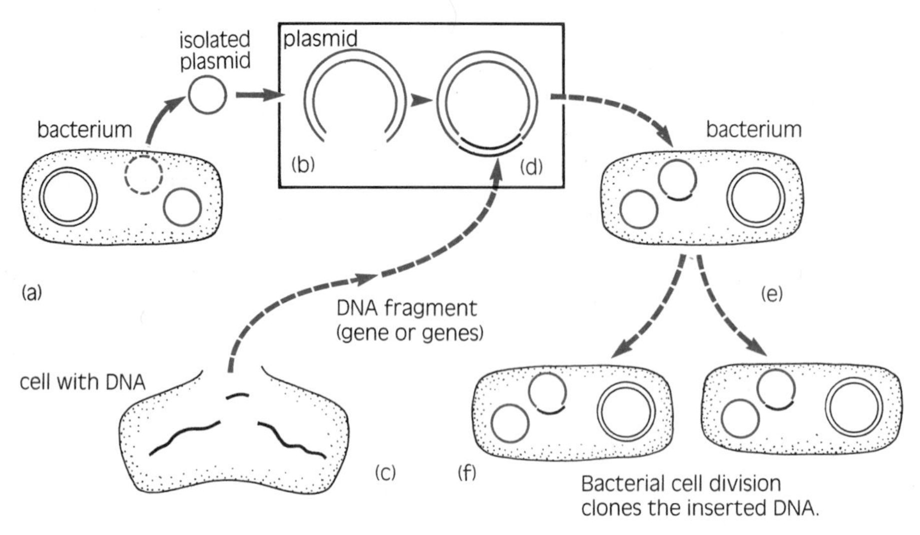

Figure 18.15
The steps involved in producing recombinant DNA.

(a) A plasmid is removed from a bacterium. (A plasmid is a small molecule of DNA found in some bacterial cells.)

(b) The plasmid is broken at one point.

(c) DNA molecules from a plant or animal cell are broken into fragments. The fragment to be used may be a specific gene or genes.

(d) The DNA fragment is spliced into the broken plasmid. The plasmid now contains recombinant DNA: a gene or genes from another organism have been added to its own genes.

(e) The plasmid is put into a bacterium.

(f) The bacterium undergoes cell division many times, making copies of itself. During cell division, the plasmid containing recombinant DNA is duplicated and a copy goes into each offspring cell. As a result, there are now many clones of the desired DNA fragment.

ACTIVITY 18D A Model of Gene Splicing

In this activity, you will model the process of gene splicing.

MATERIALS

modelling clay in at least two different colours

PROCEDURE

1. Use one colour of modelling clay to represent a human chromosome, and the other to represent a bacterial plasmid. Using the clay and a pencil and paper (to draw the outlines of cells and other structures that are part of the process), make a model to show how genes are spliced. Use arrows to indicate the order of the steps involved. Be sure to label all parts of your model. Use Figure 18.15 for guidance.
2. Compare your model with a classmate's. Are they similar? Perhaps one of you will see some way to improve your model so that it will represent the process more clearly.

DISCUSSION

1. (a) Define the term "gene splicing."
   (b) Why are spliced genes also called recombinant DNA?
2. Why are desired genes usually inserted into a bacterial plasmid for copying, instead of into the chromosome of a sexually reproducing organism?

BENEFITS OF GENE SPLICING

Recombinant DNA has been put to many uses in agriculture, industry, and medicine. In all these fields, research is continuing, and new uses for gene splicing technology are being discovered every day.

Many scientists believe that the greatest potential use for gene splicing is in agriculture. The giant mouse in the photograph at the beginning of this chapter has grown bigger as a result of the addition to its mouse genotype of a gene that produces rat growth hormone. Researchers are now working with farm animals to introduce genes that will cause a similar growth increase—resulting in higher meat or milk production. Although their success has been limited, they have shown that such a result is possible. Scientists hope that, in the future, the transfer of other beneficial genes to domestic animals will be possible. Such genes might make sheep more resistant to certain diseases, for example, or reduce the fat content of beef.

Gene splicing is also being done with plants. It has already improved the ability of some crop plants to withstand certain insect pests and herbicides (chemicals that kill plants). Scientists hope that this technology will eventually allow them to add genes to improve features such as drought resistance and crop yield. Gene splicing techniques make it possible for them to do this far more quickly than they could by selective breeding. The economic benefits of such improvements would be enormous.

Industry has also benefited from recombinant DNA technology. This is particularly true of the pharmaceutical industry, which produces medicinal drugs. Some of these drugs are now manufactured using gene splicing. For example, by inserting into bacteria the human gene that controls production of insulin, researchers have been able to produce large amounts of this drug. (Insulin, which is normally produced by cells in the pancreas, regulates the amount of sugar in the blood.) Insulin is vitally important to people with diabetes, a condition in which the pancreas does not produce insulin (Figure 18.16). Previously, it was obtained from cows or pigs. Although animal insulin is similar to human insulin, it is not identical, and some diabetic patients have allergic reactions to it. Mass production of human insulin in bacteria has reduced this problem.

DID YOU KNOW?

The insulin produced by recombinant DNA technology is called Humulin. It was the first drug produced with gene splicing techniques to be approved for human use (in 1982).

Another drug now manufactured using recombinant DNA technology is human growth hormone. Produced by the pituitary gland, this substance is needed for normal growth. Some individuals, however, do not produce enough growth hormone. These people, who have a condition known as pituitary dwarfism, do not grow to the usual human height. The condition can be treated by giving affected people growth hormone obtained from another person's pituitary gland. Until recently, the hormone could only be obtained from donors after their death, and it was in very short supply. Today, ample amounts of growth hormone are manufactured using recombinant technology, by inserting the appropriate human gene into bacterial DNA. There is now more than enough to treat all these who need it. Two other useful drugs manufactured in

Figure 18.16
This person has diabetes. She is preparing to give herself a shot of insulin. The injected insulin will help her body perform the necessary function of regulating the amount of sugar in her blood.

this way are interferon (an antiviral substance produced naturally by the human body) and the blood-clotting factor missing in hemophiliacs.

Other industries have also found uses for recombinant DNA. Genes that break down the molecules in oil have been introduced into a species of ocean-dwelling bacteria. The resulting clones of this bacteria have then helped to clean up ocean oil spills. Researchers are trying to develop micro-organisms that will break down toxic chemicals and others that will convert agricultural waste into fuel.

In medicine, scientists hope to use recombinant DNA to alter a human's genes — rather than just altering traits by using drugs. Their goal is to remove disease-causing alleles and splice normal alleles into the chromosomes of some of the person's cells. A person who might otherwise have been a pituitary dwarf, for example, would then be able to produce his or her own growth hormone. If this were done when the individual was an infant, it would not be necessary to provide the hormone as a drug. Although this sounds quite possible—just another type of recombinant DNA technology — there are many problems involved in transferring genes between humans. At present this process, known as gene therapy, is just beginning to be tried out with humans.

EXTENSION

Recombinant DNA technology has been used to make some types of plants resistant to herbicides. Do you think that the expensive and time-consuming research to discover how to do things like this is worthwhile? Would it be better to stop using herbicides? (Such chemicals are a major source of pollution.) Find out what other methods are available for controlling unwanted plants. Do you think these alternatives could replace herbicides? Write a brief report summarizing your views on this issue.

HAZARDS OF GENE SPLICING

Gene splicing became possible in the early 1970s. At that time, some scientists began to worry about the possible hazards of the technique. For example, what if genes from a cancer cell were introduced into a type of bacteria that could then infect humans with the disease? Or what if some new and dangerous life form were invented, against which humans — or other organisms — had no natural defences? At an international meeting in 1975, scientists and government officials set up safety guidelines for gene splicing research. In the years since then, an enormous amount has been learned about recombinant DNA molecules and about the bacteria being used for gene splicing. It is now known that most of the early fears about this technology were unjustified. There is little chance of accidents happening. Many of the 1975 guidelines have been relaxed. Currently, recombinant DNA research is carried on in many laboratories around the world.

The social and legal issues surrounding the use of this technology have not been as easily resolved. Suppose, for example, that a scientist or a drug company creates a new organism or a new drug by splicing genes together. Unlike machinery or other such items, the raw materials of these inventions are living things. Should the inventors own the legal right (the patent) to sell or distribute these products, as is the case for other types of inventions? Many judges and lawyers differ in opinion over such matters.

Another problem is the potential for misuse of the results of this research. Now that human growth hormone is abundant, people other than pituitary dwarfs are seeking to use it. Athletes and other individuals who would like to be taller (even though they are within the range of normal human height) want to use the drug. However, it is not yet certain whether the drug causes any negative side effects in humans. The insertion of the gene for rat growth hormone can result in damage to the mouse's reproductive system. A similar problem might result from the use of human growth hormone by people whose bodies already produce it. Negative side effects often show up only after long-term use and study of a new drug. Should pharmaceutical companies be allowed to sell human growth hormone before such information is known? Should doctors be allowed to prescribe it for anyone? Who should decide? As is true of other advances in reproductive or genetic technology, the possibility exists that recombinant DNA techniques may be misused.

DID YOU KNOW?

Industry and science have co-operated on research into gene splicing, embryo transfer, *in vitro* fertilization, and other related technologies. The development and use of these and similar procedures is called biotechnology. Canada has over 200 biotechnology firms, which employ tens of thousands of people. And the field is constantly growing!

REVIEW 18.3

1. Define the term "cloning."
2. (a) Describe Gurdon's experiment in which he cloned a toad.
   (b) What is the difference in chromosome number between an unfertilized ovum and an intestine cell?
   (c) Why do you think the ovum started to divide as soon it received the nucleus of the intestine cell?
3. What is genetic engineering?
4. What is gene splicing?
5. (a) List at least five things that are currently produced using gene splicing.
   (b) What are three other potential uses of this technique?
6. (a) Describe some of the possible hazards or disadvantages of recombinant DNA technology.
   (b) What are some ideas you have about dealing with these hazards?

18.4 The Potential for Misuse of Genetic Ideas

There is a danger that the recent advances in genetics, like all technological breakthroughs, may be misused. If we want to avoid such misuse, we need to understand the issues involved.

THE EUGENICS MOVEMENT

During the early part of this century, certain discoveries being made in genetics were wrongly interpreted and misapplied. This particular misuse was associated with a number of ideas commonly referred to as **eugenics**. Eugenics refers to the endeavour by some to improve future generations of the human population through the selection of parents.

The school of thought called eugenics was started in Britain in the 1880s, when a mathematician named Francis Galton (1822–1911) published a book on the subject. "Eugenics" comes from the Greek words *eu*, which means "good" or "well," and *gene*, which means "birth" or "origin." Galton defined eugenics as "the science which deals with all influences that improve inborn qualities of race." His ideas soon gained support and spread. Supporters of eugenics included some scientists, politicians, businesspeople, and other individuals around the world. They felt that most human characteristics were genetically determined. In other words, they thought that mental abilities and personality, as well as physical features, were inherited from the parents. Thus, they concluded that "feeblemindedness," insanity, and even poverty were inherited—as were success in business or politics.

The acceptance of Mendelian genetics in the early 1900s seemed to support these ideas. However, Mendel and others had investigated the inheritance of only simple physical traits such as plant height, seed shape, and flower colour. It was not correct of eugenicists to assume that conclusions based on these studies also applied to complex human characteristics. Also, at the time, no one had closely examined the effect of the environment on human characteristics. Compared with what we know today, the scientists of that time knew very little about genetics. Consequently, many of their conclusions were incorrect.

Despite these errors, eugenic ideas caught hold. Laws based on eugenics were passed in North America and elsewhere in the 1920s and 1930s. These laws were designed to genetically "improve" the populations within various countries. Although some of the people who originally supported eugenics had noble ideals, eugenic ideas soon became a tool of various narrow-minded individuals. Eugenics was used to justify the passing of laws that resulted in the sterilization of people in mental institutions and prohibited marriage between people of different races. Other laws severely restricted immigration to North America from countries whose citizens were considered to be "genetically inferior."

EXTENSION

Scientist, writer, and broadcaster David Suzuki, who lives for part of each year in British Columbia, has reflected on many of the topics in this chapter. His book *Inventing the Future* (Toronto: Stoddart Publishing Co., 1989) contains some of his writings on genetics. You may find the section entitled "The Great Code: Genetics and Society" to be particularly interesting.

Figure 18.17
These people are being sent to the notorious German concentration camp at Auschwitz. The Nazi government used eugenic ideas to justify killing more than 6 million people.

The misuse of genetics reached a peak in Germany in the 1930s and 1940s. During this time, Hitler and the Nazi government used eugenic ideas to maintain the "purity of the master race" and justify the killing of millions of people (Figure 18.17). They had labelled these individuals genetically "inferior" because they were mentally or physically handicapped, or because they belonged to certain religions, races, or political groups.

Since that time, society has generally become very suspicious of eugenics. We now know that *both* genes and environment influence almost every human characteristic. Studies have shown that the environment plays a very significant role in determining complex human characteristics such as intelligence or personality type. In most places, eugenic laws have long since been revoked.

The idea of race assumes that there are identifiable groups whose members tend to marry only within their group. It is difficult to define the basis for certain racial distinctions, and between 2 and 30 races have been suggested. There are no pure races. Genetic studies have shown that the vast majority of genes and alleles are shared by all races. In fact, there is more difference (in terms of shared alleles) between the individuals *within* a single race than there is between different races. Look around you at the variations in height, eye colour, hair colour, skin colour, and other features that exist even within a particular race.

A certain number of scientists and others still support the ideas proposed by eugenicists. In order to make up your own mind about the validity of such ideas, you need to understand some basic genetic ideas. This unit has introduced to you some of these ideas. But new discoveries, like those related to gene splicing, for instance, are taking place every day, and this makes understanding more difficult. Without a clear understanding of the extent and limitations of these genetic discoveries, the possibility of accepting incorrect assumptions becomes a reality. If this is the case, it could become easy once again for a few people to mislead others with their biased and narrow-minded ideas.

Figure 18.18
Dr. Lap-Chee Tsui was the leader of a medical team at Toronto's Hospital for Sick Children. In 1989, this team located the gene responsible for cystic fibrosis. Tsui says that a complete map of human genes would have allowed the discovery to take place three years earlier.

MAPPING THE GENETIC CODE

One of the new technologies you may have heard about is our ability to construct "maps" of chromosomes. These maps describe the order of nucleotides in a DNA molecule. (Chapter 15 described nucleotides, the building blocks that make up a gene.) Until recently, nucleotide mapping was a very time-consuming, expensive process. At first, only single genes could be mapped. Later, the sequences of nucleotides along entire chromosomes were mapped. In 1977, the complete sequence of nucleotides in an entire virus DNA molecule (5400 nucleotides making up nine genes) was mapped. In 1987, all 3 million nucleotides in the DNA of a bacterial species, *Escherichia coli*, were mapped.

Now it is possible to map the human genome—all the genes and their alleles present in humans. Many researchers are now involved in a massive project, the Human Genome Project, which is designed to do this. They hope to have this map completed early in the 21st century.

Such a map could provide many benefits in treating genetic diseases (Figure 18.18). For example, knowing where disease-causing genes are located on chromosomes would be an important first step in applying gene therapy. The genes must be located before disease-causing alleles can be replaced with normal ones.

However, it is important that this genetic information, when it becomes available, is not misused. Suppose that the map shows that certain alleles are found more commonly in criminals or in the mentally ill. Will we then assume that people who have these alleles must *always* display such traits? Or will we remember that the environment plays a vital role in determining how a person thinks and acts, regardless of the alleles he or she possesses? Will society seek ways of overcoming a possible genetic *tendency* towards undesirable behaviour? Or will we simply give up, and isolate individuals with certain genotypes? Will society forbid them to reproduce so that they do not pass on these alleles?

Does a world in which such policies exist sound like the plot for a science fiction movie or novel? Remember that genetic ideas have been misused in the past. Learning about these mistakes can help us avoid repeating them.

ACTIVITY 18E *Discussion*

With a partner or in a small group, discuss the benefits and drawbacks of being able to preselect—at least partially—the genotype a child will have. This can be done by selecting sperm or ova from individuals with certain desired characteristics. The sperm or ova are then joined by artificial insemination or *in vitro* fertilization. Consider the following:

- What effect might predetermining the genetic traits of offspring have for society, as well as for the parents of the children?
- What might happen if the leaders of a country or a group of people decided to make these decisions for all group members—instead of individual couples making their own decisions?
- What controls, if any, would you place on the selection of specific ova or sperm? How would you exercise these controls?

COMPUTER APPLICATION

THE HUMAN GENOME PROJECT

The goal of the Human Genome Project is to "map" the human genome. This map would record the sequence of all 3 billion nucleotides on the DNA in our 23 pairs of chromosomes.

The human body contains about 100 000 genes. (Most are 10 000 to 150 000 nucleotides long.) By 1989, the approximate location on a chromosome had been determined for about 1800 genes. However, the exact sequence and number of nucleotides has been determined for only a handful. If scientists can locate genes responsible for certain genetic diseases, researchers are more likely to find tests and cures.

Until recently, gene mapping was done by hand. It took scientists years to determine the exact sequence of nucleotides for a single gene. Now sequencing by hand or machine can be done much faster—up to 16 000 nucleotides a day. However, the work of sequencing the whole genome will not be attempted without new technology—with machines that can sequence up to 100 000, or maybe even 1 million, nucleotides a day.

Hundreds of scientists in dozens of laboratories all over the world are working on this project. The enormous volume of data it will generate can be managed only with computers. Giant databases are needed to store and retrieve information, and sophisticated software programs are needed to analyse it. In order for everyone to be able to add to, access, and use the material, it is important to have a common method and format for submitting data.

When a scientist maps a section of DNA, the information about both its location on the chromosome and its sequence of nucleotides is reported to a database. Everybody will need to use the same way of recording information. Some new techniques have made it possible for other scientists to read this information as a "recipe" and make their own copies of the DNA section.

Let's say you are a scientist working on the project. You have worked out the sequence of nucleotides for a section of chromosome 4. You send it from your computer to a database. Other scientists are searching for the allele that causes Huntington's disease, which is somewhere on chromosome 4. They call up the database on their computers, read the recipe you submitted, recreate the DNA section in their laboratories, and use it in their research.

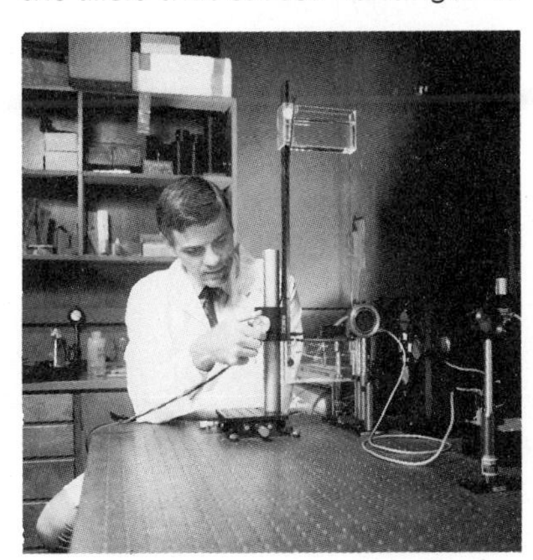

Leroy Hood of the California Institute of Technology. With his colleagues, Hood developed a machine that sequences the nucleotides in DNA. This automated DNA sequencer has greatly reduced the time and cost of gene mapping.

If the project is successful, the information gained will go far beyond locating disease-causing alleles. The gene or genes controlling each function of the human body could be known.

Thirty years from now, your doctor might obtain a profile of your genome. You could get detailed information on your genetic risks for certain diseases. You might be able to alter your lifestyle, diet, or environment to reduce or eliminate some of those risks. Or you might have the anguish of knowing that you are likely to develop a disease for which there is no treatment or cure. What if such information were obtained by a prospective employer or an insurance company? People who seem to be at risk for certain diseases might be unable to find jobs or might be charged more for life insurance.

The project is controversial, both inside and outside the scientific community. It is very expensive, and some scientists worry that time and money will be spent researching unimportant parts of the genome instead of concentrating on the search for disease-causing alleles. The project also raises ethical issues about how we use the resulting information about an individual's genetic make-up. We all must consider the possible harmful as well as beneficial uses for the vast amount of information the Human Genome Project may uncover.

REVIEW 18.4

1. Define the term "eugenics."
2. What did those who supported eugenics believe about human traits?
3. (a) Why did eugenic ideas at first seem reasonable?
   (b) What is the major error in eugenic ideas?
4. What were some of the results of the eugenics movement?
5. (a) What benefits might the Human Genome Project provide?
   (b) What ways might information from this project be misused?

CHAPTER • REVIEW

Key Ideas

- Selective breeding is one of the most widely used methods of influencing the inheritance of genetic traits.
- Seed storage, cross-pollination, and grafting are methods of controlling the reproduction of plants in order to influence their heredity.
- Artificial insemination, *in vitro* fertilization, and the use of surrogate mothers are ways of controlling reproduction in animals. These influence the inheritance of traits.
- These reproductive techniques provide numerous advantages for humans and other animals. Their use in humans also creates a number of social and legal problems.
- Genetic counselling provides information and advice for couples who face the risk of having a child with a genetic disease.
- Various methods of prenatal testing can detect the existence of certain genetic diseases in a developing fetus. These tests include amniocentesis, chorionic villus sampling, and ultrasound examination.
- Direct manipulation of genes is another way to influence genetic traits. Methods for manipulating the genes include cloning and gene splicing (the production of recombinant DNA).
- Recombinant DNA has many beneficial uses in industry, medicine, and agriculture. There are also some hazards involved in producing recombinant DNA, and a number of social and legal issues have arisen from the use of this technology.
- Eugenics is a school of thought that believes that most human characteristics are inherited. Eugenicists, therefore, think that a group of people can be improved genetically by enforcing various reproductive controls.
- The original eugenicists misinterpreted many of the results of early genetic studies. Furthermore, they lacked much of the knowledge we have today. Although eugenic ideas are still supported by some, most people no longer accept them.

Vocabulary

grafting
artificial insemination
in vitro fertilization
surrogate mother
embryo transfer
genetic counselling
prenatal diagnosis
ultrasound examination
amniocentesis
chorionic villus sampling
cloning
genetic engineering
recombinant DNA
gene splicing
eugenics

1. Construct a concept map using some or all of the terms in this vocabulary list.
2. Design a crossword puzzle using as many of the words in this list as possible.

Connections

1. Both selective breeding and gene therapy can be used to influence heredity. The two techniques, however, provide two very different kinds of influence. Explain this comment.
2. In what way are grafting and cloning similar?
3. Imagine that your uncle and one of your male cousins have hemophilia. (You do not.) Put yourself in the different roles listed below, and explain your answers.
   (a) You and your husband are considering whether or not to have children. What should you do?
   (b) You and your wife are considering whether or not to have a family. What should you do?

4. A genetic counsellor has advised a couple that because the wife is a carrier of a genetic disease, there is a high probability that any children she might have would be affected by it. What alternatives are available to the woman and her husband if they wish to have children?
5. Describe three things a rancher might do to improve the genetic quality of a herd of cattle.
6. If they had been available at the time, how might each of the following have been used to further the goals of early eugenicists: *in vitro* fertilization, artificial insemination, and surrogate motherhood?
7. Can selective breeding, artificial insemination, or *in vitro* fertilization always ensure that a particular offspring will have all of the desirable genetic traits of its parents? Explain your answer in full.

Explorations

1. Make a list of the techniques described in this chapter that are used in agriculture. Interview a farmer or rancher in your area to discover whether he or she has made use of any of these techniques to grow crops or breed domestic animals. Alternatively, you might obtain information from a representative of Agriculture Canada or the British Columbia Ministry of Agriculture to investigate which of these techniques are used in your area. Write a brief report summarizing which procedures are used, where, and how frequently.

Figure 18.19
Collecting articles related to genetic technology.

2. Maintain a class bulletin board or scrapbook on recent advances in genetics (Figure 18.19). Include magazine or newspaper articles, pamphlets, or other pieces of information on this topic that you think are interesting. How many of the new discoveries described in these articles are related to medicine and how many to agriculture or industry? How many articles view favourably the technology used? How many concentrate on the disadvantages or new difficulties caused by the technology?

3. Do research on ways gene splicing is used today to help solve environmental problems. Think about other ways in which this technology, as well as other techniques discussed in this chapter, might be used to help the environment. Write a brief report on both the current uses and your ideas. To whom might you send your proposals for consideration? You may want to pursue this matter further after reading the next unit.

Reflections

1. Look back at the notes you made in your learning journal while doing Activity 18A. Have your opinions on any of these issues changed? If so, what do you think now, and why did your opinions change?
2. If you had unlimited funds, what type of genetic research would you do—what specifically would you study? Explain the reasons for your choice.

UNIT VI

Understanding Ecology

Most people in Canada today live in large towns or cities like the one shown here. What do you think the people living in this city need to live? What do you think the people in your community need?

Imagine what would happen if the people in your community could not obtain the things they need. Without resources such as food and water, you and the other people in your community would soon die. Without a supply of energy, you would not be able to have heat and light on a cold winter night. And without large amounts of wood, metals, plastics, and other materials, you would not have a house to live in or cars, blue jeans, toothbrushes, television sets, and the other manufactured items you use every day. Human beings, like all other living things, need and use resources.

Do you know where all the resources your community uses come from? Even more important, do you know what your community does about the garbage, sewage, and pollution that are produced when resources are obtained and used? The more people there are in a place, the more resources they use—and the more garbage, sewage, and pollution they produce.

This unit is about the connections between population size, pollution, and our use of resources. Throughout the world today, the size of the human population is growing rapidly. The many people on Earth are quickly using up some resources and producing more and more pollution. What does this mean for our planet's future? What should we be doing? As you read the chapters in this unit, you will have a chance to consider these and other important questions.

CHAPTER

19 Populations and Environment

A huge swarm of locusts, such as the one in the photograph, may contain as many as 50 billion insects and cover an area of several thousand square kilometres. As the restless swarm migrates across the countryside, the locusts devour vast areas of vegetation and leave the land bare and brown. Swarms occur at unpredictable intervals, after locust numbers begin to skyrocket. Eventually the swarm breaks up as most of the locusts die or are killed. The number of locusts then returns to its previous smaller level.

Locust swarms show the capacity of organisms to increase rapidly in numbers. Although such dramatic fluctuations in population size are unusual, all populations of organisms may change slightly in size from year to year. You may have observed that populations of some animals and plants in your area are increasing. For instance, you may have noticed more pigeons or more dandelions around your school. You may also have noticed some organisms declining in numbers. You may be catching fewer of some types of fish in your favourite lake, or seeing fewer of certain types of birds at your feeder than in previous years.

In this chapter, you will examine some different patterns of population change and consider what causes these changes. By understanding the factors that affect population size, people can predict how populations may change in the future. This is important for controlling growth in populations of pest organisms, preventing declines in populations of rare animals and plants, and planning for the future of the human population.

ACTIVITY 19A Observing a Population

In this activity, you will think about changes in the size of a **population** of a particular **species** of plant or animal in your neighbourhood. (Remember: a species is a group of closely related and similar organisms that can interbreed. A population is the number of individuals of a species that occur together in a particular area at the same time.) Choose a species of plant or animal in your area. What have you noticed about the population size of your chosen species? Is it growing? Is it shrinking? Is it staying the same size? Briefly record your observations, then write down any questions you have about reasons for the change—or lack of change—in the population size of your species.

19.1 What Causes Changes in Population Size?

There may be a population of 320 seals in a seal colony (Figure 19.1), 650 cedar trees on a tree farm, 23 fleas on a dog, or 1724 people in a town. Populations have several characteristics. These include their size (the number of individuals in the population); **density** (how crowded the individuals are); sex ratio (how many males and females); and age ratio (how many young, middle-aged, and old). All these characteristics are important, but population size has the greatest impact on the environment. It is this characteristic you will be studying in this chapter.

Figure 19.1
What might cause the number of seals in this population to change?

Changes in population size can be brought about by four different processes. These processes are already familiar to you. Populations can change in size when individuals *move into an area* or **immigrate**, *move out of an area* or **emigrate,** and when individuals are born or die (see Figure 19.2). Two of these processes—immigration and birth—increase the size of a population. The other two processes—emigration and death—decrease the size of a population.

Think about the population of people that live in your community for a moment. Imagine you can travel through time and visit your community one year from today. You find that the population size has changed. There may be more people, or there may be fewer. Obviously, one or more of the four processes affecting population size will have been responsible for this change. People will have been born, or died, or moved in or out.

Now suppose you travel a year into the future and find the same number of people in your community as live there today. You might conclude that nobody has moved in or out and that nobody has been born or died during that time. This is possible, but it is more likely that all or some of the processes of change have taken place among individuals in the population. The population as a whole has remained the same because the total that were born and immigrated equalled the total of those that died and emigrated. The overall effect of all four processes on population size can be summarized in a simple equation:

Change in Population Size = (Births + Immigrants) − (Deaths + Emigrants)

During your studies of populations, keep in mind that births, deaths, and movement of individuals occur in all populations over time. Whether a population grows, shrinks, or remains the same depends on which of these processes occurs most frequently relative to the others.

Figure 19.2
The four processes that affect population size.

ACTIVITY 19B Measuring Changes in Population Size

The following activity uses beans to represent a population of organisms (Figure 19.3). The activity will show how several processes contribute to changes in population size over a period of time.

Figure 19.3
Beans can be used to illustrate the ways populations change in size.

MATERIALS

50 beans

PROCEDURE

1. Count 30 beans into a pile and set the rest aside. Create a data table with the following headings: Month, Births, Immigrants, Deaths, Emigrants, Population. Record the size of your initial population (30). As you read the account below, transfer beans between your population and the reserved pile to demonstrate the effects of the four processes described. Record the changes on your data table.
   - During one month, your population produces 20 offspring. Ten individuals die, and 5 move into the area. Count or calculate the net change in the population at the end of the month and record this figure.
   - During the next month, 5 more offspring are born, 10 individuals die, and 15 emigrate. Count or calculate the net change and record the size of the population at the end of the month.
   - In the third month, 3 offspring are produced, 8 individuals die, 7 move into the population, and 2 move out of the area. Count or calculate the final population size.
2. Draw a bar graph to compare the size of the population at the start of the experiment and at the end of each month.

DISCUSSION

1. From the figures recorded on your data table and illustrated in your bar graph, give
   (a) the biggest population size,
   (b) the smallest population size,
   (c) the average population size.
2. During the three months of observation, did the population size grow, shrink, or remain the same?
3. Suppose that for each of the next six months the number of immigrants equals the number of emigrants. Could the population
   (a) increase in size?
   (b) decrease in size?
   (c) remain the same?
   Explain your answers.

HOW POPULATIONS GROW

In some regions of the world, and in certain periods of history, human population growth has been greatly influenced by the movement of individuals from country to country. For example, between 1976 and 1981, the population of Canada grew by 1 349 000. Nearly one-third of this increase was the result of immigration (Figure 19.4). For most species, however, the effect of immigration on population growth is very small compared with the effect of births.

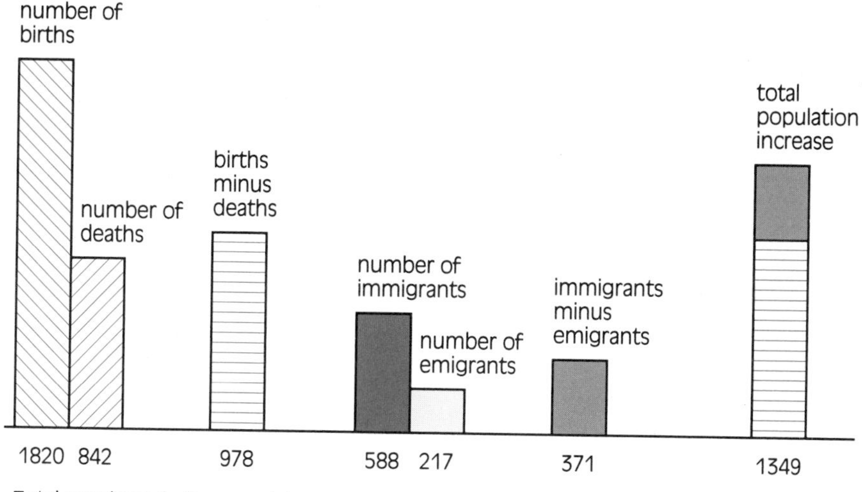

Figure 19.4
Births and immigration have increased Canada's population size.

By reproducing, all organisms can increase their numbers (Figure 19.5). When a mouse gives birth to baby mice, the size of the mouse population goes up. When dandelion seeds drift to the ground and begin to grow, the dandelion population increases. If *all* those baby mice and dandelion plants survived and reproduced in their turn, before long the face of the Earth would be covered with mice and dandelions. The fact that this does not happen indicates that there are factors limiting population growth. You will be studying some of these factors later in this chapter.

DID YOU KNOW?

The population of gulls in North America has been doubling every 12 to 15 years since the beginning of this century. This population boom is partly a result of the increased food supply provided by garbage dumps.

Figure 19.5
Reproduction is the most important means of increasing population size among animals and plants.

ACTIVITY 19C Multiplying Microbes

This activity will allow you to calculate and graph the growth of a rapidly expanding population.

MATERIALS

calculator
graph paper

PROCEDURE

1. Imagine that a scientist places a single bacterium (plural: bacteria) into a beaker and fills the beaker with all the nutrients the bacterium needs to survive. The beaker is sealed so that no more bacteria can get in, and none can get out. After 20 min, the bacterium reproduces by dividing in two. Assume that every 20 min afterwards each bacterium in the beaker divides in two. How many bacteria will be in the beaker when the scientist leaves the lab 6 h later?
2. Draw a table to record your results. You will need to keep track of the time elapsed, the number of generations, and the total number of bacteria in each generation.
3. Calculate and record the number of bacteria in the beaker every 20 min for a total of 6 h.
4. Plot the data from your table onto a graph. The horizontal axis (x-axis) will show the generations from 0 through 18. The vertical axis (y-axis) will show the total population of bacteria in each generation. Be sure to choose scales that allow you to include all of the data.

DISCUSSION

1. Describe the shape of the curve on your graph.
2. How many generations would it take to produce over a million bacteria in the population?
3. How many generations would it take to produce 5 million bacteria in the population? Ten million?
4. Which of the four processes that affect population size had the most influence here?
5. Would you expect a real population of organisms to grow in the same way? Explain why or why not. What factors might limit the final population size?
6. Why was it difficult to choose a scale for your vertical axis? How else could you plot the data to avoid this problem?

EXPONENTIAL GROWTH

The type of dramatic population growth shown by the curve in Figure 19.6 is called **exponential growth**. The population of Eurasian water milfoil in British Columbia grew in this way after the plants were introduced from Europe. In their new environment, the plants had plenty of space and nutrients. There were no diseases that affected them or animals that fed on them. Their high rate of reproduction and low rate of death soon caused their numbers to increase rapidly.

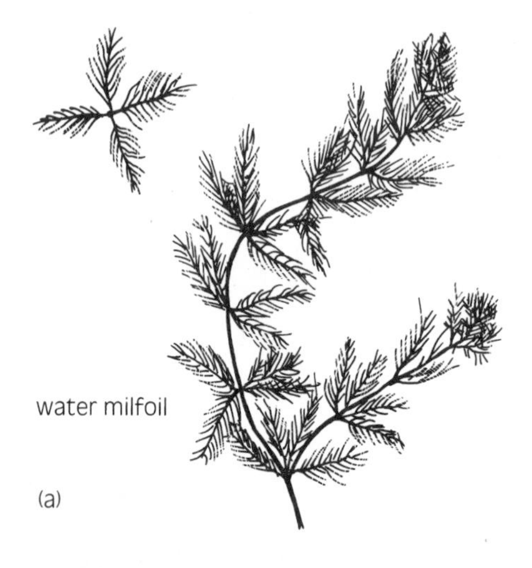

(a)

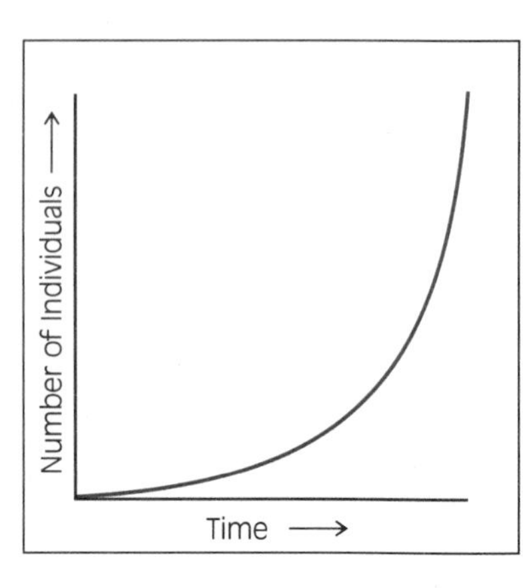

(b)

Figure 19.6
(a) The water milfoil was brought to North America from Europe. The plants spread rapidly in British Columbia and soon choked many waterways.
(b) The increase in water milfoil is an example of exponential population growth.

An exponential growth curve always has the same shape. It begins with a stage when the population grows slowly and the curve rises gently. This is followed by a stage when the population grows rapidly and the curve rises steeply. The completed graph line is often referred to as a **J-shaped curve** because it resembles the letter J.

During exponential growth, the population multiplies from generation to generation by a constant amount (for example, by two). Why, then, does the growth rate change from slow to fast? At the beginning, when the population is relatively small, the increase in numbers from one generation to the next is also small. For example, two multiplied by two produces four, four multiplied by two produces eight, and so on. But as the numbers grow bigger, the increase from one generation to the next also gets bigger. Two thousand multiplied by two produces 4000, and 4000 multiplied by two produces 8000. Although it takes many generations of doubling to increase the population size from two to 1 million, it takes only one generation more to add another million to the population!

All populations of organisms have the *potential* for exponential growth. The J-shaped pattern of population growth has been observed among micro-organisms, plants, and animals. But no populations can grow in this explosive way for very long. Imagine what might happen if they did! Long before most populations reach the size of the locust swarm you saw at the beginning of this chapter, environmental factors cause population growth to slow down or decline. In the next section, you will discover what some of these factors are and how they affect the pattern of population growth.

EXTENSION

Suppose you work for one cent on Monday, two cents on Tuesday, four cents on Wednesday, eight cents on Thursday, and so on—doubling your pay each day. Calculate how many days you must work before earning a million dollars.

REVIEW 19.1

1. What is a species?
2. What is a population?
3. What two processes lead directly to an increase in population size?
4. What two processes lead directly to a decline in population size?
5. Suppose that in January you count a population of 25 cats on your street. Six months later, you count the cats again and come up with a population of 32. Assuming both your counts were accurate, what factors might account for this increase in numbers?
6. Suppose you plot a graph of population size against time and produce a J-shaped curve. Identify two stages in this curve and describe what is happening in the population at each stage.
7. A single water plant is put into a large pond. Every day the plant population doubles in size. After 30 days the plants completely cover the surface of the pond. After what number of days would the plants cover half the surface of the pond?
8. Sketch a graph to represent the growth curve of the plant population in question 7. What kind of growth does the curve represent? What might happen to the curve after the thirtieth day? Why?

19.2 Effects of the Environment on Population Size

The photograph in Figure 19.7 was taken during the 1950s in Australia, where a population explosion of rabbits caused problems for farmers across the country. A favourable environment with lots of food and few predators had allowed rabbit numbers to grow exponentially. But environmental factors also led eventually to a population crash. Crowded conditions throughout the rabbit population helped the rapid spread of a fatal disease introduced deliberately by humans. Hundreds of thousands of rabbits died and the population fell to a much lower level.

Figure 19.7
What factors in their environment might cause more rabbits to be born or more rabbits to die?

FACTORS AFFECTING POPULATION GROWTH

Environmental factors can affect population growth by altering the **birth rate** and **death rate** (that is, the number of births and deaths over a certain period of time). Some of these factors increase or decrease in intensity as the size of the population itself changes. Consider the death rate caused by owls preying on a population of lemmings. The lemming death rate is likely to increase as the lemming population increases, either because the larger number of lemmings will attract and support more owls, or because each owl is likely to encounter lemmings more often. Whatever the reason, there is clearly a relationship between population density and population growth. Factors that are influenced by the number of individuals in a population are called **density-dependent factors.**

Other factors can affect population growth regardless of population size. For instance, a natural disaster such as a severe snowstorm may dramatically increase the death rate and reduce the population size of robins. The population density of robins clearly did not cause the snowstorm or affect the number of birds killed. Such factors, which are not affected by the size of the population, are called **density-independent factors.**

ACTIVITY 19D Predicting Changes in Population Size

The following activity will help you think about the different factors that affect population growth and predict what might happen to a sample population.

MATERIALS

graph paper

PROCEDURE

Read the paragraph below. It describes a population and its environment and asks you to create a graph that indicates how the growth of the population might change over time. (Note: You are not required to put actual figures on your graph. Your aim is to sketch the general shape of the graph and to identify probable changes in population growth.)

A small population of deer lives on a forested island with a mild climate (Figure 19.8). There are no predators on the island and no diseases. A game manager wants to find out how one environmental factor—the availability of food—affects the deer population. Once every month, the game manager ships 10 truckloads of hay to the island for the deer to feed on. Sketch a graph that shows probable changes in population size over the next 10 years: use the x-axis to show time and the y-axis to show population size. (Note: Your only options after plotting the starting point are to have the population go up, go down, or remain the same. Decide how rapidly your population may go up or down, and for how long it may stay in a stage of growth, decline, or stability.)

Figure 19.8
Deer feed on vegetation, including grass, twigs, bark, leaves, and shoots.

DISCUSSION

1. What assumptions did you make in order to produce your graph?
2. Explain each different stage in your population curve and relate each stage to events occurring in the population.
3. What other environmental factors, apart from food supply, might affect the growth of the deer population?

POPULATIONS AND RESOURCES

The availability of resources clearly affects the size of a population. In turn, the size of the population itself affects the availability of resources. For example, imagine a population of 100 rabbits living on an island (see Figure 19.9 on page 444). Now imagine a population of 1000 rabbits living on the same island. In the second population, each rabbit has, on average, 10 times less space, less food, less water, and less shelter than each rabbit in the first population—even though the amount of resources on the island is initially the same.

As any population grows in size, the organisms use more and more of the resources in their environment. With increased competition for resources, there is likely to be a decrease in the birth rate and/or an increase in the death rate. These changes slow the rate of population growth. The demand of the population on the environment then becomes more constant as the number of individuals begins to stabilize. Eventually, the death rate matches the birth rate and the population levels off. This chain of events produces an **S-shaped curve** of population growth (Figure 19.10 on page 444). Like the J-shaped curve, it starts with a stage of slow increase in numbers followed by a stage of rapid increase. In the third stage, population growth slows and eventually stops. As a result, the population size levels out or stabilizes.

Figure 19.9
(Top) Each rabbit has more of everything—space, food, water, shelter—than each rabbit in the lower illustration.

Figure 19.10
An S-shaped growth curve has three stages.

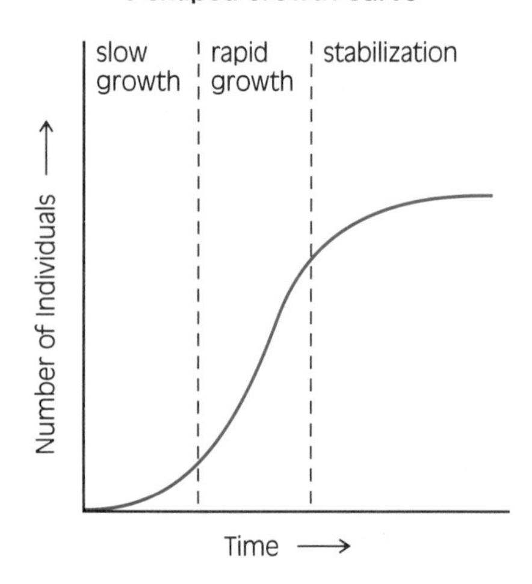

CARRYING CAPACITY

Carrying capacity refers to the maximum number of organisms that can be supported indefinitely by a particular environment. Most populations continue to fluctuate slightly above or below the carrying capacity. When the population grows *above* the carrying capacity, there are not enough resources for each individual. Without all the food, water, and shelter they need, more individuals may die and/or move away from the area. There may also be fewer births if members of the population are not well nourished enough to reproduce. These factors soon bring the numbers down. When the population drops *below* the carrying capacity, there is an excess of resources for each individual. More organisms are likely to survive and reproduce, which takes the numbers back up (Figure 19.11).

The carrying capacity of an environment is different for each species, depending on which resources and how much of them the species requires. If any factors in the environment change, then the carrying capacity, and the population size, may also change. The carrying capacity may go either up or down. For example, if bird feeders or nesting boxes are provided in a neighbourhood, the carrying capacity is increased for some species of birds and their populations may grow. If all the weeds in a neighbourhood are removed, the carrying capacity is reduced for species of butterflies that lay their eggs on the weeds, and their populations may shrink.

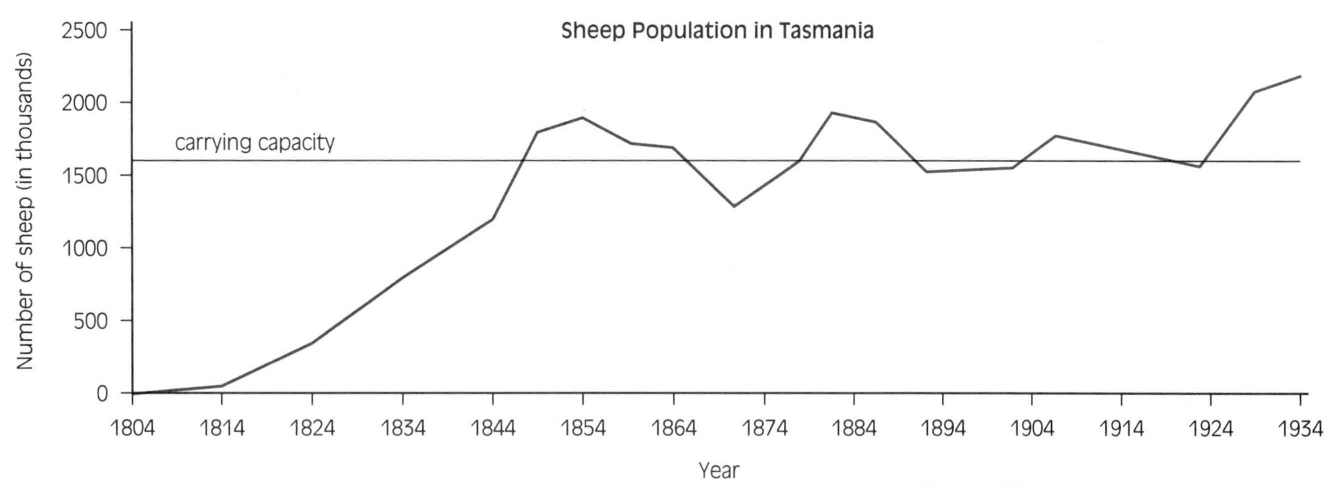

Figure 19.11
After wild sheep were introduced to the island of Tasmania in the early 1800s, the population grew in an S-shaped curve. The population reached the carrying capacity of the island in the mid-1800s, then fluctuated between 1.5 and 2 million over the next 80 years.

HUMAN POPULATION GROWTH

Scientists who study changes in the human population are called **demographers.** They analyse patterns of population growth and try to answer questions such as: Why do more people live in one place than another? Why is the birth rate high in some parts of the world and low in others? What are the impacts of the human population on the world around us? These are important questions because many alarming problems in the world today—from starvation and disease to pollution, revolution, and mass migration—may be connected with human population growth (Figure 19.12).

Examine the figures of world population in Table 19.1 on the next page. Demographers have estimated that the entire human population of the world in 5000 B.C. was only about 10 million (10 000 000) people. It took all of human history up to 1850 for the population to climb to 1 billion (1 000 000 000). Only 75 years later, in 1925, the population was 2 billion. It took 50 more years to reach 4 billion, in 1975, and only 12 more years for the world population to reach 5 billion. How many more people might be added to the population during your lifetime?

DID YOU KNOW?

In 1990, Mexico City had over 20 million residents. By the year 2000, it is expected to have another 7 million.

Figure 19.12
There are more people on the Earth today than ever before.

Table 19.1 World Population

Date	Approximate population of the Earth
5000 B.C.	10 000 000
A.D. 1	300 000 000
A.D. 1500	500 000 000
1850	1 000 000 000
1925	2 000 000 000
1962	3 000 000 000
1975	4 000 000 000
1987	5 000 000 000

ACTIVITY 19E *Illustrating Human Population Growth*

In this activity, you can get a clearer idea of the pattern of human population growth by plotting the data from Table 19.1 on a graph.

MATERIALS

graph paper

PROCEDURE

1. Use the long side of your sheet of graph paper for the x-axis. Start with the year 1500 at the left side of the axis. Divide the axis into equal 50-year intervals to the year 2000.
2. Divide the y-axis into five equal intervals, each representing 1 billion people.
3. Plot each of the population figures listed from 1500 to 1987, then join the points.

DISCUSSION

1. Does the shape of your graph most resemble a J-shaped (exponential) growth pattern or an S-shaped growth pattern?
2. Suggest two factors that might have produced this growth curve in the human population.
3. At its present rate of growth, what will the approximate human population be when you are
   (a) 20 years old?
   (b) 30 years old?
4. Do you think human population growth is subject to the same controls that affect populations of other organisms? Explain your answer.
5. Predict what might happen to human population growth over the next 100 years, giving your reasons. How might this pattern of growth affect the future of your children? How might it affect our environment?
6. How many people do you think the Earth can support? Give reasons for your answer.

ZERO POPULATION GROWTH

The global human population cannot grow indefinitely. The current total of over 5 billion is being added to at a rate of about 85 million a year. This means that *every week* there are about 1.6 million more people on our planet—that's similar to the population size of Greater Vancouver! Like populations of locusts, rabbits, or deer, at some point the human population must stabilize or decline. In order to achieve a stable population, the number of births per year must be equal to the number of deaths. This situation is called **zero population growth.**

To bring about population stability without increasing the death rate, it is necessary to reduce the birth rate. Two children per couple would be enough to replace the parents in the population when the parents die. In practice, the average number of children per couple would have to be a little more, to allow for deaths among children before they reach adulthood. Figure 19.13 compares the future population growth in Canada based on average family sizes of 2.1 children, 1.8 children, and 2.36 children per woman. Notice how seemingly small differences in family size can have large effects on population growth. (Note: In 1986, the average number of children per woman in Canadian families with children was 1.9. In many countries today, the average is more than 5 children per woman.)

Even in the unlikely event that all the people in the world agreed today to limit their family size to achieve zero population growth, the population would not stabilize right away. In many countries, about half the population is under 15 years old. These individuals are likely to have children of their own during the next 5 to 10 years, while their parents are still alive. Thus, the population would continue to grow rapidly for another generation or two even with replacement-level family size.

EXTENSION

Devise a dramatic way of letting other people know about the size and growth rate of the human population. You might plot the numbers on a large graph or chart. You might make a model, using small objects, containers of liquid, or even repetitive sounds to represent the existing population and the rate of increase. A small group might act out the numbers in some way. What scale will you use? Based on the approximate annual increase in numbers (85 million), calculate the rate of increase per month, week, day, minute, or second—whichever you choose for your display.

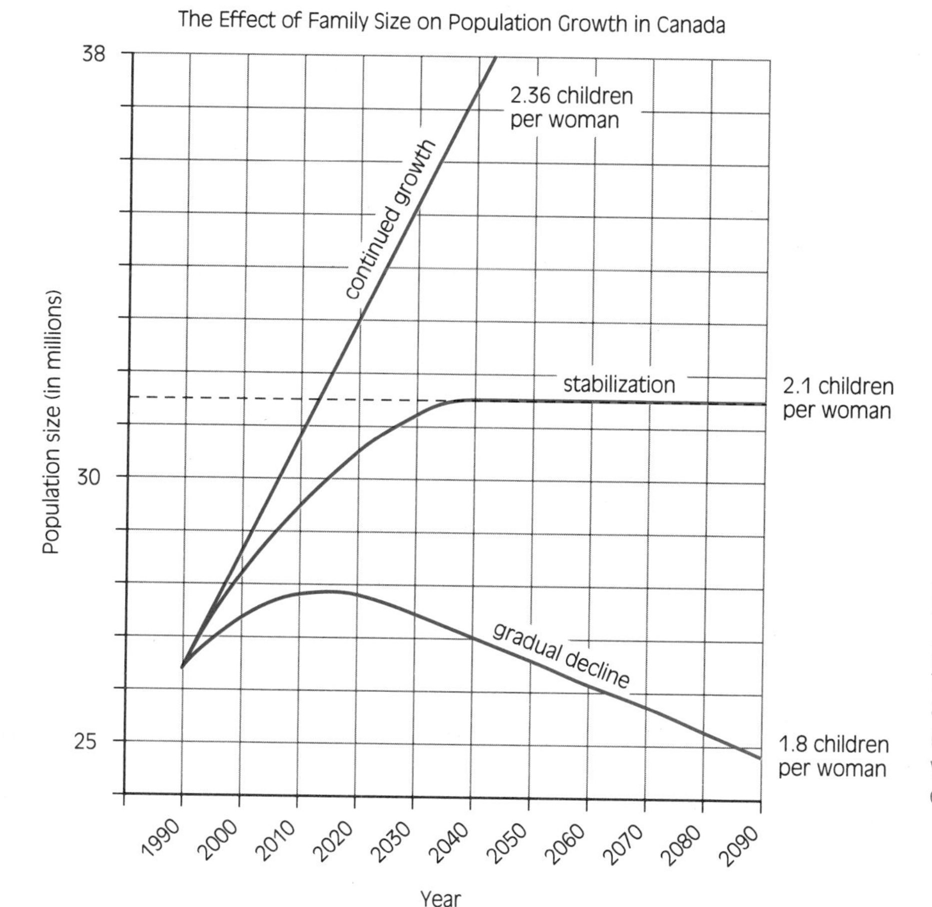

Figure 19.13
The graph shows projected populations that would result from three different family sizes, assuming a net migration of zero. In reality, immigration to Canada would increase all of these estimates.

HUMANS AND THE GLOBAL ENVIRONMENT

You have learned that rapid exponential growth may occur in populations of organisms when they have extra space or food, and when their death rate from factors such as predators or disease is reduced. The same factors have also allowed the human population to grow exponentially.

Until a little over 200 years ago, disease, food shortages, and wars kept the global human death rate more or less in balance with the birth rate. The population fluctuated or grew slowly. During the 1800s, however, rapid advances in scientific knowledge and technology brought about several important changes that greatly reduced the death rate (Figures 19.14, 19.15). Developments in agriculture and transportation allowed people to produce and distribute more and better food. Developments in medicine and hygiene — including sewer systems — greatly reduced the risk of disease. As a result of these changes, more children survived into adulthood and more people lived longer. With a lower death rate, population numbers began to climb. At the same time, people were exploring and developing more regions of the world.

Figure 19.14
Edward Jenner, an English physician, helped to lay the foundations of modern immunological science with his discovery that people could be vaccinated against smallpox.

Figure 19.15
Louis Pasteur, a French chemist, extended Jenner's work with his studies of bacteriology and infectious diseases.

In effect, we humans have increased the world's carrying capacity for our own species. Through science and technology, we have increased our food supply and living space. We have reduced or eliminated many of our predators, parasites, and diseases. Our expanding numbers are now using the resources of the world at an unprecedented rate. You will consider what impact the growing human population has on other species and the environment in the next section.

SCIENCE IN OUR WORLD

PANDEMICS AND POPULATIONS

They are too small to see, yet their effect on population size can be more devastating than war. For centuries, bacteria and viruses have played a major role in reducing the size of human populations by causing pandemics — epidemics that affect vast areas.

The best-known pandemic disease is the bubonic plague, a bacterial infection that has killed untold millions over the last 1500 years. Caused by the bacterium *Yersinia pestis*, bubonic plague is usually transmitted from rats to human beings by fleas. Humans infected by the bacteria often develop black swollen patches. Because of this, the term "Black Death" was used for the disease when it swept Europe during the 1300s.

The Black Death had a serious effect on Europe's population size. From 1347 to 1351, plague killed about 24 million people. And even though thousands fled cities, the disease still spread because no one understood the nature of the plague bacteria or the part played by rats and fleas. It was not until rats were brought under control in the 1800s that bubonic plague became less common in Europe.

In British Columbia, another disease did its part to reduce population size. During the 1860s, the coastal native Indian population was decreased by a smallpox epidemic. Two years after the smallpox virus was introduced by European traders and settlers, a third of the coastal native Indian population was wiped out and a generation of Indian leaders was lost.

The next pandemic to attack a generation was a worldwide outbreak of virulent influenza in 1918. Almost 22 million people, many of them young adults, died in less than one year. These deaths pushed population sizes around the world back to the levels of the 1800s.

Today, population levels worldwide are higher than ever before, but we also have another possible pandemic to worry about. In the 1970s, cases of the disease we now call acquired immune deficiency syndrome (AIDS) became common in Africa. In 1983, researchers identified the cause of AIDS, the human immunodeficiency virus (HIV). They also determined that HIV is a virus, spread mainly by sexual contact and intravenous drug use. Both heterosexuals and homosexuals can get it. Some infected people carry HIV for years with no obvious ill effects, whereas others develop AIDS and die very quickly.

At present, there is no vaccine and no cure for AIDS. This shows us we cannot assume that epidemic diseases are a fear of the past. Viruses in particular are astonishingly adaptable and prolific. Geneticist Joshua Lederberg has called them "our only real competitors for dominion of the planet." Undoubtedly, the size of human populations everywhere will continue to be altered by the pandemics that these "competitors" and other microscopic agents of disease can cause.

Native Indian communities such as this were ravaged by smallpox during the 1800s.

REVIEW 19.2

1. Name four environmental factors and say how each may affect either the birth rate or the death rate in a population.
2. Explain the difference between a density-dependent factor and a density-independent factor.
3. Suppose you plot a graph of population size against time, and the result is an S-shaped curve. Describe how and why the population changes in this pattern over time.
4. What happens to the relationship between resources and the individuals in a population as the population grows in size?
5. Describe what is meant by the term "carrying capacity."
6. Give an original example of how you might increase or decrease the population of a species by altering the carrying capacity of its environment.
7. A scientist placed equal numbers of two different aquatic microorganisms in a test tube of water. One species of microorganism obtains its food by photosynthesis. The second species obtains its food by eating the first. Draw a sketch to show how the populations of both species might change over time.
8. Complete the following sentences:
   (a) In the past, human population growth was kept low by . . .
   (b) Since the 1800s, the world population has grown rapidly because . . .
9. What is meant by the term "zero population growth"? How may it be achieved?

Figure 19.16
In the southwestern deserts of the United States, starlings and owls compete for nesting holes made by woodpeckers in cacti. The number of nesting holes determines the maximum population size of these two species.

19.3 Effects of Population Size on the Environment

The interactions among populations of different organisms and their environment are complex. Consider the ways in which three species of birds and one species of plant living in the California deserts affect one another (Figure 19.16). All three birds nest in holes in giant cacti. Woodpeckers use their long beaks to dig out nesting holes. Starlings and owls cannot make holes for themselves, so they nest in abandoned woodpecker holes. The starlings nest early in the breeding season while the owls are still in their wintering grounds. When the owls return to the deserts, the aggressive starlings defend their holes. Because of these relationships, an increase in the starling population has led to a decrease in the owl population. The total population size of both, in turn, depends on the number of woodpeckers—and on the number of cacti.

This example shows how the population growth of each species may depend on only one or two factors in its environment. These factors may be **biotic** (living), such as a plant, or **abiotic** (non-living), such as the soil. In turn, environmental factors are themselves affected by the size of the populations they support.

EFFECTS OF HUMAN POPULATION GROWTH ON OTHER SPECIES

The growth of the human population affects the population size of many other animals and plants, increasing some and decreasing others. For example, a large human population produces large populations of the domestic crops and livestock we need and the pets and ornamental plants we value. The growing human population also helps other species increase, by reducing their natural predators, or by providing them with more food or living space. For example, populations of rats, cockroaches, and pigeons tend to thrive in towns or cities and increase as the human population grows (Figure 19.17).

In contrast, the numbers of many animals and plants drop as the human population increases. This may happen because human activities reduce their living space or food supply. For example, when large areas of forest and marsh are cut and drained to provide room for expanding farmland and towns, organisms that depend on these habitats become scarcer (Figure 19.18). Other organisms may decline in numbers as a direct result of hunting, harvesting, or collecting by humans.

DID YOU KNOW?

At the beginning of this century, intensive hunting reduced the population of grey whales that migrate past the coast of British Columbia from several thousand to a few hundred. Protection from hunting has allowed this population to increase again to about 11 000 in the eastern Pacific.

Figure 19.17
The pigeon population grows with the human population.

Figure 19.18
Populations of frogs have fallen as wetlands in many parts of the world have been drained and polluted by people.

ACTIVITY 19F Managing Fish Populations

Figure 19.19
Is overfishing reducing fish populations off the B.C. coast?

How many fish of a particular species can be caught each year without reducing their population? This is an important question for fisheries managers (Figure 19.19). To predict the future size of fish stocks, scientists need to examine data on changes in the population size of fish. The following activity allows you to analyse data of this kind.

PROCEDURE

Examine Figure 19.20 on the next page. The map shows a commercial fishing region off the coast of British Columbia. The graph shows changes in the quantities of three kinds of fish caught there by Canadian vessels between 1955 and 1984. Use this information to answer the following questions.

→

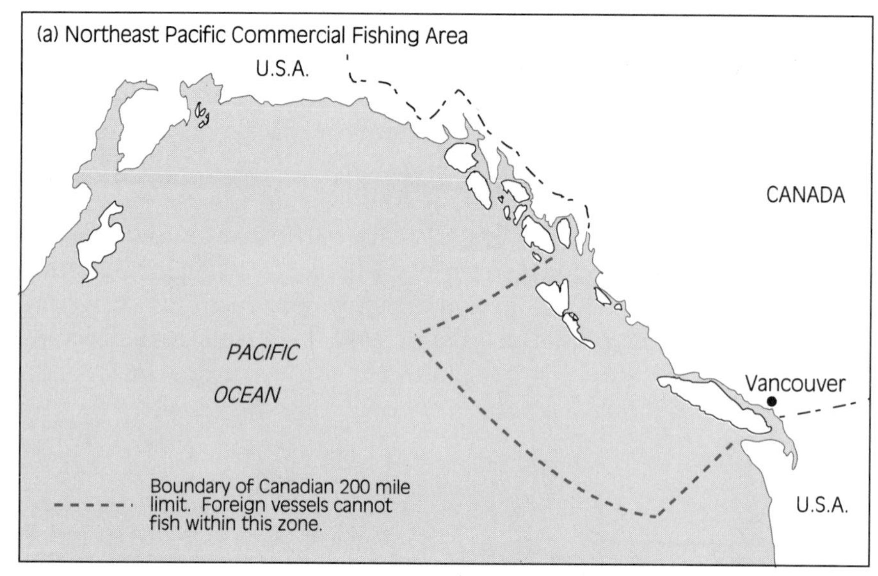

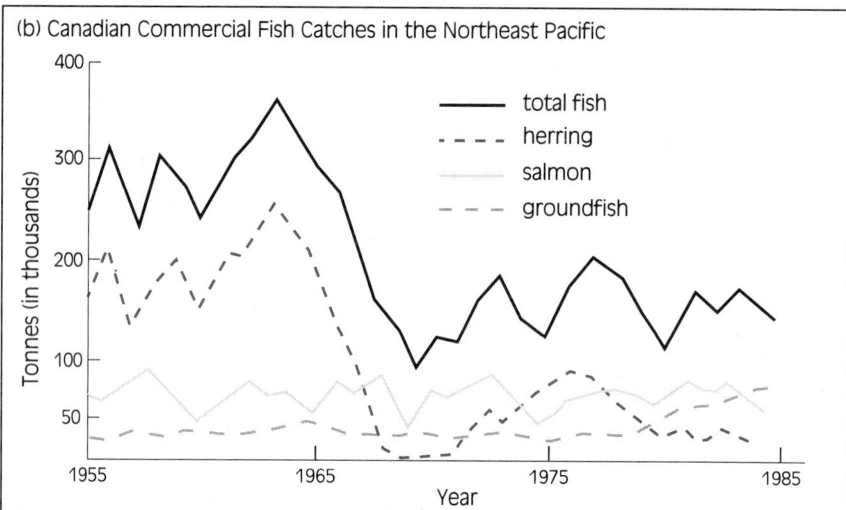

Figure 19.20
Fisheries officials consider information about catches over many years when estimating the size of fish populations.

DISCUSSION

1. From 1955 to 1967, herring were caught in much larger quantities than other species of fish. From 1968 to 1972 they were caught in smaller quantities than other fish. Suggest what might have caused this drop in the quantity of herring harvested.
2. What data would you need to support your answer to question 1?
3. How would you explain the increase in herring catches after 1972?
4. What other explanation(s) might account for the shape of the graph showing herring catches?
5. How would you explain the fact that catches of salmon and groundfish did not decline in the same way as catches of herring, but fluctuated at similar levels throughout the 30 years?
6. What recommendations would you make to ensure that populations of fish do not decline in the future?

OVERPOPULATION AND EXTINCTION

The population of any species may occasionally grow to a much larger size than its environment can support. The forest tent caterpillars shown in Figure 19.21 were part of such a population explosion. Large numbers of caterpillars stripped away the leaves from huge areas of forest in central Ontario, weakening and killing many trees. When a large population begins to damage the resources on which it depends, it is said to be in a state of **overpopulation**.

Any species in a state of overpopulation can cause long-term changes to its environment. For example, growing populations of cattle on Africa's dry grasslands overgraze the vegetation on which they feed (Figure

19.22). The herds' overgrazing and trampling prevent the vegetation from becoming re-established. The lack of plant cover increases soil erosion and slowly changes the area into a desert. This new desert cannot support the cattle or other grazing animals, such as antelope. These animals will either starve to death or move elsewhere, and with them will go the predators, such as lions, that feed on the grazing animals.

When the overpopulation of an organism damages the environment in this way, the damaged environment in turn affects the organism's future population growth. There are three possible outcomes that can follow (Figure 19.23). All of these outcomes will affect the environment and other populations.

Figure 19.21
Population explosions of forest tent caterpillars in Ontario have damaged large areas of forest.

Figure 19.22
Large populations of cattle on dry grassland have helped create a desert.

Three Outcomes of Overpopulation

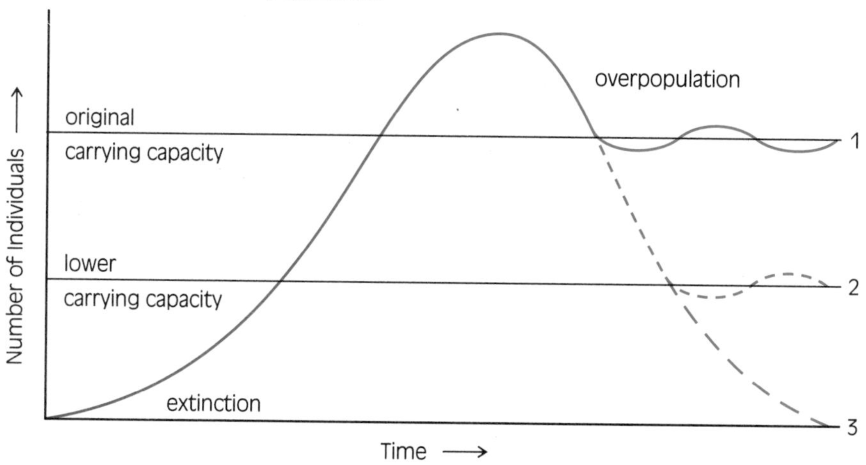

Figure 19.23
Following overpopulation, a population may: (1) stabilize at the original carrying capacity; (2) stabilize at a lower carrying capacity; or (3) become extinct.

1. If the damage to the environment is not too severe, the population may drop back to the original carrying capacity and become stabilized.

2. If the damage to the environment has reduced the carrying capacity below its original level, the population may plunge as rapidly as it grew and stabilize at a new, smaller size.
3. If the damage to the environment is so great that the organisms can no longer survive in the area, the population may leave the area, fall to a very low level, or become **extinct**—die out altogether.

HUMAN POPULATIONS AND RESOURCES

So far, you have considered all the people in the world as part of a single global human population. However, the rates of population growth and the impact of people on the environment differ greatly from region to region (Figure 19.24). The population in some regions is growing slowly or is already stable, with birth rate equal to death rate. In other regions, the birth rate is much higher than the death rate, and the population is growing rapidly. These differences in population growth rate and environmental impact occur for many reasons, including differences in political systems and cultural traditions; levels of industrialization, economic development, and education; distribution of wealth; and differences in climate, vegetation, and terrain.

Unlike populations of other animals, whose resource use is based only on population size, resource use by human populations also depends on wealth and cultural values. Wealth, usually expressed as the average income per person in a population, is a measure of the population's ability to demand, control, and use resources such as food and energy.

Figure 19.24
In some countries, each individual in the population uses a large amount of materials and energy. In other countries, people use far less of the Earth's resources.

ACTIVITY 19G Comparing World Regions

Figure 19.25
Students can represent different world regions in Activity 19G.

In this group activity, you will model the distribution of population and wealth in different regions of the world (Figure 19.25). You will consider how each of these factors affects population growth and the environment.

MATERIALS

100 beans for each group of six students
100 paper "dollars" for each group of six students

PROCEDURE

1. Work in groups of six around a table. Each person will represent one of the world regions listed in Table 19.2. Write the name of your region on a sheet of paper and place it on the table in front of you.
2. Table 19.2 lists the approximate percentage of the world's population in each region. From the pile of 100 beans, take the number equivalent to your region's share of the world population. Arrange the beans on the table beside your region's name. The beans represent the total population of your region.
3. Table 19.3 lists the approximate distribution of wealth for each region. This figure, called the Gross National Product (GNP) per person, is calculated from the total value of the region's yearly production of goods and services. From the pile of paper dollars, take the number equivalent to your region's share of the world's wealth and spread them out beside your beans.

Table 19.2 Approximate Distribution of the World's Population by Region

Region	Percentage of world population
Africa	13
Central and South America	9
Asia (excluding Japan and U.S.S.R.)	53
North America, Japan, Australia	9
U.S.S.R.	6
Europe	10
	100

Table 19.3 Approximate Distribution of the World's Wealth by Region (in Gross National Product per person)

Region	Percentage of world wealth
Africa	3
Central and South America	5
Asia (excluding Japan and U.S.S.R.)	14
North America, Japan, Australia	33
U.S.S.R.	15
Europe	30
	100

4. Calculate the ratio of wealth to population in each region by dividing the number of dollars by the number of beans. Write this number beside the name of your region. The greater the number, the greater the wealth per person in the region.

DISCUSSION

1. Take turns with members of your group to explain how your region's share of the world's population and wealth may affect the lives of people in your region. Do some research on these topics or use your existing knowledge, experience, and imagination to explain how wealth, or lack of wealth, may affect:
   - food supply
   - water supply
   - consumption of energy and materials
   - agricultural practices
   - industrial practices
   - nature conservation
   - health
   - education
   - pollution
   - population growth
2. As a group, discuss some possible solutions to the problems created by the size and demands of the population in each region. What specific actions may be taken to reduce population growth and environmental impact in the more populated, less wealthy regions? What specific actions may be taken to reduce resource consumption and environmental impact in the less populated, more wealthy regions? List your solutions.

TWO WORLDS

On the basis of wealth, the nations of the world may be divided into two groups. The populations of Canada, Australia, the United States, Japan, and countries in western Europe have relatively high levels of wealth. These and others are known as **more developed countries.** The populations living in most of the countries of Africa, Asia, and South and Central America have relatively low levels of wealth. They are known as **less developed countries** (Figure 19.26).

Figure 19.26
A map showing regions of the world.

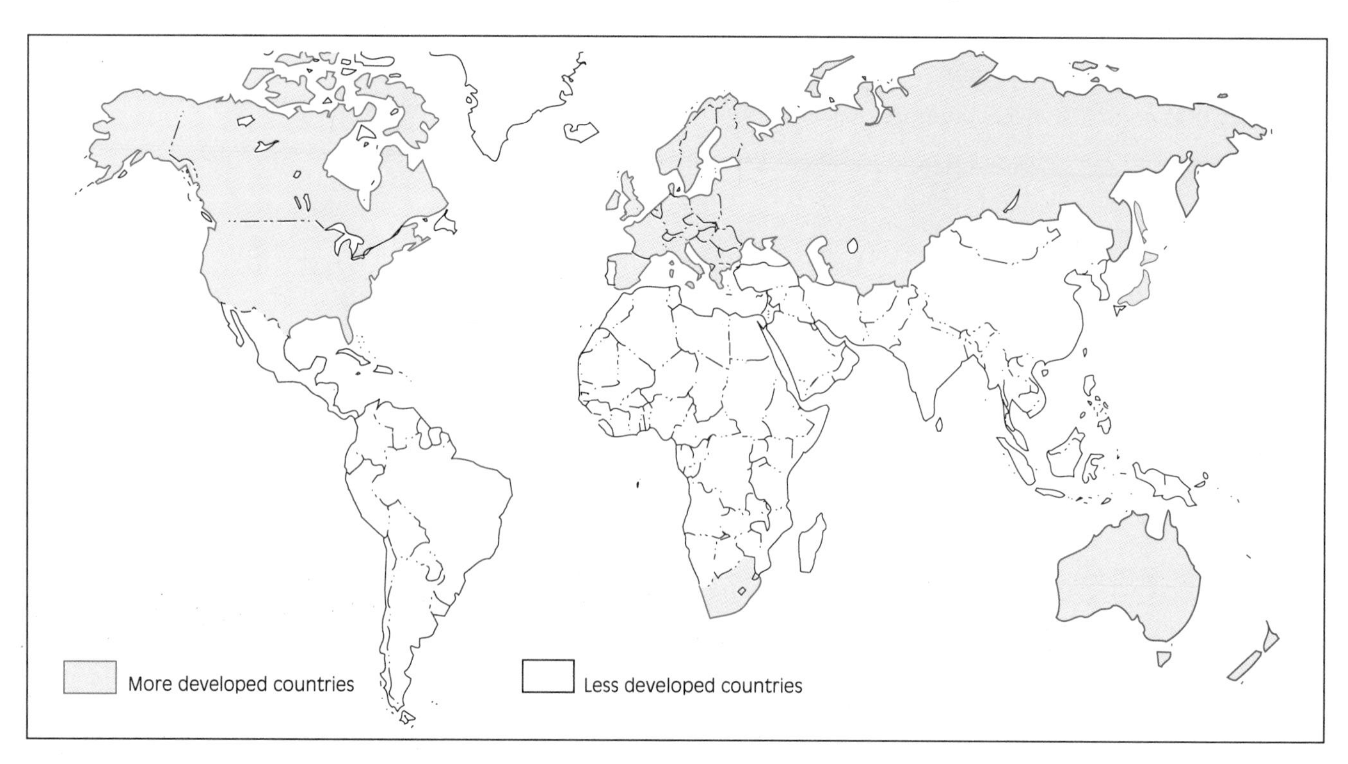

Both very high levels of wealth and extremely low levels of wealth, or poverty, can lead to environmental damage. People in the more developed countries tend to use large amounts of materials and energy for such things as housing, vehicles, electrical appliances, and a wide range of manufactured products. Each person living in one of these countries consumes, on average, about 30 times as much energy and materials as a person living in a less developed country.

Although people in less developed countries do not, on average, use large quantities of resources per person, they still can have a serious impact on their environment. For example, to increase land for growing food crops and to obtain firewood for cooking, people in less developed countries have cut down and burned large areas of woodland and forest. Added to this, the less developed regions of the world currently have the highest rates of population growth. During your lifetime, about 85 per cent of the world's total population growth is expected to occur in the less developed countries (Table 19.4).

EXTENSION

Discuss what factors might lead to higher birth rates among people in less developed countries. List possible reasons and explain how each is related to large family size.

Table 19.4 World Population in 1988 and Expected Increase by the Year 2000

Region	Population in 1988 (in millions)	Expected population in 2000 (in millions)	Estimated percentage increase
Africa	623	887	42
Central and South America	429	536	25
Asia (excluding Japan and U.S.S.R.)	2587	3170	23
North America, Japan, Australia	421	457	9
U.S.S.R.	286	311	9
Europe	497	507	2

DID YOU KNOW?

The average African woman has 6 children. At this rate, one woman can have 36 grandchildren and 216 great grandchildren. In contrast, the Canadian average is 2 children per woman, resulting in an average of 4 grandchildren and 8 great-grandchildren per woman.

The reason for considering the wealth and lifestyle of people in different regions is that these factors affect the Earth's carrying capacity. What would happen if all of the more than 5 billion people in the world lived in the way that the average Canadian lives? Would such a thing be possible? Are there enough resources on the planet to allow the present population of the world to live at our level of resource consumption? The maximum population that the world can support will depend on the demand for resources that each person makes and the impact on the environment that each person has.

REVIEW 19.3

1. Give an example to show how an increase in the population of one species may either increase or decrease the population of another species.
2. What is meant by the term "overpopulation"?
3. How might overpopulation in a species damage its abiotic environment? Give an example.
4. Explain what may happen to population growth following overpopulation in a species, and why.
5. In your own words, define the terms "more developed country" and "less developed country."
6. Give an example of one way in which a high level of wealth in a region might influence
   (a) population growth,
   (b) the environment.
7. Give an example of one way in which population growth in a region might influence
   (a) the region's level of wealth,
   (b) the environment.

CHAPTER • REVIEW

Key Ideas

- The number of organisms in a population—the population size—may increase, decrease, or remain the same. These changes in population size depend on the relative numbers of births, deaths, immigrants, and emigrants.
- A high birth rate and low death rate produce exponential growth in population size. This appears on a graph as a J-shaped curve. The global human population is now in a period of rapid exponential growth.
- A balance between the birth rate and death rate produces a stable population. When this follows a period of population growth, it appears on a graph as an S-shaped curve.
- Growing populations have an impact on both the abiotic and biotic aspects of their environment.
- The carrying capacity of an environment is the maximum number of organisms that it can support indefinitely. When a population overshoots the environment's carrying capacity, it begins to damage its resources.
- The maximum size of the human population that the Earth can support will depend on the amount of resources each person uses and the impact on the environment each person has.

Vocabulary

population
species
density
immigrate
emigrate
exponential growth
J-shaped curve
birth rate
death rate
density-dependent factor
density-independent factor
S-shaped curve
carrying capacity
demographer
zero population growth
biotic
abiotic
overpopulation
extinct
more developed country
less developed country

1. Select a term from the vocabulary list and write a question to which your term is the answer. Do the same for four more terms. Exchange your list of five questions with a partner and write in the correct answers on one another's list.
2. Use some or all of the terms from the vocabulary list to make a concept map showing the relationship between a population and its environment.

Connections

1. Two friends keep guppies in aquariums. Both began their hobby a year ago with one pair of fish each. Neither has bought any more since then. Roberta now has 10 guppies in her aquarium, whereas Miguel has 20. Suggest reasons that Miguel's fish population is larger than Roberta's. Suggest how Roberta might increase her fish population.
2. During the last century, rabbits from Britain were introduced in Australia, where rabbits had never lived before. A few rabbits escaped from a farm. Within 50 years a huge rabbit population had spread from one end of Australia to the other,

despite many efforts to control them.

(a) What type of population growth curve was shown by the rabbit population? Sketch a curve of this kind.

(b) In terms of birth rate and death rate, explain how this type of population growth is produced.

(c) What environmental factors may have led to the rabbit population growth?

(d) Suggest a method for stopping the population growth of the rabbits. How might your method affect other organisms or non-living aspects of the environment?

3. A Canadian lake had large populations of fish, waterfowl—and mosquitoes! To make life more pleasant for visitors to the lake, the area was sprayed with insecticide. Explain how the reduction in the mosquito population might affect the populations of fish and waterfowl.

4. Name two specific changes in the environment that might increase the birth rate and decrease the death rate in a population of deer.

5. Name two effects on a forest that might be caused by overpopulation of deer.

6. How has growth in the human population led to a decline in the populations of many other species? How might future declines be prevented?

7. A farmer has too many mice in the barn. What controls would you recommend to reduce the barn's carrying capacity for mice? How would each control you suggest affect the birth rate or death rate of the mice?

8. A gardener observed that the population of duckweed in her pond increased sharply in the spring and then remained at about the same level.

(a) Sketch a graph to illustrate the type of population growth curve shown by the duckweed population.

(b) Describe each stage in the graph and say what is happening to the birth rate and death rate in the population at that stage.

(c) Suggest what factors in the environment are most likely affecting the birth and death rates at each stage.

9. How might the human population achieve zero population growth?

10. Population growth is not the only factor that determines the impact of human populations on the environment. What other factor(s) are important, and how do they affect the population's impact on the environment?

Explorations

1. The population of sea otters off the coast of British Columbia became extinct early in this century (Figure 19.27). A small population of these animals was moved from Alaska to the west coast of Vancouver Island during the early 1970s. Do some research to find out how many sea otters live off the B.C. coast today, and where these populations are located. Prepare a report or display on sea otters to show whether their numbers are rising, falling, or staying about the same. What factors in their environment affect the birth rate and death rate of sea otters? Suggest what people might do to help the number of sea otters increase.

Figure 19.27
The sea otter was hunted to extinction along the B.C. coast because so many people wanted its fur.

2. Wolves in some areas of British Columbia have been killed in recent years as a control measure intended to increase the population of deer for hunters. In small groups, prepare for a formal debate by discussing what effects reducing predators might have on the environment. Prepare to defend your answer to the following question: "Should people kill predators, such as wolves, in order to increase the numbers of other species?"

3. Interview a person whose job is related to the management of populations of living things: a commercial fishing-boat operator, a game manager, a farmer, a nursery operator, someone who works in a fish hatchery, or a forester. Find out what measures they take to increase or decrease populations of animals or plants that are important to their job. For example, ask how a nursery operator reduces the numbers of insect pests and weeds, and increases the reproduction and survival of the plants raised and sold at the nursery. What impact might their actions have on populations of other organisms and on the environment? Find out whether their current management practices will allow them to produce the same or increased numbers of the animals or plants they want year after year in the future, or whether their actions may lead to a decline in their population of organisms.

4. Find out how the population of your town or city has changed since you were born. Based on these figures, what do you predict will happen to the population in the next 10 years? How will this change affect your life?

Reflections

1. In some ways, studying populations is simple. Populations can only go up, go down, or stay the same. They can only be changed by births, deaths, and movement. Do you think this knowledge helps you understand the essential reasons for the growth of the human population, and what actions are needed to plan for its future? Or do you think that there are other important factors that affect human population growth? What else would you like to know about human populations?

2. In what ways have your ideas about populations in general, and the human population in particular, changed? Why did you change your mind?

CHAPTER

20

People and Pollution

As you approach a community in British Columbia, your first view may be through an overhanging haze of yellowish or grey air. You may even smell the community before you see it. The materials that make the air hazy or smelly are a mixture of dust and exhaust fumes from gasoline-powered vehicles, homes, businesses, hospitals, pulp mills, smelters, and factories. The discoloured air is a sign that the community is producing these waste materials at a rate faster than the atmosphere can recycle or disperse them. As a result, the particles and gases build up in the surrounding air, creating pollution.

In this chapter, you will discover how pollution is produced by many everyday activities—from mining, farming, and forestry to the production of energy and the manufacture, transport, use, and disposal of every kind of product. The total amount of pollution tends to increase with population size, because more people use more resources and produce more wastes.

Some types of pollution pose direct threats to human health through the air we breathe, the water we drink, and the foods we eat. Other types of pollution threaten our survival through long-term damage to ecosystems and the climatic patterns that sustain the web of life worldwide. Through understanding these connections, we can take action to reduce or eliminate pollution and to protect and improve our environment.

ACTIVITY 20A Thinking about Pollution

In this activity, you will record some of your ideas about pollution in your learning journal. First, title a new page in your learning journal: "What pollution means to me."

Next, complete the following:
I think pollution can be defined as . . .
I think most pollution is caused by . . .
I could personally help reduce pollution by . . .

When you have completed this chapter, reread your notes and decide whether you could change or add to your ideas.

Figure 20.1
Pollution can kill living things.

20.1 Pollution: Everyone's Problem

Pollution occurs when something that is added to the environment causes harm to living things, such as plants and animals (Figure 20.1), or non-living things, such as air, water, and buildings. The "something," or **pollutant,** may be a substance, such as sulphur dioxide, or a form of energy, such as noise. Some pollutants are not harmful in themselves. For example, carbon dioxide is a substance produced by animal and plant respiration. It is naturally recycled during photosynthesis. However, carbon dioxide can do harm when added to the environment in larger than normal amounts. Other pollutants, such as mercury, can harm living things even when released in very small quantities.

Pollution is occasionally caused by natural events such as volcanic eruptions, forest or grass fires, and dust storms in desert areas. The eruption of Mount Saint Helens in 1980, for example, sent millions of tonnes of dust and gases into the atmosphere. Particles from the eruption disrupted air traffic and settled in thick layers over areas of Oregon and Washington State (Figure 20.2), damaging property, suffocating vegetation, and choking streams. But although natural events can lead to pollution, most pollution problems are caused by human activities.

Figure 20.2
Some pollution is caused by natural events. After the eruption of Mount Saint Helens, many communities were showered with ash.

HUMAN-MADE POLLUTION

Human-made pollution is not just a twentieth-century phenomenon. In some ways, our communities today are less polluted than they were in the past (Figure 20.3). In other ways, however, they are much more polluted. Today we no longer dump our garbage onto the streets—but we are producing so much more of it that sites for burying solid waste are rapidly filling up. We have greatly reduced some forms of air pollution from industries and homes—but air pollution from automobile exhausts has increased with the greater number of cars. Our drinking water today is treated to remove most harmful materials—but every week communities across Canada pour hundreds of thousands of tonnes of liquid wastes directly into their nearest body of water.

Figure 20.4 shows why pollution is a serious problem today.

◄ **Figure 20.3**
Many people in early industrial communities lived in crowded, polluted, unsanitary conditions. Garbage was dumped into streets, and chimneys poured smoke and soot into the air. Water was often contaminated by human waste.

Figure 20.4
A growing population, inefficient technology, and a wasteful lifestyle—all lead to increased pollution.
▼

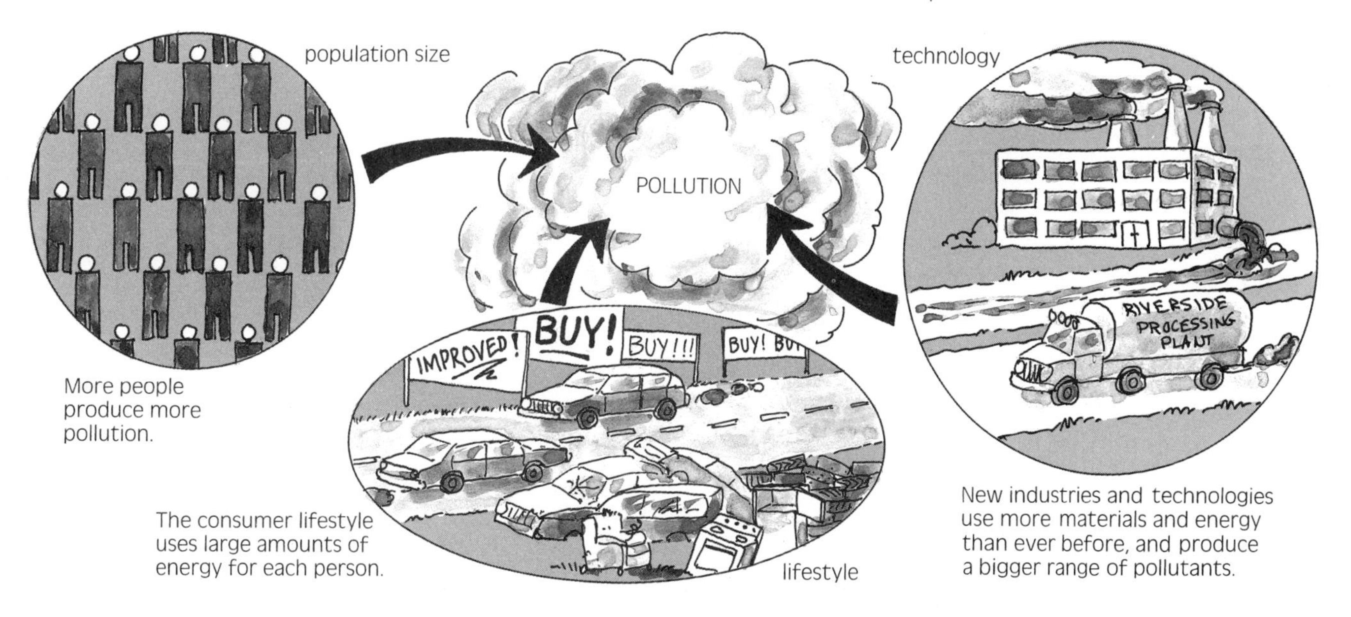

EXTENSION

Study your municipal garbage disposal system. How many truckloads of garbage are picked up in your community each week? How much material is recycled? What happens to the material that is recycled? What happens to the material that is not recycled? What impact might these disposal and recycling methods have on your community? Can the existing garbage disposal system in your community operate for the next 10 years?

GARBAGE: WHAT GOES IN MUST COME OUT

What is the stuff we call "garbage"? If you were to dig into a landfill site, you would find the torn and broken remains of magazines, toys, furniture, appliances, disposable diapers, bones, tea bags, kitchen scraps, boxes, plastic bags, containers from paint and household chemicals, running shoes, light bulbs, toothbrushes—in fact, the remains of all the things you find in a home.

Everybody is responsible for producing garbage. It may surprise you to learn that the quantity of garbage produced by most communities is about the same as the quantity of new products shipped into the communities. In other words, most things you see in the stores will sooner or later be transformed into garbage (Figure 20.5). The more you buy, the more you throw out.

Figure 20.5
The end result of shopping. A product-hungry society generates mountains of solid waste.

ACTIVITY 20B How Much Garbage Does Your School Produce?

In this activity, you will find out about the kinds and amounts of materials thrown away at your school.

PROCEDURE

1. Predict the kinds of materials thrown out by the students in your school.
2. During one lunch hour, work in groups to observe and record the types of materials thrown into garbage containers placed in your school eating area(s). List the different objects discarded and the materials they are made of—for example, paper napkin, plastic drinking straw, foil wrapper, apple core, etc. Do *not* handle the garbage.
3. Arrange with the custodial staff to collect all the garbage containers after lunch and take them to a central point. Measure or estimate the total mass and/or volume of the garbage. When you have obtained this figure, ensure that the garbage is disposed of in the normal way.

DISCUSSION

1. List any materials you found that you did not predict would be thrown away.
2. What is the total quantity of garbage produced by your school after one lunch hour?
3. Using your measurements, estimate the quantity of waste that will be produced by your school during a school day and a school year (200 days).
4. From your answer to question 3, calculate the average amount of garbage produced per student per school year. (You will need to know the total number of students in your school to answer this question.)
5. Assume all students in your town or city produce the same amount of garbage. How much garbage is produced by the students in your community per school year? (You will need to know the total number of students in your community to answer this question.)
6. Suggest some effects the garbage from your community might have on the environment.
7. Predict what might happen if your community had no garbage collection or disposal for a month.
8. Find out what your school is already doing to reduce the amount of garbage produced daily. List your ideas about how to reduce this amount even more. Can you put these ideas into action?
9. Criticize or defend the following statement: "Current garbage disposal problems are easily solved." Suggest some alternatives to present methods of dealing with garbage.

THE PROBLEM OF TOXIC WASTES

Toxic wastes are poisonous pollutants that pose special problems: even small quantities in our soil, air, and water may kill or injure living things. There are more than 30 000 chemicals used in Canada that are classified as toxic. They include arsenic, cyanide, chlorine, sulphuric acid, chloroform, mercury, ammonia, furans, copper, zinc, lead, formaldehyde, dioxins, and PCBs.

Many toxic pollutants come from household products, such as bleaches, disinfectants, herbicides, pesticides, medicines, cleaners, dyes, solvents, and battery acid (Figure 20.6). Other toxic pollutants are by-products of pulp and paper mills, mines, metal smelters, petroleum refineries, and chemical manufacturing plants.

Mining activities release toxic chemicals into the environment indirectly. For example, when sulphur-bearing rock is exposed to air and water, it reacts to form sulphuric acid. This acid may then dissolve other minerals locked in the rock, releasing toxic compounds such as arsenic, uranium, and copper.

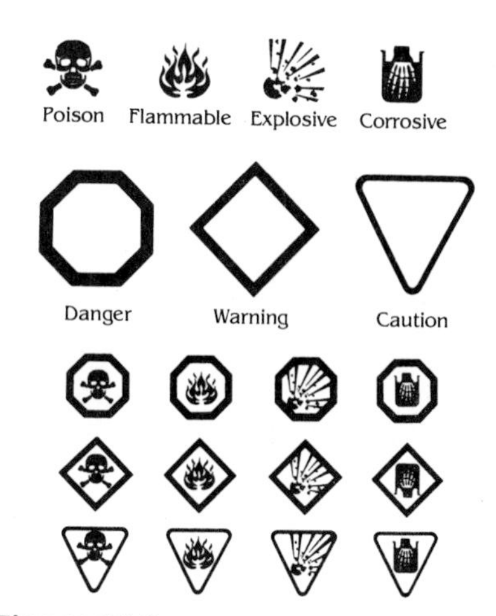

Figure 20.6
Hazardous household products have one or more of these symbols on their labels. These products must be handled with care.

The effects of toxic substances on organisms depend on many factors, including the quantity of the pollutant, the health of the organisms, and conditions in the environment, such as temperature, acidity, and the presence of other pollutants. Short-term effects, apart from death, may consist of chemical "burns" and ulcers. Longer-term effects come from the storage and build-up of toxins in the body. Predators acquire high levels of toxic compounds over time by eating other organisms that contain poisons. In this way, toxic compounds that may have been in the soil or water in low concentrations become more highly concentrated as they pass up the food chain. Among fish, birds, and mammals, high concentrations of toxic substances may lead to damage to the kidneys, liver, and nervous system, and an increase in cancer and birth defects.

ACTIVITY 20C *Toxic Pollutants*

In this activity, you will investigate the hazardous chemicals found in common household products.

PROCEDURE

1. Select one of the following products to study: batteries, drain cleaners, herbicides, paint, motor oil, pesticides, pool chemicals, polishes, antifreeze, bleach, pharmaceuticals. Do *not* handle any of these materials.
2. Find out what toxic chemicals are in the product you have selected by examining the labels and instructions, researching in your library, or writing to the manufacturer. List each chemical and record why it is used. Describe what effect each chemical might have on the biotic or abiotic environment.
3. Suggest how the problem of pollution from this product can be eliminated.

DEALING WITH TOXIC WASTES

The best way to reduce the impact of toxic pollutants is to avoid producing them wherever possible and to reduce the quantities produced. This can be done by:

- Using fewer or no toxic materials.
- Developing less toxic or non-toxic alternatives.
- Recycling and reusing harmful chemicals.

Toxic pollutants that cannot be eliminated by reduction, substitution, recycling or reuse can be collected and treated to make them less dangerous. Materials such as oil can be treated by biological methods, using micro-organisms that decompose them into simpler and less harmful substances. Other hazardous wastes can be treated by chemical reaction. For example, strong acids or bases may be mixed with solutions that neutralize them. Some toxic pollutants can be separated from other wastes by a simple physical process, such as placing absorbent filters and traps into chimneys and outflow pipes.

When toxic materials cannot be treated, they must be stored securely to isolate them from the environment. The sites for stored hazardous wastes must be carefully chosen, so that there is no risk of materials

EXTENSION

Review Chapter 13 and decide where on our planet you would *not* store hazardous wastes, and why.

escaping into the soil, air, or water. Gases and liquids are more difficult to contain than solids, as they may easily leak out through any small crack or hole in the container. This problem can be solved for some liquid and semi-liquid wastes by mixing them with substances that fix them into a solid mass.

The disposal of radioactive wastes presents particularly difficult problems (Figure 20.7). Radioactivity is impossible to detect by sight or smell. Radioactivity decays over time, but some materials remain potentially hazardous for hundreds or thousands of years. As you may have learned in Chapter 7, no completely safe methods of disposal have yet been developed. Radioactive waste packaged in concrete-encased steel drums has already been deposited on the ocean floor, but these containers may eventually corrode and begin to leak. Researchers are now looking at a way of sealing radioactive wastes deep in the Earth's crust. An alternative is to store the waste containers at nuclear power-generating stations and to continue monitoring them for leakage indefinitely. This has the advantage of retaining the used fuel until we have the technology to recycle it and obtain more energy from the fuel.

Figure 20.7
Radioactive materials must be kept isolated from the environment for hundreds or even thousands of years.

POLLUTION AND LIFESTYLE

Canada is an example of a **consumer society** — that is, Canadians buy and use, or consume, a huge amount of material goods (Figure 20.8). Although this seems perfectly normal to you, it is a lifestyle that has developed only during the last two or three generations. Many of the things that you regularly buy or use were unknown when your great-grandparents were your age. Products new to this century include nearly all electrical and battery-powered products, most aluminum products, and everything made of plastic, nylon, and various other synthetic materials.

Not only do Canadians have many more things to buy than in the past, they also have, on average, more money to spend. Table 20.1 shows the growth over 20 years in the percentage of Canadian households owning various consumer products.

Table 20.1 Percentage of Canadian Households Owning Major Retail Items

Item	Year		
	1965	1975	1985
Clothes dryer	25	48	66
Air conditioner	2	12	18
Dishwasher	3	15	37
Electric stove	69	85	92
Freezer	23	42	57
Microwave oven	n.a.	1	23
Colour television set	n.a.	53	91
Video recorder	n.a.	n.a.	23
Number of households (in millions)	5	6.7	9.07

n.a. = not available

Source: Statistics Canada

Figure 20.8
Many of the products and materials in our stores were unknown at the start of this century.

The growing demand for products does more than lead to a growing volume of garbage: it is indirectly responsible for most of the other pollution we produce. How so? Look at the data in Table 20.1. Note that in 1965, colour televisions were not yet generally available in stores — they were a new invention. Only 20 years later, 91 per cent, or about 9 out of every 10 households in Canada had a colour television set. Furthermore, the number of households itself had increased during this time — from 5 million to just over 9 million. Where did all the old black and white television sets go? And what materials and activities were needed to manufacture and sell the millions of *new* television sets (Figure 20.9)?

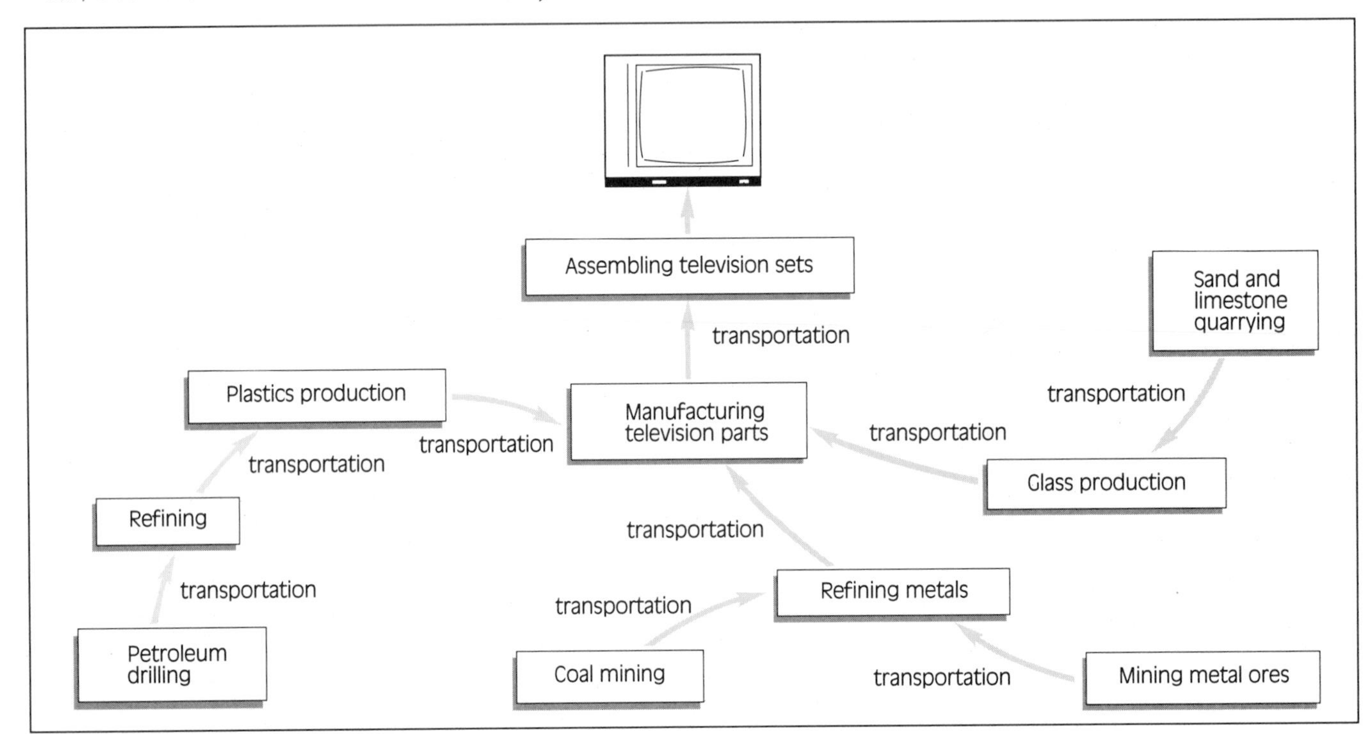

Figure 20.9
Many polluting activities are involved in making consumer products such as television sets. Only a few are shown here.

EXTENSION

Construct a topic web like that in Figure 20.9 to show some of the activities involved in producing a pair of jeans. The materials in a typical pair of jeans include cotton for the fabric, dye to colour the cotton, and metal for the zipper or buttons. During which activities are energy and materials consumed? At which stages might polluting wastes be produced?

As you can see from this example, the demands of each person in a consumer society, multiplied by the number of people in the population, add up to a huge impact on the environment. The pollution from mines, factories, refineries, mills, vehicles, and power-generating stations exists as part of a chain of events that begins with the demand by consumers for products and energy. In other words, pollution is caused not only by a growth in population size, or by technology, but also by the way in which we choose to live.

Canadians can choose to live in a way that uses less and wastes less of the world's limited supply of resources. In a **conserver society**, people would not buy things unnecessarily; they would not buy disposable or overpackaged products; they would buy products that last and that take less energy to make or use. By making these changes in lifestyle from consuming to conserving, people can directly reduce or eliminate much of the waste and pollution we now produce.

Profile

PHILIP PAUL

Philip Paul sees pollution problems through native Indian eyes. A member of the Tsartlip Band and hereditary spokesperson for the Saanich Tribal Council on Vancouver Island, he feels that Canada's first people have something important to share: an appreciation of the relationship between humans and all other parts of the natural world.

Paul's comments arise from his own experience and his study of economics, sociology, and anthropology. For many years, Paul has been talking about his people's vision and working as a leader inside and outside the native community. He served on the local Band Council for 12 years and on a national committee advising the federal government on native affairs. He recently co-ordinated the establishment of a tribal school serving over 200 students from the four Saanich bands. As well as attending classes, students spend time with native elders. They learn traditional arts, languages, and values, including respect for the Earth.

Paul believes native people had a balanced way of life before the first white settlers arrived. Two hundred years ago, resources were used on a small scale and pollution was not a problem. Now, he says, we are all beginning to realize that there is a high price to pay for the technological "progress" and consumerism that settlers brought.

Native people have always believed that the Earth is a living organism. Paul says, "It isn't a question of us and something 'out there' called the environment that we use, manage, or experiment on. Everything is interdependent: if we damage the Earth, we damage ourselves. If we deplete Earth's resources, we threaten our own survival. The next 10 years will be extremely important. We must turn around and undo what's been done to the environment."

Philip Paul, spokesperson for the Saanich Tribal Council, believes that native Indian people have something important to say about pollution problems.

As population pressures mount on the Saanich peninsula, Paul's people must increase their efforts to protect their environment. Recently, Saanich bands successfully opposed the dumping of sewage into a bay near the Tsartlip Reserve and the development of a huge marine complex near the Tsawout Reserve. They are currently involved in discussions about the management of Victoria's main garbage dump, which is located near band lands.

Paul believes that native activism is not always appreciated and that some people would like Victoria's native neighbours to be less demanding. But if he has anything to do with it, the Saanich people will continue to fight pollution. Philip Paul says they will also continue to share their vision of an interdependent world "for the good of all."

REVIEW 20.1

1. What is pollution?
2. "All pollutants are made by people." Decide whether you agree or disagree with this statement and list reasons that support your decision.
3. Give two reasons why pollution worldwide is a greater problem today than in the past.
4. List two solutions to the problem of too much garbage. Which solution do you prefer and why?
5. Name one source of toxic wastes.
6. Describe one short-term and one long-term effect of toxic pollution.
7. Give three methods of controlling and/or treating toxic wastes.
8. Figure 20.10 shows the typical composition of garbage produced in a Canadian city in the early 1990s.
   (a) What proportion of this waste could be recycled?
   (b) What proportion could be composted?
   (c) What percentage of the garbage would remain after the recycling and composting parts had been removed?
9. Explain the meaning of the terms "consumer society" and "conserver society." How does the lifestyle of each society affect the environment?
10. Suggest some ways in which you personally could reduce pollution.

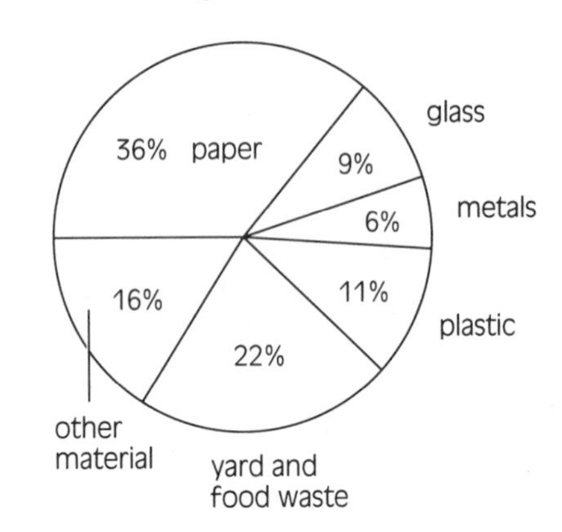

Figure 20.10
What we throw away.

Figure 20.11
Where do the pollutants from this smoking chimney go? What effects do they have?

20.2 Focus on Air

Does the air in your neighbourhood ever appear hazy? Does it sometimes have an unpleasant smell, or make your eyes or throat sting? Whether you live in the city or the country, each lungful of air you breathe is certain to contain pollutants put there by human activity (Figure 20.11).

Air pollutants can be gases, solids, or liquids. The most important polluting gases include carbon monoxide, carbon dioxide, nitrogen oxides, hydrocarbons, sulphur oxides, ozone, and CFCs (chlorofluorocarbons, also known as freon gases). Microscopic solid particles and fine mists suspended in the air are grouped together as **particulates**. They include solids such as dust, pollen, soot, and ash, and liquid aerosol sprays such as pesticides, medicines, and cosmetics.

The period of time that each pollutant remains in the atmosphere can vary from hours to years, depending on the type of pollutant and the weather conditions. Those that remain in the air for long periods may be carried by winds far from their point of origin, and can be dispersed worldwide.

Although air pollutants make up only a tiny fraction of the total volume of the atmosphere, they can cause serious health problems in humans, animals, and plants. Air pollution also increases the rate of damage from rusting, corrosion, and acidification. Some types of pollution have the potential to cause major changes to the Earth's climate.

ACTIVITY 20D Air Pollutants

Prepare a table to list various air pollutants as you read about them or discuss them in class. Beside each pollutant record the major sources, the effects, and possible means of control. Some of this information will be in the text that follows, and some you may research for yourself. Your table will be a useful summary of all that you have learned about air pollution.

AIR POLLUTION AND HEALTH

For thousands of years, smoke from fires and furnaces was the only human-made source of air pollution. Then, in the early 1800s, the development of industry and the building of fuel-powered machinery and vehicles greatly increased the amount of fuel that people could extract and burn. These developments led to a huge increase in air pollution, but almost no attempt was made to reduce it. People simply relied on the wind to carry their fumes away. Where people were concentrated in large industrial cities, the volume of air pollution they produced sometimes outgrew the capacity of the atmosphere to disperse it. Filthy air was trapped within the cities, and millions of people were forced to breathe it.

Figure 20.12
Once a common sight, the London smog has become rare since laws to control air pollution were passed.

One of the worst examples of air pollution ever recorded was the London smog of December 1952 (Figure 20.12). Smog is a mixture of fog and smoke. The 1952 smog was so dense that visibility was reduced practically to zero. People had to grope their way around the streets, and several accidentally walked off sidewalks into the River Thames. Polluted air seeping into movie theatres under doors and windows prevented those in the back rows from seeing the screen! During that month, about 4000 people died as a result of the pollution, most of them from respiratory diseases such as bronchitis and asthma.

The dramatic example of London forced people to take air pollution more seriously. Laws were soon passed in many countries to restrict the burning of fuels and the production of waste emissions. Despite these laws, air pollution is still a problem. In many cities across Canada today, newspapers report air pollution levels along with the daily weather. The **Air Quality Index** uses a standard scale to show the degree of risk to health from air pollution. It is based on the concentrations of five major air pollutants (see Table 20.2 on the next page).

Monitoring stations that measure air pollution are located in various parts of urban regions such as Greater Vancouver. Each station produces an Air Quality Index for its area, based on the pollutant with the highest concentration relative to government standards. High index values (over 50) indicate poor air quality. During such times, people with heart or lung problems are advised to reduce their physical activity and stay indoors.

Table 20.2 Major Air Pollutants

Pollutant	Description	Effect on health
Sulphur dioxide (SO_2)	Colourless gas with a pungent odour	Causes irritation of the upper respiratory tract and eyes. May lead to an increase in respiratory diseases.
Carbon monoxide (CO)	Colourless, odourless, tasteless gas	Slows reflexes when in low concentrations. May cause drowsiness and asphyxiation in higher concentrations.
Nitrogen dioxide (NO_2)	Reddish-brown gas	Causes increased risk of respiratory infection. Produces constricted air passages in people suffering from asthma.
Ozone (O_3)	Pungent-smelling, faintly bluish gas	Causes coughing and irritation to lungs and eyes.
Particulates	Microscopic solids and liquids that make the air hazy	Reduces visibility. Very fine particles can damage lungs.

ACTIVITY 20E Sources and Patterns of Air Pollution

Knowing where pollutants come from may help you to recommend ways of reducing the amount of pollution. In this activity, you will analyse data that show the sources of some major air pollutants in Canada, and investigate patterns of air pollution within Greater Vancouver.

PART I

PROCEDURE

The main gaseous and particulate pollutants in Canada are shown in Figure 20.13. Examine the data carefully and consider the sources for each pollutant.

DISCUSSION

1. List the major source of each pollutant.
2. Considering all five pollutants together, what is the biggest single source of air pollution?

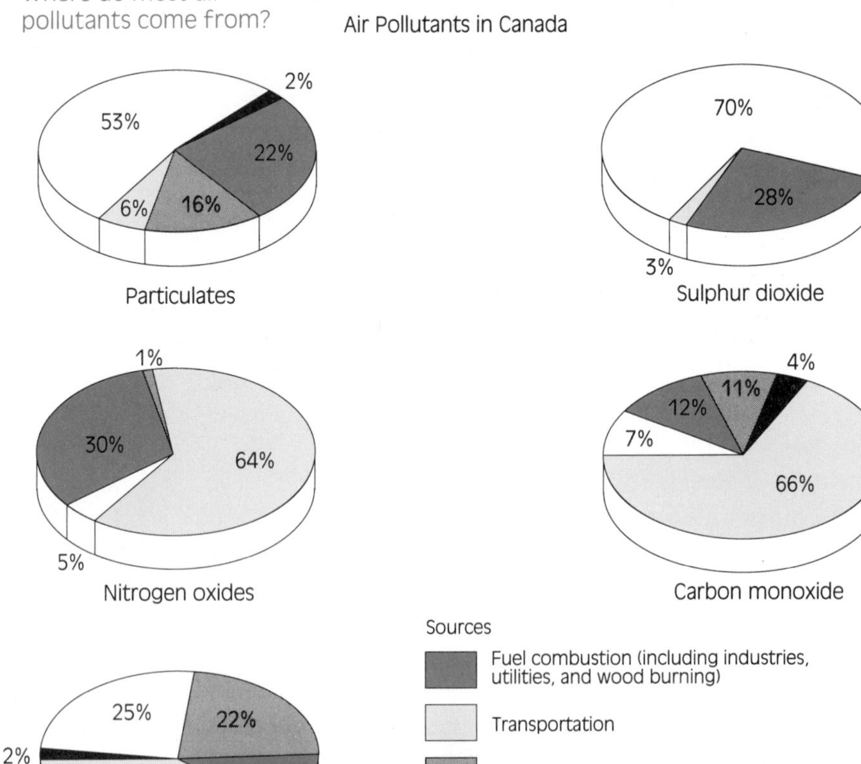

Figure 20.13 Where do most air pollutants come from?

3. What action(s) could you recommend to reduce the amount of air pollution by (a) particulates and sulphur dioxide, and (b) nitrogen oxides, carbon monoxide, and hydrocarbons?

PART II

PROCEDURE

Table 20.3 shows the Air Quality Index for three locations (Vancouver, Burnaby, and Port Moody), recorded daily over an 11-day period during August. Graph the data on a single graph, using a different colour or symbol for each location.

Table 20.3 Air Quality Index Data

Day	S	M	T	W	T	F	S	S	M	T	W
Downtown Vancouver (1)	7	12	10	9	11	15	9	5	13	13	16
North Burnaby (4)	6	22	19	12	20	25	17	3	18	20	23
Port Moody (9)	5	28	34	30	40	44	21	5	42	48	46

DISCUSSION

1. Using your graph and Figure 20.14, what conclusions can you draw about changes in air quality (a) from one location to another? (b) from one day to another?

2. Describe any patterns you see in your graph. Suggest explanations for these patterns. (HINT: Study the map in Figure 20.14, then find out about the wind patterns in the area.)

Figure 20.14
Information about air pollutants is gathered at many different sites.

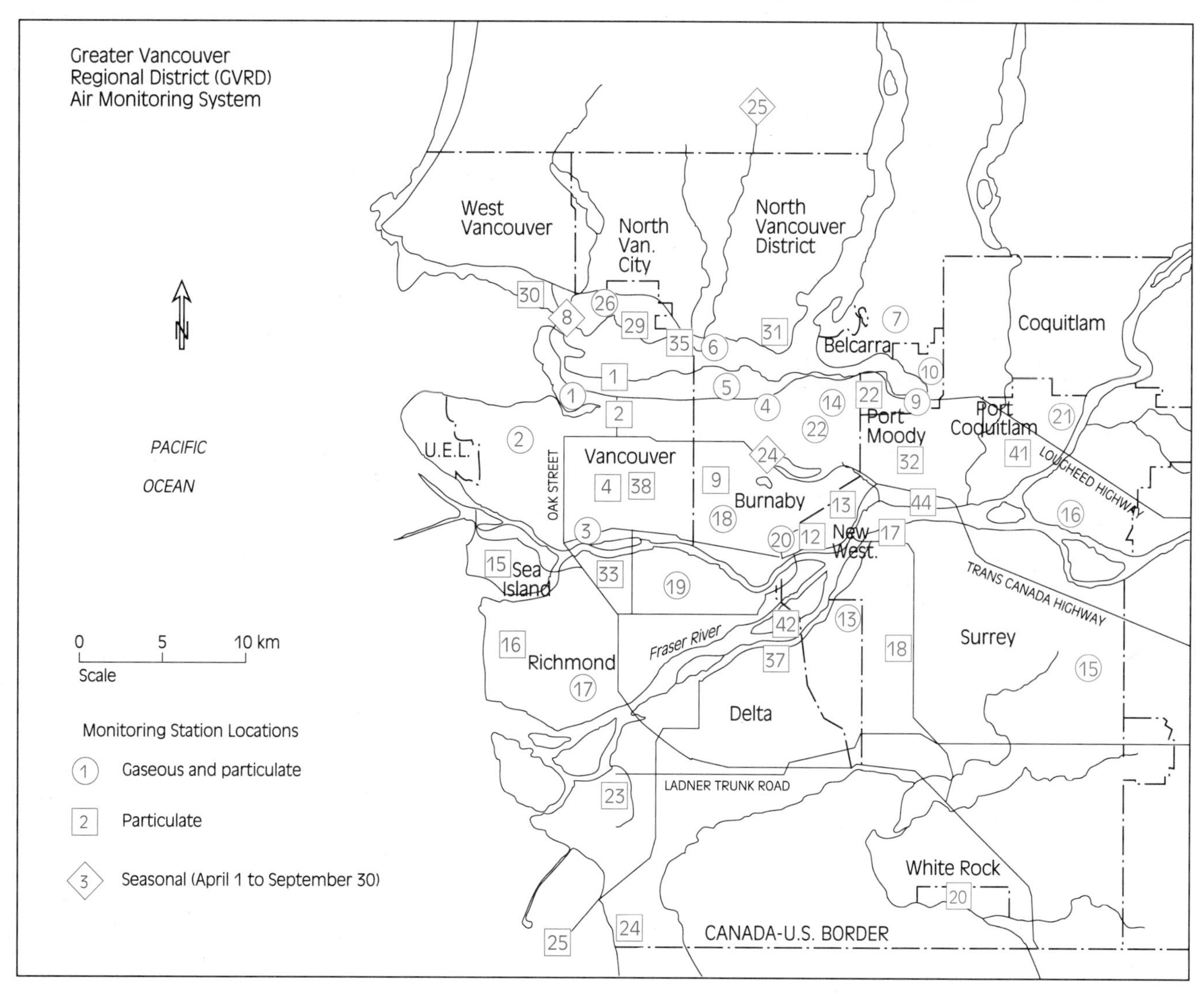

Table 20.4 Hourly Variation in CO Levels (parts per million × 10)

Hour of day	1	2	3	4	5	6	7	8	9	10	11	12	13	14	15	16	17	18	19	20	21	22	23	24
Station 1	25	19	17	13	10	9	12	22	32	34	31	28	26	28	31	36	42	50	37	34	32	31	29	29
Station 9	11	8	6	4	4	4	8	18	21	15	8	5	5	5	7	13	20	22	24	23	20	17	16	12

PART III

PROCEDURE

Table 20.4 shows hourly changes in the average levels of carbon monoxide at two monitoring stations—in downtown Vancouver (Station 1) and Port Moody (Station 9)—measured during November. Plot the data for both stations on a single graph.

DISCUSSION

1. What similarities are there in the pattern of hourly concentration of carbon monoxide at the two stations? What differences are there?
2. How would you explain the patterns you observe? (HINT: Use the information on sources of carbon monoxide from Figure 20.13.)

EXTENSION

Set up a particle trap to study air pollution in your area. Make the trap using double-sided tape or petroleum jelly on a slide. Study what you collect with a hand lens or microscope. What do the particles consist of? Which locations are more polluted? Why might pollution vary from one day to another?

OZONE: BOTH HARMFUL AND HELPFUL

You may have noticed that one of the major gases monitored for the Air Quality Index is ozone, yet you were not given a source for ozone in Figure 20.13. That is because large quantities of ozone are not the direct result of any one human activity. The ozone found near ground level is formed by the action of sunlight on air pollutants, especially the gases produced by vehicle exhausts and gasoline vapours. Because such reactions require energy from sunlight to occur, they are known as **photochemical reactions**. Ozone produced in this way is one of the main ingredients of the haze known as "photochemical smog," a particularly common health hazard over sunny, traffic-filled cities such as Mexico City, Athens, and Los Angeles.

About 90 per cent of the Earth's ozone, however, is not a pollutant. It is not caused by human activity and it is not found near ground level. It occurs between 15 and 35 km above our heads in a natural **ozone layer** that surrounds the entire planet (Figure 20.15). The photochemical reaction that produces ozone in the upper atmosphere is shown below:

$$\underset{\text{oxygen}}{3\,O_2} \xrightarrow{\text{sunlight}} \underset{\text{ozone}}{2\,O_3}$$

This reaction absorbs most of the sun's ultraviolet energy, which would otherwise penetrate to the Earth's surface. Ultraviolet rays are those that produce sunburn and some forms of skin cancer. In addition, high levels of ultraviolet light are particularly damaging to some agricultural crops and microscopic marine life, and can cause genetic mutations in most organisms. By filtering out these rays in the upper atmosphere, the ozone layer protects life on Earth.

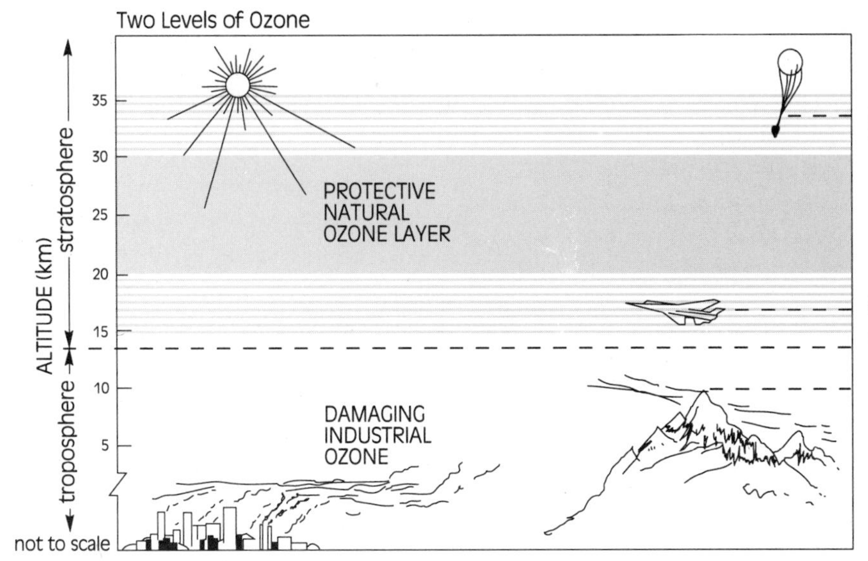

Figure 20.15
In the upper atmosphere, a layer of ozone shields us from the sun's damaging rays. Near ground level, ozone is a serious air pollutant.

THREATS TO THE OZONE LAYER

During the 1970s, scientists discovered that the amount of ozone in the upper atmosphere was declining. Beginning in the 1980s, large portions of the ozone layer over both the North Pole and the South Pole thinned out by about 50 per cent. The "holes" remained for two or three months of each year, but later filled in again (Figure 20.16). Scientists believe that the reason for this thinning involves the reaction of ozone with human-made pollutants that have been accumulating in the upper atmosphere.

Can We Repair the Sky?

Stratospheric Ozone Decreasing

Fate of Arctic Ozone Remains up in the Air

Can We Keep the Sunshine Safe?

Deadly Danger in a Spray Can

Ozone Hole's Reappearance Linked to Chlorofluorocarbons

Arctic Ozone: Signs of Chemical Destruction

Mysterious threats

Winter Ozone Gap Detected Over Artic

Ozone Holes Bode Ill for the Globe

urope to Ban CFCs by 2000

New Threats to the Sky

Research Balloons Study Thinning of Arctic Zone

Does the Ozone Hole Threaten Antarctic Life?

The Poles in Peril

The Search for Ozone-Friendly Refrigerants

Ozone Depletion Worsens

Figure 20.16
Newspaper stories during the 1980s told people about holes in the ozone layer.

Researchers identified two main ozone destroyers: high-altitude jet aircraft and a group of chemicals called CFCs (chlorofluorocarbons). High-altitude jets fly in the lower levels of the ozone layer, and their exhaust fumes react chemically with ozone molecules. But a much larger impact has come from CFCs, used in some aerosol sprays (such as deodorants and hair sprays) and in the manufacturing of refrigerators, air conditioners, and some kinds of polystyrene foam. Once released into the air, CFCs slowly diffuse upwards through the atmosphere. At high altitudes, they break apart, starting a series of reactions that destroy huge amounts of ozone (Figure 20.17).

How CFCs Break Down Ozone Molecules

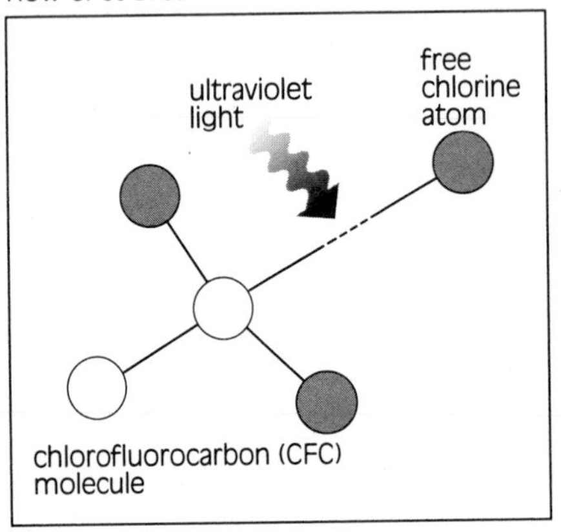

(a)

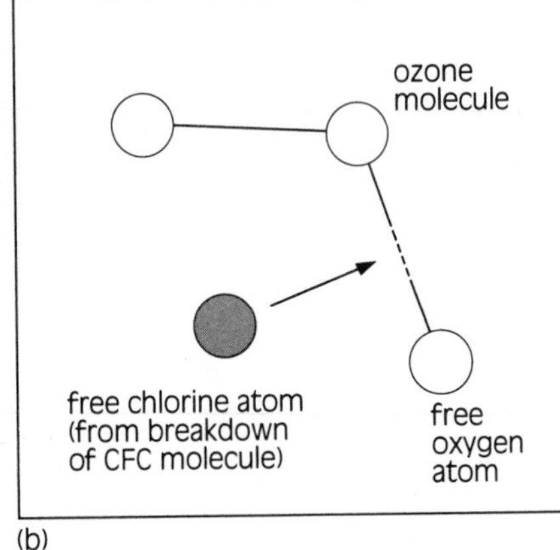

(b)

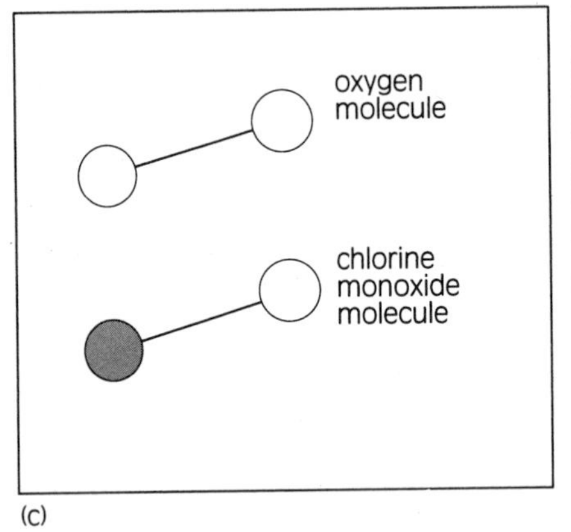

(c)

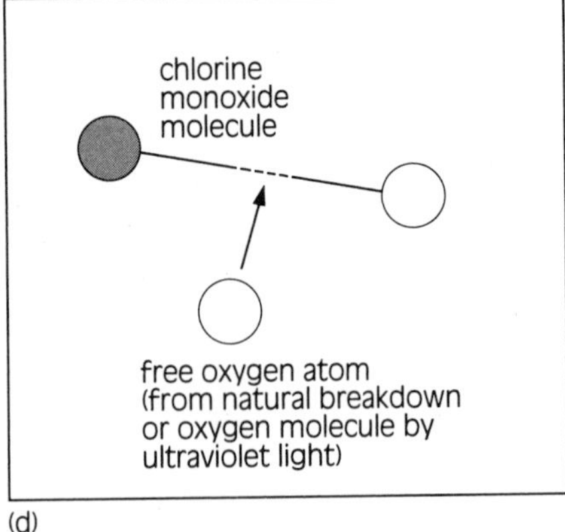

(d)

Figure 20.17
(a) Ultraviolet light breaks down a chlorofluorocarbon molecule (CFC), releasing a free chlorine atom (Cl). (b) The free chlorine atom breaks down an ozone molecule (O_3), releasing a free oxygen atom (O). (c) The free chlorine atom combines with the free oxygen atom and produces chlorine monoxide (ClO). One molecule of oxygen is also produced. (d) Meanwhile, ultraviolet light is also breaking down oxygen molecules (O_2). This produces still more free oxygen atoms. Most free oxygen atoms (O) react with oxygen molecules (O_2) to produce ozone (O_3). But some free oxygen atoms attack and break down chlorine monoxide molecules. Now the chlorine atoms are free to repeat the cycle shown in (b) and (c). In this way, the same chlorine atom can break down thousands of ozone molecules.

DID YOU KNOW?

Today's mean temperature on earth (15°C) is only a few degrees above the estimated mean temperature during the last ice age (11°C).

A HOLISTIC APPROACH TO POLLUTION

The story of ozone illustrates how substances originally thought to be harmless can interact and have unforeseen and far-reaching effects. Because huge numbers of variables can affect one another in the environment, the scientific method of isolating and controlling separate parts cannot be used to study certain environmental problems. Instead, ecosystems must be analysed as a whole. This is called a **holistic approach.**

It often takes several years for researchers to link particular pollutants to specific environmental effects. Even then, uncertainties may lead scientists to disagree about how to interpret the evidence. An example is the debate about trends in global temperature change.

Some scientists argue that average temperatures worldwide are gradually increasing (Figure 20.18). They base their conclusions on studies of the **greenhouse effect**, a term that describes how certain gases in the atmosphere work like the glass in a greenhouse to trap heat. They allow incoming sunlight to warm the Earth's surface, and then reflect some of the *outgoing* heat energy back to the Earth.

But other scientists disagree. They say that pollution has also led to increased dust and cloud in the atmosphere, which should reflect some of the *incoming* sunlight, producing a cooling effect large enough to offset the greenhouse effect.

Whatever their disagreements, however, most scientists agree that the volume of pollution now produced by people is great enough to affect ecosystems around the world. In effect, the entire planet has become a laboratory, and pollution is an experiment with unknown results.

Figure 20.18
Could human-made increases in the amount of "greenhouse gases" in the atmosphere make parts of Canada look like this during the next century?

ACTIVITY 20F *Investigating the Greenhouse Effect*

In this activity, you will find out more about the greenhouse effect. You will then present your research findings in a way that explains the connection between everyday activities and the greenhouse effect.

PROCEDURE

1. From your library or other sources, find out what pollutants are known as greenhouse gases. Find out where and how these pollutants are produced.
2. Find out what impact an increase in the greenhouse effect may have on your life.
3. Prepare something—a poster, a cartoon, a model, a short story, a play, a video—that explains how one of your activities can produce a pollutant that adds to the greenhouse effect. Explain how the greenhouse effect could in turn affect something that you do. Your presentation should answer the following questions.
   - What is the greenhouse effect?
   - What are some common activities that produce greenhouse gases?
   - What are the possible short-term and long-term effects of a specific air pollutant on (a) living things and (b) the non-living environment?
   - What impact(s) might the greenhouse effect have on the lives of Canadians?
   - What action(s) can people take to reduce their production of greenhouse gases?

REVIEW 20.2

1. Name three major air pollutants and give one or two sources of each.
2. Suggest why air in cities tends to be more polluted than air in rural or wilderness areas.
3. What precautions would you take if the Air Quality Index in your community were over 50? Why?
4. Sketch a graph to show the way in which the concentration of one specific air pollutant changes during the course of a day or a week. Explain the probable cause of the pattern you have drawn.
5. "Ozone is a pollutant." "We need ozone." Explain why both of these statements are correct.
6. Why is building taller chimneys not an effective solution to the problem of air pollution from factories?
7. Give an example of one way in which your ideas about air pollution have changed since you began your study of this subject.

20.3 Focus on Water

What happens to the waste water after you brush your teeth, take a shower, flush the toilet, wash the dishes, do the laundry, water the garden, or clean the car (see Figure 20.19)? Perhaps more important for you, what has happened to the water before it flows out of your kitchen tap and into your drinking glass?

Figure 20.19
Where does your drinking water come from? Where does your waste water go?

ACTIVITY 20G *Water Pollutants*

Prepare a table to list various water pollutants as you read about them or discuss them in class. Beside each pollutant record the major sources, the effects, and possible means of control. Some of this information will be in the text that follows, and some you may research for yourself. Your table will be a useful summary of all that you have learned about water pollution.

DID YOU KNOW?

In parts of the world where there is a shortage of water, several communities now reclaim a large percentage of their sewage, returning it directly to their drinking water supply after complete treatment.

The liquid wastes that leave your house, together with the wastes from other houses, businesses, and storm drains, are collectively known as **sewage** (Figure 20.20). It may contain a wide variety of pollutants, ranging from feces and detergents to improperly disposed-of toxic chemicals and waste oil. The way in which municipalities deal with their sewage varies greatly from one region to another. Some Canadian cities, like Victoria, B.C., pour untreated sewage directly into waterways. Others process their sewage through several stages to remove pollutants before the treated liquid is emptied into a river, lake, or sea. A sewage treatment plant basically speeds up the natural processes of water cleansing: it accomplishes in a few hours what a river takes days or weeks to do.

SEWAGE TREATMENT

Pollutants are removed from sewage by a variety of physical, biological, and chemical methods. The simplest method of sewage treatment, called primary treatment, separates solid particles from liquid. This is done by filtering the sewage through screens, or allowing the solids to settle in still tanks.

Secondary treatment systems use naturally occurring bacteria and other micro-organisms to break down solid organic matter, called sludge. To help this process, air is circulated through the sludge. The aerobic (oxygen-using) micro-organisms then multiply rapidly. They decompose the waste material into harmless products that dissolve in water. Any remaining sludge is removed and dried out. Depending on its contents, it may be burned, buried in landfill, or added to the soil as fertilizer.

Many communities combine primary and secondary treatment by leaving sewage to stand in artificial shallow pools called lagoons. Here, solids settle out and natural decomposition of waste occurs through contact with the air.

After secondary treatment, the water is usually channeled directly into a waterway. In some communities, it receives a third or tertiary treatment, which involves the addition of chemicals such as chlorine to kill any disease-causing bacteria. At this point, the water is pure enough to drink (Figure 20.21).

Figure 20.20
Pipes such as this carry sewage out of your community.

Figure 20.21
Sewage must be treated to make it safe.

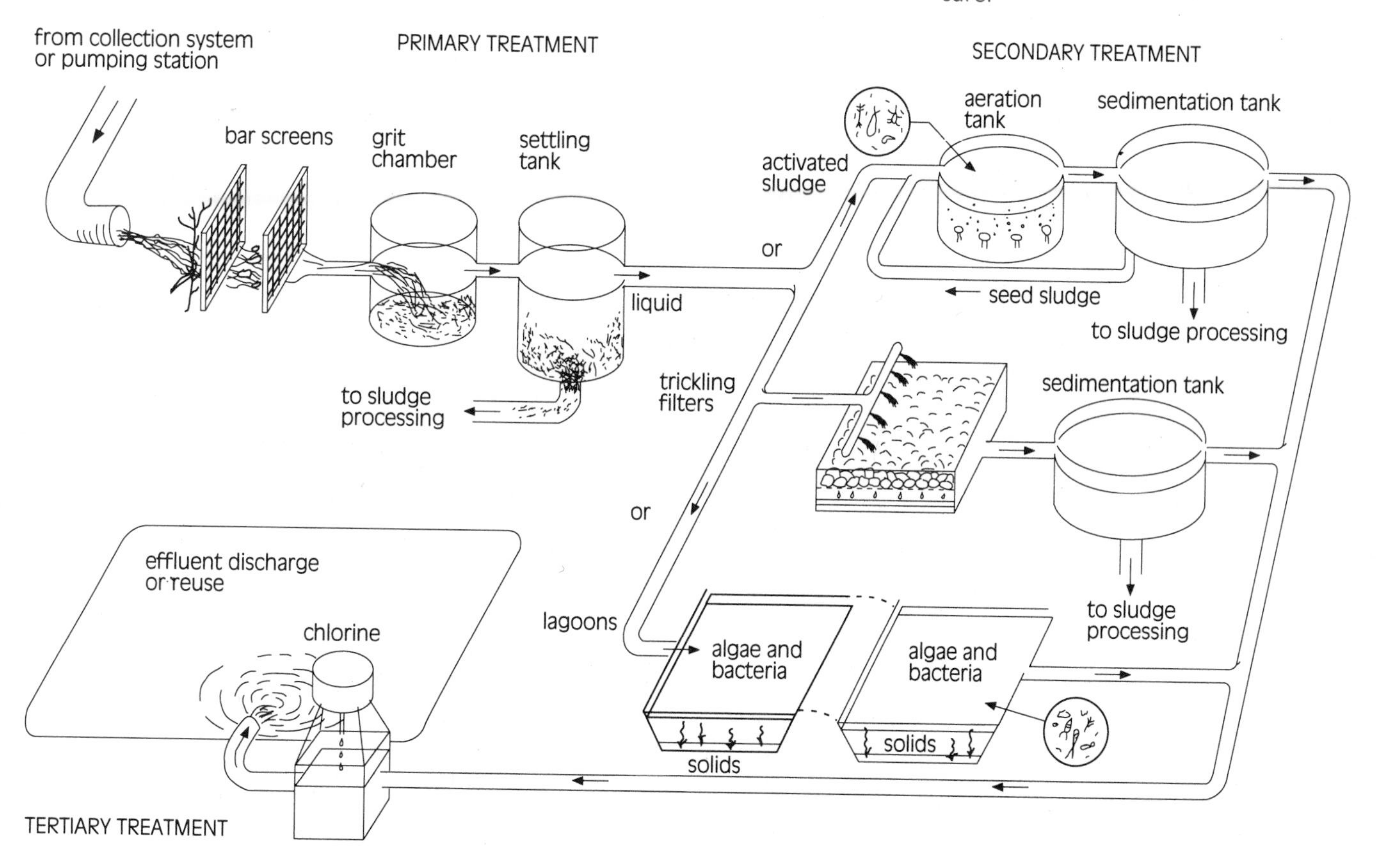

The volume of sewage produced by a community depends on the volume of water it uses. Sanitation engineers must plan to treat and dispose of a volume of sewage equal to the volume of water that is brought into the community for use by all businesses and homes. As is true for garbage, what goes in must come out.

ACTIVITY 20H *How Clean Is Your Water?*

In this activity, you will compare your drinking water with water from a nearby river, lake, or reservoir (Figure 20.22).

MATERIALS

safety goggles
sample of drinking water
sample of water from a nearby river, lake, or other local source
two test tubes
two glass Petri dishes
5 mL graduated cylinder
hotplate
pH paper, pH meter, or Universal Indicator solution
liquid soap
dropper
stirring stick

PROCEDURE

1. Read the following and prepare a table with five columns to record your observations.
2. Carefully smell each sample. Describe any odour you detect and suggest what might cause it.
3. Hold a sheet of white paper behind both samples of water and note which sample is clearer.

CAUTION!
Wear safety goggles while warming a Petri dish in case it breaks.

4. Measure 5 mL of each sample into separate clean Petri dishes. Place the dishes on a warm (not hot) hotplate and leave the water to evaporate while you carry out the next procedure. When the dishes are dry, note whether any solid material remains. Record the colour of any residues and compare the amounts left by each sample.
5. The amount of mineral matter in water determines its hardness. You can compare the hardness of your samples by measuring how easily they form suds with soap. Measure 10 mL of drinking water into a clean test tube. Add liquid soap one drop at a time. Stir or agitate the water after each drop. Record how many drops are required to form suds. Repeat this procedure with your other sample, using a second clean test tube.

CAUTION!
Hot and cold burners on a hotplate can look the same. Be careful!

6. Using the pH paper, a pH meter, or Universal Indicator solution, measure and record the approximate Ph of each sample.

Figure 20.22
Various tests can help determine water quality.

DISCUSSION

1. If you did not know the origin of each sample of water, would the results of your comparison tell you which was drinking water? Explain.
2. From your comparison, do you think water treatment affects the water's
   (a) odour?
   (b) clarity?
   (c) dissolved solids?
   (d) hardness?
   (e) pH?
   If you answered "yes," say how you think water treatment affects each characteristic.
3. There are many things found in water that your testing procedure cannot identify. Name one of these.
4. Suggest one or two other tests that you could carry out to compare the quality of drinking water and other local water.

ACID PRECIPITATION

In many parts of North America, rainfall and snowfall are no longer pure — high concentrations of pollutants in the atmosphere affect the precipitation even before it reaches the ground. **Acid rain** is produced when sulphur dioxide and nitrogen oxides react with water vapour in the air to form sulphuric and nitric acids:

$$H_2O + SO_2 \longrightarrow H_2SO_3$$
$$H_2O + NO_2 \longrightarrow H_2NO_3$$

Most of the sulphur dioxide and nitrogen oxides in the atmosphere come from the burning of fossil fuels in vehicles, metal smelters, and coal-burning power generators. These sources produce an invisible stream of pollution that drifts high in the atmosphere. The pollutants spread over a large area before eventually falling as acid precipitation (Figure 20.23), often many kilometres from the original source of pollution.

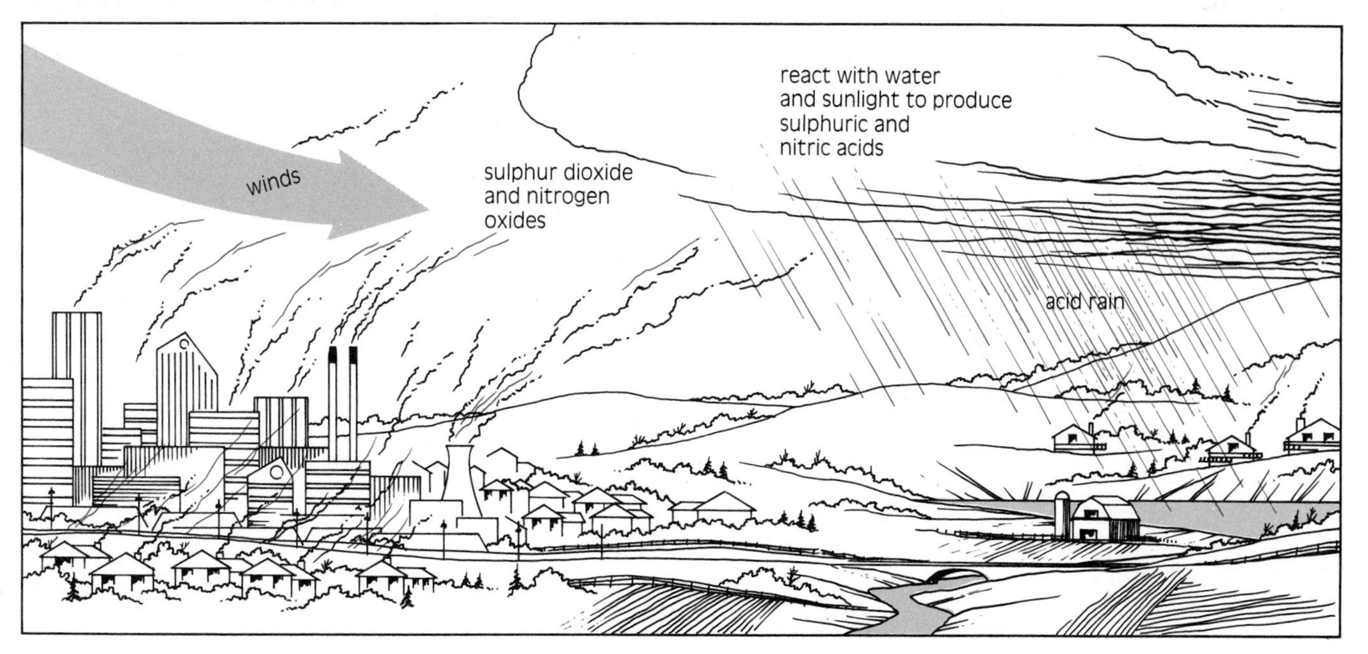

Figure 20.23
Acid rain may fall hundreds or thousands of kilometres from the original source of pollution.

In some places, the concentration of pollutants is so great that the rain is more acidic than vinegar (Figure 20.24). Acid rain has caused the gradual death of thousands of lakes and many hundreds of hectares of forests over large areas of North America and Europe. It has also damaged crops, buildings, and metal structures such as bridges and vehicles.

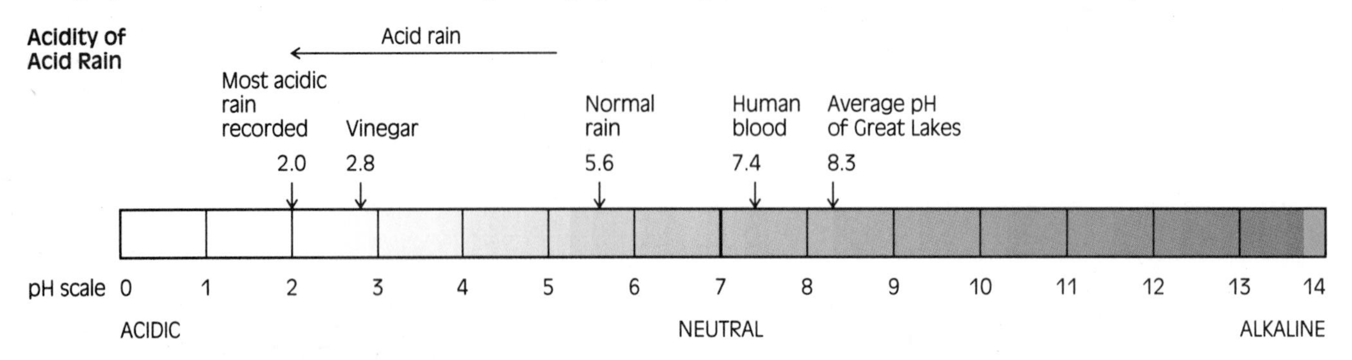

Figure 20.24
Acid rain can be more acidic than vinegar, as shown on this pH scale.

ACTIVITY 20I Finding Out about Acid Precipitation

In this activity, you will use what you have already learned about acids (see Chapter 9) and acid precipitation to answer some questions.

PROCEDURE

1. Examine the map in Figure 20.25. Note that many of the areas that are described as most sensitive to damage by acid rain are distributed downwind from the major industrial and urban centres on the continent.
2. Examine the photograph in Figure 20.26—it shows a building in Mexico that has been corroded (eaten away) by acid rain. Note that buildings made of limestone and marble are much more vulnerable to damage from acid rain than those made of granite.

DISCUSSION

1. Explain why industries that burn high-sulphur coal (such as electricity-generating stations and metal smelters) are major contributor to acid rain.

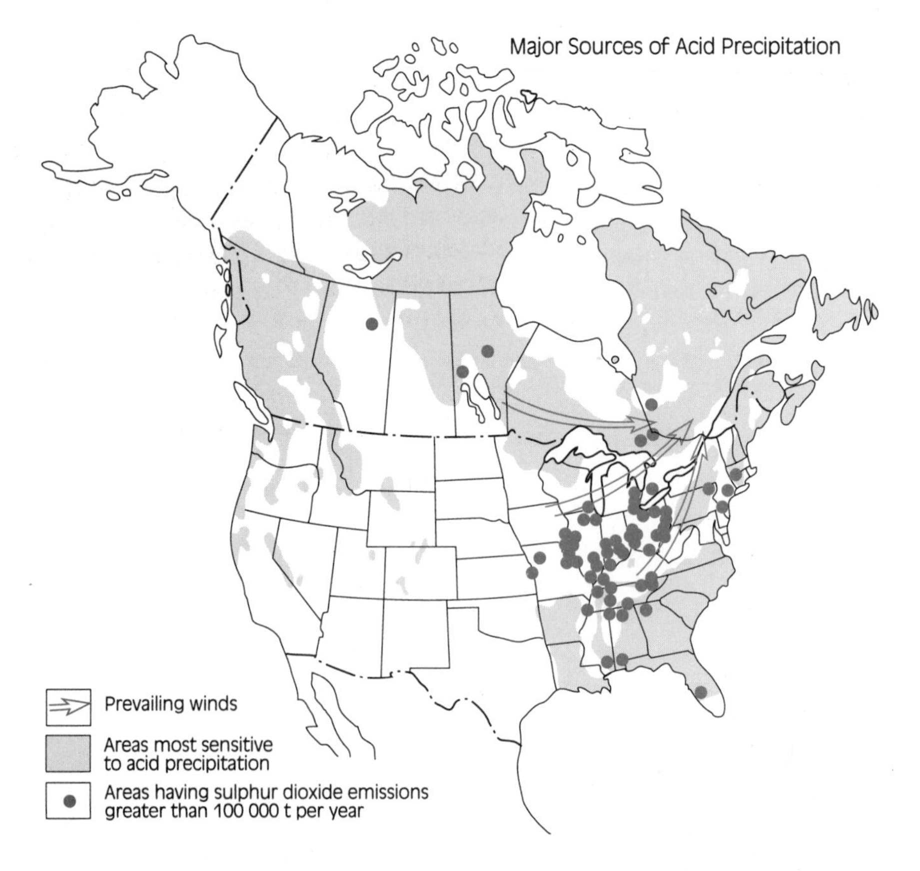

Figure 20.25
This map shows why Canada's efforts to control acid precipitation are concentrated in the east. Most major Canadian sources of sulphur dioxide are located within this area. In addition, prevailing winds carry the pollution from the United States into the eastern portions of Canada.

Figure 20.26
Acid rain threatens many of the world's cultural treasures, such as the Acropolis in Athens and this ancient Mayan palace in Uxmal, Mexico.

2. Suggest why marble ($CaCO_3$) is more vulnerable to damage from acid rain than granite—an igneous rock consisting chiefly of feldspar and quartz (SiO_2). (HINT: See Chapter 9.)
3. Lakes that are situated on bedrock consisting of limestone and sandstone suffer less damage to aquatic life from acid rain than lakes that lie over rock such as granite. Why might this be so?
4. Suggest why British Columbia has not had as much damage from acid rain as eastern Canada. Do you think this could change?

EXTENSION

Design and carry out an experiment that will allow you to observe the effect on water plants of a slow and steady increase in water acidity. (Remember to obtain your teacher's approval before proceeding with your experiment.) You can simulate acid rain using vinegar, which has a pH of about 2.8. What effects might you predict? How will you record these effects? Remember to set up a control for your experiment.

REVIEW 20.3

1. Name five major sources of water pollution in Canada.
2. Describe three processes used to remove pollutants from sewage.
3. What tests might you use to determine water quality?
4. Describe how acid rain is produced.
5. Give an example of one way in which your ideas about water pollution have changed since you began your study of this subject.

20.4 Controlling Pollution

To some people, cars are a wasteful method of moving people around. Most North Americans, however, find it difficult to imagine life without them. Whatever your opinion, automobiles are a good example of the connections that link pollution to population size, technology, and lifestyles (Figure 20.27). Understanding these connections can help you to identify the most effective methods of controlling pollution.

As you saw in Figure 20.13, automobile exhausts are one of the biggest sources of atmospheric pollutants. This is partly a result of the large number of cars on the roads (population size), partly due to the fact that burning gasoline in an internal combustion engine is a relatively inefficient way to power a car (technology), and partly a result of people's dependence on the everyday use of personal vehicles (lifestyle). But the impact of automobiles on the environment doesn't begin or end with what comes out of the exhaust pipes. The activities, materials, and energy used to manufacture cars, and to produce the fuel and roadways needed to operate them, all add up to a multitude of pollution problems.

Figure 20.27
What effect do all these vehicles have on the environment? How can we change this?

Table 20.5 lists some possible solutions for controlling the pollution produced by making and using cars. Many of these solutions have already been achieved by some automobile manufacturers and in some communities. Note that the solutions involve a variety of strategies that you have already studied. They include both technological answers (such as redesigning engines and developing alternative fuels) and lifestyle changes (such as reducing automobile use and increasing public transportation).

Table 20.5 Ways of Reducing the Pollution Caused by Automobiles

Solutions	How achieved
Reduce the quantity of materials and energy used to manufacture automobiles	• Make cars smaller • Make cars that last longer • Reduce demand for cars by providing more and better public transportation • Recycle and reuse scrapped cars and parts
Reduce the quantity of gasoline consumed by automobiles	• Improve engine efficiency to increase km/L • Maintain lower speed limit on highways (faster driving burns more fuel) • Encourage car pools • Reduce unnecessary use of cars and encourage alternatives of walking, cycling, and public transportation • Develop alternative fuels such as hydrogen, gasohol, and electricity
Reduce the quantity of oil used by automobiles	• Improve seals and valves to minimize leakage • Clean and reuse engine oil
Reduce polluting exhausts	• Improve engine design to burn fuel more completely • Improve catalytic converters that reduce emissions of hydrocarbons and lead • Use lead-free gasoline • Reduce fuel consumption (see above)

Figure 20.28
Equipment such as this can be used to detect minute quantities of pollutants.

SETTING STANDARDS

One of the questions to be answered before developing costly pollution control methods is: What levels of each pollutant are dangerous? This is not an easy question to answer because long-term exposure to low levels of some pollutants may be just as harmful as short-term exposure to high levels. Furthermore, two or more pollutants in small quantities may combine in the body to have a greater effect than either pollutant on its own. This problem is made worse by the huge number of new chemicals developed for use every year.

Modern technology allows scientists to measure a wide range of chemicals in much smaller quantities than ever before (Figure 20.28). Toxic substances are usually reported in parts per million (ppm). One part per million is equivalent to 1 mg per kg, or about one minute in two years. Our ability to detect and measure many substances is, in fact, much better developed than our understanding of the substances' long-term effects in the environment or in our bodies.

Levels of tolerance, or maximum acceptable concentrations for substances, are usually based on studies of the effect each toxic substance has on laboratory animals. Past experience shows that this is not always reliable, since some chemicals produce different results in laboratory animals than occur in humans.

Despite these difficulties, many pollution control standards have been set and enforced over the past few years. The results show that a cleaner environment can be achieved. In Canada, for example, the concentrations of several atmospheric pollutants have been reduced by installing pollution control equipment on industrial smokestacks, and by improving fuel consumption, adding catalytic converters, and using lead-free gas in automobiles. Between 1974 and 1982, the national annual average concentrations of sulphur dioxide decreased by 30 per cent; nitrous oxides dropped 25 per cent; particles fell by 24 per cent, and lead by 60 per cent.

Pollution is not only a national problem, however. It is a global issue. Pollutants produced by one country can be carried beyond borders, travelling worldwide on circulating air and water currents. Toxic chemicals can pass from organism to organism through food chains, turning up in animals and in places far from their point of origin. Through processes such as these, each part of the environment is connected to the others. Whatever we put into the environment enters a system of which humans are a part (Figure 20.29). While it is important to act locally to reduce pollution, people also need to think globally. The world is one place, in which we all live downwind and downstream from one another. Both strong national actions and international co-operation are needed to tackle the many problems of pollution.

Figure 20.29
Pollutants cannot be thrown away. On planet Earth, there is no "away."

ACTIVITY 20J Acting Locally, Thinking Globally

What evidence of pollution can you find in your community? In this activity, you will produce a report that connects pollution to the development of a local resource.

PROCEDURE

1. Consider the development of a resource in your community. How you complete your report will depend on where you live. For example, your community may be affected by a nearby factory, mill, or fish-packing plant. Or perhaps agricultural, mining, or forestry activities have an impact on your local environment. Or maybe you have a garbage or erosion problem from the development of a local tourist activity.
2. When you have chosen what to study, identify its possible impact on the environment. This impact may involve short-term or long-term damage to the ecosystem. If pollutants are involved, note how the amount of pollution compares with acceptable standards developed by the government. Describe what effect(s) the object of your study has on the air, water, or soil, and on the plant, animal, and human life in your area.
3. Analyse the problem you have identified. Suggest methods of eliminating or reducing the problem. Include solutions that involve changes in technology and those that require changes in people's attitudes and behaviour. Finally, consider the problem in your area in global terms. What effect might something done in your community have on the rest of the world?

REVIEW 20.4

1. Give an example to show how one form of pollution is affected by
   (a) population size,
   (b) technology,
   (c) lifestyle.
2. Why is it important to identify the minimum amount of a pollutant that might cause harm to human health? Why is it difficult to establish what this amount might be?
3. Give your reasons for agreeing or disagreeing with each of the following:
   (a) "We have no air pollution problem because the prevailing winds carry our air pollutants out to sea."
   (b) "A country that has no industries does not have to worry about pollution."
4. Describe one way in which technology may (a) cause pollution, (b) solve a pollution problem.

CHAPTER • REVIEW

Key Ideas

- Pollution occurs when something that may cause harm to living or non-living things is added to the environment. Pollutants may be natural or human-made.
- Most pollutants are waste products from human activities; many of them can be reduced, reused, or recycled.
- In a consumer society, the demand for products sets off a chain of activities that eventually result in increased pollution. Pollution can be decreased by a conserver lifestyle, in which resource use and waste are minimized.
- Air pollutants may be gases, solids, or liquids. Most air pollution comes from the burning of fuels.
- Pollutants may interact with one another and the environment to have far-reaching effects. Examples of this are the production of photochemical smog, the thinning of the ozone layer, the greenhouse effect, and acid rain.
- Water pollutants include human sewage and industrial waste.
- Acid rain is a major pollutant in North America.
- Pollution may be controlled by a variety of responses that society can make. Most responses involve changes in both technology and lifestyle.
- Pollution problems cannot be solved by dispersing pollutants because anything added to the environment becomes part of a global system of which humans are a part.

Vocabulary

pollution
pollutant
toxic waste
consumer society
conserver society
particulates
Air Quality Index
photochemical reaction
ozone layer
holistic approach
greenhouse effect
sewage
acid rain

1. Create a concept map that uses at least five of the terms from the vocabulary list.
2. Match each of the descriptions below to a term from the vocabulary list.
   (a) A chemical process that depends on light.
   (b) A society that reduces the individual's impact on the environment.
   (c) It helps protect you from sunburn.
   (d) It may lead to an increase in average global temperatures.

Connections

1. The volume of garbage may be reduced by incineration or by reducing the amount of packaging on consumer goods. Give one advantage and one disadvantage for each of these solutions.
2. "It is a waste of money to spend billions of dollars to reduce dioxin pollution from the manufacture

Figure 20.30
Has pollution stopped you from going to a place you once enjoyed?

of pulp and paper because these chemicals occur naturally (for example, as a result of forest fires) and so will always be present in the environment." Evaluate this statement and say why you agree or disagree with it.

3. A small town is located beside a lake. The town decides to save money by allowing untreated sewage into the lake, letting natural processes of decomposition break it down. Predict what may happen as the town grows in size.
4. The amount of air pollution in a city is particularly high in the morning and late afternoon when many people are driving to and from work. How would you solve this problem?
5. Suggest some tips that will allow individuals to reduce the amount of waste water produced in their homes.
6. Think of a place you once enjoyed that has been spoiled by pollution (Figure 20.30). What happened to it? What could have been done to prevent this?

Explorations

1. Some people believe that the problems of pollution are so great that the government should take stronger action to reduce it, including punishing polluters with heavy fines and jail sentences. In small groups, debate the proposal: "All polluters should be considered criminals." Under what circumstances would you *not* prosecute a polluter? Do you think punishment is an effective method of controlling polluters?
2. In groups, devise a game that could educate people about the connections between their lifestyle and pollution. Award bonus points for all activities that conserve materials and energy and reduce waste. Give penalties for actions that contribute to pollution. Evaluate which actions have a large impact and which have only a minor impact. Develop instructions and materials, then try out your game on another group.
3. Interview someone in your community whose work involves controlling pollution: a sanitation engineer, a refuse-disposal worker, a researcher, or a manufacturer. Find out about the pollutants their work involves. How aware are they of pollution problems? Have these problems been reduced or grown worse over the years?

Reflections

1. What are some B.C. pollution problems that have not been discussed in this chapter that you would like to know more about?
2. How can you personally bring about changes that reduce pollution?
3. Look back at the notes you made in Activity 20A. How have your ideas about pollution changed, and why?

CHAPTER

21

People and Resources

Imagine you are having dinner at a friend's home. After the meal is over, your friend helps his parents sweep all the china, cutlery, and glasses into a garbage can and carry it outside. Your friend explains that since his family has plenty of money, it is more convenient to throw out dirty dishes and use new ones the next time, rather than wash dishes every day.

An unlikely scene? Consider how many things people in Canada use for a short time, then throw away, give away, or sell, only to get more things to replace them. Did you ever wonder what it takes to produce all the things you use in your daily life? What resources are used to produce items made of glass, china, steel, or plastic? Do you think there are any limits to our ability to produce such items?

Today, the growing population means there is less of each resource, on average, for each person. There are also competing demands for the way each resource is used. For example, there are several possible uses for forests. They can be preserved as wilderness areas. They can be used to provide wood and pulp. Or they can be developed as areas for tourism and recreation.

Using a resource for one purpose often means that it cannot be used for another. Sometimes, however, compromises can be made to allow several different uses of the same resource. Who decides how resources are used? Who *should* decide? What data are needed to help make these decisions? In this chapter, you will consider these questions and others.

ACTIVITY 21A Resource Use Begins at Home

How does the behaviour of the family described in the first paragraph of this chapter relate to resource use? How is resource use connected with lifestyle, population size, and pollution? To begin your study of resource management, construct a diagram to show what you already know, or think you know, about resources. Write the word "resources" at the top of a page. Write "environment" at the bottom. Across the middle of the page, write the words "pollution," "population," and "lifestyle." Draw lines to connect each of these words to others on the page. Write one or more words beside each line to show how these things are related.

21.1 Resources and Lifestyle

Like all living organisms, people must obtain everything they use from naturally occurring parts of the environment, or **natural resources.** All the manufactured items that you see in a city begin with resources of some kind (Figure 21.1). For example, concrete is a manufactured material produced by mixing together water, sand, gravel, and powdered cement. Cement itself is made of lime and other materials obtained from natural sources such as limestone, chalk, and clay (Figure 21.2).

In the past, the term "natural resources" referred only to parts of the environment that directly provide people with raw materials or forms of energy: for example, minerals, metal ores, water, forests, and certain species of animals. Other parts that have few obvious uses, such as the atmosphere and deserts, were considered unimportant. More recently,

Figure 21.1
Huge quantities of concrete, steel, copper, glass, plastic, rubber, and other materials are needed to build a city like this. Vast amounts of gasoline, oil, and electricity are consumed each day to keep the city running. Where do all these things come from?

+

Figure 21.2
The concrete used in buildings is made from natural resources: sand, gravel, limestone, and water.

we have come to understand how all parts of the environment, living and non-living, are interconnected and equally important. Today, "natural resources" has come to mean all of nature. Thus, the atmosphere and deserts, together with the polar regions, oceans, soils, and every type of animal, plant, and micro-organism, can also be seen as resources that ultimately contribute to human well-being. For example, certain kinds of fungi that live on the roots of some trees help the trees obtain dissolved minerals. Without these fungi, the trees would not survive and people would not be able to use the trees.

Figure 21.3
What resources is this student using? Would you have guessed that someone riding a bicycle depends on coal?

USING RESOURCES EVERY DAY

People are often unaware of the connections between the products they buy or use and the resources on which these products depend (Figure 21.3). For example, the steel in a bicycle is a manufactured material produced by melting iron together with small quantities of carbon and other elements such as nickel and chromium. Both the heat used to melt the iron and the carbon added to make the steel come from coal.

In many cases, products are made using resources obtained from different places around the world. For example, your bicycle probably contains chromium from southern Africa (Figure 21.4). You may never see the chromium mines in southern Africa, but the impact of the mining activities on the environment is directly connected to every product eventually made using chromium from those mines. Every time you buy or use a product, therefore, you may have an impact on the world far beyond your own community. As a first step towards making decisions about your personal use of resources, you need to find out exactly what resources you use each day.

Figure 21.4
A chromium mine in southern Africa, where most of the world's reserve of chromium is found.

E X T E N S I O N

The Worldwatch Institute, based in Washington, D.C., gathers information about resource use and environmental issues from contacts all over the world. Find out more about the institute. Who founded it? When? Why? What does the institute hope to accomplish?

ACTIVITY 21B From Resources to Products

In this activity, you will carry out research to discover what resources are used to make everyday products. The activity may be completed by small groups of students working together.

PROCEDURE

1. Make a list of some products you use every day. If possible, work with your classmates to compile a class list of products.
2. Choose one product that many people in the class use regularly, such as a television, a tube of toothpaste, or a bed. Obtain an illustration or sample of your product.
3. Make a table with four columns, as in Table 21.1. Put the name of your product in the first column. In the second column, list all the materials that the product is made of. Include the packaging materials if the product cannot be used without them (for example, a tube of toothpaste). You may need to do some preliminary research to complete your list of materials.
4. Choose one of the materials used to make your product and find out how it is produced. Use your library or other sources and try to answer the following questions:
   - What natural resources are used to produce the material?
   - What major activities are needed to convert each natural resource into the material?

   Use your answers to complete the table.
5. Construct a flow chart that begins with natural resources and ends with the product you studied. Show the different steps needed to get from the resources to the product, including any impacts on the environment.

Table 21.1 Sample Data Table for Activity 21B

Product	What is it made of?	Major natural resources used	Related activities
Television	Steel	Iron ore Nickel Chromium Coal	Mining Smelting Transportation
	Glass	Sand (silica) Limestone Soda (sodium carbonate)	Mining Melting in furnace Transportation
	Plastic	etc.	
	Copper	etc.	

SAMPLE ONLY

DISCUSSION

1. Were certain resources used in every product studied by your classmates? If so, name them.
2. What was the average number of different resources used in each product?
3. What impact(s) might the production of your product have on the environment?
4. Choose one resource used in your group's product and imagine that it no longer exists. Would people still be able to make this product? Explain your answer.
5. Name any materials used in your product that can be reused or recycled.
6. "The Canadian lifestyle depends on mining." Explain why you agree or disagree with this statement.
7. Suggest ways in which you might alter your lifestyle to reduce your personal resource use. Suggest why you should.

REVIEW 21.1

1. What is meant by the term "natural resource"?
2. Give examples of three different types of natural resources.
3. Choose one material found in a manufactured item and describe how it is made from natural resources.
4. Every spring, the Grey family's garden chairs, made of plastic and aluminum, are cracked and worn after being left outside all winter. The Greys put the chairs out with the garbage and buy new ones.
   (a) Describe three impacts on the environment produced by this aspect of the Grey family's lifestyle. →

(b) Suggest three ways in which the Grey family could reduce their impact on the environment.

5. Describe one way in which your personal lifestyle affects a particular resource.

6. In what way(s) is human dependence on natural resources the same as that of other animals? In what way(s) is it different?

21.2 Managing Resources

DID YOU KNOW?

During the late 1980s, North American consumers spent more for food *packaging* than farmers earned selling their crops.

Our impact on the environment depends on more than just the kinds of resources we use. It also depends on the amounts we use. In Canada, we rarely experience shortages of energy or materials. We usually get all the electricity we need by flicking on a switch. We get all the water we want simply by turning on a tap. Car owners can fill their gas tanks with as much fuel as they want. We can choose from a huge number and variety of items in our stores. These common experiences may give the impression that the world's supply of resources has no limits. But clearly this is not the case. There is not an endless supply of resources. We must therefore learn how to manage the resources we have properly (Figure 21.5).

Figure 21.5
Canadians are using resources at a rapid rate. Some of these resources may be gone by the time you reach old age.

A major goal of **resource management** is to make the best use of a given resource for the maximum number of people over the longest period of time. This must be done with a minimum of damage to the environment. Without such planning, a resource may be divided unfairly among those who use it, or it may be used up too rapidly. Lack of planning can also lead to increased damage to the environment.

So who *does* manage natural resources? During the last century, there was little or no management of some resources. As long as there was a demand to be met, mining companies would dig out an ore until it was gone, then move on to the next site. Most people did not see the damage to the environment in the remote places where resources were extracted (Figure 21.6). Few thought the government should be involved. By the

Figure 21.6
The barren landscape around an open-pit mine. Poor management of resources can cause costly damage to the environment.

early 1900s, however, people began to realize that resource use affects everyone. Today, the people who make decisions about resource use include politicians, consumers, corporations that use the resources, and individuals who own land on which resources are located.

You may have already studied the management of fish and forest resources during your science courses in earlier grades. These are considered **renewable resources**, because fish and trees can grow and reproduce themselves. By balancing the rate at which we use them with the rate at which they are replaced, renewable resources can be managed to supply human needs for an indefinite period into the future.

In some ways, the problems involved with managing **non-renewable resources**, such as metals, minerals, and fossil fuels, are more straightforward (Figure 21.7). These materials exist in fixed amounts that grow smaller as the resource is used. At some point, a non-renewable resource may be reduced to such small amounts that it will no longer be available for human use.

Figure 21.7
Many non-renewable resources are obtained by digging and drilling.

ACTIVITY 21C *Managing a Non-Renewable Resource*

In this activity, you will analyse some of the factors that affect resource use by using a fictional situation as a model.

PROCEDURE

Read the paragraphs about "goc" use among the "Nujanac" people and examine Figure 21.8 on the next page. Use the information to answer the questions that follow.

Goc Use among the Nujanacs

A population of 1000 Nujanacs lives on a remote island. The most important non-renewable resource on the island is a mineral called goc. It is estimated that a total of 1 million tonnes of goc can be extracted economically from the goc mines on the island.

The Nujanacs use goc to fuel their vehicles, called mobes. Every Nujanac has one mobe. They consume, on average, 1 t of goc per mobe per year. The goc can also be used to produce a material called rikstik, which is made into fashionable clothing and household utensils. It takes 250 t of goc to make enough rikstik to supply the rikstik industry for one year at current levels of demand. →

Figure 21.8
The Nujanacs use goc to fuel their mobes and to make rikstik.

DISCUSSION

1. How much goc is used by the Nujanacs each year? At this rate, how long will the goc last?
2. Assume goc is used only as a fuel. How long will it last if
   (a) The Nujanac population remains stable?
   (b) There are twice as many Nujanacs in the population?
   (c) Each Nujanac buys a second mobe?
3. A Nujanac scientist invents a new mobe that uses 25 per cent less goc than the old models. By how many years will this increase the life of the goc mines if all Nujanacs use the new mobe?
4. Because of concern over pollution from the goc mines, the mine owners install pollution control equipment. To pay for this, they increase the price of goc. The Nujanac consumption of goc decreases by 20 per cent per year. At this reduced rate, how long will the goc last?
5. After 400 years, a Nujanac explorer finds a new underground supply of goc in a remote corner of the island. It is estimated to contain 50 000 t of goc. How many more years after this discovery will all the known reserves of goc last?
6. Suppose the Nujanacs earn more money by becoming international consultants on resource management. With their new wealth, they double the yearly demand for fashionable clothes and household utensils. How long would the original supply of goc last?
7. Predict what will happen to the rate of goc use if
   (a) another use is found for rikstik,
   (b) mobes are replaced by vehicles that use a different fuel,
   (c) there is an increase in unemployment among Nujanacs.
8. Suggest another factor that could affect the rate of goc use and explain how it might increase or decrease the rate.
9. From your analysis of goc use among the Nujanacs, complete the following general statements about resource use:
   (a) The length of time for which a non-renewable resource lasts depends on . . .
   (b) The rate at which a resource is used depends on . . .
   (c) The demand for a resource depends on . . .

CHANGING TRENDS IN RESOURCE USE

The total amount of a resource that has been measured and judged economical to extract is known as a **reserve.** This amount is used as a basis for estimating how long supplies of the resource may last. Reserves decline as the resource is extracted and used (Figure 21.9). However, reserves may be increased if additional supplies are discovered, such as a large new oilfield under the ocean floor or a mineral body in the Arctic.

Another way that a reserve may increase is if the price paid for the resource goes up, or a less expensive method of extracting the resource is developed. A mining company may close a gold mine when the expense of extracting the gold is greater than the money the company can get by selling it. If this happens, the gold left in the ground is not considered part of the reserve. But if the selling price of gold increases, or the cost of extraction is reduced, the mine may become profitable once again. The quantity of gold left in the mine will then be added to the reserve.

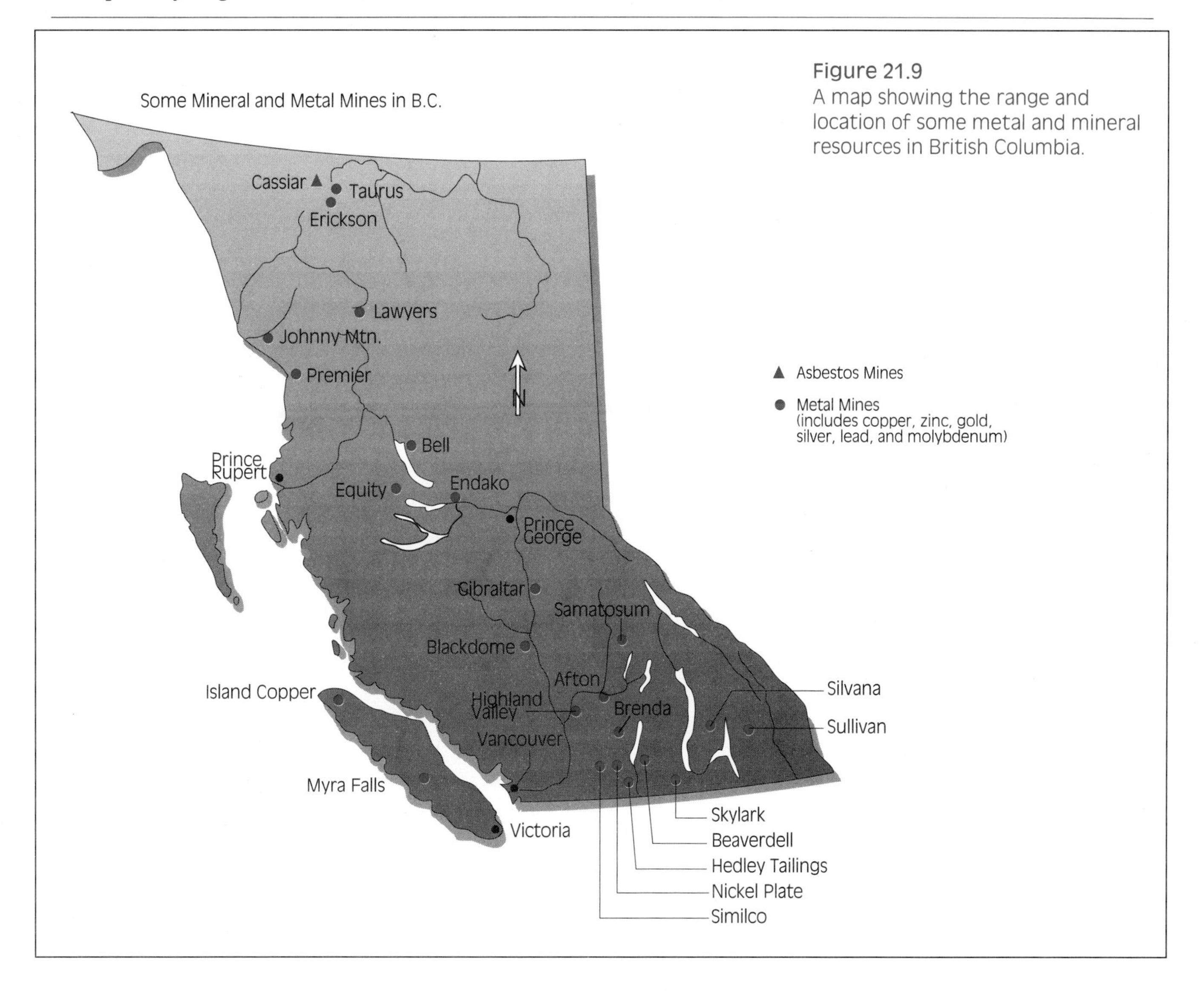

Figure 21.9
A map showing the range and location of some metal and mineral resources in British Columbia.

EXTENSION

From 1943 to 1952, Canada produced just under 7 million tonnes of salt (Figure 21.10). From 1973 to 1982, Canada's salt production grew nearly 10 times, to over 63 million tonnes. Find out what salt is used for, and suggest why there was such a huge growth in the production of this resource.

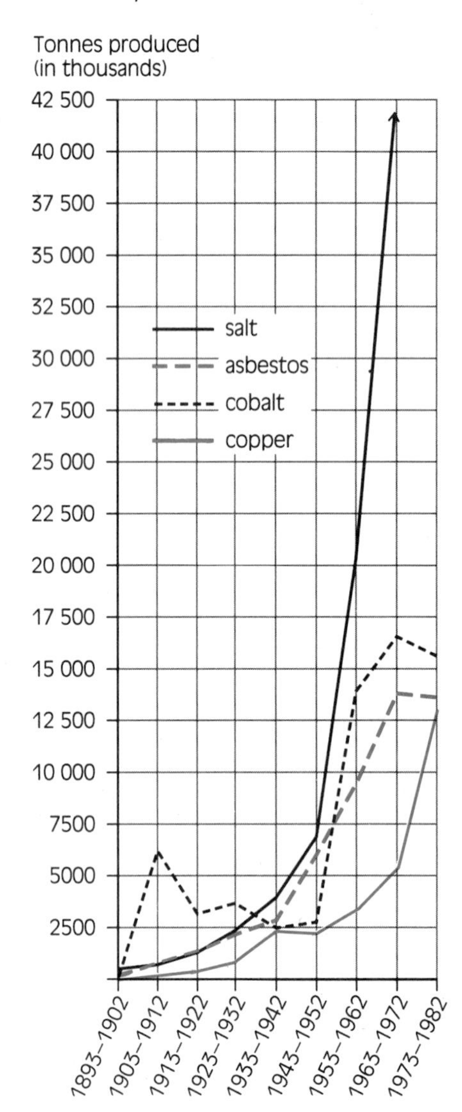

Figure 21.10
Why might the production of cobalt and asbestos be declining?

In order to predict how long a resource may last, resource analysts look at **trends**, or patterns of resource use. Data like those in Figure 21.10 are typical for most mineral and metal resources. They show a trend of increasing use since the beginning of the century, with a particularly sharp growth after the early 1940s. This increase in resource use is partly due to the increase in population and partly a result of other factors, such as those considered in Activity 21C. A rough estimate of future supplies can be made by dividing the amount in the reserve by the current rate of extraction. For example, Canada's reserve of copper is about 17 billion tonnes. The mining industry extracts about 700 million tonnes of copper per year. If extraction continues at this rate, Canada's copper reserve will be used up in just over 24 years.

Such estimates must be revised frequently. Not only may known reserves be increased over a period of a few years, but many factors can rapidly change the rate of resource use in either direction.

ACTIVITY 21D *Analysing Trends in Coal Use*

This activity will allow you to find out more about trends in B.C. coal production. Coal is used mainly as a fuel and worldwide demand for fuels is growing. Knowing this, and taking other information into account, can you forecast the future of B.C.'s coal industry?

PROCEDURE

Table 21.2 gives figures for the total coal produced in B.C. over a period of 20 years. Examine the figures, then plot them on a graph. Use the information from the table and your graph to answer the questions below.

DISCUSSION

1. What percentage increase was there in coal production between
   (a) 1970 and 1979?
   (b) 1980 and 1989?
   (c) 1970 and 1989?
2. (a) If you had been examining the data in 1979, and assumed that growth in production would remain constant, what level of coal

production would you have predicted for 1989?
  (b) What changes that you couldn't have predicted in 1979 might have produced a change in the rate of coal use?

3. Based on current trends in coal production, estimate how much coal will be produced in B.C. in 1999 and 2009.
4. Suggest two or more factors that might make your prediction
   (a) too high,
   (b) too low.
5. Examine these coal production figures for earlier years:

   1940 1 507 758 t
   1950 1 427 907 t
   1960 715 455 t
   1969 773 226 t

   Do these additional data affect your estimates in question 3? Explain.
6. What additional information might help you make a better estimate of future coal production? How might this information affect your estimates? Give examples.

Table 21.2 Coal Production in British Columbia, 1970 to 1989

Year	Quantity produced (in tonnes)	Year	Quantity produced (in tonnes)
1970	2 398 635	1980	10 823 530
1971	4 141 496	1981	11 752 621
1972	5 466 846	1982	10 645 742
1973	6 924 733	1983	11 480 298
1974	7 757 440	1984	20 739 725
1975	8 924 816	1985	22 612 810
1976	7 537 695	1986	20 836 863
1977	8 424 181	1987	22 586 852
1978	9 463 920	1988	24 813 082
1979	10 570 370	1989	25 134 198

RESOURCE USE AND CHANGING TECHNOLOGY

One of the most important factors affecting how people use resources is changes in technology — that is, the development of new materials, machinery, or processes (Figure 21.11). These developments may reduce the need for a once commonly used resource, or boost the demand for a resource that was previously little used.

Figure 21.11
Different technologies use different resources. What resources are used by steam engines and nuclear generators?

For example, large quantities of wood and coal were once used as fuel to boil water for trains powered by steam. When steam engines were replaced by diesel-driven engines, the need for wood and coal was reduced and the demand for diesel fuel increased. In many countries, diesel engines have in turn given way to engines that run on electricity. Where the electricity is produced by coal-fired generating stations, these modern trains once again depend on coal!

Before nuclear-powered electricity production was developed during the late 1950s, few people would have predicted a large demand for uranium. By the mid-1970s, however, the world demand for nuclear power, and hence for uranium, had grown rapidly. During the 1980s, the trend changed again. Demand for uranium slowed as its cost increased and as the risks associated with nuclear power became clear. Accidents at Three Mile Island in the United States in 1980, and at Chernobyl in the Soviet Union in 1985, made many people less enthusiastic about nuclear power.

RESOURCE USE AND LIFESTYLE

The demand for all kinds of metals, minerals, and fossil fuels is growing rapidly in the world today, mainly because of a change in lifestyle. Many countries whose economies were once based on farming and rural life now have economies based on industry, manufacturing, and city life. Many scientists believe our planet cannot provide enough resources to allow this trend to continue. Indeed, it would be impossible for even the current population of the world to live the kind of lifestyle typical of industrialized countries such as Canada.

Figure 21.12
Canadian imports of crude oil fell sharply after 1975. Ask some adults what they remember about the "energy crisis" that began in the mid-1970s.

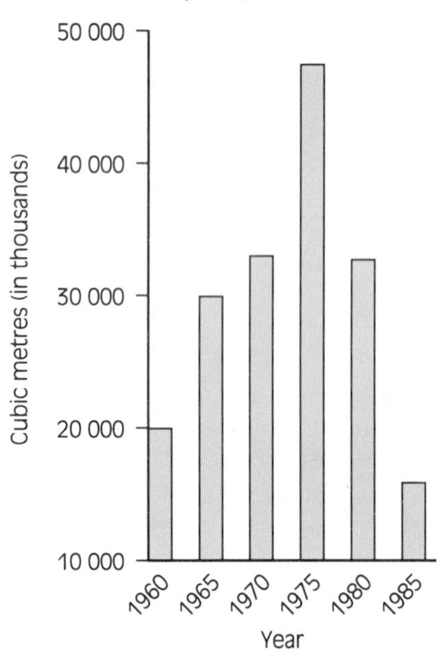

RESOURCE USE AND POLITICS

Resource use can be affected by political events as well as by population growth and changes in technology and lifestyle. This is most likely to occur when important resources are located in only a few countries. For example, the Arab countries of the Middle East and North Africa are the world's largest exporters of oil to the industrialized countries of North America, western Europe, and Asia. In 1973, when the Arab countries were at war with Israel, they decided to cut their exports of oil to countries that supported Israel. The reduced oil supply caused an "energy crisis" in the industrialized countries that depended on imported oil from the Middle East. Within 10 years, the price of oil had grown to more than 10 times its price at the start of the Arab-Israeli war. This eventually affected every country in the world. The higher cost of oil had a dramatic effect in Canada, where oil imports dropped by over half between 1975 and 1982 (Figure 21.12). During 1990, the invasion of Kuwait by Iraq raised the issue of stable worldwide oil supplies once again.

COMPUTER APPLICATION

GEOGRAPHIC INFORMATION SYSTEMS AND RESOURCE MANAGEMENT

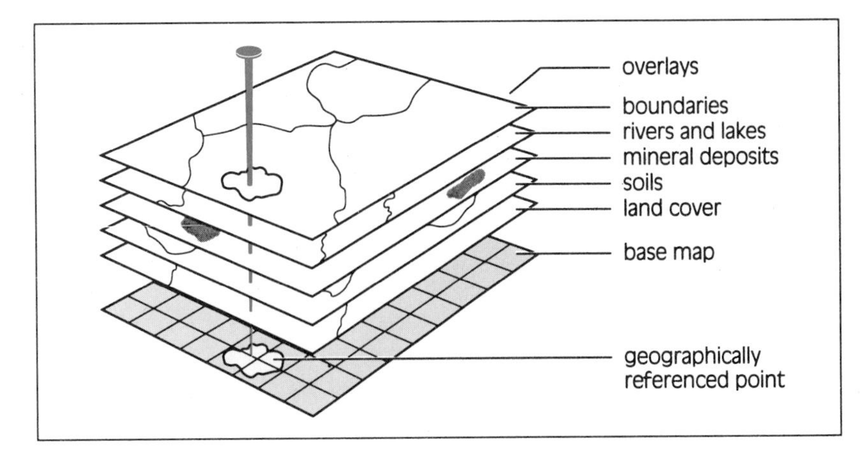

You are developing plans for a mine and want to check some field study information about water run-off that might affect fish stocks in the area, but you cannot go to the mine site. What can you do? Look at a map . . . or visit the site on your computer.

Using data about the area and a mapping program, you call up a three-dimensional model of the mine site on your computer screen. You examine the peaks, valleys, and river corridors from various angles. You call up data on fish stocks and water flow. Before long, your questions are answered. Without leaving your desk, you have paid a "visit" to the site with a geographic information system (GIS).

Geographic information systems use computer technology to store and manipulate large amounts of data relating to locations on the Earth's surface. You might visualize a GIS as a base map accompanied by several overlays. The base map provides precise geographic references (latitude and longitude), and the overlays provide data that can be analysed by computer programs. A single geographically referenced point can be studied in detail using this information.

By providing access to large amounts of data, geographic information systems are revolutionizing the management of natural resources. Today, a GIS can help resource managers examine the effect of a dam or highway on a forest, or make projections about the amount of a mineral that can be extracted from a particular site. A GIS does this by allowing the resource manager to organize, analyse, combine, and display different types of information about a location. Often this involves "mixing" data—comparing data about human population and fish stocks, or coal deposits and land cover. Any number of different databases can be combined and displayed together on a computer screen.

The display of data on screen is probably where geographic information systems depart most dramatically from the world of paper maps. Paper maps are restricted by the fact that all information is stored and displayed in a permanent form on a sheet. GIS technology has separated the storage function from the display function. While information can be printed out on a sheet, it does not have to be. As a GIS user, you can select a specific area and particular features for display on your computer screen. You can zoom in for a closer look or move back for a larger view. You can examine the area in three dimensions, choosing your angle and viewpoint. And you can always call up additional statistical or graphic information.

In the future, animation will let you "walk" or "drive" around in modelled terrain, or "fly" over it. You will even be able to see water "flowing" and wildlife "migrating." These capabilities are sure to make geographic information systems important tools for resource management in the next decade.

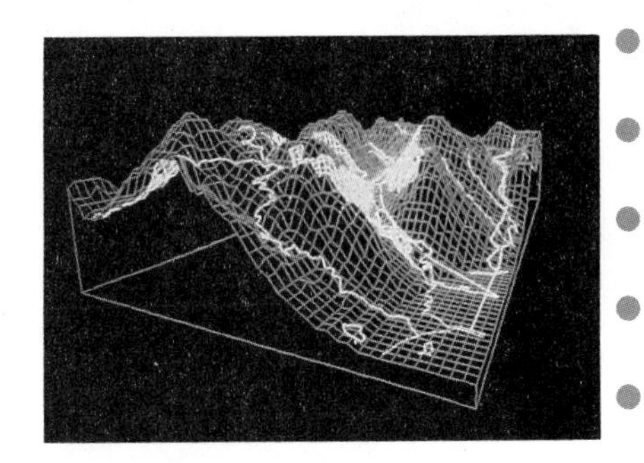

Computers have changed the way we store and display geographic information. Computerized terrain models like the one shown here are being used now to make decisions about how we use resources.

MANAGING BY CONSERVING

There are several ways in which resources can be made to last longer. You may have already considered some of them in Chapter 20. We can reduce our consumption of materials and energy. We can recycle materials such as glass, aluminum, and oil. We can reuse products rather than throwing them out as garbage. All these actions reduce both waste and the demand for natural resources, which ultimately will make the resources last longer. The strategy of managing resources by careful use is known as **conservation**.

Many new technological developments can help people conserve energy and materials (Figure 21.13). Improved home insulation materials and window construction prevent heat escaping from buildings. Better designed cars last longer and travel farther on the same amount of fuel. Furnaces, refrigerators, and other appliances use less energy than in the past by using improved design and materials to do the same job. Resources can also be saved by switching from non-renewable to renewable resources. You may have studied some of the renewable sources of energy in Chapter 6.

DID YOU KNOW?

A fluorescent light bulb uses only one-quarter of the electricity needed by an incandescent bulb and lasts 10 times as long.

Figure 21.13
Energy can be saved by new technology.

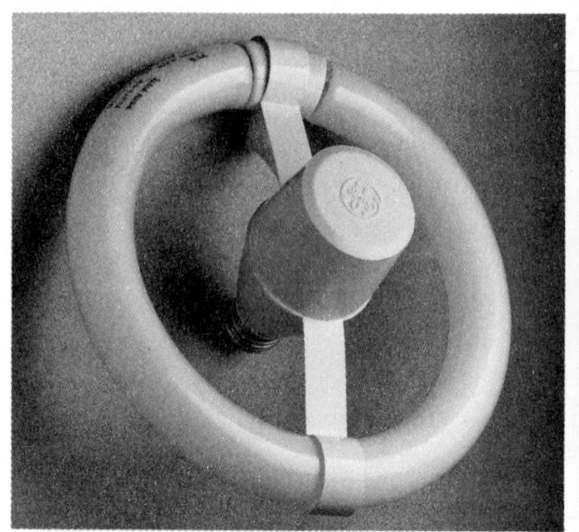

fluorescent light bulb

microwave oven

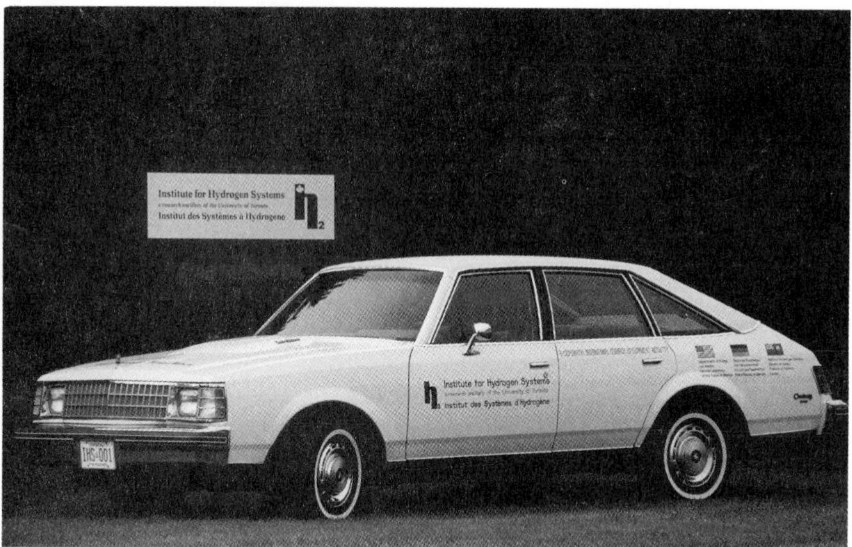

hydrogen-powered car

A scientist once compared the problem of managing shrinking resources to the problem of keeping water in a leaking bathtub. The traditional approach of governments and corporations is to meet the growing demand by searching for and extracting more and more resources. This is like adding more hot-water tanks to keep the bathtub filled. It greatly increases the amount of water consumed for the task of bathing. The approach of conservers is to recycle and reuse resources and eliminate waste. This is like plugging the leak in the bathtub. As the comparison shows, it often costs much less to make the changes needed to *save* a given quantity of a resource than it does to provide the same quantity from a new supply (Figure 21.14).

Figure 21.14
Two ways to keep a leaking bathtub filled.

ACTIVITY 21E How to Be a Conserver

How could you decrease your personal impact on resources and the environment? To find out, you can measure your use of a resource, then suggest ways of slowing down your rate of use.

PROCEDURE

1. Observe and measure the quantity of something that your family uses during one week. Choose something that is practical for you to research. It could be water, gasoline, paper, laundry detergent, electricity, aluminum, plastic, glass, or something else. Either measure the total amount used by your household during seven consecutive days (for example, the total mass of paper, the total litres of gasoline); or estimate the quantity from samples (for example, determine the quantity of water used per shower, toilet flush, and laundry cycle and multiply by the number of times the shower, toilet, and washing machine are used in a week).
2. Record your data in a table to show the material or energy studied, the total quantity used per week, and the resources used (see Table 21.3).
3. Suggest ways of reducing your use of the material or energy. This could involve a change in habits, or the use of a new product or invention that does the same job using less of what you have chosen to study in this activity. Measure or estimate the quantity of material or energy saved by your alternative.
4. Prepare a display to illustrate your impact on resources before and after you become a conserver.

DISCUSSION

1. What quantity of the material or energy you studied could your family save in one year by conserving? (Multiply your weekly saving by 52.)
2. There are about 11 million households in Canada. How much material or energy would be used each week if every household used what your family did (a) before conserving? (b) after conserving?
3. Name two effects that conserving has on the environment, based on the activities needed to produce material or energy from natural resources.

Table 21.3 Sample Data Table for Activity 21E

Material or energy	Quantity used	Natural resource(s) used
Plastic	0.5 kg per week	Oil or coal

REVIEW 21.2

1. What is a major goal of resource management?
2. How might a reserve of oil be increased? Are there limits to this increase? Explain.
3. Name three things that affect the rate of resource use and say whether each increases or decreases the rate.
4. Explain how information about trends is useful for predicting how long a non-renewable resource may last. Why must these predictions be updated regularly?
5. Explain the meaning of "conservation of resources" and give an example.

21.3 Coal: A British Columbia Resource

Before certain resources can be used, they must be extracted, transported, and processed. Activities such as these provide employment to many people in British Columbia and produce wealth for the province. But what are the effects of resource development on the environment? In this section, you will study coal as an example of how a particular non-renewable resource is obtained and used in British Columbia.

Coal is a very widespread resource, found on every continent. More than 60 countries produce a total of over 3000 million tonnes of coal every year. Most of Canada's coal is mined in Alberta and British Columbia, which together provide about three-quarters of Canada's annual coal production (Figure 21.15). At the present rate of production, there is enough coal in Canada's reserves to last over 100 years.

Figure 21.15
Most of Canada's coal is found west of Manitoba.

USES OF COAL

Most coal is burned to produce heat energy and generate electric power. Coal is the world's second biggest source of energy, only slightly behind oil. Nearly half the world's electricity is generated in coal-burning stations.

A major product from coal is coke, an almost pure form of carbon that burns with an intense, smokeless heat. Coke is a vital ingredient in metal smelters, where it is burned in furnaces to purify metals and to produce steel from iron ore. Coke is the solid product left behind after coal is heated in the absence of air in huge brick and steel ovens. (This process is called destructive distillation.) During the production of coke, gases are given off. These may be collected and used as a fuel. Coal is also a valuable source of hundreds of chemicals. Industrial chemists have used coal to make such things as plastics, explosives, dyes, tar, perfumes, and medicines (Figure 21.16).

Figure 21.16
Some uses of coal.

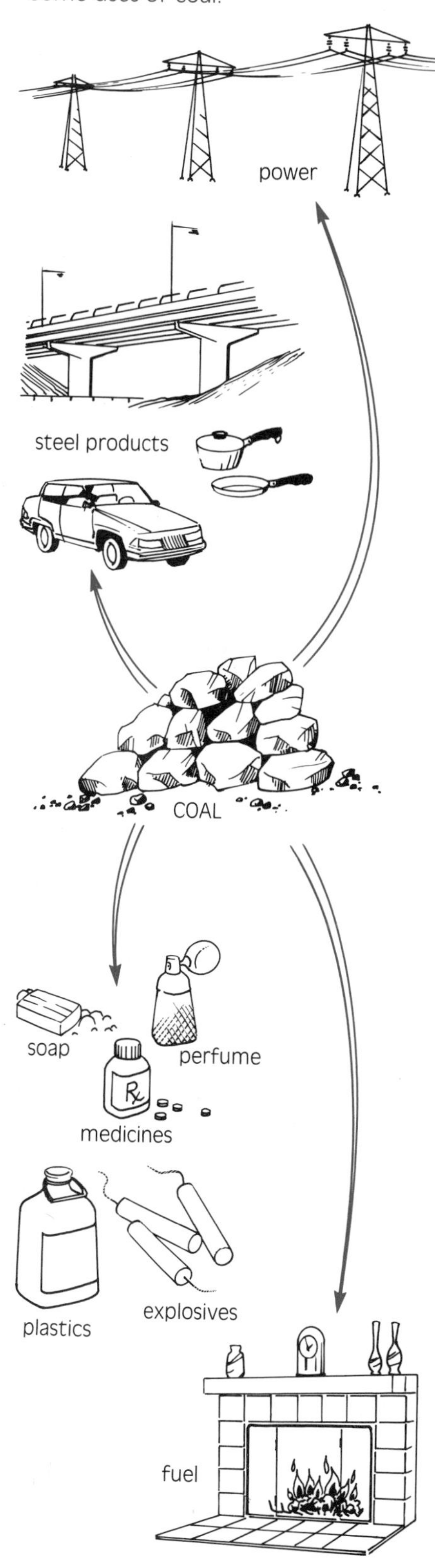

TYPES OF COAL

Coal varies in quality and composition, depending on how and when it was formed. The main ingredient in coal is carbon. The higher the carbon content of the coal, the more heat energy and the less smoke it produces when burned. Coal also contains small amounts of minerals and other elements such as sulphur. The four main types of coal are lignite, sub-bituminous, bituminous, and anthracite (see Table 21.4). In the mining industry, coal is also named according to its major use. Coal that is burned to produce heat and generate electric power (anthracite, sub-bituminous, and lignite) is known as thermal coal. Coal that is used for making coke (mainly bituminous) is called metallurgical coal.

Table 21.4 Major Types of Coal

Type of coal	Properties
Lignite	Brown to black coal with the lowest carbon content (60% to 75%). Very smoky when burned. Often contains high levels of sulphur. A big contributor to air pollution.
Sub-bituminous	Intermediate form between lignite and bituminous. Cannot be used to make coke. Mostly used in electricity generators.
Bituminous	The most common type of coal. Burns with a smoky, yellowish flame. May be used to make coke or generate electricity.
Anthracite	A black, hard coal with a high carbon content (80% to 85%). Burns slowly with a pale blue flame and produces little smoke. Best coal for home use. May be mixed in small quantities with bituminous to make coke.

ACTIVITY 21F Managing Coal Resources

In this activity, you will explore some of the issues that are involved in developing British Columbia's coal resources (Figure 21.17).

PROCEDURE

1. Read the information on coal that follows this activity, making notes about the effects of coal development under the following headings in your notebook: Mining Activities; Transportation Methods; Processing and Using Coal.

2. Make a simple flow chart to show the activities involved from the mining of a coal deposit to the production of steel and electricity.

Figure 21.17
Taking part in a debate can help you find out more about an issue.

3. Make lists of the various ways in which the extraction, transportation, and use of coal may affect (a) land, (b) water, (c) air, (d) plants and animals, (e) people. You may wish to do this with a partner.

4. "The development of B.C.'s coal resources has more advantages than disadvantages for the province." Take part in a debate with a partner; one of you can agree with the statement and the other can disagree. First, review the information in your notes, then present your argument to your partner.

DID YOU KNOW?

Some scientists have attempted to obtain energy from coal without mining it! This is done by burning the coal deposit underground. Air passed through the deposit keeps it burning. Water piped through the underground fire is converted to steam and used to generate electricity. Attempts so far have not overcome the problems of controlling the combustion and producing consistent energy supplies.

MINING COAL

The methods used to get coal out of the ground vary with the size of the coal deposit, its depth beneath the surface, and the location. The two major types of mining are underground mining and surface mining.

Coal that is deep underground is usually reached by tunnels cut into the ground. Vertical or inclined shafts are sunk into coal deposits, and horizontal tunnels may be dug outwards from the shafts into the coal. The coal is removed by large mining machines that grind it into workable pieces. It is then loaded onto conveyer belts or shuttle cars and brought to the surface.

In British Columbia, nearly all coal is removed by a surface mining technique called open-pit mining (Figure 21.18). This method is used to extract thick beds of coal that lie underneath a thin covering of rock and earth known as overburden. After the topsoil is carefully scraped away, the remaining overburden is blasted apart and the coal is extracted by large mobile excavators (Figure 21.19).

Mined coal usually includes a proportion of other substances, such as shales and other rocks. These are separated from the coal in special washers. Most coal cleaning is based on the fact that coal is less dense than rock. Devices that use water, air, heavy liquid, or vibration are all used to separate coal from rock. On average, nearly six tonnes of rock waste are produced for every tonne of coal mined in British Columbia.

Without proper management, a mine can scar the landscape, alter drainage patterns, and increase erosion. In addition, mining activities can pollute waterways with silt, coal dust, and toxic materials from exposed rock. With suitable planning, however, a mining operation can

Figure 21.18
An open-pit mine in British Columbia.

Figure 21.19
Equipment used to dig and lift coal from the surface in an open-pit mine.

Figure 21.20
A mine site that has been rehabilitated.

avoid these results. Drainage and erosion problems can be minimized by carefully controlling how the excavation alters the slope of the pit walls. The overburden that is removed can be piled securely to prevent run-off or erosion. Some of this overburden may later be used to backfill the pit, or it may be used in nearby road construction or other building projects. When the excavated coal is washed, the problem of polluting streams and rivers with waste water can be prevented by cleaning and recycling the water.

Although the process of open-pit mining cannot avoid removing soil and vegetation and digging out tonnes of rock, this use of the land is a temporary one. A mining operation may last from 10 years to 30 years or more, but eventually the coal in a particular location will be gone or the mine will no longer be economical to operate. When a mine is closed, the land can be restored for other uses (Figure 21.20). Any remaining coal, or other toxic materials, is first covered by rock. Topsoil is then returned to the surface, and the area is contoured and planted. The restored area might be used for parkland, farmland, or tree farming.

TRANSPORTING COAL

How would you get coal from the mines to the places where it is processed or used? Some ways are shown in Figure 21.21. Overland, the obvious ways are by truck or railway car. For overseas transport, cargo ships are used. Coal is also shipped through inland waterways by barge. The choice of transport depends partly on the locations of the mine and the customer and partly on the cost of each method. Because of its bulk, the cost of transporting coal is often a large proportion of the total price paid by the consumer. Mining companies are therefore always looking for the most economical routes and methods of transportation.

The major impact on the environment from transporting coal comes from such road, rail, and port building activities, and from the pollutants

Figure 21.21
Methods of transporting coal.

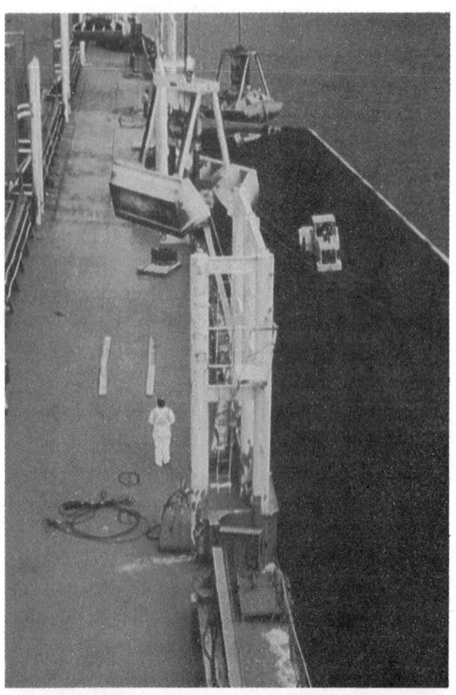

produced by using the fuel that powers the transporting vehicles. In other respects, coal is a less hazardous material to transport than, say, oil. A spill of coal on the land or at sea does less damage to living things and is easier to remove than a spill of liquids or gases.

In some areas of the world, coal is shipped "by wire." To be more precise, the energy in coal is shipped by wire. The coal is burned in an electricity-generating station at the coal-mine site, and the electric power is transmitted to users over extra-high voltage wires. Sending coal "by wire" eliminates the need for transporting coal, and may be the cheapest source of large-scale electricity production after hydroelectricity.

Coal may also be converted into gas at the coal mine. The gas is then transported to users by pipeline. Pipelines have also been used to transport a slurry (semi-fluid mixture) of coal and water up to 440 km.

USING COAL

The production and use of B.C.'s coal is an example of the international nature of resource management. As you have learned, coal is used mainly as a fuel in electricity-generating stations and metal smelters (Figure 21.22). But British Columbia produces its electricity in hydro stations, and does not have a large metal-processing industry. Most of the coal mined in this province is used by customers outside the province. At the end of the 1980s, about 90 per cent of the coal produced in British Columbia was metallurgical coal, shipped to Japan to make coke for steel mills.

Because little coal is used in the province, it may seem that the impact of the coal industry on our environment is limited to mining and transportation activities. However, pollution from burning coal produces changes in the atmosphere that affect the entire planet. Like other fossil fuels, coal releases polluting gases and particulates into the air when burned. Moreover, coal is the "dirtiest" of the fossil fuels. It releases 20 per cent more oxides of carbon than oil does and 75 per cent more than natural gas. In addition, burning coal produces oxides of sulphur and nitrogen. You may have already studied the connections among these pollutants and their contribution to poor air quality, acid rain, and the greenhouse effect.

The impact of pollution from burning coal can be reduced by several management strategies. These include:

- improving the efficiency with which coal is burned and converted into electricity
- reducing the waste of electrical energy by consumers
- increasing the production of electricity by alternative, non-polluting, renewable resources such as wind and solar power
- reducing the demand for new steel and recycling more scrapped steel, which takes less energy to process

All these steps have the additional effect of reducing the demand for coal, which reduces the impact of mining and transportation activities and lengthens the life of coal reserves.

DID YOU KNOW?

Steel produced from scrap reduces air pollution by 85 per cent and cuts water pollution by 76 per cent compared with steel produced from iron ore. Furthermore, it eliminates mining wastes altogether.

(a)

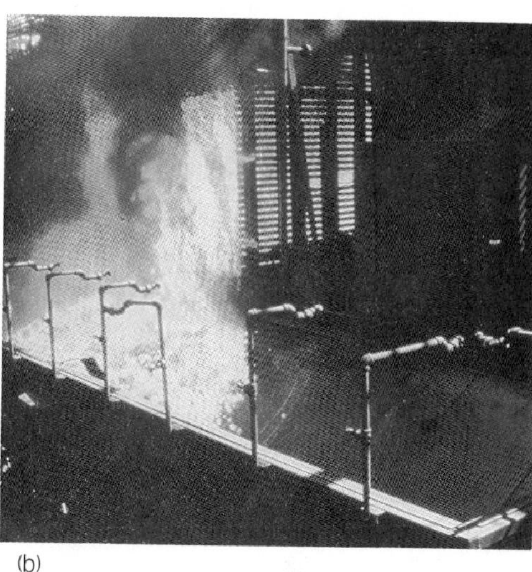

(b)

Figure 21.22
(a) A coal-burning power station.
(b) A coal-burning metal smelter.

REVIEW 21.3

1. What is the main difference between metallurgical coal and thermal coal?
2. How is coke made and what is it used for?
3. List some of the impacts of mining activities on the environment. How can these impacts be reduced?
4. What factors help determine the method used to transport coal from the mine to the user?
5. How is coal mining in British Columbia connected with
   (a) car manufacturing in Japan?
   (b) the greenhouse effect?
6. List three questions that resource managers might want answered to help them determine how to develop a non-renewable resource.
7. "By exporting its coal, British Columbia exports its biggest pollution problems." Explain why you agree or disagree with this statement.
8. What effects might it have on you personally and on the province as a whole if there were
   (a) more coal mining in B.C?
   (b) less coal mining in B.C.?

EXTENSION

What happens to mining communities after all the coal in an area has been mined? Will they become abandoned "ghost towns"? Investigate the history of mining towns in British Columbia. Find out when coal mines now operating in the province are predicted to close down. Do B.C. government departments, or the mining companies, have plans for the people who will no longer be employed when that happens?

21.4 Making Decisions about Resource Use

Suppose one evening you hear the following news broadcast on your television.

News Announcer: "The government recently announced that a new mine will be opening soon in a remote part of northwestern B.C. A spokesperson for Global Resources Limited, the company that will operate the mine, said that it will create about 500 jobs in the area. The company plans to begin mining gold, silver, copper, and cobalt early in 1998. The project includes a 100-km access road from the highway that runs through northwestern B.C. into the Yukon. The government is now reviewing the company's Environmental Impact Assessment on the mining operation. Here's how two residents from a nearby community reacted to the government's announcement."

Figure 21.23
What does news about resource use and the environment mean to you?

First Resident: "It's great news! The mine will bring people and money into the area and help our community to grow. It's the kind of sustainable development we need. Without development, small communities like this will die out. This company runs a good operation and it's never caused any pollution or damage to the environment, so I support it."

Second Resident: "A mine near here will ruin our way of life. We don't have any problems now, and don't need more people here just for the sake of growth. You can't keep mining forever. Sure there'll be jobs for a few years, but then the mine will close and what will happen to people then? This company has polluted other areas, and I'm against it."

What is *your* reaction to the news and to the views of the two residents? Why do people often disagree about resource development? In this section, you will learn how a knowledge of science can help you to evaluate media reports about environmental issues and resource conflicts (Figure 21.23).

MANY VIEWS: YOU CHOOSE

Debates about the environment are rarely straightforward. This is partly because the environment itself is complex and made up of many interacting parts. Even scientists disagree about the impact particular activities have on the air, water, or soil. But disagreements also occur because of different cultural, economic, and political concerns. Added to this, individuals can be influenced by a particular **bias**, or preference. Bias can make it difficult to judge a particular situation fairly.

Before making your own decisions about an issue, it is important to identify facts and opinions, as well as the biases of the various people involved. You also need to identify the different *kinds* of concerns and issues that people may disagree about. For example, what exactly are the two residents in the broadcast described above disagreeing about? If you analyse each resident's statement, you can identify at least three different kinds of issues:

1. **Value issues.** These issues are about what people consider important — the things they believe people should or shouldn't do. The two residents disagree about developing the mine because they value different things. The first resident values growth and expansion of the community. The second resident values the existing way of life in the community. Science cannot directly help decide issues based

on personal values. However, information from science can be used to predict the possible outcomes of different choices that have been made based on values.

2. **Conceptual issues.** These issues involve disagreements about the meaning of a particular term or concept. For example, what is meant by the term "sustainable development"? The first resident states that the mining operation will sustain growth in the community. The second resident states: "You can't keep mining forever." In other words, the extraction of minerals cannot be sustained indefinitely. Conceptual issues may be resolved by clearly defining each concept.
3. **Empirical issues.** These issues depend on data gathered from observations or experiments, and on the interpretation of these data. Disagreements based on empirical issues may therefore be settled by obtaining more scientific information. For example, the two residents disagree about whether the company has caused pollution in the past. If the second resident can provide data showing examples of polluting activity by the company, the first resident can no longer claim the company "has never caused any pollution."

DID YOU KNOW?

"Sustainable development" can be defined as development that meets the needs of the present without risking the ability of future generations to meet their own needs.

EVALUATING MEDIA REPORTS

How can you tell if a news report is biased? One way is to examine accounts of the same event from several different sources. Do the accounts all agree about the facts of the event, or do different accounts present different "facts"? If you find major differences among news stories, you need to discover why such differences occur (Figure 21.24).

To begin with, knowing who produced a particular news report, and for whom it was produced, can give you clues about possible bias. You would expect an oil spill to be reported differently in a newsletter produced by an oil company and one produced by a citizens' group called "Coalition to Save Our Beaches." A report may include certain facts and leave out others in order to present a particular point of view to the reader.

Less obvious is the unconscious bias that comes from a reporter's cultural and political background. For example, during the 1970s, many people in less developed nations felt that the image of their countries was distorted in reports by journalists from the more developed nations. They believed that western journalists were ignorant of many of the problems faced by less developed nations. As a result of this concern, the United Nations Educational, Scientific and Cultural Organization (UNESCO) passed a resolution in 1980 calling for a New World Information Order. This would include more training and technology for journalists from less developed nations, and a code of responsibility to be adopted by all journalists. Some more developed countries opposed the resolution. They argued that it was an attempt to control journalists and muzzle a free press.

EXTENSION

Reports of environmental issues generally occupy much less space in the media than reports of political, sporting, or entertainment news. Write a brief essay explaining why you think this is so.

Figure 21.24
News reporters may portray the same event in different ways.

Because we live in a society that values a free press, different groups can present different accounts of the same event. And whenever different groups present different accounts, you must be the judge. You must be aware of your own biases and try to keep an open mind. An understanding of the scientific background, and of scientific processes, can help you focus on the issues and ask appropriate questions.

ACTIVITY 21G *Analysing the Media*

This activity involves evaluating the way media reports present information on a particular issue (Figure 21.25).

PROCEDURE

1. Choose a specific issue related to the proposed development of a resource in B.C. Collect several articles or advertisements that report on the proposal. Look for accounts in newspapers, magazines, and a variety of other publications. If possible, search for reports from local, provincial, national, and international sources.
2. Carefully read each report and prepare a presentation giving your evaluation of the reporting. Your presentation may be a talk, a display, or a written report.

Base your evaluation on:
- the content of each article (How much factual information does the article include? How much opinion? What issues are presented?)
- the author's bias (Who wrote the article? What does the author do for a living? Who published the article?)
- the author's selection of facts supporting his or her point of view (Which facts are emphasized? Which facts are ignored?)
- the persuasiveness of the reporting (How is the argument presented?)

Conclude with an account of how your personal views on the issue were influenced by what you read.

Figure 21.25
Media reports of an issue or event need to be evaluated.

CAREER

ENVIRONMENTAL CONSULTANT

Sliding down a cliff with a portable computer sounds like something from an adventure film, doesn't it? But to Betsy Gordon, a wildlife biologist and environmental consultant, it's all part of a day's work.

What does your work involve?
I work with an environmental consulting firm. The firm's clients are developers of projects such as mines or pulp mills, and government agencies reviewing environmental aspects of these projects. Before a company can get government approval to start an industrial operation, it has to prepare an extensive report called an Environmental Impact Assessment, or EIA. An EIA is a scientific study of the way the proposed development might affect the surrounding environment.

The consulting firm I work with assembles teams of scientists that go into the field and gather data for EIAs. Our field trips vary in length from a few days to several weeks; we try to cover each season. Typically we examine water quality and flow, aquatic insects, fish species and habitat, wildlife, terrain, soils, and vegetation. I do most of the terrestrial biology.

There's a lot of discussion between the scientists, the development company, and different levels of government before and after the EIAs are submitted. Often a monitoring program is set up for the duration of the project.

What's the best part of your work?
I enjoy so many things about it. Each area that I study is geographically and environmentally different: I like the variety. I also like the challenge of reporting. I have to think carefully before I give an opinion, and I have to do research in the scientific literature as well as on the site.

Nobody told me this in school, but science is political! Big decisions and millions of dollars can rest on the scientific opinions contained in an Environmental Impact Assessment. I have to be able to defend my position. But the best part of my work is when I know my assessment of an area really made a difference, when both people and the environment benefitted. I believe we can have economically *and* environmentally sound development. And I'm in a position to help that happen.

How did you get into this work?
I've always been interested in the natural world, so I took a degree in Wildlife Biology at the University of Guelph. The program offered a broad combination of courses in botany, zoology, ecology, and wildlife management.

It wasn't easy to find jobs after I graduated in 1975. Many firms and government departments expected biologists to look at a single species or group in isolation. They wanted specialists in fisheries, mammalogy, or botany. Also, the standard approach to assessing the impact of development on an area has been to look only at "valuable" wildlife species like big game and fish, and not at smaller species like frogs or insects, or the plant species that animals depend on. But diversity is so important! The stability and sustainability of an area is directly related to its diversity.

Now, in the early 1990s, the "environment industry" is booming. It wasn't that way in the early 1980s—the bottom dropped out of the natural resources industries, and I had to find other work for two years. As well as environmental consulting, I now write, edit, illustrate, and do desktop publishing for a variety of organizations.

The fact that I've had to practise diversity myself is interesting. I remember a counsellor in high school telling me I should expect to retrain twice in my career. I didn't understand that then, but I do now. In fact, I think high school graduates today should expect to retrain at least four times.

What trends do you see in your work?
I see more environmental consultants with a holistic or overall systems approach working to help resolve environmental and development issues. I also see more women enjoying the challenges and rewards of a career in the natural sciences.

Betsy Gordon is shown here collecting water samples from a creek that might be affected by a mine development.

REVIEW 21.4

1. List three or more reasons why people may disagree about environmental issues.

2. "The facts in this government report show what will happen." "This development won't degrade the environment." "We shouldn't go ahead with this development." Say which of these statements deals with
   (a) values,
   (b) concepts,
   (c) empirical issues.

3. Give an example of media bias and explain why it occurs.

CHAPTER • REVIEW

Key Ideas

- A resource is any part of the environment that contributes to an organism's survival and well-being. The rate at which humans use resources depends on population size, technology, and lifestyle.
- Resources must be managed to make the best use of them for the greatest number of people over the longest period of time. This must be done with a minimum impact on the environment. Conservation is a way of managing resources by careful use and elimination of waste.
- Renewable resources can last indefinitely if well managed. Non-renewable resources will not last indefinitely. How long a non-renewable resource will last depends on the quantity of the resource and the rate at which it is used.
- Predictions about the use of resources in the future are based in part on the analysis of trends. The rate of resource use may rapidly increase or decrease as a result of developments in technology, lifestyle, politics, or economy.
- The development of a resource may have impacts on the environment during several stages, such as extraction or harvesting, transportation, processing, and use.
- Media reports of conflicts over the use of natural resources must be carefully analysed. Analysis helps us understand why opinions differ and what biases exist.

Vocabulary

natural resource
resource management
renewable resource
non-renewable resource
reserve
trend
conservation
bias
value issue
conceptual issue
empirical issue

1. Arrange as many terms as possible from the vocabulary list into a crossword puzzle and write clues for each term.

2. Create a concept map using at least five terms from the vocabulary list.

Connections

1. Jonas goes to school on a skateboard, Susan travels by bicycle, and Lorne drives a car. List the resources used by each student. Choose one resource and make a flow chart showing how the resource becomes the final product used in one of the methods of transportation.

2. Predict whether each of the following is likely to increase or decrease the rate of use of oil, and explain your answers.
   (a) an increase in the size of Canada's population
   (b) an increase in the sale of automobiles
   (c) an increase in unemployment
   (d) adoption of a conserver lifestyle

3. Scientists estimate that the non-renewable resource that a particular country uses for fuel will be used up in 20 years, at current rates of use. Suggest at least three things that can be done to ensure that the population will continue to have a source of fuel after 20 years.

4. Over a billion people in the world now cook their meals over small fires of wood or other fuels. It has been said that providing these people with electricity from coal-burning stations will help reduce the amount of air pollution. What data would you need before you could decide whether you agree or disagree with this idea?

5. "Natural resources are things people use. If we can't use something, it has no value." Eval-

uate this statement and say why you agree or disagree with it.

6. Two cities have an air pollution problem caused by automobiles being driven to and from the downtown area (Figure 21.26). One city proposes to solve the problem using improved technology. It will require vehicle engines to have devices that reduce the production of pollutants. The other city proposes to solve the problem by encouraging a change in lifestyle. It will ban vehicles that carry fewer than two people. Compare these two solutions, giving one advantage and one disadvantage of each.

7. Suppose that coal has been discovered in two different locations. One is in a remote area of the Arctic, hundreds of kilometres from the nearest community. The other is in one corner of a large provincial park situated only 20 km from the city where the coal is needed. Which source of coal would you recommend developing and why?

8. Write two or three sentences describing an oil spill, as it might be written by a naturalist who helped rescue birds harmed by the oil. Write a contrasting description of the same event from the point of view of a writer who believes that the harm to birds is an unavoidable result of our need for oil. Finally, write what you believe would be a "balanced" account of the event.

Explorations

1. What difference would it make to your life if there were no petroleum? Slightly over 100 years ago, very little petroleum was used in Canada. One hundred years from now, it is likely that there will be little or no petroleum left in the world's reserves. Yet petroleum is so important in today's society that it is difficult to imagine life without it. List some of the ways you now use petroleum or products made from it, then write a story about a Canadian student in the year 2090—perhaps your own great-grandchild. Describe how he or she might manage living in a world without petroleum.

2. You are stranded on a desert island. List six natural resources you would want on the island (list A). Now suppose you could only have two of those resources. Which two would it be? Make a second list (list B). Compare your two lists with your classmates' lists. Help create a class chart to show how often each resource was mentioned in list A and in list B. Take part in a class discussion based on these results. Consider the differences between resources people want and those they need.

3. Choose a natural resource and find out how many different uses it has. Present your findings in the form of a flow chart.

4. Product packaging or advertisements may contain a lot of information about the product, such as the materials it is made of and where it is made. In recent years, some companies have added information about the product's impact on the environment, such as whether it is biodegradable, uses CFCs, has a low energy consumption, or involves practices or materials that avoid killing wildlife. Select a product and design an advertisement or package for it that gives the consumer an important piece of information about the connection between the product and the environment.

Figure 21.26
How can we solve pollution problems caused by automobiles?

Reflections

1. Do you think education is enough to make people take action to solve the problems of overpopulation, pollution, and resource depletion?

2. How and why have your ideas changed since reading this chapter?

Appendix A
SI Units of Measurement

The International System of Units — the metric system — is usually referred to as the SI. SI comes from the French name *Le Système internationale d'unités*.

Table A.1 shows some common SI units. It includes all the SI base units used in this book. Sometimes SI units are combined to measure other quantities. For example, speed can be measured in kilometres per hour (km/h), and energy use can be measured in kilowatt hours (kW•h).

Table A.2 shows all the SI prefixes. You can see that all the prefixes are related to each other by multiples of 10. The SI prefixes may be combined with the base units in Table A.1 — for example, kilometre (km), milligram (mg), or microcoulomb (μC).

Table A.1 Some Common SI Symbols, Units, and Quantities

Symbol	Unit	Quantity
A	ampere	electric current
a	year	time
Bq	becquerel	radioactivity
C	coulomb	electric charge
d	day	time
g	gram	mass
Gy	gray	absorbed dose (radiation)
h	hour	time
ha	hectare	area
Hz	hertz	frequency
J	joule	energy, work
L	litre	volume
m	metre	length
min	minute	time
N	newton	force
Pa	pascal	pressure, stress
s	second	time
Sv	sievert	dose equivalent (radiation)
t	tonne, metric ton	mass
V	volt	electric potential
W	watt	power
Ω	ohm	electric resistance
°C	degree Celsius	Celsius temperature

Note: one tonne (1 t) = one megagram (1 Mg)
one hectare (1 ha) = one square hectometre (1 hm^2)

Table A.2 SI Prefixes

Prefix	Symbol	Factor by which the base unit is multiplied	Example
exa	E	10^{18} = 1 000 000 000 000 000 000	
peta	P	10^{15} = 1 000 000 000 000 000	
tera	T	10^{12} = 1 000 000 000 000	
giga	G	10^{9} = 1 000 000 000	
mega	M	10^{6} = 1 000 000	10^{6} m = 1 Mm
kilo	k	10^{3} = 1 000	10^{3} m = 1 km
hecto	h	10^{2} = 100	
deca	da	10^{1} = 10	
		10^{0} = 1	m
deci	d	10^{-1} = 0.1	
centi	c	10^{-2} = 0.01	10^{-2} m = 1 cm
milli	m	10^{-3} = 0.001	10^{-3} m = 1 mm
micro	μ	10^{-6} = 0.000 001	10^{-6} m = 1 μm
nano	n	10^{-9} = 0.000 000 001	
pico	p	10^{-12} = 0.000 000 000 001	
femto	f	10^{-15} = 0.000 000 000 000 001	
atto	a	10^{-18} = 0.000 000 000 000 000 001	

Appendix B

The Modern Periodic Table of Elements

Source: *Understanding Chemistry*, 1988 (modified)

1 +1 **H** 1.008 Hydrogen

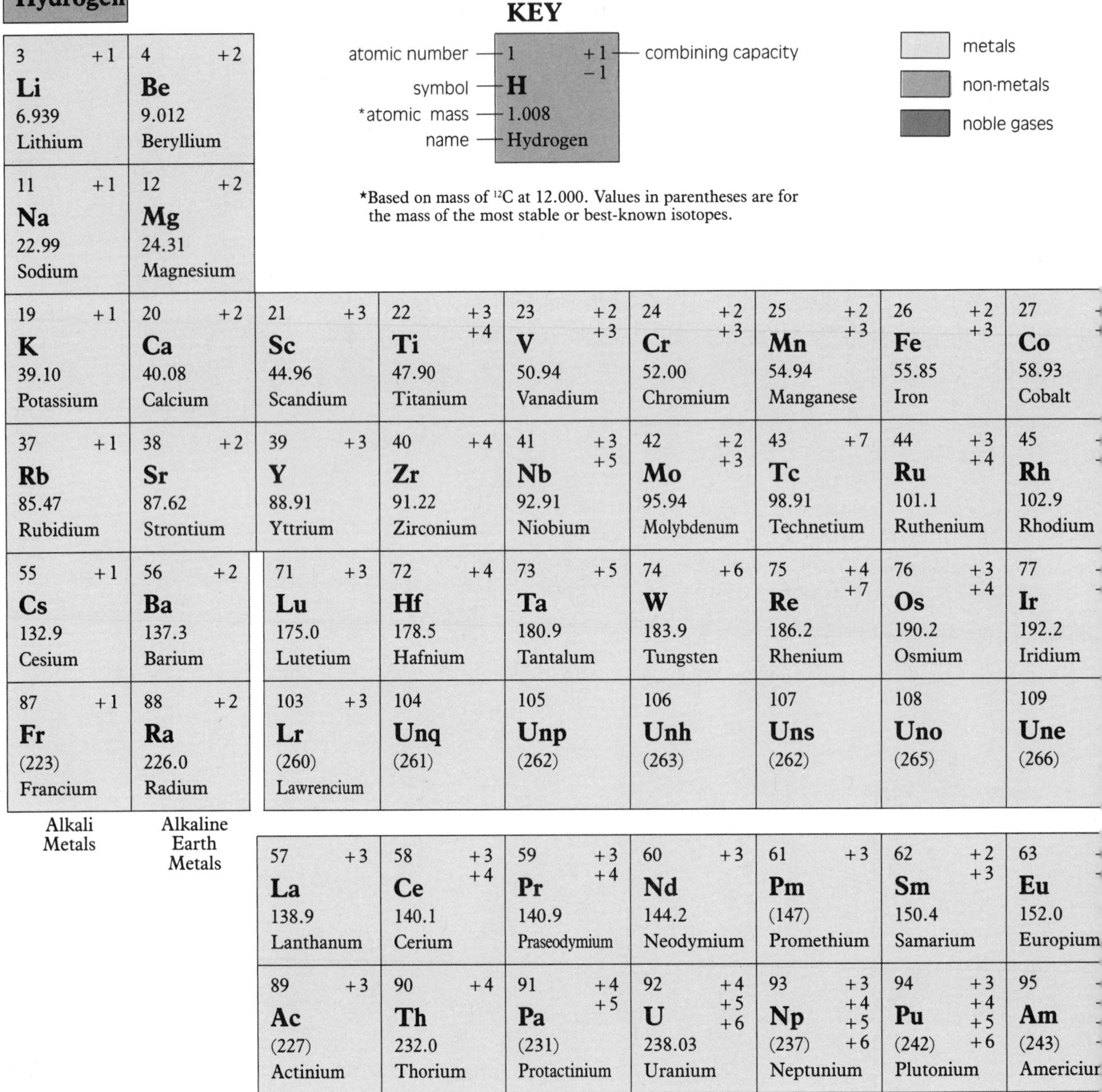

KEY

atomic number — 1 +1 −1 — combining capacity
symbol — **H**
*atomic mass — 1.008
name — Hydrogen

metals
non-metals
noble gases

*Based on mass of ^{12}C at 12.000. Values in parentheses are for the mass of the most stable or best-known isotopes.

3 +1 **Li** 6.939 Lithium	4 +2 **Be** 9.012 Beryllium							
11 +1 **Na** 22.99 Sodium	12 +2 **Mg** 24.31 Magnesium							
19 +1 **K** 39.10 Potassium	20 +2 **Ca** 40.08 Calcium	21 +3 **Sc** 44.96 Scandium	22 +3 +4 **Ti** 47.90 Titanium	23 +2 +3 **V** 50.94 Vanadium	24 +2 +3 **Cr** 52.00 Chromium	25 +2 +3 **Mn** 54.94 Manganese	26 +2 +3 **Fe** 55.85 Iron	27 **Co** 58.93 Cobalt
37 +1 **Rb** 85.47 Rubidium	38 +2 **Sr** 87.62 Strontium	39 +3 **Y** 88.91 Yttrium	40 +4 **Zr** 91.22 Zirconium	41 +3 +5 **Nb** 92.91 Niobium	42 +2 +3 **Mo** 95.94 Molybdenum	43 +7 **Tc** 98.91 Technetium	44 +3 +4 **Ru** 101.1 Ruthenium	45 **Rh** 102.9 Rhodium
55 +1 **Cs** 132.9 Cesium	56 +2 **Ba** 137.3 Barium	71 +3 **Lu** 175.0 Lutetium	72 +4 **Hf** 178.5 Hafnium	73 +5 **Ta** 180.9 Tantalum	74 +6 **W** 183.9 Tungsten	75 +4 +7 **Re** 186.2 Rhenium	76 +3 +4 **Os** 190.2 Osmium	77 **Ir** 192.2 Iridium
87 +1 **Fr** (223) Francium	88 +2 **Ra** 226.0 Radium	103 +3 **Lr** (260) Lawrencium	104 **Unq** (261)	105 **Unp** (262)	106 **Unh** (263)	107 **Uns** (262)	108 **Uno** (265)	109 **Une** (266)
Alkali Metals	Alkaline Earth Metals							

57 +3 **La** 138.9 Lanthanum	58 +3 +4 **Ce** 140.1 Cerium	59 +3 +4 **Pr** 140.9 Praseodymium	60 +3 **Nd** 144.2 Neodymium	61 +3 **Pm** (147) Promethium	62 +2 +3 **Sm** 150.4 Samarium	63 **Eu** 152.0 Europium
89 +3 **Ac** (227) Actinium	90 +4 **Th** 232.0 Thorium	91 +4 +5 **Pa** (231) Protactinium	92 +4 +5 +6 **U** 238.03 Uranium	93 +3 +4 +5 +6 **Np** (237) Neptunium	94 +3 +4 +5 +6 **Pu** (242) Plutonium	95 **Am** (243) Americiur

								2 0 **He** 4.003 Helium
			5 +3 **B** 10.81 Boron	6 +4 **C** 12.01 Carbon	7 −3 **N** 14.01 Nitrogen	8 −2 **O** 16.00 Oxygen	9 −1 **F** 19.00 Fluorine	10 0 **Ne** 20.18 Neon
			13 +3 **Al** 26.98 Aluminum	14 +4 **Si** 28.09 Silicon	15 −3 **P** 30.97 Phosphorus	16 −2 +4 +6 **S** 32.06 Sulphur	17 −1 **Cl** 35.45 Chlorine	18 0 **Ar** 39.95 Argon
8 +2 +3 **Ni** 8.71 Nickel	29 +1 +2 **Cu** 63.55 Copper	30 +2 **Zn** 65.38 Zinc	31 +3 **Ga** 69.72 Gallium	32 +4 **Ge** 72.59 Germanium	33 −3 +5 **As** 74.92 Arsenic	34 −2 +4 +6 **Se** 78.96 Selenium	35 −1 **Br** 79.91 Bromine	36 0 **Kr** 83.80 Krypton
6 +2 +4 **Pd** 06.4 Palladium	47 +1 **Ag** 107.9 Silver	48 +2 **Cd** 112.4 Cadmium	49 +3 **In** 114.8 Indium	50 +2 +4 **Sn** 118.7 Tin	51 −3 +5 **Sb** 121.8 Antimony	52 −2 +4 +6 **Te** 127.6 Tellurium	53 −1 **I** 126.9 Iodine	54 0 **Xe** 131.3 Xenon
8 +2 +4 **Pt** 95.1 Platinum	79 +1 +3 **Au** 197.0 Gold	80 +1 +2 **Hg** 200.6 Mercury	81 +1 +3 **Tl** 204.4 Thallium	82 +2 +4 **Pb** 207.2 Lead	83 −3 +5 **Bi** 209.0 Bismuth	84 +2 +4 **Po** (209) Polonium	85 −1 **At** (210) Astatine	86 0 **Rn** (222) Radon
							Halogens	Noble Gases

4 +3 **Gd** 57.3 Gadolinium	65 +3 +4 **Tb** 158.9 Terbium	66 +3 **Dy** 162.5 Dysprosium	67 +3 **Ho** 164.9 Holmium	68 +3 **Er** 167.3 Erbium	69 +2 +3 **Tm** 168.9 Thulium	70 +2 +3 **Yb** 173.0 Ytterbium
6 +3 **Cm** 247) Curium	97 +3 +4 **Bk** (247) Berkelium	98 +3 **Cf** (251) Californium	99 +3 **Es** (252) Einsteinium	100 +3 **Fm** (257) Fermium	101 +2 +3 **Md** (258) Mendelevium	102 +2 +3 **No** (259) Nobelium

Appendix C
List of Elements

Element	Symbol	Atomic number	Atomic mass*
Actinium	Ac	89	(227)
Aluminum	Al	13	26.98
Americium	Am	95	(243)
Antimony	Sb	51	121.8
Argon	Ar	18	39.95
Arsenic	As	33	74.92
Astatine	At	85	(210)
Barium	Ba	56	137.3
Berkelium	Bk	97	(247)
Beryllium	Be	4	9.012
Bismuth	Bi	83	209.0
Boron	B	5	10.81
Bromine	Br	35	79.91
Cadmium	Cd	48	112.4
Calcium	Ca	20	40.08
Californium	Cf	98	(251)
Carbon	C	6	12.011
Cerium	Ce	58	140.1
Cesium	Cs	55	132.9
Chlorine	Cl	17	35.45
Chromium	Cr	24	52.00
Cobalt	Co	27	58.93
Copper	Cu	29	63.55
Curium	Cm	96	(247)
Dysprosium	Dy	66	162.5
Einsteinium	Es	99	(252)
Erbium	Er	68	167.3
Europium	Eu	63	152.0
Fermium	Fm	100	(257)
Fluorine	F	9	19.00
Francium	Fr	87	(223)
Gadolinium	Gd	64	157.3
Gallium	Ga	31	69.72
Germanium	Ge	32	72.59
Gold	Au	79	197.0
Hafnium	Hf	72	178.5
Helium	He	2	4.003
Holmium	Ho	67	164.9
Hydrogen	H	1	1.008
Indium	In	49	114.8
Iodine	I	53	126.9
Iridium	Ir	77	192.2
Iron	Fe	26	55.85
Krypton	Kr	36	83.80
Lanthanum	La	57	138.9
Lawrencium	Lw	103	(260)
Lead	Pb	82	207.2
Lithium	Li	3	6.939
Lutetium	Lu	71	175.0
Magnesium	Mg	12	24.31

Element	Symbol	Atomic number	Atomic mass*
Manganese	Mn	25	54.94
Mendelevium	Md	101	(258)
Mercury	Hg	80	200.6
Molybdenum	Mo	42	95.94
Neodymium	Nd	60	144.2
Neon	Ne	10	20.18
Neptunium	Np	93	(237)
Nickel	Ni	28	58.71
Niobium	Nb	41	92.91
Nitrogen	N	7	14.01
Nobelium	No	102	(254)
Osmium	Os	76	190.2
Oxygen	O	8	16.0
Palladium	Pd	46	106.4
Phosphorus	P	15	30.97
Platinum	Pt	78	195.1
Plutonium	Pu	94	(242)
Polonium	Po	84	(209)
Potassium	K	19	39.100
Praseodymium	Pr	59	140.90
Promethium	Pm	61	(147)
Protactinium	Pa	91	(231)
Radium	Ra	88	(226)
Radon	Rn	86	(222)
Rhenium	Re	75	186.20
Rhodium	Rh	45	102.90
Rubidium	Rb	37	85.47
Ruthenium	Ru	44	101.1
Samarium	Sm	62	150.4
Scandium	Sc	21	44.96
Selenium	Se	34	78.96
Silicon	Si	14	28.09
Silver	Ag	47	107.9
Sodium	Na	11	22.99
Strontium	Sr	38	87.62
Sulphur	S	16	32.060
Tantalum	Ta	73	180.90
Technetium	Tc	43	98.91
Tellurium	Te	52	127.6
Terbium	Tb	65	158.9
Thallium	Tl	81	204.4
Thorium	Th	90	232.0
Thulium	Tm	69	168.9
Tin	Sn	50	118.7
Titanium	Ti	22	47.90
Tungsten	W	74	183.9
Uranium	U	92	238.03
Vanadium	V	23	50.94
Xenon	Xe	54	131.3
Ytterbium	Yb	70	173.0
Yttrium	Y	39	88.91
Zinc	Zn	30	65.38
Zirconium	Zr	40	91.22
—	—	104	(261)
—	—	105	(262)

*Based on mass of ^{12}C at 12.000. Values in parentheses are for the mass of the most stable or best-known isotopes.

Appendix D
Geological Time Scale

Age in millions of years	Era	Period	Geological event	Biological event
0.004 1 1.5–2	CENOZOIC (Recent life)	QUATERNARY	Great Lakes formed	Humans appear
26 34 55		TERTIARY		
			Rocky Mountains formed	First primates
65	MESOZOIC (Middle life)	CRETACEOUS		Age of the dinosaurs
136		JURASSIC		First mammals; First birds
190–195		TRIASSIC		
225	PALEOZOIC (Ancient life)	PERMIAN	Appalachian Mountains formed	
280		PENNSYL VANIAN		First reptiles
		MISSISSIPPIAN		
345		DEVONIAN		First insects
395		SILURIAN		
430–440		ORDOVICIAN		First fish
500		CAMBRIAN	Burgess Shale deposited	Invertebrates flourish
570	PRECAMBRIAN – PROTEROZOIC	HADRYNIAN		Very simple organisms
995		HELIKIAN		
		APHEBIAN		
2600	PRECAMBRIAN – ARCHEAN			

Glossary

A

abiotic: Having to do with non-living things.

absolute age: In geology, the amount of time in years that has passed since the occurrence of a particular event.

acid: A compound that tastes sour, turns blue litmus paper red, corrodes active metals and produces hydrogen gas, conducts electricity when in solution, and is capable of combining with a base to form a salt.

acid rain: Rain produced when oxides of nitrogen and sulphur dissolve in atmospheric water vapour to produce nitric, sulphurous, and sulphuric acids.

activity: The number of nuclei of atoms in a sample that undergo radioactive decay each second. Activity is measured in becquerels.

aftershock: A secondary earthquake occurring some time after the main shock.

Air Quality Index: An index used to indicate the degree of risk to health from air pollution; based on five major air pollutants.

aircraft: A vehicle that travels through the air.

alkali metal: Any of these metals: lithium, sodium, potassium, rubidium, cesium, and francium.

alleles: The different forms of a gene.

alpha radiation: A type of radiation resulting from the emission of helium nuclei from the nuclei of atoms.

alternating current (AC): Electric current that flows first in one direction through a circuit and then in the other and that varies its direction at regular intervals.

altitude: The height of an object above sea level, measured in metres or kilometres.

ammeter: A device that measures electric current.

ammonite: An extinct group of marine animals. They are related to the modern nautilus and the modern squid.

amniocentesis: A prenatal test in which a sample of amniotic fluid containing fetal cells is removed from the female and examined for the presence of various genetic diseases.

ampere (A): The unit of electric current. One ampere equals one coulomb of charge per second.

amplitude: The height or strength of a wave.

anaphase: The third stage of mitosis, in which each pair of chromatids splits at its centromere to form two new chromosomes. These new chromosomes move to opposite ends of the cell, forming two genetically identical groups.

anode: A positively charged electrode; the electrode in an electrolysis cell that loses an electron.

antibiotic: A substance that kills (or slows the growth of) specific types of living cells.

antibody: A type of protein molecule produced by certain cells, which binds with a specific antigen and deactivates it.

antigen: A molecule carried on the surface of a pathogen, which stimulates the production of the corresponding antibodies.

artificial insemination: A process in which a special tube is used to insert sperm into a female's reproductive system at the time she ovulates.

asexual reproduction: Reproduction in which only one parent is involved, in which all the offspring are identical to each other and to the parent, and for which specialized reproductive cells are not usually necessary.

ash: Fine-grained particles of solid lava that have been sprayed out of a volcanic vent.

atmosphere: The envelope of gases approximately 100 km thick that surrounds the Earth.

atom: The smallest particle of an element that can exist by itself.

atomic mass: The total mass of the protons, neutrons, and electrons that make up an atom.

atomic number: The number of protons found in the nucleus of an atom.

atomic theory: A theory stating that matter is made up of tiny particles called atoms, that each atom has a positive nucleus surrounded by negative electrons, and that atoms can combine to form molecules.

attract: To draw to oneself or to cause to come near as a result of an electric or magnetic force.

B

bacterium (plural: bacteria): A single-celled organism that lacks a nucleus and most other types of membrane-bound organelles.

balanced equation: A chemical equation that has the same number of atoms of each element on each side of the equation.

base: A compound that feels slippery in solution, turns red litmus paper blue, conducts electricity when in solution, and is capable of combining with an acid to form a salt.
battery: Two or more chemical cells connected together.
becquerel (Bq): Unit used to measure the activity of a radioactive substance. One becquerel equals one radioactive decay of a nucleus per second.
beta radiation: A type of radiation resulting from the emission of high-energy electrons from the nuclei of atoms.
bias: A preference for one thing over another based on preconceived ideas.
binary fission: A method of asexual reproduction in which the reproducing cell or organism splits into two equal parts.
biotic: Having to do with living things.
birth rate: The rate at which new individuals are added to a population by the process of reproduction.
body tube: The part of a compound microscope that holds the objective and ocular lenses.
booster rockets: Rocket engines used only to help launch a vehicle into space.
branch circuit: A circuit that supplies electricity to some household outlets and lights.
budding: A method of asexual reproduction in which the offspring develops as a bud on the parent until it drops off and becomes independent.

C

Canadarm: Another name for the Remote Manipulator System.
cancer: A disease in which uncontrolled cell division results in the growth of malignant tumours in the body.
carrying capacity: The maximum size of population that can be supported indefinitely by a given amount of resources.
cathode: A negatively charged electrode; the electrode in an electrolysis cell that gains an electron.
cell: The basic structural and functional unit of life. All living things are composed of one or more cells.
cell cycle: The life cycle of a cell.
cell division: The division of the cell into two halves by the processes of mitosis and cytokinesis. It is the method by which cells reproduce.
cell membrane: The thin, flexible material that surrounds the cytoplasm of every cell.
cell reproduction: The method by which cells reproduce. The cell divides into two halves by the processes of mitosis and cytokinesis.
cell theory: A theory about the nature of living organisms, which states three ideas: (1) all living things are made up of cells, (2) cells are the basic functional units of life, and (3) all cells come from pre-existing cells.
cell wall: A layer of non-living material that serves as the external boundary of plant cells. Some types of bacteria also have cell walls.
cellular respiration: In biology, the process in which energy-rich molecules are broken down to release their energy in a form that can be used to drive chemical reactions within cells. Cellular respiration involves a complex series of chemical reactions.
centrioles: A pair of organelles located near the nucleus in animal cells. The function of centrioles is uncertain, but they are associated with the spindle during mitosis and meiosis.
centromere: A region of the chromosome where paired chromatids are joined.
cervix: The ring of muscle that surrounds the mouth of the uterus.
CFCs: Gases used to propel liquids from aerosol cans, or liquids used as refrigerants and solvents. Also known as *c*hloro*f*luoro*c*arbons or freon gases. CFCs are associated with the thinning of the ozone layer.
chain reaction: A nuclear reaction in which some of the products of the reaction go on to cause more reactions to occur.
chemical cell: An apparatus that generates electric current by means of chemical reactions.
chemical change: The same as a chemical reaction.
chemical equation: A summary of a chemical reaction using chemical formulas for reactants and products.
chemical reaction: A change that produces one or more new substances with different properties from those of the starting materials.
chert: Rock composed of submicroscopic quartz grains.
chlorofluorocarbons: See CFCs.
chlorophyll: A substance found in chloroplasts that enables the cell to trap energy from the sun and convert it into chemical energy, which can then be used to drive the chemical reactions of photosynthesis.
chloroplast: An organelle that contains chlorophyll; it is the site where photosynthesis occurs. Chloroplasts are found in plant cells as well as in the cells of some protists and bacteria.
chorionic villus (plural: villi)**:** A finger-like extension of the chorion, the outermost membrane surrounding the fetus. The chorionic villi extend into the placenta and become part of it.
chorionic villus sampling: A prenatal test in which a sample of fetal cells is obtained from the chorionic villi and examined for the presence of various genetic diseases.

chromatid: Each of the two identical strands of a duplicated chromosome, which are joined together at the centromere.

chromosome: A thread-like structure found in the nucleus of cells. Each chromosome is made up of proteins and a single DNA molecule.

circuit breaker: A switch that turns off automatically if the current flowing through it exceeds a preset limit.

circuit overload: A condition that exists when too many appliances are connected to the same circuit and are used at the same time, thus causing more current to flow in the circuit than the circuit can carry.

clone: The group of identical offspring organisms formed as a result of asexual reproduction by a single organism. "Clone" can also refer to any one of these offspring.

cloning: Producing one or more genetically identical copies of an individual from a single cell.

coarse adjustment knob: A focusing knob that produces large changes in the working distance on a microscope.

codominance: A situation in which none of the alleles of a particular gene is dominant over the other, but both contribute equally to the phenotype.

coefficient: A number placed in front of a formula in a balanced equation. The number multiplies all atoms in the formula that follows.

coke: An almost pure form of carbon, made by heating bituminous coal in the absence of air.

combining capacity: The number of electrons an atom can lose, gain, or share to form chemical bonds.

compound: A substance made up of two or more elements chemically combined according to a definite formula.

compound microscope: A microscope that contains more than one lens or lens system.

conceptual issue: An issue that involves disagreement about the meaning of a particular term or idea.

conductor: A material in which electrons can move easily.

conservation: Managing resources by careful use and reducing or eliminating waste.

conserver society: A society in which people use a minimum amount of resources and create minimum waste.

consumer society: A society based on the increasing use of materials and energy.

control rod: A cadmium or boron rod that can be inserted into a reactor to absorb neutrons released in the fission reaction, thus slowing the rate of fission.

convection current: Current of moving material that is driven by a heat source. Wind and ocean currents are examples of convection currents.

conventional current: A flow of electric current from a positively charged object to a negatively charged one.

core: The centremost region of an object. The Earth's core is approximately 7500 km in diameter and consists of a dense, liquid outer core and a denser, solid inner core.

corpus luteum: A mass of cells formed from the remains of the follicle after ovulation. It secretes estrogen and progesterone.

coulomb (C): The unit of electric charge. One coulomb equals 6.24×10^{18} electrons or protons.

covalent bond: A chemical bond in which two or more atoms share one or more pairs of electrons.

covalent compound: A compound that is formed when non-metallic atoms share electrons and form covalent bonds.

Cowper's gland: A gland that produces one component of the seminal fluid. It empties its secretions into the urethra.

critical mass: The smallest mass in which a fission chain reaction can be sustained.

crosscutting rule: A rule stating that any event that disturbs a rock is always younger than the rock itself.

crust: Comparatively thin, outermost, solid layer of the Earth.

cytokinesis: The process in which the cytoplasm of a cell divides into two roughly equal halves.

cytoplasm: The fluid found outside the nuclear membrane, but inside the cell membrane, of every cell.

D

death rate: The rate at which individuals in a population die.

decay product: The nucleus produced in a radioactive decay.

decomposition: A chemical reaction in which a compound is broken down into two or more elements or simpler compounds.

demographer: A person who studies the characteristics of human populations such as growth rate and sex ratio.

density (ecological)**:** In populations, the number of individuals in a particular area.

density (physical)**:** The mass of one unit of volume of a substance, expressed as the ratio of mass to volume. Water has a density of 1.0 g/cm^3.

density-dependent factor: A factor, such as food supply or disease, that affects a population's growth and depends on population density.

density-independent factor: A factor, such as fire, flood, or drought, that affects a population's growth but is not dependent on population density.

deposition: The laying down of sediment such as gravel, sand, or mud.

depth of field: The thickness of the image that is in focus at any one time when a specimen is viewed with a microscope.

diaphragm: In microscopy, a device that controls the amount of light shining up into the body tube of a microscope. Making the opening of the diaphragm larger increases the brightness, whereas making the opening smaller reduces glare.

diffusion: The movement of molecules from an area where they are in higher concentration to an area where they are in lower concentration.

dike: A fracture that has been filled with upwelling magma which has since solidified.

diploid: Having pairs of homologous chromosomes.

direct current (DC): Current that flows in only one direction in a circuit.

distribution station: A facility that receives electricity from a substation (usually at 60 kV) and distributes the electricity to the streets of a community at a lower voltage, typically 25 kV.

DNA (*d*eoxyribo*n*ucleic *a*cid)**:** The genetic material of the cell, made up of four different types of nucleotides arranged in a chain. DNA is the major component of the chromosomes in the nucleus.

dominant: In genetics, referring to an allele that is expressed whenever it is present. A dominant allele masks the effects of any other allele of that gene.

double replacement: A chemical reaction in which elements in different compounds replace each other or exchange places.

E

earth scientists: Geologists and others who study the Earth's surface, crust, and interior.

earthquake: Violent vibrations on the surface of the Earth, usually caused by sudden movement along a fault.

efficiency: The ability of a device to transform electrical energy to another desired form of energy. Efficiency equals the ratio of useful energy obtained from a device to the energy used to operate the device.

elastic: The property of returning to original form after a force is removed.

elastic rebound theory: A theory describing the periodic storing up and sudden release of elastic energy in the Earth. The energy is released suddenly by movement on a fault, causing an earthquake.

electric circuit: A complete pathway for electric current, from the source and back again.

electric current: The flow of electrically charged particles through a conductor.

electric generator: Any device that can transform mechanical energy into electrical energy.

electric meter: A device used by an electric utility company to measure and record the energy used in a building.

electric motor: A device that can transform electrical energy into mechanical energy.

electrical potential energy: Stored energy that results from the attraction or repulsion of charged objects.

electrochemistry: The study of chemical reactions that either use or produce electricity.

electrolysis: The decomposition of a compound caused by the passage of electricity.

electrolyte: A substance that dissolves in water to form a solution that conducts electricity.

electrolytic cell: An apparatus used for performing electrolysis.

electromagnet: A magnet that produces a magnetic field when an electric current flows through it. An electromagnet is usually constructed by winding a coil of wire around a core made of a magnetic material such as soft iron.

electromagnetic radiation: A type of radiation that consists of an electric and magnetic disturbance that spreads through space. Light is an example of electromagnetic radiation.

electromagnetic spectrum: The range of electromagnetic radiation from low frequency and long wavelength to high frequency and short wavelength.

electron: A negatively charged subatomic particle found outside the nucleus of an atom.

electron microscope: A microscope that uses a beam of electrons to produce a magnified image of a specimen.

electroplating: The coating of an object with a thin film of metal deposited from an electrolytic solution by electrolysis.

electroscope: A device that detects the presence of static electric charge on an object.

element: A pure substance that cannot be broken down into simpler substances by ordinary chemical reactions.

embryo: A young, developing organism before it has been born or hatched. In humans, the developing offspring is called an embryo from the time cell division begins until about 8 to 10 weeks after the egg has been fertilized.

embryo transfer: The removal of embryos from one female animal—usually with particularly desirable genetic traits—and their placement in the uteruses of several other females, where the embryos can develop successfully until birth.

emigrate: To move out of a population to another area.

empirical issue: An issue relying on data from experiments and observation.

endothermic reaction: A chemical reaction that absorbs energy from its surroundings.

Energuide label: A label attached to an electrical appliance stating the number of kilowatt hours of energy per month the appliance will use.
energy: The ability to do work. The unit of energy is the joule.
enzyme: A type of protein that speeds up the rate of a chemical reaction. Different enzymes speed up different reactions.
epicentre: A location on the surface of the Earth directly above the point (focus) where an earthquake occurs.
epididymis: The long, coiled tube that lies on top of each testis. It stores sperm and secretes a substance that causes the sperm to actively move along the reproductive tract.
era: A large division of geological time, made up of one or more periods.
erosion: Transportation of weathered rock fragments from one place to another. Gravity, wind, water, and glaciers are the main agents of erosion.
erosion surface: A rock surface that is or has been exposed to weathering through the action of water, ice, or wind.
estrogen: A female sex hormone secreted by the follicle cells of the ovary.
eugenics: A school of thought, or movement, that has the goal of improving the inherited characteristics of humans by controlling reproduction. The movement developed in the late 1800s and was particularly prevalent during the first half of this century.
exothermic reaction: A chemical reaction that releases energy to its surroundings.
exponential growth: An increase produced when multiplying numbers again and again by a constant amount (for example, by two: $2 \times 2 \times 2 \times 2 \ldots$).
extinct: No longer in existence.
extinction: The permanent disappearance of a species from the Earth as a result of environmental events or human actions.

F

family: In the periodic table, a vertical column of elements that have similar chemical and physical properties and similar electron arrangements.
fault: A crack in a rock body along which there has been some movement.
fertilization: The process in which a male and a female gamete fuse to form a zygote.
fetus: In many organisms, the name given to a young, developing organism before it is born or hatched, but after all the internal organs have begun to function to some extent and bone formation has begun.
field of view: The area of the slide that a person can see when looking into the microscope.
fine adjustment knob: A focusing knob that produces small changes in the working distance on a microscope.
fission: The process by which a large nucleus splits into two pieces of roughly equal mass, accompanied by the release of very large amounts of energy.
fissure: A crack or fracture that has separated to form a gap.
focus (geological)**:** The actual location along a fault, usually at some depth, where earthquake energy originates.
fold: The bending and tilting of rock layers.
follicle: The covering of cells that surrounds each ovum in a female's ovary.
formula: A shorthand representation of the composition of a substance, indicating the elements present and the number of atoms of each of these elements present in the substance.
fossil: Evidence of ancient life preserved in the Earth's crust, often in the form of shells, bones, imprints, or traces.
fossil fuel: A fuel that has been formed by the decomposition of once-living material over millions of years.
fracture: A rough, uneven break in a body of rock or a mineral.
fragmentation: A type of asexual reproduction in which large or small fragments of an organism break off and develop into new organisms.
frequency: The number of cycles of alternating current per second, measured in hertz.
friction: The force that resists the relative movement between sliding surfaces.
fuel cell: A chemical cell that continuously transforms the chemical energy of fuel supplied to the cell into electrical energy.
fuse: A device that contains a small piece of wire that will melt if the current flowing through this wire exceeds a preset maximum current.
fusion: The joining of two small nuclei of atoms to make a larger nucleus, accompanied by the release of very large amounts of energy.

G

galaxy: A huge system of stars that travel as a group in the universe. Our own galaxy is called the Milky Way; it consists of hundreds of billions of stars.
galvanometer: A very sensitive ammeter with a needle centred so that small electric currents flowing through the meter in either direction can be detected.
gamete: A reproductive cell of a sexually reproducing organism. Gametes are haploid.
gamma radiation: Electromagnetic radiation emitted from the nuclei of atoms.
gastropod: A class of mollusc that includes snails.

Geiger counter: A device that detects radiation.
gene: A segment of a chromosome, made up of a particular combination of nucleotides, which codes for a specific enzyme. Each chromosome contains many different genes.
gene splicing: The process of producing recombinant DNA. Desired genes are removed from an individual and spliced into bacterial plasmids (or into the chromosomes of other asexually reproducing organisms) so that they can be copied many times during cell division.
gene therapy: The alteration of an individual's genes by removing disease-causing alleles and replacing them with alleles that produce a disease-free phenotype.
genetic counselling: A process in which couples who are at risk of having a child with a genetic disease receive information about the disease and the risk of their child's inheriting it.
genetic engineering: The alteration of the genetic material of an organism by the addition of genes or the substitution of certain alleles for other ones.
genetics: The field of science that deals with heredity.
genome: All the genes and their alleles present in a population.
genotype: The set of alleles possessed by an individual for a given trait or traits.
geographic pole: One of the points through which the Earth's axis passes.
geographic north: The direction from a place to the geographic North Pole. Also known as true north.
geological time scale: A kind of calendar of Earth history, divided into intervals called eras and periods (See Appendix D).
geology: A branch of science dealing with the study of the Earth.
geomagnetic pole: One of two places on the Earth's surface where the Earth's magnetic field is the most concentrated.
geophysics: The science that studies the structure and behaviour of the Earth.
geosynchronous orbit: The orbit of a satellite that remains in the same location above the Earth's surface.
grafting: Attaching buds or twigs from a plant with desirable genetic traits to the established trunk or roots of another plant.
granite: A plutonic (igneous) rock formed from the slow cooling of underground magma.
gravity: The force of attraction between any two objects that have mass.
gray (Gy): A measure of the energy transferred to a material by a dose of radiation. A dose of one gray delivers one joule of energy per kilogram of material absorbing the radiation.
greenhouse effect: An increase in the amount of heat trapped in the Earth's atmosphere caused by an increase in the levels of gases such as carbon dioxide.
greenhouse gases: Gases that absorb heat that would otherwise dissipate from the atmosphere.
ground fault circuit interrupter (GFCI): A special type of circuit breaker that shuts off if too much current flows in a circuit or if there is an extremely small difference between the current in the hot and neutral wires in a circuit.
ground rod: A long metal rod that is driven deep into the ground to electrically connect a circuit to the ground.
ground terminal: Terminal that electrically connects appliances to the ground.

H

half-life: The amount of time required for half the nuclei in a sample of a radioactive isotope to decay.
halogen: Any of these reactive nonmetals: fluorine, chlorine, bromine, iodine, and astatine.
haploid: Having only one chromosome of each type or form.
heredity: The passing on (transmitting) of features from parents to offspring.
hertz (Hz): The unit of frequency. One hertz is equal to one cycle per second.
heterozygous: Having different alleles for a particular gene.
high-risk earthquake zone: An area on the surface of the Earth where there is a high likelihood that earthquakes will occur.
holistic approach: A way of studying environmental problems by considering whole ecosystems rather than parts of ecosystems.
homologous: Describes chromosomes that are similar in overall appearance and made up of the same kind of genes, thus forming a pair.
homozygous: Having the same alleles for a particular gene.
hormone: A substance, secreted by a gland or other structure located in one part of the body, that regulates the activity of a particular organ or tissue in another part of the body.
hot wire: A wire that has a different voltage from that of the ground.
hybridization: The crossing of individuals of different species; or, in breeding experiments, the crossing of individuals that show different forms of the same trait.
hydrosphere: A discontinuous layer of water found on the Earth. The hydrosphere includes all bodies of surface water and ice.
hypothesis: An "educated guess" that predicts the solution to a research problem.

I

igneous: Referring to a family of rocks that were once molten. These include both volcanic and plutonic rocks.

image: A representation or picture of an object that resembles the object in many details, but may be larger or smaller than the object.

immigrate: To move into a population from another area.

***in vitro* fertilization:** The union of egg and sperm outside the body in a Petri dish. After several days, one (or more) of the resulting embryos is inserted into the uterus, to develop until birth.

incomplete dominance: A situation in which none of the alleles of a particular gene is completely dominant over the other(s). The resulting phenotype shows some mixture of the alleles that are present in the genotype.

index fossil: A fossil known to have lived at a particular time in the Earth's history. It indicates the age of the rock in which it is found. (Also known as a guide fossil.)

indicator: A substance used to detect the presence of another material. An acid-base indicator shows the presence of an acid or a base by changing colour.

induced charge: An electric charge that results from the presence of a charged object nearby, rather than from direct contact with a charged object.

inertia: The property of a mass that resists a change in motion. Stationary objects tend to stay in place and moving objects tend to keep moving.

insulator: A material in which electrons do not move easily.

intensity: The observable effects or damage caused by earthquakes. Earthquake intensity usually tends to decrease with distance from the epicentre. It is commonly measured on the Mercalli scale.

interphase: The phase of the cell cycle between cell divisions. During this phase, the cell grows and the chromosomes are duplicated (except in the interphase period between meiosis I and II).

interstitial cells: Cells scattered throughout the tissue surrounding the seminiferous tubules in the testes. The interstitial cells produce testosterone.

ion: An atom or a group of atoms that is positively or negatively charged as a result of losing or gaining one or more electrons.

ionic bond: The attraction between positive and negative ions.

ionic compound: A compound consisting of positive and negative ions held together by ionic bonds.

ionizing radiation: Radiation from radioactive isotopes, short-wavelength ultraviolet radiation, and X-ray radiation that has enough energy to ionize atoms.

isotope: One of two or more atoms of the same element that have the same number of protons, and therefore the same atomic number, but that have different mass numbers as a result of having different numbers of neutrons in the nucleus.

J

J-shaped curve: The curve on a graph produced by plotting exponential growth in numbers against time.

joule (J): The unit of work or energy. One joule equals one newton metre, or one watt second.

K

karyotype: A visual arrangement of an individual's chromosomes.

kilowatt hour (kW•h)**:** A non-metric unit of energy that corresponds to 3 600 000 J. One kilowatt hour equals the product of the power used (in kilowatts) and the time of use (in hours).

kinetic energy: Energy of motion.

L

lava: Molten rock on the Earth's surface. Also, the same material after it has solidified.

launcher: A device that carries a payload into space.

law of conservation of mass: A law stating that the total mass of all of the reactants in a chemical reaction is equal to the total mass of all of the products.

law of superposition: A law stating that in an undisturbed sequence of rock, the oldest layer is at the bottom and the youngest is at the top.

leaching: The movement of dissolved materials through soil.

lens: A device, often made of glass, that has at least one curved surface, and that changes the path of light travelling through it.

less developed country: A nation with a relatively low level of wealth and technology.

light microscope: A microscope that uses light to produce a magnified image of a specimen.

limestone: A sedimentary rock consisting mainly of calcium carbonate. The calcium carbonate may have originally been in the form of shells, rock fragments, or mineral grains.

liquefy: To change from a solid to a liquid state.

lithosphere: A rigid layer of the Earth consisting of part of the upper mantle and the crust.

low Earth orbit: An orbit just above the Earth's atmosphere.

M

magma: Molten rock below the Earth's surface.

magnet: An object that can attract a piece of iron as a result of the object's magnetic field.

magnetic: Any material that is affected by a magnetic field.

magnetic domain: A group of atoms within a magnetic material whose individual magnetic fields are lined up in the same direction.

magnetic field: A region in space where magnetic effects can be detected.
magnetic lines of force: Lines drawn from the north pole of a magnet to the south pole to represent a magnetic field. The closer the lines, the stronger the magnetic field.
magnetic north: North as indicated by a magnetic compass.
magnetic poles: The regions of a magnet where the magnetic field is strongest and where the magnetic lines of force are concentrated.
magnification: The extent to which an image is enlarged.
magnify: To cause something to appear larger than it really is.
magnitude: A measure of the total energy released by an earthquake. It is commonly measured on the Richter scale.
main circuit breaker: The single large circuit breaker that connects the main electricity supply wires to the service panel in a building. This breaker can shut off all electricity to the service panel.
mantle: Thick (2900 km), semi-solid layer of the Earth found between the crust and outer core.
marble: A metamorphic rock formed from the action of heat, pressure, and/or hot solutions on limestone.
mass number: The total number of protons and neutrons found in the nucleus of an atom.
medium (plural: media)**:** The material through which energy is carried.
meiosis: The process that produces cells with one-half the original number of chromosomes as a result of the occurrence of two separate nuclear divisions.
meiosis I: The first division of the nucleus in meiosis, in which pairs of homologous chromosomes separate and the number of chromosomes in each group of chromosomes is halved.
meiosis II: The second and final division of the nucleus in meiosis, in which the chromatids separate, producing haploid cells.
menopause: The period during which follicles in a woman's ovaries gradually cease developing. The production of estrogen and progesterone declines and, as a result, menstrual cycles stop.
menstrual cycle: The cycle of changes that occur in the reproductive organs of a female's body in response to changing levels of specific hormones. Events in this cycle include ovulation and menstruation.
menstruation: The process by which the thickened lining of the uterus leaves a female's body through the vagina. The four or five days during which this lining is lost (mainly as blood) is called the menstrual period.
Mercalli scale: A scale of earthquake intensity based on observation of the effect produced by ground motion. The full name is the Modified Mercalli Scale of Earthquake Intensity.
metallurgical coal: Coal used mainly for making coke, which is burned in metal smelters.
metamorphic: Referring to rock changed by the action of pressure, heat, and/or chemical solutions.
metamorphism: Process of changing rocks through heat, pressure, and/or the addition of chemical solutions.
metaphase: The second stage of mitosis, in which the joined chromatids move to the centre of the cell.
microgravity: The situation where objects in orbit are falling toward the Earth and thus seem weightless.
micro-organism: A living thing that is so small that it can be seen only with the aid of a microscope.
microscope: An instrument that produces enlarged images of specimens that are too small to be seen with the human eye alone.
microscopist: An individual who is skilled with the use of a microscope.
mineral: A naturally occurring chemical compound that makes up a rock or part of a rock.
mitochondria (singular: mitochondrion)**:** The organelles where most of the reactions that make up cellular respiration occur.
mitosis: The process in which the material within the nucleus of a cell is divided into genetically identical halves and two new nuclei are formed.
Mobile Servicing Centre (MSC): A sliding platform on the main frame of the proposed space station *Freedom*, containing robotic devices; it is Canada's contribution to the space station.
model: A representation made of a process or material, intended to help us visualize or understand it.
moderator: A substance used in a nuclear reactor to slow neutrons down so that they will be absorbed by nuclei, thus increasing the likelihood of further fission.
molecule: A particle consisting of two or more atoms joined together.
more developed country: A nation with a relatively high level of wealth and technology.
multiple alleles: Three or more different forms of a particular gene in the population.
mutagen: A substance or agent that causes changes (called mutations) in the genetic code.
mutation: A change in the hereditary material of a cell, resulting from a change in the order of the nucleotides in one or more chromosomes.

N

natural background radiation: The total amount of radiation from naturally occurring radioactive isotopes in the environment and from space.
natural resource: A naturally occurring part of the environment that supplies raw materials or energy, or in other ways contributes to human survival and well-being.

negative charge: The type of charge that results when an object contains more electrons than protons. A negative charge is developed on vinyl plastic when it is rubbed with wool, or on amber when it is rubbed with fur.

neutral: Carrying the same number of electrons as protons; having no net electric charge.

neutral wire: A wire that enters a house and is connected to the ground so that there is no voltage difference between the wire and the ground.

neutralization: A chemical reaction in which an acid and a base combine to produce a salt and water.

neutron: An uncharged subatomic particle found in the nucleus of an atom.

noble gas: Any one of the unreactive elements, including helium, neon, argon, krypton, xenon, and radon.

non-renewable energy resource (physical): An energy resource that cannot be replaced during a human lifetime.

non-renewable resource (ecological): A resource that cannot be replaced after it is used (for example, coal).

north pole: The end of a magnet that is attracted towards the northern part of the Earth. The north pole of a magnet attracts the south pole of a compass.

nosepiece: The part of a compound microscope to which the objective lenses are attached.

nuclear equation: In physics, a symbolic statement that shows the nuclei and particles at the beginning and end of a nuclear process.

nuclear force: In physics, a short-range force that acts within nuclei to hold the protons and neutrons of the nucleus together.

nuclear membrane: In biology, the membrane that surrounds the nucleus of cells, separating it from the cytoplasm.

nuclear reactor: In physics, a structure designed for the controlled release of heat energy from a sustained fission chain reaction.

nuclear transmutation: In physics, a change in the type of nucleus caused by a nuclear process.

nucleus (plural: nuclei) (physical): The very dense, positively charged part of an atom. Nuclei of different atoms consist of varying numbers of protons and neutrons.

nucleus (plural: nuclei) (biological): The control centre of the cell. The nucleus contains the cell's genetic material, which directs the production of the enzymes needed to carry out the chemical reactions that are the cell's life functions.

O

objective lens: The lens or lens system in a compound microscope that is nearest the specimen (at the lower end of the body tube). It produces a magnified image of the specimen.

ocular lens: The lens or lens system in a compound microscope that is nearest the eye of the viewer (near the top of the body tube), also called the ocular. The ocular magnifies the image produced by the objective lens.

ohm (Ω): The unit of electrical resistance. One ohm equals one volt per ampere.

Ohm's law: A law stating that the current flowing through a material is proportional to the voltage applied to it.

open-pit mining: Removing surface rock and digging a large pit to extract resources that lie just below the Earth's surface.

orbit (physics): According to Bohr, the circular path followed by an electron as it revolves around the nucleus.

orbit (space): The regular path of a spacecraft or a natural satellite around a planet, such as Earth, or other large object.

organelle: The term used for the many different types of structures found within the cytoplasm of cells. Each type of organelle serves a different function and has a different structure. Each organelle is usually surrounded by its own membrane.

organic: Matter that is or once was a part of living organisms. Some fossil shells and charcoal are examples of organic matter.

osmosis: The movement of water molecules through a selectively permeable membrane from an area where they are in higher concentration to an area where they are in lower concentration.

ovary: The female organ that produces the gametes (ova). Human females have a pair of ovaries.

overburden: The rock and earth removed in order to reach minerals and fossil fuels.

overpopulation: A condition that occurs when a population is so large that it damages the environment on which it depends.

oviduct: The tube that conducts ova from the abdominal cavity near the ovary to the uterus. It is also called a Fallopian tube.

ovulation: The process in which an ovum is released from its follicle and from the ovary into the abdominal cavity.

ovum (plural: ova): The mature reproductive cell, or gamete, of females. These cells are also called egg cells.

ozone layer: A 20 km thick band of ozone in the upper atmosphere that absorbs ultraviolet light from the sun.

P

P (primary) wave: (Also known as a compression wave.) A type of earthquake wave in which the particles in the medium vibrate parallel to the wave direction. It is the first wave to arrive at a seismographic station.

paleontologist: A scientist who studies ancient life through the examination of fossil remains. Also known as a paleobiologist.
parallel circuit: A circuit in which there is more than one pathway for electric current. If cells are connected in parallel, the positive terminals of all the cells are connected together, and so are all the negative terminals.
parent material: A radioactive isotope of an element that undergoes radioactive decay.
parent nucleus: The original nucleus of an atom in a radioactive decay.
particulates: Microscopic bits of solids or liquids suspended in the air.
pathogen: A disease-causing agent, such as a bacterium, protist, or virus.
payload: Any satellite, piloted vehicle, or cargo launched into space.
pedigree: A chart or family-tree diagram showing the pattern of inheritance of a single trait within a family.
penis: The male reproductive organ that transports sperm to the female during sexual intercourse.
period (chemical): A horizontal row of elements in the periodic table. (See Appendix B.)
period (geological): Part of the standard, worldwide geological time scale. (See Appendix D.)
periodic table: A chart of the elements, arranged according to their atomic numbers. Elements with similar physical and chemical properties and similar electron arrangements are in the same column. (See Appendix B.)
permanent magnet: A magnetic material in which the magnetic fields of most of the magnetic domains are aligned in the same direction and retain this alignment for a long period of time, thus allowing the material to remain a magnet for a long time.
phenotype: The physical appearance of an individual.
photochemical reaction: A chemical reaction that occurs in the presence of light.
photosynthesis: The process in which the sun's energy is used to produce energy-rich molecules, such as glucose. Photosynthesis involves a complex series of chemical reactions.
photovoltaic cell: A device that converts the energy of light directly into electrical energy. Also called a solar cell.
physical change: A change in which the chemical composition of the starting materials does not change and no new substance is formed.
piezoelectric effect: The production of electricity by subjecting certain types of crystal to stress.
piloted spacecraft: A spacecraft with people on board.
placenta: A structure composed partly of the membranes that surround the embryo and partly of tissue from the wall of the uterus. The developing offspring receives oxygen and nutrients and gets rid of waste materials through the placenta.
plasmid: In many bacterial cells, a small, second DNA molecule that carries genes responsible for a few less important traits. Plasmids are duplicated during cell division.
plate: Large, mobile section of the Earth's lithosphere.
plate boundary: The region where adjoining plates meet. Volcanic and earthquake activity are common in these areas.
plate tectonics theory: A theory describing the movement of and interaction between large segments of the Earth's lithosphere.
plutonic rock: An igneous rock formed from magma cooled beneath the Earth's surface. Also called intrusive rock.
polarized: Having terminals that must be connected in a particular way within a circuit.
polarized plug: A plug that has one prong that is wider than the other and that can be inserted into an outlet in only one way; reduces the risk of electrocution, and required for the safe use of some appliances.
pollutant: Substance or energy that causes pollution.
pollution: Damage to the living or non-living environment caused by the addition of certain substances or energy.
polyatomic ion: A group of atoms that carries a charge and remains bonded together in most chemical reactions.
population: All the individuals of a species that occur together in a particular area at a particular time.
positive charge: The type of charge that results when an object contains more protons than electrons. A positive charge is developed on acetate plastic when it is rubbed with cotton, or on glass when it is rubbed with silk.
potential energy: Energy that is stored, such as the energy in a stretched spring.
power: The rate at which energy is produced, absorbed, or transferred; measured in watts.
prenatal diagnosis: Obtaining information, through various tests, about the health of a fetus.
primary cell: A cell that uses chemical energy to produce its own electricity and that can only be used once.
primary sexual characteristics: The parts of the reproductive system that are directly involved in the production of gametes, in fertilization, or in the development of an embryo.
probability: Chance or likelihood.
process skill: A procedure, such as observing or measuring, that is used in scientific research.
product: The substance formed in a chemical reaction.
progesterone: A female sex hormone secreted by the corpus luteum.

property: An observable characteristic of matter that can be used to distinguish one substance from another.

prophase: The first stage of mitosis, in which the chromosomes become visible, the spindle forms, and the nuclear membrane breaks down.

prostate: A gland that produces one component of the seminal fluid. It empties its secretions into the urethra.

protist: A type of organism, usually small, and in most cases consisting of a single cell.

proton: A positively charged subatomic particle found in the nucleus of an atom.

puberty: The period of an organism's life in which the reproductive organs become functional and the secondary sexual characteristics develop.

Punnett square: A grid for determining the possible genotypes of offspring resulting from a cross between two parents with particular genotypes.

R

radiation: A general term that is applied to electromagnetic energy or to particles emitted from the nuclei of atoms.

radioactive: Emitting radiation; said of atoms that emit radiation from their nuclei.

radioactive decay: The process whereby the nuclei of radioactive parent isotopes emit alpha, beta, or gamma radiation to form decay products.

radioactivity: The property of some elements whose nuclei spontaneously release radiation.

radiometric age: Time since a mineral grain last crystallized.

radiometric dating: Determining the age of a material by calculating the percentage of the radioactive atoms which have decayed.

reactant: The starting material in a chemical reaction.

reactor meltdown: The destruction of a nuclear reactor due to loss of control of the energy-releasing fission reaction within the reactor.

recessive: In genetics, referring to an allele that is not expressed when a dominant allele for the gene is also present.

recombinant DNA: Genes from different types of organisms that have been spliced together.

recycle: To process waste materials, such as glass, paper, or aluminum, so that they can be used again.

reduce: To cut down on consumption of materials and energy.

relative age: The age of one event compared to another, that is, whether one of the events is older or younger than the other.

Remote Manipulator System: The robotic arm designed and built in Canada and used on the space shuttle.

remote sensing: Making observations from a distance using an imaging device, such as a camera.

renewable energy resource (physical): An energy resource that constantly replenishes itself for continuous use.

renewable resource (ecological): A resource that can be continually replaced by natural processes (for example, fish).

repel: To push away from oneself or force apart as a result of an electric or magnetic force.

reserve: The total amount of a resource that has been measured and is judged economical to extract.

resistance: A measure of how hard it is for an electric current to flow through a material. Resistance, measured in ohms, is equal to the voltage, measured in volts, divided by current, measured in amperes ($R = V/I$).

resistor: A device made of a material that restricts the flow of current in a circuit.

resolving power: The ability of an optical instrument or eye to show separate objects as distinct. It is also called the resolution of the instrument or eye.

resource management: Making the best use of resources for the greatest number of people over the longest period of time, while minimizing the impact on the environment.

reuse: To use an item such as a glass bottle or plastic container over and over again instead of throwing it out after a single use.

Richter scale: A scale of earthquake magnitude, based on seismograph recordings, that is a measure of the amount of energy released from an earthquake.

right-hand rule: A rule stating that if the thumb of the right hand is pointed in the direction of the conventional current flow in a wire, the fingers will curl in the direction of the magnetic lines of force around the wire.

rock cycle: Over long periods of time, any piece of the Earth's crust may be transformed into any of the three families of rock—igneous, metamorphic, or sedimentary—depending on conditions. Thus each family is linked to the others in a cycle.

rock families: The classification system of rocks based on the manner in which they were formed. The three families are called igneous, sedimentary, and metamorphic rocks.

rocket engine: A device in which fuel is burned rapidly to create gases that leave through a nozzle, exerting a thrust on the device or vehicle; it can operate in space because it has its own supply of fuel and an oxygen-containing chemical.

S

S (secondary or shear) wave: An earthquake wave in which particles in the medium vibrate perpendicular to the wave direction, similar to the shaking of a rope. Since the S wave travels more slowly than a P wave, it is the second wave to arrive.

S-shaped curve: The shape of a population growth curve in which a period of rapid growth is followed by a slowing of the growth rate and finally by stabilization of numbers.

salt: A compound formed by the reaction of an acid and a base.

sandstone: A rock composed of sand grains cemented together.

satellite: An object that travels in an orbit around another object.

schematic diagram: A diagram that uses simplified symbols to represent the parts of electric circuits.

science: Both a body of knowledge and an organized process for gaining more knowledge about the natural world.

scientific method: A problem-solving approach that uses science process skills in a logical order: (1) identify the problem, (2) make a hypothesis, (3) design and perform an experiment to test the hypothesis, (4) interpret and analyse results, (5) communicate procedures, observations, and conclusions to other scientists.

scrotum: The sac that contains the pair of testes.

secondary cell: A cell that needs to be charged to produce an electric current and that can be recharged.

secondary sexual characteristics: Features that develop during puberty and serve to distinguish the sexes, but that do not play a direct role in the production of gametes, fertilization, or the development of an embryo.

sediment: Rock, gravel, sand, mud, clay, or shell fragments which have settled out of either air or water. They may later form layers of sedimentary rock.

sedimentary rock: Rock formed from the compaction of layers of sediment.

seismogram: The graphic record made by a seismograph.

seismograph: A mechanical instrument used to measure ground motion graphically; usually measures motion from earthquakes.

seismology: A branch of science devoted to the study of earthquakes.

seismometer: An electronic instrument used to measure ground motion.

selective breeding: Allowing or encouraging certain individual plants and animals with desirable characteristics to reproduce in order to increase the chance of obtaining offspring with these characteristics.

selectively permeable: Something (for example, a membrane) that permits some, but not all, materials to pass through it.

semen: A thick fluid composed of sperm and seminal fluid.

seminal fluid: A fluid made up of secretions from several glands of the male reproductive tract. It protects and nourishes sperm and assists their movement.

seminal vesicle: A gland that produces one component of the seminal fluid. It empties its secretions into the vas deferentia.

seminiferous tubules: Thread-like tubes, coiled and tightly packed within the testes, in which the sperm are produced and stored.

series circuit: A circuit in which there is only one pathway for electric current. If cells are connected in series, the positive terminal of one cell is connected to the negative terminal of the next cell.

series-parallel circuit: A circuit that contains both series and parallel connections.

service panel: The electrical panel from which the branch circuits are distributed.

sewage: Liquid or semi-liquid waste carried away from houses and other buildings.

sex chromosomes: The chromosomes that determine the sex of an individual, such as the X and Y chromosomes of humans.

sex-linked trait: A trait controlled by a gene located on one of the sex chromosomes. As a result, the trait is much more common in one sex than the other.

sexual reproduction: Reproduction that requires the participation of two parents, each one contributing a gamete. The two gametes fuse to form the first cell of an offspring organism.

shale: A sedimentary rock made from mud or clay.

short circuit: A condition that exists when bare wires touch and thus provide an undesired low-resistance pathway for electricity.

sievert (Sv): A measure of the effect of a dose of radiation on living things.

simple microscope: A microscope that contains only one lens or lens system.

single replacement: A chemical reaction in which one element replaces another element in a compound.

slide: A rectangular piece of transparent glass or plastic used to hold a specimen that is being examined with a microscope.

solar cell: A commonly used name for a photovoltaic cell.

solar system: Our sun (which is an average-size star) and everything influenced by its force of gravity, such as the planets, comets, moons, and asteroids.

solid: Matter with a definite shape, volume, and strength.

south pole: The end of a magnet that is attracted towards the southern part of the Earth. The south pole of a magnet attracts the north pole of a compass.

space: The region above the Earth's atmosphere.

spacecraft: A vehicle designed to travel in space where there is no air.

space probe: A spacecraft with many instruments on board that is sent to explore our solar system.

space station: A large spacecraft in which humans live and work, designed to stay in space for many years.

species: A group of organisms. Individual members of a species are able to interbreed with one another.

spectroscope: An instrument that separates light into coloured bands of a spectrum.

sperm: A short form of spermatozoon or spermatozoa.

spermatozoon (plural: spermatozoa): The mature reproductive cell, or gamete, of males. These cells are also called sperm.

spindle: A system of tiny fibres that assists the movement of chromosomes during mitosis and meiosis.

spinoff: An application or a benefit to society that is an indirect result of space exploration.

spore formation: A method of asexual reproduction in which specialized cells called spores (each genetically identical to the parent) are formed and then dispersed via wind or water or in other ways.

stage: The platform on which a specimen is placed for viewing. It is located below the objective lens of a compound microscope.

standard geologic column: A vertical diagram that shows the subdivisions of geological time and a general view of the rock layers and fossils.

static electric charge: An excess of positive or negative charges that stay in place on an object for a relatively long period of time.

strata: Layers of sedimentary rock.

structure: In geology, the features of a rock mass usually produced after its original formation.

subatomic particle: A particle that is smaller than an atom. This term is usually applied to particles that make up atoms, including protons, neutrons, and electrons.

sub-station: A facility that receives electricity from a long-distance power transmission line at voltages up to 500 kV and transfers it to distribution stations at a lower voltage, typically 60 kV.

superconductor: A substance that offers almost no resistance to the flow of electric current.

surrogate mother: A female who carries an embryo until birth for another female. The embryo can either have been removed from the other female or have been produced by *in vitro* fertilization. A human surrogate mother is a woman who has made a contract to have a child for a couple, with the child being genetically related to at least one member of the couple.

synthesis: A chemical reaction in which two or more elements or compounds combine to form a single compound.

T

technology: The application of scientific knowledge.

telophase: The fourth and last stage of mitosis, in which nuclear membranes form around the groups of chromosomes at each end of the cell. The chromosomes uncoil and the spindle disappears.

terminal: The electrical connection to an object such as a chemical cell.

testis (plural: testes): The male organ that produces the gametes (sperm). Human males have a pair of testes.

testosterone: A male sex hormone secreted by the interstitial cells of the testes.

theory: A well-accepted hypothesis that is supported by repeated observation and experimentation.

thermal coal: Coal used mainly to produce heat in electricity-generating stations.

thermocouple: A device that makes use of the thermoelectric effect to produce electricity.

thermoelectric effect: The production of electricity due to the difference in temperature between junctions of two different metals.

thrust: The force that causes an airplane, rocket, or spacecraft to move.

toxic waste: Poisonous material that may cause harm even in small amounts.

trait: Another word for characteristic, or feature.

transformer: A device that can raise or lower the voltage of alternating-current electricity.

tremor: An earthquake with small ground motion.

trend: Pattern of resource use showing changes in rate of use.

trilobite: An extinct class of animals (arthropods) related to modern crabs and shrimp.

tsunami: A large sea wave, usually triggered by an earthquake on the sea floor.

tumour: An abnormal clump of cells formed when damaged cells reproduce too quickly.

U

ultrasound examination: A prenatal test in which high-frequency sound waves are projected through the mother's body. There the waves bounce off the structures inside to give a "picture" of the fetus.

unpiloted spacecraft: A spacecraft with no people on aboard.

urethra: The tube, connected to both the vas deferentia and the bladder, which runs through the penis to provide an exit from the body. It transports both semen and urine (at different times).

uterus: The female organ within which the offspring develops before birth. It is also called the womb.

V

vaccine: A substance used to prevent disease. Manufactured from a weakened strain of a specific pathogen, a vaccine is injected into an organism to stimulate the production of antibodies, which will give the organism immunity to later infections of the same pathogen.

vacuole: An organelle found in the cells of most organisms. Vacuoles serve a variety of functions, including storing nutrients, wastes, or water.

vagina: The elastic, muscular tube leading from the cervix out of the female's body.

value issue: An issue that involves personal beliefs and values.

vas deferens (plural: vas deferentia): The long tube in which sperm travel through the abdominal cavity from each epididymis to the urethra.

vector: An organism that carries a disease-causing agent from an infected organism to a healthy one. Insects, dogs, foxes, and other mammals are the vectors for various diseases.

vegetative reproduction: A method of asexual reproduction in plants, in which an offspring develops from a part of the plant other than the flower, such as from a stem or leaf.

velocity: Speed in a given direction, found by dividing distance by time; for example, 40 km/h towards the north.

virus: A non-living unit that can enter the cell of an organism and then make use of this cell's internal processes and structures to reproduce itself. Infection by viruses usually results in illness.

visible spectrum (plural: spectra): That portion of the electromagnetic spectrum that includes visible light (red, orange, yellow, green, blue, indigo, and violet).

volcanic rock: Rock that has formed from magma reaching the Earth's surface to become lava or ash.

volt: The unit of voltage. One volt equals one joule per coulomb.

voltage: The amount of electrical potential energy per coulomb of charge transferred.

voltmeter: A device that measures voltage.

W

watt (W): The unit of power. One watt equals one joule per second.

wavelength: The distance from a point on one wave to the same position on the next wave.

weathering: The physical or chemical breakdown of rock into fragments such as mud, sand, or gravel.

weightlessness: The sensation of floating in space caused either by continuously falling toward the Earth while in orbit or by being located extremely far from the force of gravity of any large body.

work: The product of the force used to move an object and the distance the object is moved. Work is measured in joules.

working distance: In microscopy, the distance between the objective lens and the specimen being examined.

X

X chromosome: One of the two types of sex chromosomes in humans and many other types of organisms. Female humans have two X chromosomes.

Y

Y chromosome: One of the two types of sex chromosomes in humans and other types of organisms. Only males have a Y chromosome.

Z

zero population growth: A stable human population in which the number of births each year is equal to the number of deaths.

zygote: The diploid cell formed by the fusion of a male and a female gamete during fertilization.

Index

D

E

L

M

N

O

Photo Credits

All photos not credited here are courtesy of the authors or of Wiley photo library files.

To the Student, page xiii: Al Harvey. Chap. 1 opening: NASA. Fig. 1.2: Al Harvey. Fig. 1.5: (a) The Metropolitan Museum of Art, Gift of Edward S. Harkness; (b) Royal Ontario Museum; (c) Bell Canada Telephone Historical Collections. Fig. 1.6: Al Harvey. Profile: P. Rona, University of Victoria. Fig. 1.8: The Bettmann Archive.

UNIT I
Unit opener: NASA. Chap. 2 opening: NASA. Fig. 2.2: Canada Centre for Remote Sensing. Fig. 2.3: NASA. Fig. 2.4: Esther Goddard/AIP Niels Bohr Library. Fig. 2.5: NASA. Fig. 2.8: NASA. Fig. 2.12: Environment Canada. Fig. 2.13: Landsat-5/SPOT Imagery processed and supplied by MacDonald Dettwiler, courtesy of the Canada Centre for Remote Sensing, © CNES. Fig. 2.14: Andrew Antenna Company. Fig. 2.16: NASA. Fig. 2.17(a), (b), (c): NASA. Fig. 2.18: NASA. Fig. 2.19: NASA. Fig. 2.20: NASA. Fig. 2.21(a), (b): NASA. Fig. 2.22: NASA. Profile: Canadian Space Agency. Fig. 2.23: NASA. Fig. 2.24: SPAR Aerospace/Canadian Nuclear Association. Fig. 2.25: D. Bullard.

UNIT II
Unit opener: B.C. Hydro. Chap. 3 opening: Bill Ivy. Profile: AIP/Niels Bohr Library. Fig. 3.9: Ontario Science Centre. Chap. 4 opening: Jack Finch, Fairbanks, Alaska. Fig. 4.20: Aluminum Company of Canada, Kitimat, B.C. Fig. 5.8: CSA and Consumer and Corporate Affairs Canada. Career: Kathy Gibson. Chap. 6 opening: B.C. Hydro. Fig. 6.3(b): B.C. Hydro. Fig. 6.4(a): B.C. Hydro. Fig. 6.5: B.C. Hydro; Pacific Gas and Energy, San Francisco CA.; Nova Scotia Power Corporation. Fig. 6.7: (a) NASA; (b) Arco Solar Inc. Science in Our World: Instructional Media Centre, Simon Fraser University. Science in Our World: Colorado Superconductor Inc. Fig. 6.17: Rosemary Wurts. Computer Application: B.C. Hydro Control Centre. Chap. 7 opening: The Deutches Museum, Munich. Fig. 7.1: Ontario Hydro. Fig. 7.2: Wayne Sheridan. Fig. 7.18: Toronto East General Hospital. Profile: AIP/Niels Bohr Library. Fig. 7.21: British Columbia Institute of Technology. Fig. 7.22: AIP/Niels Bohr Library. Fig. 7.25: Ontario Hydro. Science in Our World: Canapress Photo Service.

UNIT III
Unit opener: Ken Shelton, U.S. Fish and Wildlife Service. Chap. 8 opening: Paul Thompson. Fig. 8.1: Fisher Collection, Fisher Scientific Company. Fig. 8.3: Fisher Collection, Fisher Scientific Company. Fig. 8.4: Smith Collection, Center for the History of Chemistry. Fig. 8.5: Fisher Collection, Fisher Scientific Company. Profile: Fisher Collection, Fisher Scientific Company. Fig. 8.11: AIP/Niels Bohr Library. Fig. 8.15: Smith Collection, Center for the History of Chemistry. Fig. 8.21: OPC, Inc. Fig. 8.24: Canapress Photo Service. Chap. 9 opening: Steven Groer. Fig. 9.2: Robert J. Capece. Fig. 9.3: D. Bullard. Fig. 9.4: Vancouver General Hospital. Fig. 9.5: Robert J. Capece. Fig. 9.6: (a) International Salt Company, Clarks Summit PA; (b) Don McCoy/Rainbow. Fig. 9.7: OPC, Inc. Computer Application: R.H. Czerneda. Fig. 9.23: D. Bullard. Chap. 10 opening: S. T. Fox. Fig. 10.1: J. D. Rising. Fig. 10.2: Sharon Baillie. Fig. 10.3: (a) Sharon Baillie; (b) OPC, Inc.; (c) Sharon Baillie. Fig. 10.9: D. Bullard. Fig. 10.12: Gary Kaustinen and Peter Senkowski. Fig. 10.19: Jane L. Forbes. Career: R. H. Czerneda. Fig. 11.6: S. T. Fox. Fig. 11.11: Kennecott Coppery Refinery. Fig. 11.15: D. Bullard.

UNIT IV
Unit opener: R. E. Wallace, U.S. Geological Survey. Chap. 12 opening: Canadian Museum of Civilization/National Museums of Canada. Fig. 12.1: (a) TV Ontario;

(b) MacMillan Bloedel; (c) Imperial Oil Limited; (d) Rosemary Wurts. Fig. 12.2: (a) Automobile Quarterly, Inc.; (b) MacMillan Bloedel; (d) Ontario Hydro. Fig. 12.3: Mildred McPhee. Fig. 12.4: (a) Kraft/ Explorer/Science Source; (b) K. Wayne Savigny. Fig. 12.6: (a) D. Grose; (b), (c) Carlo Giovanella. Fig. 12.8: (b), (c) W. R. Danner; (d) Carlo Giovanella. Fig. 12.11: (a), (b) Carlo Giovanella. Science in Our World: *Fossils of the Burgess Shale*, Miscellaneous Report 43, Geological Survey of Canada; Energy, Mines and Resources Canada. Reproduced with the permission of the Minister of Supply and Services Canada, 1991. Career: R. H. Czerneda. Fig. 12.24: Vancouver Public Library files. Fig. 12.28(a), (b), (c): NASA. Chap. 13 opening: *Vancouver Sun*. Fig. 13.4: Alberni Valley Museum. Profile: Ontario Science Centre. Figs. 13.9 and 13.17: *Canadian Earthquakes—1985–86*, Geological Survey of Canada Paper 88–14; Energy, Mines and Resources Canada. Reproduced with the permission of the Minister of Supply and Services Canada, 1990. Fig. 13.10: University of Washington, Bellingham WA, March 28, 1964. Fig. 13.16: *Canadian Geographic*. Fig. 13.18: *Geos*, Energy, Mines and Resources Canada. Fig. 13.19: B. Atwater, Uses/*Geos*, Energy Mines and Resources Canada. Fig. 13.20: *The Loma Prieta Earthquake of October 17, 1989*, U.S. Geological Survey, 1990. Computer Application: Geological Survey of Canada. Fig. 13.21: Al Harvey. Appendix D: *Fossils of the Burgess Shale*, Geological Survey of Canada Miscellaneous Report 43; Energy, Mines and Resources Canada. Reproduced with the permission of the Minister of Supply and Services Canada, 1991.

UNIT V

Unit opener: Sharon Baillie. Chap. 14 opening: National Archives. Fig. 14.3(b): Al Harvey. Fig. 14:5: The Bettmann Archive. Fig. 14.6: Smith Collection, Center for the History of Chemistry. Fig. 14.7(a): Carl Zeiss Canada Limited. Fig. 14.8: National Library of Medicine. Fig. 14.9: National Library of Medicine. Fig. 14.10: (a) Tony Williams; (b) MPR Teltech Ltd.; (c) RCMP Forensic Laboratories. Fig. 14.22: Al Harvey. Fig. 14.23(a), (b), (c): Vancouver General Hospital, Dept. of Pathology. Fig. 14.24: Simon Fraser University. Chap. 15 opening: J. W. Schopf. Fig. 15.1: Paul Bradbury, British Columbia Institute of Technology. Fig. 15.2: B.C. Research Corporation. Career: Kathleen Gibson. Fig. 15.11: Finn Larsen. Fig. 15.16: Province of British Columbia. Fig. 15.17: Health and Welfare Canada. Fig. 15.18: J. Lai-Fook. Fig. 15.22(b): Carolina Biological Supply Company. Fig. 15.28: World Health Organization/J. Wickett. Fig. 15.30: The Kidney Foundation of Canada. Chap. 16 opening: Ontario Science Centre. Fig. 16.1: National Film Board/John Flanders. Fig. 16.3: (a) Ontario Science Centre; (b) D. I. Gailbraith et al., Life Probe. Fig. 16.7: Al Harvey. Fig. 16.8: Al Harvey. Profile: Quadra Logics. Fig. 16.11: D. I. Gailbraith et al., Life Probe. Fig. 16.12: Carolina Biological Supply Company. Fig. 16.13: Valan Photos/Fred Bauendam. Fig. 16.14: H. M. Lang. Fig. 16.15: Heilman Photography/John Colwell. Fig. 16.20: Frank Norman. Chap. 17 opening: *Vancouver Sun*. Fig. 17.4: National Library of Medicine. Fig. 17.5: Al Harvey. Science in Our World: Nancy Wexler. Fig. 17.14: Ontario Science Centre. Chap. 18 opening: R. L. Brinster/Peter Arnold, Inc. Fig. 18.3: Metropolitan Toronto Zoo. Fig. 18.5: Metropolitan Toronto Zoo. Fig. 18.6: Cincinnati Zoo and Botanical Garden. Fig. 18.7: New Westminster News/Brian Giebelhaus. Fig. 18.10(b): Vancouver General Hospital. Fig. 18.11(a): Sheila Camer. Fig. 18.16: Crystal Image Photography. Fig. 18.17: UPI/Bettmann. Fig. 18.18: The Hospital for Sick Children. Computer Application: California Institute of Technology. Fig. 18.19: Al Harvey.

UNIT VI

Unit opener: Province of British Columbia. Chap. 19 opening: UPI/

Bettmann. Fig. 19.1: Province of British Columbia. Fig. 19.3: Al Harvey. Fig. 19.7: Ontario Science Centre. Fig. 19.8: Province of British Columbia. Fig. 19.12: Mahdu Ranadive. Fig. 19.14: National Library of Medicine. Fig. 19.15: National Library of Medicine. Science in Our World: Courtesy of the Royal British Columbia Museum, Victoria, B.C. Fig. 19.17: Bill Hurst. Fig. 19.18: Bill Ivy. Fig. 19.22: CIDA/Roger Lemoyne. Fig. 19.24: (a) UNICEF; (b) Rosemary Wurts. Fig. 19.25: Al Harvey. Fig. 19.27: Finn Larsen. Chap. 20 opening: *Vancouver Sun*. Fig. 20.1: SCC Photocentre. Fig. 20.2: National Geographic Society/Robert Madden. Fig. 20.3: Metropolitan Toronto Reference Library. Fig. 20.5: (top centre) Al Harvey; (top right, bottom left) Rosemary Wurts. Fig. 20.8: Al Harvey. Profile: *Monday Magazine*, Victoria/Lawrence McLagan. Fig. 20.11: H. M. Lambert/Miller Services. Fig. 20.12: D. Bullard. Fig. 20.20: *Calgary Herald*/Bill Herriot. Page 480: Al Harvey. Fig. 20.26: Mexican National Tourist Council. Fig. 20.27: Take Stock Photography Inc./Rick Rudnicki. Fig. 20.28: Sayed Sattar. Page 485: NASA. Page 487: Bill Ivy. Fig. 21.1: The British Columbia Pavilion Corporation. Fig. 21.2(top): Carlo Giovanella/UBC Dept. of Geology; (bottom): Rosemary Wurts. Fig. 21.3: Al Harvey. Fig. 21.4: Ken Graff/Union Carbide Corporation. Fig. 21.6: B.C. Coal. Fig. 21.7: (left) Government of Saskatchewan; (right) Ontario Hydro. Fig. 21.11: (left) B.C. Rail; (right) Ontario Hydro. Computer Application: Crown Lands Communications Branch, B.C. Fig. 21.13: (right) Institute for Hydrogen Systems; (bottom left) Toshiba of Canada Ltd.; (top left) B.C. Hydro. Fig. 21.17: Al Harvey. Fig. 21.18: B.C. Ministry of Energy, Mines and Petroleum Resources. Fig. 21.19: B.C. Ministry of Energy, Mines and Petroleum Resources. Fig. 21.20: B.C. Coal. Fig. 21.21 (top and bottom): B.C. Ministry of Energy, Mines and Petroleum Resources. Fig. 21.22(a), (b): The Coal Association of Canada. Fig. 21.24: L. T. Webster. Fig. 21.25: Al Harvey. Career: Betsy Gordon. Fig. 21.26: Miller Comstock/Armstrong Roberts.